Lecture Notes in Computer Science 1664

Edited by G. Goos, J. Hartmanis and J. van Leeuwen

Springer

Berlin
Heidelberg
New York
Barcelona
Hong Kong
London
Milan
Paris
Singapore
Tokyo

Jos C.M. Baeten   Sjouke Mauw (Eds.)

CONCUR'99
Concurrency Theory

10th International Conference
Eindhoven, The Netherlands, August 24-27, 1999
Proceedings

Springer

Series Editors

Gerhard Goos, Karlsruhe University, Germany
Juris Hartmanis, Cornell University, NY, USA
Jan van Leeuwen, Utrecht University, The Netherlands

Volume Editors

Jos C.M. Baeten
Sjouke Mauw
Eindhoven University of Technology, Department of Computing Science
P.O. Box 513, 5600 MB Eindhoven, The Netherlands
E-mail: {josb,sjouke}@win.tue.nl

Cataloging-in-Publication data applied for

Die Deutsche Bibliothek - CIP-Einheitsaufnahme

Concurrency theory : 10th international conference ; proceedings / CONCUR '99, Eindhoven, The Netherlands, August 24 - 27, 1999. Jos C. M. Baeten ; Sjouke Mauw (ed.). - Berlin ; Heidelberg ; New York ; Barcelona ; Hong Kong ; London ; Milan ; Paris ; Singapore ; Tokyo : Springer, 1999
(Lecture notes in computer science ; Vol. 1664)
ISBN 3-540-66425-4

CR Subject Classification (1998): F.3, F.1, D.3, D.1, C.2

ISSN 0302-9743
ISBN 3-540-66425-4 Springer-Verlag Berlin Heidelberg New York

Typesetting: Camera-ready by author
SPIN: 10704232 06/3142 – 5 4 3 2 1 0 Printed on acid-free paper

Preface

This volume contains the proceedings of the 10th International Conference on Concurrency Theory (CONCUR'99) held in Eindhoven, The Netherlands, 24-27 August 1999.

The purpose of the CONCUR conferences is to bring together researchers, developers and students in order to advance the theory of concurrency and promote its applications. Interest in this topic is continuously growing, as a consequence of the importance and ubiquity of concurrent systems and their applications, and of the scientific relevance of their foundations. The scope of CONCUR'99 covers all areas of semantics, logics and verification techniques for concurrent systems. A list of specific topics includes (but is not limited to) concurrency-related aspects of: models of computation and semantic domains, process algebras, Petri nets, event structures, real-time systems, hybrid systems, stochastic systems, decidability, model-checking, verification techniques, refinement techniques, term and graph rewriting, distributed programming, logic constraint programming, object-oriented programming, typing systems and algorithms, case studies, and tools and environments for programming and verification.

The first two CONCUR conferences were held in Amsterdam (NL) in 1990 and 1991, the following ones in Stony Brook (USA), Hildesheim (D), Uppsala (S), Philadelphia (USA), Pisa (I), Warsaw (PL) and Nice (F). The proceedings have appeared in Springer LNCS, as Volumes 458, 527, 630, 715, 836, 962, 1119, 1243, and 1466.

Of the 91 regular papers submitted this year, 32 were accepted for presentation at the conference and are included in the present volume. Apart from these, the conference included four invited presentations, by Rance Cleaveland (State University of New York at Stony Brook, USA), Javier Esparza (Technische Universität München, D), Rob van Glabbeek (Stanford University, USA) and Catuscia Palamidessi (Pennsylvania State University, USA), and three invited tutorials, by Petr Jančar (Technical University of Ostrava, CZ), Nils Klarlund (AT&T Labs Research, USA) and Jan Tretmans (Universiteit Twente, NL).

We want to thank all members of the program committee, and their subreferees, for selecting the papers to be presented.

Special thanks are due to the local organization committee, chaired by Jan Friso Groote. Dragan Bošnački arranged the tool demonstrations, André Engels was webmaster, Kees Middelburg was in charge of the tutorials, and Martijn Oostdijk took care of the submission software (written by Vladimiro Sassone). Local arrangements, and help with registration, were provided by Marcella de Rooij, Desiree Meijers, and Anne-Meta Oversteegen.

The conference had three satellite events, all held on 23 August 1999. These were PROBMIV'99 (Workshop on Probabilistic Methods in Verification), EXPRESS'99 (6th International Workshop on Expressiveness in Concurrency), and VFM (Symposium on Visual Formal Methods). We thank Eindhoven University of Technology for hosting the event and providing many facilities. We thank our sponsors IPA (Institute of Programming Research and Algorithmics, NL), Philips Research Eindhoven, and EESI (Eindhoven Embedded Systems Institute).

Eindhoven
June 1999

Jos Baeten and Sjouke Mauw
Program Committee Co-chairs

CONCUR Steering Committee

Jos Baeten (Chair, Eindhoven, NL)
Eike Best (Oldenburg, D)
Kim Larsen (Aalborg, DK)
Ugo Montanari (Pisa, I)
Scott Smolka (Stony Brook, USA)
Pierre Wolper (Liège, B)

CONCUR'99 Program Committee

Ralph Back (Åbo Akademi University, SF)
Jos Baeten (Eindhoven University of Technology, NL, Co-chair)
Jan Bergstra (Universiteit van Amsterdam, NL)
Manfred Broy (Technische Universität München, D)
Rocco De Nicola (Universitá di Firenze, I)
Andrew Gordon (Microsoft Research, UK)
Roberto Gorrieri (University of Bologna, I)
Tom Henzinger (University of California, USA)
Bengt Jonsson (Uppsala University, S)
Maciej Koutny (University of Newcastle, UK)
Nancy Lynch (MIT, USA)
Sjouke Mauw (Eindhoven University of Technology, NL, Co-chair)
Arend Rensink (University of Twente, NL)
Philippe Schnoebelen (Ecole Normale Supérieure de Cachan, F)
Robert de Simone (INRIA, F)
P.S. Thiagarajan (SPIC Mathematical Institute, IN)
David Walker (Oxford University, UK)
Glynn Winskel (University of Århus, DK)

Referees

Parosh Aziz Abdulla
Luca Aceto
Alessandro Aldini
Luca de Alfaro
Roberto Amadio
Henrik Reif Andersen
Suzana Andova
Ralph Back
Jos Baeten
Christel Baier
Twan Basten
Marek A. Bednarczyk
Michael von der Beeck
Saddek Bensalem
Jan Bergstra
Marco Bernardo
G. Berry
Eike Best
Michele Boreale
Andrzej Borzyszkowski
Victor Bos
Dragan Bošnački
Ahmed Bouajjani
Gerard Boudol
Mario Bravetti
Max Breitling
Antonio Brogi
Stephen Brookes
Nadia Busi
Martin Büchi
P. Caspi
Flavio Corradini
Jordi Cortadella
Pedro R. D'Argenio
Silvano Dal-Zilio
Mads Dam
Philippe Darondeau
Alexandre David
Anuj Dawar
Conrado Daws
Raymond Devillers
André Engels
Emanuela Fachini
G. Ferrari
Fabrice Le Fessant
Hans Fleischhack
Cedric Fournet
Daniel Fridlender
Fabio Gadducci
Mauro Gaspari
Simon Gay
Thomas Gehrke
Herman Geuvers
Rob van Glabbeek
Stefania Gnesi
Andrew Gordon
Roberto Gorrieri
Jan Friso Groote
Irène Guessarian
Dilian Gurov
Paul Hankin
Boudewijn Haverkort
Matthew Hennessy
Thomas Henzinger
Holger Hermanns
Thomas T. Hildebrandt
Tony Hoare
Leszek Holenderski
Kohei Honda
Furio Honsell
Benjamin Horowitz
Franz Huber
Michaela Huhn
Kees Huizing
Thomas Hune
Anna Ingólfsdóttir
Daniel Jackson
Petr Jančar
Ryszard Janicki
Alan Jeffrey
Bengt Jonsson
Marcin Jurdzinski
Joost-Pieter Katoen
Claude Kirchner
Hanna Klaudel
Maciej Koutny
Dexter Kozen
S Krishnan
Ingolf Krueger
Ruurd Kuiper
K. Narayan Kumar
Orna Kupferman
Marta Kwiatkowska
Anna Labella
Rom Langerak
Izak van Langevelde
François Laroussinie
Alexander Lavrov
Francesca Levi
Xuandong Li
Johan Lilius
Jorn Lind-Nielsen
Michele Loreti
Victor Luchangco
Bas Luttik
Nancy Lynch
Davide Marchignoli
A. Mateescu
Sjouke Mauw
Richard Mayr
Kees Middelburg
Leonid Mikhajlov
Anna Mikhajlova
Faron Moller
Rémi Morin
Laurent Mounier
Olaf Mueller
Madhavan Mukund
Gustaf Naeser
Rajagopal Nagarajan
Margherita Napoli
Wolfgang Naraschewski
Monica Nesi
Uwe Nestmann
Rocco De Nicola
Peter Niebert
Mogens Nielsen
Jan Nyström
Vincent van Oostrom
Giuseppe Pappalardo
Joachim Parrow
Adriano Peron

Paul Pettersson
Anna Philippou
Jan Philipps
Claudine Picaronny
Sophie Pinchinat
Marco Pistore
A. Podelski
Jaco van de Pol
Ivan Porres
John Power
Sanjiva Prasad
Leonor Prensa-Nieto
Corrado Priami
Rosario Pugliese
Paola Quaglia
Mauno Rönkkö
R. Ramanujam
J-F. Raskin
Julian Rathke
Andreas Rausch
Antonio Ravara
Michel Reniers
Arend Rensink
Christian Retore
Judi Romijn
Bill Roscoe
Giuseppe Rosolini
Brigitte Rozoy
Theo C. Ruys
Jeff Sanders
Robert Sandner
Davide Sangiorgi
Birgit Schieder
Vincent Schmitt
Philippe Schnoebelen
Claus Schröter
Bernhard Schätz
Roberto Segala
Peter Sewell
Robert de Simone
Arne Skou
Konrad Slind
Scott Smolka
Katharina Spies
Thomas Stauner
Jason Steggles
Colin Stirling
Laurent Théry
P. S. Thiagarajan
Luca Trevisan
Yaroslav S. Usenko
Mandana Vaziri
Betti Venneri
Bjorn Victor
Patrick Viry
Marc Voorhoeve
Farn Wang
Heike Wehrheim
Jack van Wijk
Tim Willemse
Poul Frederick Williams
Glynn Winskel
J. von Wright
Alex Yakovlev
Wang Yi
Nobuko Yoshida
Sergio Yovine
Mark van der Zwaag

Table of Contents

Invited Lectures

Tutorials

Accepted Papers

Temporal Process Logic*

Rance Cleaveland

Dept. of Comp. Sci., SUNY at Stony Brook, Stony Brook, NY 11794-4400, USA
rance@cs.sunysb.edu

Abstract of Invited Talk

Research in the specification and verification of concurrent systems falls into two general categories. The *temporal logic* school advocates temporal logic as a language for formulating system requirements, with the semantics of the logic being used as a basis for determining whether or not a system is correct. The *process-algebraic* community focuses on the use of "higher-level" system descriptions as specifications of "lower-level" ones, with a refinement relation being used to determine whether an implementation conforms to a specification. From a user's perspective, the approaches offer different benefits and drawbacks. Temporal logic supports "scenario-based" specifications, since formulas may be given that focus on single aspects of system behavior. On the other hand, temporal logic specifications suffer from a lack of compositionality, since the language of specifications differs from the system description language. In contrast, compositional specification is the hallmark of process algebraic reasoning, but at the expense of requiring what some view as overly detailed specifications. Although much research has studied the connections between the temporal logic and process algebra, a truly uniform formalism that combines the advantages of the two approaches has yet to emerge.

In my talk I present preliminary results obtained by Gerald Lüttgen, of ICASE, and me on the development of such a formalism. Our approach features a process-algebra-inspired notation that enriches traditional process algebras by allowing linear-time temporal formulas to be embedded in system descriptions. We show how the combined formalism may be given a uniform operational semantics in Plotkin's Structural Operational Semantics (SOS) style, and we define a refinement relation based on Denicola/Hennessy testing and discuss its congruence properties. We then demonstrate that traditional temporal-logic-style arguments about system correctness can be naturally captured via refinement; we also illustrate how the combination of logical and system operators allows users to define systems in which some "components" remain specified only as formulas.

* Research supported by NSF grants CCR-9257963, CCR-9505562 and CCR-9804091, AFOSR grant F49620-95-1-0508, and ARO grant P-38682-MA.

An Unfolding Algorithm for Synchronous Products of Transition Systems*

Javier Esparza and Stefan Römer

Institut für Informatik, Technische Universität München
e-mail: {esparza,roemer}@in.tum.de

Abstract. The unfolding method, initially introduced for systems modelled by Petri nets, is applied to synchronous products of transition systems, a model introduced by Arnold [2]. An unfolding procedure is provided which exploits the product structure of the model. Its performance is evaluated on a set of benchmarks.

1 Introduction

The unfolding method is a partial order approach to the verification of concurrent systems introduced by McMillan in his Ph. D. Thesis [6]. A finite state system, modelled as a Petri net, is *unfolded* to yield an equivalent acyclic net with a simpler structure. This net is usually infinite, and so in general it cannot be used for automatic verification. However, it is possible to construct a *complete finite prefix* of it containing as much information as the infinite net itself: Loosely speaking, this prefix already contains all the reachable states of the system. The prefix is usually far smaller than the state space, and often smaller than a BDD representation of it, and it can be used as input for efficient verification algorithms. A rather complete bibliography on the unfolding method, containing over 60 papers on semantics, algorithms, and applications is accessible online [1].

The thesis of this paper is that the unfolding method is applicable to any model of concurrency for which a notion of 'events occurring independently from each other' can be defined, and not only to Petri nets—as is often assumed. We provide evidence in favour of this thesis by applying the method to *synchronous products of labelled transition systems.* In this model, introduced by Arnold in [2], a system consists of a tuple of communicating sequential components. The communication discipline, formalised by means of so-called synchronisation vectors, is very general, and contains as special cases the communication mechanisms of process algebras like CCS and CSP.

Readers acquainted with both Arnold's and the Petri net model will probably think that our task is not very difficult, and they are right. It is indeed straightforward to give synchronous products of transition systems a Petri net semantics, and then apply the usual machinery. But we go a bit further: We show that the

* Work partially supported by the Teilprojekt A3 SAM of the Sonderforschungsbereich 342 "Werkzeuge und Methoden für die Nutzung paralleler Rechnerarchitekturen".

additional structure of Arnold's model with respect to Petri nets—the fact that we are given a decomposition of the system into sequential components—can be used to simplify the unfolding method. More precisely, in a former paper by Vogler and the authors [4], we showed that the key to an efficient algorithm for the construction of a complete finite prefix is to find a mathematical object called a *total adequate order*, and provided such an order for systems modelled by Petri nets[1]. In this paper we present a new total adequate order for synchronous products of labelled transition systems. The proof of adequacy for this new order is simpler than the proof of [4].

In a second part of the paper we describe an efficient implementation of the algorithm , and compare it with the algorithm of [4] on a set of benchmarks.

Very recently, further evidence for the wide applicability of unfoldings has been provided by Langerak and Brinksma in [5]. Independently from us, they have applied the unfolding technique to a CSP-like process algebra, a model even further away from Petri nets than ours. A brief discussion of the relation to our work can be found in the conclusions.

The paper is organised as follows. Section 2 introduces synchronous products of transition systems following [2], and Section 3 gives them a partial order semantics based on unfoldings. Section 4 describes an algorithm to construct a complete finite prefix. Section 5 discusses how to efficiently implement it. Section 6 discusses the performance of the new algorithm.

2 Synchronous Products of Transition Systems

In this section we introduce Arnold's model and its standard interleaving semantics. Notations follow [2] with very few minor changes.

2.1 Labelled Transition Systems

A *labelled transition system* is a tuple $\mathcal{A} = \langle S, T, \alpha, \beta, \lambda \rangle$, where S is a set of *states*, T is a set of *transitions*, $\alpha, \beta : T \to S$ are the *source* and *target* mappings, and $\lambda: T \to A$ is a labelling mapping assigning to each transition a letter from an alphabet A. We assume that A contains a special label ϵ, and that for each state $s \in S$ there is a transition ϵ_s such that $\alpha(\epsilon_s) = s = \beta(\epsilon_s)$, and $\lambda(\epsilon_s) = \epsilon$. Moreover, no other transitions are labelled by ϵ. Transitions labelled by ϵ are called *idle* transitions in the sequel.

We use a graphical representation for labelled transition systems. States are represented by circles, and a transition t with $\alpha(t) = s$, $\beta(t) = s'$, and $\lambda(t) = a$ is represented by an arrow leading from s to s' labelled by $t : a$. Idle transitions are not represented. Figure 1 shows two labelled transition systems.

[1] More exactly, systems modelled by 1-safe Petri nets, i.e., Petri nets whose places can hold at most one token.

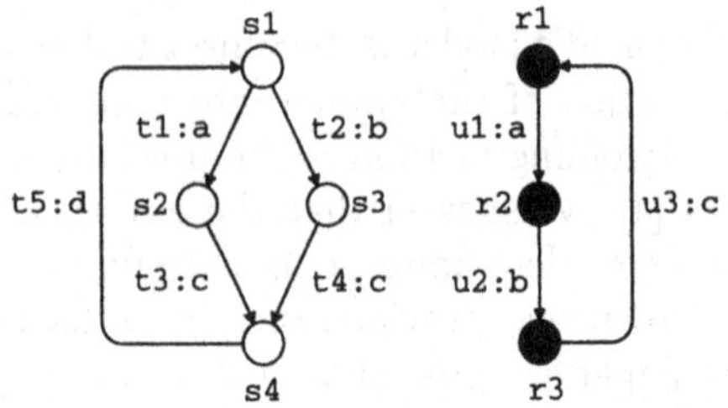

Fig. 1. Two labelled transition systems

2.2 Synchronous Products

Let $\mathcal{A}_1, \ldots, \mathcal{A}_n$ be labelled transition systems, where $\mathcal{A}_i = \langle S_i, T_i, \alpha_i, \beta_i, \lambda_i \rangle$, and λ_i labels each transition of T_i with an element of an alphabet A_i. We assume for convenience that the sets S_i and T_i are pairwise disjoint. A subset I of $(A_1 \times \ldots \times A_n) \setminus (\epsilon, \ldots, \epsilon)$ is called a *synchronisation constraint*, and the elements of I are called *synchronisation vectors*. Loosely speaking, these vectors indicate which transitions of $\mathcal{A}_1, \ldots, \mathcal{A}_n$ must synchronise. The tuple $\mathbf{A} = \langle \mathcal{A}_1, \ldots, \mathcal{A}_n, I \rangle$ is called the *synchronous product* of the $\mathcal{A}_i$ under I.

As running example we use $\mathbf{A} = \langle \mathcal{A}_1, \mathcal{A}_2, I \rangle$, where $\mathcal{A}_1, \mathcal{A}_2$ are the two labelled transition systems of Figure 1, and I contains the following synchronisation vectors:

$$\langle a, \epsilon \rangle, \langle b, \epsilon \rangle, \langle d, \epsilon \rangle, \langle \epsilon, a \rangle, \langle \epsilon, c \rangle, \langle c, b \rangle$$

I.e., c-labelled transitions of $\mathcal{A}_1$ must synchronise with b-labelled transitions of $\mathcal{A}_2$. The other transitions do not synchronise.

The interleaving semantics of $\mathbf{A}$ is the labelled transition system $\mathcal{A}_{int} = \langle S, T, \alpha, \beta, \lambda \rangle$, where $\lambda : T \to I$, and

$$\begin{aligned} S &= S_1 \times \ldots \times S_n \\ T &= \{\langle t_1, \ldots, t_n \rangle \mid \langle \lambda_1(t_1), \ldots, \lambda_n(t_n) \rangle \in I\} \\ \alpha(\langle t_1, \ldots, t_n \rangle) &= \langle \alpha_1(t_1), \ldots, \alpha_n(t_n) \rangle \\ \beta(\langle t_1, \ldots, t_n \rangle) &= \langle \beta_1(t_1), \ldots, \beta_n(t_n) \rangle \\ \lambda(\langle t_1, \ldots, t_n \rangle) &= \langle \lambda_1(t_1), \ldots, \lambda_n(t_n) \rangle \end{aligned}$$

The elements of S and T are called *global states* and global transitions, respectively.

If each of the $\mathcal{A}_i$ has a distinguished initial state is_i, then the initial state of $\mathbf{A}$ is the tuple $\mathbf{is} = \langle is_1, \ldots, is_n \rangle$, and $\mathbf{A}$ with $\mathbf{is}$ as initial state is denoted by $\langle \mathbf{A}, \mathbf{is} \rangle$. The set of *reachable* global states is then the set of global states reachable from $\mathbf{is}$. For our running example we take $\mathbf{is} = \langle s_1, r_1 \rangle$.

We introduce a notation that will help us to later define the unfolding of $\mathbf{A}$. Given a global transition $\mathbf{t} = \langle t_1, \ldots, t_n \rangle$ of $\mathbf{A}$, we define

$$\begin{aligned} {}^\bullet\mathbf{t} &= \{\alpha_i(t_i) \mid 1 \leq i \leq n \text{ and } \lambda_i(t_i) \neq \epsilon\} \\ \mathbf{t}^\bullet &= \{\beta_i(t_i) \mid 1 \leq i \leq n \text{ and } \lambda_i(t_i) \neq \epsilon\} \end{aligned}$$

Loosely speaking, ${}^\bullet\mathbf{t}$ contains the sources of the non-idle transitions of $\mathbf{t}$, and $\mathbf{t}^\bullet$ their targets.

3 Unfolding of a Synchronous Product

In [2], synchronous products are only given an interleaving semantics. In this section we give them a partial order semantics based on the notion of unfolding of a synchronous product, and show its compatibility with the interleaving semantics. We introduce a number of standard notions about Petri nets, but sometimes our definitions are not completely formalised. The reader interested in rigorous definitions is referred to [4].

3.1 Petri nets

As usual, a *net* consists of a set of *places*, graphically represented by circles, a set of *transitions*, graphically represented as boxes, and a flow relation assigning to each place (transition) a set of input and a set of output transitions (places). The flow relation is graphically represented by arrows leading from places to transitions and from transitions to places. In order to avoid confusions between the transitions of a transition system and the transitions of a Petri net, we call the latter *events* in the sequel. Places and events are called *nodes*; given a node x, the set of input and output nodes of x is denoted by ${}^\bullet x$ and $x^\bullet$, respectively. A place of a net can hold tokens, and a mapping assigning to each place a number of tokens is called a *marking*. If, at a given marking, all the input places of an event hold at least one token, then the event can *occur*, which leads to a new marking obtained by removing one token from each input place and adding one token to each output place. An *occurrence sequence* is a sequence of events that can occur in the order specified by the sequence.

A synchronous product can be associated a Petri net as follows: Take a place for each state of each component, and an event for each global transition; add an arc from s to $\mathbf{t}$ if $s \in {}^\bullet\mathbf{t}$, and from $\mathbf{t}$ to s if $s \in \mathbf{t}^\bullet$; put a token in the initial state of each component, and no tokens elsewhere. The unfolding of a synchronous product can be defined as the unfolding of its associated Petri net, but in the rest of the section we give a direct definition.

3.2 Ocurrence nets

Given two nodes x and y of a net, we say that x is *causally related* to y, denoted by $x \leq y$, if there is a (possibly empty) path of arrows from x to y. We say that x and y are in *conflict*, denoted by $x \# y$, if there is a place z, different from x and y, from which one can reach x and y, exiting x by different arrows. Finally, we say that x and y are *concurrent*, denoted by $x \, co \, y$, if neither $x \leq y$ nor $y \leq x$ nor $x \# y$ hold. *Occurrence nets* are those satisfying the following three properties:

- the net, seen as a graph, has no cycles;
- every place has at most one input event;
- no node is in self-conflict, i.e., $x \# x$ holds for no x.

The nets of Figure 2 and Figure 3 are occurrence nets.

Occurrence nets can be infinite. We restrict ourselves to those in which every event has at least one input place, and in which the arrows cannot be followed backward infinitely from any point (this is called *well-foundedness*). It follows that by following the arrows backward we eventually reach a place without predecessors. These are the *minimal places* of the net.

We associate to an occurrence net a default *initial marking*, in which the minimal places carry exactly one token, and the other places no tokens. It is easy to see that all the markings reachable from the initial marking also put at most one token on a place. Therefore, we represent reachable markings as sets of places.

3.3 Branching processes

Given a synchronous product of transition systems, we associate to it a set of *labelled* occurrence nets, called the *branching processes* of $\mathbf{A}$. The places[2] of these nets are labelled with states of the components of $\mathbf{A}$, and their events are labelled with global transitions. The places and events of the branching processes are all taken from two sets $\mathcal{P}$ and $\mathcal{E}$, inductively defined as follows:

- $\bot \in \mathcal{E}$, where $\bot$ is a special symbol;
- if $e \in \mathcal{E}$, then $(s, e) \in \mathcal{P}$ for every $s \in S_1 \cup \ldots \cup S_n$;
- if $X \subseteq \mathcal{P}$, then $(\mathbf{t}, X) \in \mathcal{E}$ for every $\mathbf{t} \in T$.

In our definition of branching process (see below) we make consistent use of these names: The label of a place (s, e) is s, and its unique input event is e. Places $(s, \bot)$ are those having no input event, i.e., the special symbol $\bot$ is used for the minimal places of the occurrence net. Similarly, the label of an event $(\mathbf{t}, X)$ is $\mathbf{t}$, and its set of input places is X. The advantage of this scheme is that a branching process is completely determined by its sets of places and events. In the sequel, we make use of this and represent a branching process as a pair (P, E).

Definition 1. *The set of* finite branching processes of $\langle \mathbf{A}, \mathbf{is} \rangle$, *where* $\mathbf{is} = \langle is_1, \ldots, is_n \rangle$, *is inductively defined as follows:*

- $(\{(is_1, \bot), \ldots, (is_n, \bot)\}, \emptyset)$ *is a branching process of* $\langle \mathbf{A}, \mathbf{is} \rangle$.
- *If* (P, E) *is a branching process,* $\mathbf{t}$ *is a global transition, and* $X \subseteq P$ *is a co-set labelled by* ${}^\bullet\mathbf{t}$*, then*

$$(P \cup \{(s, e) | s \in \mathbf{t}^\bullet\}\ ,\ E \cup \{e\})$$

is also a branching process of $\langle \mathbf{A}, \mathbf{is} \rangle$, *where* $e = (\mathbf{t}, X)$. *If* $e \notin E$, *then* e *is called a* possible extension *of* (P, E).

We denote the set of possible extensions of a branching process BP *by* $PE(BP)$.

[2] In some papers (including [4]), the name *conditions* is used instead of places.

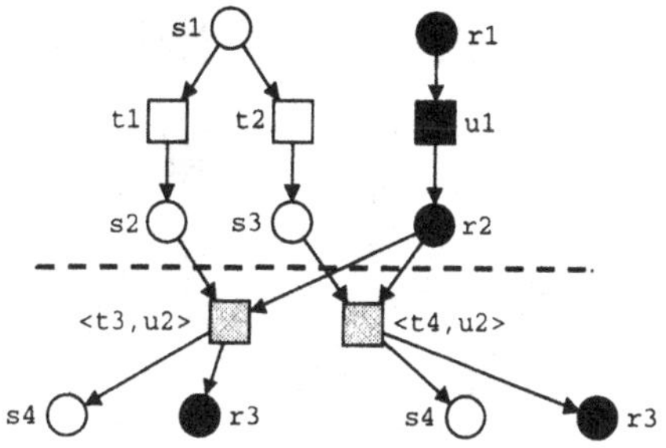

Fig. 2. A branching process of $\langle \mathbf{A}, \mathbf{is} \rangle$

A place of the form $(s, \bot)$ or (s, e) such that $s \in S_i$ is called an *i-place*. An event of the form $(\mathbf{t}, X)$ such that $\mathbf{t}(i)$ is not an idle transition is called an *i-event*. Observe that an event can be both an i-event and a j-event for $i \neq j$ (in this case we say that $\mathcal{A}_i$ and $\mathcal{A}_j$ *synchronize* in e), but a place cannot, since the states of the different components are disjoint by assumption.

Figure 2 shows a finite branching process of our running example (above the dashed line), together with its two possible extensions (below that line). 1-nodes are white, 2-nodes are dark grey, and events that are both 1- and 2 events are light grey. The labels of events have been simplified for clarity: We write t instead of $\langle t, \epsilon \rangle$, and u instead of $\langle \epsilon, u \rangle$.

The set of branching processes of $\langle \mathbf{A}, \mathbf{is} \rangle$ is obtained by declaring that the union of any finite or infinite set of branching processes is also a branching process, where union of branching processes is defined componentwise on places and events. Since branching processes are closed under union, there is a unique maximal branching process. We call it the *unfolding* of $\langle \mathbf{A}, \mathbf{is} \rangle$. The unfolding of our running example is an infinite occurrence net. Figure 3 shows an initial part. Events and places have been assigned identificators that will be used in the examples.

The following Proposition is easy to prove by structural induction on branching processes:

Proposition 1. *Two i-nodes of a branching process are either causally related or in conflict.*

For instance, in Figure 3 all white and light grey nodes are causally related or in conflict.

3.4 Configurations and cuts

For our purposes, the most interesting property of occurrence nets is that their sets of occurrence sequences and reachable markings can be easily characterised in graph-theoretic terms using the notions of configuration and cut.

Definition 2. *A* configuration *of an occurrence net is a set of events C satisfying the two following properties: C is causally closed, i.e., if $e \in C$ and $e' < e$ then $e' \in C$, and C is conflict-free, i.e., no two events of C are in conflict.*

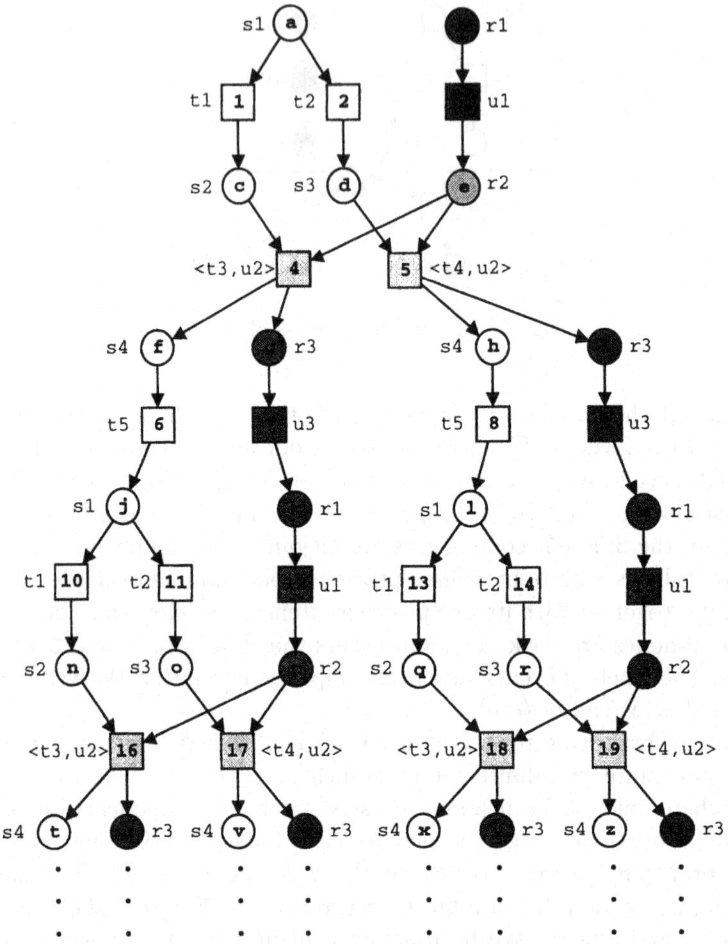

Fig. 3. The unfolding of $\langle \mathbf{A}, \mathbf{is} \rangle$

In Figure 3, $\{1,3,4,6\}$ is a configuration, and $\{1,4\}$ (not causally closed) or $\{1,2\}$ (not conflict-free) are not.

It is easy to prove that a set of events is a configuration if and only if there is an occurrence sequence of the net (from the default initial marking) containing each event from the set exactly once, and no further events. This occurrence sequence is not necessarily unique. For instance, for the configuration $\{1,3,4,6\}$ there are two occurrence sequences like 1 3 4 6 or 3 1 4 6. However, all occurrence sequences corresponding to the same configuration lead to the same reachable marking. For example, the two sequences above lead to the marking $\{j, g\}$.

Definition 3. *A* cut *is a set of places* **c** *satisfying the two following properties:* **c** *is a co-set, i.e., any two elements of* **c** *are concurrent, and* **c** *is maximal, i.e., it is not properly included in any other co-set.*

It is easy to prove that the reachable markings of an occurrence net coincide with its cuts. We can assign to a configuration C the marking reached by any of the occurrence sequences mentioned above. This marking is a cut, and it is easy to prove that it is equal to $(Min \cup {}^\bullet C) \setminus C^\bullet$, where *Min* denotes the set of minimal places of the branching process.

The following Proposition can also be easily proved by structural induction on branching processes.

Proposition 2. *A cut* $\mathbf{c}$ *of a branching process contains exactly one* i*-place for each component* $\mathcal{A}_i$.

This result allows us to use the notation $\mathbf{c} = \langle p_1, \ldots, p_n \rangle$ for cuts. Since the place p_i is labelled by some state $s_i \in S_i$, the tuple $\langle s_1, \ldots, s_n \rangle$ is a reachable global state of $\langle \mathbf{A}, \mathbf{is} \rangle$. The global state corresponding to the cut of a configuration C is denoted by $GState(C)$.

We take as partial order semantics of $\langle \mathbf{A}, \mathbf{is} \rangle$ its unfolding. The relationship between the interleaving and partial order semantics of $\langle \mathbf{A}, \mathbf{is} \rangle$ is given by the following result:

Theorem 1. *Let* $\langle \mathbf{A}, \mathbf{is} \rangle$ *be a synchronous product of transition systems.*

(a) Let C *be a configuration of a branching process of* $\langle \mathbf{A}, \mathbf{is} \rangle$. *There is a state* $\mathbf{s}$ *of* $\mathcal{A}_{int}$, *reachable from* $\mathbf{is}$, *such that: (1)* $\mathbf{s} = GState(C)$, *and (2) for every configuration* $C \cup \{e\}$ $(e \notin C)$ *there is a transition* $\mathbf{t}$ *of* $\mathcal{A}_{int}$ *such that* $\alpha(\mathbf{t}) = GState(C)$ *and* $\beta(\mathbf{t}) = GState(C \cup \{e\})$.

(b) Let $\mathbf{s}$ *be a state of* $\mathcal{A}_{int}$, *reachable from* $\mathbf{is}$. *There is a configuration* C *of the unfolding of* $\langle \mathbf{A}, \mathbf{is} \rangle$ *such that: (1)* $GState(C) = \mathbf{s}$, *and (2) for every transition* $\mathbf{t}$ *of* $\mathcal{A}_{int}$ *such that* $\alpha(\mathbf{t}) = \mathbf{s}$ *and* $\beta(\mathbf{t}) = s'$ *there exists a configuration* $C \cup \{e\}$ $(e \notin C)$ *such that* e *is labelled by* $\mathbf{t}$, *and* $GState(C \cup \{e\}) = \mathbf{s}'$.

Informally, (a) means that the information a branching process has about $\mathcal{A}_{int}$ is correct, while (b) means that the unfolding has complete information about $\mathcal{A}_{int}$ (actually, the unfolding also contains "true concurrency" information).

4 Constructing a complete finite prefix

We say that a branching process of $\langle \mathbf{A}, \mathbf{is} \rangle$ is *complete* if it contains complete information about $\mathcal{A}_{int}$, i.e., if condition (b) of Theorem 1, which is always fulfilled by the unfolding, also holds for it.[3] The important fact is that *finite* complete prefixes exist, the main reason being that the number of global states of $\langle \mathbf{A}, \mathbf{is} \rangle$ is finite. For instance, the prefix of Figure 3 containing the places $\{a, \ldots, k, n, o, p\}$ and the events $\{1, \ldots, 7, 10, 11, 12\}$ can be shown to be a complete prefix.

[3] In fact, it is easy to see that a complete prefix contains as much information as the unfolding itself, in the sense that given a complete prefix there is a unique unfolding containing it.

In [4] an algorithm is presented for the construction of a complete finite prefix, which improves on a previous construction presented in [6]. The algorithm makes use of a so-called ***adequate order*** on the configurations of the unfolding. Different adequate orders lead to different versions of the algorithm, and also to different complete prefixes. Total adequate orders are particularly nice, since they lead to complete prefixes which, loosely speaking, are guaranteed not to be larger than the transition system $\mathcal{A}_{int}$[4]. In [4] a total adequate order for the unfoldings of Petri nets is presented. In this section we recall the algorithm of [4], and then present a total adequate order for the unfoldings of synchronous products of transition systems. The additional structure of a synchronous product with respect to a Petri net leads to a simpler order, with a simpler proof of adequacy.

4.1 The algorithm

Given a configuration C of the unfolding, we denote by $C \oplus E$ the set $C \cup E$, under the condition that $C \cup E$ is a configuration satisfying $C \cap E = \emptyset$. We say that $C \oplus E$ is an *extension* of C, and that E is a *suffix* of $C \oplus E$. Obviously, if $C \subseteq C'$ then there is a suffix E of C' such that $C \oplus E = C'$.

Now, let C_1 and C_2 be two finite configurations leading to the same global state, i.e. $GState(C_1) = \mathbf{s} = GState(C_2)$. The 'continuations' of the unfolding from the cuts corresponding to C_1 and C_2 (the nodes lying below these cuts) are isomorphic (see [4] for a more formal description). For example, in Figure 3 the configurations $\{1,3,4\}$ and $\{2,3,5\}$ lead to the cuts $\langle f,g\rangle$ and $\langle h,i\rangle$, which correspond to the global state $\langle s_4, r_3\rangle$. Loosely speaking, the continuations from these cuts contain the nodes below f,g and h,i, respectively (f,g and h,i included). This isomorphism, say I, induces a mapping from the extensions of C_1 onto the extensions of C_2, which maps $C_1 \oplus E$ onto $C_2 \oplus I(E)$. For example, $\{1,3,4,7,12\}$ is mapped onto $\{2,3,5,9,15\}$.

The intuitive idea behind the algorithm is to avoid computing isomorphic continuations, since one representative suffices. However, a correct formalisation is not easily achieved. It requires the following three basic notions:

Definition 4. *A partial order $\prec$ on the finite configurations of the unfolding is* adequate *if:*

- *it is well-founded,*
- *it refines the inclusion order, i.e. $C_1 \subset C_2$ implies $C_1 \prec C_2$, and*
- *it is preserved by finite extensions, i.e. if $C_1 \prec C_2$ and $GState(C_1) = GState(C_2)$, then the isomorphism I above satisfies $C_1 \oplus E \prec C_2 \oplus I(E)$ for all finite extensions $C_1 \oplus E$ of C_1.*

Definition 5. *The local configuration $[e]$ associated to an event e of a branching process is the set of events e' such that $e' \leq e$.*[5]

[4] For a more precise statement see [4].
[5] It is immediate to prove that $[e]$ is a configuration.

Definition 6. *Let $\prec$ be an adequate order on the configurations of the unfolding, and let BP be a branching process containing an event e. The event e is a* cut-off event *of BP (with respect to $\prec$) if BP contains a local configuration $[e']$ such that $GState([e]) = GState([e'])$, and $[e'] \prec [e]$.*

The algorithm is in fact a family of algorithms: each adequate order $\prec$ leads to a different member of the family. It computes a branching process, and whenever it identifies a cut-off event it takes care of not extending the process behind it.

input: a synchronous product $\langle \mathbf{A}, \mathbf{is} \rangle$, where $\mathbf{is} = \langle is_1, \ldots, is_n \rangle$.
output: a complete finite prefix of the unfolding of $\langle \mathbf{A}, \mathbf{is} \rangle$.

```
begin
    bp := ({(is_1, ⊥), ..., (is_n, ⊥)}, ∅);
    pe := PE(bp);
    cut-off := ∅;
    while pe ≠ ∅ do
        choose e = (t, X) in pe such that [e] is minimal with respect to ≺;
        if [e] ∩ cut-off = ∅ then
            extend bp with the event e and with a place (s, e)
                for every output place s of t;
            pe := PE(bp);
            if GState([e]) = GState([e']) for some event e' of bp then
                cut-off := cut-off ∪ {e}
            endif
        else pe := pe \ {e}
        endif
    endwhile;
    return bp
end
```

One of the main results of [4] states that this algorithm is correct if $\prec$ is an adequate order. The order $\prec$ need not be total, but, loosely speaking, total orders lead to more cut-off events, and so to smaller prefixes. In fact, totality is a sufficient condition for the output of the algorithm to be at most as large as the interleaving semantics $\mathcal{A}_{int}$. Weaker conditions achieve the same effect (the order need only be total among configurations with the same associated global state, a fact exploited in [5]), but we do not need them here.

5 Adequate orders for the unfolding of a synchronous product

In this section we introduce a total adequate order on the configurations of the unfolding of a synchronous product. The order is simpler to define and to prove adequate than the order introduced in [4] for systems modelled by Petri nets.

5.1 Local views

Our adequate order is based on the notion of *local view* of a configuration. Given a finite configuration C, we define its *projection* $C|_i$ *onto* A_i as its set of i-events. If we take $C = \{2, 3, 5, 8, 9, 13\}$ in Figure 3, then we have $C|_1 = \{2, 5, 8, 13\}$ and $C|_2 = \{3, 5, 9\}$. The events of $C|_i$ are totally ordered by the causal relation $<$. This is so because i-events are either causally related or in conflict (Proposition 1), and the events of $C|_i$ are not in conflict because they belong to a configuration. We define:

Definition 7. *Let C be a configuration, and let $e_1 < e_2 < \ldots < e_{k_i}$ be the result of ordering $C|_i$ with respect to $<$. The* i-view *of a configuration C, denoted by $V_i(C)$, is the sequence $\mathbf{t}_1\mathbf{t}_2\ldots\mathbf{t}_{k_i}$, where $\mathbf{t}_j$ is the global transition labelling the event e_j. We denote by $\mathbf{V}(C) = \langle V_1(C), \ldots, V_n(C)\rangle$ the n-tuple of local views of a configuration.*

Intuitively, $V_i(C)$ is the history of the computation as seen by the i-th component. In our example we have $2 < 5 < 8 < 13$ for $C|_1$ and $3 < 5 < 9$ for $C|_2$. Furthermore, $V_1(C) = t_2\langle t_4, u_2\rangle t_5 t_1$ and $V_2(C) = u_1\langle t_4, u_2\rangle u_3$.

The definition of local view can be extended without problems to suffixes of configurations, for instance to the set $\{8, 9\}$: We have then $V_1(\{8, 9\}) = \langle t_5, \epsilon\rangle$ and $V_2(\{8, 9\}) = \langle \epsilon, u_3\rangle$. In particular, for an event $e = (\mathbf{t}, X)$ we have that $V_i(\{e\})$ is the empty sequence if $\mathbf{t}(i)$ is an idle transition, and $V_i(\{e\}) = \mathbf{t}$ otherwise.

The following result will be crucial:

Theorem 2. *The mapping* $\mathbf{V}$ *is injective.*

Proof. Let $C_1 = C_2$ be two configurations such that $\mathbf{V}(C_1) = \mathbf{V}(C_2)$. We prove $C_1 = C_2$ by showing $C_1 = C$ and $C_2 = C$, where $C = C_1 \cap C_2$. By symmetry it suffices to prove $C_1 = C$. We proceed by contradiction.

Assume $C \neq C_1$. Then C can be extended by an event $e_1 \in C_1 \setminus C$. We prove $e_1 \in C_2$, a contradiction to $C = C_1 \cap C_2$. Let $e_1 = (\mathbf{t}, X_1)$, where $\mathbf{t} = \langle t_1, \ldots, t_n\rangle$. Since $\mathbf{t} \neq (\epsilon, \ldots, \epsilon)$ by the definition of global transition, some component of $\mathbf{t}$, say t_i, satisfies $\lambda_i(t_i) \neq \epsilon$. By the definition of local view, $V_i(C) \cdot \mathbf{t}$ is a prefix of $V_i(C_1)$, and, since $\mathbf{V}(C_1) = \mathbf{V}(C_2)$ holds by assumption, also a prefix of $V_i(C_2)$. So C can be extended by an event $e_2 \in C_2$ such that $e_2 = (\mathbf{t}, X_2)$ for some *co*-set X_2. We prove:

- X_1 and X_2 are both labelled by ${}^\bullet\mathbf{t}$. Follows immediately from $e_1 = (\mathbf{t}, X_1)$ and $e_2 = (\mathbf{t}, X_2)$.
- Each place of $X_1 \cup X_2$ carries a different label. Since both e_1 and e_2 extend the same configuration C, we have that $X_1 \cup X_2$ is a *co*-set. Since every *co*-set can be extended to a cut, we can apply Proposition 4.

It follows $X_1 = X_2$, which implies $e_1 = e_2$. So $e_1 \in C_2$, and we are done.

In words, Theorem 2 states that a configuration is characterised by its tuple of local views. If we let $\mathbf{T}^*$ be the set of n-tuples whose elements are sequences of global transitions, i.e., $\mathbf{T}^* = (T^*)^n$, then a tuple of local views is an element of $\mathbf{T}^*$. By Theorem 2, an order $\prec$ on $\mathbf{T}^*$ induces an order on configurations:

$$C_1 \prec_C C_2 \text{ if and only if } \mathbf{V}(C_1) \prec \mathbf{V}(C_2)$$

Moreover, if $\prec$ is total, then $\prec_C$ is total.

5.2 From orders on local views to adequate orders

We identify sufficient conditions for an order $\prec$ on $\mathbf{T}^*$ to induce an *adequate* total order on configurations. We need to introduce some definitions. The concatenation of two elements $\sigma, \tau \in \mathbf{T}^*$ is defined componentwise, and denoted by $\sigma \cdot \tau$. The partial order $\sqsubseteq$ on $\mathbf{T}^*$ is defined as follows: $\sigma \sqsubseteq \tau$ if there exists σ' such that $\tau = \sigma \cdot \sigma'$. In other words, $\sigma \sqsubseteq \tau$ if each component of σ is a prefix of the corresponding component of τ.

We start with the following two observations, which follow easily from the definitions.

Proposition 3. *(1) If $C_1 \subseteq C_2$ then $\mathbf{V}(C_1) \sqsubseteq \mathbf{V}(C_2)$.*
(2) $\mathbf{V}(C \oplus E) = \mathbf{V}(C) \cdot \mathbf{V}(E)$.

Let us illustrate this result with configurations from Figure 3. Let $C_1 = \{2,3,5,8\}$ and $C_2 = \{2,3,5,8,9,13\}$. We have

$$\mathbf{V}(C_1) = \{t_2\langle t_4, u_2\rangle t_5, u_1\langle t_4, u_2\rangle\}$$
$$\mathbf{V}(C_2) = \{t_2\langle t_4, u_2\rangle t_5 t_1, u_1\langle t_4, u_2\rangle u_3\}$$

Furthermore, we have $C_2 = C_1 \oplus E$, where $E = \{9, 13\}$, and $\mathbf{V}(E) = \{t_1, u_3\}$, and indeed $\mathbf{V}(C_2) = \mathbf{V}(C_1) \cdot \mathbf{V}(E)$.

We can now obtain sufficient conditions for the induced order $\prec_C$ to be adequate and total:

Lemma 1. *Let $\prec$ be an order on $\mathbf{T}^*$ satisfying the following conditions:*
(1) $\prec$ is well-founded;
(2) $\prec$ refines $\sqsubseteq$, i.e. $\sigma \sqsubseteq \tau$ implies $\sigma \prec \tau$;
(3) $\prec$ is preserved by concatenation, i.e., if $\sigma \prec \tau$ then $\sigma \cdot \sigma' \prec \tau \cdot \sigma'$ for every $\sigma' \in \mathbf{T}^*$;
(4) $\prec$ is a total order.
Then the induced order $\prec_C$ is a total adequate order.

Proof. We prove that $\prec_C$ satisfies the properties of a total adequate order:

(a) $\prec_C$ is well-founded. $C_1 \succ_C C_2 \succ_C \ldots$ implies $\mathbf{V}(C_1) \succ \mathbf{V}(C_2) \succ \ldots$, contradicting the well-foundedness of $\prec$.
(b) If $C_1 \subseteq C_2$ then $C_1 \prec_C C_2$. By Proposition 3(1), $\mathbf{V}(C_1) \sqsubseteq \mathbf{V}(C_2)$. By (2), $\mathbf{V}(C_1) \prec \mathbf{V}(C_2)$. By the definition of $\prec_C$, $C_1 \prec_C C_2$.
(c) If $C_1 \prec_C C_2$ then $C_1 \oplus E \prec_C C_2 \oplus E$. If $C_1 \prec_C C_2$ then $\mathbf{V}(C_1) \prec \mathbf{V}(C_2)$. By (3), $\mathbf{V}(C_1) \cdot \mathbf{V}(E) \prec \mathbf{V}(C_2) \cdot \mathbf{V}(E)$. By Proposition 3(2), $\mathbf{V}(C_1 \oplus E) \prec \mathbf{V}(C_2 \oplus E)$. By the definition of $\prec_C$, $C_1 \oplus E \prec C_2 \oplus E$.
(d) $\prec_C$ is total. Immediate from (4) and the definition of $\prec_C$.

5.3 Orders on $\mathbf{T}^*$ inducing adequate orders

We describe in this section two total orders on $\mathbf{T}^*$ satisfying conditions (1)–(4) of Lemma 1. We start with an arbitrary total order $\prec_T$ on T, and use the following three auxiliary orders on T^*:

- the *size* order: σ is smaller than τ if $|\sigma| < |\tau|$;
- the *lexicographic* order: σ is smaller than τ if σ is lexicographically smaller than τ with respect to $\prec_T$.
- the *silex* (size-lexicographic) order: σ is smaller than τ if $|\sigma| < |\tau|$ or if $|\sigma| = |\tau|$ and σ is lexicographically smaller than τ.

Let us first consider the case $n = 1$, i.e, $\mathbf{A}$ contains only one component. We have then $\mathbf{T}^* = T^*$, i.e., we look for an order on sequences of global transitions satisfying (1)–(4). It is immediate to see that the silex order does the job: the order $\sqsubseteq$ is in this case the prefix order on sequences, and the concatenation operation is just the ordinary concatenation of sequences.

The silex order can be extended to an arbitrary number n of components in two different ways:

Definition 8. *Let σ, τ be elements of $\mathbf{T}^*$. We say $\sigma \prec_1 \tau$ if there is an index $1 \leq i \leq n$ such that $\sigma(j) = \tau(j)$ for all $1 \leq j < i$, and $\sigma(i)$ is smaller than $\tau(i)$ with respect to the silex order on sequences. We say $\sigma \prec_2 \tau$ if*
(a) there is an index $1 \leq i \leq n$ such that $|\sigma(j)| = |\tau(j)|$ for all $1 \leq j < i$, and $|\sigma(i)| < |\tau(i)|$, or
(b) $|\sigma(i)| = |\tau(i)|$ for all $1 \leq i \leq n$, and there is an index i such that $\sigma(j) = \tau(j)$ for all $1 \leq j < i$ and $\sigma(i)$ is lexicographically smaller than $\tau(i)$.

It is only a small exercise to prove that $\prec_1$ and $\prec_2$ satisfy conditions (1)–(4):

Theorem 3. *The orders $\prec_1$ and $\prec_2$ satisfy conditions (1)–(4) of Lemma 1. Therefore, they induce total adequate orders on configurations.*

Proof. Let us prove condition (3) for the order $\prec_2$, the others being similar or simpler. Assume $\sigma \prec_2 \tau$. We prove $\sigma \cdot \sigma \prec_2 \tau \cdot \sigma$. Let $\sigma(i)$ be the first component of σ such that $\sigma(i)$ is smaller than $\tau(i)$ with respect to the silex order. Consider two cases:

- There is an index $1 \leq i \leq n$ such that $|\sigma(j)| = |\tau(j)|$ for all $1 \leq j < i$ and $|\sigma(i)| < |\tau(i)|$. Then $|\sigma(j)\sigma'(j)| = |\tau(j)\sigma'(j)|$ for all $1 \leq j < i$ and $|\sigma(i)\sigma'(i)| < |\tau(i)\sigma'(i)|$. Hence $\sigma \cdot \sigma \prec_2 \tau \cdot \sigma$.
- $|\sigma(i)| = |\tau(i)|$ for all $1 \leq i \leq n$, and there is an index i such that $\sigma(j) = \tau(j)$ for all $1 \leq j < i$, and $\sigma(i)$ is lexicographically smaller than $\tau(i)$. Then $|\sigma(i)\sigma'(i)| = |\tau(i)\sigma'(i)|$ for all $1 \leq i \leq n$, and there is an index i such that $\sigma(j)\sigma'(j) = \tau(j)\sigma'(j)$ for all $1 \leq j < i$ and $\sigma(i)\sigma'(i)$ is lexicographically smaller than $\tau(i)\sigma'(i)$. Hence $\sigma \cdot \sigma \prec_2 \tau \cdot \sigma$.

This concludes the proof of adequacy of the two orders $\prec_1$ and $\prec_2$. The proof consists of Theorem 2, Lemma 1, and Theorem 3. The latter two have very simple proofs, only Theorem 2 requires a bit of ingenuity.

Which of the two orders is more suitable for an implementation is a question of efficiency, and is discussed—together with other implementation points—in the next section.

6 Efficient implementation of the complete finite prefix algorithm

The algorithm presented in Section 4.1 is hopefully easy to understand. However, it is still far too abstract. It leaves the choice of the order $\prec$ open, and it does not explain how to compute the functions *PE* and *GState*, nor how to compute a minimal event with respect to $\prec$. In the algorithm of [4] the computation of the functions and the minimal event involved expensive forward and backward global searches in branching processes. The additional structure of synchronous products allows to compute *GState* and minimal events using new procedures, described in Sections 6.2 and 6.1, respectively. In Section 6.3 we also describe how to speed-up the computation of *PE*; however, in this case the improvement does not exploit the structure of synchronous products, and can be used for Petri net systems as well.

In the sequel, the abstract algorithm of the last section is called 'the algorithm'. The concrete algorithm using the procedures just mentioned is called 'our implementation'.

6.1 Computing a minimal event

In order to determine the minimal event, our implementation maintains a queue of possible extensions sorted according to $\prec_C$. So we need a procedure to decide for two given configurations $[e_1], [e_2]$ whether $[e_1] \prec_C [e_2]$ or $[e_2] \prec_C [e_1]$. For both $\prec = \prec_1$ and $\prec = \prec_2$ we face a trade-off between time and space. The fastest procedure is to attach to each event e in the queue the whole vector $\mathbf{V}([e])$, which leads to a high memory overhead. The most economic procedure in memory terms is to recompute $\mathbf{V}([e])$ whenever it is needed by means of a backward search, a much slower solution. In our implementation we adopt an intermediate solution: We attach to each event e in the queue the integer vector $\langle |V_1([e])|, \ldots, |V_n([e])| \rangle$.

Once this design choice has been made, the order $\prec_2$ becomes superior to $\prec_1$. With $\prec = \prec_2$, the vectors $\mathbf{V}([e])$ and $\mathbf{V}([e'])$ have to be computed only if the integer vectors attached to e and e' coincide, which is rarely the case. With $\prec_C = \prec_1$, we have to compute $V_1([e])$ and $V_1([e'])$ if the first components of the integer vectors are equal; we have to compute $V_2([e])$ and $V_2([e'])$ if $V_1([e]) = V_1([e'])$ and the second components of the integer vectors are equal, and so on.

6.2 Computing *GState*$([e])$

Whenever the current branching process is extended with a new event e, the state *GState*$([e])$ has to be computed in order to determine if e is a cut-off event or not. For that, we first compute the cut corresponding to $[e]$; the labels of the conditions of this cut are *GState*$([e])$. Recall that the cut corresponding to $[e]$ is given by $(Min \cup [e]^\bullet) \setminus {}^\bullet[e]$, which provides a procedure to compute it. However, since it is too costly to store $[e]$ for each event e, the procedure involves computing the events preceding e.

The additional structure of synchronous products allows to easily compute the cut of $[e]$ from the cuts of the immediate predecessors of e, i.e., of the input events of e's input conditions. Let us start with a definition and a lemma:

Definition 9. *Let $p = (s, e)$ be an i-place of a branching process. The* depth *$d(p)$ of p is recursively defined as follows:*

- *If $e = \bot$, then $d(p) = 0$;*
- *If $e = (\mathbf{t}, X)$, then let p' be the unique i-place of X; define $d(p) = d(p') + 1$.*

Lemma 2. *Let $C_1, \ldots, C_k$ be configurations such that $C = C_1 \cup \ldots \cup C_k$ is also a configuration. Let $\mathbf{c}_i$ be the cut corresponding to C_i, and let $\mathbf{c}$ be the cut corresponding to C. For every $1 \leq j \leq n$, $\mathbf{c}(j)$ is the unique condition of the set $\{\mathbf{c}_1(j), \ldots, \mathbf{c}_k(j)\}$ having maximal depth.*

Proof. Since all the elements of $\{\mathbf{c}_1(j), \ldots, \mathbf{c}_k(j)\}$ are j-places, they are causally related or in conflict (Proposition 1). Since C is a configuration, they cannot be in conflict, and so they are all causally ordered. It follows that they all have different depths (notice that not all of $\mathbf{c}_1(j), \ldots, \mathbf{c}_k(j)$ have to be different, but of course all elements of $\{\mathbf{c}_1(j), \ldots, \mathbf{c}_k(j)\}$ are different by definition of set). So $\mathbf{c}(j)$ is well defined. We prove that $\mathbf{c}(j)$ belongs to the cut of C, i.e., that $\mathbf{c}(j) \in (\mathit{Min} \cup C^\bullet) \setminus {}^\bullet C$.

Assume without loss of generality that $\mathbf{c}(j) = \mathbf{c}_1(j)$. Then we have $\mathbf{c}(j) \in (\mathit{Min} \cup C_1^\bullet) \setminus {}^\bullet C_1$. So $\mathbf{c}(j) \in (\mathit{Min} \cup C_1^\bullet)$, and so $\mathbf{c}(j) \in (\mathit{Min} \cup C^\bullet)$. It remains to prove $\mathbf{c}(j) \notin {}^\bullet C$. Assume the contrary. Then there exists an index i such that $\mathbf{c}(j) \in {}^\bullet C_i$. It follows that the depth of $\mathbf{c}_i(j)$ must be greater than the depth of $\mathbf{c}(j)$, a contradiction.

We can now compute the cut of an event e as follows:

Proposition 4. *Let $e = (\mathbf{t}, X)$ be an event, and let $e_1, \ldots, e_k$ be its immediate predecessors. The cut of $[e]$ can be computed in two steps as follows:*

- *Compute the cut of $[e_1] \cup \ldots \cup [e_k]$ using Lemma 2; let $\mathbf{c}$ be this cut;*
- *For each output place p of e: If p is an i-place then replace the i-place of $\mathbf{c}$ by p.*

Proof. Observe that the output places of e belong to the cut of $[e]$. The rest follows easily from Lemma 2 and the definitions.

Let us apply this Proposition to compute the cut of [16] in Figure 3. The immediate predecessors of event 16 are events 10 and 12. Their corresponding cuts are $\langle n, g \rangle$ and $\langle f, p \rangle$. We have $d(n) = 4, d(g) = 2$ and $d(f) = 2, d(p) = 4$. So the cut of $[10] \cup [12]$ is $\langle n, p \rangle$. Now, the second step says to replace n by t and p by u. So the final result is $\langle t, u \rangle$. The fact that this is also the set of output places of event 16 is a coincidence.

In order to apply Proposition 4, our implementation has to compute the depth of each place of the current branching process. Fortunately, this leads to no time overhead. Recall that in order to decide if $[e_1] \prec_C [e_2]$ we attach to each event e the vector $\langle |V_1([e])|, \ldots, |V_n([e])| \rangle$. It follows immediately from the definitions that the depth of an i-place with input event e is equal to $|V_i([e])|$.

6.3 Computing $PE(BP)$

The computation of $PE(BP)$ is the most time consuming part of the algorithm. The computation is performed by considering each global transition $\mathbf{t} \in T$ in turn, and computing the possible extensions of BP of the form $(\mathbf{t}, X)$. So the problem consists of finding all $X \subseteq P$ such that (a) X is labelled by ${}^\bullet\mathbf{t}$, and (b) X is a *co*-set. Since the places of BP can be easily indexed according to the states they are labelled with, we search among all sets X satisfying (a) for those satisfying also (b). The implementation stores the *co*-relation of the places contained in the current branching process. Therefore, whenever the process is extended by a new event e, it is necessary to compute the places of the process that are in *co*-relation with the output places of e (notice that these output places themselves build a *co*-set).

A first procedure to compute this set of places applies the definition of the concurrency relation. Take the set of all places of the branching process, and perform the following steps:

(1) remove all places which are causally related with $e^\bullet$, by iteratively computing e's immediate predecessors, their immediate predecessors and so on; mark along the way all the places having more than one successor;
(2) remove the successors of the marked places (not already removed in (1)) ; these are the places in conflict with $e^\bullet$;
(3) give as output the remaining set of places.

To illustrate this procedure, assume that the current branching process is the prefix of Figure 3 containing events $1, 2, \ldots 11$, and that event 12 is the new event. Step (1) removes $\{k, g, c, a, e, b\}$, and marks a and e. Step (2) removes $\{d, h, i, l, m\}$. Step (3) yields $\{f, j, n, o\}$.

In the worst case, these steps require to visit all nodes of the current branching process, and since they have to be carried out whenever a new event e is added, the cost can be high. In the rest of the section we give a more efficient procedure.

Proposition 5. *Let $e = (\mathbf{t}, X)$ be a possible extension of a branching process (P, E). Let p be an output place of e, and let $p' \in P$ be an arbitrary place. $p\ co\ p'$ holds if and only if p' is an output place of e different from p, or $x\ co\ p'$ for every $x \in X$.*

Proof. If $p = p'$ then we are done, and so we consider only the case $p \neq p'$. Since e is a possible extension, $p < p'$ cannot hold, and so we have $p\ co\ p' \iff \neg(p' < p \vee p \# p')$. So it suffices to prove:

$$(p' < p \vee p \# p') \iff (p' \notin e^\bullet) \wedge (\exists x \in X. x \leq p' \vee p' \leq x \vee x \# p')$$

($\Longrightarrow$) We prove four statements:

- $p' < p \Rightarrow p' \notin e^\bullet$. Obvious, because no two output places of e are causally related.
- $p' < p \Rightarrow \exists x \in X. p' \leq x$. Since p has e as unique input event, the path from p' to p must necessarily contain e, and so it must also contain some input place of e, i.e., some element of X.

- $p\#p' \Rightarrow p' \notin e^\bullet$. Obvious, because no two output places of e are in conflict.
- $p\#p' \Rightarrow \exists x \in X.x \leq p' \vee x\#p'$. Since $p\#p'$ there exist two paths from a condition p'' to p and p' sharing only p''. If $p'' \in X$, then we have $p'' < p'$, and by taking $x = p''$ we are done. If $p'' \notin X$, then the path from p'' to p contains some element x of X, and so $x\#p'$.

($\Leftarrow$) We consider three cases:

- $p' \notin e^\bullet \wedge \exists x \in X.x \leq p'$. Then there exist two paths from x to p and p' sharing only x. So $p\#p'$.
- $p' \notin e^\bullet \wedge \exists x \in X.p' \leq x$. Then, since $x < p$, we have $p' \leq p$.
- $p' \notin e^\bullet \wedge \exists x \in X.x\#p'$. Since $x < p$ and $x\#p'$, we have $p\#p'$.

If we assume that the *co*-relation is updated whenever a new event is added to the current branching process (P, E), then at the point of adding a new event $e = (\mathbf{t}, X)$ we can assume that we already know whether $x \; co \; p'$ holds or not for every $x \in X$ and every $p' \in P$. Updating the relation is now a simple matter. The following procedure takes care of it.

Procedure Update($(P, E), co, e = (\mathbf{t}, X)$)
begin
    places := P;
        for every $p \in P$ **do**
            for every $x \in X$ **do**
                if $\neg(x \; co \; p)$ **then** *places* := *places* $\setminus \{p\}$ **endif**
            endfor
        endfor;
    $co := co \cup (e^\bullet \times e^\bullet) \cup (e^\bullet \times places) \cup (places \times e^\bullet)$
end

The operations in the procedure can be efficiently implemented using a bitvector $co(p)$ for each place p.

There is also an obvious improvement concerning recomputations of $PE(BP)$. The algorithm computes $PE(BP)$, extends β by one event, say e, and recomputes $PE(BP)$. This is very inefficient, since numerous possible extensions may be recomputed again and again. In fact, the only new possible extensions after the addition of e are those having e as immediate predecessor. When the first event of the queue of possible extensions is added to the current branching process, only new extensions having this event as immediate predecessor are computed and inserted in the queue.

7 Experimental results

The abstract algorithm of section 4.1 was originally introduced in [4] for systems modelled by Petri nets. The same paper contained performance measures of an implementation, called `Imp1` in the sequel. Since synchronous products can be given a Petri net semantics, as sketched in Section 3, `Imp1` can also be applied to

System	Synch.Prod. *Comp.*	*Trans.*	Imp1 *Events*	Cut-offs	Time	Imp12 Time	Imp2 *Events*	cut-offs	Time
DPH(7)	15	121	40672	21427	623.79	117.57	19306	9693	22.39
ELEVATOR(4)	7	939	16935	7337	96.03	24.32	16935	7337	25.42
KEY(3)	8	133	6940	2921	16.38	3.57	7187	3032	2.44
MMGT(3)	7	172	5841	2529	7.88	2.61	5841	2529	2.18
Q(1)	18	194	8402	1173	44.34	12.67	8030	1125	10.21
RING(24)	48	264	12745	1082	152.42	33.90	10722	1082	34.70
RW(12)	25	313	49177	45069	69.95	22.61	49177	45069	83.54
BUFFER(240)	240	241	28921	1	7098.06	1980.78	28921	1	34.81
CYCLIC(1000)	2000	5999	8996	1001	1372.24	1338.36	8996	1001	63.83
SENTST(2000)	2005	2030	2191	40	311.81	186.65	2030	40	8.33

Table 1. Experimental results

synchronous products. So it is possible to compare the performances of `Imp1` and the implementation of Section 6, called `Imp2` in the sequel. The main differences between `Imp1` and `Imp2` are

(a) `Imp1` uses the adequate order of [4], while `Imp2` uses $\prec_2$;
(b) `Imp1` computes the concurrency relation by the three-step procedure described at the beginning of Section 6.3, while `Imp2` uses the *Update* procedure;
(c) `Imp1` computes *Marking*([*e*]) (the equivalent of *GState*([*e*]) in [4]) by means of a backward search, while `Imp2` uses the procedure derived from Prop. 4.

The differences (a) and (c) are inherent to the change of model: Petri nets for `Imp1`, and synchronous products for `Imp2`. On the contrary, the difference (b) is accidental: When `Imp1` was programmed, we had not found the *Update* procedure. So it makes sense to consider a third implementation, `Imp12`, which coincides with `Imp1` on (a) and (c), and with `Imp2` on (b).

We have chosen a set of benchmarks compiled by Corbett in [3]; for a description of the systems the reader is referred to [3] and [7]. All benchmarks are scalable. Table 1 displays the results of the experiments for some representative cases. The experiments were carried out on a Sun Ultra 60 (295 MHz UltraSPARC-II) with 640 MB RAM using Solaris 2.7. The displayed data are the number of components and the number of global transitions of the product, the number of events of the complete prefix, the number of cut-off events, and the computation time (in seconds). The size of the unfoldings for `Imp1` and `Imp12` is always the same, since they both use the same adequate order. The benchmarks above the double horizontal line have a large ratio cut-offs/events, corresponding to wide but shallow prefixes, while those below have a small ratio, corresponding to narrow and deep prefixes. The results indicate that `Imp2` is indeed more efficient than `Imp1`. A closer look, and a comparison with `Imp12`, indicates that:

- For large cut-off ratios, the speed-up factor lies between 3 and 5, and it is due to the new procedure for the computation of the concurrency relation.

- For small cut-off ratios, the speed-up factor is of 1 to 2 orders of magnitude, and it is due to the new order, and to the new procedure for computing *GState*.

These provisory conclusions still need to be tested on more examples.

8 Conclusions

We have adapted the unfolding technique to Arnold's synchronous products of transition systems. The fact that a synchronous product consists of a fixed number of communicating sequential components has been used to simplify the unfolding procedure. We have obtained adequate orders simple to define and simple to prove correct.

We mentioned in the introduction that Langerak and Brinksma have applied the unfolding technique to a CSP-like process algebra [5]. The algebra has more modelling power than synchronous products; in particular, it is able to model nested parallelism, which synchronous products cannot. The price to pay is a more complicated adequate order than $\prec_1$ or $\prec_2$, although simpler than the order of [4] for Petri nets. Together with ours, Langerak and Brinksma's paper gives strong evidence that the unfolding technique can be applied to any model of concurrency allowing for a notion of independent actions.

We have presented an efficient implementation of the abstract algorithm for the construction of a complete finite prefix, which improves on the implementation of [4]. The speed-ups can reach two orders of magnitude in very favourable cases. A speed-up factor of at least 3 to 5 is achieved in nearly all cases.

Acknowledgements

Many thanks to Walter Vogler and Rom Langerak for helpful remarks and discussions.

References

1. Bibliography on net unfoldings. Accessible at `http://wwwbrauer.informatik.tu-muenchen.de/~esparza`.
2. A. Arnold. *Finite Transition Systems*. Prentice-Hall, 1992.
3. J.C. Corbett. Evaluating Deadlock Detection Methods for Concurrent Software. In T. Ostrand, editor, *Proceedings of the 1994 International Symposium on Software Testing and Analysis*, 204–215, 1994.
4. J. Esparza, S. Römer und W. Vogler: An Improvement of McMillan's Unfolding Algorithm. In T. Margaria und B. Steffen, editors, *Proceedings of TACAS'96*, LNCS 1055, 87–106, 1996.
5. R. Langerak and E. Brinksma. A Complete Finite Prefix for Process Algebra. To appear in Proceedings of CAV '99, 1999.
6. K. McMillan. *Symbolic Model Checking: An Approach to the State Explosion Problem*. Kluwer, 1993.
7. S. Melzer. Verification of Distributed Systems Using Linear and Constraint Programming. Ph. D. Thesis, Technical University of Munich, 1998 (in German).

Petri Nets, Configuration Structures and Higher Dimensional Automata

R.J. van Glabbeek*

Computer Science Department, Stanford University
Stanford, CA 94305-9045, USA.
http://theory.stanford.edu/~rvg/
rvg@cs.stanford.edu

In this talk, translations between several models of concurrent systems are reviewed c.q. proposed. The models considered capture causality, branching time, and their interplay, and these features are preserved by the translations. To the extent that the models are intertranslatable, this yields support for the point of view that they are all different representations of the same phenomena. The translations can then be applied to reformulate any issue that arises in the context of one model into one expressed in another model, which might be more suitable for analysing that issue. To the extent that the models are not intertranslatable, my investigations are aimed at classifying them w.r.t. their expressiveness in modelling phenomena in concurrency. The results are summarised in the figure at the end of this paper.

Starting point is the work of NIELSEN, PLOTKIN & WINSKEL [17], in which safe Petri nets are translated, through the intermediate stages of occurrence nets, prime event structures with a binary conflict relation, and their families of configurations, into a class of Scott domains.

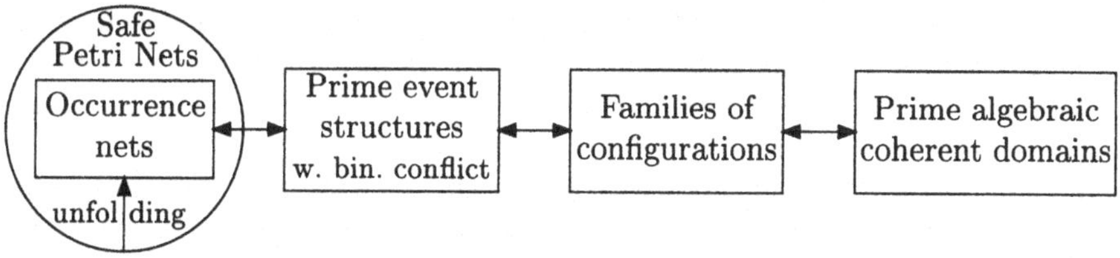

1 From nets to configurations

In VAN GLABBEEK & PLOTKIN [10] extensions of the translations above to unsafe Petri nets have been studied. For this purpose two different schools of thought in interpreting the causal behaviour of nets needed to be distinguished, which we called the *individual* and *collective token* philosophy. Their difference is illustrated by the following net. According to the individual token philosophy, A

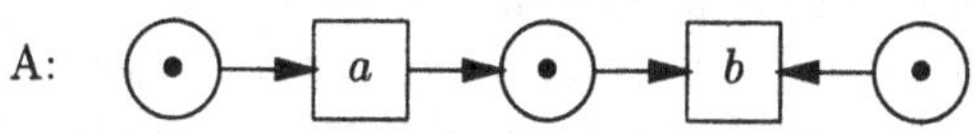

has an execution in which the action b causally depends on a, whereas according to the collective token philosophy, a and b are always causally independent.

* This work was supported by ONR under grant number N00014-92-J-1974.

In MESEGUER, MONTANARI & SASSONE [15], the unfolding from [17], translating safe nets into the subclass of occurrence nets, is extended to arbitrary nets, while preserving the individual token interpretation. It follows that under this interpretation prime event structures are expressive enough to represent all processes expressible by Petri nets.

Under the collective token interpretation there turn out to be nets whose causal behaviour cannot be faithfully represented by a prime event structure. Representative examples are the two nets below, modelling what I often call

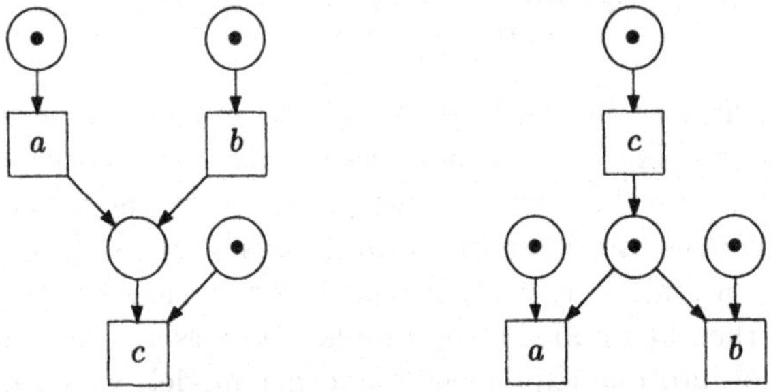

disjunctive causality and *resolvable conflict*, respectively.

In WINSKEL [25] a more general notion of event structure is proposed, extending the prime event structures with a binary conflict relation from [17], along with matching generalisations of the families of configurations and Scott domains. These event structures capture disjunctive causality, but not resolvable conflict.

The families of configurations of Winskel's event structures were introduced merely to facilitate the construction of the Scott domains associated to these event structures. In VAN GLABBEEK & GOLTZ [9] we found it convenient to use such families as a model of concurrency in its own right. In this context the families were called *configuration structures.* A configuration structure can be given by a set of *events*, modelling occurrences of actions the represented system may perform, possibly a *labelling function*, associating actions to events, and a collection of sets of events, the *configurations*, modelling the states of the represented system, and satisfying a number of closure conditions. A configuration represents the state in which the events it contains have occurred. The closure conditions ensure that each configuration structure can be regarded as the family of configurations of an event structure.

In [10] we proposed to drop the closure conditions, thereby obtaining a more general model of concurrency, capturing both disjunctive causality and resolvable conflict. The resulting configuration structures are, up to isomorphism, the *extensional Chu spaces* of GUPTA & PRATT [13], but equipped with a slightly different computational interpretation. Through suitable translations we showed that these configuration structures are equally expressive as general Petri nets without self-loops. Such nets are called *pure.* To this end we defined a *1-occurrence net* to be a Petri net in which each transition can fire at most once, and we showed how any (pure) Petri net can be converted into a (pure) 1-occurrence net, using a construction we called *1-unfolding.* We argued that this conversion preserves essential features of the represented system like causality and branching time. It may convert a finite net into an infinite one, however. The translations between pure 1-occurrence nets and configuration structures take the transitions of the net to be the events of the configuration structure and vice versa; this way a configuration structure can be fully recovered from its Petri net representation. Our translations also extend the correspondence between flow nets and flow event structures proposed in BOUDOL [3].

ST-configuration structures are a further generalisation of configuration structures in which the configurations may contain certain events 'partially' (in case they are currently being executed). They are (a mild generalisation of) what are called *local event structures* in HOOGERS, KLEIJN & THIAGARAJAN [14]. In forthcoming work, Gordon Plotkin and I extend the translations between pure nets and configuration structures to translations between arbitrary Petri nets and ST-configuration structures, thus showing that also these models are equally expressive. The same was done, using a different construction, for general Petri nets without autoconcurrency in [14]. We also propose a matching generalisation of the model of event structures.

2 Scott domains versus process graphs

In [17] a "curious mismatch" is observed between the domains that result from translating nets or event structures, and the ones originally studied by SCOTT [22]. Although mathematically of the same nature, a domain that arises through the translations of [17] represents a single concurrent system, namely the same one represented by the Petri net or event structure it originated from. In domain theory, on the other hand, processes show up at best as the *elements* of a domain. Thus the use of domains to represent concurrent systems is novel in [17].

In most models of concurrency, attention is restricted to *discrete* processes, i.e. processes that can perform only finitely many actions in a finite time. Petri nets are commonly interpreted to represent discrete processes—this comes with the common definitions of the *firing rule.* On prime event structures the axiom of *finite causes* restricts attention to the structures representing discrete systems, and in Winskel's general event structures discreteness is obtained by the way the notion of a configuration of an event structure is defined. A Scott domain is a partially ordered set, satisfying certain conditions. The *finite* elements in such a domain are the ones that dominate only finitely many other elements. A discrete Scott domain (resulting from translating a discrete event structure) has the property that its infinitary part is redundant, in the sense that it can be recovered in full from its finitary part (the partial suborder of its finite elements). The finitary part of a domain can, without loss of information, be trivially represented, and is often displayed, as an unlabelled rooted graph. Therefore I argue that the correspondence between event structures and domains proposed in [17], and generalised to all event structures in [25], can equivalently, or maybe better, be regarded as a correspondence between event structures and a class of unlabelled transition systems or *process graphs.* As remarked in [7], this correspondence can trivially be extended to *labelled* event structures and transitions systems; the latter are easier to label than domains. It follows immediately that process graphs, or labelled transition systems, are at least as capable of expressing causality as labelled event structures.

The computational interpretation of domains, inherited from that of event structures, naturally applies to the process graphs corresponding with those domains. These graphs capture causality through confluence of squares of transitions. This computational interpretation can be extended to process graphs that do not correspond to event structures or Scott domains. It can be seen as an enrichment of the classical interpretation of process graphs. Just like any process graph can be unfolded into a tree, while preserving its interleaving interpretation, I propose a causality respecting unfolding of arbitrary process graphs into so-called *history preserving* ones, which preserves transition squares. History preserving process graphs generalise the Scott domains originating from the general event structures of [25]. They can also model phenomena like resolvable conflict that are not expressible by these event structures.

Several brands of transition systems enriched with some auxiliary structure to capture causality have been proposed as models of concurrency, cf. the *asynchronous transition systems* of SHIELDS [23] and BEDNARCZYK [2], the *behaviour structures* of RABINOVICH & TRAKHTENBROT [21], the *concurrent transition systems* of STARK [24] and DROSTE, [5] and the *transition systems with independence* of NIELSEN & WINSKEL [26]. In each of these cases the added structure does not fundamentally increase their expressiveness: after a suitable behaviour-preserving unfolding, the causalities expressed by this added structure are completely determined by the underlying transition system, which always forms a history preserving process graph.

Event automata, studied by PINNA & POIGNÉ [19], fit between configuration structures and ST-configuration structures. Through appropriate translations these can be shown to be equally expressive as the so-called *configuration-deterministic* process graphs. Graphs which are not configuration deterministic do not correspond to nets or event-oriented models. Interestingly, translating back and forth between event automata and process graphs may repeatedly increase the number of events of the event automaton representation of the system in question, namely by spitting events into subevents that occur in disconnected parts of the system representation. Hence these translations cannot be expressed as reflexions or coreflexions in a suitable categorical framework.

3 Higher dimensional automata

The concurrent interpretation of process graphs allows one to think of squares and cubes as being "filled in". PRATT [20] proposes a *geometric model of concurrency*, refining this approach by not necessarily filling in *all* squares and cubes, but explicitly filling in only those that one wants to represent concurrency. Alternative formalisations of this idea appear in VAN GLABBEEK [8], GOUBAULT & JENSEN [11] and CATTANI & SASSONE [4]. Although the resulting model of *higher dimensional automata* is more complicated than that of plain automata or process graphs, it is more expressive as well. The Petri net below, for in-

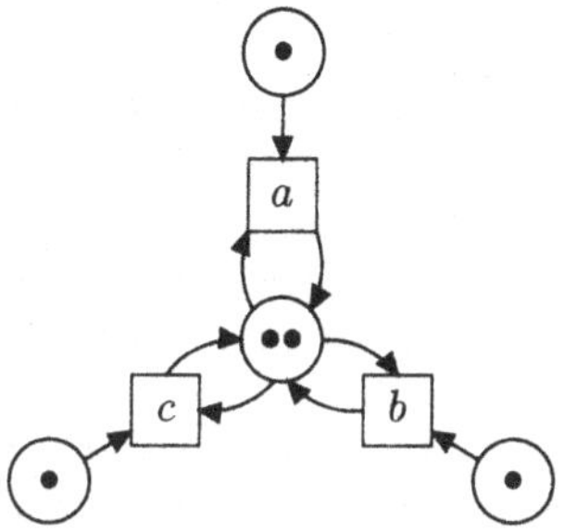

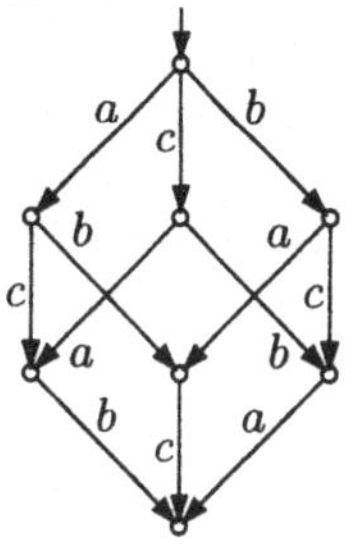

stance, is expressible by a higher dimensional automaton in the form of a cube, as displayed above, of which all 6 sides are filled in (representing the 6 possible concurrent firings of two transitions, either before or after the third one fires), but the interior is not. The causal behaviour of this system cannot be represented by a process graph, and hence neither by an event automaton, nor by a pure Petri net. Process graphs and the mentioned transition systems with extra structure to capture causality can be regarded as one- and two-dimensional automata, respectively.

A representation of higher dimensional automata in which the names of both events and actions are incorporated, is given by *labelled step transition systems* (LSTSs), see also BADOUEL [1]. These naturally unfold into (alternative representations of) ST-configuration structures. Petri nets translate to LSTSs by taking their marking graphs; LSTSs translate to higher dimensional automata as in [8] by forgetting event names (but remembering their labels).

EHRENFEUCHT & ROZENBERG [6] characterised which process graphs can be obtained as the marking graphs of a safe nets. Likewise, MUKUND [16] characterised which LSTSs can be obtained as the step marking graphs of general Petri nets. Both papers also yield translations back from (step) transition systems to nets, but only for systems in the characterised class. More general translations from LSTSs to nets can be obtained through unfolding.

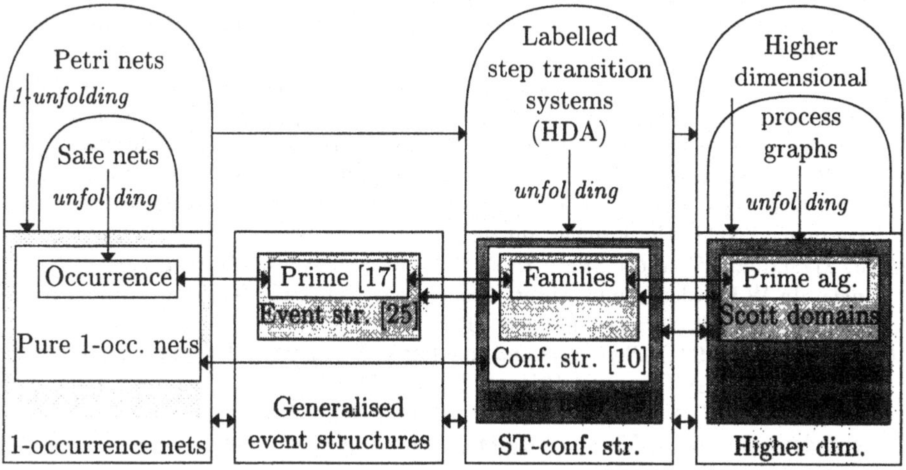

References

1. E. BADOUEL (1996): *Splitting of actions, higher-dimensional automata, and net synthesis*. Technical Report RR-3490, Inria, France.
2. M. BEDNARCZYK (1987): *Categories of asynchronous systems*. PhD thesis, Computer Science, University of Sussex, Brighton.
3. G. BOUDOL (1990): *Flow event structures and flow nets*. In I. Guessarian, editor: *Semantics of Systems of Concurrent Processes, Proceedings LITP Spring School on Theoretical Computer Science*, La Roche Posay, France, LNCS 469, Springer, pp. 62–95.
4. G.L. CATTANI & V. SASSONE (1996): *Higher dimensional transition systems*. In Proceedings 11th Annual IEEE Symposium on *Logic in Computer Science* (LICS 96), New Brunswick, USA, IEEE Computer Society Press, pp. 55–62.
5. M. DROSTE (1992): *Concurrent automata and domains*. International Journal of Foundations of Computer Science 3(4), pp. 389–418.
6. A. EHRENFEUCHT & G. ROZENBERG (1990): *Partial 2-structures*. Acta Informatica 27(4), pp. 315–368.
7. R.J. VAN GLABBEEK (1988): *An operational non-interleaved process graph semantics of CCSP* (abstract). In E.-R. Olderog, U. Goltz & R.J. van Glabbeek, editors: *Combining compositionality and concurrency, summary of a GMD-workshop, Königswinter, March 1988*, Arbeitspapiere der GMD 320, pp. 18–19.
8. R.J. VAN GLABBEEK (1991): *Bisimulations for higher dimensional automata*. Email message, July 7, '91. Available at `http://theory.stanford.edu/~rvg/hda`.
9. R.J. VAN GLABBEEK & U. GOLTZ (1990): *Refinement of actions in causality based models*. In J.W. de Bakker, W.P. de Roever & G. Rozenberg, editors: *Proceedings REX Workshop on Stepwise Refinement of Distributed Systems: Models, Formalism, Correctness*, Mook, The Netherlands, May/June 1989, LNCS 430, Springer, pp. 267–300.
10. R.J. VAN GLABBEEK & G.D. PLOTKIN (1995): *Configuration structures (extended abstract)*. In D. Kozen, editor: Proceedings 10th Annual IEEE Symposium on *Logic in Computer Science* (LICS 95), San Diego, USA, IEEE Computer Society Press, pp. 199–209.
11. E. GOUBAULT & T. JENSEN (1992): *Homology of higher dimensional automata*. In W.R. Cleaveland, editor: *Proceedings CONCUR 92*, Stony Brook, NY, USA, LNCS 630, Springer, pp. 254–268.
12. J. GUNAWARDENA (1992): *Causal automata*. Theoretical Computer Science 101, pp. 265–288.
13. V. GUPTA & V.R. PRATT (1993): *Gates accept concurrent behavior*. In *Proc. 34th Ann. IEEE Symp. on Foundations of Comp. Sci.*, pp. 62–71. More material on Chu spaces can be found at `http://boole.stanford.edu/chuguide.html`.
14. P.W. HOOGERS, H.C.M. KLEIJN & P.S. THIAGARAJAN (1993): *Local event structures and Petri nets*. In E. Best, editor: *Proceedings CONCUR 93*, Hildesheim, Germany, LNCS 715, Springer, pp. 462–476.
15. J. MESEGUER, U. MONTANARI & V. SASSONE (1992): *On the semantics of Petri nets*. In W.R. Cleaveland, editor: *Proceedings CONCUR 92*, Stony Brook, NY, USA, LNCS 630, Springer, pp. 286–301.
16. M. MUKUND (1992): *Petri nets and step transition systems*. International Journal of Foundations of Computer Science 3(4), pp. 443–478.
17. M. NIELSEN, G.D. PLOTKIN & G. WINSKEL (1981): *Petri nets, event structures and domains, part I*. Theoretical Computer Science 13(1), pp. 85–108.

18. M. NIELSEN, G. ROZENBERG & P.S. THIAGARAJAN (1992): *Elementary transition systems.* *Theoretical Computer Science* 96, pp. 3–33.
19. G.M. PINNA & A. POIGNÉ (1995): *On the nature of events: another perspective in concurrency.* *Theoretical Computer Science* 138(2), pp. 425–454.
20. V.R. PRATT (1991): *Modeling concurrency with geometry.* In *Proc. 18th Ann. ACM Symposium on Principles of Programming Languages*, pp. 311–322.
21. A. RABINOVICH & B.A. TRAKHTENBROT (1988): *Behavior structures and nets.* *Fundamenta Informaticae* 11(4), pp. 357–404.
22. D. SCOTT (1970): *Outline of a mathematical theory of computation.* In *Proceedings of the 4th Annual Princeton Conference on Information Sciences and Systems*, pp. 169–176.
23. M.W. SHIELDS (1985): *Concurrent machines.* *The Computer Journal* 28(5), pp. 449–465.
24. E.W. STARK (1989): *Concurrent transition systems.* *Theoretical Computer Science* 64, pp. 221–269.
25. G. WINSKEL (1987): *Event structures.* In W. Brauer, W. Reisig & G. Rozenberg, editors: *Petri Nets: Applications and Relationships to Other Models of Concurrency, Advances in Petri Nets 1986, Part II; Proceedings of an Advanced Course*, Bad Honnef, September 1986, LNCS 255, Springer, pp. 325–392.
26. G. WINSKEL & M. NIELSEN (1995): *Models for concurrency.* In S. Abramsky, D.M. Gabbay & T.S.E. Maibaum, editors: *Handbook of Logic in Computer Science*, volume 4: Semantic Modelling, chapter 1. Oxford University Press.

Expressiveness and Distributed Implementation of Concurrent Calculi with Link Mobility

Catuscia Palamidessi

The Pennsylvania State University
catuscia@cse.psu.edu   http://www.cse.psu.edu/~catuscia

Abstract

The π-calculus [6] has introduced in concurrency the concept of link mobility, namely the possibility of communicating values which can afterwards be used as communication means (i.e. channels). Since the original work on the π-calculus, many variants and related paradigms have been introduced, including the asynchronous π-calculus [1, 4, 5], the π-calculus with input-guarded choice [8], the π-calculus with internal communication [11], the Fusion Calculus [10], and the Join Calculus [2, 3]. In general, these variants introduce restrictions that allow for a simpler formal treatment, and/or a more direct modeling of some of the features of distributed systems (like asynchronous communication).

Some recent results [7, 9] suggest that the expressive power of these variants can be very different when distribution constraints are taken into consideration. In this talk, I will focus on the relative expressiveness of some of these variants, and discuss possible approaches to their distributed implementation.

References

1. G. Boudol. Asynchrony and the π-calculus. Technical Report 1702, INRIA, Sophia-Antipolis, 1992.
2. C. Fournet and G. Gonthier. The Reflexive Chemical Abstract Machine and the Join-Calculus. In *Proc. of the 23rd Annual Symposium on Principles of Programming Languages (POPL)*, pages 372–385, 1996.
3. C. Fournet, G. Gonthier, J.-J. Lévy, Luc Maranget, and D. Rémy. A Calculus of Mobile Agents. In U. Montanari, editor, *Proc. of the 7th International Conference on Concurrency Theory (CONCUR)*, volume 1119 of *Lecture Notes in Computer Science*, pages 406–421, 1996.
4. K. Honda and M. Tokoro. An object calculus for asynchronous communication. In P. America, editor, *Proc. of ECOOP 91*, volume 512 of *Lecture Notes in Computer Science*, pages 133–147. Springer-Verlag, 1991.
5. K. Honda and M. Tokoro. On asynchronous communication semantics. In M. Tokoro, O. Nierstrasz, and P. Wegner, editors, *Proc. of Object-Based Concurrent Computing*, volume 612 of *Lecture Notes in Computer Science*, pages 21–51. Springer-Verlag, 1992.
6. R. Milner, J. Parrow, and D. Walker. A calculus of mobile processes, Part I/II. *Information and Computation*, 100(1):1–77, 1992.

7. U. Nestmann. What Is a 'Good' Encoding of Guarded Choice? In C. Palamidessi and J. Parrow, editors, *Proc. of EXPRESS'97*, volume 7 of *Electronic Notes in Theoretical Computer Science*, 1997. Extended version accepted for publication on *Information and Computation*.
8. U. Nestmann and B.C. Pierce. Decoding choice encodings. In U. Montanari, editor, *Proc. of the 7th International Conference on Concurrency Theory (CONCUR)*, volume 1119 of *Lecture Notes in Computer Science*, pages 179–194. Springer-Verlag, 1996.
9. C. Palamidessi. Comparing the expressive power of the Synchronous and the Asynchronous π-calculus. In *Proc. of the 24th ACM Symposium on Principles of Programming Languages (POPL)*, pages 256–265, ACM, 1997.
10. J. Parrow and B. Victor. The Fusion Calculus: Expressiveness and Symmetry in Mobile Processes. In *Proc. of the Thirteenth Annual Symposium on Logic in Computer Science (LICS)*, pages 176–185, IEEE Computer Society Press, 1998.
11. D. Sangiorgi. π-calculus, internal mobility, and agent-passing calculi. *Theoretical Computer Science* 167(2):235–274, 1996.

Techniques for Decidability and Undecidability of Bisimilarity

Petr Jančar[1*] and Faron Moller[2**]

[1] Technical University of Ostrava, Czech Republic
Petr.Jancar@vsb.cz
[2] Uppsala University, Sweden
fm@csd.uu.se

Abstract. In this tutorial we describe general approaches to deciding bisimilarity between vertices of (infinite) directed edge-labelled graphs. The approaches are based on a systematic search following the definition of bisimilarity. We outline (in decreasing levels of detail) how the search is modified to solve the problem for finite graphs, BPP graphs, BPA graphs, normed PA graphs, and normed PDA graphs. We complete this by showing the technique used in the case of graphs generated by one-counter machines. Finally, we demonstrate a general reduction strategy for proving undecidability, which we apply in the case of graphs generated by state-extended BPP (a restricted form of labelled Petri nets).

1 Bisimulation Equivalence

The ***bisimulation game*** is played on a "board" which consists of a (generally infinite) directed edge-labelled multigraph (several edges can lead between two vertices), simply called a ***graph*** in the following. We assume that this graph is labelled from a *finite* set of labels, and that it is ***finite-branching***: that every vertex has finite out-degree. In particular, it is ***image-finite***: for every vertex E and every label a, the set $succ_a(E) = \{F : E \stackrel{a}{\to} F\}$ is finite. We also asume that these successor sets are effectively constructible.

A game is defined by two vertices E_0 and F_0 of a graph, as well as a predetermined time limit $n \in \mathbb{N} \cup \{\omega\}$ (where $\mathbb{N} = \{0, 1, 2, 3, \ldots\}$); we specify such a game by $G_n(E_0, F_0)$. We have two ***players*** competing in the game whom we refer to as **Alice** (the *"Attacker"*) and **Bob** (the *"Bisimulator"*). Their individual goals are as follows:

1. **Alice** wants to show that E_0 and F_0 are in some sense "different".
2. **Bob** wants to show that E_0 and F_0 are in the same sense "the same".

* This paper was written during a visit by the first author to Uppsala University supported by a grant from the Swedish STINT Fellowship Programme. He is also partially supported by the Grant Agency of the Czech Republic, Grant No. 201/97/0456.

** The second author is supported by Swedish TFR grants No. 221-98-103 *"Verification of Infinite State Automata"* and 221-97-275 *"Games for Processes"*.

The sense in which two vertices are deemed to be the same is given by the rules for playing the game. A ***play*** of the game is a sequence of pairs $(E_0, F_0)\,(E_1, F_1)\cdots$ of length $\leq(1{+}n)$, with the next pair in the sequence after (E_i, F_i) arising as follows:

1. **Alice** chooses an edge $E_i \overset{a}{\rightarrow} E_{i+1}$ or $F_i \overset{a}{\rightarrow} F_{i+1}$;
2. **Bob** chooses a matching edge $F_i \overset{a}{\rightarrow} F_{i+1}$ or $E_i \overset{a}{\rightarrow} E_{i+1}$.

Alice is thus acting as an attacker, trying to choose an edge leading out of one of the vertices which she believes cannot be matched by any edge (with the same label) leading out of the other vertex; **Bob** on the other hand is defending his thesis that the vertices are equal, that any edge leading out of either of the vertices has a matching edge leading out of the other vertex. **Alice** wins a play of the game if **Bob** ever gets stuck (that is, if he cannot respond to a move by **Alice**); and **Bob** wins any play of length n (that is, any "timed-out" play in which the players have exchanged n moves) as well as any play in which **Alice** finds herself with no move possible (that is, if there are no edges leading out of either of the two specified vertices).

We are interested in knowing if **Bob** has a winning (i.e., defending) strategy for the game $G_n(E_0, F_0)$, that is, if he is able to win *any* play of the game regardless of the moves made by **Alice**. To this end, we make the following definitions. For $n{\in}\mathbb{N}$, we say that E_0 and F_0 are ***n-game equivalent*** (written $E_0 \sim_n F_0$) iff **Bob** has a winning strategy for the game $G_n(E_0, F_0)$; and we say that E_0 and F_0 are ***game equivalent*** (written $E_0 \sim F_0$) iff **Bob** has a winning strategy for the game $G_\omega(E_0, F_0)$. The relation $\sim$ is referred to as ***bisimulation equivalence***, or ***bisimilarity***. Before proceeding, it is worth recording the following straightforward facts.

Fact 1 *$\sim_n$ and $\sim$* are *equivalence (reflexive, symmetric, transitive) relations.*

Fact 2 $\sim_0 \;\supseteq\; \sim_1 \;\supseteq\; \sim_2 \;\supseteq\; \sim_3 \;\supseteq\; \cdots \;\supseteq\; \sim$.

The following fact gives us an *inductive* characterisation of the finite-game equivalences.

Fact 3 ("Stratified Bisimilarity")

1. $E \sim_0 F$ *for all* E, F.
2. $E \sim_{n+1} F$ *iff*
   (a) *if* $E \overset{a}{\rightarrow} E'$ *then* $F \overset{a}{\rightarrow} F'$ *with* $E' \sim_n F'$;
   (b) *if* $F \overset{a}{\rightarrow} F'$ *then* $E \overset{a}{\rightarrow} E'$ *with* $E' \sim_n F'$.

Fact 3 can be used as the basis of a recursive algorithm for determining if $E \sim_n F$, that is, if **Bob** has a winning strategy for $G_n(E, F)$. Less obvious is how to devise an algorithm for determining if $E \sim F$. As a start, the following fact gives us a *coinductive* characterisation of the infinite-game equivalence.

Fact 4 ("Bisimilarity")

$\sim$ *is the* ***largest*** *relation* $\equiv$ *satisfying: if* $E \equiv F$ *then*

1. *if* $E \xrightarrow{a} E'$ *then* $F \xrightarrow{a} F'$ *with* $E' \equiv F'$;
2. *if* $F \xrightarrow{a} F'$ *then* $E \xrightarrow{a} E'$ *with* $E' \equiv F'$.

The characterisation given in Fact 4 does not in general give rise to any effective procedure for determining if $E \sim F$. However, given our assumption of image-finiteness, we can use the following fact to exploit the finite-game characterisation of Fact 3 to verify that $E \not\sim F$.

Fact 5 *For* ***image-finite*** *graphs,* $\sim \; = \bigcap_{n \in \mathbb{N}} \sim_n$.

Hence in the case of image-finite graphs, the non-equivalence problem $E \not\sim F$ is semi-decidable: we simply use Fact 3 to find the smallest n such that $E \not\sim_n F$.

We can be more explicit as to what constitutes a winning strategy for **Bob** for the game $G_\omega(E_0, F_0)$; this is a set $\mathcal{B}$ of pairs of vertices containing (E_0, F_0) which satisfies the following property:

For every pair $(E, F) \in \mathcal{B}$,

1. if $E \xrightarrow{a} E'$ then $F \xrightarrow{a} F'$ with $(E', F') \in \mathcal{B}$;
2. if $F \xrightarrow{a} F'$ then $E \xrightarrow{a} E'$ with $(E', F') \in \mathcal{B}$.

That such a set constitutes a winning strategy for **Bob** is clear: he merely uses this set to choose matching edges, maintaining the invariant that the pair offered to **Alice** is in the set $\mathcal{B}$. Furthermore, if **Bob** has a winning strategy for the game $G_\omega(E_0, F_0)$, then this strategy can be represented by such a set: we merely take the collection of all pairs which appear after every exchange of moves during any and all plays in which **Bob** uses this strategy. A set $\mathcal{B}$ which satisfies the above property is referred to as a ***bisimulation relation***. Thus $\sim$ is the maximal bisimulation and the following is apparent.

Fact 6 $\sim$ *is the union of all bisimulation relations (winning strategies for* ***Bob****).*

2 Decidability

Deciding whether or not $E_0 \sim F_0$ amounts to deciding whether or not there is a bisimulation relation containing the pair (E_0, F_0). A straightforward idea for tackling this problem is to employ some systematic search for such a potential bisimulation. In Section 2.1 we describe a general procedure which arises naturally from the definition of a bisimulation, and demonstrate its (modified) use in various contexts, particularly in Section 2.2 for classes of graphs determined by various process algebras. In Section 2.3 we outline a different approach, exemplified on the class of graphs generated by one-counter machines.

2.1 Expansion trees

Given two sets $B \neq \emptyset$ and A of pairs of vertices, A is called an ***expansion*** of B iff it is a minimal set (wrt inclusion) satisfying the following property:

For every pair $(E, F) \in B$,
1. if $E \xrightarrow{a} E'$ then $F \xrightarrow{a} F'$ with $(E', F') \in A$;
2. if $F \xrightarrow{a} F'$ then $E \xrightarrow{a} E'$ with $(E', F') \in A$.

Note that a nonempty set *fails* to have an expansion precisely when it contains a pair (E', F') such that $E' \not\sim_1 F'$; and it has (the single) expansion $\emptyset$ iff it contains only pairs of vertices with out-degree zero. Note further that, due to our image-finiteness assumption, a finite set has only finitely many expansions, all of which are finite.

Comparing this definition to that of a bisimulation relation, we observe that a bisimulation relation (containing) B must contain some expansion A of B. The following fact then becomes apparent.

Fact 7 *If $A \subseteq B$ and A is an expansion of B then B is a bisimulation relation, and hence $B \subseteq \sim$.*

More generally, to decide if $E_0 \sim F_0$, that is, if there is a bisimulation B containing the pair (E_0, F_0), we might try to expand $A_0 = \{(E_0, F_0)\}$ recursively (find an expansion A_{k+1} of A_k for each $k = 0, 1, 2, \ldots$) and hope to arrive at a set $A_n \subseteq B = \bigcup_{k<n} A_k$; in this case, some $A \subseteq B$ would be an expansion of B, so Fact 7 would give us that $B \subseteq \sim$, in particular, that $E_0 \sim F_0$.

Following this idea, we adapt from Hirshfeld [8] the idea of an ***expansion tree***, which is a (generally infinite) tree whose nodes are (labelled by) sets of pairs of vertices, in which the children of a node are precisely the (finitely many) expansions of that node. The empty node is an example of a ***leaf*** (it has no successors) and is deemed to be ***successful***; all other leaves are ***unsuccessful***. We say that a ***branch*** (a full path through the tree) is ***successful*** iff it is infinite or finishes with a successful node; otherwise it is ***unsuccessful*** (it finishes with an unsuccessful node). We observe that the union of the nodes along a successful branch constitutes a bisimulation, and the following fact thus becomes apparent.

Fact 8 *$E_0 \sim F_0$ iff the expansion tree rooted at $\{(E_0, F_0)\}$ has a successful branch.*

Note that in the case $E_0 \not\sim F_0$, the expansion tree rooted at (E_0, F_0) is necessarily finite, and hence semidecidability of nonbisimilarity once again becomes apparent.

In general, we can search for a (finite) successful branch using breadth-first search (recall that the tree is finite branching). This is surely sufficient in the case of finite acyclic graphs, where the expansion trees are always finite; the search of the expansion tree rooted at $\{(E_0, F_0)\}$ either finds a successful leaf (in which case $E_0 \sim F_0$) or terminates having found only unsuccessful leaves (in which case $E_0 \not\sim F_0$).

Example 1: ***Finite acyclic graphs.***
In this example, we have a finite acyclic graph containing vertices X, A

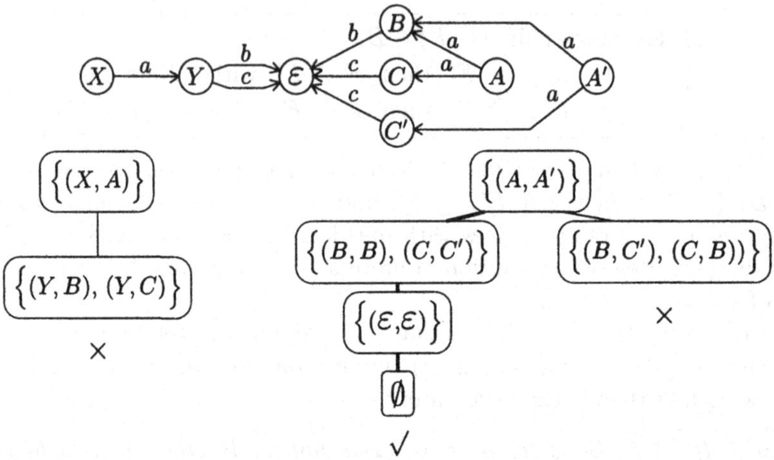

and A' (amongst others), and we demonstrate that $X \not\sim A$ and $A \sim A'$, by building the expansion trees rooted at $\{(X, A)\}$ and $\{(A, A')\}$, respectively. The first tree has only one leaf, which is unsuccessful, while the second tree has two leaves, one of which is successful (the other not).

The naïve procedure described above is of course of little use if the expansion tree of interest has infinitely many nodes. To handle such cases, we shall devise techniques for modifying the tree during construction. What we aim for is an effective construction of a tree which is still finite branching, still has only finite nodes, and such that Fact 8 is maintained: that our pair of vertices is bisimilar iff there is a successful branch in this tree. Finally, we further aim to achieve the ***finite witness property***: if there *are* successful branches, then at least one of these is finite. This property will ensure decidability, as the breadth-first search must either find this finite witness or terminate having discovered all branches are unsuccessful.

The modifications are accomplished through the use of ***rules*** which alter the definition of the children of a node in the tree and are ***safe***, meaning that they maintain Fact 8. To suggest such rules, we first observe the following.

Fact 9 *For any node $A \neq \emptyset$ and any $n \in \mathbb{N}$,*

$$A \subseteq \sim_{n+1} \quad \textit{iff} \quad A \textit{ has a child } C \subseteq \sim_n.$$

As a consequence, $A \subseteq \sim$ iff A has a child $C \subseteq \sim$.

This fact implies Fact 8 and hence any rule which respects it maintains safeness. Furthermore, the least n such that $E \not\sim_n F$ for some pair (E, F) in a node gives an upper bound on the depth of the subtree rooted at this node; in other words, if the subtree rooted at B is of depth n, then $E \sim_n F$ for all (E, F) in B. Therefore, any application of the following "abstract" rule maintains Fact 9 (and is hence safe); it can only diminish the size of the tree. (By $A^{\Uparrow}$ we mean the union of all ancestor nodes to A.)

Omitting Rule: We can omit from a node A any pair (E, F) whenever there is $B \subseteq A^{\Uparrow}$ such that $\forall n \in \mathbb{N}$, if $B \subseteq \sim_n$ then $E \sim_n F$.

We call this rule *abstract* since it is in general not effectively computable. A *concrete*, that is, effectively computable, instance of this rule, based on the fact that the relations $\sim_n$ are equivalences, is provided by the following.

Equivalence Rule: Omit from node A the pair (E, F) if it belongs to the least equivalence containing $A^{\Uparrow}$.

This rule, for example, allows us to omit reflexive pairs (E, E) from expansion sets, as well as pairs (E, F) which have appeared in some ancestor node either as (E, F) or symmetrically as (F, E). However, there are only finitely many non-reflexive pairs in the equivalence generated by the finitely-many ancestor pairs; these pairs can be easily computed, showing that this rule is effective. As such, the rule is not too powerful; essentially, it only allows us to handle finite graphs.

Example 2: ***Finite graphs.***
In this example, we have a graph similar to that of Example 1 but with

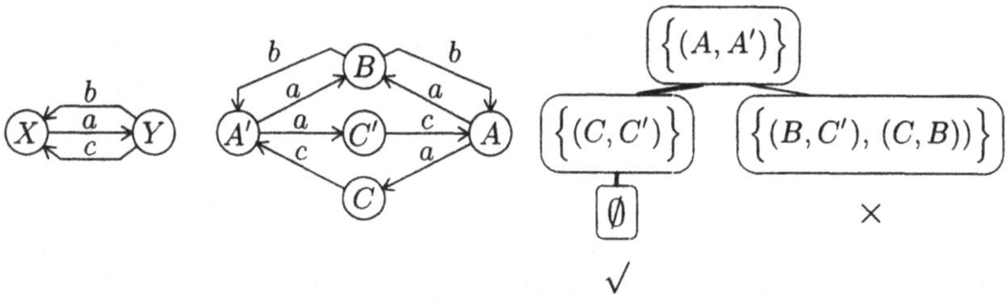

cycles introduced. Using our newly-introduced simplification rule, we get the same expansion tree as in Example 1 to show that $X \not\sim A$, and the given expansion tree to show that $A \sim A'$.

Note that this is by no means the most efficient algorithm for this problem; better algorithms based on partition refinement have been devised [24, 17]. We merely present this as a basic application of the general method (which we shall apply to various classes of infinite graphs).

Finally, note that the union of the nodes along a successful branch need no longer be a bisimulation; it is now a set B which is guaranteed to have an expansion A such that for all $n \in \mathbb{N}$, $B \subseteq \sim_n$ implies $A \subseteq \sim_n$ (which implies that $B \subseteq \sim$). Sets with this property are also the outcomes (when the input pair *is* bisimilar) of the algorithms outlined further in the following sections.

2.2 Process Algebras

In this section we consider bisimulation equivalence over graphs generated by (subalgebras of) the process algebra PA. A ***PA graph*** is defined by a finite set of ***productions*** of the form $X \stackrel{a}{\to} \alpha$, where X ranges over a finite set V of ***variables*** (there is at least one production for each $X \in V$), a ranges over a finite set Σ of ***labels***, and α ranges over the terms of the free algebra over V generated by a non-commutative associative operator "$\cdot$" (representing *sequential composition*) and a commutative associative operator "$|$" (representing *parallel composition*); we take ε as the empty term. We shall usually drop the operator "$\cdot$", thus representing the sequential composition of terms by concatenation. The ***vertices*** are given by the terms of the algebra, and the ***edges*** are specified by the production rules as extended to terms as follows: if $\alpha \stackrel{a}{\to} \beta$ then $\alpha{\cdot}\gamma \stackrel{a}{\to} \beta{\cdot}\gamma$ and $\alpha|\gamma \stackrel{a}{\to} \beta|\gamma$ (recall that $|$ is commutative). The subcases of ***BPA graphs*** (***Basic Process Algebra***) and ***BPP graphs*** (***Basic Parallel Processes***) are specified by taking terms from the free algebra over V generated by only one of these operators, "$\cdot$" in the case of BPA and "$|$" in the case of BPP.

As there are only finitely many productions, the graphs are quickly seen to be finitely-branching (and hence image-finite); and taking the algebraic structure of terms into consideration, we can readily verify (by induction) that the relations $\sim_n$ (and hence also $\sim$) are not just equivalences over terms but congruences with respect to the two operators. Hence we can immediately introduce the following safe abstract rule, which is a special instance of the **Omitting Rule**.

> **Congruence Rule**: Omit from node A the pair (α, β) if it belongs to the least congruence containing $A^{\Uparrow}$.

Again, the rule is called abstract since we do not claim its effectiveness in general. However, restricting ourselves to BPP, we notice that the terms can be viewed as nonnegative integer vectors (elements of $\mathbb{N}^k$ for $k = |V|$), and we thus work with finitely generated commutative semigroups, FGCSs (monoids, in fact). Decidability of the word problem for FGCSs (its EXPSPACE-completeness is shown in [19]) implies that the **Congruence Rule** is effective for BPP. Moreover, the result that *every* congruence on a FGCS is finitely generated [25] (so we cannot have an infinite sequence of strictly increasing congruences) shows that applying the rule makes (all branches of) our tree finite.

However, this is not how decidability was originally proved by Christensen, Hirshfeld and Moller [3]. The ingeniously simple rule underlying this algorithm is given next. We suppose (by symmetry) that all pairs (α, β) in nodes are ordered lexicographically: $\alpha <_L \beta$; note that $<_L$ is a well ordering. The rule is then as follows.

> **BPP Rule**: Do not include a pair $(\alpha', \beta|\gamma)$ in a node A when some (α, β) is in $A^{\Uparrow}$; instead consider the pair $(\alpha', \alpha|\gamma)$ (or the symmetric pair, depending on $<_L$).

The rule is obviously effective; it gives rise to a finite chain of considerations (for ever smaller, in $<_L$, right-hand sides) and finishes either by including a pair which has "no bigger" right-hand side than any pair in $A^{\Uparrow}$, or by including nothing (due to reflexivity). Dickson's Lemma [6] (that every infinite sequence in $\mathbb{N}^k$ has an infinite increasing subsequence) then guarantees the finiteness of every branch (every infinite branch is transformed into a finite successful one) and hence the finiteness of the whole tree. Furthermore, the safeness of this rule is guaranteed by the congruence property: if $\alpha \sim_n \beta$, then $\alpha|\gamma \sim_n \beta|\gamma$, so $\alpha' \sim_n \alpha|\gamma$ iff $\alpha' \sim_n \beta|\gamma$.

Remark Basing on Dickson's Lemma only, we cannot provide a primitive recursive bound for the length of branches; nevertheless it is well possible that a detailed analysis of the special case for BPP would reveal an elementary bound (bounded number of exponentiations).

Remark Note that this idea (of lexicographic ordering together with Dickson's Lemma) has been used by Hirshfeld [7] to give a short self-contained proof that *every* congruence on a FGCS is finitely generated.

In fact, knowing that $\sim$ is a finitely generated congruence gives rise to another *semidecision* procedure for $\alpha \sim \beta$: "guess" (i.e., systematically search for) a finite set B of pairs including (α, β), and an expansion A of B together with the proof that A is included in the least congruence containing B. This semidecision procedure can be combined with the semidecision procedure for the negative question described earlier to yield a decision algorithm.

Unlike the case for BPP, for BPA algebras (noncommutative semigroups) we have non-finitely generated congruences (and the word problem is undecidable; see references in [19]). For example, in the free semigroup generated by $\{a, b\}$, the infinite sequence $(bab, baab), (baab, baaab), \ldots, (ba^ib, ba^{i+1}b), \ldots$ has the property that every element does not belong to the least congruence generated by the predecessing elements. Nevertheless, Christensen, Hüttel and Stirling [4] show that bisimulation equivalence on any BPA graph *is* a finitely generated congruence.

Given this fact, we can simply use the idea of combining two semidecision procedures as above. Nevertheless, we give a collection of safe and effective rules for modifying the expansion tree which illustrate the use of certain decompositions, that is, replacing "large" pairs by "smaller" ones. The rules will be instances of the next abstract rule.

Replacing Rule: To a node A, we can add a new sibling node where we replace a pair (E, F) by a (finite) set S of pairs when there is $B \subseteq A^{\Uparrow}$ such that for all $n \in \mathbb{N}$, if $B \subseteq \sim_n$ and $S \subseteq \sim_n$ then $E \sim_n F$.

This rule is readily seen to be safe. Furthermore, if each of two siblings arises from the other through the application of this rule, then we can omit one of them; in this sense, the **BPP Rule** given above is a special instance of this rule.

Before giving our rules, we make the following technical definitions. We say that a term (sequence of variables) α is ***normed*** iff there is a path in the graph from α to ε; we denote by norm(α) the length of a shortest such path. If α is not normed, then we say that it is ***unnormed***, and we note that $\alpha \sim \alpha\beta$ whenever α is unnormed. Hence we can assume that every term is of the form α or αX where α is normed and X is unnormed.

BPA Rules:

1. If $(X\alpha, Y\beta)$ is in A and some $(X\alpha', Y\beta')$ is in $A^{\Uparrow}$, then we create a sibling node for A containing (α,α') and (β,β') instead of $(X\alpha, Y\beta)$.
2. If $(X\alpha, Y\beta)$ is in A where X and Y are normed, then we create sibling nodes containing ("decomposition pairs") $(X, Y\gamma)$, $(\gamma\alpha, \beta)$ instead, where norm(X) = norm$(Y\gamma)$ (there are only finitely many of these).

The symmetric rules, and the use of the **Equivalence Rule**, are supposed implicitly; the safeness is readily verified. Since the rules apply recursively to all newly-created sibling nodes, the finite-branching of our tree can be cast in doubt. But we can demonstrate that the (modified) tree *is* finitely branching by using the following measure of ***size***. We take size$(\alpha,\beta) = \max\{\text{xnorm}(\alpha), \text{xnorm}(\beta)\}$ where for normed γ and unnormed X, xnorm(γ) = xnorm(γX) = norm(γ). There are only finitely many pairs with size $\leq n$ (for any n), and we observe that the size of any pair in any node created due to A is bounded by the maximum size in $A \cup A^{\Uparrow}$. In this sense we replace "large" pairs by "smaller" ones.

The finite-witness property is then guaranteed as follows. Suppose there is an infinite branch but no finite successful branch. Then we could extract an infinite sequence of pairs $(X\alpha_1, Y\beta_1), (X\alpha_2, Y\beta_2), \ldots$ with $\alpha_i \not\sim \alpha_j$ or $\beta_i \not\sim \beta_j$ for each $i \neq j$ which are non-decomposable in the above sense, but for which $X\alpha_i \sim Y\beta_i$ for each i; this leads to a contradiction [4] (due to image-finiteness).

Remark Again, this idea provides no reasonable complexity bound. However, the problem is studied in more detail by Burkart, Caucal and Steffen [2], where the algorithm is modified to operate within an elementary bound.

Remark We have exploited certain decomposition properties for BPA. In the normed case, both for BPP and BPA, a unique decomposition (into "prime" processes as defined by Milner and Moller [20]) is guaranteed, which Hirshfeld, Jerrum and Moller [10, 11] exploited to give (nontrivial) proofs that polynomial-time algorithms exist for these cases.

Finally, for the case of PA, the answer is known only in the normed case. Hirshfeld and Jerrum [9] use, in principle, the idea of the modified expansion tree, having found sufficient rules for replacing "large" pairs by "smaller" ones which induce them. They furthermore establish a bound for the "largeness" of the resulting pairs, which results in an upper bound on the complexity. The proof makes substantial use of the decomposition properties of BPA and BPP; but the main challenge is to handle "mixed pairs" $(\alpha\cdot\alpha', \beta|\beta')$; it turns out that such (bisimilar) pairs have a surprisingly rich structure. Furthermore, it is not clear if $\sim$ is a finitely generated congruence in the case of (normed) PA. The nodes of the tree in [9] are no longer just finite sets of pairs, but they can also contain "schemes" representing infinite sets.

Remark The authors of [9] do not expect their technique to be directly generalized for the whole class of (unnormed) PA graphs. They illustrate our still insufficient insight by noting that a polynomial algorithm as well as undecidability are still possible in this case.

2.3 State-extended Process Algebras

Richer classes of graphs have been studied, arising from process algebras by adding a finite-state control. Formally, the productions are of the form $pX \stackrel{a}{\to} q\alpha$ where p and q denote states of the finite control. The extension to terms is as follows: if $p\alpha \stackrel{a}{\to} q\beta$ then $p(\alpha{\cdot}\gamma) \stackrel{a}{\to} q(\beta{\cdot}\gamma)$ and $p(\alpha|\gamma) \stackrel{a}{\to} q(\beta|\gamma)$.

State-extended BPA graphs are also called ***PDA graphs*** (from ***push-down automata***). Though language equivalent to BPA, they constitute a richer family of graphs, and to solve the decidability question for bisimilarity needs more insight. Stirling [27] proves the decidability in the *normed* case; a PDA vertex $p\alpha$ is deemed to be normed iff for any $q\beta$ which is reachable from $p\alpha$ there is some $q'\varepsilon$ reachable from $q\beta$.

Stirling also invents a way how to replace "large" pairs $(p\alpha, q\beta)$ by "smaller" ones. To this aim he introduces special stack symbols (constants) which enable to finitely describe (infinite) "regular behaviours" of the (suffix of the) stack and enable a "congruence property" (wrt "stacking"). He shows that it is sufficient to have finitely many such constants but gives no bound on their number, so no bound for the complexity is provided.

Remark Very close to this topic is the long-standing open problem of the decidability of language equivalence for deterministic pushdown automata, which has only recently been solved (positively) by Sénizergues [26]. The full version of his proof is more than 70 pages, and still longer is his announced proof for decidability of bisimilarity for the whole class of PDA graphs. Based on insight got from Sénizergues' proof and on the techniques which we have sketched above, Stirling provides (in May 1999) a much shorter proof (approx. 20 pages) of DPDA-equivalence, and conjectures that his "structural" technique will yield a shorter proof of decidability of bisimilarity as well.

Here we describe in more detail the case of graphs generated by one-counter machines. It is a subcase of PDA, where all considered graph vertices are of the form $pXX \cdots XZ$, and the productions are of the appropriately restricted form; here the number of symbols X is interpreted as the value of the counter while Z is a special "bottom-of-stack" symbol allowing to test for zero. We use the more natural notation $p(m)$ to denote the vertex pX^mZ.

A decidability proof is given by Jančar [13]; however, in view of the above Remark, it is (probably) subsumed by more general results. Nevertheless the technique in [13] is different from what we mentioned so far, and fairly simple to describe. Moreover, it is more amenable for yielding a complexity bound in this (sub)case. Performing again a systematic search, the algorithm constructs (in the case when the given pair *is* bisimilar) a finite representation of a (generally infinite) bisimulation relation (not using the "large-replace-by-small" approach).

Remark This approach could be again illustrated on BPP. Here $\sim$ is a congruence and every congruence on a finitely generated commutative semigroup is semilinear [5] (see [7] for a short proof). The property of a semilinear relation being a bisimulation is easily seen to be expressible in Presburger arithmetic.

Hence due to its decidability (see, e.g., [23]), the semidecidability of bisimilarity for BPP can be shown as follows: systematically generate descriptions of all semilinear relations and check for each of them if it is a bisimulation containing the given pair.

Considering (pairs of) one-counter processes (i.e., vertices in a respective graph), a conceptual advantage is that we have a natural presentation in terms of "colouring" the set $\mathbb{N} \times \mathbb{N}$ (viewed as the integral grid in the first quadrant of the two-dimensional plane). By a ***colouring*** with colours from a finite set C we mean a function $c : \mathbb{N} \times \mathbb{N} \to C$. Given a one-counter machine M (that is, the respective collection of PDA productions) with control state set Q, by a ***colouring c related to M*** we mean any colouring arising as the product $c = \prod_{p,q \in Q} c_{(p,q)}$ where $c_{(p,q)} : \mathbb{N}\times\mathbb{N} \to \{black, white\}$. Such a colouring c ***represents*** the relation $\mathcal{R}_c$ on the set of vertices of the respective graph as follows: $p(m)\ \mathcal{R}_c\ q(n)$ iff $c_{(p,q)}(m,n){=}black$.

We distinguish the colouring c^M representing bisimulation equivalence:

$$c^M = \prod_{p,q \in Q} c^M_{(p,q)}$$

where $c^M_{(p,q)}(i,j) = black$ iff $p(i) \sim q(j)$ and $c^M_{(p,q)}(i,j) = white$ iff $p(i) \not\sim q(j)$. The key fact here is the following.

Fact 10 *For a one-counter machine M, the colouring c^M is regular.*

Here by a ***regular colouring*** we mean a colouring arising from a periodic "background" colouring by changing the colours in finitely many belts (whose slopes are nonnegative rational or ∞) so that each belt colouring is periodic as well, and by final recolouring of an initial square. The following pictures illustrate these notions.

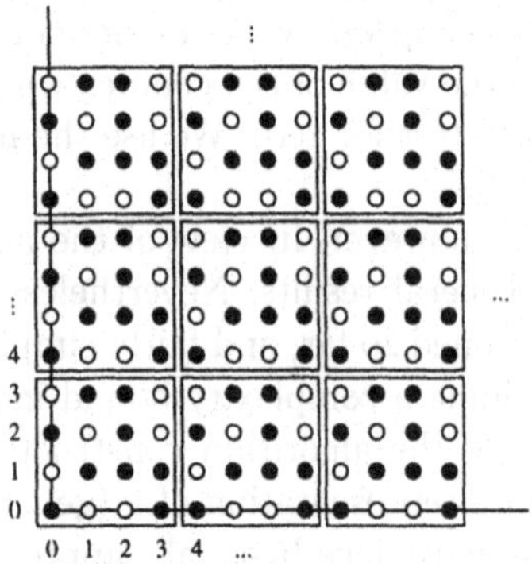

A periodic colouring.

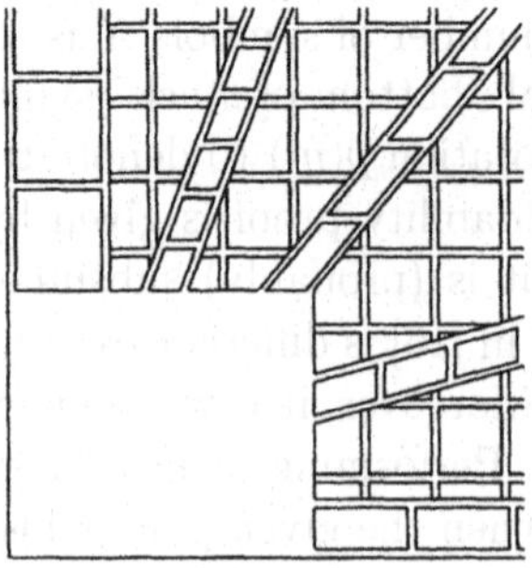
Scheme of a regular colouring.

Therefore, knowing Fact 10, we can generate all (descriptions of) regular colourings c (related to M) and check for each of them whether $\mathcal{R}_c$ is a bisimulation relation containing the given pair (this is straightforward since it is sufficient to check just finitely many pairs, due to the periodicity). This gives a semidecision procedure for bisimilarity. (It can be presented as a decision procedure as in [16] where a similar technique for simulation is used.)

The idea behind the technical details showing Fact 10 can be sketched as follows. If $p(m)$ and $q(n)$ are bisimilar then their distances (i.e., lengths of shortest sequences) to certain vertices with 0 as the counter value must be the same. Such distances are, in principle, linear in the counter value, and the constraint of the mentioned equality yields the linear belts. Periodicity (inside the belts and on the background) then follows from some simple observations.

Remark As mentioned above, a complexity bound can be achieved by a more detailed analysis; nevertheless, it was not done in [13].

3 Undecidability

We certainly cannot expect bisimulation equivalence to be decidable for a class of graphs representing systems which are able to faithfully model universal devices, in particular Minsky machines [21], which are simple straight-line programs which operate on nonnegative counters. It is sufficient for us to assume just two counters, initialized to 0, since the halting problem is undecidable even in this simple case. Formally, a ***Minsky machine*** M is a sequence of labelled instructions

$$\mathtt{X}_0 : \mathrm{comm}_0\,; \quad \mathtt{X}_1 : \mathrm{comm}_1\,; \quad \cdots \quad \mathtt{X}_{n-1} : \mathrm{comm}_{n-1}\,; \quad \mathtt{X}_n : \mathrm{comm}_n$$

where each of the first n instructions is either of the form

$$\mathtt{X}_\ell : \mathtt{c}_0 := \mathtt{c}_0{+}1;\ \mathtt{goto}\ \mathtt{X}_j \qquad \text{or} \qquad \mathtt{X}_\ell : \mathtt{c}_1 := \mathtt{c}_1{+}1;\ \mathtt{goto}\ \mathtt{X}_j$$

or of the form

$$\begin{array}{lcl} \mathtt{X}_\ell : \mathtt{if}\ \mathtt{c}_0 = 0\ \mathtt{then}\ \mathtt{goto}\ \mathtt{X}_j & \text{or} & \mathtt{X}_\ell : \mathtt{if}\ \mathtt{c}_1 = 0\ \mathtt{then}\ \mathtt{goto}\ \mathtt{X}_j \\ \quad\ \mathtt{else}\ \mathtt{c}_0 := \mathtt{c}_0{-}1;\ \mathtt{goto}\ \mathtt{X}_k & & \quad\ \mathtt{else}\ \mathtt{c}_1 := \mathtt{c}_1{-}1;\ \mathtt{goto}\ \mathtt{X}_k \end{array}$$

The machine M starts executing with the value 0 in the counters $\mathtt{c}_0$ and $\mathtt{c}_1$ and the control at label $\mathtt{X}_0$. When the control is at label $\mathtt{X}_\ell$ ($0 \le \ell < n$), the machine executes instruction comm_ℓ, appropriately modifying the contents of the respective counter and transferring the control to the appropriate label as directed by the instruction. The machine halts if and when the control reaches the `halt` instruction at label $\mathtt{X}_n$.

State-extended PA can easily model Minsky machines, as we can use the parallel composition of two sequential terms $(XX \ldots XZ)|(YY \ldots YZ)$ to model the two counters in the obvious way. So a straightforward reduction from the halting problem shows undecidability of bisimilarity for state-extended PA. Recall that there is no such reduction for (normed) state-extended BPA, that is, for (normed) PDA, due to the decidability in that case, which we discussed in Section 2.3. However, it turns out that ***state-extended BPP***—which we further denote ***MSA*** (***multiset automata***)—allow for such a reduction, even in the normed case, though they cannot faithfully model Minsky machines. They represent a subclass of labelled place/transition Petri nets and thus they lack the ability to test-for-zero. In the rest of this section, we outline the reduction, thus demonstrating undecidability of bisimilarity for MSA.

The idea is to construct an MSA which "weakly" models a given Minsky machine, and then take two slightly modified copies. The modifications guarantee that whenever **Alice** "cheats" (i.e., makes a "zero" move while the counter of the modelled Minsky machine is positive), **Bob** can punish her by reaching a pair of equal (and hence trivially bisimilar) vertices.

To the Minsky machine M as presented above, we define the (description of the) MSA graph as follows.

- The set of labels is $\Sigma = \{i, d, z, h\}$.
- The set of control states is $Q = \{p_0, p_1, \ldots, p_{n-1}, p_n, q_0, q_1, \ldots, q_{n-1}, q_n\}$.
- The set of variables is $V = \{Z, 0, 1\}$.
- For each machine instruction

$$\mathtt{X}_\ell : \mathtt{c}_b := \mathtt{c}_b{+}1;\ \mathtt{goto}\ \mathtt{X}_j$$

  we have the productions

$$p_\ell Z \xrightarrow{i} p_j(b|Z) \qquad \text{and} \qquad q_\ell Z \xrightarrow{i} q_j(b|Z).$$

- For each machine instruction

$$\begin{aligned}\mathtt{X}_\ell : \mathtt{if}\ \mathtt{c}_b = 0\ \mathtt{then\ goto}\ \mathtt{X}_j \\ \mathtt{else}\ \mathtt{c}_b := \mathtt{c}_b{-}1;\ \mathtt{goto}\ \mathtt{X}_k\end{aligned}$$

  we have the productions

$$\begin{array}{lll} p_\ell b \xrightarrow{d} p_k & p_\ell Z \xrightarrow{z} p_j Z & p_\ell b \xrightarrow{z} q_j b \\ q_\ell b \xrightarrow{d} q_k & q_\ell Z \xrightarrow{z} q_j Z & q_\ell b \xrightarrow{z} p_j b \end{array}$$

- We have the one final production

$$p_n Z \xrightarrow{h} p_n$$

The MSA graph reflects the (computation of the) machine M in the following sense.

- When M is at the command labelled X_ℓ with the values x and y in its counters, this is reflected by the "p-vertex" $p_\ell(0^x|1^y|Z)$ as well as the "q-vertex" $q_\ell(0^x|1^y|Z)$ (where 0^x represents $0|0|\ldots|0$ with x occurrences of 0).
- If this command is an increment, then the associated p-vertex has only one outgoing edge, labelled by i (for "increment"), leading to the p-vertex reflecting the state of M upon executing the increment command; similarly for the associated q-vertex.
- If this command is a successful test for zero (that is, the relevant counter has the value 0), then the associated p-vertex (as well as the associated q-vertex) has only one outgoing edge, labelled by z (for "zero"), leading to the p-vertex (q-vertex) reflecting the state of M upon executing the respective command.

- If this command is a decrement (that is, a failed test for zero), then the associated p-vertex (as well as the associated q-vertex) has three outgoing edges, exactly one of which is labelled d (for "decrement") which again leads to the p-vertex (q-vertex) reflecting the state of M upon executing the command.
- In this last instance, the additional two outgoing edges, both labelled by z, reflect the weakness of MSA (or, more generally, Petri nets) in their inability to test for zero (a weakness which works in their favour with respect to several important positive decidability results such as the reachability problem [18]). One of the two edges represents an "honest cheating", and arises from the rule $p_\ell Z \xrightarrow{z} p_j Z$ ($q_\ell Z \xrightarrow{z} q_j Z$), while the other edge represents "knowingly cheating", and arises from the rule $p_\ell b \xrightarrow{z} q_j b$ ($q_\ell b \xrightarrow{z} p_j b$); in this final case, the edge leads to the domain of the other copy weakly modelling M.

Fact 11 $p_0 Z \sim q_0 Z$ *iff the Minsky machine M does not halt.*

To see this, we note that if M halts, then a winning strategy for **Alice** in the bisimulation game would be to mimic the behaviour of M in either of the two copies (in p-vertices or q-vertices). **Bob**'s only option in response would be to do the same in the other copy. Upon termination, the pair of vertices reached in the game will be $p_n(0^x|1^y|Z)$ and $q_n(0^x|1^y|Z)$ for some values x and y. **Alice** may then choose the edge $p_n(0^x|1^y|Z) \xrightarrow{h} p_n(0^x|1^y)$ which cannot be matched by **Bob**. Hence $p_0 Z$ and $q_0 Z$ are not bisimilar.

On the other hand, if M fails to halt, then a winning strategy for **Bob** would be to mimic **Alice**'s moves for as long as **Alice** mimics M, and to cheat knowingly or honestly, respectively, in the instance that **Alice** cheats honestly or knowingly, respectively, so as to arrive at the situation where the two vertices are identical; from here **Bob** can copy every move of **Alice** verbatim. Hence $p_0 Z$ and $q_0 Z$ are bisimilar.

The above MSA graph can be made normed (in the same sense as for PDA) by adding a new label e along with the following productions (one for each $\ell = 0, 1, \ldots, n$ and each $X = Z, 0, 1$).

$$p_\ell X \xrightarrow{e} p_\ell \qquad p_\ell X \xrightarrow{e} q_\ell \qquad q_\ell X \xrightarrow{e} q_\ell \qquad q_\ell X \xrightarrow{e} p_\ell$$

This guarantees that every vertex has an outgoing path to some vertex $r\varepsilon$, and continues to allow **Bob** to produce a pair of identical vertices if **Alice** chooses one of these non-M-mimicking edges. The same argument can then be made to show that $p_0 Z$ and $q_0 Z$ are bisimilar exactly when the Minsky machine M does not halt. We thus arrive at our result.

Fact 12 *Bisimilarity is undecidable over the class of normed MSA graphs.*

4 Additional Remarks

The undecidability proof in the previous section stems from the proof for labelled Petri nets given by Jančar [12] which was adapted for MSA by Moller [22] (there referred to as PPDA, *parallel pushdown automata*). We thus have undecidability of bisimulation equivalence over a very restricted class of Petri nets: those with only two unbounded places and a minimal degree of nondeterminism. Decidability results for various subcases have been obtained in [12, 15].

The study of the decidability of bisimilarity over various classes of graphs became a popular topic in concurrency theory with the appearance of the first, then-surprising, proof of the decidability result for normed BPA by Baeten, Bergstra, Klop [1]. Since then there has been a great deal of research on the topic, and we have only touched on a fragment of it in this tutorial paper. In particular, we have not discussed *weak* bisimilarity here, which seems much less amenable for obtaining decidability results. (Some partial results are obtained in [14]). For further references in the area, we refer to the survey paper [22].

References

1. J.C.M. Baeten, J.A. Bergstra and J.W. Klop (1993). Decidability of bisimulation equivalence for processes generating context-free languages. *Journal of the ACM* **40**:653–682. (Preliminary version in the proceedings of PARLE'87, *Lecture Notes in Computer Science* **259**:94–113, 1987.)
2. O. Burkart, D. Caucal and B. Steffen (1995). An elementary decision procedure for arbitrary context-free processes. Proceedings of MFCS'95. *Lecture Notes in Computer Science* **969**:423–433.
3. S. Christensen, Y. Hirshfeld and F. Moller (1993). Bisimulation equivalence is decidable for basic parallel processes. Proceedings of CONCUR'93, *Lecture Notes in Computer Science* **715**:143–157.
4. S. Christensen, H. Hüttel and C. Stirling (1995). Bisimulation equivalence is decidable for all context-free processes. *Information and Computation* **121**(2):143–148. (Preliminary version in the proceedings of CONCUR'92, *Lecture Notes in Computer Science* **630**:138–147, 1992.)
5. S. Eilenberg and M.P. Schützenberger (1969). Rational sets in commutative monoids. *Journal of Algebra* **13**:173–191.
6. L.E. Dickson (1913). Finiteness of the odd perfect and primitive abundant numbers with distinct factors. *American Journal of Mathematics* **35**:413–422.
7. Y. Hirshfeld (1994). Congruences in commutative semigroups. Research report ECS-LFCS-94-291, Department of Computer Science, University of Edinburgh.
8. Y. Hirshfeld (1997). Bisimulation trees and the decidability of weak bisimulations. *Electronic Notes in Theoretical Computer Science* **5**, http://www.elsevier.nl/locate/entcs/volume5.html.
9. Y. Hirshfeld and M. Jerrum (1999). Bisimulation equivalence is decidable for normed Process Algebra. Proceedings of ICALP'99, to appear, *Lecture Notes in Computer Science* **1644**.
10. Y. Hirshfeld, M. Jerrum and F. Moller (1996). A polynomial algorithm for deciding bisimilarity of normed context-free processes. *Theoretical Computer Science* **158**:143–159. (Preliminary version in the proceedings of FOCS'94:623–631, 1994.)
11. Y. Hirshfeld, M. Jerrum and F. Moller (1996). A polynomial algorithm for deciding bisimulation equivalence of normed basic parallel processes. *Mathematical Structures in Computer Science* **6**:251–259.

12. P. Jančar (1995). Undecidability of bisimilarity for Petri nets and related problems. *Theoretical Computer Science* **148**:281–301. (Preliminary version in the proceedings of STACS'94, *Lecture Notes in Computer Science* **775**:581–592, 1994.)
13. P. Jančar (1997). Bisimulation equivalence is decidable for one-counter processes. Proceedings of ICALP'97, *Lecture Notes in Computer Science* **1256**:549-559. (To appear in *Information and Computation.*)
14. P. Jančar, A. Kučera and R. Mayr (1998). Deciding bisimulation-like equivalences with finite-state processes. Proceedings of ICALP'98, *Lecture Notes in Computer Science* **1443**:200–211. (To appear in *Theoretical Computer Science.*)
15. P. Jančar, J. Esparza and F. Moller (1999). Petri nets and regular processes. *Journal of Computer and System Sciences* (to appear). (Research report 162, Computing Science Department, Uppsala University.
16. P. Jančar and F. Moller (1999). Simulation of one-counter nets via colouring. Research report 159, Computing Science Department, Uppsala University.
17. P.C. Kanellakis and S.A. Smolka. CCS expressions, finite-state processes and three problems of equivalence. *Information and Computation* **86**:43–68, 1990.
18. E. Mayr (1984). An algorithm for the general Petri net reachability problem. *SIAM Journal of Computing* **13**:441–460.
19. E.W. Mayr and A.R. Meyer (1982). The complexity of the word problems for commutative semigroups and polynomial ideals. *Advances in Mathematics* **46**:305–329.
20. R. Milner and F. Moller (1993). Unique decomposition of processes. *Theoretical Computer Science* **107**:357–363. (Preliminary version in the *Bulletin of the EATCS* **41**:226–232, 1990.)
21. M. Minsky (1967). **Computation: Finite and Infinite Machines**. Prentice-Hall.
22. F. Moller (1996). Infinite results. Proceedings of CONCUR'96, *Lecture Notes in Computer Science* **1119**:195–216.
23. D.C. Oppen (1978). A $2^{2^{2^{p(n)}}}$ upper bound on the complexity of Presburger arithmetic. *Journal of Computer and System Science* **16**:323–332.
24. R. Paige and R.E. Tarjan. Three partition refinement algorithms. *SIAM Journal on Computing* **16**:937–989, 1987.
25. L. Redei (1965). **The theory of finitely generated commutative semigroups.** Oxford University Press.
26. G. Sénizergues (1997). The Equivalence Problem for Deterministic Pushdown Automata is Decidable. Proceedings of ICALP'97, *Lecture Notes in Computer Science* **1256**:671–681.
27. C. Stirling (1998). Decidability of bisimulation equivalence for normed pushdown processes. *Theoretical Computer Science* **195**:113–131. (Preliminary version in the proceedings of CONCUR'96, *Lecture Notes in Computer Science* **1119**:217–232, 1996.)

Testing Concurrent Systems : A Formal Approach

Jan Tretmans

University of Twente *
Faculty of Computer Science, Formal Methods and Tools research group
P.O. Box 217, 7500 AE Enschede, The Netherlands
email: tretmans@cs.utwente.nl

Abstract. This paper discusses the use of formal methods in testing of concurrent systems. It is argued that formal methods and testing can be mutually profitable and useful. A framework for testing based on formal specifications is presented. This framework is elaborated for labelled transition systems, providing formal definitions of conformance, test execution and test derivation. A test derivation algorithm is given and its tool implementation is briefly discussed.

1 Introduction

During the last decades much theoretical research in computing science has been devoted to formal methods. This research has resulted in many formal languages and in verification techniques, supported by prototype tools, to verify properties of high-level, formal system descriptions. Although these methods are based on sound mathematical theories, there are not many systems developed nowadays for which correctness is completely formally verified using these methods.

On the other hand, the current practice of checking correctness of computing systems is based on a more informal and pragmatic approach. Testing is usually the predominant technique, where an implementation is subjected to a number of tests which have been obtained in an ad-hoc or heuristic manner. A formal, underlying theory for testing is mostly lacking.

The combination of testing and formal methods is not very often made. Sometimes it is claimed that formally verifying computer programs would make testing superfluous, and that, from a formal point of view, testing is inferior as a way of assessing correctness. Also, some people cannot imagine how the practical, operational, and 'dirty-hands' approach of testing could be combined with the mathematical and 'clean' way of verification using formal methods. Moreover, the classical biases against the use of formal verification methods, such as that formal methods are not practical, that they are not applicable to any real system

* This research is supported by the Dutch Technology Foundation STW under project STW TIF.4111: *Côte de Resyste* – COnformance TEsting of REactive SYSTEms; URL: http://fmt.cs.utwente.nl/CdR.

but very simple toy systems, and that they require a profound mathematical training, do not help in making test engineers adopt formal methods.

Fortunately, views are changing. Academic research on testing is increasing, and even the most formal verifyer admits that a formally verified system should still be tested. (Because: Who verified the compiler? And the operating system? And who verified the verifyer?). On the other hand, formal methods are used in more and more software projects, in particular for safety critical systems, and also the view that a formal specification can be beneficial during testing is getting more support.

The aim of this paper is to strengthen this process of changing views. To that purpose, this paper discusses how testing can be performed based on formal specifications, and how advantage can be obtained in terms of precision, clarity and consistency of the testing process by adopting this formal approach. Also, it will be shown how the use of formal methods helps automating the testing process, in particular the automated derivation of tests from formal specifications. The discussion about testing and formal methods will support the following claims: (*i*) formal methods and testing are a perfect couple; (*ii*) testing and formal verification are both necessary; (*iii*) a formally verified specification is a good starting point for testing; (*iv*) formal testing is a good starting point for introducing formal methods in software development.

The structure of this paper is as follows. In the next section we start with some informal discussion on classical software testing, see, e.g., [38, 3]. Section 3 then discusses a formal, generic framework for testing with formal methods. Section 4 makes this framework more specific by instantiating it for the formalism of labelled transition systems. Section 5 discusses tool support and a concrete application of testing a simple protocol based on the labelled transition system testing theory. Finally, section 6 comes back to the claims made above and discusses some open issues.

The intention of this paper is to give an idea about how testing and formal methods can be mutually beneficial. A complete overview of formal approaches to testing is outside the scope of this paper. Other approaches exist, e.g., [21] for Abstract Data Type testing, and certainly other instantiations of the generic framework of section 3 are possible, e.g., with Finite-State Machines (Mealy machines) [7, 35]. Also it is not the intention to give a complete and precise overview of testing for labelled transition systems. However, the branch of testing theory, which is elaborated in section 4, is shown to be a realistic and practically applicable approach in section 5. Moreover, many pointers to the literature are provided which allow to explore alternatives and to study further details.

2 Software Testing

What is testing? Testing is an operational way to check the correctness of a system implementation by means of experimenting with it. Tests are applied to the implementation under test in a controlled environment, and, based on observations made during the execution of the tests, a verdict about the correct functioning of the implementation is given. The correctness criterion that is to be tested is given by the system specification; the specification is the basis for testing.

Testing is an important technique to increase confidence in the quality of a computing system. In almost any software development trajectory some form of testing is included.

Sorts of testing There are many different kinds of testing. In the first place, different aspects of system behaviour can be tested: Does the system have the intended functionality and does it comply with its functional specification (functional tests or conformance tests)? Does the system work as fast as required (performance tests)? How does the system react if its environment shows unexpected or strange behaviour (robustness tests)? Can the system cope with heavy loads (stress testing)? How long can we rely on the correct functioning of the system (reliability tests)? What is the availability of the system (availability tests)?

Moreover, testing can be applied at different levels of abstraction and for different levels of (sub-)systems: individual functions, modules, combinations of modules, subsystems and complete systems can all be tested.

Another distinction can be made according to the parties or persons performing (or responsible for) testing. In this dimension there are, for example, system developer tests, factory acceptance tests, user acceptance tests, operational acceptance tests, and third party (independent) tests, e.g., for certification.

A very common distinction is the one between black box and white box testing. In black box testing, or functional testing, only the outside of the system under test is known to the tester. In white box testing, also the internal structure of the system is known and this knowledge can be used by the tester. Naturally, the distinction between black and white box testing leads to many gradations of grey box testing, e.g., when the module structure of a system is known, but not the code of each module.

In this paper, we concentrate on black box, functional testing, also called *conformance testing*. We do not care about the level of (sub-)systems or who is performing the testing. Key points are that there is a system implementation exhibiting behaviour and that there is a specification. The specification is a prescription of what the system should do; the goal of testing is to check, by means of testing, whether the implemented system indeed satisfies this prescription. In particular, the rest of this paper will consider a conformance testing process based on specifications which are given in a formal notation.

Confusion of tongues Sometimes the term testing is also used for performing static checks on the program code, e.g., checking declarations of variables using a static checker, or code inspections. This kind of testing is then denoted by *static testing*. However, we restrict to *dynamic testing*, i.e., testing consisting of really executing the implemented system, as described above. Another broader use of the term testing is to include *monitoring*. Monitoring is then called *passive testing* as opposed to *active testing* as described above, where the tester has active control over the test environment, and a set of predefined tests is executed. A third extension of the term testing, sometimes made, is to include all checking activities in the whole software development trajectory, e.g., reviews and inspections.

The testing process In the conformance testing process there are two main phases: *test generation* and *test execution*. Test generation involves analysis of the specification and determination of which functionalities will be tested, determining how these can be tested, and developing and specifying test scripts. Test execution involves the development of a test environment in which the test scripts can be executed, the actual execution of the test scripts and analysis of the execution results and the assignment of a verdict about the well-functioning of the implementation under test.

Other important activities in the testing process are test management and test maintenance. In particular, test maintenance is often underestimated. It involves recording and documenting the test scripts, test environments, used test tools, relating test sets to versions of specifications and implementations, with the aim of making the testing process repeatable and reusable, in particular for regression testing. Regression testing is the re-testing of unmodified functionality in case of a modification of the system. It is one of the most expensive (and thus often deliberately neglected) aspects of testing.

Test automation Testing is a difficult, expensive, time-consuming and labour-intensive process. Moreover, testing is (should be) repeated each time a system is modified. Hence, testing would be an ideal candidate for automation.

The main class of commercially available test tools are *record & playback* tools (capture and replay tools) which support the test execution process. Record & playback tools are able to record user actions at a (graphical) user interface, such as keyboard and mouse actions, in order to replay these actions at a later point in time. In this way a recorded test can be replayed several times, which may be advantageous during regression testing.

For the test generation phase there are tools which are able to generate large amounts of input test data. However, these tools are mainly used for performance and stress tests and hence, are outside the scope of this paper. Some tools exist that are able to generate a set of tests with the same structure based on a template of a test case by only varying the input parameters in this template. In the area of communication protocol testing there exist some (prototype) test tools that can (semi-) automatically generate test cases for conformance testing from a formal specification. Some of these tools will be briefly described in section 5.

To relate test cases to the requirements that they test, standard requirements management tools can be used, but such tools are not specific for testing. The main functionality of such tools is to relate high level system requirements to (lower level) sub-system requirements and to relate requirements to test cases.

A kind of test tools which are used during test execution, but which (should) influence test generation, are code coverage tools. Code coverage tools calculate the percentage of the system code executed during test execution according to some criterion, e.g., "all paths", "all statements", or "all definition-usage combinations" of variables. They give an indication about the completeness of a set of tests. Note that this notion of completeness refers to the implemented code (white box testing); it does not say anything about the extent to which the requirements or the specification were covered.

3 Formal Framework for Testing

In section 2 the software testing process was described from a traditional perspective. Conformance testing was introduced as a kind of testing where the behaviour of a system is systematically tested with respect to the system's specification of functional behaviour.

In this section a framework is presented for the use of formal methods in conformance testing [10, 44, 32]. The framework can be used for testing of an implementation with respect to a formal specification of its functional behaviour. It introduces, at a high level of abstraction, the concepts used in a formal conformance testing process and it defines a structure which allows to reason about testing in a formal way. The most important part of this is to link the informal world of implementations, tests and experiments with the formal world of specifications and models. To this extent the framework introduces the concepts of conformance, i.e., functional correctness, testing, sound and exhaustive test suites, and test derivation. All these concepts are introduced at a generic level; sections 4 and 5 will show how to instantiate and apply these concepts.

Conformance For talking about conformance we need implementations and specifications. The specifications are formal, so a universe of formal specifications denoted *SPECS* is assumed. Implementations are the systems that we are going to test, henceforth they will be called IUT, implementation under test, and the class of all IUT's is denoted by *IMPS*. So, conformance could be introduced by having a relation **conforms-to** $\subseteq$ *IMPS* $\times$ *SPECS* with IUT **conforms-to** s expressing that IUT is a correct implementation of specification s.

However, unlike specifications, implementations under test are real, physical objects, such as pieces of hardware or software; they are treated as black boxes exhibiting behaviour and interacting with their environment, but not amenable to formal reasoning. This makes it difficult to give a formal definition of **conforms-to** which should be our aim in a formal testing framework. In order to reason formally about implementations, we make the assumption that any real implementation IUT $\in$ *IMPS* can be modelled by a formal object $i_{\text{IUT}} \in MODS$, where *MODS* is referred to as the universe of models. This assumption is referred to as the ***test hypothesis*** [6]. Note that the test hypothesis only assumes that a model i_{IUT} exists, but not that it is known a priori.

Thus the test hypothesis allows to reason about implementations as if they were formal objects, and, consequently, to express conformance by a formal relation between models of implementations and specifications. Such a relation is called an *implementation relation* **imp** $\subseteq$ *MODS* $\times$ *SPECS* [10, 32]. Implementation IUT $\in$ *IMPS* is said to be correct with respect to $s \in$ *SPECS*,

IUT **conforms-to** s, if and only if the model $i_{\text{IUT}} \in MODS$ of IUT is **imp**-related to s: i_{IUT} **imp** s.

Observation and testing The behaviour of an implementation under test is investigated by performing experiments on the implementation and observing the reactions that the implementation produces to these experiments. The specification of such an experiment is called a *test case*, and the process of applying a test to an implementation under test is called *test execution*.

Let test cases be formally expressed as elements of a domain $TESTS$. Then test execution requires an operational procedure to execute and apply a test case $t \in TESTS$ to an implementation under test IUT $\in IMPS$. This operational procedure is denoted by $\text{EXEC}(t, \text{IUT})$. During test execution a number of observations will be made, e.g., occurring events will be logged, or the response of the implementation to a particular stimulus will be recorded. Let (the formal interpretation of) these observations be given in a domain of observations OBS, then test execution $\text{EXEC}(t, \text{IUT})$ will lead to a subset of OBS. Note that EXEC is not a formal concept; it captures the action of "pushing the button" to let t run with IUT. Also note that $\text{EXEC}(t, \text{IUT})$ may involve multiple runs of t and IUT, e.g., in case nondeterminism is involved.

Again, since $\text{EXEC}(t, \text{IUT})$ corresponds to the physical execution of a test case, we have to model this process of test execution in our formal domain to allow formal reasoning about it. This is done by introducing an observation function $obs : TESTS \times MODS \to \mathcal{P}(OBS)$. So, $obs(t, i_{\text{IUT}})$ formally models the real test execution $\text{EXEC}(t, \text{IUT})$.

In the context of an *observational framework* consisting of $TESTS$, OBS, EXEC and obs, it can now be stated more precisely what is meant by the test hypothesis:

$$\forall \text{IUT} \in IMPS\ \exists i_{\text{IUT}} \in MODS\ \forall t \in TESTS : \ \text{EXEC}(t, \text{IUT}) = obs(t, i_{\text{IUT}}) \quad (1)$$

This could be paraphrased as follows: for all real implementations that we are testing, it is assumed that there is a model, such that if we would put the IUT and the model in black boxes and would perform all possible experiments defined in $TESTS$, then we would not be able to distinguish between the real IUT and the model. Actually, this notion of testing is analogous to the ideas underlying testing equivalences [15, 14], which will be elaborated for transition systems in section 4.

Usually, we like to interpret observations of test execution in terms of being right or wrong. So we introduce a family of *verdict functions* $\nu_t : \mathcal{P}(OBS) \to \{\textbf{fail}, \textbf{pass}\}$ which allows to introduce the following abbreviation:

$$\text{IUT } \textbf{passes } t \quad \Longleftrightarrow_{\text{def}} \quad \nu_t(\text{EXEC}(t, \text{IUT})) = \textbf{pass} \quad (2)$$

This is easily extended to a *test suite* $T \subseteq TESTS$: IUT **passes** $T \Leftrightarrow \forall t \in T$: IUT **passes** t. Moreover, an implementation fails test suite T if it does not pass: IUT **fails** $T \Leftrightarrow$ IUT $\not\textbf{passes}$ T.

Conformance testing Conformance testing involves assessing, by means of testing, whether an implementation conforms, with respect to implementation relation **imp**, to its specification. Hence, the notions of conformance, expressed by **imp**, and of test execution, expressed by EXEC, have to be linked in such a way that from test execution an indication about conformance is obtained. So, ideally, we would like to have a test suite T_s such that for a given specification s

$$\text{IUT } \mathbf{conforms\text{-}to} \ s \iff \text{IUT } \mathbf{passes} \ T_s \tag{3}$$

A test suite with this property is called *complete*; it can distinguish exactly between all conforming and non-conforming implementations. Unfortunately, this is a very strong requirement for practical testing: complete test suites are usually infinite, and consequently not practically executable. Hence, usually a weaker requirement on test suites is posed: they should be *sound*, which means that all correct implementations, and possibly some incorrect implementations, will pass them; or, in other words, any detected erroneous implementation is indeed non-conforming, but not the other way around. Soundness corresponds to the left-to-right implication in (3). The right-to-left implication is called *exhaustiveness*; it means that all non-conforming implementations will be detected.

To show soundness (or exhaustiveness) for a particular test suite we have to use the formal models of implementations and test execution:

$$\forall i \in MODS: \quad i \ \mathbf{imp} \ s \iff \forall t \in T: \ \nu_t(obs(t,i)) = \mathbf{pass} \tag{4}$$

Once (4) has been shown it follows that

$$\begin{array}{ll} & \text{IUT } \mathbf{passes} \ T \\ \text{iff} & (* \text{ definition } \mathbf{passes} \ T \ *) \\ & \forall t \in T: \ \text{IUT } \mathbf{passes} \ t \\ \text{iff} & (* \text{ definition } \mathbf{passes} \ t \ *) \\ & \forall t \in T: \ \nu_t(\text{EXEC}(t, \text{IUT})) = \mathbf{pass} \\ \text{iff} & (* \text{ test hypothesis (1) } *) \\ & \forall t \in T: \ \nu_t(obs(t, i_{\text{IUT}})) = \mathbf{pass} \\ \text{iff} & (* \text{ completeness on models (4) applied to } i_{\text{IUT}} \ *) \\ & i_{\text{IUT}} \ \mathbf{imp} \ s \\ \text{iff} & (* \text{ definition of conformance } *) \\ & \text{IUT } \mathbf{conforms\text{-}to} \ s \end{array}$$

So, if the completeness property has been proved on the level of models and if there is ground to assume that the test hypothesis holds, then conformance of an implementation with respect to its specification can be decided by means of a testing procedure.

Now, of course, an important activity is to devise algorithms which produce sound and/or complete test suites from a specification given an implementation relation. This activity is known as *test derivation*. It can be seen as a function $der_{\mathbf{imp}} : SPECS \to \mathcal{P}(TESTS)$. Following the requirement on soundness of test suites, such a function should only produce sound test suites for any specification $s \in SPECS$, so the test suite $der_{\mathbf{imp}}(s)$ should satisfy the left-to-right implication of (4).

Extensions Some extensions to and refinements of the formal testing framework can be made. Two of them are mentioned here. The first one concerns the *test architecture* [44, 32]. A test architecture defines the environment in which an implementation is tested. It gives an abstract view of how the tester communicates with the IUT. Usually, an IUT is embedded in a test context, which is there when the IUT is tested, but which is not the object of testing. In order to formally reason about testing in context, the test context must be formally modelled. Sometimes, the term SUT – system under test – is then used to denote the implementation with its test context, whereas IUT is used to denote the bare implementation without its context.

The second extension is the introduction of *coverage* within the formal framework. The coverage of a test suite can be introduced by assigning to each erroneous implementation that is detected by a test suite a value and subsequently integrating all values. This can be combined with a stochastic view on erroneous implementations and a probabilistic view on test execution [9, 26].

4 Labelled Transition Systems

One of the formalisms studied in the realm of conformance testing is that of *labelled transition systems.* A labelled transition system is a structure consisting of states with transitions, labelled with actions, between them. The formalism of labelled transition systems can be used for modelling the behaviour of processes, such as specifications, implementations and tests, and it serves as a semantical model for various formal languages, e.g., ACP [5], CCS [37], and CSP [28]. Also (most parts of) the semantics of standardized languages like LOTOS [30] and SDL [12], and of the modelling language PROMELA [29] can be expressed in labelled transition systems. We assume the basic definitions of labelled transition systems to be familiar; they can be found in many of the given references, e.g., in [45] the definitions are given in the same notation as they are used here (however, we will not consider internal actions τ in this section).

This section instantiates the generic, formal testing framework of section 3 with labelled transition systems. This means that the formal domains *SPECS*, *MODS* and *TESTS* will now consist of (some kind of) transition systems. In particular, it will be shown how the **ioco**-testing theory based on inputs, outputs and repetitive quiescence fits within the testing framework [45].

Traditionally, for labelled transition systems the term testing theory does not refer to conformance testing. Instead of starting with a specification to find a test suite that characterizes the class of its conforming implementations, these testing theories aim at defining implementation relations, given a class of tests: a transition system p is equivalent to a system q if any test case in the class leads to the same observations with p as with q (or more generally, p relates to q if for all possible tests, the observations made of p are related in some sense to the observations made of q). In terms of an observational framework as introduced in section 3, an implementation relation **imp** is defined by

$$p \ \mathbf{imp} \ q \quad \Longleftrightarrow_{\text{def}} \quad \forall t \in \mathit{TESTS} : \ \mathit{obs}(t,p) \otimes \mathit{obs}(t,q) \qquad (5)$$

Many different relations can be defined by variations of the class of tests $TESTS$, the way observations obs are obtained, and the required relation between observations $\otimes$ [15, 1, 14, 40, 23, 24].

Once an implementation relation has been defined, conformance testing involves finding a test derivation algorithm such that test suites can be derived from a specification which are sound, and, in some sense, minimal. Conformance testing for labelled transition systems has been studied especially in the context of testing communication protocols with the language LOTOS, e.g., [11, 8, 41, 49, 34, 47, 45, 27].

For the discussion of the **ioco**-testing theory both kinds of testing theory are used: firstly, the implementation relation **ioco** is defined following the principle of (5); secondly, test derivation from specifications for **ioco** is investigated resulting in a sound and exhaustive test derivation algorithm.

In the remainder of this section we will successively instantiate all the ingredients of the formal testing framework of section 3 for **ioco**-based testing. These include $SPECS$, $IMPS$, $MODS$, **imp**, $TESTS$, OBS, ν_t, EXEC, obs and $der_{\mathbf{imp}}$. The description of the different concepts will be done semi-formally; full technical details can be found in [45]. The next section, section 5, will briefly discuss the use of this **ioco**-testing theory for building of software tools and for testing some simple communication protocol implementations based on LOTOS and PROMELA specifications.

Specifications For specifications we allow to use labelled transition systems, or any formal language with a labelled transition system semantics. We require that the actions of the transition system are known and can be partitioned into inputs and outputs, denoted by L_I and L_U, respectively. However, we do not impose any restrictions on inputs or outputs. For $\mathcal{LTS}(L)$ the class of labelled transition systems over action alphabet L, $SPECS := \mathcal{LTS}(L_I \cup L_U)$.

Implementations and their models We assume implementations to be modelled by a special class of transition systems called *input-output transition systems*, which, inspired by Input/Output Automata (IOA) [36], have the property that any input action is always enabled in any state. For $\mathcal{IOTS}(L_I, L_U)$ the class of input-output transition systems with inputs in L_I and outputs in L_U, $MODS := \mathcal{IOTS}(L_I, L_U)$.

For $IMPS$ we allow any computer system or program which can be modelled as an input-output transition system, i.e., a system which has distinct inputs and outputs, where inputs can be mapped 1:1 on L_I and outputs on L_U, and where inputs can always occur.

Implementation relation The implementation relation is instantiated with the relation $\mathbf{ioco} \subseteq \mathcal{IOTS}(L_I, L_U) \times \mathcal{LTS}(L_I \cup L_U)$, which is briefly discussed here.

The relation **ioco** inherits many ideas from other relations defined in the literature. Its roots are in the theory of testing equivalence and preorders [15, 14], where testing preorder on transitions systems is defined following (5) using transition systems as tests, traces and completed traces of the synchronized parallel composition of t and p as observations, and inclusion of observations as comparison criterion. Three developments, which build on these testing preorders, are of importance for **ioco**.

Firstly, a relation with more discriminating power than testing preorder was defined in [40] by having more powerful testers which can detect not only the occurrence of actions but also the absence of actions, i.e., refusals. We follow [33] in modelling the observation of a refusal by adding a special label $\theta \notin L$ to observers: $\mathit{TESTS} = \mathcal{LTS}(L \cup \{\theta\})$. While observing a process, a transition labelled with θ can only occur if no other transition is possible. In this way the observer knows that the process under observation cannot perform the other actions it offers. This is modelled using a parallel operator $\rceil\!|$ which is the usual synchronized parallel composition operator extended with the following inference rule to cope with the refusal-detecting features of θ:

$$u \xrightarrow{\theta} u', \quad \forall a \in L: \; u \xrightarrow{a}\!\!\!\!\!/\;\; \text{or} \;\; p \xrightarrow{a}\!\!\!\!\!/ \qquad \vdash \qquad u\rceil\!|\,p \xrightarrow{\theta} u'\rceil\!|\,p$$

The implementation relation defined in this way is called *refusal preorder*.

A second development was the definition of a weaker implementation relation **conf** that is strongly related to testing preorder [11]. It is a modification of testing preorder by restricting all observations to only those traces that are contained in the specification s. This restriction is in particular used in conformance testing. It makes testing a lot easier: only traces of the specification have to be considered, not the huge complement of this set, i.e., the traces not explicitly specified. In other words, **conf** requires that an implementation does what it should do, not that it does not do what it is not allowed to do. Several test generation algorithms have been developed for the relation **conf** [41, 49], among which the canonical tester theory [8], corresponding tools have been implemented [17, 2], and extensions have been studied [34, 16].

The third development of importance for **ioco** was the application of the principles of testing preorder to Input/Output Automata in [42]. It was shown that testing preorder coincides with quiescent trace preorder introduced in [46] when requiring that inputs are always enabled.

The relation **ioco** inherits from all these developments. The definition of **ioco** follows the principles of testing preorder (5) with tests that can also detect the refusal of actions as in refusal preorder. Outputs and always enabled inputs are distinguished analogous to IOA, and, moreover, a restriction is made to only the traces of the specification as in **conf**. The resulting relation **ioco** can be defined semi-formally as follows.

Let $i \in \mathcal{IOTS}(L_I, L_U)$, $s \in \mathcal{LTS}(L_I \cup L_U)$ then

$$i \;\mathbf{ioco}\; s \quad \Longleftrightarrow_{\text{def}} \quad \forall \sigma \in \mathit{Straces}(s): \quad \mathit{out}(\, i \;\mathbf{after}\; \sigma\,) \;\subseteq\; \mathit{out}(\, s \;\mathbf{after}\; \sigma\,)$$

where

- $p\;\mathbf{after}\;\sigma$ is the set of states in which transition system p can be after having executed the trace σ.

- $out(\,p\,\mathbf{after}\,\sigma\,)$ is the set of output actions which may occur in some state of $p\,\mathbf{after}\,\sigma$. Additionally, the special action δ, indicating *quiescence*, may occur if there is a *quiescent state* in $p\,\mathbf{after}\,\sigma$.
- A state p is *quiescent*, denoted by $p \xrightarrow{\delta} p$, if no output action can occur: $\forall x \in L_U : p \stackrel{x}{\not\longrightarrow}$.
- $Straces(s)$ are the *suspension traces* of specification s, i.e., the traces in which the special action δ may occur beside normal input and output actions.

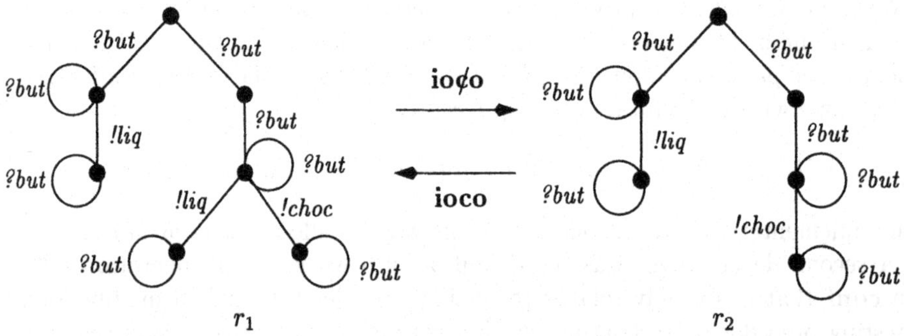

Fig. 1. (Non-)**ioco**-related input-output transition systems.

The relation **ioco** is chosen as implementation relation in our framework: **imp** := **ioco**. Informally, an implementation i is **ioco**-correct with respect to the specification s if i can never produce an output which could not have been produced by s in the same situation, i.e., after the same suspension trace. Moreover, i may only be quiescent, i.e., produce no output at all, if s can do so.

Example 1. Figure 1 gives two input-output transition systems with $L_I = \{\,?but\}$ and $L_U = \{\,!liq,\,!choc\}$ and their **ioco**-relation.
r_1 **io**$\not{c}$**o** r_2 since $out(\,r_1\ \mathbf{after}\ ?but\cdot\delta\cdot ?but\,) = \{\,!liq,\,!choc\}$,
while $out(\,r_2\ \mathbf{after}\ ?but\cdot\delta\cdot ?but\,) = \{\,!choc\}$.

For more details about the relation **ioco**, argumentation for its use, and for more generic definitions we refer to [45]. New developments have led to a variant of **ioco**, called **mioco**, were explicit communication channels for actions are distinguished. Moreover, this **mioco**-theory allows to include all testing-based implementation relations, including refusal preorder, testing preorder, trace preorder, quiescent trace preorder and different variants of **ioco** and **mioco**, in a single lattice [27, 25].

Tests Also *TESTS* is instantiated with transition systems, but this time we add an extra label θ, as in [33], to model the detection of refusals, in particular the detection of the refusal of all outputs, i.e., quiescence. Moreover, we restrict tests to deterministic transition systems with finite behaviour, so that any test execution is always finite and ends in a terminal state of the test case. We will denote these terminal states as either **pass** or **fail**. Finally, we require that for each non-terminal state s of a test case either $init(s) = \{a\}$ for some $a \in L_I$, or $init(s) = L_U \cup \{\theta\}$; $init(t)$ is the set of initial actions of t: $init(t) = \{a | t \xrightarrow{a}\}$. So, the behaviour of a test case is described by a (finite) tree where in each state either one specific input action can occur, or all outputs together with the special action θ. The special label $\theta \notin L \cup \{\delta\}$ will be used in a test case to detect quiescent states of an implementation, so it can be thought of as the communicating counterpart of a δ-action. It will usually be implemented by a kind of time-out.

Example 2. Figure 2 gives an example of a test case t.

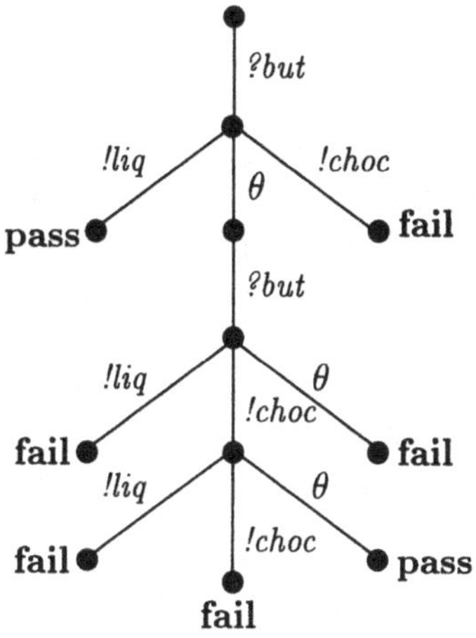

Fig. 2. A test case t

Observations Observations are logs of actions, i.e., traces over $L \cup \{\theta\}$: $OBS := (L \cup \{\theta\})^*$.

Observation function The observation function *obs* is defined by the synchronized parallel composition of t and i ending in a final state of t:

$$obs(t,i) \;=_{\text{def}}\; \{\, \sigma \in (L \cup \theta)^* \mid t]|i \xRightarrow{\sigma} t']|i',\ t' = \textbf{pass} \text{ or } t' = \textbf{fail} \,\}$$

Example 3. For r_1 (figure 1) there are three observations with t of figure 2:

$$t]|r_1 \xRightarrow{?but\cdot !liq} \textbf{pass}]|r_1'$$
$$t]|r_1 \xRightarrow{?but\cdot\theta\cdot ?but\cdot !liq} \textbf{fail}]|r_1''$$
$$t]|r_1 \xRightarrow{?but\cdot\theta\cdot ?but\cdot !choc\cdot\theta} \textbf{pass}]|r_1'''$$

where r_1', r_1'', and r_1''' are the leaves of r_1 from left to right.

Verdicts The verdict assigned to a set of observations $O \subseteq OBS$ is **pass** if all traces in O lead to the terminal state **pass** of the test case:

$$\nu_t(O) \quad =_{\text{def}} \quad \begin{cases} \textbf{pass} & \text{if } \forall \sigma \in O : t \overset{\sigma}{\Longrightarrow} \textbf{pass} \\ \textbf{fail} & \text{otherwise} \end{cases}$$

Example 4. Continuing example 3 we have that, since the terminal state of t for the second run is **fail**, the verdict for r_1 is **fail**. Similarly, it can be checked that the verdict for r_2 is **pass**.

Test execution Test execution EXEC(t, IUT) should be correctly implemented, i.e., it should be implemented such that it correctly reflects the semantics as expressed by $obs(t, i_{\text{IUT}})$ and establishes the test hypothesis.

Test derivation The following algorithm specifies the derivation of test cases from a labelled transition system specification for the implementation relation **ioco**. The test cases are denoted using a process-algebraic notation: ";" denotes action prefix; "+" denotes choice; "Σ" denotes generalized choice. Moreover, for S a set of states, S **after** a denotes the set of states which can be reached from any state in S via action a.

Algorithm – **ioco** *test derivation:* Let s be a specification with initial state s_0. Let S be a non-empty set of states, with initially $S = \{s_0\}$. Then a test case t is obtained from S by a finite number of recursive applications of one of the following three nondeterministic choices:

1. (* terminate the test case *)
   $t :=$ **pass**
2. (* give a next input to the implementation *)
   $t := a \; ; \; t'$, if S **after** $a \neq \emptyset$
   where $a \in L_I$, and t' is obtained by recursively applying the algorithm for $S' = S$ **after** a.
3. (* check the next output of the implementation *)
   $$\begin{aligned} t := \; & \Sigma \{ x \; ; \; \textbf{fail} \mid x \in L_U, \; x \notin out(S) \} \\ & + \Sigma \{ \theta \; ; \; \textbf{fail} \mid \delta \notin out(S) \} \\ & + \Sigma \{ x \; ; \; t_x \mid x \in L_U, \; x \in out(S) \} \\ & + \Sigma \{ \theta \; ; \; t_\theta \mid \delta \in out(S) \} \end{aligned}$$
   where t_x and t_θ are obtained by recursively applying the algorithm for S **after** x and S **after** δ, respectively.

Given a specification $s \in \mathcal{LTS}(L_I \cup L_U)$, this algorithm was proved in [45] to produce only sound test cases, i.e., test cases which never produce **fail** while testing an **ioco**-conforming implementation. Formally, let *der* be any function satisfying the (nondeterministic) algorithm, then the following holds

$$\forall i \in \mathcal{IOTS}(L_I, L_U) : \; i \; \textbf{ioco} \; s \implies \forall t \in der(s) : \nu_t(obs(t, i)) = \textbf{pass}$$

Moreover, it was shown in [45] that any non-conforming implementation can always be detected by a test case generated with this algorithm, i.e., let T_s be the set of all test cases which can be generated by the algorithm from s, then

$$\forall i \in \mathcal{IOTS}(L_I, L_U) : i \textbf{ ioco } s \iff \forall t \in T_s : \nu_t(obs(t,i)) = \textbf{pass}$$

Example 5. Using the **ioco**-test derivation algorithm the test case t of figure 2 can be derived from specification r_2 in figure 1. This is consistent with figure 1 and example 4: $r_1 \; \mathbf{io\not{c}o} \; r_2$, $r_2 \textbf{ ioco } r_2$ (**ioco** is reflexive), and indeed $\nu_t(obs(t, r_1)) = \textbf{fail}$, and $\nu_t(obs(t, r_2)) = \textbf{pass}$. So, test case t can be used to detect that r_1 is not **ioco**-correct with respect to r_2.

5 Tools and an Application

The algorithm for **ioco**-test derivation has a wider applicability than candy machines. Different tools have been built which implement, more or less strictly, this algorithm. These include TVEDA [39,13], TGV [18] and TORX [4].

TVEDA is a tool which is able to generate test cases in TTCN [31, part 3] from single-process SDL specifications. Actually, it is interesting to note that the test generation algorithm of TVEDA was not based on the algorithm for **ioco**-test derivation but on the intuition and heuristics of experienced test case developers at France Telecom CNET. Only careful analysis afterwards showed that this algorithm generates test cases for an implementation relation which was called "R_1" in [39] and which is almost the same as **ioco**.

The tool TGV generates tests in TTCN from LOTOS or SDL specifications. It implements a test derivation algorithm for **ioco** with an unfair extension for finite-state divergences. Moreover, it allows test purposes to be specified by means of automata, which makes it possible to identify the parts of a specification which are interesting from a testing point of view.

Whereas TVEDA and TGV only support the test derivation process by deriving test suites and expressing them in TTCN, the tool TORX combines **ioco**-test derivation and test execution in an integrated manner. This approach, where test derivation and test execution occur simultaneously, is called *on-the-fly testing*. Instead of deriving a complete test case, the test derivation process only derives the next test event from the specification and this test event is immediately executed. While executing a test case, only the necessary part of the test case is considered: the test case is derived *lazily* (cf. lazy evaluation of functional languages). This can reduce the effort needed for deriving a test case, see also [48].

TORX is currently able to derive test cases from LOTOS and PROMELA specifications, but since its implementation uses the OPEN/CÆSAR interface [20] for traversing through a labelled transition system, the tool can be easily extended to any formalism with transition system semantics for which there is an OPEN/CÆSAR interface implementation available.

A simple experiment was conducted to show the viability and the practical applicability of the **ioco** testing theory and the tool TORX [4]. For this experiment a simple protocol, *the Conference Protocol*, was considered [19]. The

Conference Protocol resembles a "chatbox". It offers to users the possibility to join a group, to chat with the members of the group, and to leave the group. It is implemented on top of the UDP protocol from the TCP/IP protocol suite.

Specifications in LOTOS and in PROMELA were developed for the Conference Protocol. An implementation in the C programming language was developed, too. From this implementation 27 (erroneous) mutants were derived. Moreover, for benchmarking, an SDL specification was developed from which 13 TTCN test cases were generated using the tool AUTOLINK which is part of the SDL tool set TAU [43].

The 28 different implementations were tested with respect to the LOTOS and PROMELA specifications using TORX with the on-the-fly approach. All 25 **ioco**-incorrect mutants could be detected, based on the LOTOS as well as on the PROMELA specification. The length of the test run, i.e., the number of test events before the defect was detected, varied between 2 and 498 test events. Two mutants, although differing from the specification, were **ioco**-correct, and indeed no errors were found in these implementations. (These implementations differed in traces not explicitly contained in the specification, i.e., traces σ, with $\sigma \notin \mathit{Straces}(s)$, cf. the definition of **ioco** in section 4).

While testing the **ioco**-correct implementations based on the LOTOS specification, we were able to execute test runs consisting of 28,000 test events without finding a discrepancy between implementation and specification. Then the infamous message "out of memory" occurred while consuming 1.4 Gb. of memory. Since our PROMELA implementation in TORX inherits the state-space exploration algorithm from the very efficient model checker SPIN [29], much longer test runs could be made with PROMELA: 450,000 test events using 400 Mb.

Using the 13 SDL-derived test cases, 5 erroneous mutants slipped through the testing procedure: they obtained a verdict **pass**. Although this experiment was certainly not significant enough for a fair comparison between the tool TORX and the commercial tool AUTOLINK, we dare conclude that the **ioco**-based test theory as implemented in TORX constitutes a sound, feasible, and practically applicable approach for conformance testing based on formal methods.

6 Concluding Remarks

We have shown in this paper how formal methods can be used in conformance testing. It can be concluded that the use of formal methods in testing has many advantages. These advantages include

- a formal, thus more precise and less ambiguous specification of what should be tested;
- formal preciseness and clarity in the properties that are being tested;
- formal reasoning about the validity of tests; and
- algorithmic generation of test cases, with the potential of automated test case derivation.

The first advantage is already present in the testing process even if the testing process itself is not formal. Analysis of practical testing processes shows that most of the problems encountered are not due to the testing process itself but to unclear, imprecise and ambiguous specifications. Formalizing these specifications helps in reducing testing problems even without any formal testing. This is also one of the main conclusions of testing in the BOS-project where specifications were written in Z and PROMELA and testing was performed systematically based on these formal specifications, but using manual, conventional techniques without any formal derivation steps [22].

The third advantage addresses another practical testing problem, viz. that the occurrence of a **fail** verdict does not always point to an error in the implementation. In many cases, sometimes up to 50%, the error is due to an erroneous test case. Formal reasoning about conformance and about the validity of test cases may help to alleviate this problem.

The second advantage opens ways to combine verification and testing in a systematic and precise way. Some properties of a system may be verified while others are tested.

The fourth advantage has the largest economic implications. By automation the testing effort in software projects, which may currently take up to 40% of software development costs, may be reduced significantly. And this can be a good starting point for the introduction of formal methods in software development: most likely, more people will invest in using formal methods if test cases are for free once a formal system specification has been developed.

Formal verification does not make testing superfluous, nor does testing make formal verification superfluous. They are complementary techniques for analysis and checking of correctness of systems. While verification aims at proving properties about systems by formal manipulation on a mathematical model of the system, testing is performed by exercising the real, executing implementation (or an executable simulation model). Verification can give certainty about satisfaction of a required property, but this certainty only applies to the model of the system: any verification is only as good as the validity of the system model. Testing, being based on observing only a small subset of all possible instances of system behaviour, can never be complete: testing can only show the presence of errors, not their absence. But since testing can be applied to the real implementation, it is useful in those cases when a valid and reliable model of the system is difficult to build due to complexity, when the complete system is a combination of formal parts and parts which cannot be formally modelled (e.g., physical devices), when the model is proprietary (e.g., third party testing), or when the validity of a constructed model is to be checked with respect to the physical implementation. Moreover, testing based on a formal specification only makes sense if this specification can be assumed to be valid, i.e., has been sufficiently verified.

A crucial point both in formal verification and in formal testing is the link with the non-formal reality. In verification this occurs when a model of reality is built for which informal arguments are given that it is a valid modelling of reality. Subsequently, reasoning occurs completely in the formal domain under the assumption that the formal results will also apply to reality if the model is valid. In formal testing the link with reality is established using the test hypothesis. Here, a model is assumed to exist in a particular formal domain. It is not necessary that this model is available (then we could perform formal verification), nor that we will ever be able to develop it. Moreover, it is assumed that the way of doing experiments on the real system is modelled in a valid way by the formal function *obs*. This incorporates, among others, that test cases are assumed to be correctly implemented. Whether in formal testing or in verification, somewhere the link to the non-formal reality has to be made. It is important to be aware of the assumptions on which this is based, so that results are interpreted in the right context and with the appropriate precautions.

The formal testing framework of section 3 and its instantiation in section 4 provide a good basis for testing with formal methods. But they also point to some open problems. One of the most important ones is the problem of *test selection*. The algorithm for **ioco** test derivation, and many other similar algorithms, allow to derive infinitely many sound test cases. But which ones shall be selected and executed? Can test suites be compared with respect to their error detecting capabilities? Can measures be assigned to test suites expressing their quality? Can the quality of an implementation passing a particular test suite be quantified? To these questions there are not many usable answers, yet. Solutions can be sought by defining coverage measures, fault models, quantifying test hypotheses, etc. [6, 32, 39, 9].

Acknowledgement

Numerous people, in particular the participants in the ISO/ITU-T standardization group on "Formal Methods in Conformance Testing", the partners in the *Côte de Resyste* research project, testing engineers at CMG The Hague B.V. and at CMG Finance B.V., and the members of the Formal Methods and Tools group at the University of Twente, contributed to the developments described in this paper by means of stimulating discussions or commenting on earlier papers, for which I am grateful. Joost Katoen, René de Vries, Axel Belinfante and Jan Feenstra are thanked for proof-reading.

References

1. S. Abramsky. Observational equivalence as a testing equivalence. *Theoretical Computer Science*, 53(3):225–241, 1987.
2. R. Alderden. COOPER, the compositional construction of a canonical tester. In S.T. Vuong, editor, *FORTE'89*, pages 13–17. North-Holland, 1990.
3. B. Beizer. *Software Testing Techniques*. Van Nostrand Reinhold, 1990.
4. A. Belinfante, J. Feenstra, R.G. de Vries, J. Tretmans, N. Goga, L. Feijs, S. Mauw, and L. Heerink. Formal test automation: A simple experiment. In G. Csopaki, S. Dibuz, and K. Tarnay, editors, 12th *Int. Workshop on Testing of Communicating Systems*. Kluwer Academic Publishers, 1999.

5. J.A. Bergstra and J.W. Klop. Algebra of communicating processes with abstraction. *Theoretical Computer Science*, 37(1):77–121, 1985.
6. G. Bernot. Testing against formal specifications: A theoretical view. In S. Abramsky and T. S. E. Maibaum, editors, *TAPSOFT'91, Volume 2*, pages 99–119. Lecture Notes in Computer Science 494, Springer-Verlag, 1991.
7. B. S. Bosik and M. Ü. Uyar. Finite state machine based formal methods in protocol conformance testing: From theory to implementation. *Computer Networks and ISDN Systems*, 22(1):7–33, 1991.
8. E. Brinksma. A theory for the derivation of tests. In S. Aggarwal and K. Sabnani, editors, *Protocol Specification, Testing, and Verification VIII*, pages 63–74. North-Holland, 1988.
9. E. Brinksma. On the coverage of partial validations. In M. Nivat, C.M.I. Rattray, T. Rus, and G. Scollo, editors, *AMAST'93*, pages 247–254. BCS-FACS Workshops in Computing Series, Springer-Verlag, 1993.
10. E. Brinksma, R. Alderden, R. Langerak, J. van de Lagemaat, and J. Tretmans. A formal approach to conformance testing. In J. de Meer, L. Mackert, and W. Effelsberg, editors, *Second Int. Workshop on Protocol Test Systems*, pages 349–363. North-Holland, 1990.
11. E. Brinksma, G. Scollo, and C. Steenbergen. LOTOS specifications, their implementations and their tests. In G. von Bochmann and B. Sarikaya, editors, *Protocol Specification, Testing, and Verification VI*, pages 349–360. North-Holland, 1987.
12. CCITT. *Specification and Description Language (SDL)*. Recommendation Z.100. ITU-T General Secretariat, Geneve, Switzerland, 1992.
13. M. Clatin. Manuel d'utilisation de TVEDA V3. Manual LAA/EIA/EVP/109, France Télécom CNET LAA/EIA/EVP, Lannion, France, 1996.
14. R. De Nicola. Extensional equivalences for transition systems. *Acta Informatica*, 24:211–237, 1987.
15. R. De Nicola and M.C.B. Hennessy. Testing equivalences for processes. *Theoretical Computer Science*, 34:83–133, 1984.
16. K. Drira. The refusal graph: a tradeoff between verification and test. In O. Rafiq, editor, *Sixth Int. Workshop on Protocol Test Systems*, pages 297–312. North-Holland, 1994.
17. H. Eertink. The implementation of a test derivation algorithm. Memorandum INF-87-36, University of Twente, Enschede, The Netherlands, 1987.
18. J.-C. Fernandez, C. Jard, T. Jéron, and C. Viho. An experiment in automatic generation of test suites for protocols with verification technology. *Science of Computer Programming - Special Issue on COST247, Verification and Validation Methods for Formal Descriptions*, 29(1–2):123–146, 1997.
19. L. Ferreira Pires. Protocol implementation: Manual for practical exercises 1995/1996. Lecture notes, University of Twente, Enschede, The Netherlands, August 1995.
20. H. Garavel. OPEN/CÆSAR: An open software architecture for verification, simulation, and testing. In B. Steffen, editor, *Fourth Int. Workshop on Tools and Algorithms for the Construction and Analysis of Systems (TACAS'98)*, pages 68–84. Lecture Notes in Computer Science 1384, Springer-Verlag, 1998.
21. M.-C. Gaudel. Testing can be formal, too. In P.D. Mosses, M. Nielsen, and M.I. Schwartzbach, editors, *TAPSOFT'95: Theory and Practice of Software Development*, pages 82–96. Lecture Notes in Computer Science 915, Springer-Verlag, 1995.
22. W. Geurts, K. Wijbrans, and J. Tretmans. Testing and formal methods — BOS project case study. In *EuroSTAR'98: 6th European Int. Conference on Software*

Testing, Analysis & Review, pages 215–229, Munich, Germany, November 30 – December 1 1998.

23. R.J. van Glabbeek. The linear time – branching time spectrum. In J.C.M. Baeten and J.W. Klop, editors, *CONCUR'90*, Lecture Notes in Computer Science 458, pages 278–297. Springer-Verlag, 1990.
24. R.J. van Glabbeek. The linear time – branching time spectrum II (The semantics of sequential systems with silent moves). In E. Best, editor, *CONCUR'93*, Lecture Notes in Computer Science 715, pages 66–81. Springer-Verlag, 1993.
25. L. Heerink. *Ins and Outs in Refusal Testing.* PhD thesis, University of Twente, Enschede, The Netherlands, 1998.
26. L. Heerink and J. Tretmans. Formal methods in conformance testing: A probabilistic refinement. In B. Baumgarten, H.-J. Burkhardt, and A. Giessler, editors, *Ninth Int. Workshop on Testing of Communicating Systems*, pages 261–276. Chapman & Hall, 1996.
27. L. Heerink and J. Tretmans. Refusal testing for classes of transition systems with inputs and outputs. In T. Mizuno, N. Shiratori, T. Higashino, and A. Togashi, editors, *Formal Desciption Techniques and Protocol Specification, Testing and Verification FORTE X /PSTV XVII '97*, pages 23–38. Chapman & Hall, 1997.
28. C.A.R. Hoare. *Communicating Sequential Processes.* Prentice-Hall, 1985.
29. G. J. Holzmann. *Design and Validation of Computer Protocols.* Prentice-Hall Inc., 1991.
30. ISO. *Information Processing Systems, Open Systems Interconnection, LOTOS - A Formal Description Technique Based on the Temporal Ordering of Observational Behaviour.* International Standard IS-8807. ISO, Geneve, 1989.
31. ISO. *Information Technology, Open Systems Interconnection, Conformance Testing Methodology and Framework.* International Standard IS-9646. ISO, Geneve, 1991. Also: CCITT X.290–X.294.
32. ISO/IEC JTC1/SC21 WG7, ITU-T SG 10/Q.8. *Information Retrieval, Transfer and Management for OSI; Framework: Formal Methods in Conformance Testing.* Committee Draft CD 13245-1, ITU-T proposed recommendation Z.500. ISO – ITU-T, Geneve, 1996.
33. R. Langerak. A testing theory for LOTOS using deadlock detection. In E. Brinksma, G. Scollo, and C. A. Vissers, editors, *Protocol Specification, Testing, and Verification IX*, pages 87–98. North-Holland, 1990.
34. G. Leduc. A framework based on implementation relations for implementing LOTOS specifications. *Computer Networks and ISDN Systems*, 25(1):23–41, 1992.
35. D. Lee and M. Yannakakis. Principles and methods for testing finite state machines. *The Proceedings of the IEEE*, August 1996.
36. N.A. Lynch and M.R. Tuttle. An introduction to Input/Output Automata. *CWI Quarterly*, 2(3):219–246, 1989.
37. R. Milner. *Communication and Concurrency.* Prentice-Hall, 1989.
38. G.J. Myers. *The Art of Software Testing.* John Wiley & Sons Inc, 1979.
39. M. Phalippou. *Relations d'Implantation et Hypothèses de Test sur des Automates à Entrées et Sorties.* PhD thesis, L'Université de Bordeaux I, France, 1994.
40. I. Phillips. Refusal testing. *Theoretical Computer Science*, 50(2):241–284, 1987.
41. D. H. Pitt and D. Freestone. The derivation of conformance tests from LOTOS specifications. *IEEE Transactions on Software Engineering*, 16(12):1337–1343, 1990.
42. R. Segala. Quiescence, fairness, testing, and the notion of implementation. In E. Best, editor, *CONCUR'93*, pages 324–338. Lecture Notes in Computer Science 715, Springer-Verlag, 1993.

43. Telelogic. TAU *SDL Tool Set Documentation.* Telelogic AB, Malmö, Sweden, 1998.
44. J. Tretmans. A formal approach to conformance testing. In O. Rafiq, editor, *Sixth Int. Workshop on Protocol Test Systems*, number C-19 in IFIP Transactions, pages 257–276. North-Holland, 1994.
45. J. Tretmans. Test generation with inputs, outputs and repetitive quiescence. *Software—Concepts and Tools*, 17(3):103–120, 1996.
46. F. Vaandrager. On the relationship between process algebra and Input/Output Automata. In *Logic in Computer Science*, pages 387–398. Sixth Annual IEEE Symposium, IEEE Computer Society Press, 1991.
47. L. Verhaard, J. Tretmans, P. Kars, and E. Brinksma. On asynchronous testing. In G. von Bochmann, R. Dssouli, and A. Das, editors, *Fifth Int. Workshop on Protocol Test Systems*, IFIP Transactions. North-Holland, 1993.
48. R.G. de Vries and J. Tretmans. On-the-Fly Conformance Testing using SPIN. In G. Holzmann, E. Najm, and A. Serhrouchni, editors, *Fourth Workshop on Automata Theoretic Verification with the* SPIN *Model Checker*, ENST 98 S 002, pages 115–128, Paris, France, November 2 1998. Ecole Nationale Supérieure des Télécommunications. Also to appear in Software Tools for Technology Transfer.
49. C. D. Wezeman. The CO-OP method for compositional derivation of conformance testers. In E. Brinksma, G. Scollo, and C. A. Vissers, editors, *Protocol Specification, Testing, and Verification IX*, pages 145–158. North-Holland, 1990.

Computing Minimum and Maximum Reachability Times in Probabilistic Systems*

Luca de Alfaro

Department of Electrical Engineering and Computer Sciences,
University of California at Berkeley. Email: `dealfaro@eecs.berkeley.edu`

Abstract. A Markov decision process is a generalization of a Markov chain in which both probabilistic and nondeterministic choice coexist. Given a Markov decision process with costs associated with the transitions and a set of target states, the *stochastic shortest path* problem consists in computing the minimum expected cost of a control strategy that guarantees to reach the target. In this paper, we consider the classes of stochastic shortest path problems in which the costs are all non-negative, or all non-positive. Previously, these two classes of problems could be solved only under the assumption that the policies that minimize or maximize the expected cost also lead to the target with probability 1. This assumption does not necessarily hold for Markov decision processes that arise as model for distributed probabilistic systems. We present efficient methods for solving these two classes of problems without relying on additional assumptions. The methods are based on algorithms to transform the original problems into problems that satisfy the required assumptions. The methods lead to the efficient solution of two basic problems in the analysis of the reliability and performance of partially-specified systems: the computation of the minimum (or maximum) probability of reaching a target set, and the computation of the minimum (or maximum) expected time to reach the set.

1 Introduction

Markov decision processes are generalizations of Markov chains in which probabilistic choice coexists with nondeterministic choice [2]. Several models of distributed probabilistic systems are based either on Markov decision processes [1, 15] or on closely related formalisms, such as the *concurrent Markov chains* of [23], the *probabilistic automata* of [22, 25], and the *timed probabilistic automata* of [21]. Several models based on process algebras are also closely related to Markov decision processes [18, 14, 25]. In these proposals, probability enables the modeling of phenomena related to reliability and performance, while nondeterminism has been used to model concurrency [23, 19, 21], inputs [21], imprecise knowledge of the transition probabilities [8, 9], and in general any behavior for which probabilistic information is not known.

* This research was supported in part by the NSF CAREER award CCR-9501708, by the DARPA (NASA Ames) grant NAG2-1214, by the DARPA (Wright-Patterson AFB) grant F33615-98-C-3614, by the ARO MURI grant DAAH-04-96-1-0341, and by the Gigascale Silicon Research Center.

A Markov decision process (MDP) consists of a set of states; with each state is associated a set of possible actions. At every state, the choice of the next action is nondeterministic; once chosen, the action determines the transition probability distribution for the successor state. In order to quantify the probabilistic properties of an MDP, the concept of *policy* is introduced [11], related to the *schedulers* of [23, 19] and to the *adversaries* of [22, 21]. A policy is a criterion for selecting the actions during a behavior of the system; once the policy is fixed, the MDP is reduced to a conventional stochastic process. A simple way to introduce time in these models is to associate with each pair consisting of state and of a related action the time (or the expected time) spent at the state when the action is selected [13, 21, 9]. One of the basic questions we can ask about the timing behavior of such a system is the expected time needed to reach a given set of target states from a specified starting state. Being able to answer this question opens the way to the automated verification of systems properties such as expected time to failure, expected task completion time, and several others. Since the system model includes nondeterminism, the answer to this expected time question consists not in a single value, but rather in a range of values comprised between a minimum and a maximum, depending on whether the policy in use hastens or delays the reaching of the target. This paper is concerned with the question of how to compute these minimum and maximum values.

The problem of computing the maximum and minimum reachability times can be reduced to the *stochastic shortest path* (SSP) problem [12, 11]. In the statement of the SSP problem, with each state-action pair is associated a real-valued cost; the SSP problem consists in computing the minimum expected cost incurred to reach a set of target states. Hence, to compute the minimum (resp. maximum) reachability time, it suffices to equate the cost to the time (resp. to the time multiplied by -1) and to solve the resulting SSP problem. However, previous solutions to the SSP problem rely on assumptions that do not necessarily hold for the SSP problems obtained by the above reduction. In particular, previous solutions require that the target set can be reached with probability 1 from every state, and that either (a) every policy that does not lead to the target with probability 1 yields infinite expected total cost, or (b) the policies that minimize or maximize the expected total cost also lead to the target with probability 1 [4, 3]. Under either one of these assumptions, the goal of reaching the target can be disregarded in the solution of the optimization problem, and the SSP problem can be solved by determining the policy that minimizes the total cost. If the starting and target states are part of a formal specification, or if the time associated with state-action pairs can be 0, as in [13, 21, 8, 9], these assumptions do not hold in general, and new solution methods are required.

The aim of this paper is to present methods for solving the SSP problem that rely on the assumptions that the costs are all non-negative, or all non-positive. We call the SSP problems that satisfy these assumptions the *non-negative* and *non-positive* SSP problems. Solving these SSP problems suffices for solving the original problem about the maximum and minimum reachability times. Furthermore, we show that the proposed solution methods can be applied to the efficient computation of the maximum and minimum probability of reaching a target set of states.

The minimum expected cost to reach a set of target states is well defined only if the target can be reached with probability 1. The first step in the solution of the SSP problem consists thus in computing the set of states from which the target set can be reached with probability 1. This problem can be solved in polynomial time by a reduction to linear programming [11]. In this paper we present a more efficient algorithm, that solves the problem in time quadratic in the size of the MDP, and that does not require numerical computation. The algorithm, originating from [8], is related to an algorithm for solving two-person reachability games presented in [10].

Once we have determined the states from which the target set cannot be reached with probability 1, we present two methods for solving the SSP problem on the remaining states. First, we show that non-negative and non-positive SSP problems can be solved using linear programming over the extended field $\mathbb{R} \cup \{\pm\infty\}$. Second, we present translation algorithms that transform non-negative and non-positive SSP problems into SSP problems that satisfy the assumptions previously considered in the literature [4, 3]. This enables the use of several well-known techniques for the solution of non-negative and non-positive SSP problems, such as value iteration methods, and methods based on learning and sample path analysis (see [4, 3] again). The translation algorithms have strongly-polynomial time complexity in the size of the MDP being translated. As the algorithms never increase and often reduce the size of the MDPs, they also perform a beneficial pre-conditioning prior to the application of numerical solution methods.

Finally, we apply the algorithms presented in this paper to the computation of the minimum and maximum probability of reaching a set of target states. The computation of the minimum reachability probability is useful for determining lower bounds for the probability of reaching desirable system configurations, or of accomplishing tasks from given starting points. The computation of the maximum reachability probability is one of the basic problems in probabilistic verification: aside from being of interest in its own right, it is at the basis of the algorithms for the determination of the maximum and minimum probability with which a linear-time temporal logic formula holds over an MDP [6, 7, 1]. While the maximum reachability probability can be computed with the algorithms of [6], the proposed approach minimizes the size of the numerical problem to be solved.

2 Preliminaries

A *Markov decision process* (MDP) is a generalization of a Markov chain in which nondeterministic choice coexists with probabilistic one. Markov decision processes are closely related to the *probabilistic automata* of [20], to the *concurrent Markov chains* of [23], and to the *simple probabilistic automata* of [22, 21]. To present their definition, given a countable set C we denote by $\mathcal{D}(C)$ the set of probability distributions over C, i.e. the set of functions $f : C \mapsto [0, 1]$ such that $\sum_{x \in C} f(x) = 1$. Given a distribution $f \in \mathcal{D}(C)$, we indicate by $\mathit{Support}(f) = \{x \in C \mid f(x) > 0\}$.

An MDP $\mathcal{M} = (S, \mathit{Acts}, A, p)$ consists of the following components:

- a finite set S of states;
- a finite set *Acts* of actions;

- a function $A : S \mapsto 2^{Acts}$ that associates with each $s \in S$ a finite set $A(s) \subseteq Acts$ of actions available at s;
- a function $p : S \times Acts \mapsto \mathcal{D}(S)$ that associates with each $s, t \in S$ and $a \in A(s)$ the probability $p(s,a)(t)$ of a transition from s to t when action a is selected.

A *path* of the MDP $\mathcal{M}$ is an infinite sequence $\omega : s_0, a_0, s_1, a_1, \ldots$ of alternating states and actions, such that $s_i \in S$, $a_i \in A(s_i)$ and $p(s_i, a_i)(s_{i+1}) > 0$ for all $i \geq 0$. For $i \geq 0$, the sequence is constructed by iterating a two-phase selection process. First, an action $a_i \in A(s_i)$ is selected nondeterministically; second, the successor state s_{i+1} is chosen according to the probability distribution $p(s_i, a)$. Given a path $\omega : s_0, a_0, s_1, a_1, \ldots$ and $k \geq 0$, we denote by $X_k(\omega)$, $Y_k(\omega)$ its k-th state s_k and its k-th action a_k, respectively. Given a state $s \in S$ and an action $a \in A(s)$ for s, we also denote by $dest(s,a) = \{t \in S \mid p(s,a)(t) > 0\}$ the set of possible successors of s when a is selected.

To be able to talk about the probability of system behaviors, we need to specify the criteria with which the actions are chosen. To this end, we use the concept of *policy* [11], closely related to the adversaries of [22, 21] and to the schedulers of [23, 19]. A policy η is a mapping $\eta : S^+ \mapsto \mathcal{D}(Acts)$, which associates with each finite sequence of states $s_0, s_1, \ldots, s_n \in S^+$ and each $a \in A(s_n)$ the probability $\eta(s_0, \ldots, s_n)(a)$ of choosing a after following the sequence of states $s_0, \ldots, s_n$. We require that $\eta(s_0, \ldots, s_n)(a) > 0$ implies $a \in A(s_n)$: a policy can choose only among the actions that are available at the state where the choice is made. We indicate with *Pol* the set of all policies. We say that a policy η is *memoryless* if $\eta(s_0, \ldots, s_n)(a) = \eta(s_n)(a)$ for all sequences of states $s_0, \ldots, s_n \in S^+$ and all $a \in A(s)$.

For every state $s \in S$, we denote by Ω_s the set of paths having s as initial state, and we let $\mathcal{B}_s \subseteq 2^{\Omega_s}$ be the σ-algebra of *measurable* subsets of Ω_s, following the classical definition of [17]. Under policy η the probability of following a finite path prefix $s_0 a_0 s_1 a_1 \cdots s_n$ is $\prod_{i=0}^{n-1} p(s_i, a_i)(s_{i+1})\, \eta(s_0 \cdots s_i)(a_i)$. These probabilities for prefixes give rise to a unique probability measure on $\mathcal{B}_s$. We write $\Pr_s^\eta(\mathcal{A})$ to denote the probability of event $\mathcal{A}$ in Ω_s under policy η, and $\mathrm{E}_s^\eta\{f\}$ to denote the expectation of the random function f from state s under policy η.

2.1 The stochastic shortest path problem

An instance $\Pi = (S, Acts, A, p, R, c, g)$ of the *stochastic shortest path* problem consists of an MDP $(S, Acts, A, p)$, together with the additional components R, c and g:

- $R \subseteq S$ is the the set of *destination states;*
- $c : S \times Acts \mapsto \mathbb{R}$ is the *running cost function,* that associates with each state $s \in S \setminus R$ and each action $a \in A(s)$ the cost $c(s,a)$;
- $g : R \mapsto \mathbb{R}$ is the *terminal cost function,* that associates to each $s \in R$ its terminal cost $g(s)$.

We say that an instance of the SSP problem is *non-negative* (resp. *non-positive*) if $c(s,a) \geq 0$ (resp. $c(s,a) \leq 0$) for all $s \in S$ and $a \in A(s)$; note that the sign of g is not relevant for this definition.

The SSP problem consists in determining the minimum cost of reaching R when following a policy that reaches R with probability 1, provided such a policy exists.

Precisely, let $T_R(\omega) = \min\{k \mid X_k(\omega) \in R\}$ be the position of first visit of a path in R. For all $s \in S$ we denote by $Prp(s) = \{\eta \in Pol \mid \Pr_s^\eta(T_R < \infty) = 1\}$ the set of policies that lead from s to R with probability 1; these policies are the *proper policies* for s. Given a state $s \in S$, the *cost* v_s^η of a policy η is defined by

$$v_s^\eta = \mathrm{E}_s^\eta\left\{g(X_{T_R}) + \sum_{k=0}^{T_R-1} c(X_k, Y_k)\right\} . \tag{1}$$

A policy η is *optimal* if $v_s^\eta = v_s^*$ for all $s \in S \setminus R$. With this notation, the SSP problem consists in:

1. determining the set $Q = \{s \in S \setminus R \mid Prp(s) \neq \emptyset\}$ of states having at least one proper policy;
2. computing the minimum cost $v_s^* = \inf_{\eta \in Prp(s)} v_s^\eta$ of a proper policy at all $s \in Q$.

Usually, the SSP problem is considered to consist only in the second question, and the existence of at least one proper policy for each state is stated as an assumption. However, when the SSP problem is used to compute the minimum or maximum reachability times between an initial state and a set of target states that are part of a reliability of performance specification, we cannot assume that the target set can be reached from the initial state with probability 1. Hence, in Section 2.3 we present an algorithm to solve also this first question. In addition, we will characterize the optimal policies for non-negative and non-positive SSP problems.

SSP problem and reachability time. In a *timed probabilistic system*, the timing behavior of an MDP $(S, Acts, A, p)$ is specified by means of a function $time : S \times Acts \mapsto \mathbb{R}^+$ that associates with each $s \in S$ and $a \in A(s)$ the expected amount of time $time(s,a)$ spent at state s when action a is selected [9]. Given a set R of target states, to compute the minimum (resp. maximum) expected time to reach R it suffices to solve an SSP problem having cost functions defined by $c(s,a) = time(s,a)$ (resp. $c(s,a) = -time(s,a)$) and $g(s) = 0$, for all $s \in S$ and $a \in A(a)$. The minimum (resp. maximum) expected time to reach R from $s \in S \setminus R$ is then given by v_s^* (resp. $-v_s^*$).

2.2 End components

The algorithms that we present to solve the classes of SSP problems rely on the notion of *end component* [8]. End components are the analogous concept in Markov decision processes of the closed recurrent classes of Markov chains [17]: they represent the set of states and actions that can be repeated infinitely often along a path with non-zero probability. Related sets of states have been used for solving optimization problems on MDPs [7]. Given an MDP $\mathcal{M} = (S, Acts, A, p)$, a *sub-MDP* is a pair (C, D), where $C \subseteq S$ is a subset of states and $D : S \mapsto Acts$ is a function that associates to each $s \in S$ a subset $D(s) \subseteq A(s)$ of actions. A sub-MDP (C, D) is an *end component* if the following conditions hold:

- *Closure:* for all $s \in C$, $a \in D(s)$, and $t \in S$, if $p(s,a)(t) > 0$ then $t \in C$.

- *Connectivity:* Let $E = \{(s,t) \in C \times C \mid \exists a \in D(s) \,.\, p(s,a)(t) > 0\}$; then, the graph (C, E) is strongly connected.

We say that an end component (C, D) is contained in a sub-MDP (C', D') if

$$\{(s,a) \mid s \in C \wedge a \in D(s)\} \subseteq \{(s,a) \mid s \in C' \wedge a \in D'(s)\} .$$

We say that an end component (C, D) is *maximal* in a sub-MDP (C', D') if there is no other end component (C'', D'') contained in (C', D') that properly contains (C, D). We denote by $Mec(C', D')$ the set of maximal end components of (C', D'). It is not difficult to see that, given a sub-MDP (C, D), the set $Mec(C, D)$ can be computed in time polynomial in $|C| + \sum_{s \in C} |D(s)|$ using simple graph algorithms; an algorithm to do so is given in [8, §3]. Given a path ω, denote by $\mathit{InfS}(\omega) = \{s \in S \mid \stackrel{\infty}{\exists} k \,.\, X_k(\omega) = s\}$ the set of states visited infinitely often by ω, where $\stackrel{\infty}{\exists}$ is a shorthand for "there are infinitely many distinct". Also, define $\mathit{InfA}(\omega) : S \mapsto 2^{Acts}$ by $\{a \in A(s) \mid \stackrel{\infty}{\exists} k \,.\, X_k(\omega) = s \wedge Y_k(\omega) = a\}$ for all $s \in S$. The following theorem summarizes the basic property of end components [8].

Theorem 1 *For all $s \in S$ and all $\eta \in Pol$, we have*

$$\Pr_s^\eta\big((\mathit{InfS}(\omega), \mathit{InfA}(\omega)) \textit{is an end component}\big) = 1 .$$

2.3 Computing the set of states having proper policies

As a first step in the solution of the SSP problem, we must compute the set

$$Reach(R) = \big\{s \in S \mid \exists \eta \in Pol \,.\, \Pr_s^\eta(T_R < \infty) = 1\big\}$$

consisting of the states having at least one proper policy. This problem can be solved by reducing it to several well-known dynamic programming problems, such as the *maximum average reward* problem [11] or the *maximum reachability probability* problem [6]. However, these reductions yield algorithms that are based on linear programming, and their time complexity is only weakly polynomial, i.e. it depends on the size of the bit strings encoding the probability values in the input description of the problem. We present here an algorithm that solves the problem in time quadratic in the size of the MDP, and that does not require any numerical computation. The algorithm is originally from [8], and is related to an algorithm for solving reachability problems in two-person games presented in [10]. The algorithm is also reminiscent of an algorithm independently proposed in [24]. To present the algorithm, given two subsets $X, Y \subseteq S$ of states we define the predicate $APre(Y, X)$ so that for all $s \in S$,

$$s \models APre(X, Y) \quad \textit{iff} \quad \exists a \in A(s) \,.\, \big(dest(s,a) \subseteq Y \wedge dest(s,a) \cap X \neq \emptyset\big) .$$

Given a subset R of target states, we compute $Reach(R)$ by the following μ-calculus expression:

$$Reach(R) = \nu Y \,.\, \mu X \,.\, (APre(Y, X) \vee R) , \qquad (2)$$

where we have used the slightly improper notation of denoting by R a predicate that holds exactly for the states in R. The algorithm (2) can be understood as follows. Denoting by Y_k the value of the set Y computed at iteration $k \geq 0$, we have initially $Y_0 = S$. At the end of the first iteration, we have $Y_1 = S \setminus C_0$, where C_0 is the subset of states of S that cannot reach R. At the end of the second iteration, we have $Y_2 = Y_1 \setminus C_1$, where C_1 is the set of states that cannot reach R without risking to enter C_0. In general, at the end of iteration $k > 0$, we have $S_k = S_{k-1} \setminus C_{k-1}$, where C_{k-1} consists of the states that cannot reach R without risking to enter $\bigcup_{i=0}^{k-2} C_i$. Given an MDP $\mathcal{M} = (S, A, p)$, define its graph size $|\mathcal{M}|$ by

$$|\mathcal{M}| = \sum_{s \in S} \sum_{a \in A(s)} \left| \mathit{Support}(p(s,a)) \right| .$$

The following theorem summarizes the results about this algorithm.

Theorem 2 *Given an MDP $\mathcal{M} = (S, A, p)$ and a set $R \subseteq S$ of target states, relation (2) correctly computes Reach(R) in time quadratic in $|\mathcal{M}|$.*

Once the set $\mathit{Reach}(R)$ has been computed, we can replace the original SSP problem $(S, \mathit{Acts}, A, p, R, c, g)$ with a new problem $(Q, \mathit{Acts}, A', p', R, c', g')$, where $Q = \mathit{Reach}(R)$, where p', c', g' are the restrictions of p, c, g to Q, and where for all $s \in Q$ we let $A'(s) = \{a \in A(s) \mid \mathit{dest}(s,a) \subseteq Q\}$. To avoid a change of notation, in the following we denote an instance of the SSP problem again by $(S, \mathit{Acts}, A, p, R, c, g)$, but we assume that $\mathit{Reach}(R) = S$. This is equivalent to assuming that the above reduction has been made already.

3 Solving Non-Negative SSP Problems

The class of SSP problems that is most closely related to the non-negative class, and for which solution methods have been presented in the literature, is discussed in [4, 3]. There, it is shown that the SSP problem can be solved under the additional assumption that, for all $s \in S$, there is a *proper* policy that minimizes the total cost (1). An example of SSP problem in which this assumption does not hold is depicted in Figure 1. Clearly, the policy that minimizes (1) is the policy η_1 that always chooses action a at s_3; this policy leads to the expected cost $v_{s_1}^{\eta_1} = 1$. However, this policy is not proper, and it is easy to see that for every proper policy η it is $v_{s_1}^{\eta} = 3$.

To understand why the iterative approaches such as value iteration cannot be applied immediately to this problem, let $n = |S \setminus R|$, and denote with $v = [v_s]_{s \in S \setminus R} \in \mathbb{R}^n$ a vector of real numbers indexed by the states of $S \setminus R$. Define the Bellman operator $L : \mathbb{R}^n \mapsto \mathbb{R}^n$ on the space of v by

$$[L(v)]_s = \min_{a \in A(s)} \Big[c(s,a) + \sum_{t \in S \setminus U} p(s,a)(t)\, v_t + \sum_{t \in R} p(s,a)(t)\, g(t) \Big] \qquad s \in S \setminus R\,, \tag{3}$$

where $[L(v)]_s$ denotes the s-components of vector $L(v)$. Given an initial vector v^0, the value iteration method computes the sequence of vectors $v^0, v^1, v^2, \ldots$ by $v^{k+1} = L(v^k)$, for $k \geq 0$, and returns as answer $\lim_{k \to \infty} v^k$, provided the limit exists. The initial vector v^0 represents an initial (often arbitrary) estimate for the minimum ex-

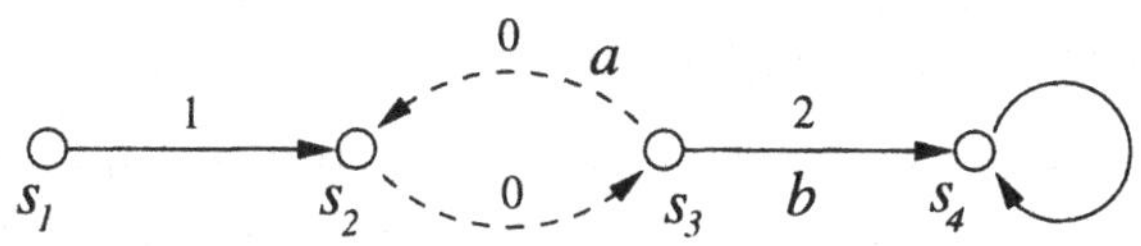

Fig. 1. An instance of SSP problem. The target set is $R = \{s_4\}$, and the terminal cost is $g(s_4) = 0$. States are represented as nodes of a graph, and actions as edges. We have indicated only the actions a and b corresponding to state s_3, where $A(s_3) = \{a, b\}$. In this example, all actions (including a and b) are deterministic, i.e. they lead to only one destination state. The actions are labeled with their cost c. The two actions having cost 0 have been indicated with dashed lines. A larger instance of SSP problem is presented in Figure 3.

pected reachability cost; each iteration of the Bellman operator L is aimed at improving the estimate. Clearly, the answer returned by the value iteration procedure is a fixpoint of L. However, in non-negative SSP problems the Bellman operator L may admit more than one fixpoint: for example, in the SSP problem of Figure 1, for $x \geq 0$ all vectors

$$v(x) = [v_1, v_2, v_3] = [3, 2, 2] - x[1, 1, 1] \tag{4}$$

satisfy $v = L(v)$. If L admits more than one fixpoint, the sequence $v^0, v^1, v^2, \ldots$ can converge to any one of them, depending on the value of the initial estimate v^0. In the example of Figure 1, starting from the initial vector $[0, 0, 0]$, the value iteration method converges to the fixpoint $[1, 0, 0]$. However, we will prove that the solution of the SSP problem corresponds to the largest fixpoint, which in this case is $[3, 2, 2]$. The fact that the Bellman operator does not necessarily admit a unique fixpoint in non-negative SSP problems not only prevents a direct application of value iteration methods, but also blocks the line of analysis of [4] for the solution based on linear programming.

We present two approaches to the solution of non-negative SSP problems. The first approach is based on the observation that the difficulties in solving non-negative SSP problems stem from the presence in the SSP problem of end components consisting of state-action pairs having 0 cost. If we remove these components, we obtain an equivalent problem whose Bellman operator has a unique fixpoint; the problem can then be solved using any of several methods that have been developed for SSP problems, including linear programming and value iteration. This approach has two advantages. First, it enables to exploit in the solution of the SSP problem many numerical techniques that have been devised to handle large-sized problems. Second, the algorithm that removes the end components often achieves a reduction of the size of the problem.

The second approach consists in reducing the SSP problem directly to linear programming: since the solution of the linear programming problem corresponds to the greatest fixpoint of the Bellman operator, as we will show, it also corresponds to the solution of the SSP problem. The correctness proof of this second approach relies on an analysis of the first approach.

3.1 Eliminating 0-cost end components

A *0-cost* end component is an end component (C, D) such that $c(s, a) = 0$ for all $s \in C$ and all $a \in D(s)$. As we will show (see Theorem 3), the lack of uniqueness of the fixpoint is due to the presence of 0-cost end components in the MDP. In a 0-cost component (C, D), by selecting at each $s \in C$ the actions in $D(s)$ uniformly at random, we can go from any state of C to any other state of C with probability 1 while incurring cost 0. Hence, the states of a 0-cost end component are equivalent from the point of view of the minimum cost to the target. The following algorithm exploits this fact to eliminate the 0-cost end components of an MDP by replacing them with single states. The algorithm opens the way to the use of iterative methods based on the Bellman operator for the solution of the non-negative SSP problem.

Algorithm 1 (eliminating 0-cost end components)
Input: SSP problem $\Pi = (S, \mathit{Acts}, A, p, R, c, g)$.
Output: SSP problem $\widehat{\Pi} = \mathit{ElimEC}(\Pi) = (\widehat{S}, \mathit{Acts}, \widehat{A}, \widehat{p}, R, \widehat{c}, \widehat{g})$.
Method: For each $s \in S \setminus R$, let $D(s) = \{a \in A(s) \mid c(s, a) = 0\}$, and let $\{(B_1, D_1), \ldots, (B_n, D_n)\} = \mathit{Mec}(S \setminus R, D)$ be the set of 0-cost maximal end components that lie outside R. Define $\widehat{S} = S \cup \{\widehat{s}_1, \ldots, \widehat{s}_n\} \setminus \bigcup_{i=1}^n B_i$, where $\widehat{s}_1, \ldots, \widehat{s}_n$ are new states. The action sets associated with the states are defined by:

$$s \in S \setminus \textstyle\bigcup_{i=1}^n B_i : \qquad \widehat{A}(s) = \{\langle s, a\rangle \mid a \in A(s)\}$$

$$1 \le i \le n : \qquad \widehat{A}(\widehat{s}_i) = \big\{\langle s, a\rangle \mid s \in B_i \wedge a \in A(s) \setminus D_i(s)\big\} .$$

For $s \in \widehat{S}$, $t \in S \setminus \bigcup_{i=1}^n B_i$ and $\langle u, a\rangle \in \widehat{A}(s)$, the transition probabilities are defined by $\widehat{p}(s, \langle u, a\rangle)(t) = p(u, a)(t)$ and $\widehat{p}(s, \langle u, a\rangle)(\widehat{s}_i) = \sum_{t \in B_i} p(u, a)(t)$. For $s \in \widehat{S}$ and $\langle u, a\rangle \in \widehat{A}(s)$ we let $\widehat{c}(s, \langle u, a\rangle) = c(u, a)$; for $s \in R$ we let $\widehat{g}(s) = g(s)$. ∎

The algorithm replaces each 0-cost end component (B_i, D_i) with a single new state $\widehat{s}_i$, for $1 \le i \le n$. The actions associated with $\widehat{s}_i$ consist in all the pairs $\langle t, a\rangle$ such that $s \in C_i$ and $a \in A(s)$ is an action not belonging to the end component. Intuitively, taking action $\langle s, a\rangle$ at $\widehat{s}_i$ corresponds to taking action a from s, possibly leaving C_i. The transition probabilities and costs of the corresponding actions are unchanged, except that the probability of a transition to $\widehat{s}_i$ is equal to the probability of a transition into C_i in the original system, for $1 \le i \le n$. The result of applying Algorithm 1 to the instance of SSP depicted in Figure 1 is illustrated in Figure 2. The (maximal) end component formed by states s_2, s_3 together with the 0-cost actions has been replaced by the single state $\hat{s}_1$. Figure 3 depicts another example of application of Algorithm 1. The algorithm computes the 0-cost end components $(B_1, D_1), (B_2, D_2)$, where the first end component is given by $B_1 = \{s_3, s_4, s_7\}$ and $D_1(s_3) = \{d\}$, $D_1(s_4) = \{k\}$, $D_1(s_7) = \{j\}$, and the second one by $B_2 = \{s_5, s_6\}$ and $D_2(s_5) = \{f\}$, $D_2(s_6) = \{g\}$. The algorithm replaces these end components with the two new states $\widehat{s}_1$ and $\widehat{s}_2$. This example illustrates the potential reduction of the state-space of the system.

Once the 0-cost end components have been eliminated, the next lemma shows that the reduced problem satisfies the following two assumptions:

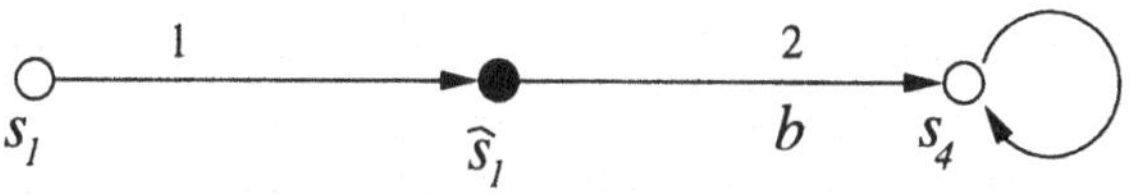

Fig. 2. Result of applying Algorithm 1 to the instance of SSP problem depicted in Figure 2. The new state $\widehat{s}_1$ introduced by the algorithm is drawn as a filled circle.

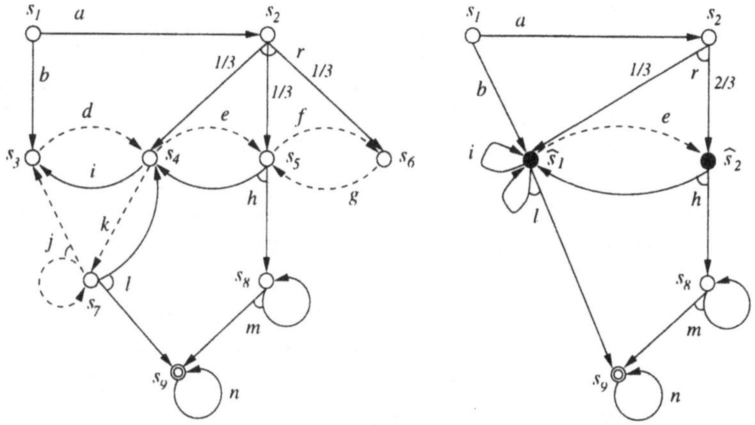

Fig. 3. An instance of SSP problem (left), and the result of applying Algorithm 1 to it (right). Here, not all actions are deterministic, and we depict actions that can lead to more than one destination by "bundles" of edges. To simplify the diagrams, we have indicated only the transition probabilities corresponding to action r, and we have omitted all costs. The actions that have cost 0 have been represented by dashed edges. The target set is $R = \{s_9\}$. The new states $\widehat{s}_1$ and $\widehat{s}_2$ that have been introduced to replace the zero-cost end components are indicated by filled circles.

SSP-1 For all $s \in S$, we have $\mathit{Prp}(s) \neq \emptyset$.
SSP-2 For all $s \in S$ and $\eta \notin \mathit{Prp}(s)$, we have $v_s^\eta = \infty$.

Lemma 1 *Consider an instance Π of non-negative SSP problem such that there is at least one proper policy for each state, and let $\widehat{\Pi} = \mathit{ElimEC}(\Pi)$. Then, $\widehat{\Pi}$ satisfies assumptions SSP-1 and SSP-2.*

Proof. By hypothesis (or more accurately, by the algorithm presented in Section 2.3), $\widehat{\Pi}$ satisfies SSP-1. By Theorem 1, the set of states and actions that are repeated infinitely often along a path is an end component. Hence, if all 0-cost end components have been eliminated, with probability 1 a path that does not reach R has infinite cost, showing that $\widehat{\Pi}$ satisfies SSP-2. ∎

The class of SSP problems that satisfies assumptions SSP-1 and SSP-2 has been studied in depth in the literature. In particular, it is known that the Bellman operator admits a unique fixpoint for this class of problems, and that there exist optimal policies that are memoryless [4]. Moreover, such problems can be solved using value-iteration and policy-iteration methods, which converge to the solution [4]. Other refined iterative methods for the solutions of this class of problems are presented in [5]. As hinted by Lemma 1, the uniqueness of the fixpoint of the Bellman operator is related to the presence of 0-cost end components.

Theorem 3 *Given a non-negative instance $(S, Acts, A, p, R, c, g)$ of SSP problem, the Bellman operator L admits a unique fixpoint iff there is no 0-cost end component (C, D) with $C \subseteq S \setminus R$.*

Proof. In one direction, assume that a non-negative instance of SSP problem does not contain any 0-cost end component. Reasoning as for Lemma 1, we have that assumption SSP-2 holds. If assumption SSP-1 also holds, then the uniqueness of the fixpoint follows from [4]. If assumption SSP-1 does not hold, then assumption SSP-2 ensures that the fixpoint of the Bellman operator diverges to $+\infty$ on the states where there is no proper policy. This, together with the analysis of [4] for the states where there are proper policies, ensures again the uniqueness of the fixpoint. Conversely, if there is a 0-cost end component in a non-negative SSP problem, then we can obtain multiple fixpoints of the Bellman operator by selecting one such end component, and by setting the value of the fixpoint there to any negative value, as done in (4). ∎

The following theorem relates the solutions of the SSP problems Π and $\widehat{\Pi}$, and it enables the (trivial) derivation of a solution for Π from a solution for $\widehat{\Pi}$.

Theorem 4 *Consider an instance Π of non-negative SSP such that there is at least one proper policy for each state, and let $\widehat{\Pi} = ElimEC(\Pi)$. Let also $B_1, \dots, B_n$ be the 0-cost end components that are replaced by states $\widehat{s}_1, \dots, \widehat{s}_n$. Denoting by v^* (resp. $\widehat{v}^*$) the solution of the SSP problems on Π (resp. $\widehat{\Pi}$), we have $v^*_s = \widehat{v}^*_s$ for $s \in S \setminus \bigcup_{i=1}^n B_i$, and $v^*_s = \widehat{v}^*_{\widehat{s}_i}$ for $s \in B_i$, $1 \leq i \leq n$.*

Even though it might appear intuitively plausible that eliminating the 0-cost end components should not modify the solution of the SSP problem, the proof of the above theorem is somewhat involved; it can be found in [8]. The same analysis also leads to the following result.

Corollary 1 *Non-negative SSP problems admit memoryless optimal policies.*

3.2 Linear programming

The second approach is given by the following theorem.

Theorem 5 *Consider an instance Π of non-negative SSP such that there is at least one proper policy for each state. Then, the solution v^* of the SSP problem is the largest fixpoint of operator L defined in (3). Moreover, the following linear programming problem has v^* as unique solution:*

$$\textit{Maximize} \sum_{s \in S \setminus R} v_s \quad \textit{subject to} \quad v_s \leq c(s,a) + \sum_{t \in S \setminus R} p(s,a)(t)\, v_t + \sum_{t \in R} p(s,a)(t)\, g(t)$$

for all $s \in S \setminus R$ and $a \in A(s)$.

Proof. The theorem is proved by showing first that every fixpoint of the Bellman operator (3) is no greater (componentwise) than the solution of the SSP problem. Next, we use the relationship between Π and $\widehat{\Pi} = \mathit{ElimEC}(\Pi)$ to show that one of the fixpoints is equal to the solution of the SSP problem; this implies that the solution of the SSP problem is the largest fixpoint. Finally, it can be shown that the linear programming problem converges to the largest fixpoint, and thus to the solution of the SSP problem. The details can be found in [8]. ∎

4 Solving Non-Positive SSP Problems

Consider an instance $\Pi = (S, \mathit{Acts}, A, p, R, c, g)$ of non-positive SSP problem, and assume that $S = \mathit{Reach}(R)$, i.e. that for every state there is a proper policy. Unlike in the non-negative case, it is possible that $v_s^* = -\infty$ for some $s \in S \setminus R$, and the first step towards the solution of non-negative SSP problems consists in determining the set of states from which the minimum cost diverges to $-\infty$. This can be done with the following algorithm.

Algorithm 2

Input: A non-positive SSP problem $\Pi = (S, \mathit{Acts}, A, p, Rc, g)$, with $\mathit{Reach}(R) = S$.
Output: The subset $\mathit{Diverge}(\Pi) = \{s \mid v_s^* = -\infty\}$.
Method: Let $\mathcal{L} := \{(C, D) \in \mathit{Mec}(S \setminus R, A) \mid \exists s \in C \,.\, \exists a \in D(s) \,.\, c(s,a) < 0\}$ be the set of end components outside R that have at least one strictly negative state-action pair, and let $C = \bigcup_{(C,D) \in \mathcal{L}} C$ be the union of their states.
Let $C_\infty = \mu X \,.\, (\neg R \wedge (\mathit{APre}(S, X) \vee C))$ be the set of states that can reach C without entering R.
Return: C_∞. ∎

Theorem 6 *For an instance Π of non-positive SSP such that $S = \mathit{Reach}(R)$, we have that $v_s^* = -\infty$ iff $s \in \mathit{Diverge}(\Pi)$.*

Proof. From a state $s \in \mathit{Diverge}(\Pi)$, we can reach with positive probability an end component in $\mathcal{L}$. Once there, we can stay in the end component arbitrarily long, accumulating an arbitrarily large amount of negative cost, before proceeding to the target. Hence, we have $v_s^* = -\infty$. The details can be found in [8]. The proof of the converse, i.e., that if $s \notin \mathit{Diverge}(\Pi)$ then $v_s^* > -\infty$, will be given in Section 4.1. ∎

Once the set $\mathit{Diverge}(\Pi)$ has been computed, it remains to compute v_s^* for $s \in S \setminus (R \cup \mathit{Diverge}(\Pi))$. To this end, we first reduce the SSP problem by eliminating the states in $\mathit{Diverge}(\Pi)$. We define a new instance of SSP $\widetilde{\Pi} = \mathit{Converge}(\Pi) = (\widetilde{S}, \mathit{Acts}, \widetilde{A}, \widetilde{p}, R, \widetilde{c}, \widetilde{g})$, where $\widetilde{S} = S \setminus \mathit{Diverge}(\Pi)$, where $\widetilde{p}, \widetilde{c}, \widetilde{g}$ are the restrictions of p, c, g to $\widetilde{S}$, and where for all $s \in \widetilde{S}$ we let $\widetilde{A}(s) = A(s)$. The reduced non-positive SSP problem can then be solved in three ways: by eliminating the 0-cost end components, by linear programming, and by value iteration.

4.1 Eliminating 0-cost components

The first method for solving the reduced problem consists in eliminating the 0-cost end components using Algorithm 1 to compute $\widehat{\Pi} = \mathit{ElimEC}(\widetilde{\Pi})$. The following theorem asserts that $\widehat{\Pi}$ satisfies conditions SSP-1 and SSP-2: hence, the SSP instance $\widehat{\Pi}$ can be solved with the methods presented in [4, 5, 3].

Theorem 7 *The non-positive SSP instance $\widehat{\Pi} = \mathit{ElimEC}(\widetilde{\Pi})$ satisfies conditions SSP-1 and SSP-2. Moreover, let $B_1, \ldots, B_n$ be the 0-cost end components that are replaced by states $\widehat{s}_1, \ldots, \widehat{s}_n$. Denoting by $\widetilde{v}^*$ (resp. $\widehat{v}^*$) the solution of the SSP problems on $\widetilde{\Pi}$ (resp. $\widehat{\Pi}$), we have $\widetilde{v}^*_s = \widehat{v}^*_s$ for $s \in \widetilde{S} \setminus \bigcup_{i=1}^n B_i$, and $\widetilde{v}^*_s = \widehat{v}^*_{\widehat{s}_i}$ for $s \in B_i$, $1 \leq i \leq n$.*

Proof. Since the costs are non-positive, the cost from a state never diverges to $+\infty$. Hence, by Theorem 1, a non-positive instance satisfies condition SSP-2 iff there are no end components entirely outside of the target R. To see that this condition holds for $\widehat{\Pi}$, note that the end components containing some negative cost have been eliminated by Algorithm 2, while those consisting entirely of 0-cost state-action pairs have been eliminated by Algorithm 1. The second part of the result is proved in an analogous way to Theorem 4, and the proof can be found in [8]. ∎

Theorem 7 also leads to the second part of Theorem 6. If $s \notin \mathit{Diverge}(\Pi)$, then $s \in \widetilde{S}$. The fact that assumptions SSP 1 and SSP 2 hold for $\widehat{\Pi}$, together with the results of [4], ensures then that $v^*_s > -\infty$.

Theorem 8 *An instance of non-negative SSP problem Π admits memoryless optimal (proper) policies iff $\mathit{Diverge}(\Pi) = \emptyset$. In any case, there is always a (possibly non memoryless) optimal proper policy.*

Proof. To see that if $\mathit{Diverge}(\Pi) \neq \emptyset$, then there are no memoryless optimal policies, refer to Algorithm 2. Since under a memoryless policy the MDP behaves like a Markov chain, under a memoryless proper policy each path stays for a finite expected amount of time in the end components in $\mathcal{L}$ before reaching R, so that $v^\eta_s > -\infty$ for all $s \in S \setminus R$. On the other hand, there is a (non-memoryless) policy such that, once we reach an end component in $\mathcal{L}$, we stay for infinite expected time in the end component (accumulating an infinite expected cost) before reaching the target with probability 1. The proof that if $\mathit{Diverge}(\Pi) = \emptyset$ there are memoryless optimal policies can be found in [8]. ∎

4.2 Linear programming

Reasoning as in the proof of Theorem 5, it is possible to show that the solution of the SSP problem corresponds to the largest fixpoint of the Bellman operator. The solution can thus be computed by linear programming.

Theorem 9 *Consider an instance Π of non-positive SSP problem such that $\Pi = \mathit{Converge}(\Pi)$, and such that there is at least one proper policy for each state. Then, the solution v^* of the SSP problem is the largest fixpoint of operator L of (3). Moreover, the following linear programming problem has v^* as unique solution:*

$$\textit{Maximize} \sum_{s \in S \setminus R} v_s \quad \textit{subject to} \quad v_s \leq c(s,a) + \sum_{t \in S \setminus R} p(s,a)(t)\, v_t + \sum_{t \in R} p(s,a)(t)\, g(t)$$

for all $s \in S \setminus R$ *and all* $a \in A(s)$.

4.3 Value iteration

The third way to solve the reduced problem is by value iteration. Convergence to the solution of the SSP problem can be ensured simply by using an initial estimate $\boldsymbol{v}^0$ that is identically 0.

Theorem 10 *Consider an instance Π of non-positive SSP such that $\Pi = \mathit{Converge}(\Pi)$, and such that there is at least one proper policy for each state. Then, the solution of the SSP problem is given by* $\lim_{k\to\infty} L^k(\mathbf{0})$, *where* $\mathbf{0}$ *is the vector all whose entries are 0.*

Proof. The theorem follows from the fact that, in a non-positive SSP problem, all fixpoints of the Bellman operator are componentwise smaller or equal to $\mathbf{0}$. Since the solution computed by Theorem 10 is the largest such fixpoint, by Theorem 9 it is also the solution of the SSP problem. ∎

5 Maximum and Minimum Reachability Probabilities

An instance $\Lambda = (S, \mathit{Acts}, A, p, T)$ of the *maximum* or *minimum reachability* problems consists of an MDP $\Pi = (S, \mathit{Acts}, A, p)$ together with a destination set T. The maximum and minimum reachability probability problems consists in determining, for all $s \in S$, the values

$$u_s^+ = \sup_{\eta \in \mathit{Pol}} \Pr_s^\eta(\exists k \,.\, X_k \in T) \qquad u_s^- = \inf_{\eta \in \mathit{Pol}} \Pr_s^\eta(\exists k \,.\, X_k \in T)\,.$$

Let $Z \subseteq S$ be the subset of states that cannot reach T (so that $u_s^+ = 0$ for $s \in Z$). From [6], we know that the maximum reachability probability can be solved using a linear programming problem on the set of variables $\{u_s \mid s \in S \setminus (T \cup Z)\}$. Here, we show how our results on the SSP problem can be used to improve the efficiency of that solution, as well as to solve the minimum reachability probability problem.

Maximum reachability probability. To reduce the maximum reachability probability problem to the SSP problem, we construct from the instance Λ an SSP instance $\Pi = \mathit{Ssp}^+(\Lambda) = (S, \mathit{Acts}, A, p, R, c, g)$, where $R := \mathit{Reach}(T) \cup Z$, the cost c is identically 0, and the terminal cost is defined by $g(s) = -1$ for $s \in \mathit{Reach}(T)$, and $g(s) = 0$ for $s \in Z$. Note that Λ is both a non-negative and a non-positive instance of SSP problem. The following theorem relates the two problems.

Theorem 11 *If $\Pi = \mathit{Ssp}^+(\Lambda)$, then $u_s^+ = -v_s^*$ for all $s \in S \setminus R$, where u_s^+ is computed on Λ and v_s^* on Π.*

Proof. Since every state of $S \setminus R$ can reach R, we have that $Reach(R) = S$, so that every state of S has a proper policy. From a memoryless optimal policy η_s for the SSP problem, we can construct a policy η_r that coincides with η_s on $S \setminus R$ such that $-v_s^{\eta_s} = u_s^{\eta_r}$, yielding $-v_s^* \leq u_s^+$ for all $s \in S \setminus T$. In the other direction, consider a memoryless policy η_r optimal for reachability (we know from [8] that such a policy exists). We have $u_s^{\eta_r} = -v_s^{\eta_r}$ for all $s \in S \setminus T$. Moreover, η_r is proper, since every state of $S \setminus R$ can reach T with positive probability. This yields the reverse inequality $-v_s^* \geq u_s^+$ for all $s \in S \setminus R$, and hence the result. ∎

Note that we have used algorithm (2) to reduce the size of the set of states on which the maximum reachability probability must be determined, from $S \setminus (T \cup Z)$ to $S \setminus (Reach(T) \cup Z)$. Theorem 11 opens the way to the application of Algorithm 1 for the solution of maximum reachability probability problems. Since the running cost c is identically 0, the algorithm eliminates all end components of the MDP that lie completely outside of $Reach(T) \cup Z$, achieving a further potential reduction in the size of the problem.

Minimum Reachability Probability. Let $\{(C_1, D_1), \ldots, (C_n, D_n)\} = Mec(S \setminus T, A)$ be the set of maximal end components lying outside T, and let $C = \bigcup_{i=1}^{n} C_i$ be the union of their states. Clearly, from $Z \cup C$ the minimum probability of reaching T is 0. Moreover, the MDP does not have any end component completely contained in $S \setminus (T \cup Z \cup C)$. From the instance $\Lambda = (S, Acts, A, p, T)$ we construct an SSP instance $\Pi = Ssp^-(\Lambda) = (S, Acts, A, p, R, c, g)$, where $R := T \cup Z \cup C$, the cost c is identically 0, and the terminal cost is defined by $g(s) = 0$ for $s \in Z \cup C$, and $g(s) = 1$ for $s \in T$. The following theorem relates the two problems, and it enables the computation of the minimum probability of reaching the target.

Theorem 12 *If $\Pi = Ssp^-(\Lambda)$, then $u_s^- = v_s^*$ for all $s \in S$, where u_s^- is computed on Λ and v_s^* is computed on Π.*

Proof. The proof of the theorem follows from the fact that all policies of Π are proper, and from the observation that from a policy η_s of Π, we can easily obtain a policy η_r for Λ such that $u_s^{\eta_r} = v_s^{\eta_s}$ for all $s \in S$, and vice versa. ∎

In this case, Algorithm 1 cannot be used to reduce the size of the problem, since there are no end components in $S \setminus R$. The reduction has been effected in a more direct way by adding the set C to the set of target states of the SSP problem.

Optimal policies. The maximum and minimum reachability problems admit memoryless optimal policies. This result is proved in [8].

References

1. A. Bianco and L. de Alfaro. Model checking of probabilistic and nondeterministic systems. In *Found. of Software Tech. and Theor. Comp. Sci.*, volume 1026 of *Lect. Notes in Comp. Sci.*, pages 499–513. Springer-Verlag, 1995.
2. R.E. Bellman. *Dynamic Programming*. Princeton University Press, 1957.

3. D.P. Bertsekas. *Dynamic Programming and Optimal Control.* Athena Scientific, 1995. Volumes I and II.
4. D.P. Bertsekas and J.N. Tsitsiklis. An analysis of stochastic shortest path problems. *Math. of Op. Res.*, 16(3):580–595, 1991.
5. D.P. Bertsekas and J.N. Tsitsiklis. *Neuro-Dynamic Programming.* Athena Scientific, 1996.
6. C. Courcoubetis and M. Yannakakis. Markov decision processes and regular events. In *Proc. 17th Int. Colloq. Aut. Lang. Prog.*, volume 443 of *Lect. Notes in Comp. Sci.*, pages 336–349. Springer-Verlag, 1990.
7. C. Courcoubetis and M. Yannakakis. The complexity of probabilistic verification. *J. ACM*, 42(4):857–907, 1995.
8. L. de Alfaro. *Formal Verification of Probabilistic Systems.* PhD thesis, Stanford University, 1997. Technical Report STAN-CS-TR-98-1601.
9. L. de Alfaro. Stochastic transition systems. In *CONCUR'98: Concurrency Theory. 9th Int. Conf.*, Lect. Notes in Comp. Sci. Springer-Verlag, 1998.
10. L. de Alfaro, T.A. Henzinger, and O. Kupferman. Concurrent reachability games. In *Proc. 39th IEEE Symp. Found. of Comp. Sci.*, 1998.
11. C. Derman. *Finite State Markovian Decision Processes.* Academic Press, 1970. cut?
12. J.H. Eaton and L.A. Zadeh. Optimal pursuit strategies in discrete-state probabilistic systems. *J. of Basic Engineering*, pages 23–29, 1962.
13. H. Hansson. *Time and Probabilities in Formal Design of Distributed Systems.* Real-Time Safety Critical Systems Series. Elsevier, 1994.
14. B. Jonsson and K.G. Larsen. Specification and refinement of probabilistic processes. In *Proc. 6th IEEE Symp. Logic in Comp. Sci.*, pages 266–277, 1991.
15. C. Baier and M. Kwiatkowska. Model checking for a probabilistic branching time logic with fairness. *Distributed Computing,* vol. 11, May 1998.
16. M.Z. Kwiatkowska, G. Norman, D. Parker, and R. Segala. Symbolic model checking of concurrent probabilistic systems using MTBDDS and simplex. Technical Report CSR-99-1, University of Birmingham, 1999.
17. J.G. Kemeny, J.L. Snell, and A.W. Knapp. *Denumerable Markov Chains.* D. Van Nostrand Company, 1966.
18. K.G. Larsen and A. Skou. Bisimulation through probabilistic testing (preliminary report). In *Proc. 16th ACM Symp. Princ. of Prog. Lang.*, pages 344–352, 1989.
19. A. Pnueli and L. Zuck. Probabilistic verification by tableaux. In *Proc. First IEEE Symp. Logic in Comp. Sci.*, pages 322–331, 1986.
20. M.O. Rabin. Probabilistic automata. *Information and Computation*, 6:230–245, 1963.
21. R. Segala. *Modeling and Verification of Randomized Distributed Real-Time Systems.* PhD thesis, MIT, 1995. Technical Report MIT/LCS/TR-676.
22. R. Segala and N.A. Lynch. Probabilistic simulations for probabilistic processes. In *CONCUR'94: Concurrency Theory. 5th Int. Conf.*, volume 836 of *Lect. Notes in Comp. Sci.*, pages 481–496. Springer-Verlag, 1994.
23. M.Y. Vardi. Automatic verification of probabilistic concurrent finite-state systems. In *Proc. 26th IEEE Symp. Found. of Comp. Sci.*, pages 327–338, 1985.
24. M. Vardi. Infinite games against nature. Unpublished manuscript, 1995.
25. S.-H. Wu, S.A. Smolka, and E.W. Stark. Composition and behaviors of probabilistic I/O automata. In Springer-Verlag, editor, *CONCUR'94: Concurrency Theory. 5th Int. Conf.*, Lect. Notes in Comp. Sci., pages 511–528, 1994.

Automating Modular Verification*

Rajeev Alur[1], Luca de Alfaro[2], Thomas A. Henzinger[3], and Freddy Y.C. Mang[2]

[1] Department of Computer and Information Science, University of Pennsylvania, and Bell Laboratories, Lucent Technologies. Email: alur@cis.upenn.edu

[2] Department of Electrical Engineering and Computer Sciences, University of California at Berkeley. Email: {dealfaro, fmang}@eecs.berkeley.edu

[3] Department of Electrical Engineering and Computer Sciences, University of California, Berkeley, and Max-Planck Institute for Computer Science, Saarbrücken. Email: tah@eecs.berkeley.edu

Abstract. Modular techniques for automatic verification attempt to overcome the state-explosion problem by exploiting the modular structure naturally present in many system designs. Unlike other tasks in the verification of finite-state systems, current modular techniques rely heavily on user guidance. In particular, the user is typically required to construct module abstractions that are neither too detailed as to render insufficient benefits in state exploration, nor too coarse as to invalidate the desired system properties. In this paper, we construct abstract modules automatically, using reachability and controllability information about the concrete modules. This allows us to leverage automatic verification techniques by applying them in layers: first we compute on the state spaces of system components, then we use the results for constructing abstractions, and finally we compute on the abstract state space of the system. Our experimental results indicate that if reachability and controllability information is used in the construction of abstractions, the resulting abstract modules are often significantly smaller than the concrete modules and can drastically reduce the space and time requirements for verification.

1 Introduction

The single largest obstacle to the use of automatic methods in system verification is the state-explosion problem, which is the exponential increase in the number of system states caused by a linear increase in the number of system components or variables. Modular verification techniques attempt to overcome the state-explosion problem by exploiting the modular structure naturally present in most system designs. The basic idea is to analyze each module of the system separately, perhaps together with an environment that represents a simplified model of the rest of the system; the results obtained for the individual modules are then combined into a single result about the compound system. Unlike other tasks in the verification of finite-state systems, which have been

* This research was supported in part by the NSF CAREER award CCR-9734115, by the NSF CAREER award CCR-9501708, by the DARPA (NASA Ames) grant NAG2-1214, by the DARPA (Wright-Patterson AFB) grant F33615-98-C-3614, by the ARO MURI grant DAAH-04-96-1-0341, and by the Gigascale Silicon Research Center.

largely automated, current modular verification techniques still rely heavily on user guidance. Aside from deciding how to break up a system into modules, the user also has to specify the environment in which to study each module, which is usually a difficult task. In this paper, we present an approach to modular verification that is almost entirely automatic, leaving to the user only the task of specifying which variables of a module should be relevant to the other modules.

For each concrete module, we erase some variables to construct an abstract module, which has a smaller state space; the abstract module is then used to replace the concrete module in the verification process. If this approach is pursued naively, typically one of two things happens. Either one abstracts only variables that do not influence the property to be verified, which is certainly prudent but more often than not leads to insufficient savings, or one abstracts variables that do influence the desired property, in which case the abstract module may violate the property even though the concrete module does not. We take the second route, but use additional information about the concrete module in order to construct more useful abstractions than could be achieved by simply erasing variables. In the most basic variation of our method, we use reachability information about the concrete module when erasing variables to construct an abstraction. In a more advanced variation, we also use controllability information about the concrete module with respect to the desired property. In all cases, the additional information we use can be obtained fully automatically by looking only at individual modules and the property to be verified —there is no need to involve the compound system. Our experimental results indicate that the use of reachability and controllability information can lead to dramatic improvements in verification: the resulting module abstractions are often much smaller than the concrete modules yet still preserve the desired property.

Our model of computation is that of transition systems defined over finite sets of state variables. We describe systems as the parallel composition of one or more modules. A *module* $P = (\mathcal{V}_P, I_P, T_P)$ consists of a set $\mathcal{V}_P$ of variables, partitioned into *input* and *output variables*, an *initial predicate* I_P over $\mathcal{V}_P$ defining the initial states of P, and a *transition predicate* T_P over $\mathcal{V}_P \cup \mathcal{V}'_P$ defining the possible state transitions of P in terms of their source states (over $\mathcal{V}_P$) and destination states (over $\mathcal{V}'_P = \{x' \mid x \in \mathcal{V}_P\}$). We consider systems consisting of *non-blocking* modules, in which every state has a successor, regardless of the inputs to the module. The semantics of parallel composition is conjunction: $P \,||\, Q = (\mathcal{V}_P \cup \mathcal{V}_Q, I_P \wedge I_Q, T_P \wedge T_Q)$. For the sake of simplicity, in this paper we focus on Moore modules, for which the outputs during a transition depend only on the source state of the transition. Our approach can be adapted with only minor modifications to Mealy-type modules, such as the Reactive Modules of [1]. We consider the verification of invariance properties. An invariance property for the module P is specified by an *invariant predicate* φ over $\mathcal{V}_P$. The module P *satisfies* the invariant predicate φ, written $P \models \Box\varphi$, if P never leaves the set of states defined by φ.

Consider a system $P \,||\, Q$ consisting of two modules P and Q, and a desired invariant predicate φ for $P \,||\, Q$. To check if $P \,||\, Q \models \Box\varphi$ without constructing the global state space of $P \,||\, Q$, we can remove a subset $\mathcal{W}_P \subseteq \mathcal{V}_P$ of the variables of P and a subset $\mathcal{W}_Q \subseteq \mathcal{V}_Q$ of the variables of Q. Formally, the abstract module $(\exists\, \mathcal{W}_P . P) = (\mathcal{V}_P \backslash \mathcal{W}_P, \exists \mathcal{W}_P \,.\, I_P, \exists \mathcal{W}_P \exists \mathcal{W}'_P \,.\, T_P)$ is constructed by existentially quantifying the removed variables in the initial and transition predicates; we say that $(\exists \mathcal{W}_P.P)$ is obtained by *erasing* from P the variables in $\mathcal{W}_P$. Then we can attempt to use the following standard inference rule:

$$\frac{(\exists \mathcal{W}_P.P) \,\|\, (\exists \mathcal{W}_Q.Q) \models \Box\varphi}{P \,\|\, Q \models \Box\varphi} \tag{1}$$

This rule is sound, because every reachable state of the concrete system $P \,\|\, Q$ corresponds to a reachable state of the abstract system $(\exists \mathcal{W}_P.P) \,\|\, (\exists \mathcal{W}_Q.Q)$. The efficiency advantage of the rule stems from the fact that the premise involves fewer variables than the conclusion, reducing the size of the state space to be explored. However, the premise may fail even though the conclusion holds, because there may be many reachable states of the abstract system that do not correspond to reachable states of the concrete system. In fact, it is often impossible to choose suitable, reasonable large sets $\mathcal{W}_P$ and $\mathcal{W}_Q$, because modular designs aggregate naturally within each module only closely interdependent variables. By erasing such dependencies between variables, the number of transitions of the abstract system grows quickly to the point of violating all but trivial invariants. Our goal is to confine this growth in abstract transitions by utilizing additional information about the component modules P and Q.

More precisely, a state s of P can be written as a pair $s = (s_a, s_w)$, where s_a is a state over the set $\mathcal{V}_P \backslash \mathcal{W}_P$ of variables, and s_w is a state over the set $\mathcal{W}_P$ of erased variables. The abstract module $(\exists \mathcal{W}_P.P)$ contains a transition from source state s_a to destination state s'_a iff the concrete module P contains a transition from (s_a, s_w) to (s'_a, s'_w) for some s_w and s'_w. As a first improvement, we can include a transition from s_a to s'_a in the abstract module only if, for some s_w and s'_w, there is a transition from (s_a, s_w) to (s'_a, s'_w) in the concrete module *and* the state (s_a, s_w) is reachable in the concrete module. This is because it is certainly not useful to include abstract transitions that have no reachable concrete counterparts. To this end, we compute a predicate R_P over $\mathcal{V}_P$ that defines the reachable states of P. The predicate R_P can be computed using standard state-space exploration (symbolic or enumerative). Our experiments based on symbolic methods indicate that this computation is efficient, since the module P is considered in isolation. From the predicate R_P we construct the module $(P \,\&\, R_P) = (\mathcal{V}_P, I_P, T_P \wedge R_P)$, which is like P, except that it allows only transitions from reachable states. After erasing the variables in $\mathcal{W}_P$, we obtain the abstract module $(\exists \mathcal{W}_P.(P \,\&\, R_P))$. In a similar way, we compute the reachability predicate R_Q for Q and construct the abstract module $(\exists \mathcal{W}_Q.(Q \,\&\, R_Q))$. To complete the verification process, we then use the following rule:

$$\frac{(\exists \mathcal{W}_P.(P \,\&\, R_P)) \,\|\, (\exists \mathcal{W}_Q.(Q \,\&\, R_Q)) \models \Box\varphi}{P \,\|\, Q \models \Box\varphi} \tag{2}$$

Since the systems $P \,\|\, Q$ and $(P \,\&\, R_P) \,\|\, (Q \,\&\, R_Q)$ have the same reachable states, rule (2) is sound. As we shall see, unlike the simplistic rule (1), the improved rule (2) can often be successfully applied even when the sets $\mathcal{W}_P$ and $\mathcal{W}_Q$ include variables that contribute to ensure the invariant φ. Yet the savings in checking the premise of rule (2) are just as great as those for checking the premise of the earlier rule (1), because the same sets of variables are erased. In other words, $(\exists \mathcal{W}_P.(P \,\&\, R_P)) \,\|\, (\exists \mathcal{W}_Q.(Q \,\&\, R_Q))$ is a more accurate but no more detailed abstraction of $P \,\|\, Q$ than is $(\exists \mathcal{W}_P.P) \,\|\, (\exists \mathcal{W}_Q.Q)$. In our experiments we shall obtain dramatic results by applying rule (2) with the simple heuristics of erasing those variables that are not involved in the communication between P and Q. While reachability information is often used in algorithmic verification, the novelty of rule (2) consists in the use of such information for the modular construction of abstractions.

The effectiveness of a rule such as (1) or (2) is directly related to the number of variables that can be erased in a successful application of the rule. Rule (2) improves on rule (1) by using *reachability* information about the individual modules in the construction of the abstractions, which usually permits the erasure of more variables. It is possible to further improve on the rule (2) by using, in addition to reachability information, also information about the *controllability* of the individual modules with respect to the specification $\Box\varphi$. This improvement is based on the following observation. The predicate R_P used in (2) defines the reachable states of P *when P is in a completely general environment.* However, the module P may exhibit anomalous behaviors in a completely general environment; in particular, more states may be reachable under a completely general environment than under the specific environment provided by Q. Of course, we do not want to compute the reachable states of P when P is composed with Q: doing so would require the exploration of the state space of the global system $P \,\|\, Q$, which is exactly what our modular verification rules try to avoid. To study the module P under a suitable confining environment, while still avoiding the exploration of the global state space, we consider the module P *in the most general environment E that ensures the invariant φ*; that is, E is the least restrictive module such that $P \,\|\, E \models \Box\varphi$. In practice, we need not construct E explicitly, but compute only the predicate D_P that defines the set of reachable states of $P \,\|\, E$. Since E is more restrictive than the completely general environment, the predicate D_P is stronger than R_P, and the implication $D_P \to R_P$ holds. The algorithm for computing D_P follows from the standard game-theoretic algorithm for computing the set of states of the module P that are controllable with respect to the invariant φ; it can be implemented symbolically or enumeratively, with a time complexity that is linear in the size of the state space of P [4]. This leads to the following modular verification rule:

$$\frac{\begin{array}{c}(I_P \wedge I_Q) \to (D_P \wedge D_Q)\\ P \,\|\, (\exists \mathcal{W}_Q.(Q \,\&\, D_Q)) \models \Box D_P\\ Q \,\|\, (\exists \mathcal{W}_P.(P \,\&\, D_P)) \models \Box D_Q\end{array}}{P \,\|\, Q \models \Box\varphi} \qquad (3)$$

where $\mathcal{W}_P \subseteq \mathcal{V}_P$ and $\mathcal{W}_Q \subseteq \mathcal{V}_Q$. The soundness of this rule depends on an inductive argument, and it will be proved in detail in the paper. Essentially, the first premise ensures that the modules P and Q are initially in states satisfying $D_P \wedge D_Q$. The second premise shows that, as long as Q does not leave the set defined by D_Q, the module P will not leave the set defined by D_P; the third premise is symmetrical. As the implications $D_P \to \varphi$ and $D_Q \to \varphi$ hold, the three premises lead to the conclusion. The rule is in fact closely related to inductive forms of assume-guarantee reasoning [16, 3, 1, 14]. The use of the stronger predicates D_P and D_Q in the second and third premises of the rule (3) potentially enables the erasure of more variables compared to the earlier rule (2). However, in rule (3) this erasure can take place only on one side of the parallel composition operator or, in the case of multi-module systems, for all modules but one.

While automatic approaches to the construction of abstractions for model checking have been proposed, for example, in [13,9,10,7], these approaches do not exploit reachability and controllability information in a modular fashion. In particular, instead of the standard principle "first abstract, then model check the abstraction," our approach follows the more refined principle "first model check the components, then use this information to abstract, then model check the compound abstraction." In this way, our modular verification rules are doubly geared towards automatic verification methods: state-space exploration is used both to compute the reachability and controllability predicates, and to check all temporal premises (those which contain the $\models$ operator). It is worth pointing out that nontemporal premises would result in rules that are considerably less powerful. For example, suppressing variable erasures, the temporal premise $(P \,\&\, R_P) \,\|\, (Q \,\&\, R_Q) \models \Box\varphi$ of rule (2) is weaker than the two nontemporal premises $I_P \wedge I_Q \rightarrow \varphi$ and $\varphi \wedge R_P \wedge T_P \wedge R_Q \wedge T_Q \rightarrow \varphi'$ would be (here, φ' results from φ by replacing all variables with their primed versions). Similarly, the second premise of rule (3) is weaker than the two nontemporal premises $I_P \wedge I_Q \rightarrow D_Q \wedge D_P$ and $D_P \wedge T_P \wedge D_Q \wedge T_Q \rightarrow D'_P$ would be. It is easy to find examples where our temporal premises apply, but their nontemporal counterparts do not.

The outline of the paper is as follows. After introducing preliminary definitions in Section 2, we develop the technical details of the proposed modular verification rules in Section 3. The verification rules have been implemented on top of the MOCHA model checker [2], using BDD-based fixpoint algorithms for the computation of the reachability and controllability predicates. In Section 4 we discuss the implementation of the verification rules, and we describe the *script language* we devised in order to be able to experiment efficiently with various modular verification techniques. In Section 5 we present experimental results for three examples: a demarcation protocol used to maintain the consistency between distributed databases [5], a token-ring arbiter, and a sliding-window protocol for data communication [11]. We conclude the paper with some insights gathered in the course of the experimentation with the proposed verification rules.

2 Modules

Given a set $\mathcal{V}$ of typed variables with finite domain, a state s over $\mathcal{V}$ is an assignment for $\mathcal{V}$ that assigns to each $x \in \mathcal{V}$ a value $s[\![x]\!]$. We also denote by $\mathcal{V}' = \{x' \mid x \in \mathcal{V}\}$ the set obtained by priming each variable in $\mathcal{V}$. Given a predicate H over $\mathcal{V}$, we denote by H' the predicate obtained by replacing in H every $x \in \mathcal{V}$ with $x' \in \mathcal{V}'$. Given a set A and an element x, we often write $A \backslash x$ for $A \backslash \{x\}$, when this generates no confusion. A module $P = (\mathcal{C}_P, \mathcal{E}_P, I_P, T_P)$ consists of the following components:

1. A (finite) set $\mathcal{C}_P$ of *controlled variables,* each with finite domain, consisting of the variables whose values can be accessed and modified by P.
2. A (finite) set $\mathcal{E}_P$ of *external variables,* each with finite domain, consisting of the variables whose values can be accessed, but not modified, by P.

3. A *transition predicate* T_P over $\mathcal{C}_P \cup \mathcal{E}_P \cup \mathcal{C}'_P$.
4. An *initial predicate* I_P over $\mathcal{C}_P$.

We denote by $\mathcal{V}_P = \mathcal{C}_P \cup \mathcal{E}_P$ the set of variables mentioned by the module. Given a state s over $\mathcal{V}_P$, we write $s \models I_P$ if I_P is satisfied under the variable interpretation specified by s. Given two states s, s' over $\mathcal{V}_P$, we write $(s, s') \models T_P$ if predicate T_P is satisfied by the interpretation that assigns to $x \in \mathcal{V}_P$ the value $s[\![x]\!]$, and to $x' \in \mathcal{V}'_P$ the value $s'[\![x]\!]$. A module P is *non-blocking* if the predicate I_P is satisfiable, i.e., if the module has at least one initial state, and if the assertion $\forall \mathcal{V}_P \,.\, \exists \mathcal{C}'_P \,.\, T_P$ holds, so that every state has a successor. A *trace* of module P is a finite sequence of states $s_0, s_1, s_2, \ldots s_n \in \mathit{States}(\mathcal{V}_P)$, where $n \geq 0$ and $(s_k, s_{k+1}) \models T_P$ for all $0 \leq k < n$; the trace is *initial* if $s_0 \models I_P$. Two modules P and Q are *composable* if $\mathcal{C}_P \cap \mathcal{C}_Q = \emptyset$; in this case, their *parallel composition* $P \,||\, Q$ is defined as:

$$P \,||\, Q = \left(\mathcal{C}_P \cup \mathcal{C}_Q, (\mathcal{E}_P \cup \mathcal{E}_Q) \backslash (\mathcal{C}_P \cup \mathcal{C}_Q), I_P \wedge I_Q, T_P \wedge T_Q\right) .$$

Given a module P and a predicate H over $\mathcal{V}_P$, we denote by

$$(P \,\&\, H) = \left(\mathcal{C}_P, \mathcal{E}_P, I_P \wedge H, T_P \wedge H\right)$$

the module like P, except that only transitions from states that satisfy H are allowed. Given a module P and a set $\mathcal{W}$ of variables, we let

$$(\exists \mathcal{W}.P) = \left(\mathcal{C}_P \backslash \mathcal{W}, \mathcal{E}_P \backslash \mathcal{W}, \exists \mathcal{W} \,.\, I_P, \exists \mathcal{W}, \mathcal{W}' \,.\, T_P\right)$$

be the module obtained by *erasing* the variables $\mathcal{W}$ in P. Note that the module $(P \,\&\, H)$ can be blocking even if module P is non-blocking. On the other hand, the parallel composition of non-blocking modules is non-blocking, and a module obtained from a non-blocking module by erasing variables is also non-blocking.

A state of a module P is *reachable* if it appears in some initial trace of P. We denote by *Reach*(P) the predicate defining the reachable states of P; this predicate can be compute using standard state-space exploration techniques [8]. Given a module P and a predicate φ, the relation $P \models \Box\varphi$ holds iff the implication *Reach*$(P) \rightarrow \varphi$ is valid. In this paper, we present modular techniques for verifying whether the relation $P_1 \,||\, \cdots \,||\, P_n \models \Box\varphi$ holds, where $P_1, P_2, \ldots, P_n$ are composable modules, for $n > 0$, and where φ is defined over the set of variables $\bigcup_{i=1}^{n} \mathcal{V}_{P_i}$. This verification problem is known as the *invariant verification problem,* and it is one of the most basic problems in formal verification.

3 Modular Rules for Invariant Verification

In this section, we present three modular rules for the verification of invariants; the rules are presented in order of increasing sophistication, and of increasing ability of successfully erasing variables. The first rule is a standard rule based on the construction of abstract modules:

$$\frac{(\exists \mathcal{W}_1.P_1) \,||\, \cdots \,||\, (\exists \mathcal{W}_n.P_n) \models \Box\varphi}{P_1 \,||\, \cdots \,||\, P_n \models \Box\varphi} \tag{4}$$

The second rule is derived from the above rule, by using in the construction of the abstract modules also information about the reachable states of the concrete modules. The third rule constructs the abstract modules using both reachability and controllability information about the concrete modules.

3.1 Reachability-based abstractions

In order to improve the ability of rule (4) to successfully erase variables, we construct the abstract modules using reachability information about the concrete modules. Hence, we formulate the following modular verification rule:

$$\frac{(\exists \mathcal{W}_1.(P_1 \,\&\, \mathit{Reach}(P_1))) \parallel \cdots \parallel (\exists \mathcal{W}_n.(P_n \,\&\, \mathit{Reach}(P_n))) \models \Box\varphi}{P_1 \parallel \cdots \parallel P_n \models \Box\varphi} \tag{5}$$

This rule is sound. The rule is also complete, since whenever the conclusion holds, the premise also does, with the choice $\mathcal{W}_1 = \cdots = \mathcal{W}_n = \emptyset$. Our experiments indicated that rule (5) is often surprisingly effective in enabling the successful erasure of variables, leading to dramatic savings in the space and time requirements of verification. We illustrate this with an example.

Example 1 This example is a simplified version of the token-ring example presented in Section 5. Consider a system composed of two modules P and Q that circulate a token through a 4-phase handshake protocol. The module P has controlled variables $\mathcal{C}_P = \{\mathit{grant}_1, \mathit{ack}_1, x_1, y_1, c_1\}$ and external variables $\mathcal{E}_P = \{\mathit{grant}_2, \mathit{ack}_2\}$. All variables are boolean, except for c_1 that has domain $\{0, 1, 2, 3\}$. The module Q is defined similarly, except that the subscripts 1 and 2 are exchanged. Intuitively, grant_2 and ack_1 form the handshake that passes a token from Q to P. Once the token arrives into P, it is stored first in x_1, then in y_1. The handshake variables grant_1 and ack_2 are used to pass the token back to Q. The variable c_1 is an auxiliary variable that records the number of tokens in P. The initial condition of P is $I_P : \neg \mathit{ack}_1 \wedge \neg \mathit{grant}_1 \wedge x_1 \wedge \neg y_1 \wedge (c_1 = 0)$; the initial condition of Q is $I_Q : \neg \mathit{ack}_2 \wedge \neg \mathit{grant}_2 \wedge \neg x_2 \wedge \neg y_2 \wedge (c_2 = 0)$, so that the token is initially in x_1. We present the transition predicate of P in guarded-commands notation, with the convention that the values of the variables not mentioned in the assignments are not modified, and that the command to be executed is chosen nondeterministically among those whose guards are true:

$$\begin{array}{lll}
[\!] \; \mathit{grant}_2 \wedge \neg \mathit{ack}_1 \wedge \neg x_1 & \longrightarrow & \mathit{ack}_1' := \mathrm{T};\ x_1' := \mathrm{T};\ c_1' := (c_1 + 1) \bmod 4 \\
[\!] \; \neg \mathit{grant}_2 \wedge \mathit{ack}_1 & \longrightarrow & \mathit{ack}_1' := \mathrm{F} \\
[\!] \; x_1 \wedge \neg y_1 & \longrightarrow & x_1' := \mathrm{F};\ y_1' := \mathrm{T} \\
[\!] \; \neg \mathit{grant}_1 \wedge \neg \mathit{ack}_2 \wedge y_1 & \longrightarrow & \mathit{grant}_1' := \mathrm{T};\ y_1' := \mathrm{F};\ c_1' := (c_1 - 1) \bmod 4 \\
[\!] \; \mathit{grant}_1 \wedge \mathit{ack}_2 & \longrightarrow & \mathit{grant}_1' := \mathrm{F} \\
[\!] \; \mathrm{T} & \longrightarrow &
\end{array}$$

The transition predicate of Q is identical, except that the subscripts 1 and 2 are exchanged. The invariant is $\varphi : [(c_1 + c_2) \bmod 4 < 2]$, and states that there is at most one token. To verify that $P \parallel Q \models \Box\varphi$, we can apply rule (5) with sets of erased variables $\mathcal{W}_P = \{x_1, y_1\}$ and $\mathcal{W}_Q = \{x_2, y_2\}$. Hence, we are able to erase all the variables that are not used for communication, and that do not appear in the invariant. The intuition is that, once the value of c_1 is known, the predicate

$$Reach(P) : (c_1 = 0 \wedge \neg x_1 \wedge \neg y_1) \vee (c_1 = 1 \wedge (x_1 \not\equiv y_1)) \vee (c_1 = 2 \wedge x_1 \wedge x_2)$$

provides sufficient information about the possible values of the erased variables x_1 and y_1 to enable an accurate computation of the successor states. In contrast, rule (4) does not enable the erasure of any variables. ∎

3.2 Controllability and reachability-based abstractions

Consider an instance $P_1 \| \cdots \| P_n \models \Box\varphi$ of the invariant verification problem, for $n \geq 1$. As mentioned in the introduction, the predicate $Reach(P_i)$ defines the reachable states of module P_i *when the module P_i is in a completely arbitrary environment,* for $1 \leq i \leq n$. However, a module may have many more reachable states when composed with a completely arbitrary environment, than when composed with the other modules of the system. To obtain more precise predicates, we consider the states of P_i that are reachable under the *most general environment under which P_i satisfies the specification* $\Box\varphi$, for $1 \leq i \leq n$. The idea is that, if the system has been properly designed, then the actual environment of P_i is a special case of this most general environment.

An *environment* for a module P is a non-blocking module E composable with P. Given a module P and a predicate φ, we denote by $Envs(P)$ the set of all environments of P, and we let $Envs_\varphi(P) = \{E \in Envs(P) \mid P \| E \models \Box\varphi\}$ the set of environments of P under which the specification $\Box\varphi$ holds. We define

$$CR(P, \varphi) = \bigvee_{E \in Envs_\varphi(P)} \exists(\mathcal{V}_E \backslash \mathcal{V}_P) \,.\, Reach(P \| E)$$

with the convention that $CR(P, \varphi) = \mathrm{F}$ if $Envs_\varphi(P) = \emptyset$. The predicate $CR(P, \varphi)$ defines the set of states of P that can be reached when P is composed with an environment under which $\Box\varphi$ holds. Denote by $\mathcal{V}_\varphi$ the variables occurring in φ. The following proposition gives some additional properties of the predicate $CR(P, \varphi)$.

Proposition 1 *Given a non-blocking module P and a predicate φ, the following assertions hold.*

1. *There is an environment $E \in Envs_\varphi(P)$ with $\mathcal{V}_E = \mathcal{V}_P \cup \mathcal{V}_\varphi$ such that $CR(P, \varphi) \equiv \exists(\mathcal{V}_\varphi \backslash \mathcal{V}_P) \,.\, Reach(P \| E)$.*
2. *The implications $CR(P, \varphi) \to \exists(\mathcal{V}_\varphi \backslash \mathcal{V}_P) \,.\, \varphi$ and $CR(P, \varphi) \to Reach(P)$ hold.*

Regarding the second assertion, note that in the introduction we implicitly assumed $\mathcal{V}_\varphi \subseteq \mathcal{V}_{P_i}$ for $1 \leq i \leq n$ for the sake of simplicity, while here we are only assuming the weaker $\mathcal{V}_\varphi \subseteq \bigcup_{i=1}^{n} \mathcal{V}_{P_i}$. We can then formulate the verification rule:

$$\frac{\begin{array}{l} \bigwedge_{i=1}^{n} I_{P_i} \;\to\; \bigwedge_{i=1}^{n} CR(P_i, \varphi) \\ P_i \| \left(\|_{j \in \{1,\ldots,n\} \backslash i} (\exists \mathcal{W}_j . (P_j \,\&\, CR(P_j, \varphi)))\right) \models \Box CR(P_i, \varphi) \quad 1 \leq i \leq n \end{array}}{P_1 \| \cdots \| P_n \models \Box\varphi} \qquad (6)$$

In the second premise of this rule, for $1 \leq i \leq n$, we cannot erase variables of P_i. In fact, the predicate $CR(P_i, \varphi)$ on the right hand side of $\models$ involves most of the variables in P_i, preventing their erasure. In the experiments described in Section 5, the systems were composed of two modules, and rule (5) performed better than rule (6), since in rule (5) the variables could be erased in both the composing modules. In systems composed of many modules, it is conceivable that the advantage derived from using the stronger predicates of rule (6) in all modules but one, thus possibly erasing more variables, outweighs the disadvantage of not being able to erase variables in one of the modules.

Proposition 2 *Rule (6) is sound. If $P_1, \ldots, P_n$ are non-blocking, rule (6) is also complete: if the conclusion holds, then the premises also hold for $\mathcal{W}_1 = \cdots = \mathcal{W}_n = \emptyset$.*

Proof. It suffices to consider the case $\mathcal{W}_1 = \cdots = \mathcal{W}_n = \emptyset$. To show that the rule is sound, we assume that its premises hold, and we prove by induction on $k \geq 0$ that, if $s_0, s_1, \ldots, s_k$ is an initial trace of $P_1 \| \cdots \| P_n$, then $s_i \models CR(P_j, \varphi)$ for all $0 \leq i \leq k$ and $1 \leq j \leq n$. The base case follows from the first premise of (6). For the induction step, assume that the assertion holds for k, and consider the assertion for $k+1$ for any j, with $1 \leq j \leq n$. The trace $s_0, s_1, \ldots, s_k, s_{k+1}$ is an initial trace of $P_j \| \left(\|_{l \in \{1,\ldots,n\} \setminus j} \left(P_j \,\&\, CR(P_j, \varphi)\right)\right)$ Hence, we have that $s_{k+1} \models CR(P_j, \varphi)$, completing the induction step. From $\mathcal{V}_\varphi \subseteq \bigcup_{i=1}^n \mathcal{V}_{P_i}$ and from Proposition 1, part 2, we have that the implication $(\bigwedge_{i=1}^n CR(P_i, \varphi)) \rightarrow \varphi$ holds. This implication, together with the conclusion of the induction proof, leads to the desired result. The completeness of the rule follows by noticing that if $P_1 \| \cdots \| P_n \models \Box\varphi$, then by definition of $CR(\cdot, \varphi)$ we have $P_1 \| \cdots \| P_n \models \Box(CR(P_1, \varphi) \wedge \cdots \wedge CR(P_n, \varphi))$. ∎

To compute the predicate $CR(P, \varphi)$ given P and φ, we proceed in two steps. First, we compute the predicate $Ctr(P, \varphi)$ defining the set of states from which P is *controllable* with respect to the safety property $\Box\varphi$. The predicate $Ctr(P, \varphi)$ can be computed with a standard controllability algorithm [17,4,15].

Algorithm 1
Input: Module P and predicate φ.
Output: Predicate $Ctr(P, \varphi)$ over $\mathcal{V}_P$.

Initialization: Let $\mathcal{F} = \mathcal{V}_\varphi \backslash \mathcal{V}_P$ and $U_0 = \exists \mathcal{F} \,.\, \varphi$.
Repeat: For $k \geq 0$, let $U_{k+1} = U_k \wedge \exists(\mathcal{E}'_P \cup \mathcal{F}') \,.\, \forall \mathcal{C}'_P \,.\, (T_P \rightarrow (U'_k \wedge \varphi'))$.
Until: $U_{k+1} \equiv U_k$.
Return: U_k.

The algorithm computes a sequence $U_0, U_1, U_2, \ldots$ of increasingly strong predicates. For $k \geq 0$, predicate U_k defines the states from which it is possible to control P to satisfy predicate φ for at least $k+1$ steps; note that the implication $U_k \rightarrow \exists \mathcal{F} \,.\, \varphi$ holds for $k \geq 0$. At each iteration $k \geq 0$, the algorithm lets U_{k+1} define the set of states from which the environment can choose the next value for the external variables, so that for all choice of the controlled variables, the successor states of the transitions satisfy U_k. The following algorithm computes the predicate $CR(P, \varphi)$, using the previous algorithm as a subroutine.

Algorithm 2
Input: Module P and predicate φ.
Output: Predicate $CR(P, \varphi)$ over $\mathcal{V}_P$.

Initialization: Let $\mathcal{F} = \mathcal{V}_\varphi \backslash \mathcal{V}_P$, and $V_0 = I_P \wedge \exists \mathcal{F} \,.\, \forall \mathcal{C}_P \,.\, \big(I_P \rightarrow (Ctr(P, \varphi) \wedge \varphi)\big)$.
Repeat: For $k \geq 0$, let

$$V'_{k+1} = V'_k \vee \exists \mathcal{V}_P \,.\, \big[V_k \wedge T_P \wedge \exists \mathcal{F}' \,.\, \forall \mathcal{C}'_P \,.\, \big(T_P \rightarrow (Ctr'(P, \varphi) \wedge \varphi)\big)\big] \,.$$

Until: $V_{k+1} \equiv V_k$.
Return: V_k.

For each $k \geq 0$, the predicate V_k over $\mathcal{V}_P$ defines the set of states of P that can be reached in k or less steps when P is composed with an environment E such that $P \,||\, E \models \Box\varphi$. To understand how this predicate is computed, note that the predicate $\forall \mathcal{C}_P \,.\, (I_P \rightarrow (Ctr(P, \varphi) \wedge \varphi))$ defines the set of initial valuations for the variables in $\mathcal{E}_P \cup \mathcal{F}$ that are *safe for the environment:* if one such valuation is chosen by the environment, the system will start in a controllable state that satisfies φ, regardless of the valuation for the controlled variables in $\mathcal{C}_P$ chosen by the module P. The iteration step follows a similar idea. If V_k defines the set of current states, then the formula $K_1 : \exists \mathcal{V}_P \,.\, (V_k \wedge T_P)$ over $\mathcal{C}'_P$ defines the valuations for the controlled variables that can be chosen by P for the following state. The environment must choose a valuation for the variables in $\mathcal{E}'_P \cup \mathcal{F}'$ that ensures that, regardless of the valuation for $\mathcal{C}'_P$ chosen by the module, the successor state satisfies $Ctr'(P, \varphi) \wedge \varphi$. If V_k defines the set of current states, the set of such valuations for $\mathcal{E}'_P \cup \mathcal{F}'$ is defined by the formula

$$K_2 : \exists \mathcal{V}_P \,.\, \forall \mathcal{C}'_P \,.\, \big((V_k \wedge T_P) \rightarrow (Ctr'(P, \varphi) \wedge \varphi)\big).$$

It is then easy to see that the iteration step of Algorithm 2 can be written simply as $V'_{k+1} = K_1 \wedge \exists \mathcal{F}' \,.\, K_2$, so that K_1 constrains the next valuation of the controlled variables, and $\exists \mathcal{F}' \,.\, K_2$ constrains the next valuation of the external variables. Algorithms 1 and 2 can be implemented enumeratively or symbolically, and they have running time linear in $|States(\mathcal{V}_P \cup \mathcal{V}_\varphi)|$. In the next example, we see how rule (6) can enable the erasure of variables that could not be erased with rule (5).

Example 2 Consider the verification problem $P_1 \,||\, P_2 \models \Box\varphi$, where the invariant is $\varphi : \neg z_1 \wedge \neg z_2$. The modules have variables $\mathcal{C}_{P_i} = \{x_i, y_i, z_i\}$ and $\mathcal{E}_{P_i} = \{x_{2-i}, z_{2-i}\}$, for $1 \leq i \leq 2$; all the variables are boolean. Module P_1 has initial predicate I_{P_1} : $\neg x_1 \wedge \neg y_1 \wedge \neg z_1$, and has transition predicate $T_{P_1} : [x'_1 \equiv z_2] \wedge [(\neg x_1 \wedge \neg x_2) \rightarrow (y'_1 \equiv y_1)] \wedge [\neg y_1 \rightarrow (z'_1 \equiv z_1)]$. Module P_2 is defined in a symmetrical fashion. Informally, module P_1 behaves as follows. Initially, all variables are false. At each step, the new value for x_1 is the old value of z_2. If $x_1 \vee x_2$ holds, then y_1 can change value; otherwise, it retains its previous value. If y_1 is true, then z_1 can change value; otherwise, it retains its previous value. It is easy to check that $P_1 \,||\, P_2 \models \Box\varphi$ holds.

Consider module P_1. The states where $z_1 = \mathrm{T}$ or $z_2 = \mathrm{T}$ are obviously not controllable. The states where $y_1 = \mathrm{T}$ are also not controllable, since from these states module P_1 can reach a state where $z_1 = \mathrm{T}$ regardless of the values of the external variables x_2 and z_2. Likewise, the states where $x_1 = \mathrm{T}$ or $x_2 = \mathrm{T}$ are not controllable, since from these states the module can reach a state where $y_1 = \text{T}$ regardless of the values of the external variables. The only controllable (and reachable) state of P_1 is thus defined by the predicate $CR(P_1, \varphi) : \neg x_1 \wedge \neg y_1 \wedge \neg z_1 \wedge \neg x_2 \wedge \neg z_2$. Predicate $CR(P_2, \varphi)$ is defined in a symmetrical fashion. The reachability predicates are given simply by $Reach(P_1) : \text{T}$ and $Reach(P_2) : \text{T}$.

Rule (6) can be applied by taking $\mathcal{W}_1 = \mathcal{W}_2 = \{y_1, y_2\}$. In fact, the composite module $P_1 \parallel (\exists \mathcal{W}_2.(P_2 \,\&\, CR(P_2, \varphi)))$ admits only the initial traces consisting of repetitions of the state $[x_1 = \text{F}, y_1 = \text{F}, z_1 = \text{F}, x_2 = \text{F}, z_2 = \text{F}]$. This shows that the first premise holds; the case for the second premise is symmetrical. On the other hand, no variable can be successfully erased using rule (5). In fact, if we erase variable y_2, then the right hand side exhibits the initial trace s_0, s_1, where $s_0 : [x_1 = \text{F}, y_1 = \text{F}, z_1 = \text{F}, x_2 = \text{F}, z_2 = \text{F}]$ and $s_1 : [x_1 = \text{F}, y_1 = \text{F}, z_1 = \text{F}, x_2 = \text{F}, z_2 = \text{T}]$. This trace is possible because the state $t_0 : [x_1 = \text{F}, z_1 = \text{F}, x_2 = \text{F}, y_2 = \text{T}, z_2 = \text{F}]$ over $\mathcal{V}_{P_2}$ is reachable, and hence it satisfies $Reach(P_2)$, and agrees with s_0 on the shared variables. The trace is then a consequence of the transition from t_0 to $t_1 : [x_1 = \text{F}, z_1 = \text{F}, x_2 = \text{F}, y_2 = \text{T}, z_2 = \text{T}]$ in P_2. A similar argument shows that it is not possible to erase the variable x_2. ∎

4 Implementation of the Verification Rules

We have implemented the algorithms described in this paper in the verification tool MOCHA [2]. MOCHA is an interactive verification environment and it enables, among other things, the verification of invariants using both enumerative and symbolic techniques; for the latter, it relies on the BDD package and image computation engine provided by VIS [6], which we used in our implementation.

One important technique we use in the implementation of the rules is that, instead of computing the abstract modules explicitly, we compute them *implicitly*. The idea is as follows: suppose we are computing the reachable states of $(\exists \mathcal{W}_P.P) \parallel (\exists \mathcal{W}_Q.Q)$. A straight-forward algorithm would be to first compute the two abstract modules, and then compute the reachable states of their composition. This is very inefficient in terms of the usage of space. Transition relations are usually presented as a list of conjuncts rather than as a single, larger conjunct. The explicit computation of the abstract modules would imply conjoining all the transition relations and building a monolithic one: if represented as a BDD, such a monolithic conjunct would often be prohibitively large. Instead, we quantify away the erased variables of the abstract modules only when necessary, as for example in the computation of the reachable states. For instance, we use the following symbolic algorithm to compute the reachable states of the parallel composition of two abstract modules:

Algorithm 3
Input: Modules P and Q, and variables $\mathcal{W}_P \subseteq \mathcal{V}_P \backslash \mathcal{C}_Q$ and $\mathcal{W}_Q \subseteq \mathcal{V}_Q \backslash \mathcal{C}_P$.
Output: $Reach((\exists \mathcal{W}_P.P) \parallel (\exists \mathcal{W}_Q.Q))$.

Initialization: Let $U_0 = \exists(\mathcal{W}_P \cup \mathcal{W}_Q) \,.\, (I_P \wedge I_Q)$.
Repeat For $k \geq 0$, let $U'_{k+1} = U'_k \vee \exists(\mathcal{V}_P \cup \mathcal{V}_Q \cup \mathcal{W}'_P \cup \mathcal{W}'_Q) \,.\, (U_k \wedge T_P \wedge T_Q)$.
Until $U_{k+1} \equiv U_k$.
Return: U_k.

In the body of the loop, we rely on the early quantification algorithm in VIS to keep the intermediate BDDs small. With this scheme, a monolithic transition relation is never built. In particular, our implementation represents abstract modules as pairs consisting of a concrete module and of a list of variables that have been erased from it; such pairs are called *extended modules.*

In order to experiment with the verification rules proposed in this paper, we implemented a simple script language, called `sl`, built on top of MOCHA and based on the Tcl/Tk API. The algorithms and methodologies described in this paper provide the theoretical basis of the commands provided by `sl`. The verification rules proposed in this paper can be implemented as `sl` scripts, and the language `sl` provides invaluable flexibility for experimenting with alternative forms of the rules. An example of script is the following, which verifies the correctness of the *demarcation protocol* using rule (5) (the demarcation protocol is described in Section 5.1).

```
read_module   demarc.rm
sl_em         P Q Spec
sl_reach      phi    em_Spec s
sl_reach      rp     em_P s
sl_restrict   Prest  rp em_P
sl_erase      Pabs   Prest  P/xw P/xr P/req1 P/grant1 P/req2 \
                            P/grant2 P/xlupd1 P/xlupd2 P/busy
sl_reach      rq     em_Q s
sl_restrict   Qrest  rq em_Q
sl_erase      Qabs   Qrest  Q/xw Q/xr Q/req1 Q/grant1 Q/req2 \
                            Q/grant2 Q/xlupd1 Q/xlupd2 Q/busy
sl_compose    Rabs   Pabs Qabs
sl_checkinv   Rabs   phi s
```

The command `read_module` parses the file `demarc.rm`, containing the declarations of the modules `P` and `Q`, composing the protocol, and `Spec`, whose reachable states constitute the invariant. The command `sl_em P Q Spec` builds the extended modules `em_P`, `em_Q`, and `em_Spec` from `P`, `Q`, and `Spec`; of course, these extended modules have empty sets of erased variables. The command `sl_reach phi em_Spec s` computes the predicate `phi` $=$ *Reach*(`em_Spec`). The parameter `s` of this and other commands means "silent", i.e., no diagnostic information is printed. The rest of the script checks that `em_P` $\|$ `em_Q` $\models \Box$`phi` using rule (5). First, the commands `sl_reach` and `sl_restrict` are used to compute `rp` $=$ *Reach*(`em_P`) and `Prest` $=$ (`em_P` & `rp`). Then, the command `sl_erase` erases a specified list of variables from `Prest`, producing the extended module `Pabs`. As discussed earlier, the command `sl_erase` performs no actual computation, but simply adds the specified variables to the list of erased variables. The extended module `Qabs` is constructed in an analogous fashion. Finally, the command `sl_compose` composes `Pabs` and `Qabs` into a single extended module `Rabs`, which is checked against the specification $\Box$`phi` by command `sl_checkinv`.

Apart from these commands, we also have implemented commands including `sl_wcontr` and `sl_contrreach`, which together compute the predicate $CR(P, \varphi)$ given a module P and a predicate φ.

5 Experimental Results

To demonstrate the effectiveness of the proposed approach to modular verification, we compare the time and memory requirements of global state-space exploration with those of rule (5) and rule (6). We do not compare our approach with other modular verification approaches, since these approaches involve user intervention for the construction of the environments. By manually constructing the environments or the abstractions it is possible to improve on our results.

We consider three examples: a demarcation protocol used in distributed databases, a token-ring arbiter, and a sliding-window protocol for data communication. All experiments have been run on a 233 MHz Pentium® II PC with 128MB memory running Linux. We report the memory usage by giving the maximum number of BDD nodes used in any fixpoint computation or predicate; this is essentially the maximum number of BDD nodes used at any single time during verification. We also report the total CPU time; this time does not include swap activity (swap activity was in any case very limited for all examples reported). The automatic variable reordering heuristics of MOCHA were enabled during the experiments. We remark that differences in time or memory usage of up to a factor of 2 are not significant, since they can easily be produced by a variation in the automatic choice of variable ordering.

5.1 Demarcation protocol

The *demarcation protocol* is a distributed protocol aimed at maintaining numerical constraints between data residing in distributed copies of a database, while minimizing the communication requirements [5]. We consider an instance of the protocol that ensures that two databases, residing at sites 1 and 2, never sell more than the maximum available number of seats m aboard a plane. The variables x_1 and x_2 indicate the number of seats that have been sold at sites 1 and 2. Each site can both sell seats, and receive seats returned due to cancellations. In order to minimize the communication between two sites, each site $i = 1, 2$ maintains a variable xl_i indicating the maximum number of seats it can sell autonomously. If a site wishes to sell more seats than this limit allows, the site can send a request to the other site for more seats. Depending on the number of unsold seats, the other site has the option of rejecting the request, or of granting it in part or in full.

We model each site $i = 1, 2$ by a module P_i; the specification is $\Box[(x_1 \leq xl_1) \wedge (x_2 \leq xl_2) \wedge (xl_1 + xl_2 \leq m)]$. Each of P_1 and P_2 controls 20 variables, of which 8 are used for communication with the other module or appear in the invariant, and 12 are internal. Rule (5) enable the erasure of 9 of these 12 variables in each of P_1 and P_2; all of these variables are in the cone of influence of the specification. The table below compares the time and space requirements of global state space exploration with those of rules (5) and (6), for various values of m. To check the robustness of rule (5) against changes in the system model, we also wrote an alternative, somewhat more complex model for the demarcation protocol. For $m = 4$, the verification of the alternative model required 136156 BDD nodes and 2009 seconds with the global approach, and 18720 BDD nodes and 211 seconds with rule (5).

	Global		Rule (5)		Rule (6)	
m	BDD nodes	seconds	BDD nodes	seconds	BDD nodes	seconds
4	20881	97	2847	25	8695	75
6	64345	439	3338	40	20953	218
8	179364	1671	8367	81	43915	517
10	633102	8707	10475	112	65410	1878
12	space-out	—	15923	174	93295	1980
14	space-out	—	22205	300	145676	3913

5.2 Token ring arbiter

The second example is a synchronous token-ring arbiter. It involves a ring of m stations, around which a single *token* is passed unidirectionally through four-phase handshake protocols. The invariant states that there is at most one token present in the stations. A straightforward invariant would involve nearly all the variables in the system, and be rather tedious to write. Hence, we introduce *observer modules* that observe the number of tokens in the system. To enable the decomposition of the ring into two modules P_1 and P_2 representing the half-rings, we introduce two such observers, one for each half. We were able to erase all the variables used for the internal communications and state of the half-rings, even though these variables clearly belong to the cone of influence of the invariant. Each half ring controls $1 + 5m/2$ variables; of these, all but 4 could be erased. Below we compare the performance of global state-space exploration and of rules (5) and (6).

	Global		Rule (5)		Rule (6)	
m	BDD nodes	seconds	BDD nodes	seconds	BDD nodes	seconds
16	657	8	979	7	608	8
20	466	10	1619	9	308	12
24	1138	22	1297	26	473	20
28	1300	39	3486	24	519	29
32	1187	110	3190	143	772	143
36	1323	611	8230	242	1346	195

5.3 Sliding window protocol

Our last example is a classical sliding windows protocol from [11], whose encoding is taken from the MOCHA distribution. The protocol uses send and receive windows of size m, and it is composed of a sender module and a receiver module. Our invariant states essentially that the windows are not over-run by the protocols. In both the sender and the receiver, roughly half of the variables not used for communication with the other module can be erased when applying our modular approach. The comparison between the performance of global state-space exploration and rules (5) and (6) is presented below.

	Global		Rule (5)		Rule (6)	
m	BDD nodes	seconds	BDD nodes	seconds	BDD nodes	seconds
3	8992	35	776	12	2443	33
4	11831	99	1723	41	3740	42
5	36359	1911	3843	84	8503	105
6	94684	4994	7048	156	18316	500
7	95667	2630	8282	513	22289	771
8	space-out	—	26611	1582	47605	6245

5.4 Discussion

The experimental results indicate that the proposed approach leads to a considerable reduction in the time and space requirements for the verification process.

In the examples we considered, we identified which variables could be erased in the application of rule (5) by a simple trial-and-error process. We can automate this process by providing, for each module P, a list $\{x_1, \ldots, x_k\} \subseteq \mathcal{C}_P$ of variables of P that are not part of the specification, and that are not accessed by other modules. We list first the variables that are more likely to be successfully erased: those that are more "internal" to the module, and that interact with fewer other variables. We then apply rule (5) successively with the sets of erased variables $\{x_1, \ldots, x_k\}$, $\{x_1, \ldots, x_{k-1}\}$, $\{x_1, \ldots, x_{k-2}\}$, ..., until the rule succeeds. This process is efficient in practice. In fact, the more variables are erased, the smaller is the state space of the abstract modules: hence if too many variables are erased, the rule will fail in a fraction of the time required for a successful proof.

In the three examples considered, the stronger reachability predicates used to construct the abstract modules in rule (6) did not enable the erasure of any additional variable. In the demarcation protocol and in the sliding window protocol examples, the ability of rule (5) to erase variables on both sides of the parallel composition operator led to superior results compared with rule (6). In the token ring arbiter example, module P_i has many more reachable states in a completely general environment than in an environment compatible with the specification, for $i = 1, 2$. Hence, the predicates $\mathit{Reach}(P_i)$ are much weaker (and take more time and space to compute) than the predicates $\mathit{CR}(P_i, \varphi)$, for $i = 1, 2$. For this reason, rule (6) performs better than rule (5) in this example.

If the premise of rule (5) does not hold, we can construct automatically a trace over the variables in $\bigcup_{i=1}^{n} (\mathcal{V}_{P_i} \setminus \mathcal{W}_i)$, leading to a state that does not satisfy φ. This trace is a trace over a partial set of system variables, and it does not necessarily correspond to a counterexample to the conclusion. If the first premise of rule (6) does not hold, then using facts about controllability we can reconstruct automatically a counterexample trace over the complete set of system variables. On the other hand, if the second premise of rule (6) does not hold for some $1 \leq i \leq n$, then we obtain a trace over a partial set of system variables that leads to a state t_i where the predicate $\mathit{CR}(P_i, \varphi)$ does not hold. From t_i, using facts about controllability we can again construct a trace over the complete set of system variables that leads to a state where φ does not hold. When confronted with a trace over a partial set of variables, we have taken the naïve approach of selectively un-erasing some variables in the premises, until either the premises became valid, or the design error could be identified.

References

1. R. Alur and T.A. Henzinger. Reactive modules. In *Proc. 11th IEEE Symp. Logic in Comp. Sci.*, 1996.
2. R. Alur, T.A. Henzinger, F.Y.C. Mang, S. Qadeer, S.K. Rajamani, and S. Tasiran. Mocha: modularity in model checking. In *Computer Aided Verification*, LNCS 1427, pages 521–525. Springer-Verlag, 1998.
3. Martín Abadi and Leslie Lamport. Conjoining specifications. *ACM Trans. Prog. Lang. Sys.*, 17(3):507–534, 1995.
4. C. Beeri. On the membership problem for functional and multivalued dependencies in relational databases. *ACM Transactions on Database Systems*, 5:241–259, 1980.
5. D. Barbara and H. Garcia-Molina. The demarcation protocol: a technique for maintaining linear arithmetic constraints in distributed database systems. In *EDBT'92: 3rd International Conference on Extending Database Technology*, LNCS 580, pages 373–388. Springer-Verlag, 1992.
6. R. Brayton, G. Hachtel, A. Sangiovanni-Vincentelli, F. Somenzi, A. Aziz, S. Cheng, S. Edwards, S. Khatri, Y. Kukimoto, A. Pardo, S. Qadeer, R. Ranjan, S. Sarwary, T. Shiple, G. Swamy, and T. Villa. VIS: A system for verification and synthesis. In *Computer Aided Verification*, LNCS 1102, pages 428–432. Springer-Verlag, 1996.
7. P. Cousot and R. Cousot. Refining model checking by abstract interpretation *Automated Software Engineering Journal*, 6(1):69–95, 1999.
8. E.M. Clarke, E.A. Emerson, and A.P. Sistla. Automatic verification of finite state concurrent systems using temporal logic. In *Proc. 10th ACM Symp. Princ. of Prog. Lang.*, 1983.
9. D. Dams. *Abstract Interpretation and Partition Refinement for Model Checking*. PhD thesis, Technical University of Eindhoven, 1996.
10. S. Graf and H. Saïdi. Construction of abstract state graphs with PVS. In *Computer Aided Verification*, LNCS. Springer-Verlag, 1997.
11. G.J. Holzman. *Design and Validation of Computer Protocols*. Prentice Hall, 1991.
12. T.A. Henzinger, S. Qadeer, S.K. Rajamani, and S. Tasiran. An assume-guarantee rule for checking simulation. In *Proceedings of the Second International Conference on Formal Methods in Computer-Aided Design (FMCAD 1998)*, LNCS 1522, pages 421–432. Springer-Verlag, 1998.
13. R.P. Kurshan. *Computer-aided Verification of Coordinating Processes: The Automata-Theoretic Approach*. Princeton University Press, 1994.
14. K.L. McMillan. A compositional rule for hardware design refinement. In *Computer Aided Verification*, LNCS 1254, pages 24–35. Springer-Verlag, 1997.
15. P.J. Ramadge and W.M. Wonham. Supervisory control of a class of discrete-event processes. *SIAM Journal of Control and Optimization*, 25:206–230, 1987.
16. E.W. Stark. A proof technique for rely/guarantee properties. In *Proc. of 5th Conference on Foundations of Software Technology and Theoretical Computer Science*, LNCS 206, pages 369–391. Springer-Verlag, 1985.
17. J.W. Thatcher and J.B. Wright. Generalized finite-automata theory with an application to a decision problem of second-order logic. *Mathematical Systems Theory*, 2:57–81, 1968.

"Next" Heuristic for On-the-Fly Model Checking

Rajeev Alur* and Bow-Yaw Wang**

Abstract. We present a new heuristic for on-the-fly enumerative invariant verification. The heuristic is based on a construct for temporal scaling, called next, that compresses a sequence of transitions leading to a given target set into a single metatransition. First, we give an on-the-fly algorithm to search a process expression built using the constructs of hiding, parallel composition, and temporal scaling. Second, we show that as long the target set Θ of transitions includes all transitions that access variables shared with the environment, the process next Θ for P and P are equivalent according to the weak-simulation equivalence. As a result, to search the product of given processes, we can cluster processes into groups with as little communication among them as possible, and compose the groups only after applying appropriate hiding and temporal scaling operators. Applying this process recursively gives an expression that has multiple nested applications of next, and has potentially much fewer states than the original product. We report on an implementation, and show significant reductions for a tree-structured parity computer and a ring-structured leader-election protocol.

1 Introduction

Model checking [7, 24, 9, 16] has proved to be a useful technique for automatic debugging of high-level designs of hardware and protocols. Model checking requires search of the global state-space of the design, and since the number of global states increases exponentially with the size of the description, model checking tools must employ a variety of heuristics to battle this so-called ***state explosion problem***. In this paper, we present a new heuristic for enumerative verification of invariant requirements. Before we describe our approach, let us briefly review two techniques, on-the-fly search of reachable state-space, as implemented in tools such as SPIN [16] and Murφ [11], and reduction techniques based on process equivalences, as implemented in tools such as *Concurrency workbench* [10] and CADP [13].

In on-the-fly model checking, the system is described as a collection of communicating processes. The global state-transition graph of the system is explored on demand starting from initial states using a systematic search algorithm. Only the states, and not the transitions, are stored in a global table, and as soon as

* Department of Computer and Information Science, University of Pennsylvania, and Bell Laboratories, Lucent Technologies. Email: `alur@cis.upenn.edu`. Supported in part by NSF CAREER award CCR-9734115 and by the DARPA grant NAG2-1214.

** Department of Computer and Information Science, University of Pennsylvania. Email: `bywang@saul.cis.upenn.edu`. Supported by the DARPA grant NAG2-1214.

a violation of the invariant is encountered, a counter-example is reported to the user. A variety of heuristics have been shown to be effective in restricting the search to a subset of the global state-space. Sample heuristics are reductions using symmetries [12, 8, 17] and partial orders [22, 25, 14].

In the so-called compositional approaches, the system is described using a richer language (e.g. CCS [21]) that supports operators such as parallel composition and hiding. Compositional techniques rely on the definition of a process equivalence (e.g. weak bisimulation) and algorithms to reduce a process with respect to this equivalence [19, 23, 10, 4, 6]. Starting with a complex expression, subexpressions can then be replaced by equivalent smaller ones in a bottom-up fashion. The order in which processes are composed reflects the connectivity, and a judicious use of hiding allows greater reductions by making more transitions internal at each step. For well-structured designs such as trees and rings, this approach can yield significant reductions in principle. However, it does not guarantee early detection of violations which is extremely important in typical applications of model checking. Furthermore, processing of subexpressions can be expensive. First, minimization with respect to an equivalence is typically $O(n \log n)$ or $O(n^2)$, where n is the number of states in the unreduced graph. While minimization can be replaced by less expensive heuristic reduction algorithms, as far as we know, all the tools explicitly build the state-transition graphs for the subexpressions. Second, a subexpression, when considered on its own, has lot more states than when considered in the context of the whole expression. Heuristic approaches to abstract the context by assumptions during reduction have been proposed [18], but offer only a partial remedy to this problem.

In this paper, we propose an on-the-fly search algorithm that takes into account the hierarchical structure of architecture of the system. The reduction performed by the algorithm on subexpressions compresses internal transitions, and preserves the weak-simulation equivalence. While the basic reduction strategy is simple and well known, the novelty lies in applying it on demand in a hierarchical manner.

We develop our search algorithm in the more general setting of explicitly structured transition relations. Our process model, in addition to the standard operations of *hiding* and *parallel composition*, employs a construct for hierarchically structuring transition relations. For a process P and a set Θ of target transitions, the process next Θ for P is obtained by compressing a sequence of transitions of P ending in a Θ-transition into a single metatransition. The next operator is a temporal scaling operator inspired by notions of multiform time in synchronous languages [5, 15], and was introduced in the language *reactive modules* [1]. The parallel composition and temporal scaling can be mixed freely giving complex processes. Note that applications of next can cause significant reductions in the global state-space by ruling out transient states. For instance, the expression (next Θ for $P \,||$ next Σ for Q) can have much fewer reachable states than $P||Q$: to explore the state-space of $P||Q$, at every step either P or Q is allowed to take a step, while to explore the expression (next Θ for $P \,||$ next Σ for Q), once

P is chosen, it is executed until it takes a Θ-transition, and once Q is scheduled, it is executed until it takes a Σ-transition.

The first challenge is to design an on-the-fly algorithm to search a process expression built using parallel composition and temporal scaling. Such an algorithm is presented in Section 3. It generates global states only on demand, and can do early reporting of violations of invariants. To avoid recomputation, it stores in the global hash-table, along with the global states, transient states generated during processing of next operator. It is guaranteed to visit every reachable state of the given expression, and visits no more, and typically much less, than the reachable states of the flattened expression obtained by removing the applications of next. The running time is linear in the number of states (more precisely, the transitions) visited by the algorithm, and there is basically no overhead in applying the reduction.

Then, we proceed to identify restriction under which the answer to the invariant verification problem is preserved by the application of next. Along the lines of well known results concerning compressing internal transitions (cf. [6]), we establish that if Θ includes those transitions that read or write shared variables then the processes P and next Θ for P are equivalent according to the *weak-simulation* relation. Since weak-simulation is shown to be a congruence with respect to all the operators in our language, we can substitute freely and repeatedly any subexpression with next applied to it. Thus, given a problem to search the composition of a collection of processes, we can heuristically build an expression that clusters the given processes to limit the shared variables across clusters, and apply hiding and temporal scaling to individual clusters before composing them. This process is applied recursively to maximize the reduction. The resulting expression is then searched using the algorithm of Section 3.

The heuristic is demonstrated on a couple of benchmark examples using a prototype implementation. The first example concerns a tree-structured system which computes the parity of inputs presented at the leaves. This example is ideally suited for our reduction: while the number of reachable states of the global system increases exponentially with the size of the tree, the number of reachable states of the reduced search increases almost linearly. The second example concerns a standard leader election algorithm in a ring network. While the reduction is less dramatic, it is still significant for both space and time. The experimental results also support our analysis that the running time is proportional to the number of states visited.

The rest of the paper is organized as follows. Section 2 presents our process model. Section 3 gives the search algorithm for process expressions employing next and parallel composition. Section 4 presents the heuristic application of next and justifies its correctness using the theory of simulation equivalences. Section 5 gives experimental results. Finally, in Section 6, we conclude with comparison with other heuristics.

2 Process Model

We start with the definition of processes. Our process model uses interleaving semantics, and read-shared exclusive-write variables. The model is a simplification of *reactive modules* [1].

Given a set X of typed variables, a *state* over X is a function mapping variables to their values. For a state s, define $s[v/x]$ to be the state obtained from s by replacing the value of x by v.

A process is defined by the set of its variables, rules for initializing the variables, and rules for updating the variables. The variables of a process P are partitioned into three classes: *private* variables that cannot be read or written by other processes, *interface* variables that are written only by P, but can be read by other processes, and *external* variables that can only be read by P, and are written by other processes. Thus, interface and external variables are used for communication, and are called *observable* variables. The process controls its private and interface variables, and its environment controls the external variables. Once the variables of a process are specified, the state space of the process is determined. A state is also partitioned into different components as the variables are (e.g., controlled state and external state). The initialization specifies initial controlled states, and the transition relation specifies how to change the controlled state as a function of the current state.

Definition 1. *A* process P *is a tuple* (X, I, T) *where*

- $X = (X_p, X_i, X_e)$ *is the (typed) variable declaration.* X_p, X_i, X_e *represent the sets of* private variables, interface variables *and* external variables *respectively. Define the* controlled variables $X_c = X_p \cup X_i$ *and the* observable variables $X_o = X_i \cup X_e$;
- Q_c *is the set of* controlled states *over* X_c *and* Q_e *is the set of* external states *over* X_e. $Q = Q_c \times Q_e$ *is the set of states. We also define* Q_o *to be the set of* observable states *over* X_o;
- $I \subseteq Q_c$ *is the set of* initial states;
- $T \subseteq Q_c \times Q_e \times Q_c$ *is the* transition relation. *We use the notation* $q \xrightarrow{e} q'$ *for* $(q, e, q') \in T$. ■

A state q can be updated in two ways. If the process makes its move, it changes its controlled state according to the transition relation. If the environment makes its move, it can change the external state to any arbitrary value.

Definition 2. *Let* $P = ((X_p, X_i, X_e), I, T)$ *be a process, and* q, q' *be states. Then* q' *is a* successor *of* q, *written* $q \to q'$, *if*

- $q[X_e] = q'[X_e]$ *and* $q[X_c] \xrightarrow{q[X_e]} q'[X_c]$; *or*
- $q[X_c] = q'[X_c]$.

An execution *of* P *is a sequence of states in* Q^*, $q_0 q_1 \cdots q_n$, *where* $q_0[X_c] \in I$ *and* $q_i \to q_{i+1}$ *for* $0 \leq i < n$.

The reachable set *of* P, *denoted* $\mathcal{R}(P)$, *contains states* q *such that there is an execution ending in* q. ■

In order to support structured descriptions, we would like to build complex processes from simpler ones. Three constructs, hide H in P, $P||P'$ and next Θ for P for building new processes are defined in our model as follows. The hide operator makes interface variables inaccessible to other processes.

Definition 3. *Let* $P = ((X_p, X_i, X_e), I, T)$ *be a process and* $H \subseteq X_i$. *Define the process* hide H in P *to be* $((X_p \cup H, X_i \setminus H, X_e), I, T)$. ■

The parallel composition operator allows to combine two processes into a single one. The composition is defined only when the controlled variables of the two processes are disjoint. Intuitively, a step of $P||Q$ is taken by either P or Q but not both.

Definition 4. *Let* $P' = ((X'_p, X'_i, X'_e), I', T')$ *and* $P'' = ((X''_p, X''_i, X''_e), I'', T'')$ *be processes where* $X'_c \cap X''_c = \emptyset$. *The* composition *$P$ of P' and P'', denoted $P'||P''$, is defined as follows.*

- $X_p = X'_p \cup X''_p$;
- $X_i = X'_i \cup X''_i$;
- $X_e = (X'_e \cup X''_e) \setminus X_i$;
- $I = I' \times I''$;
- $T \subseteq Q_c \times Q_e \times Q_c$ *where* $(q, e, r) \in T$ *if*
  - $q[X'_c] \stackrel{(e \cup q)[X'_e]}{\longrightarrow} r[X'_c]$ *in* T' *and* $q[X''_c] = r[X''_c]$; *or*
  - $q[X''_c] \stackrel{(e \cup q)[X''_e]}{\longrightarrow} r[X''_c]$ *in* T'' *and* $q[X'_c] = r[X'_c]$. ■

Finally, we introduce the next operator. The next operator is a *temporal* construct where new transitions are created by grouping a number of lower-level transitions. A transition in next Θ for P is a sequence of transitions in P where the last transition is in Θ. The external state is assumed to stay fixed during the sequence. By applying the next operator, a sequence of small steps becomes a big step. The next operator is inspired by the notions of multi-form time in synchronous languages [15], and was introduced in the language *reactive modules* [1].

Definition 5. *Let* $P = (X, I, T)$ *be a process and* Θ *a subset of* T. *Define the process* next Θ for P *to be* (X, I, T') *where* $(q, e, r) \in T'$ *if there are states* $q_0 = q, q_1, \ldots, q_n = r$ *in* Q_c *such that* $(q_i, e, q_{i+1}) \in T \setminus \Theta$ *for* $0 \le i < n-1$ *and* $(q_{n-1}, e, q_n) \in \Theta$. ■

3 Search Algorithm

In this section, we will develop an on-the-fly search algorithm to explore reachable states of processes formed using the applications of parallel composition, hiding, and next. The applications of next nested with parallel composition lead to hierarchically structured transitions. The search algorithm proceeds recursively in the structure of the process. In order to present the algorithm, we distinguish atomic processes from composite ones. A process is called a *composite process* if it uses composition, hide or next operator. Otherwise, it is called an *atomic process.*

Definition 6. *A* process expression *is defined recursively as follows.*

$$M \equiv P \mid M \| M \mid \textsf{hide } H \textsf{ in } M \mid \textsf{next } \Theta \textsf{ for } M$$

where P is an atomic process, H is a set of variables and Θ is a set of transitions. ■

We will assume that a suitable syntax for atomic processes, such as guarded commands, has been chosen. We also assume that any atomic process appears in a process expression at most once. The algorithm uses the following pre-defined functions:

- The function *GetInitialStates(M)* returns the set of initial states of any process expression M.
- For an atomic process P and a state q in P, *AtomicGetNext(P, q)* returns the set of successors of q in P.
- The function *BelongsTo(s, t, Θ)* tests whether the transition from state s to state t is in Θ.

Since initial states are preserved by hide and next constructs, they can be computed by ignoring these operators. The function *AtomicGetNext(P, q)* is used to handle the basic case and is straightforward to implement as well.

Given a process expression, the algorithm of Figure 1 traverses reachable states in a depth-first manner. Since initial states are reachable, it calls the depth-first search subroutine on initial states iteratively. The algorithm uses a global hash table to store states. The depth-first search routine then uses the key function $GetNext(M, s)$ to compute the set of successors of s in process M. The function $GetNext(M, s)$ proceeds in a top-down fashion. Given a global state s, $GetNext(M, s)$ will return global states that are unvisited successors to s. It uses the pre-defined function $AtomicGetNext(P, s)$ for the atomic case. For the composition, it computes the successors of each component and returns their union. When it encounters a process defined by next operator, it generates the successors of the current state and checks the transition. If a target transition is found, the function adds new states to its returned result. Otherwise, it generates the successors of the new state repeatedly until a target transition is found.

To illustrate how the algorithm works, consider the processes P and Q shown in figure 2. Let P' and Q' denote next Θ for P and next Σ for Q respectively, where Θ and Σ consist of thick transitions (i.e. $\Theta = \{(0,1),(3,4)\}$ and $\Sigma = \{(B,C)\}$). The product $P \| Q$ is also shown in the figure, and has 15 states. The product $P' \| Q'$ has much fewer states, namely, 6, shown as filled circles. This illustrates that the application of next rules out many transient states, and can potentially save space. The challenge is to explore the expression in a top-down manner computing metatransitions only on demand.

```
 1  proc Main(M) ≡
 2     tbl := NewHashTable();
 3     foreach s ∈ GetInitialStates(M) do
 4        if s ∉ tbl then Insert(tbl, s); DFS(M, s); fi
 5     od.
 6  proc DFS(M, s) ≡
 7     foreach s' ∈ GetNext(M, s) do DFS(M, s') od.
 8  funct GetNext(M, s) ≡
 9     if M ≡ P
10        then result := ∅;
11             foreach s' ∈ AtomicGetNext(P, s[P]) do
12                if s[s'/X_P] ∉ tbl
13                   then Insert(result, s[s'/X_P]);
14                        Insert(tbl, s[s'/X_P])
15                fi
16             od;
17             return result
18        elsif M ≡ hide H in M'
19             then return GetNext(M', s)
20        elsif M ≡ M_1∥M_2
21             then return GetNext(M_1, s) ∪ GetNext(M_2, s);
22        elsif M ≡ next Θ for M'
23             then result := ∅;
24                  foreach s' ∈ GetNext(M', s) do
25                     if BelongsTo(s, s', Θ)
26                        then Insert(result, s')
27                        else Insert(result, GetNext(M, s'))
28                     fi;
29                  od
30                  return result
31        fi.
```

Fig. 1. Search Algorithm

Suppose the algorithm is exploring reachable states of $P'\|Q'$. First, it visits the initial state $0A$. After the first invocation of $GetNext(P'\|Q', 0A)$, it returns the states $1A$, $4A$ and $0C$ and the hash table contains these states as well as light gray states such as $2A$ and $3A$. It is crucial that the transient states such as $2A$ are stored in the hash-table to avoid recomputation.

Next, the algorithm searches from $1A$. It ignores successor state $2A$ since it is in the hash table, returns state $1C$ and puts state $1B$ in the hash table. Then, it tries to explore state $4A$. It in turn returns state $4C$ and puts state $4B$ in the table. When the algorithm processes the state $0C$, it adds states $2C$ and $3C$ to the hash table. However, since states $1C$ and $4C$ have been visited, it will ignore them. At the end, the algorithm returns black states, puts shaded states in the hash table. Note that the algorithm never visits the states $2B$ and $3B$, and thus, does less work than required to search $P\|Q$.

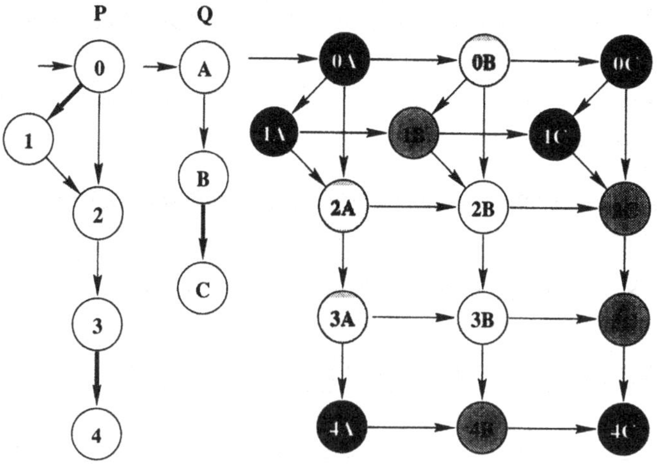

Fig. 2. Traversal Example

To analyze the algorithm, given a process expression M, define $Strip(M)$ to be the process expression obtained by removing all hide and next operators in M. More formally, we have

Definition 7. *Let M be a process expression. Define $Strip(M)$ recursively as follows.*

$$
\begin{aligned}
Strip(P) &= P \\
Strip(\text{hide } H \text{ in } M) &= Strip(M) \\
Strip(\text{next } \Theta \text{ for } M) &= Strip(M) \\
Strip(M_1 || M_2) &= Strip(M_1) || Strip(M_2)
\end{aligned}
$$

where P is an atomic process, H is a set of variables and Θ is a set of transitions.

For any process expression M, consider two reachable sets $\mathcal{R}(M)$ and $\mathcal{R}(Strip(M))$. Clearly, for each state $s \in \mathcal{R}(M)$, $s \in \mathcal{R}(Strip(M))$ because $Strip(M)$ contains all transitions of atomic processes and hence each transition in M is a sequence of transitions in $Strip(M)$. On the other hand, if M contains the next construct, some of the intermediate states may be reachable in $Strip(M)$ but not M. Hence $\mathcal{R}(M)$ and $\mathcal{R}(Strip(M))$ are the two extremes of the state spaces one can hope to search. Since we are interested in checking invariants of the process, we would like the algorithm to visit all reachable states of M. Besides, the number of states visited by the algorithm shouldn't exceed those of $Strip(M)$. In figure 2, one can see that $\mathcal{R}(P'||Q')$, *tbl* and $\mathcal{R}(Strip(P'||Q'))$ are black states, shaded states and all states respectively.

The following theorem thus shows the correctness and the space requirements of our algorithm.

Theorem 1. *Let M be an arbitrary process expression. Then at the end of the procedure $Main(M)$,*

- $\mathcal{R}(M) \subseteq tbl$;
- $tbl \subseteq \mathcal{R}(Strip(M))$.

To prove theorem 1, we begin by defining an auxiliary transition relation parameterized by process expressions.

Definition 8. *Let M be a process expression, s, s' be states of M, and N be a subexpression of M. Define $s \longrightarrow_N s'$ as follows.*

$$\frac{s[X^M \setminus X^P] = s'[X^M \setminus X^P] \quad s[X_c^P] \stackrel{s[X_e^P]}{\longrightarrow} s'[X_c^P] \in T^P}{s \longrightarrow_P s'}$$

$$\frac{s \longrightarrow_{M_1} s'}{s \longrightarrow_{M_1 \| M_2} s'} \qquad \frac{s \longrightarrow_{M_2} s'}{s \longrightarrow_{M_1 \| M_2} s'}$$

$$\frac{s \longrightarrow_M s'}{s \longrightarrow_{\textsf{hide } H \textsf{ in } M} s'}$$

$$\frac{s \longrightarrow_N s' \quad s[X_c^N] \stackrel{s[X_e^N]}{\longrightarrow} s'[X_c^N] \in \Theta}{s \longrightarrow_{\textsf{next } \Theta \textsf{ for } N} s'}$$

$$\frac{s \longrightarrow_N s'' \quad s[X_c^N] \stackrel{s[X_e^N]}{\longrightarrow} s''[X_c^N] \notin \Theta \quad s'' \longrightarrow_{\textsf{next } \Theta \textsf{ for } N} s'}{s \longrightarrow_{\textsf{next } \Theta \textsf{ for } N} s'}$$ ■

Intuitively, $s \longrightarrow_N s'$ denotes the state changes from s to s' by taking a transition of the subexpression N.

Definition 9. *Let M be a process expression, $s, s' \in Q^M$, N be a subexpression of M, and $s \longrightarrow_N s'$. Define* Expand$(s \longrightarrow_N s')$ *to be the set of paths*

$$\{s_0 = s \longrightarrow_{P_0} s_1 \cdots s_{n-1} \longrightarrow_{P_{n-1}} s_n = s' \mid$$
$$s_i \longrightarrow_{P_i} s_{i+1} \textit{ appears at the leaves of a derivation for } s \longrightarrow_N s'\}$$

■

Intuitively, Expand$(s \longrightarrow_N s')$ contains paths of atomic steps from s to s'.

Lemma 1. *Let M be a process expression, N be a subexpression of M, t, t' be states of M, and tbl be the content of the table before $GetNext(N,t)$ is invoked. If $t \longrightarrow_N t'$, then*

- $t' \in GetNext(N,t)$; *or*
- *For all paths* $s_0 = t \longrightarrow_{P_0} s_1 \cdots s_{n-1} \longrightarrow_{P_{n-1}} s_n \in Expand(t \longrightarrow_N t')$, *there is an i, $0 < i \leq n$ such that $s_i \in tbl$.*

The above lemma is used to establish that if $t \longrightarrow_N t'$ then t' is added to the hash-table eventually, and allows us to conclude that the algorithm visits all the reachable states of the input expression.

Since the hash table is used, each state in *tbl* is processed precisely once. The cost of processing a state is proportional to its outdegree, and this gives a bound on the running time of the algorithm. Thus, the saving in space comes at no extra overhead in time.

It is straightforward to modify the search algorithm to check invariants of process expressions. However, it should be noted that our algorithm cannot be used to check path properties. For instance, in the example of Figure 2, the global product $P'||Q'$ has a transition from 1A to 4A, but this transition will not be discovered by the algorithm.

4 Applying Next as a heuristic

Suppose we wish to check an invariant of the process expression $M = P||Q$. Can we find transition sets Θ and Σ, and search instead, the process expression $M' = \textsf{next}\ \Theta\ \textsf{for}\ P \,||\, \textsf{next}\ \Sigma\ \textsf{for}\ Q$. As seen already, searching M' can be more efficient than searching M, and we can use the algorithm from previous section. In this section, we would like to find conditions under which searching M' suffices to solve the original problem.

Notice that application of next compresses a sequence of transitions into a single transition. Intuitively, as long as the compressed transitions are *local*, and do not access shared variables, such a transformation should be transparent to the remaining processes. The fact that sequences of invisible transitions can be compressed preserving weak equivalences is well known in theory and tools based on standard process algebras (see, for instance, [6]). We adopt this concept to our framework in the sequel. In particular, we show that under appropriate restrictions, next Θ for P is equivalent to P under the definition of Milner's *weak-simulation equivalence*. This allows us, then, to apply next to subexpressions repeatedly in a recursive manner.

We start with the definition of observable transition. If a transition doesn't change the values of interface variables, it is called *unobservable*. From other processes' point of view, unobservable transitions are not interesting. This motivates the following definition of observable transition relation.

Definition 10. *Let $P = ((X_p, X_i, X_e), I, T)$ be a process, $q, q' \in Q_c$ and $e \in Q_e$. Then $q \xrightarrow{e}_w q'$ is an* observable transition *if there are $q_0 = q, q_1, \ldots, q_n = q' \in Q_c$ such that*

- *for all $0 \leq i < n$. $q_i[X_i] = q_0[X_i]$;*
- *for all $0 \leq i < n$. $q_i \xrightarrow{e} q_{i+1}$.* ∎

Now we define the simulation equivalence.

Definition 11. *Let $P = ((X_p, X_i, X_e), I, T)$ and $P' = ((X_p', X_i, X_e), I', T')$ be two processes, $q \in Q_c$, and $q' \in Q_c'$. A* (weak) simulation *relation $\preceq \subseteq Q_c \times Q_c'$ satisfies*

$$q \preceq q' \text{ implies } q[X_i] = q'[X_i] \text{ and}$$

$$\text{for all } r \in Q_c, \text{ for all } e \in Q_e, \text{ if } q \xrightarrow{e} r \text{ then there exists } r' \in Q_c' \text{ such that } q' \xrightarrow{e}_w r' \text{ and } r \preceq r'.$$

P is (weakly) simulated by P' *if there is a simulation* $\preceq$ *such that for all* $q \in I$, *there exists* $q' \in I'$ *with* $q \preceq q'$. $P \cong P'$ *if* $P \preceq P'$ *and* $P' \preceq P$. ■

Intuitively, if two processes are simulated by each other, one can perform the other's visible behaviors and vice versa. More specifically, if one is able to change its interface variables under certain environment, the other is able to do the same. Therefore, if an observable state is reachable for one process, it is reachable in the other too. Observe that if P and Q are simulation-equivalent, then to check invariant concerning observable states of P, it suffices to check that invariant of Q. Now we proceed to establish simulation-equivalence between a process and its next-abstraction provided the compressed transitions do not access shared variables.

Definition 12. *Let* $P = ((X_p, X_i, X_e), I, T)$ *be a process,* $q, q' \in Q_c$. *Then* $(q, e, q') \in T$ *is called a* write-visible transition *if* $q[X_i] \neq q'[X_i]$. $(q, e, q') \in T$ *is called a* read-visible transition *if there exists* $e' \in Q_e$, $(q, e', q') \notin T$.

A transition in T *is called* visible *if it is write-visible or read-visible. The set of visible transitions in* T *is denoted by* T_v. ■

The next technical lemma is needed when we prove the theorem.

Lemma 2. *Let* $P = (X, I, T)$ *be a process and* $T_v \subseteq \Theta \subseteq T$. *Define the relation* $S \subseteq Q_c \times Q_c \times Q_e$ *as follows.*

$$S = \{\ (q, q', e) \mid \exists q_0, \ldots, q_n . q_0 = q' \wedge q_n = q \wedge \forall 0 \leq i < n.(q_i, e, q_{i+1}) \in T \setminus \Theta\ \}.$$

Then for any $q \in \mathcal{R}(P)$, *there exists* $q' \in \mathcal{R}(\mathsf{next}\ \Theta\ \mathsf{for}\ P)$ *with* $(q, q', q[X_e]) \in S$.

Theorem 2. *Let* $P = (X, I, T)$ *be a process and* $T_v \subseteq \Theta \subseteq T$. *Then* $P \cong \mathsf{next}\ \Theta\ \mathsf{for}\ P$.

Given a process P, we now define its reduced version $\mathsf{NEXT}\ P$ to be $\mathsf{next}\ T_v\ \mathsf{for}\ P$. By theorem 2, $P \cong \mathsf{NEXT}\ P$. To be able to replace a process subexpression by application of NEXT freely and repeatedly, we must establish that the simulation equivalence is a congruence for the operators in our language:

Theorem 3. *Let* P, P' *be processes and* $P \preceq P'$. *Then*

- $P \| Q \preceq P' \| Q$ *for any* Q;
- $\mathsf{hide}\ H\ \mathsf{in}\ P \preceq \mathsf{hide}\ H\ \mathsf{in}\ P'$;
- $\mathsf{NEXT}\ P \preceq \mathsf{NEXT}\ P'$.

Using the search algorithm shown in Section 3, one can check invariants φ on observable variables more efficiently as follows. Consider any process expression M, we construct a new process M' by substituting subexpressions E in M by $\mathsf{NEXT}\ E$. This process can be applied repeatedly. Then every observable state of M will be visited by $Main(M')$. Thus, invariant verification of M reduces to the invariant verification of M'.

Consider a process expression $M \equiv M_1 \| \cdots \| M_n$ where each M_i is an atomic process P. We want to construct in a heuristic fashion an expression M' such

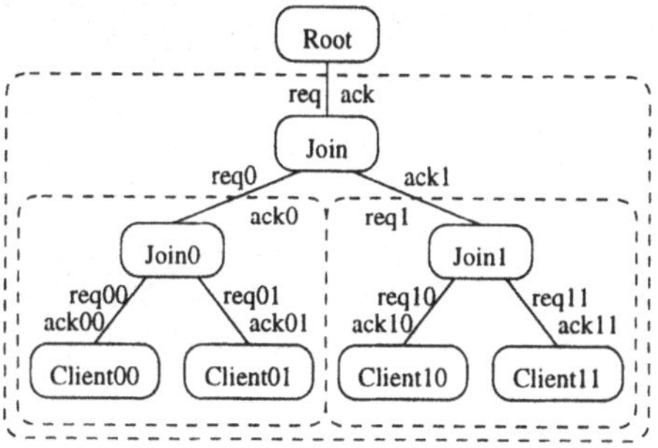

Fig. 3. Tree-structured Parity Computer

that M' involves applications of hide and NEXT to subexpressions of M with $Strip(M') = M$. By the previous sections, any substitution will yield the same answer but the effectiveness of reduction may differ by cases. How to find a suitable partition of M to apply this technique becomes an interesting problem on its own.

In principle, one would like to divide the system into parts so that the communication between any two parts is minimal. When a system is built with certain interconnection structure (like trees, rings, or grids etc), these structures should be exploited in verification. For instance, one can reduce the number of states by applying the next construct to subtrees in a tree-like structure. One can use the technique recursively on each subtrees and obtain even more efficient result. Note that hiding interface variables whenever possible is crucial since our reduction cannot ignore write-visible transitions.

5 Implementation and Experiments

We report two case-studies to test our theory. In our experiments, we use a syntax similar to MOCHA to specify modules [2]. The experiments were done using a prototype implementation in O'Caml. Details of the examples are available to the readers at `www.cis.upenn.edu/~bywang/concur99.html`.

The first example models a parity-computer with a binary tree structure. The system consists of several *Client* modules as leaves and *Join* modules as nodes. The interconnection between these modules and *Root* module is shown in Figure 3 for 4-client configuration.

Instead of searching

$$\text{Root} \parallel \text{Join} \parallel \text{Join0} \parallel \text{Join1} \parallel \text{Client00} \parallel \text{Client01} \parallel \text{Client10} \parallel \text{Client11}$$

we can naturally cluster modules as shown in Figure 3. The expression we search is

$$\text{Root} \parallel \text{NEXT hide } [\text{ Join} \parallel \begin{matrix} \text{NEXT hide (Join0} \parallel \text{Client00} \parallel \text{Client01)} \\ \text{NEXT hide (Join1} \parallel \text{Client10} \parallel \text{Client11)} \end{matrix}].$$

Num. of clients	Without next		With next	
	States	Time	States	Time
3	202	0.37s	189	0.36s
4	1748	1.05s	515	0.39s
5	6994	4.46s	2017	0.77s
6	69614	55.22s	2633	0.94s
7	320094	324.73s	3531	1.32s
8	2782634	4449.12s	4751	1.79s

Fig. 4. Reduction for the parity tree computer

For readability the argument variables to hide are not shown above, but should be obvious from context: only the variables communicating with the parent need to be observable at each level of the subtree. Notice that for the top-level expression, a step is either a step of the module *Root*, or it is a sequence of steps of the tree rooted at *Join* until either *req* is updated or *ack* is read. A step of the tree rooted at *Join* is either a step of the module *Join*, or a sequence of steps of the subtree rooted at *Join0* until *req0* or *ack0* is accessed, or a sequence of steps of the subtree rooted at *Join1* until *req1* or *ack1* is accessed. Since each *Join* node communicates with parent only after it has received and processed requests from its children, we get excellent reduction. In the table of Figure 4, we show the number of states visited, together with the CPU time required by the algorithm for different number of clients. Observe that while the number of states grows exponentially without any reduction, our heuristic scales almost linearly.

Our second example concerns the standard leader election protocol in an asynchronous ring of cells. The protocol used here is taken from [20], and has been used as a benchmark to study how model checking algorithms scale. Figure 5 shows four cells connected together and the partition used for applications of next. The results are reported in the table of Figure 6. While not as spectacular as the parity-tree-computer, observe the significant reduction in space and time.

6 Conclusions

We have presented a new heuristic for model checking based hierarchically structured transition relations. The heuristic potentially combines advantages of the on-the-fly search with early detection of violations and compositional reduction techniques based on process equivalence.

While the construct next has been studied previously in the context of symbolic search [3], this is the first study that establishes conditions under which next construct can be introduced without changing the answer to the original verification problem. The transformation of Section 3 can be used for improving efficiency of symbolic model checking as well. The reduction offered by our method differs with examples. The reduction is maximum when processes can

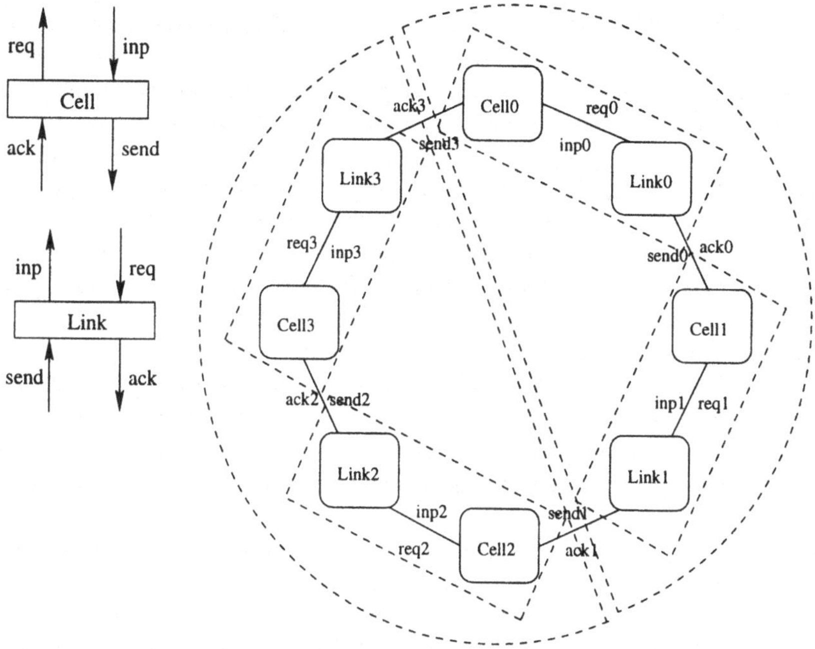

Fig. 5. Leader Module

Num. of clients	Without next		With next	
	States	Time	States	Time
3	10847	7.64s	4401	1.71s
4	195017	210.15s	26430	10.79s
5	2235421	4351.20s	230569	115.21s

Fig. 6. Reduction for the leader election in ring

be clustered so that when one cluster communicates with another, it is only in a small number of different internal states.

Partial-order reduction has been successfully combined with on-the-fly search (see, for example, [22, 25, 14]). It may appear that our reduction is a simple form of partial-order reduction, but on more scrutiny, the two techniques are incomparable, and we believe them to be compatible. Partial-order reduction does not exploit the hierarchical structure, but uses independence among transitions to restrict the number of interleavings. It is easy to construct examples in which our method works better than the partial-order reduction method, and vice versa. On the examples reported in this paper, the software SPIN gives much more impressive reductions on the leader election protocol, but our method performs better for the parity computer. The modeling languages of SPIN and MOCHA are quite different to make any meaningful comparison in the absolute number of states for each configuration, so the comparison is based on the rates of growth with the scaling parameter. More information about the comparison can be found along with the code of our examples (`www.cis.upenn.edu/~bywang/concur99.html`).

Another popular heuristic for improving on-the-fly search is reductions using symmetry [12, 8, 17]. Again, our method is incomparable to and compatible with symmetry reduction. In particular, note that in our examples, if we change the individual clients so that they behave slightly differently from each other, or the different join nodes compute different functions, there will be no significant impact on the reduction obtained. In the parity computer example, if we change $Join1$ to its complement, the experiment shows that our heuristic can still achieve the same reduction for the eight clients system.

The process-algebraic tools such as *Concurrency workbench* [10] and CADP [13] implement the reductions based on weak-simulation (or even the less expensive reduction to compress internal transitions). We have already discussed our motivation to explore on-the-fly algorithms that implement the reduction in a top-down manner.

In this paper, we have used next only to compress invisible transitions. However, our search algorithm can be used effectively in other contexts. First, next can be supplied an argument that contains only a subset of the visible transitions. This would perform a more efficient search at a coarser granularity of atomicity. Note that, while this is similar in spirit to the *atomic* construct in the language of SPIN to cluster a sequence of atomic transitions, it gives a high-level and hierarchical construct. Second, next can be used to implement the so-called "run-to-completion" semantics in which transitions are executed until there are no more enabled transitions.

References

1. R. Alur and T.A. Henzinger. Reactive modules. In *Proceedings of the 11th IEEE Symposium on Logic in Computer Science*, pages 207–218, 1996.
2. R. Alur, T. Henzinger, F. Mang, S. Qadeer, S. Rajamani, and S. Tasiran. MOCHA: Modularity in model checking. In *Proceedings of the 10th International Conference on Computer Aided Verification*, LNCS 1427, pages 516–520. Springer-Verlag, 1998.
3. R. Alur, T. Henzinger, and S. Rajamani. Symbolic exploration of transition hierarchies. In *Proceedings of the Fourth International Conference on Tools and Algorithms for the Construction and Analysis of Systems*, LNCS 1384, pages 330–344. Springer-Verlag, 1998.
4. A. Bouajjani, J.C. Fernandez, and N. Halbwachs. Minimal model generation. In *Computer-Aided Verification, 2nd International Conference, CAV'90*, LNCS 531, pages 197–203. Springer-Verlag, 1990.
5. G. Berry and G. Gonthier. The synchronous programming language ESTEREL: design, semantics, implementation. Technical Report 842, INRIA, 1988.
6. G. Boudol, V. Roy, R. de Simone, and D. Vergamini. Process calculi, from theory to practice: verification tools. In *Automatic Verification Methods for Finite State Systems*, LNCS 407, pages 1–10. 1987.

7. E.M. Clarke and E.A. Emerson. Design and synthesis of synchronization skeletons using branching time temporal logic. In *Proc. Workshop on Logic of Programs*, LNCS 131, pages 52–71. Springer-Verlag, 1981.
8. E.M. Clarke, T. Filkorn, and S. Jha. Exploiting symmetry in temporal-logic model checking. In *Computer-Aided Verification, Fifth International Conference, CAV'95*, LNCS 697, pages 450–462. Springer-Verlag, 1993.
9. E.M. Clarke and R.P. Kurshan. Computer-aided verification. *IEEE Spectrum*, 33(6):61–67, 1996.
10. R. Cleaveland, J. Parrow, and B. Steffen. The concurrency workbench: A semantics-based tool for the verification of finite-state systems. *ACM Trans. on Programming Languages and Systems*, 15(1):36–72, 1993.
11. D.L. Dill, A.J. Drexler, A.J. Hu, and C.H. Yang. Protocol verification as a hardware design aid. In *IEEE International Conference on Computer Design: VLSI in Computers and Processors*, pages 522–525, 1992.
12. E.A. Emerson and A.P. Sistla. Symmetry and model checking. In *Computer-Aided Verification, Fifth International Conference, CAV'95*, LNCS 697, pages 463–478. Springer-Verlag, 1993.
13. J. Fernandez, H. Garavel, A. Kerbrat, R. Mateescu, L. Mounier, and M. Sighireanu. CADP: A protocol validation and verification toolbox. In *Proceedings of the Eighth International Conference on Computer-Aided Verification*, LNCS 1102. Springer-Verlag, 1996.
14. P. Godefroid. Using partial orders to improve automatic verification methods. In E.M. Clarke and R.P. Kurshan, editors, *Computer-Aided Verification, 2nd International Conference, CAV'90*, LNCS 531, pages 176–185. Springer-Verlag, 1990.
15. N. Halbwachs. *Synchronous Programming of Reactive Systems*. Kluwer Academic Publishers, 1993.
16. G.J. Holzmann. The model checker spin. *IEEE Trans. on Software Engineering*, 23(5):279–295, 1997.
17. C.N. Ip and D.L. Dill. Verifying systems with replicated components in murφ. In *Proceedings of the Eighth International Conference on Computer Aided Verification*, LNCS 1102. Springer-Verlag, 1996.
18. J.-P. Krimm and L. Mounier. Compositional state space generation of lotos programs. 1997.
19. P. Kanellakis and S.A. Smolka. CCS expressions, finite state processes, and three problems of equivalence. *Information and Computation*, 86(1):43–68, 1990.
20. N.A. Lynch. *Distributed algorithms*. Morgan Kaufmann, 1996.
21. R. Milner. *A Calculus of Communicating Systems*. LNCS 92. Springer-Verlag, 1980.
22. D. Peled. Combining partial order reductions with on-the-fly model-checking. In *Computer Aided Verification, Proc. 6th Int. Conference*, LNCS 818. Springer-Verlag, 1994.
23. R. Paige and R.E. Tarjan. Three partition-refinement algorithms. *SIAM Journal on Computing*, 16(6):973–989, 1987.
24. J.P. Queille and J. Sifakis. Specification and verification of concurrent programs in CESAR. In *Proceedings of the Fifth International Symposium on Programming*, LNCS 137, pages 195–220. Springer-Verlag, 1982.
25. A. Valmari. A stubborn attack on state explosion. *Formal Methods in System Design*, 1:297–322, 1992.

Model Checking of Message Sequence Charts

Rajeev Alur* and Mihalis Yannakakis**

Abstract. Scenario-based specifications such as message sequence charts (MSC) offer an intuitive and visual way of describing design requirements. Such specifications focus on message exchanges among communicating entities in distributed software systems. Structured specifications such as MSC-graphs and Hierarchical MSC-graphs (HMSC) allow convenient expression of multiple scenarios, and can be viewed as an early *model* of the system. In this paper, we present a comprehensive study of the problem of verifying whether this model satisfies a temporal requirement given by an automaton, by developing algorithms for the different cases along with matching lower bounds.
When the model is given as an MSC, model checking can be done by constructing a suitable automaton for the linearizations of the partial order specified by the MSC, and the problem is coNP-complete. When the model is given by an MSC-graph, we consider two possible semantics depending on the *synchronous* or *asynchronous* interpretation of concatenating two MSCs. For synchronous model checking of MSC-graphs and HMSCs, we present algorithms whose time complexity is proportional to the product of the size of the description and the cost of processing MSCs at individual vertices. Under the asynchronous interpretation, we prove undecidability of the model checking problem. We, then, identify a natural requirement of *boundedness*, give algorithms to check boundedness, and establish asynchronous model checking to be PSPACE-complete for bounded MSC-graphs and EXPSPACE-complete for bounded HMSCs.

1 Introduction

Message sequence charts (MSCs), and related formalisms such as time sequence diagrams, message flow diagrams, and object interaction diagrams, are a popular visual formalism for documenting design requirements for concurrent systems such as telecommunications software [22, 18]. MSCs are often used in the first attempts to formalize design requirements for a new system and its protocols. MSCs represent typical execution scenarios, providing examples of either normal or exceptional executions ('sunny day' or 'rainy day' scenarios) of the proposed system. The clear graphical layout of an MSC immediately gives an intuitive understanding of the intended system behavior.

In the simplest form, an MSC depicts the desired exchange of messages, and corresponds to a single (partial-order) execution of the system. In recent

* Department of Computer and Information Science, University of Pennsylvania, and Bell Laboratories, Lucent Technologies. Email: `alur@cis.upenn.edu`. Supported in part by NSF CAREER award CCR-9734115 and by the DARPA grant NAG2-1214.
** Bell Laboratories, Lucent Technologies. Email: `mihalis@research.bell-labs.com`.

years, a variety of features have been introduced so that a designer can specify multiple scenarios conveniently. In particular, *MSC-graphs* allow MSCs to be combined using operations such as choice, concatenation, and repetition. *Hierarchical MSCs* (HMSC), also called *high-level MSCs*, allow improved structuring of such graphs by introducing abstraction and sharing. All these features are incorporated in an international standard, called Z.120, promoted by ITU [22]. MSCs or similar formalisms are increasingly being used by designers for specifying requirements. Such specifications are naturally compatible with object-oriented design methods, and are being supported by almost all the modern software engineering methodologies such as SDL [21], ROOM [19] and UML [3].

We believe that scenario-based requirements will play an increasingly prominent role in design of software systems that require communication among distributed agents. Requirements expressed using MSCs (or HMSCs) have a formal semantics, and hence, can be subjected to analysis. Since MSCs are used at a very early stage of design, any errors revealed during their analysis have a high pay-off. This has already motivated development of algorithms for detecting race conditions and timing conflicts [1], pattern matching [15], and detecting non-local choice [4], and tools such as uBET [1, 11]. In this paper, inspired by the success of model checking in debugging of high-level hardware and software designs [5, 6, 10], we develop a methodology and algorithms for model checking of scenario-based requirements.

It is worth noting that the traditional high-level model for concurrent systems has been communicating state machines. Both communicating state machines and HMSCs can be viewed as specifying sets of behaviors, but the two offer dual views; the former is a parallel composition of sequential machines, while the latter is a sequential composition of concurrent executions. Analyzing communicating state machines is known to be computationally expensive—PSPACE or worse, and in spite of the remarkable progress in developing heuristics, still remains the main bottleneck in application of model checking. Consequently, translating MSC-based specifications to communicating state machines, as suggested in previous approaches [13, 9], may not lead to the most efficient procedures. Also there is a difference in expressive power between the two formalisms in general. The problem of analyzing HMSCs is interesting and important in its own right, and is investigated in this paper.

We formalize the model checking problem using the automata-theoretic approach to formal verification [20, 10, 12]. The system under design is described by an MSC, or a MSC-graph, or an HMSC, in which the individual events are labeled with symbols from an alphabet Σ. The semantics of the system is a language of strings over Σ. The specification is described by an automaton over Σ whose language consists of the undesirable behaviors, and model checking corresponds to checking if the intersection of the two languages is empty. When the system is described by an MSC-graph, or an HMSC, the choice for the associated language depends on the interpretation of the concatenation of two MSCs. We consider two natural choices: in the *synchronous* concatenation of two MSCs M_1 and M_2, any event in M_2 is assumed to happen after all the events in M_1;

while the *asynchronous* interpretation corresponds to concatenating two MSCs process by process.

An MSC M specifies a partial ordering of the events it contains, and the model checking problem for M can be solved by constructing an automaton that accepts all possible linearizations of the partial order and checking this automaton against the given specification property automaton. We establish the model checking problem for MSCs to be coNP-complete. For model checking of MSC-graphs under the synchronous interpretation of concatenation, we replace each vertex of the MSC-graph by an automaton that accepts all the linearizations of the associated MSC, construct the product with the specification automaton, and check for emptiness. For HMSCs, a similar strategy reduces the model checking problem to a problem for *hierarchical state machines*, and then, we employ the efficient algorithms of [2] for searching the hierarchical structure without flattening it. The resulting complexity for both MSC-graphs and HMSCs is proportional to the size of the system description times the complexity of model checking individual MSCs. In both cases, the model checking problem is proved to be coNP-complete.

Under the asynchronous interpretation for concatenation, the model checking problem for MSC-graphs turns out to be undecidable. The problem can be traced to descriptions which allow unbounded drift between the processes and whose correct implementation requires potentially unbounded buffers. We identify a subclass of *bounded* graphs which rule out such problems and entail decidability. We give an algorithm to check if an MSC-graph or an HMSC is bounded. The boundedness requirement is similar (though not identical) to the condition identified in [4] to avoid process divergence. The algorithm in [4] to detect divergence is exponential in the number of vertices for flat MSC-graphs (and its straightforward extension to HMSCs is doubly exponential). Our algorithm for checking boundedness extends also to process divergence and is exponential in the number of processes, but linear in the size of the MSC-graph or HMSC. We show the problem of checking boundedness (and process divergence) to be coNP-complete. Finally, we establish that the asynchronous model checking problem is PSPACE-complete for bounded MSC-graphs, and EXPSPACE-complete for HMSCs. In particular, for asynchronous model checking of HMSCs, the flattening of the hierarchy is unavoidable in the worst case (unlike the synchronous model checking and the testing of boundedness, where the flattening can be avoided).

2 Message Sequence Charts

A sample message sequence chart is shown in Figure 1. Vertical lines in the chart correspond to asynchronous processes or autonomous agents. Messages exchanged between these processes are represented by arrows. The tail of each arrow corresponds to the event of sending a message, while the head corresponds to its receipt. Arrows can be drawn either horizontally or sloping downwards, but not upwards. Each arrow is labeled with a message identifier. We proceed to define MSCs formally. It is worth noting that our definitions capture the spirit

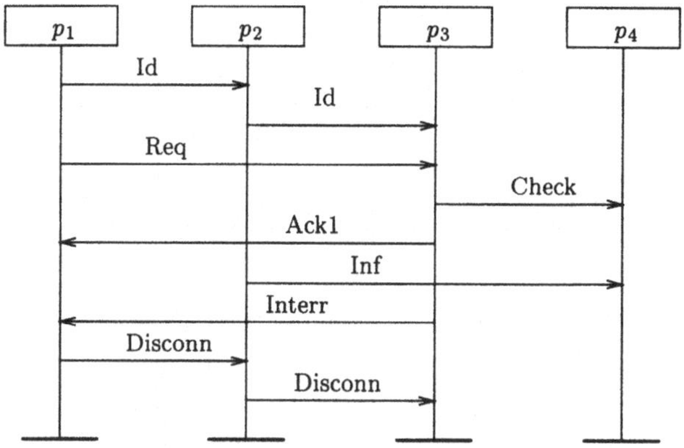

Fig. 1. A sample message sequence chart

of the standard Z.120, but differ in details and focus only on a subset of the features for the sake of clarity and simplicity.

2.1 Formalization

Formally, a message sequence chart M is a labeled directed acyclic graph with the following components [1]:

- *Processes:* A finite set P of processes.
- *Events:* A finite set E of events that is partitioned into two sets: a set S of send events and a set R of receive events.
- *Process Labels:* A labeling function g that maps each event in E to a process in P. The set of events belonging to a process p is denoted by E_p.
- *Send-receive Edges:* A bijection map $f : S \mapsto R$ that associates each send event s with a unique receive event $f(s)$ and each receive event r with a unique send event $f^{-1}(r)$.
- *Visual Order:* For every process p there is a local total order $<_p$ over the events E_p which corresponds to the order in which the events are displayed.

The local visual orders, together with the send-receive edges, define the relation

$$< \; = \; [\cup_p <_p \; \cup \{ \; (s, f(s)) \mid s \in S \; \}]^*.$$

The relation $<$ is a partial order over E since send-receive edges cannot go upwards in the chart. This formalization provides a simple, but precise, way to treat MSCs as mathematical objects. There are many alternative formalizations, for instance, via translation to process algebras [16]. Furthermore, the above formalization assumes that the ordering of the receipts of messages at a process coincides with the visual order. Depending on the underlying communication architecture, we may wish to employ alternative orderings, but this choice does not affect the complexity of the problems studied in this paper.

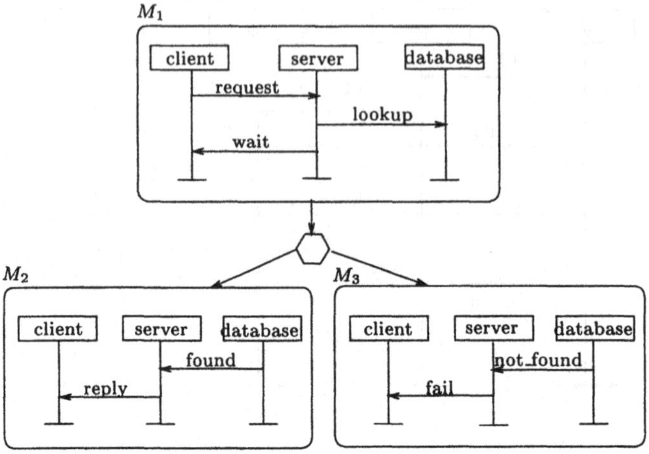

Fig. 2. A sample MSC graph G

2.2 MSC-graphs

A natural way to structure multiple scenarios is to employ graphs whose nodes are MSCs. An *MSC-graph* is a graph whose nodes are labeled with MSCs, and whose edges correspond to concatenation of MSCs. A sample MSC graph is depicted in Figure 2. The first node corresponds to a scenario M_1 in which the server initiates a database lookup to process a client request. The scenario M_1 is followed by either scenario M_2 or by the scenario M_3. The scenario M_2 corresponds to a positive response from the database, while the scenario M_3 models a negative response from the database. The hexagonal box is called a *condition* in the MSC standard, and is used to indicate a *choice* or *branching* in MSC graphs. For the purpose of this paper, conditions will be uninterpreted, and hence, can be ignored in the formalization. Formally, an MSC-graph G consists of a set V of vertices, a binary relation $\rightarrow$ over V, an initial vertex v^I, a terminal vertex v^T, and a labeling function μ that maps each vertex v to an MSC. The paths that start at the initial vertex and end at the terminal vertex represent the finite executions of the system modeled by the MSC-graph, while the infinite executions are represented by all the infinite paths starting at the initial vertex. Note that the definition can be modified to allow multiple terminal vertices without affecting any of the complexity bounds in this paper.

2.3 Hierarchical MSCs

Hierarchical MSCs (HMSC) (also called *high-level MSCs*) offer an improved structuring mechanism. Consider a sample HMSC shown in Figure 3. It is like an MSC-graph, and has three nodes M_i, M_b, and M_f. The nodes M_i and M_f are MSCs as in an MSC-graph, but the node M_b is labeled by another MSC-graph

G. Thus, the node M_b is like a *superstate* in hierarchical state-machines such as Statecharts [8]. The MSC M_i depicts the sequence of messages for initialization. As seen earlier, the MSC-graph G depicts the sequence of messages for processing individual requests. After completing one request, either G gets repeated, or the system terminates after executing the termination sequence of messages depicted in M_f. Note that the structure of G is not visible at the top level. In a typical graphical interface, the graph G itself can be viewed by clicking onto the node M_b.

More generally, a hierarchical MSC consists of a graph whose nodes are either MSCs or are labeled with another hierarchical MSC. Thus, the definition allows nesting of graphs, provided the nesting is finite. In other words, the definition of HMSCs cannot be mutually recursive: if a node of an HMSC M is labeled with another HMSC M', then a node of M' cannot be labeled with M (or any other HMSC that refers to M). Another important aspect of the definition is that different nodes can be labeled with the same HMSC. For instance, once we have defined the request-processing scenario G (Figure 2) it can be used multiple times, possibly in different contexts, just like a function in a traditional programming language. This allows reuse and sharing, and leads to succinct representation of complex scenarios.

Formally, a *Hierarchical MSC* is a tuple $H = (N, B, v^I, v^T, \mu, E)$, where

- N is a finite set of nodes.
- B is a finite set of boxes (or supernodes).
- $v^I \in N \cup B$ is the initial node or box.
- $v^T \in N \cup B$ is the terminal node or box.
- μ is a labeling function that maps each node in N to an MSC, and each box in B to another (already defined) HMSC.
- $E \subseteq (N \cup B) \times (N \cup B)$ is the set of edges that connect nodes and boxes to each other.

The meaning of an HMSC H is defined by recursively substituting each box by the corresponding HMSC to obtain an MSC-graph. For an HMSC $H = (N, B, v^I, v^T, \mu, E)$, the flattened MSC-graph H^F is defined as follows. For each box b, let b^F be the MSC-graph $(\mu(b))^F$ obtained by flattening $\mu(b)$. The MSC-graph H^F has following components:

Vertices. Every node of H is a vertex of H^F. For a box b of H, for every vertex v of b^F, the pair (b, v) is a vertex of H^F.

Initial vertex. If $v^I \in N$ then the initial vertex of H^F is v^I. If $v^I \in B$, then if the initial vertex of $(v^I)^F$ is v then the pair (v^I, v) is the initial vertex of H^F.

Terminal vertex. If $v^T \in N$ then the terminal vertex of H^F is v^T. If $v^T \in B$, then if the terminal vertex of $(v^T)^F$ is v then the pair (v^T, v) is the terminal vertex of H^F.

Labeling with MSCs. For a node u of H, the label of u in H^F is same as the label $\mu(u)$. For a box b of H, for every vertex v of b^F, the label of (b, v) in H^F is same as the label of v in b^F.

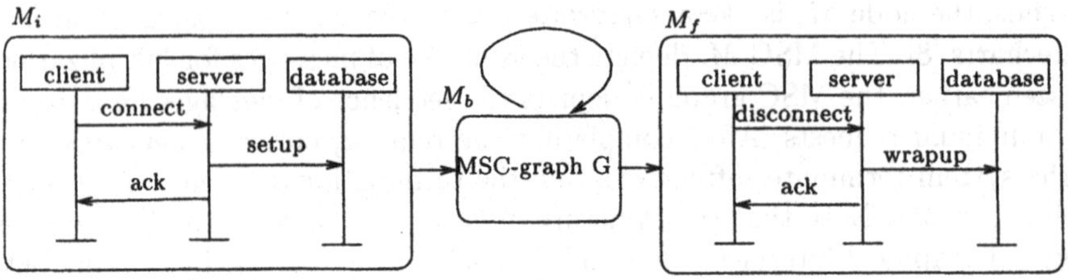

Fig. 3. A sample hierarchical MSC graph

Edges. For an edge (u, v) of H, (u', v') is an edge of H^F, where if $u \in N$ then $u' = u$ else $u' = (u, u'')$ for the terminal vertex u'' of u^F, and if $v \in N$ then $v' = v$ else $v' = (v, v'')$ for the initial vertex v'' of v^F.

Thus, the vertices in the flattened graph are tuples whose last component is a node, and remaining components are boxes that specify the context. The MSC labeling the last component determines the label of a vertex. Note that the number of components in a vertex is bounded by the nesting depth of the description, and the number of vertices can be exponential in the nesting depth.

In our definition, a box can be entered only at its entry vertex, and can be exited only at its terminal vertex. This choice is only for the sake of simplicity of presentation, and we can allow edges connecting to and from specific vertices inside a box without a significant penalty on the complexity of algorithms.

3 Model Checking of MSCs

To formalize the model checking problem, given an MSC M with event-set E, we introduce another component in the MSC-specification, namely, labeling of events in E with symbols in a given alphabet. For an alphabet Σ, a Σ-labeled MSC is a pair (M, ℓ), where M is an MSC and ℓ is a function from E to Σ. A Σ-labeled MSC can be viewed as a *partially-ordered multiset* (POMSET) [17].

Consider a Σ-labeled MSC M with events E and labeling ℓ. Recall that the MSC specifies a partial ordering of the events in E. If we consider all possible linearizations of this partial order, and map each ordering to a string over Σ by replacing each event e by its associated symbol $\ell(e)$, the resulting set of strings is called the *language of* M, and is denoted $L(M)$. Alternatively, we can label the messages in M, or we can label both the events and the messages. Such a choice would not affect the complexity of the model checking algorithms.

For a Σ-labeled MSC M, the language $L(M)$ represents the possible executions of the system. The requirement can be specified by an automaton A over Σ which accepts all the undesirable executions: the system M satisfies the specification A iff the intersection $L(M) \cap L(A)$ is empty. The *model checking problem* for MSCs is, then, given a Σ-labeled MSC M, and an automaton A over Σ, determine whether or not $L(M) \cap L(A)$ is empty.

Let M be an MSC with event set E and partial order $<$. To solve the model checking problem, we can construct an automaton A_M that accepts $L(M)$ using the standard technique of extracting global states from a partial order as follows. A *cut* c is a subset of E that is closed with respect to $<$: if $e \in c$ and $e' < e$ then $e' \in c$. Since all the events of a single process are linearly ordered, a cut can be specified by a tuple that gives the maximal event of each process. The states of the automaton A_M correspond to the cuts. The empty cut is the initial state, and the cut with all the events is the final state. If the cut d equals the cut c plus a single event e, then there is an edge from c to d on the symbol $\ell(e)$. It is easy to verify that the automaton A_M accepts the language $L(M)$. The size of A_M corresponds to the number of cuts, and is bounded by n^k, if M has n events and k processes. The model checking problem with respect to a specification automaton A can now be reduced to a reachability problem over the product of A_M and A.

Theorem 1. *Given a Σ-labeled MSC M with n events and k processes, and an automaton A of size m, the model checking problem (M, A) can be solved in time $O(m \cdot n^k)$, and is coNP-complete* [1].

4 Model Checking of MSC-graphs

For an alphabet Σ, a Σ-labeled MSC-graph G is a graph $(V, \rightarrow, v^I, v^T, \mu)$, where μ maps each vertex to a Σ-labeled MSC. To define the model checking problem for such graphs, we must associate a language with each graph. First, let us note that there is no unique interpretation of the concatenation. As an example, consider the concatenation of two MSCs M_1 and M_2 depicted in Figure 4. Under the *synchronous* interpretation, all the events in the MSC M_1 finish before any event in the MSC M_2 occurs. Thus, the event r_2 is guaranteed to occur before the event s_3. The *asynchronous* interpretation corresponds to concatenating the two MSCs process by process. Thus, the event s_3 will happen after the event s_2, but has no causal relationship to the event r_2. The partial orders of events resulting from these two interpretations are shown in Figure 4.

The synchronous interpretation is closer to the visual structure of the MSC-graph and may be closer to the behavior of the system that the designer of the MSC-graph has in mind. However it has a high implementation cost to enforce (some additional messages must be introduced to ensure that processes do not commence to execute M_2 unless all the events in M_1 have occurred). The asynchronous interpretation is advocated by the standard Z.120. It has no implementation overhead, but it introduces potentially unbounded configurations. We will study both possibilities.

[1] The proofs are omitted due to lack of space. To obtain the full version that includes the proofs, please contact the authors.

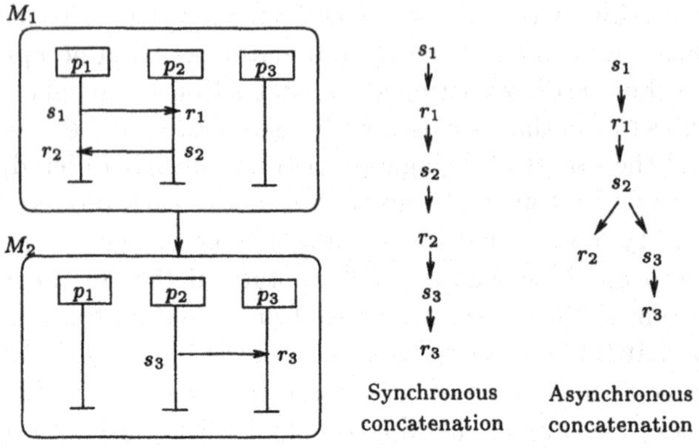

Fig. 4. Two interpretations of concatenation

4.1 Synchronous concatenation

In the synchronous interpretation, the language of concatenation of two MSCs is the concatenation of languages of the component MSCs. For an MSC-graph $G = (V, \rightarrow, v^I, v^T, \mu)$, a *path* is a sequence $\rho = v_0 v_1 \dots v_n$ such that $v_i \rightarrow v_{i+1}$ for $0 \leq i < n$. An *accepting path* is a path $v_0 v_1 \dots v_n$ such that $v_0 = v^I$ and $v_n = v^T$. In a Σ-labeled MSC-graph G, each vertex is mapped to a Σ-labeled MSC, and thus, has a language associated with it. The language of G is obtained by considering concatenation of languages of vertices along accepting paths. Formally, given a Σ-labeled MSC-graph $G = (V, \rightarrow, v^I, v^T, \mu)$, the *synchronous-language* $L^s(G)$ is the set of strings $\sigma_0 \cdot \sigma_1 \cdots \sigma_n$ such that there exists an accepting path $v_0 v_1 \dots v_n$ in G with $\sigma_i \in L(\mu(v_i))$ for $0 \leq i \leq n$.

In the synchronous model checking problem, we are given a Σ-labeled MSC-graph G, and an automaton A over Σ, and we wish to determine whether or not the intersection $L^s(G) \cap L(A)$ is empty. To solve the problem, we construct an automaton A^s_G that accepts the language $L^s(G)$ as follows. Replace each node v of G by the automaton $A_{\mu(v)}$ that accepts the language corresponding to the MSC-label of v. An edge from a vertex u to a vertex v is replaced by edges that ensure concatenation of the languages of $A_{\mu(u)}$ and $A_{\mu(v)}$ (concatenation of automata is a standard operation, and the details are omitted here). If each of the MSCs labeling the vertices of G has at most n events and k processes, then each of the individual automata has at most n^k states. If G has m vertices then A^s_G has at most $m \cdot n^k$ states.

Theorem 2. *Given a Σ-labeled MSC-graph G with m vertices, each of which is labeled with an MSC with at most n events and k processes, and an automaton A of size a, the synchronous model checking problem (G, A) can be solved in time $O(m \cdot a \cdot n^k)$, and is coNP-complete.*

4.2 Asynchronous concatenation

The asynchronous concatenation of two MSCs gives another MSC. Let $M_1 = (P_1, E_1, g_1, f_1, \{<_p^1 | \; p \in P_1\})$ and $M_2 = (P_2, E_2, g_2, f_2, \{<_p^2 | \; p \in P_2\})$ be two MSCs. *The asynchronous concatenation* of M_1 and M_2 is the MSC M defined by

- The set of processes is the union $P_1 \cup P_2$.
- Assuming the two event sets E_1 and E_2 are disjoint, the set of events is the union $E_1 \cup E_2$.
- The process labels stay unchanged: for $e \in E_1$, $g(e) = g_1(e)$ and for $e \in E_2$, $g(e) = g_2(e)$.
- The send-receive edges are unchanged: for $e \in S_1$, $f(e) = f_1(e)$ and for $e \in S_2$, $f(e) = f_2(e)$.
- For $p \in P_1 \setminus P_2$, $<_p$ equals $<_p^1$, and for $p \in P_2 \setminus P_1$, $<_p$ equals $<_p^2$. The ordering of events belonging to a common process $p \in P_1 \cap P_2$ is the concatenation of the component orderings: $<_p$ equals $<_p^1 \cup <_p^2 \cup E_{1_p} \times E_{2_p}$.

The asynchronous concatenation operation extends to Σ-labeled MSCs. To associate a language with a Σ-labeled MSC-graph G under the asynchronous interpretation, we can associate an MSC with each path by asynchronously concatenating MSCs corresponding to individual vertices. The language of the graph is the union of the languages of all such MSCs associated with the accepting paths. Formally, given a Σ-labeled MSC-graph $G = (V, \to, v^I, v^T, \mu)$, given a path $\rho = v_0 v_1 \ldots v_n$, the Σ-labeled MSC $\mu(v_0) \cdot \mu(v_1) \cdots \mu(v_n)$ is denoted M_ρ. The *asynchronous-language* $L^a(G)$ is the set

$$\{L(M_\rho) \mid \rho \text{ is an accepting path in } G\}.$$

Under the asynchronous interpretation, the language of a graph need not be regular. For instance, consider an MSC M containing a single send-receive edge: send-event s by process p_1 followed by a receive-event r by process p_2. In the MSC-graph M^*, under the asynchronous interpretation, process p_1 can send arbitrarily many messages to process p_2 before any message is actually received by process p_2. A key property that contributes to the complexity is the following: the language of the asynchronous concatenation of two MSCs with no processes in common is the shuffle of the languages of the components. This can be exploited to encode computations of Turing machines as shown below. This result strengthens the result in [15], where the intersection of two MSC graphs is shown to be undecidable.

Theorem 3. *The asynchronous model checking problem (G, A) for MSC-graphs is undecidable.*

4.3 Bounded MSC-graphs

Given an MSC M with set P of processes, define the *communication graph* H_M of M to be the graph with P as its vertices and with an arc from process p

to process q if p sends a message to q in M. Given an MSC-graph G and a subset S of its vertices, the communication graph H_S of S is the union of the communication graphs of the MSCs corresponding to the vertices in S: the set of vertices of H_S is the set P of all the processes, and there is an arc from process p to process q if p sends a message to q in the MSC $\mu(v)$ for some $v \in S$. For a set S of vertices, we denote by P_S the set of processes that send or receive a message in the MSC of some vertex in S, and call them the active processes of the set S. We call an MSC-graph *bounded* if for every cycle ρ of G, the subgraph of the communication graph H_ρ induced by the set P_ρ of active processes of the cycle is strongly connected. In other words, communication graph H_ρ on all the processes consists of one nontrivial strongly connected component and isolated nodes corresponding to processes that are inactive throughout the cycle.

We proceed to establish that the asynchronous model checking problem for bounded MSC-graphs is decidable. Given a bounded MSC-graph G, we wish to construct an automaton that generates the asynchronous language of G. Basically, the automaton traverses a path in G, and generates a linearization of the MSC obtained by concatenating the MSCs labeling the nodes on the traversed path. Such linearization can be generated by letting, at every step, one of the processes execute its next step. Due to the asynchronous nature of concatenation, the processes can drift, that is, even before all the events in the MSC corresponding to one node are executed, some processes may proceed to the next node. If we could show that processes can drift apart only by a finite distance, say, bounded by the number of nodes in the graph, then it would follow that it suffices for the automaton to remember only a finite suffix of the path. Unfortunately, this does not hold. For example, the processes may be partitioned into two disjoint sets Q and Q' such that all the processes in Q' "overtake" all the processes in Q, and proceed to execute a cycle, possibly multiple times, in which all the processes in Q are inactive. Furthermore, the processes in Q may traverse paths in which all processes in Q' are inactive while processes in Q are active, thus, imposing constraints on what the processes in Q should do in future. In the sequel, we will show that remembering only a finite amount of information suffices even if unbounded intervals of the path are of relevance. To get some intuition for the detailed construction, consider the scenario just described in which processes in Q', after overtaking Q, traverse a path that alternates between intervals in which only processes in Q are active and intervals in which only processes in Q' are active. First, due to the definition of boundedness, the number of such alternations is bounded (otherwise, there would be a cycle with two nontrivial strongly connected components in the communication graph). Second, while the individual intervals can be unbounded, it suffices to remember only the end-points of each interval (in fact, only the end-points of the intervals in which only processes in Q are active). It is worth noting that the construction would be simpler if we had used a weaker definition of boundedness which would require the communication graph of each cycle to be a single strongly connected component. However, allowing inactive processes in cycles seems important to us.

Let G be a bounded MSC-graph. Consider a path $\rho = v_0, v_1, \ldots$ through the graph, and its corresponding MSC M_ρ. Consider some linearization of M_ρ and a prefix σ of the linearization. That is, σ is the set of events executed up to some point in time. We can partition the nodes of ρ into three classes with respect to σ as follows. A node is a *past* node if all the events of the MSC of that node are already executed in the prefix σ, a *present* node if some but not all the events of the MSC of that node are executed, and a *future* node if no events corresponding to that node are executed yet. Since the MSC of each node contains at least one event (this can be assumed without loss of generality), each node of the path gets classified uniquely. Note that a node of G may occur more than once in the path and different occurrences may be classified differently.

Lemma 1. *Consider a subpath of ρ from node v_i to v_j such that the MSC-graph contains a "back" arc from v_j to v_i. Then either (i) all nodes of ρ from v_i to v_j (inclusive) are past, or (ii) all nodes of ρ from v_i to v_j (inclusive) are future, or (iii) there is a process p whose last executed step and next unexecuted step are both from the nodes $v_i, \ldots v_j$.*

We define a *configuration* as a tuple consisting of the following components:

1. A sequence of (not necessarily distinct) nodes $u_1, \ldots, u_t$ of the MSC-graph G, such that no node occurs more than k times, where k is the number of processes.
2. A mapping from each process p to one of the nodes u_i in the sequence and to a position in the process line of p in the MSC $\mu(u_i)$ (of course if p is not active in u_i, then this last part is vacuous).
3. For every $i = 1, \ldots t-1$, a bit b_i corresponding to the pair $[u_i, u_{i+1}]$.

Given a path ρ and a prefix σ of a linearization of it, we can define a configuration as follows.

1. The sequence of nodes $u_1, \ldots u_t$ consists of all the present nodes, those past nodes that are adjacent to future nodes (if the first and last node of a contiguous segment of past nodes are occurrences of the same node of the MSC graph, then we only need to keep one copy of the node), and the last node of the path ρ if it is a past node; these past nodes will be needed later to fill in the future nodes consistently. It follows from the lemma that every node v of the MSC-graph occurs at most k times in the recorded sequence u_i: If a node v occurs $k+1$ times in the path ρ, then all k processes must be executing steps in the previous k intervals from v to v, and therefore all nodes of the path ρ up to and including the $(k+1)$th last occurrence of v must be past nodes and are not selected.
2. For the second component of the configuration, map every process p to the node of the path ρ that contains the last step executed by p, and to the corresponding event of the MSC, if the node is one of the selected nodes u_i; if the node is not among the u_i's, then map p to the earliest subsequent u_i. We call this node the current node of p.

3. For the third component, if the subpath of ρ between u_i and u_{i+1} does not contain any future nodes, then set $b_i = 0$; if it contains some future nodes, then set $b_i = 1$. Note in the latter case that the subpath of ρ consists in fact entirely of a sequence of future nodes bordered by u_i and u_{i+1} (which are past or present nodes).

It is clear from the above derivation of a configuration from a partial execution that it satisfies several consistency and nonredundancy conditions. We call a configuration *legal* if it satisfies the following conditions. The mapping of the processes in the second component of a configuration induces a cut of the MSC formed by the concatenation of the MSCs $\mu(u_i)$ of the sequence of nodes u_i in the first component; i.e. if a node u_i contains a message from process p to process q, and process p is mapped before u_i or at u_i before the sending of the message, then process q is mapped before u_i or at u_i before the reception of the message. Based on the mapping of the processes we can classify the selected nodes u_i of the configuration as past, present or future. Then every node u_i of the sequence is past or present; if a node u_i is past then either $b_i = 1$ or $b_{i-1} = 1$; furthermore, if $b_{i-1} = 0$, then u_{i-1} and u_i are not occurrences of the same node of the MSC graph. If (u_i, u_{i+1}) is not an arc of the graph, then either $b_i = 1$ and there is a path in the MSC graph from u_i to u_{i+1} using only future nodes (i.e. nodes of the graphs whose MSCs involve only processes that are mapped at or before u_i), or $b_i = 0$ and there is a path from u_i to u_{i+1} using only past nodes (i.e. nodes of the graph whose MSCs involve only processes that are mapped after u_i). An obvious upper bound on the number of possible configurations is $(km)!2^{km}(mnk)^k$. The following lemma gives a better upper bound on the number of legal configurations.

Lemma 2. *The number of legal configurations is no more than* $2^{(k-1)m} \cdot (mnk)^k$, *where* k *is the number of processes,* m *is the number of vertices of* G *and* n *is the maximum number of events in a basic MSC of a vertex.*

To solve the asynchronous model checking problem, given a bounded Σ-labeled MSC-graph G, we construct an automaton A_G^a that accepts the language $L^a(G)$. The states of A_G^a are all the legal configurations. The initial state is the configuration with one node u_0, the initial node of the MSC-graph, and all processes are mapped to it, at the beginning of their process lines. The accepting state is the configuration with one node u_T, the terminal node of the MSC-graph, and all processes mapped to it at the end of their process lines. There are transitions representing the update of the configuration by execution of a single event. In addition we have ϵ−transitions that allow the addition of new nodes in the middle or the end of the sequence, the removal of nodes that are not needed any more, the advancement of a process (once it is finished with the steps of a node) and so forth.

In practice of course we will construct the automaton on the fly, generating states as needed. The automaton A_G^a which accepts the linearizations of the MSC-graph G has size at most $O(2^{km} \cdot (mnk)^k)$. This leads to the following bound.

Theorem 4. *Given a bounded Σ-labeled MSC-graph G on k processes with m vertices, each of which is labeled with an MSC with at most n events, and an automaton A of size a, the asynchronous model checking problem (G, A) can be solved in time $O(a \cdot 2^{km} \cdot (mnk)^k)$.*

A precise bound on the complexity is PSPACE:

Theorem 5. *The asynchronous model checking problem (G, A) for bounded MSC-graphs is* PSPACE*-complete. Furthermore, the* PSPACE*-hardness holds even if we bound the number of processes and the number of events in individual MSCs, and even for a fixed property.*

Finally, we address the problem of determining if a given MSC-graph is bounded. For an MSC-graph with process set P, a subset Q of processes is said to be a *witness* for unboundedness if there exists a cycle ρ such that in the MSC M_ρ, no process in Q sends a message to a process in $P \setminus Q$, and there is a process of Q and a process of $P \setminus Q$ that are active (perform some step) in ρ. Verify that if G is not bounded, then some subset must be a witness to the unboundedness.

Whether a given set Q of processes is a witness for unboundedness can be checked in linear time as follows. Remove from G all vertices v such that in the MSC $\mu(v)$ some process in Q sends a message to some process in $P \setminus Q$, and let G' be the resulting graph. Find the strongly connected components of G'. If for some strong component C of G' the corresponding set P_C of active processes intersects both Q and $P \setminus Q$, then Q is a witness for unboundedness, and G is not bounded.

Theorem 6. *Given an MSC-graph G on k processes with m vertices, each of which is labeled with an MSC with at most n events, checking whether G is bounded can be solved in time $O(m \cdot n \cdot 2^k)$, and is coNP-complete.*

A requirement similar to boundedness was identified in [4] in the context of *process divergence*, a situation in which a process sends a message an unbounded number of times ahead of a receiving process (thus requiring unbounded buffers). The condition for absence of divergence is that for every cycle ρ of the MSC-graph G, the transitive closure of the communication graph H_ρ is symmetric. This is equivalent to the requirement that every weakly connected component of H_ρ be strongly connected (thus, every bounded MSC-graph is divergence-free, but not necessarily vice-versa.) The algorithm given in [4] for process-divergence requires checking each cycle, and thus, is exponential in the number of vertices in G, and no lower bound was given. We can use the same approach as for boundedness to give an algorithm for process divergence that is exponential only in the number of processes, and we can also show a lower bound along the same lines.

Theorem 7. *Given an MSC-graph G on k processes with m vertices, each of which is labeled with an MSC with at most n events, checking G for process divergence can be solved in time $O(m \cdot n \cdot 2^k)$, and is coNP-complete.*

5 Model Checking of HMSCs

For an alphabet Σ, a Σ-labeled HMSC $H = (N, B, v^I, v^T, \mu, E)$ is like an HMSC, where μ maps nodes to Σ-labeled MSCs. By flattening a Σ-labeled HMSC H, we obtained a Σ-labeled MSC-graph H^F. Depending on whether the interpretation of concatenation is synchronous or asynchronous, we get two languages associated with H: the synchronous language $L^s(H)$ and the asynchronous language $L^a(H)$.

In the synchronous model checking problem for HMSCs, we are given a Σ-labeled HMSC H, and an automaton A over Σ, and we wish to decide if $L(A) \cap L^s(H)$ is empty. For this purpose, we translate H into a *hierarchical Kripke structure* [2] by replacing each atomic node v in H, and recursively in every HMSC associated with the boxes of H, by the automaton $A_{\mu(v)}$, and replace edges by the edges that ensure concatenation of the languages. The resulting hierarchical Kripke structure A^s_H captures the language $L^s(H)$, and model checking of H reduces to model checking of A^s_H, which can be solved using the algorithms of [2] without flattening the hierarchy.

Theorem 8. *Given a Σ-labeled HMSC H of size m, each of which nodes is labeled with an MSC with at most n events and k processes, and an automaton A of size a, the synchronous model checking problem (H, A) can be solved in time $O(m \cdot a^2 \cdot n^k)$, and is coNP-complete.*

The asynchronous model checking problem for HMSCs is, given a Σ-labeled HMSC H and an automaton A, determine if $L(A) \cap L^a(H)$ is empty. Since the problem is undecidable even for MSC-graphs, we will consider only bounded HMSCs: an HMSC is bounded if the flattened MSC-graph H^F is bounded. The asynchronous model checking problem (H, A) can be solved by first constructing the MSC-graph H^F, and then using the model checking algorithm for the bounded MSC-graphs. If the size of H is m, and its nesting depth is d, the size of H^F is $O(m^d)$. If each of the MSCs has at most n events and k processes, and A has a vertices, the resulting time bound for model checking is $O(a \cdot 2^{m^d k} \cdot (m^d nk)^k)$. A precise bound on the complexity is exponential-space:

Theorem 9. *The asynchronous model checking problem (H, A) for bounded HMSCs is* EXPSPACE*-complete.*

To determine if a given HMSC is bounded or not, for every process-set Q, we need to check if Q is a witness for unboundedness. This reduces to detecting cycles of a specific form in the hierarchical graph, and using the algorithms described in [2], can be done in linear time. A similar algorithm can be used for checking process divergence.

Theorem 10. *Given an HMSC H of size m, each of which is labeled with an MSC with at most n events and k processes, checking whether H is bounded can be solved in time $O(m \cdot n \cdot 2^k)$, and is coNP-complete.*

Acknowledgements. We thank Anca Muscholl for helpful comments.

References

1. R. Alur, G.J. Holzmann, and D. Peled. An analyzer for message sequence charts. *Software Concepts and Tools*, 17(2):70–77, 1996.
2. R. Alur and M. Yannakakis. Model checking of hierarchical state machines. In *Proc. Sixth ACM FSE*, 175–188, 1998.
3. G. Booch, I. Jacobson, and J. Rumbaugh. *Unified Modeling Language User Guide*. Addison Wesley, 1997.
4. H. Ben-Abdallah and S. Leue. Syntactic detection of process divergence and non-local choice in message sequence charts. In *Proc. of TACAS*. 1997.
5. E.M. Clarke and E.A. Emerson. Design and synthesis of synchronization skeletons using branching time temporal logic. In *Proc. Workshop on Logic of Programs*, LNCS 131, pages 52–71, 1981.
6. E.M. Clarke and R.P. Kurshan. Computer-aided verification. *IEEE Spectrum*, 33(6):61–67, 1996.
7. J. Feigenbaum, J. A. Kahn, and C. Lund. Complexity results for pomset languages. In *Proc. CAV*, 1991.
8. D. Harel. Statecharts: A visual formalism for complex systems. *Science of Computer Programming*, 8:231–274, 1987.
9. G.J. Holzmann. Early fault detection tools. *Software Concepts and Tools*, 17(2):63-69, 1996.
10. G.J. Holzmann. The model checker spin. *IEEE Trans. on Software Engineering*, 23(5):279–295, 1997.
11. G.J. Holzmann, D.A. Peled, and M.H. Redberg. Design tools for for requirements engineering. *Lucent Bell Labs Technical Journal*, 2(1):86–95, 1997.
12. R.P. Kurshan. *Computer-aided Verification of Coordinating Processes: the automata-theoretic approach*. Princeton University Press, 1994.
13. P. Ladkin and S. Leue. Interpreting message flow graphs. *Formal Aspects of Computing*, 3, 1994.
14. V. Levin, and D. Peled. Verification of message sequence charts via template matching. In *Proc. TAPSOFT*, 1997.
15. A. Muscholl, D. Peled, and Z. Su. Deciding properties of message sequence charts. In *Found. of Software Science and Computation Structures*, 1998.
16. S. Mauw and M.A. Reniers. An algebraic semantics of basic message sequence charts. *Computer Journal*, 37, 1994.
17. V.R. Pratt. Modeling concurrency with partial orders. *International Journal of Parallel Programming*, 15(1), 1986.
18. E. Rudolph, P. Graubmann, and J. Gabowski. Tutorial on message sequence charts. In *Computer Networks and ISDN Systems*, volume 28. 1996.
19. B. Selic, G. Gullekson, and P.T. Ward. *Real-time object oriented modeling and design*. J. Wiley, 1994.
20. M.Y. Vardi and P. Wolper. An automata-theoretic approach to automatic program verification. In *Proc. First LICS*, pages 332–344, 1986.
21. CCITT Specification and Description Language (SDL). ITU-T, 1994.
22. Message Sequence Charts (MSC'96). ITU-T, 1996.

Synthesis of Large Concurrent Programs via Pairwise Composition

Paul C. Attie *

School of Computer Science, Florida International University, Miami, FL, USA
http://www.cs.fiu.edu/scspage/professor/Attie.html

Abstract. We present a tractable method for synthesizing arbitrarily large concurrent programs from specifications expressed in temporal logic. Our method does not explicitly construct the global state transition diagram of the program to be synthesized, and thereby avoids *state explosion*. Instead, it constructs a state transition diagram for each pair of component processes (of the program) that interact. This "pair-program" embodies all possible interactions of the two processes. Our method proceeds in two steps. First, we construct a pair-program for every pair of "connected" processes, and analyze these pair-programs for desired correctness properties. We then take the "pair processes" of the pair-programs, and "compose" them in a certain way to synthesize the large concurrent program. We establish a "large model" theorem which shows that the synthesized large program inherits correctness properties from the pair-programs.

1 Introduction

We exhibit a method of automatically synthesizing a concurrent program consisting of K sequential processes executing in parallel, from a temporal logic specification, where K is an arbitrarily large natural number. Previous synthesis methods [1, 9, 11, 14–16] all rely on some form of exhaustive state space search, and thus suffer from the *state explosion problem*: synthesizing a concurrent program consisting of K sequential processes, each with about N local states, requires building the global state transition diagram of size at least N^K, in general. We show how to synthesize a large concurrent program by only constructing the product of small numbers of processes, and in particular, the product of a pair of processes, thereby avoiding the exponential complexity in K.

Our method is a significant improvement over the previous literature. For example, the solutions synthesized in [11] and [14] for the mutual exclusion problem were only for two processes; consideration of just three processes made the problem infeasible for hand computation. Also, the examples given in [15, 16] are reactive modules containing only two single-bit variables. Therefore, we are able to overcome the severe limitations previously imposed by state explosion on the applicability of automatic temporal logic synthesis methods.

* Supported in part by NSF CAREER Grant CCR-9702616 and AFOSR Grant F49620-96-1-0221

A crucial aspect of our method is its soundness: what correctness properties of the pair-programs are preserved by the synthesized program? We show that any formula of the branching time temporal logic ACTL [13] that is expressed over two processes, and contains no nexttime operator, is preserved. In particular, propositional invariants and some temporal leads-to properties of any pair-program also hold of the synthesized large program. (A temporal leads-to property has the following form: if condition 1 holds now, then condition 2 eventually holds. ACTL can express temporal leads-to if condition 1 is purely propositional.)

This paper extends the work of [4], in two important ways: (1) it eliminates the requirement that all pair-programs be isomorphic to each other, which in effect constrains the synthesized program to contain only one type of interaction amongst its component processes, and (2) it extends the set of correctness properties that are preserved from propositional invariants and propositional temporal leads-to properties (i.e., leads-to properties where both conditions are purely propositional) to formulae that can contain arbitrary nesting of temporal modalities. Our examples will demonstrate the utility of this greater generality.

The rest of the paper is as follows. Section 2 presents our model of concurrent computation and Section 3 discusses temporal logic. Section 4 presents the synthesis method, and Section 5 establishes the method's soundness. In Sections 6 and 7 we synthesize solutions to the readers-writers and two-phase commit problems respectively. Section 8 discusses further work and concludes.

2 Model of Concurrent Computation

A concurrent program $P = P_1 \| \cdots \| P_K$ consists of a finite number of fixed sequential processes $P_1, \ldots, P_K$ running in parallel. With every process P_i, we associate a single, unique index, namely i. Two processes are *similar* if and only if one can be obtained from the other by swapping their indices. Intuitively, this corresponds to concurrent algorithms where a single "generic" indexed piece of code gives the code body for all processes.

We use the *synchronization skeleton* model of [11]. The synchronization skeleton of a process P_i is a state-machine where each state represents a region of code that performs some sequential computation and each arc represents a conditional transition (between different regions of sequential code) used to enforce synchronization constraints. For example, a node labeled C_i may represent the critical section of P_i. While in C_i, P_i may increment a single variable, or it may perform an extensive series of updates on a large database. In general, the internal structure and intended application of the regions of sequential code are unspecified in the synchronization skeleton. The abstraction to synchronization skeletons thus eliminates all steps of the sequential computation from consideration.

Formally, the synchronization skeleton of each process P_i is a directed graph where each node s_i is a unique *local state* of P_i, and each arc has a label of the form $\oplus_{\ell \in [1:n]} B_\ell \to A_\ell$,[1] where each $B_\ell \to A_\ell$ is a guarded command [7],

[1] $[1:n]$ denotes the integers from 1 to n inclusive.

and $\oplus$ is guarded command "disjunction." For example, in Figure 2 the arc of process WP_j from N_j to T_j is labeled with $N_k \vee C_k \rightarrow \textit{skip} \oplus T_k \rightarrow x_{jk} := k$. Roughly, the operational semantics of $\oplus_{\ell \in [1:n]} B_\ell \rightarrow A_\ell$ is that if one of the B_ℓ evaluates to true, then the corresponding body A_ℓ can be executed. If none of the B_ℓ evaluates to true, then the command "blocks," i.e., waits until one of the B_ℓ holds.[2] Each node must have at least one outgoing arc, i.e., a skeleton contains no "dead ends," and two nodes are connected by at most one arc in each direction. A *global state* is a tuple of the form $(s_1, \ldots, s_K, v_1, \ldots, v_m)$ where each s_i is the current local state of P_i, and $v_1, \ldots, v_m$ is a list giving the current values of all the shared variables, $x_1, \ldots, x_m$ (we assume these are ordered in a fixed way, so that $v_1, \ldots, v_m$ specifies a unique value for each shared variable). A guard B is a predicate on states, and a body A is a parallel assignment statement that updates the values of the shared variables. If B is omitted from a command, it is interpreted as *true*, and we write the command as A. If A is omitted, the shared variables are unaltered, and we write the command as B.

We model parallelism in the usual way by the nondeterministic interleaving of the "atomic" transitions of the individual synchronization skeletons of the processes P_i. Hence, at each step of the computation, some process with an "enabled" arc is nondeterministically selected to be executed next. Assume that the current state is $s = (s_1, \ldots, s_i, \ldots, s_K, v_1, \ldots, v_m)$ and that P_i contains an arc from s_i to s'_i labeled by the command $B \rightarrow A$. If B is true in s, then a permissible next state is $(s_1, \ldots, s'_i, \ldots, s_K, v'_1, \ldots, v'_m)$ where $v'_1, \ldots, v'_m$ is the list of updated values for the shared variables produced by executing A in state s. The arc from s_i to s'_i is said to be *enabled* in state s. An arc that is not enabled is *disabled*, or *blocked*. A *(computation) path* is any sequence of states where each successive pair of states is related by the above next-state relation.

3 Temporal Logic

CTL* is a propositional branching time temporal logic [10] whose formulae are built up from atomic propositions, propositional connectives, the universal (A) and existential (E) path quantifiers, and the linear-time modalities nexttime (by process j) X_j, and strong until U. The logic CTL [11] results from restricting CTL* so that every linear-time modality is paired with a path quantifier, and vice-versa. The logic ACTL [13] results from CTL by restricting negation to propositions, and eliminating the existential path quantifier. The linear-time temporal logic PTL [14] results from removing the path quantifiers from CTL*.

Formally, we define the semantics of CTL* formulae with respect to a (K-process) structure $M = (S, R_{i_1}, \ldots, R_{i_K})$ consisting of

- S, a countable set of states. Each state is a mapping from the set $\mathcal{AP}$ of atomic propositions into $\{\textit{true}, \textit{false}\}$, and
- $R_i \subseteq S \times \{i\} \times S$, a binary relation on S giving the transitions of process i.

Here $\mathcal{AP} = \{\mathcal{AP}_{i_1}, \ldots, \mathcal{AP}_{i_K}\}$, where $\mathcal{AP}_i$ is the set of atomic propositions that "belong" to process i. Other processes can read propositions in $\mathcal{AP}_i$, but only

[2] This interpretation was proposed by [8].

process i can modify these propositions (which collectively define the local state of process i). We define the logic ACTL$^-$ to be ACTL without the AX_j modality, and the logic ACTL$^-_{ij}$ to be ACTL$^-$ where the atomic propositions are drawn only from $\mathcal{AP}_i \cup \mathcal{AP}_j$.

Let $R = R_{i_1} \cup \cdots \cup R_{i_K}$. A *path* is a sequence of states $(s_1, s_2 \ldots)$ such that $\forall i : (s_i, s_{i+1}) \in R$, and a *fullpath* is a maximal path. $M, s_1 \models f$ (respectively $M, \pi \models f$) means that f is true in structure M at state s_1 (respectively of fullpath π). Also, $M, S \models f$ means $\forall s \in S : M, s \models f$, where S is a set of states. For the full definition of $\models$, see [10, 13]. For example, $M, s_1 \models \mathsf{A}f$ iff for every fullpath $\pi = (s_1, s_2, \ldots)$ in M: $M, \pi \models f$; and $M, \pi \models f \mathsf{U} g$ iff there exists i such that $M, \pi^i \models g$ and for all $j \in [1 : (i-1)]$: $M, \pi^j \models f$ (π^i is the suffix starting at the i'th state of π).

We also introduce some additional modalities as abbreviations: $\mathsf{F}f$ (eventually) for $[true \mathsf{U} f]$, $\mathsf{G}f$ (always) for $\neg\mathsf{F}\neg f$, $[f \mathsf{U}_\mathsf{w} g]$ (weak until) for $[f \mathsf{U} g] \vee \mathsf{G}f$, $\overset{\infty}{\mathsf{F}} f$ (infinitely often) for $\mathsf{GF}f$, and $\overset{\infty}{\mathsf{G}} f$ (eventually always) for $\mathsf{FG}f$. We refer the reader to [10] for details in general, and to [13] for details of ACTL.

To guarantee liveness properties of the synthesized program, we use a form of weak fairness. Fairness is usually specified as a linear-time logic (i.e., PTL) formula Φ, and a fullpath is fair iff it satisfies Φ. To state correctness properties under the assumption of fairness, we relativize satisfaction ($\models$) so that only fair fullpaths are considered. The resulting notion of satisfaction, $\models_\Phi$, is defined by [12] as follows: $M, s_1 \models_\Phi \mathsf{A}f$ iff for every Φ-fair fullpath $\pi = (s_1, s_2, \ldots)$ in M: $M, \pi \models f$. Effectively, path quantification is only over the paths that satisfy Φ.

4 The Synthesis Method

We aim to synthesize a large concurrent program $P_{i_1} \| \cdots \| P_{i_K}$ without explicitly generating its global state transition diagram of size exponential in the number of processes K. The specification for a large concurrent program consists of:

1. a binary, irreflexive "interconnection" relation $I_c \subseteq \{i_1, \ldots, i_K\} \times \{i_1, \ldots, i_K\}$ over the set $\{i_1, \ldots, i_K\}$ of process indices, and
2. a mapping *spec* which maps each pair $(i, j) \in I$ to a formula of ACTL$^-_{ij}$ (we use $spec_{ij}$ rather than $spec((i, j))$ to denote this "pair-specification").

We use I to denote the pair $(I_c, spec)$ and abuse terminology by sometimes referring to I as the interconnection relation. Given a specification I, we synthesize an *I-program* $P^I = (S_I^0, P_{i_1}^I \| \ldots \| P_{i_K}^I)$ as follows:

1. For every pair of process indices $(i, j) \in I$, synthesize a *pair-program* $(S_{ij}^0, P_i^j \| P_j^i)$ using $spec_{ij}$ as the specification.
2. "Compose" all the pair-programs to produce P^I.

Since our focus in this article is on avoiding state-explosion, we shall not explicitly address step 1 above. Any synthesis method that produces concurrent programs in the synchronization skeleton notation can be used, e.g., [2, 3, 11].

S_{ij}^0 is the set of initial states, and P_i^j, P_j^i are the synchronization skeletons for processes i, j, in the pair-program $(S_{ij}^0, P_i^j \| P_j^i)$. We refer to the component processes of a pair-program as ***pair-processes.*** Note that P_i^j and P_j^i interact by reading each other's local state and by reading/writing a set (call it $\mathcal{SH}_{ij}$) of shared variables.[3] S_I^0 is the set of initial states, and P_i^I is the synchronization skeleton for process i, in the I-program $(S_I^0, P_{i_1}^I \| \ldots \| P_{i_K}^I)$. We refer to the component processes P_i^I of an I-program as *I-processes.* We say that P_i^I and P_j^I are ***neighbors*** when $(i,j) \in I$. We require that every process has at least one neighbor: $\forall i \in \{i_1, \ldots, i_K\} : (\exists j : (i,j) \in I)$. We also define $I(i) = \{j \mid (i,j) \in I\}$.

$spec_{ij}$ is the specification for the pair-program $(S_{ij}^0, P_i^j \| P_j^i)$, and defines the interaction of processes i and j. Thus, $spec_{ij}$ is (initially) interpreted and verified over the structure induced by $(S_{ij}^0, P_i^j \| P_j^i)$ *executing in isolation.* Once $(S_{ij}^0, P_i^j \| P_j^i)$ has been composed with all the other pair-programs to yield the I-program, we will show that $spec_{ij}$ also holds for the I-program. Unlike [4], $spec_{ij}$ and $spec_{k\ell}$ (where $\{k,\ell\} \neq \{i,j\}$) can be completely different formulae, whereas in [4] these formulae had to be "similar," i.e., one was obtained from the other by substituting process indices.

Our synthesis method requires that the pair-programs induce the same *local structure* on all common processes. That is, for pair-programs $(S_{ij}^0, P_i^j \| P_j^i)$ and $(S_{ik}^0, P_i^k \| P_k^i)$, we require $\hat{P}_i^j = \hat{P}_i^k$, where $\hat{P}_i^j, \hat{P}_i^k$ result from removing all arc labels from P_i^j, P_i^k respectively.[4] We assume, in the sequel, that this condition holds. Also, all results quoted from [4] have been reverified to hold in our setting, i.e., when the similarity assumptions of [4] are dropped.

We compose pair-programs as follows. Consider first $I = \{(i,j), (j,k), (k,i)\}$, i.e., three pairwise interconnected processes i, j, k, With respect to process i, the proper interaction (i.e., that required to satisfy $spec_{ij}$) between process i and process j is captured by the commands that label the arcs of P_i^j. Likewise, the proper interaction between process i and process k is captured by the arc labels of P_i^k. Hence, in the three-process program P^I (consisting of processes i, j, k), the proper interaction for process i with processes j and k is captured as follows: when process i traverses an arc, the command which labels that arc in P_i^j is executed "simultaneously" with the command which labels the corresponding arc in P_i^k. For example, taking as our specification the mutual exclusion problem, if process i executes a mutual exclusion protocol with respect to both processes j and k, then, when process i enters its critical section, both processes j and k must be outside their own critical sections.

Based on the above, we determine that the synchronization skeleton for process i in P^I (call it P_i^I) has the same basic graph structure as P_i^j and P_i^k, and an arc label in P^I is a "conjunction" of the labels of the corresponding arcs in P_i^j and P_i^k.

---

[3] The shared variable sets of different pair-programs are disjoint: $\mathcal{SH}_{ij} \cap \mathcal{SH}_{i'j'} = \emptyset$ if $\{i,j\} \neq \{i',j'\}$.

[4] Contrast this with the much more restrictive "process similarity assumption" of [4] which requires that P_i^j can be obtained from $P_{i'}^{j'}$ by substituting i for i' and j for j'. In effect, <u>all</u> processes must have isomorphic local structure <u>and</u> isomorphic arc labels. Thus, all pair-programs are isomorphic—Proposition 6.2.1 of [4].

Generalizing to an arbitrary interconnection relation I, P_i^I has the same basic graph structure as P_i^j, ($\hat{P}_i^I = \hat{P}_i^j$), and an arc label in P_i^I is a "conjunction" of the labels of the corresponding arcs in $P_i^{j_1}, \ldots, P_i^{j_n}$, where $\{j_1, \ldots, j_n\} = I(i)$ are all the neighbors of process i.

We now make some technical definitions. A node (i.e., local state) of P_i^j, P_i^I is a mapping of $\mathcal{AP}_i$ to $\{true, false\}$. We refer to such nodes as i-states. A state of the pair-program $(S_{ij}^0, P_i^j \,\|\, P_j^i)$ is a tuple $(s_i, s_j, v_{ij}^1, \ldots, v_{ij}^m)$ where s_i, s_j are i-states, j-states, respectively, and $v_{ij}^1, \ldots, v_{ij}^m$ give the values of all the variables in $\mathcal{SH}_{ij}$. We refer to states of $(S_{ij}^0, P_i^j \,\|\, P_j^i)$ as ij-states. An ij-state s_{ij} inherits the assignments defined by its component i- and j-states: $s_{ij}(p_i) = s_i(p_i)$, $s_{ij}(p_j) = s_j(p_j)$, where $s_{ij} = (s_i, s_j, v_{ij}^1, \ldots, v_{ij}^m)$, and p_i, p_j are arbitrary atomic propositions in $\mathcal{AP}_i$, $\mathcal{AP}_j$, respectively. A state of $(S_I^0, P_{i_1}^I \,\|\, \ldots \,\|\, P_{i_K}^I)$ is a tuple $(s_{i_1}, \ldots, s_{i_K}, v^1, \ldots, v^n)$, where s_i, $(i \in \{i_1, \ldots, i_K\})$ is an i-state and $v^1, \ldots, v^n$ give the values of all the shared variables of the I-program (i.e., those in $\bigcup_{(i,j)\in I} \mathcal{SH}_{ij}$). We refer to states of an I-program as I-states. An I-state s inherits the assignments defined by its component i-states $(i \in \{i_1, \ldots, i_K\})$: $s(p_i) = s_i(p_i)$, where $s = (s_{i_1}, \ldots, s_{i_K}, v^1, \ldots, v^n)$, and p_i is any atomic proposition in $\mathcal{AP}_i$ $(i \in \{i_1, \ldots, i_K\})$. If $J \subseteq I$, then define J-program, J-state exactly like I-program, I-state resp. but using interconnection relation J instead of I.

The state-to-formula operator $\{\!|s_i|\!\}$ takes an i-state s_i as argument and returns a propositional formula: $\{\!|s_i|\!\} = (\bigwedge_{s_i(p_i)=true} p_i) \wedge (\bigwedge_{s_i(p_i)=false} \neg p_i)$, where p_i ranges over the members of $\mathcal{AP}_i$. $\{\!|s_i|\!\}$ characterizes s_i in that $s_i \models \{\!|s_i|\!\}$, and $s_i' \not\models \{\!|s_i|\!\}$ for all $s_i' \neq s_i$. $\{\!|s_{ij}|\!\}$ is defined similarly (but note that the variables in $\mathcal{SH}_{ij}$ must be accounted for). We define the *state projection operator* $\uparrow$. This operator has several variants. First, we define projection onto a single process from both I-states and ij-states: if $s = (s_{i_1}, \ldots, s_{i_K}, v^1, \ldots, v^n)$, then $s{\uparrow}i = s_i$, and if $s_{ij} = (s_i, s_j, v_{ij}^1, \ldots, v_{ij}^m)$, then $s_{ij}{\uparrow}i = s_i$. Next we define projection of an I-state onto a pair-program: if $s = (s_{i_1}, \ldots, s_{i_K}, v^1, \ldots, v^n)$, then $s{\uparrow}ij = (s_i, s_j, v_{ij}^1, \ldots, v_{ij}^m)$, where $v_{ij}^1, \ldots, v_{ij}^m$ are those values from $v^1, \ldots, v^n$ that denote values of variables in $\mathcal{SH}_{ij}$. $s{\uparrow}ij$ is well defined only when $i\,I\,j$ (i.e., $(i, j) \in I$). Finally, we define projection of an I-state onto a J-program. If $s = (s_{i_1}, \ldots, s_{i_K}, v^1, \ldots, v^n)$, then $s{\uparrow}J = (s_{j_1}, \ldots, s_{j_L}, v_J^1, \ldots, v_J^m)$, where $\{j_1, \ldots, j_L\}$ is the domain of J, and $v_J^1, \ldots, v_J^m$ are those values from $v^1, \ldots, v^n$ that denote values of variables in $\bigcup_{(i,j)\in J} \mathcal{SH}_{ij}$. $s{\uparrow}J$ is well defined only when $J \subseteq I$.

Let π be a computation path of P^I. Then, the *path-projection* of π onto $J \subseteq I$ (denoted $\pi{\uparrow}J$) is obtained as follows. Replace every state s along π by $s{\uparrow}J$, and then remove all transitions in π that are not by some process in J, coalescing the source and target states of all such transitions (which must be the same, since if $s \xrightarrow{i} t$ and $i \notin \{j_1, \ldots, j_L\}$, then $s{\uparrow}J = t{\uparrow}J$).

The above discussion leads to the following definition for our synthesis method, which derives an I-process P_i^I of the I-program $(S_I^0, P_{i_1}^I \,\|\, \ldots \,\|\, P_{i_K}^I)$ from the pair-processes $\{P_i^j \mid j \in I(i)\}$ of the pair-programs $\{(S_{ij}^0, P_i^j \,\|\, P_j^i) \mid j \in I(i)\}$:

Definition 1 (Pairwise Synthesis). *An I-*process P_i^I $(i \in \{i_1, \ldots, i_K\})$ *is derived from the pair-processes* P_i^j, $j \in I(i)$, *as follows:*

P_i^I *contains an arc from* s_i *to* t_i *with label* $\otimes_{j \in I(i)} \oplus_{\ell \in [1:n_j]} B_{i,\ell}^j \to A_{i,\ell}^j$
iff
$\forall j \in I(i) : P_i^j$ *contains an arc from* s_i *to* t_i *with label* $\oplus_{\ell \in [1:n_j]} B_{i,\ell}^j \to A_{i,\ell}^j$.
The initial state set S_I^0 *of the I-program is derived from the pair-program initial state sets* S_{ij}^0, $(i,j) \in I$, *as follows:*

$$S_I^0 = \{s \mid \forall (i,j) \in I : s{\uparrow}ij \in S_{ij}^0\}.$$

Here $\otimes$ is guarded command "conjunction." The operational semantics of $B_1 \to A_1 \otimes B_2 \to A_2$ is that if both the guards B_1, B_2 evaluate to true, then the bodies A_1, A_2 can be executed in parallel. If at least one of B_1, B_2 evaluates to false, then the command "blocks," i.e., waits until both of B_1, B_2 evaluate to true. See [4] for complete definitions of $\oplus, \otimes$. Note that S_I^0 consists of exactly those I-states whose "projections" onto all the pairs in I give initial states of the corresponding pair-program. We assume that the initial-state sets of all the pair-programs are so that there is at least one such I-state, and so S_I^0 is nonempty.

Definition 1 is, in effect, a *syntactic transformation* that can be carried out in linear time and space (in both $(S_{ij}^0, P_i^j \,\|\, P_j^i)$ and I). In particular, we avoid explicitly constructing the global state transition diagram of $(S_I^0, P_{i_1}^I \,\| \ldots \|\, P_{i_K}^I)$, which is of size exponential in $K = |\{i_1, \ldots, i_K\}|$.

5 Soundness of the Synthesis Method

Let $M_{ij} = (S_{ij}^0, S_{ij}, R_{ij})$ and $M_I = (S_I^0, S_I, R_I)$ be the global state transition diagrams of $(S_{ij}^0, P_i^j \,\|\, P_j^i)$, $(S_I^0, P_{i_1}^I \,\| \ldots \|\, P_{i_K}^I)$, respectively. S_{ij}^0, S_I^0 are the sets of initial states of M_{ij}, M_I respectively, and S_{ij}, S_I are the sets of all states of M_{ij}, M_I respectively, and $R_{ij} \subseteq S_{ij} \times \{i,j\} \times S_{ij}$, $R_I \subseteq S_I \times \{i_1, \ldots, i_K\} \times S_I$, are the sets of transitions of M_{ij}, M_I respectively. The technical definitions of M_{ij}, M_I in terms of $(S_{ij}^0, P_i^j \,\|\, P_j^i), (S_I^0, P_{i_1}^I \,\| \ldots \|\, P_{i_K}^I)$ are straightforward and are omitted (Section 2 describes the relevant operational semantics). M_{ij} and M_I can be interpreted as ACTL structures. M_{ij} gives the semantics of $(S_{ij}^0, P_i^j \,\|\, P_j^i)$ *executing in isolation*, and M_I gives the semantics of $(S_I^0, P_{i_1}^I \,\| \ldots \|\, P_{i_K}^I)$. Our main soundness result below (the large model theorem) relates the ACTL formulae that hold in M_I to those that hold in M_{ij}. We characterize transitions in M_I as compositions of transitions in all the relevant M_{ij}:

Lemma 1. [4] *For all I-states* $s, t \in S_I$ *and* $i \in \{i_1, \ldots, i_K\}$, $s \xrightarrow{i} t \in R_I$ iff :

$$\forall j \in I(i) : s{\uparrow}ij \xrightarrow{i} t{\uparrow}ij \in R_{ij} \text{ and}$$
$$\forall j,k \in \{i_1, \ldots, i_K\} - \{i\}, j\,I\,k : s{\uparrow}jk = t{\uparrow}jk.$$

Lemma 2. [4] *Let* $J \subseteq I$. *If* π *is a path in* M_I, *then* $\pi{\uparrow}J$ *is a path in* M_J.

In particular, when $J = \{(i,j)\}$, Lemma 2 forms the basis for our soundness proof, since it relates computations of the synthesized I-program to computations of the pair-programs.

5.1 Deadlock-Freedom and the Wait-For-Graph

The *wait-for-graph* in a particular I-state s contains as nodes every I-process, and every arc whose start state is a component of s. These arcs have an outgoing edge to every I-process which blocks them.

Definition 2 (Wait-For-Graph $W_I(s)$). *Let s be an arbitrary I-state. The* wait-for-graph *$W_I(s)$ of s is a directed bipartite graph, where*

1. *the nodes of $W_I(s)$ are*
   (a) *the I-processes $\{P_i^I \mid i \in \{i_1, \ldots, i_K\}\}$, and*
   (b) *the arcs $\{a_i^I \mid i \in \{i_1, \ldots, i_K\}$ and $a_i^I \in P_i^I$ and $s{\uparrow}i = a_i^I.start\}$*
2. *there is an edge from P_i^I to every node of the form a_i^I in $W_I(s)$, and*
3. *there is an edge from a_i^I to P_j^I in $W_I(s)$ if and only if $(i,j) \in I$ and $a_i^I \in W_I(s)$ and $s{\uparrow}ij(a_i^I.guard_j) = false$.*

Here $a_i^I.guard_j$ is the conjunct of the guard of arc a_i^I which references the state shared by P_i and P_j (in effect, $\mathcal{AP}_j$ and $\mathcal{SH}_{ij}$). We characterize a deadlock as the occurrence in the wait-for-graph of a *supercycle*:

Definition 3 (Supercycle). *SC is a* supercycle *in $W_I(s)$ if and only if:*

1. *SC is nonempty,*
2. *if $P_i^I \in SC$ then for all a_i^I such that $a_i^I \in W_I(s)$, $P_i^I \longrightarrow a_i^I \in SC$, and*
3. *if $a_i^I \in SC$ then there exists P_j^I such that $a_i^I \longrightarrow P_j^I \in W_I(s)$ and $a_i^I \longrightarrow P_j^I \in SC$.*

Note that this definition implies that SC is a subgraph of $W_I(s)$. In [4], we give a criterion, the *wait-for-graph assumption*, which is evaluated over the product of a small number of processes, thereby avoiding state-explosion. We show there that if the wait-for-graph assumption holds, then $W_I(s)$ cannot contain a supercycle for any reachable state s of M_I. Furthermore, if $W_I(s)$ does not contain a supercycle, then, in state s, there exists at least one enabled arc. These results extend to the setting of this paper.

5.2 Liveness

To assure liveness properties of the synthesized I-program, we assume a form of weak fairness. Let $CL(f)$ be the set of all subformulae of f, including f itself. Let ex_i be an assertion that is true along a transition in a structure iff that transition results from executing process i. Let en_i hold in an I-state s iff P_i^I has some arc that is enabled in s. Our fairness criterion is the conjunction of *weak blocking fairness* and *weak eventuality fairness* (given below) and is defined as a formula of the linear time temporal logic PTL [14].

Definition 4 (Sometimes-Blocking, blk_i^j, blk_i). *An i-state s_i is* sometimes-blocking *in M_{ij} if and only if:*

$$\exists s_{ij}^0 \in S_{ij}^0 : M_{ij}, s_{ij}^0 \models \mathsf{EF}(\ \{\!|s_i|\!\} \wedge (\exists a_j^i \in P_j^i : (\{\!|a_j^i.start|\!\} \wedge \neg a_j^i.guard))\ ).$$

Also, $blk_i^j \stackrel{\mathrm{df}}{=} (\bigvee \{\!|s_i|\!\} : s_i$ is sometimes-blocking in $M_{ij})$, and $blk_i \stackrel{\mathrm{df}}{=} \bigvee_{j \in I(i)} blk_i^j$.

Thus, a sometimes-blocking state is an i-state s_i such that there exists a reachable ij-state s_{ij} in M_{ij} satisfying $s_{ij}{\uparrow}i = s_i$ and in which P_i blocks some arc a^i_j of P_j. $a^i_j.start$ is the start state of a^i_j, and $a^i_j.guard$ is its guard.

Definition 5 (Weak Blocking Fairness Φ_b).

$$\Phi_b \stackrel{\text{df}}{=} \bigwedge_{i \in \{i_1,\ldots,i_K\}} \overset{\infty}{\mathsf{G}}(blk_i \wedge en_i) \Rightarrow \overset{\infty}{\mathsf{F}} ex_i.$$

Weak blocking fairness requires that a process that is continuously enabled and in a sometimes-blocking state is eventually executed.

Definition 6 (Pending Eventuality, pnd_{ij}). *Let $(i,j) \in I$. An ij-state s_{ij} has a* pending eventuality *if and only if:*

$$\exists f_{ij} \in CL(spec_{ij}) : M_{ij}, s_{ij} \models \neg f_{ij} \wedge \mathsf{AF} f_{ij}.$$

Also, $pnd_{ij} \stackrel{\text{df}}{=} (\bigvee \{\!|s_{ij}|\!\} : s_{ij}$ *has a pending eventuality*$)$.

In other words, s_{ij} has a pending eventuality if there is a subformula of the pair-specification $spec_{ij}$ which does not hold in s_{ij}, but is guaranteed to eventually hold along every fullpath of M_{ij} that starts in s_{ij}.

Definition 7 (Weak Eventuality Fairness Φ_ℓ).

$$\Phi_\ell \stackrel{\text{df}}{=} \bigwedge_{(i,j)\in I} (\overset{\infty}{\mathsf{G}} en_i \vee \overset{\infty}{\mathsf{G}} en_j) \wedge \overset{\infty}{\mathsf{G}} pnd_{ij} \Rightarrow \overset{\infty}{\mathsf{F}}(ex_i \vee ex_j).$$

Weak eventuality fairness requires that if an eventuality is continuously pending, and one of P^I_i or P^I_j is continuously enabled, then eventually one of them will be executed. Our overall fairness notion Φ is then the conjunction of weak blocking and weak eventuality fairness: $\Phi \stackrel{\text{df}}{=} \Phi_b \wedge \Phi_\ell$.

Definition 8 (Liveness Condition). *For every $(i,j) \in I$:*

$$M_{ij}, S^0_{ij} \models \mathsf{AGA}(\mathsf{G} ex_i \Rightarrow \overset{\infty}{\mathsf{G}} aen_j),$$

where $aen_j \stackrel{\text{df}}{=} \forall a^i_j \in P^i_j : (\{\!|a^i_j.start|\!\} \Rightarrow a^i_j.guard)$.

aen_j means that every arc of P^i_j whose start state is a component of the current global state s is also enabled in s. The liveness condition requires, for every pair-program $(S^0_{ij}, P^j_i \| P^i_j)$, when executing in isolation, that if P^j_i can execute continuously along some path, then there exists a suffix of that path along which P^j_i does not block any arc of P^i_j. Given the liveness condition and the absence of deadlocks and the use of Φ-fair scheduling, we can show that one of P^I_i or P^I_j is guaranteed to be executed from any state of the I-program whose ij-projection has a pending eventuality.

Lemma 3 (Progress). *Let $(i,j) \in I$, and let s be an arbitrary reachable I-state. If*

1. *the liveness condition holds, and*
2. *for every reachable I-state u, $W_I(u)$ is supercycle-free, and*
3. *$M_{ij}, s{\uparrow}ij \models \neg h_{ij} \wedge \mathsf{AF} h_{ij}$ for some $h_{ij} \in CL(spec_{ij})$, then*

$$M_I, s \models_\Phi \mathsf{AF}(ex_i \vee ex_j).$$

5.3 The Large Model Theorem

The large model theorem establishes the soundness of our synthesis method. It states that any subformula of pair-specification $spec_{ij}$ which holds in the ij-projection of an I-state s also holds in s itself. That is, correctness properties satisfied by a pair-program executing in isolation also hold in the I-program.

Theorem 1 (Large Model). *Suppose $M_{ij}, S^0_{ij} \models spec_{ij}$ for some $(i,j) \in I$. Let $f_{ij} \in CL(spec_{ij})$, and let s be an arbitrary reachable I-state. If the liveness condition holds, and $W_I(u)$ is supercycle-free for every reachable I-state u, then*

$$M_{ij}, s{\upharpoonright}ij \models f_{ij} \text{ implies } M_I, s \models_\Phi f_{ij}.$$

The correctness properties we are usually interested in are those that hold in all initial states. The large model corollary states that if all pair-specifications hold in all initial states of their respective pair-programs, then all pair-specifications also hold in all initial states of the I-program. The "spatial modality" $\bigwedge_{ij}$ quantifies over all pairs $(i,j) \in I$: $\bigwedge_{ij} spec_{ij}$ is equivalent to $\bigwedge_{(i,j)\in I} spec_{ij}$.

Corollary 1 (Large Model). *If the liveness condition holds, and $W_I(u)$ is supercycle-free for every reachable I-state u, then*

$$(\forall (i,j) \in I : M_{ij}, S^0_{ij} \models spec_{ij}) \text{ implies } M_I, S^0_I \models_\Phi \bigwedge_{ij} spec_{ij}.$$

6 Example—Readers Writers

In the readers-writers problem [6] a set of reader processes and a set of writer processes contend for access to a shared file. Mutual exclusion of access between readers and writers, and also between two writers, is required. Also, all requests by writers for access must eventually be granted ("absence of starvation"), and a writer's request takes priority over a reader's request. We specify the readers-writers problem in ACTL as follows:

Local structure of both readers and writers (P_i is a reader or a writer):

N_i: P_i is initially in its noncritical region

$\mathsf{AG}(N_i \Rightarrow (\mathsf{AX}_i T_i \wedge \mathsf{EX}_i T_i)) \wedge \mathsf{AG}(T_i \Rightarrow \mathsf{AX}_i C_i) \wedge \mathsf{AG}(C_i \Rightarrow (\mathsf{AX}_i N_i \wedge \mathsf{EX}_i N_i))$: P_i moves from N_i to T_i to C_i and back to N_i. Furthermore, P_i can always move from N_i to T_i and from C_i to N_i

$\mathsf{AG}((N_i \equiv \neg(T_i \vee C_i)) \wedge (T_i \equiv \neg(N_i \vee C_i)) \wedge (C_i \equiv \neg(N_i \vee T_i)))$: P_i is always in exactly one of the states N_i (noncritical), T_i (trying), or C_i (critical)

Reader-writer pair-specification (RP_i is reader, WP_j is writer):

Local structure: The above local structure specification for both RP_i and WP_j

$\mathsf{AG}(T_i \Rightarrow \mathsf{AF}(C_i \vee \neg N_j))$: absence of starvation for readers provided no writer requests access

$\mathsf{AG}(T_j \Rightarrow \mathsf{AF} C_j)$: absence of starvation for writers

$\mathsf{AG}((T_i \wedge T_j) \Rightarrow \mathsf{A}[T_i \mathsf{U} C_j])$: priority of writers over readers for outstanding requests to enter the critical region

$\mathsf{AG}(\neg(C_i \wedge C_j))$: mutual exclusion of access between a reader and a writer

Writer-writer pair-specification (WP_j, WP_k *are writers*):
Local structure: The above local structure specification for WP_j and WP_k
$\mathsf{AG}(T_j \Rightarrow \mathsf{AF}C_j) \wedge \mathsf{AG}(T_k \Rightarrow \mathsf{AF}C_k)$: absence of starvation for writers
$\mathsf{AG}(\neg(C_j \wedge C_k))$: mutual exclusion of access between two writers

Interconnection Relation I: Let K_R, K_W be the desired number of readers, writers respectively. Then I is given by $RW \cup WW$, where $RW = \{(RP_i, WP_j) \mid i \in [1{:}K_R], j \in [1{:}K_W]\}$ gives the interconnection between readers and writers, and $WW = \{(WP_j, WP_k) \mid j, k \in [1{:}K_W], j \neq k\}$ gives the interconnection between writers and writers. There is no interconnection between readers and readers.

For each pair-specification, we synthesize a pair-program satisfying it, using the synthesis method of [11]. Figures 1, 2 display the pair-programs for the reader-writer pair-specification, writer-writer pair-specification, respectively. Finally, we apply Definition 1 to synthesize the I-program with K_R readers and K_W writers, which is shown in Figure 3. Correctness of the I-program follows immediately from Corollary 1, since the conjunction of the pair-specifications gives us the desired correctness properties (formulae of the forms $\mathsf{AG}(p_i \Rightarrow \mathsf{AX}_i q_i)$, $\mathsf{AG}(p_i \Rightarrow \mathsf{EX}_i q_i)$ are not in ACTL^-_{ij}, but were shown to be preserved in [4], and the proof given there still applies here).

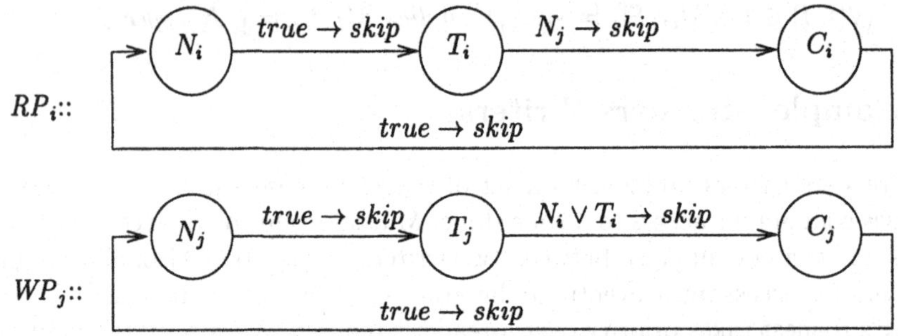

Fig. 1. Reader-writer pair-program $RP_i \parallel WP_j$.

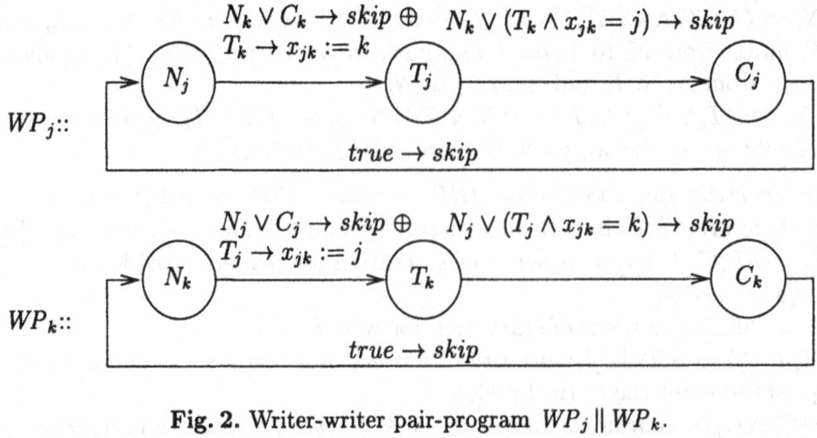

Fig. 2. Writer-writer pair-program $WP_j \parallel WP_k$.

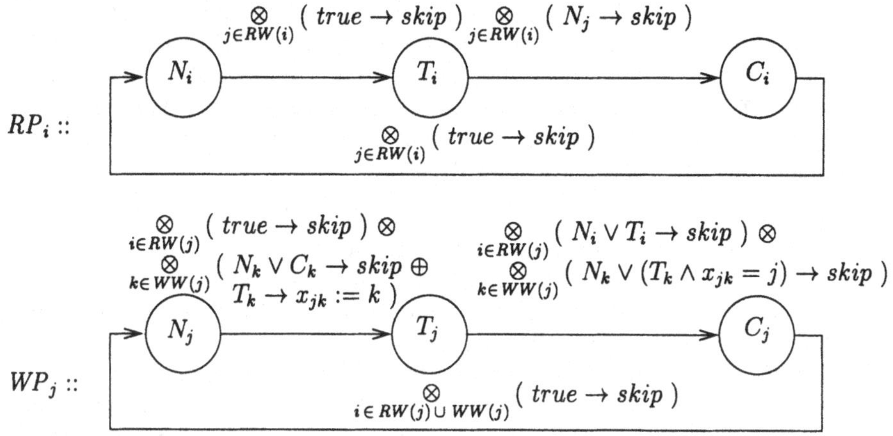

Fig. 3. Many readers, many writers program $(\|_{1\le i\le K_R} RP_i) \| (\|_{1\le j\le K_W} WP_j)$.

7 Example—Two Phase Commit

Our second example is a ring-based (non fault-tolerant) two-phase commit protocol $P^I = P_0^I \,\|\, P_1^I \,\|\, \cdots \,\|\, P_{n-1}^I$, where I specifies a ring. P_0^I is the *coordinator*, and $P_i^I, 1 \le i < n$ are the participants: each participant represents a transaction. The protocol proceeds in two cycles around the ring. The coordinator initiates the first cycle, in which each participant decides to either submit its transaction or unilaterally abort. P_i^I can submit only after it observes that P_{i-1}^I has submitted. After the first cycle, the coordinator observes the state of P_{n-1}^I. If P_{n-1}^I has submitted, that means that all participants have submitted, and so the coordinator decides commit. If P_{n-1}^I has aborted, that means that some participant P_i^I unilaterally aborted, thereby causing all participants $P_j^I, i < j \le n-1$ to abort. In that case, the coordinator decides abort. The second cycle relays the coordinators decision around the ring. The participant processes are all similar to each other, but not similar to the coordinator. Hence, there are three dissimilar pair-programs to consider: $P_{n-1}^0 \,\|\, P_0^{n-1}$, $P_0^1 \,\|\, P_1^0$, and $P_{i-1}^i \,\|\, P_i^{i-1}$ (which is replicated for each i from 2 to $n-1$ inclusive). The pair-specifications are as follows. For brevity, we omit the obvious local structure specifications (see the previous section for an example of these). The formula $f \to g$ abbreviates $\mathsf{A}[(f \Rightarrow \mathsf{AF}g)\mathsf{U_w}g]$, which means that if f holds at some point along a fullpath, then g holds at some (possibly different) point. There is no ordering on the times at which f and g hold. $f \rightsquigarrow g$ abbreviates temporal leads to: $\mathsf{AG}[f \Rightarrow \mathsf{AF}g]$.

Pair-specification for $P_{n-1}^0 \,\|\, P_0^{n-1}$:
$cm_0 \to sb_{n-1}$: the coordinator decides commit only if participant $n-1$ submits
$\mathsf{AF}(cm_0 \vee ab_0)$: the coordinator eventually decides

Pair-specification for $P_0^1 \,\|\, P_1^0$:
$\mathsf{AF}(cm_0 \vee ab_0)$: the coordinator eventually decides
$cm_1 \to cm_0$: participant 1 commits only if the coordinator decides commit

$ab_0 \rightarrow ab_1$: if the coordinator decides abort, then participant 1 aborts
$\mathsf{AG}(\neg cm_1 \vee \neg ab_1) \wedge \mathsf{AG}(cm_1 \Rightarrow \mathsf{AG}cm_1) \wedge \mathsf{AG}(ab_1 \Rightarrow \mathsf{AG}ab_1)$: participant 1 does not both commit and abort, and does not change its decision once made
$\mathsf{AG}(st_1 \Rightarrow \mathsf{EX}_1 ab_1)$: participant 1 can abort unilaterally from it's starting state
$\mathsf{AG}[sb_1 \Rightarrow \mathsf{A}[sb_1 \mathsf{U}(sb_1 \wedge (cm_0 \vee ab_0))]]$: once participant 1 submits, it does not decide until the coordinator first decides

Pair-specification for $P^i_{i-1} \,\|\, P^{i-1}_i$, *for* $2 \leq i \leq n-1$:
$sb_i \rightarrow sb_{i-1}$: participant i submits only if participant $i-1$ submits
$cm_i \rightarrow cm_{i-1}$: participant i commits only if participant $i-1$ commits
$(cm_{i-1} \wedge sb_i) \leadsto cm_i$: if participant i submits and participant $i-1$ commits, then participant i eventually commits
$ab_{i-1} \rightarrow ab_i$: if participant $i-1$ aborts, then so does participant i
$\mathsf{AG}(\neg cm_i \vee \neg ab_i) \wedge \mathsf{AG}(cm_i \Rightarrow \mathsf{AG}cm_i) \wedge \mathsf{AG}(ab_i \Rightarrow \mathsf{AG}ab_i)$: participant i does not both commit and abort, and does not change its decision once made
$\mathsf{AG}[sb_i \Rightarrow \mathsf{A}[sb_i \mathsf{U}(sb_i \wedge (cm_{i-1} \vee ab_{i-1}))]]$: once participant i submits, it does not decide until participant $i-1$ first decides
$\mathsf{AG}(st_i \Rightarrow \mathsf{EX}_i ab_i)$: participant i can abort unilaterally from it's starting state

The pair-programs synthesized from the above pair-specifications are given in Figures 4, 5, and 6, respectively, where $term_i \equiv cm_i \vee ab_i$, and an incoming arrow with no source indicates an initial local state. They satisfy the liveness condition and the wait-for-graph assumption, and so Theorem 1 is applicable. The synthesized two phase commit protocol P^I is given in Figure 7. We establish the correctness of P^I by the following deductive argument:

1.	$cm_0 \rightarrow sb_{n-1}$	LMT
2.	$\bigwedge_{2 \leq i < n}(sb_i \rightarrow sb_{i-1})$	LMT
3.	$cm_0 \rightarrow \bigwedge_{1 \leq i < n} sb_i$	1, 2
4.	$\bigwedge_{1 \leq i < n}(cm_i \rightarrow cm_{i-1})$	LMT
5.	$\bigwedge_{0 \leq i < n}(cm_i \rightarrow (\bigwedge_{1 \leq j < n} sb_j))$	3, 4
6.	$\bigwedge_{1 \leq i < n}((cm_{i-1} \wedge sb_i) \leadsto cm_i)$	LMT
7.	$\bigwedge_{0 \leq i < n} \mathsf{AG}(\neg cm_i \vee \neg ab_i) \wedge \mathsf{AG}(cm_i \Rightarrow \mathsf{AG}cm_i)$	LMT
8.	$\bigwedge_{1 \leq i < n} \mathsf{AG}[sb_i \Rightarrow \mathsf{A}[sb_i \mathsf{U}(sb_i \wedge (cm_{i-1} \vee ab_{i-1}))]]$	LMT
9.	$\bigwedge_{0 \leq i < n-1}(cm_i \rightarrow \mathsf{A}[sb_{i+1} \mathsf{U}(sb_{i+1} \wedge cm_i)])$	5, 7, 8
10.	$\bigwedge_{1 \leq i < n-1}((cm_{i-1} \wedge sb_i) \rightarrow (cm_i \wedge sb_{i+1}))$	6, 9
11.	$cm_0 \rightarrow \bigwedge_{1 \leq i < n} cm_i$	3,6,7,9,10

The formulae in the above proof hold in all initial states of M_I, the global state transition diagram of P^I. The notation LMT means that the formula is a conjunct of the pair-specifications, and then we used Theorem 1 to deduce that the formula also holds in M_I (i.e., for the I-program). A notation of some formula numbers means that the formula was deduced from preceding formulae using an appropriate CTL deductive system [10]. Formula 11 gives a correctness property of two phase commit: if the coordinator commits, then so does every participant. Likewise, we establish $ab_0 \rightarrow \bigwedge_{1 \leq i < n} ab_i$—if the coordinator aborts, then so does every participant. Finally, we establish $\mathsf{AF}(cm_0 \vee ab_0)$—the coordinator eventually decides—directly from the pair-specification for $P^1_0 \,\|\, P^0_1$ using

Theorem 1. Note that $\bigwedge_{1 \leq i < n} \mathsf{AG}(st_i \Rightarrow \mathsf{EX}_i ab_i)$—every participant can abort unilaterally—also holds in M_I.

The deductive argument we used to establish $cm_0 \rightarrow \bigwedge_{1 \leq i < n} cm_i$ required only five deductive steps (lines 3, 5, 9, 10, and 11). A completely manual correctness argument for a two-phase commit protocol would be much longer. Our vision is that the large model theorem, in combination with automatic synthesis or model checking [5] methods for verifying the correctness of pair-programs, performs most of the work in establishing behavioral properties of the synthesized I-program. Then, the use of a deductive system provides us with the flexibility needed to deduce the final desired correctness properties.

Finally, we note the significant use of nested temporal modalities in both the above pair-specifications and the deductive proof (recall that the ACTL formula $f \rightarrow g$ is really an abbreviation for $\mathsf{A}[(f \Rightarrow \mathsf{AF}g)\mathsf{U_w}g]$, which nests AF inside $\mathsf{AU_w}$). This would not have been possible in the framework of [4].

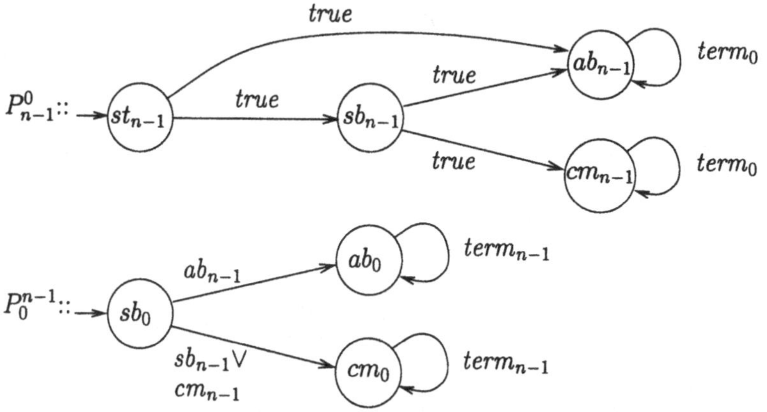

Fig. 4. Pair program $P^0_{n-1} \| P^{n-1}_0$.

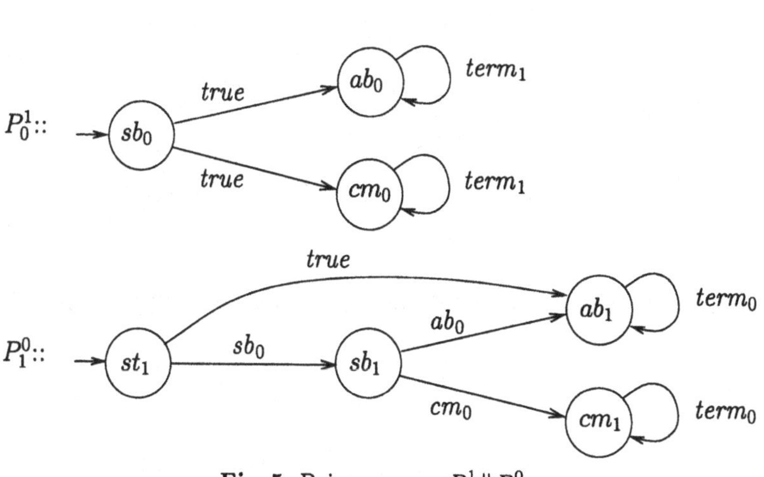

Fig. 5. Pair program $P^1_0 \| P^0_1$.

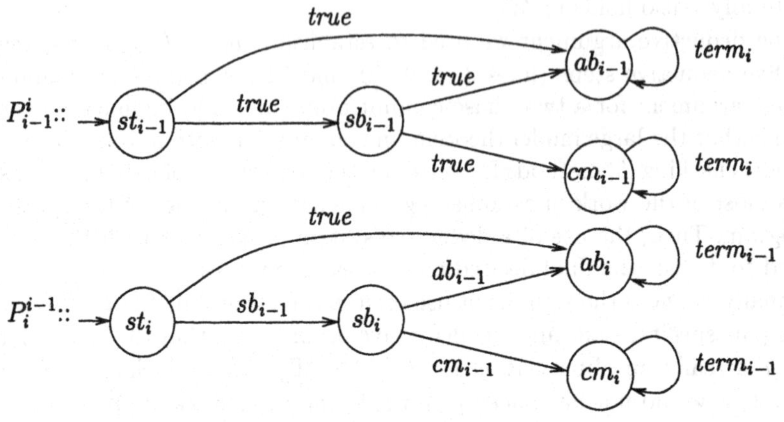

Fig. 6. Pair program $P^i_{i-1} \| P^{i-1}_i$.

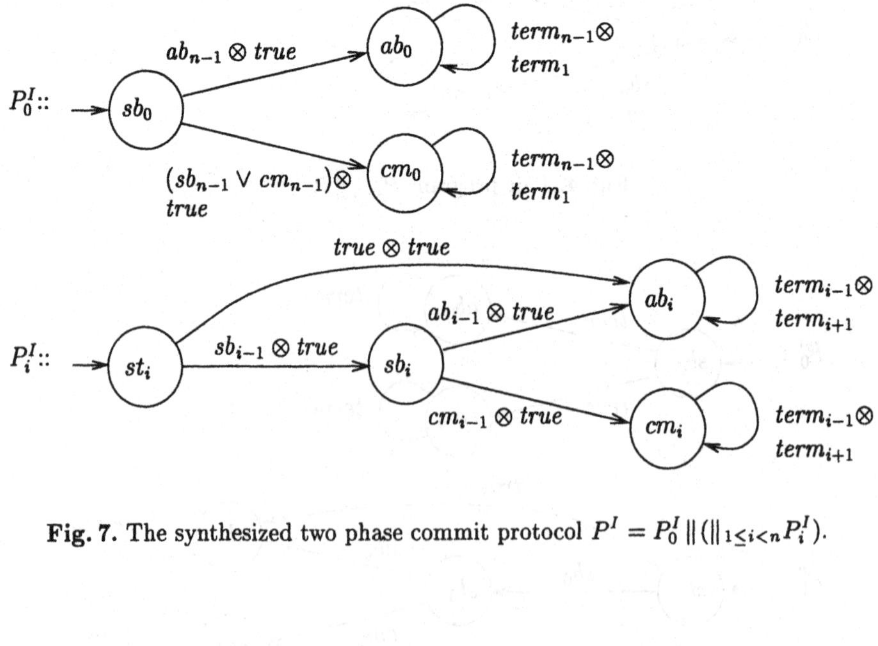

Fig. 7. The synthesized two phase commit protocol $P^I = P^I_0 \| (\|_{1 \leq i < n} P^I_i)$.

8 Conclusions and Further Work

We presented a synthesis method that deals with an arbitrary number of component processes without incurring the exponential overhead due to state explosion. Our method applies to any process interconnection scheme, does not make any assumption of similarity among the component processes, and preserves all pairwise and nexttime-free formulae of ACTL. We note that the method of implementing the synthesized programs on realistic distributed systems which was proposed in [4] is also applicable to the programs that our new method produces.

Further work includes dealing with fault-tolerance and real-time, and extending the method to a more expressive notation where the nodes of a synchronization skeleton denote sets of local states rather than individual local states.

References

1. A. Anuchitanukul and Z. Manna. Realizability and synthesis of reactive modules. In *Proc. 6th Intl. CAV Conference*, volume 818 of *LNCS*. Springer-Verlag, 1994.
2. A. Arora, P. C. Attie, and E. A. Emerson. Synthesis of fault-tolerant concurrent systems. In *Proc. 17'th Annual ACM Symposium on Principles of Distributed Computing*, pages 173 - 182, June 1998.
3. P. C. Attie and E. A. Emerson. Synthesis of concurrent systems for an atomic read/atomic write model of computation (extended abstract). In *Proc. 15'th ACM Symposium on Principles of Distributed Computing*, pages 111 - 120, May 1996.
4. P. C. Attie and E. A. Emerson. Synthesis of concurrent systems with many similar processes. *ACM Trans. Program. Lang. Syst.*, 20(1):51–115, January 1998.
5. E. M. Clarke, E. A. Emerson, and P. Sistla. Automatic verification of finite-state concurrent systems using temporal logic specifications. *ACM Trans. Program. Lang. Syst.*, 8(2):244–263, April 1986.
6. P.J. Courtois, H. Heymans, and D.L. Parnas. Concurrent control with readers and writers. *Communications of the ACM*, 14(10):667 - 668, 1971.
7. E. W. Dijkstra. *A Discipline of Programming*. Prentice-Hall Inc., 1976.
8. E. W. Dijkstra. *Selected Writings on Computing: A Personal Perspective*, pages 188–199. Springer-Verlag, New York, 1982.
9. D.L. Dill and H. Wong-Toi. Synthesizing processes and schedulers from temporal specifications. In *2'nd Intl. CAV Conference*, *LNCS* vol. 531. Springer-Verlag, 1990.
10. E. A. Emerson. Temporal and modal logic. In *Handbook of Theoretical Computer Science*, volume B. The MIT Press/Elsevier, Cambridge, Mass., 1990.
11. E. A. Emerson and E. M. Clarke. Using branching time temporal logic to synthesize synchronization skeletons. *Sci. Comput. Program.*, 2:241 - 266, 1982.
12. E. A. Emerson and C. Lei. Modalities for model checking: Branching time logic strikes back. *Sci. Comput. Program.*, 8:275–306, 1987.
13. O. Grumberg and D.E. Long. Model checking and modular verification. *ACM Trans. Program. Lang. Syst.*, 16(3):843–871, May 1994.
14. Z. Manna and P. Wolper. Synthesis of communicating processes from temporal logic specifications. *ACM Trans. Program. Lang. Syst.*, 6(1):68–93, January 1984.
15. A. Pnueli and R. Rosner. On the synthesis of a reactive module. In *Proc. 16th ACM Symposium on Principles of Programming Languages*, New York, 1989. ACM.
16. A. Pnueli and R. Rosner. On the synthesis of asynchronous reactive modules. In *Proc. 16th ICALP*, volume 372 of *LNCS*, Berlin, 1989. Springer-Verlag.

Approximate Symbolic Model Checking of Continuous-Time Markov Chains*

(extended abstract)

Christel Baier[a], Joost-Pieter Katoen[b,c], and Holger Hermanns[c]

[a]Lehrstuhl für Praktische Informatik II, University of Mannheim
68131 Mannheim, Germany

[b]Lehrstuhl für Informatik 7, University of Erlangen-Nürnberg
Martensstraße 3, 91058 Erlangen, Germany

[c]Systems Validation Centre, University of Twente
P.O. Box 217, 7500 AE Enschede, The Netherlands

Abstract. This paper presents a symbolic model checking algorithm for continuous-time Markov chains for an extension of the continuous stochastic logic **CSL** of Aziz et al [1]. The considered logic contains a time-bounded until-operator and a novel operator to express steady-state probabilities. We show that the model checking problem for this logic reduces to a system of linear equations (for unbounded until and the steady state-operator) and a Volterra integral equation system for time-bounded until. We propose a symbolic approximate method for solving the integrals using MTDDs (multi-terminal decision diagrams), a generalisation of MTBDDs. These new structures are suitable for numerical integration using quadrature formulas based on equally-spaced abscissas, like trapezoidal, Simpson and Romberg integration schemes.

1 Introduction

The mechanised verification of a given (usually) finite-state model against a property expressed in some temporal logic is known as *model checking.* For probabilistic systems, transition systems where branching is governed by discrete probability distributions, qualitative and quantitative model checking algorithms have been investigated extensively [2, 5–7, 11, 14–16, 19, 23, 27]. In a qualitative setting it is checked whether a property holds with probability 0 or 1; in a quantitative setting it is typically verified whether the probability for a certain property meets given lower- or upper-bounds. For discrete-time systems, the quantitative approach has been investigated quite thoroughly: model checking algorithms have been developed for fully probabilistic transition systems [2, 5, 14, 19], like discrete-time Markov chains or generative transition systems, as well as for probabilistic systems that contain non-determinism [6–8, 16].

* The first and second author are sponsored by the DAAD-Project AZ 313-ARC-XII-98/38 on stochastic modelling and verification.

In this paper we consider real-time probabilistic systems, that is, we consider the model checking problem for continuous-time Markov chains (CTMCs) that are at the basis of contemporary performance evaluation and reliability analysis methodologies. A branching-time logic called *continuous-time stochastic logic* (**CSL**) is used to express properties over CTMCs. This logic is an extension of the (equally named) logic by Aziz et al [1] with an operator to reason about steady-state probabilities: e.g. the formula $\mathcal{S}_{\geqslant p}(\Phi)$ asserts that the steady-state probability for a Φ-state is at least p, for $p \in [0,1]$. Apart from the usual path-formulas like next and until, a time-bounded until $\mathcal{U}^{\leqslant t}$, for t a non-negative real, is incorporated, together with standard derivatives, such as a time-bounded eventually $\diamond^{\leqslant t}$. The usual path quantifiers $\forall$ and $\exists$ are replaced by the probabilistic operator $\mathcal{P}_{\bowtie p}(.)$ for comparison operator $\bowtie$ and $p \in [0,1]$. For instance, $\mathcal{P}_{<0.001}(\diamond^{\leqslant 4} error)$ asserts that the probability for a system error within 4 time-units is less than 10^{-3}.

The model checking problem for **CSL** is known to be decidable [1] (for rational time bounds), but to the best of our knowledge no algorithms have been considered yet to verify CTMCs automatically, let alone symbolically. This paper investigates which numerical methods can be adapted to "model check" **CSL**-formulas over CTMCs as models. We show that next and (unbounded) until-formulas can be treated similarly as in the discrete-time probabilistic setting. Checking steady-state probability-properties reduces to solving a linear equation system combined with standard graph analysis methods, while checking the time-bounded until reduces to solving a (recursive) Volterra integral equation system. These integrals are characterised as least fixed points of appropriate higher-order functions, and can thus be approximated by an iterative approach.

One of the major reasons for the success of model checking tools in practice is the efficient way to cope with the state-space explosion problem. A prominent technique is to adopt a compact representation of state spaces using (reduced ordered) *binary decision diagrams*, BDDs for short [9]. This paper follows this line by proposing an alternative variant, referred to as *multi-terminal decision diagrams* (MTDDs), that is suited for the necessary real-time probability calculations. MTDDs are a novel generalisation of multi-terminal binary decision diagrams (MTBDDs [12], also called algebraic decision diagrams [3]), variants of BDDs that can efficiently deal with real matrices. MTBDDs (and MTDDs) allow arbitrary real numbers in the terminal nodes instead of just 0 and 1 (like in BDDs). Whereas MTBDDs are defined on boolean variables, MTDDs allow both boolean and real variables. This generalisation is suitable for numerical integration — needed for time-bounded until — using quadrature formulas based on equally-spaced abscissas (i.e. interval points). This includes well-known methods like trapezoidal, Simpson and Romberg integration schemes [24]. Due to their suitability for numerical integration, the potential application of MTDDs is much wider than model checking CTMCs. For the other temporal operators in **CSL** we show that slight modifications of the MTBDD-approach for discrete-time probabilistic systems [5] can be adopted.

The paper introduces MTDDs, defines appropriate operators on them and presents a symbolic model checking algorithm for **CSL** using these structures. Although it is difficult to obtain precise estimates for the time complexity of model checking using MTDDs (as with BDDs and MTBDDs), the success of (MT)BDD-based model checkers for large-scale examples (for BDDs [10] and for MTBDDs [18, 20]) provides sufficient evidence to investigate MTDDs for our setting. For instance, [18] reports experimental results of the computation of steady-state probabilities for discrete-time Markov chains of over 10^{27} states.

Organisation of the paper. Section 2 introduces the necessary concepts of CTMCs. Section 3 presents the logic **CSL** and provides some useful characterisations of **CSL**-formulas that facilitate a model checking procedure. Section 4 introduces MTDDs, describes how CTMCs can be encoded as MTDDs, and presents several operators on these structures. Section 5 presents the approximative symbolic model checking algorithm. Finally, Section 6 concludes the paper. A (small) running example is used throughout the paper to illustrate the key concepts.

2 Continuous-time Markov chains

Basic definitions. Let AP be a fixed, finite set of atomic propositions. A (labelled) *continuous-time Markov chain* (CTMC for short) is a tuple $\mathcal{M} = (S, \mathbf{Q}, L)$ where S is a finite set of *states*, $\mathbf{Q} : S \times S \to \mathbb{R}_{\geqslant 0}$ the *generator matrix*[1], and $L : S \to 2^{AP}$ the *labelling* function which assigns to each state $s \in S$ the set $L(s)$ of atomic propositions $a \in AP$ that are valid in s.

Intuitively, $\mathbf{Q}(s, s')$ specifies that the probability of moving from state s to s' within t time-units (for positive t) is $1 - e^{-\mathbf{Q}(s,s')\cdot t}$, an exponential distribution with rate $\mathbf{Q}(s, s')$. If $\mathbf{Q}(s, s') > 0$ for more than one state s', a competition between the transitions is assumed to exist, known as the *race condition*. Let $\mathbf{E}(s) = \sum_{s' \in S} \mathbf{Q}(s, s')$, the total rate at which any transition emanating from state s is taken. This rate is the reciprocal of the mean sojourn time in s. More precisely, $\mathbf{E}(s)$ specifies that the probability of leaving s within t time-units (for positive t) is $1 - e^{-\mathbf{E}(s)\cdot t}$, due to the fact that the minimum of exponential distributions (competing in a race) is characterised by the sum of their rates. Consequently, the probability of moving from state s to s' by a single transition, denoted $\mathbf{P}(s, s')$, is determined by the probability that the delay of going from s to s' finishes before the delays of other outgoing edges from s; formally, $\mathbf{P}(s, s') = \mathbf{Q}(s, s')/\mathbf{E}(s)$ (except if s is an absorbing state, i.e. if $\mathbf{E}(s) = 0$; in this case we define $\mathbf{P}(s, s') = 0$). Remark that the matrix $\mathbf{P}$ describes an embedded discrete time Markov chain. (For a more extensive treatment of CTMCs see [25].)

[1] Whereas usually the diagonal elements are defined as $\mathbf{Q}(s, s) = -\sum_{s' \neq s} \mathbf{Q}(s, s')$ we allow self-loops. This does not affect the transient and steady state behaviour of the chain, but allows the standard interpretation of the next-state operator of the logic.

Example 1. As a running example we consider $AP = \{a, b\}$, $S = \{s_0, \ldots, s_3\}$ with $L(s_0) = \varnothing$, $L(s_1) = \{a\}$, $L(s_2) = \{b\}$ and $L(s_3) = \{a, b\}$. The details of the CTMC are:

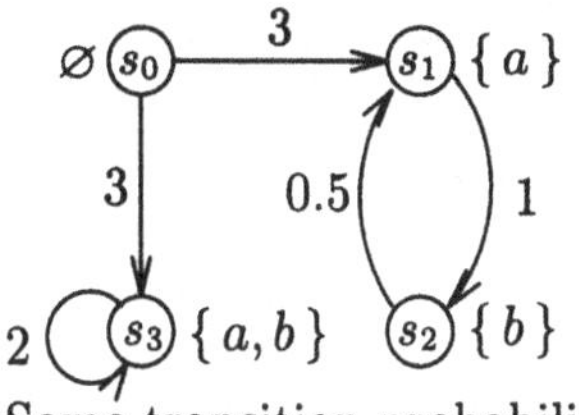

$$\mathbf{Q} = \begin{pmatrix} 0 & 3 & 0 & 3 \\ 0 & 0 & 1 & 0 \\ 0 & 0.5 & 0 & 0 \\ 0 & 0 & 0 & 2 \end{pmatrix} \text{ and } \mathbf{E} = \begin{pmatrix} 6 \\ 1 \\ 0.5 \\ 2 \end{pmatrix}.$$

Some transition probabilities are $\mathbf{P}(s_0, s_3) = \mathbf{P}(s_0, s_1) = \frac{1}{2}$ and $\mathbf{P}(s_1, s_2) = 1$.

A *path* is a (finite or infinite) sequence $s_0, t_0, s_1, t_1, s_2, t_2, \ldots$, written as

$$\sigma = s_0 \xrightarrow{t_0} s_1 \xrightarrow{t_1} s_2 \xrightarrow{t_2} \ldots,$$

with for natural i, $s_i \in S$ and $t_i \in \mathbb{R}_{>0}$ such that $\mathbf{Q}(s_i, s_{i+1}) > 0$, if σ is infinite. Otherwise, if $\sigma = s_0 \xrightarrow{t_0} \ldots \xrightarrow{t_{l-1}} s_l$ is finite, we require that s_l is absorbing, and $\mathbf{Q}(s_i, s_{i+1}) > 0$ for all $i < l$. For σ a path, $t \in \mathbb{R}_{>0}$ and natural i let $\sigma[i] = s_i$, the i-th state of σ, $\delta(\sigma, i) = t_i$, the time spent in state s_i, and $\sigma(t) = s_0$ for $t < t_0$, and $\sigma(t) = \sigma[i]$ where i is the smallest index i with $t \leqslant \sum_{0 \leqslant j \leqslant i} t_j$, otherwise. (For σ a finite path with absorbing state s_l, $\sigma[i]$ and $\delta(\sigma, i)$ are only defined for $i \leqslant l$, $\delta(\sigma, l) = \infty$, and $\sigma(t) = s_l$ for $t > t_1 + \ldots + t_{l-1}$.) Let *Path*$(s)$ denote the set of paths in $\mathcal{M}$ starting in s, and *Reach*(s) the set of states reachable from s.

Borel space. Let $s_0, \ldots, s_k \in S$ with $\mathbf{Q}(s_i, s_{i+1}) > 0$, $(0 \leqslant i < k)$, and $I_0, \ldots, I_{k-1}$ non-empty intervals in $\mathbb{R}_{\geqslant 0}$. Then, $C(s_0, I_0, \ldots, I_{k-1}, s_k)$ denotes the *cylinder set* consisting of all paths $\sigma \in Path(s_0)$ such that $\sigma[i] = s_i$ $(i \leqslant k)$, and $\delta(\sigma, i) \in I_i$ $(i < k)$. Let $\mathcal{F}(Path(s))$ be the smallest σ-algebra on $Path(s)$ which contains all sets $C(s, I_0, \ldots, I_{k-1}, s_k)$ where $s_0, \ldots, s_k$ ranges over all sequences of states such that $s = s_0$ and $\mathbf{Q}(s_i, s_{i+1}) > 0$ $(0 \leqslant i < k)$ and $I_0, \ldots, I_{k-1}$ ranges over all sequences of non-empty intervals in $\mathbb{R}_{\geqslant 0}$. The probability measure Pr on $\mathcal{F}(Path(s))$ is the unique measure defined by induction on k by $\Pr(C(s_0)) = 1$ and for $k \geqslant 0$:

$$\Pr(C(s_0, \ldots, s_k, I', s')) = \Pr(C(s_0, \ldots, s_k)) \cdot \mathbf{P}(s_k, s') \cdot \left(e^{-\mathbf{E}(s_k) \cdot a} - e^{-\mathbf{E}(s_k) \cdot b}\right)$$

where $a = \inf I'$ and $b = \sup I'$. (For $b = \infty$ and $\lambda > 0$ let $e^{-\lambda \cdot \infty} = 0$.)

3 The continuous stochastic logic CSL

Syntax. CSL is a branching-time, CTL-like temporal logic where the state-formulas are interpreted over states of a CTMC. It adopts operators of PCTL [19], like a time-bounded until operator and a probabilistic operator asserting that the probability for a certain event meets given bounds. We treat a variant of the (equally named) logic of [1] with, for reasons of simplicity, an unnested time-bounded until operator plus a novel steady-state probability operator.

Definition 1. *For $a \in AP$, $p \in [0,1]$ and $\bowtie \in \{\leqslant, <, \geqslant, >\}$, the state-formulas of* **CSL** *are defined by the grammar*

$$\Phi ::= \mathrm{tt} \;\Big|\; a \;\Big|\; \Phi \wedge \Phi \;\Big|\; \neg\Phi \;\Big|\; \mathcal{S}_{\bowtie p}(\Phi) \;\Big|\; \mathcal{P}_{\bowtie p}(\varphi)$$

where for $t \in I\!R_{\geqslant 0}$ path-formulas are defined by

$$\varphi ::= X\Phi \;\Big|\; \Phi\,\mathcal{U}\,\Phi \;\Big|\; \Phi\,\mathcal{U}^{\leqslant t}\,\Phi.$$

The other boolean connectives are derived in the usual way, i.e. $\mathrm{ff} = \neg\mathrm{tt}$, $\Phi_1 \vee \Phi_2 = \neg(\neg\Phi_1 \wedge \neg\Phi_2)$, and $\Phi_1 \rightarrow \Phi_2 = \neg\Phi_1 \vee \Phi_2$. The intended meaning of the temporal operators $\mathcal{U}$ ("until") and X ("next step") is standard. The temporal operator $\mathcal{U}^{\leqslant t}$ is the real-time variant of $\mathcal{U}$; $\Phi_1\,\mathcal{U}^{\leqslant t}\,\Phi_2$ asserts that $\Phi_1\,\mathcal{U}\,\Phi_2$ will be satisfied in the time interval $[0,t]$; i.e. there is some $x \in [0,t]$ such that Φ_1 continuously holds during the interval $[0,x[$ and Φ_2 becomes true at time instant x. The state formula $\mathcal{S}_{\bowtie p}(\Phi)$ asserts that the steady-state probability for a Φ-state falls in the interval $I_{\bowtie p} = \{q \in [0,1] \mid q \bowtie p\}$. $\mathcal{P}_{\bowtie p}(\varphi)$ asserts that the probability measure of the paths satisfying φ falls in the interval $I_{\bowtie p}$.

Temporal operators like $\Diamond$, $\Box$ and their real-time variants $\Diamond^{\leqslant t}$ or $\Box^{\leqslant t}$ can be derived, e.g. $\mathcal{P}_{\bowtie p}(\Diamond^{\leqslant t}\,\Phi) = \mathcal{P}_{\bowtie p}(\mathrm{tt}\,\mathcal{U}^{\leqslant t}\,\Phi)$ and $\mathcal{P}_{\geqslant p}(\Box\,\Phi) = \mathcal{P}_{\leqslant 1-p}(\Diamond\,\neg\Phi)$. For example, $\mathcal{P}_{\geqslant 0.99}(\Box\;(\mathit{req} \rightarrow \mathcal{P}_{\geqslant 1}(\Diamond^{\leqslant 5}\,\mathit{resp})))$ asserts that there is a probability of at least 99% that every request will be responded within the next 5 time-units.

Semantics. The state-formulas are interpreted over the states of a CTMC. Let $\mathcal{M} = (S, \mathbf{Q}, L)$ with proposition labels in AP. The definition of the satisfaction relation $\models\, \subseteq S \times$ **CSL** is as follows. Let $\mathit{Sat}(\Phi) = \{s \in S \mid s \models \Phi\}$.

$s \models \mathrm{tt}$	for all $s \in S$	$s \models \Phi_1 \wedge \Phi_2$	iff $s \models \Phi_i, i{=}1,2$
$s \models a$	iff $a \in L(s)$	$s \models \mathcal{S}_{\bowtie p}(\Phi)$	iff $\pi_{\mathit{Sat}(\Phi)}(s) \in I_{\bowtie p}$
$s \models \neg\Phi$	iff $s \not\models \Phi$	$s \models \mathcal{P}_{\bowtie p}(\varphi)$	iff $\mathit{Prob}(s,\varphi) \in I_{\bowtie p}$.

Here, $\pi_{S'}(s)$ denotes the steady-state probability for $S' \subseteq S$ wrt. state s, i.e.

$$\pi_{S'}(s) \;=\; \lim_{t \to \infty} \Pr\{\sigma \in \mathit{Path}(s) \mid \sigma(t) \in S'\}.$$

The limit exists, a consequence of S being finite [25]. Obviously, $\pi_{S'}(s) = \sum_{s' \in S'} \pi_{s'}(s)$, where we write $\pi_{s'}(s)$ instead of $\pi_{\{s'\}}(s)$. We let $\pi_{\emptyset}(s) = 0$. $\mathit{Prob}(s,\varphi)$ denotes the prob. measure of all paths $\sigma \in \mathit{Path}(s)$ satisfying φ, i.e.

$$\mathit{Prob}(s,\varphi) \;=\; \Pr\{\sigma \in \mathit{Path}(s) \mid \sigma \models \varphi\}.$$

The fact that, for each state s, the set $\{\sigma \in \mathit{Path}(s) \mid \sigma \models \varphi\}$ is measurable, follows by easy verification. The satisfaction relation (also denoted $\models$) for the path-formulas is defined as usual:

$$\begin{array}{ll}
\sigma \models X\Phi & \text{iff } \sigma[1] \text{ is defined and } \sigma[1] \models \Phi \\
\sigma \models \Phi_1\,\mathcal{U}\,\Phi_2 & \text{iff } \exists k \geqslant 0.\ (\sigma[k] \models \Phi_2 \wedge \forall 0 \leqslant i < k.\, \sigma[i] \models \Phi_1) \\
\sigma \models \Phi_1\,\mathcal{U}^{\leqslant t}\,\Phi_2 & \text{iff } \exists x \in [0,t].\ (\sigma(x) \models \Phi_2 \wedge \forall y \in [0,x[.\, \sigma(y) \models \Phi_1).
\end{array}$$

In the remainder of this section we present alternative characterisations for $\pi_{S'}(s)$ and $Prob(s, \varphi)$ that will serve as a basis for our model checking algorithm. Since the derivation of these characterisations from the theory of CTMCs and DTMCs is not much involved, proofs are omitted.

Computing steady-state probabilities. It is well known that the steady state probabilities exist for arbitrary CTMCs. For a strongly connected CTMC $\mathcal{M}$ and (non-absorbing) state s', the steady state probability $\pi_S(s')$ can be obtained by solving a *linear equation system* [25], i.e.

$$\pi_S(s') = \frac{\pi'_S(s')/\mathbf{E}(s')}{\sum_{s \in S} \pi'_S(s)/\mathbf{E}(s)}$$

where $\pi'_S(s'')$ satisfies the linear equation system

$$\pi'_S(s'') = \sum_{s \in S} \mathbf{P}(s, s'') \cdot \pi'_S(s) \text{ such that } \sum_{s \in S} \pi'_S(s) = 1.$$

For the general case we reformulate this as follows. Let G be the underlying directed graph of $\mathcal{M}$ where vertices represent states and where there is an edge from s to s' iff $\mathbf{Q}(s, s') > 0$. Sub-graph B is a *bottom strongly connected component* (bscc) of G if it is a strongly connected component such that for any $s \in B$, $Reach(s) \subseteq B$. We have $\pi_{s'}(s) = 0$ iff s' does not occur in any bscc reachable from s. Let B be a bscc of G with $Reach(s) \cap B \neq \varnothing$ (or equivalently, $B \subseteq Reach(s)$) and assume that a_B is an atomic proposition such that $a_B \in L(s)$ iff $s \in B$. Then $\Diamond a_B$ is a path-formula in **CSL** and $Prob(s, \Diamond B) = Prob(s, \Diamond a_B)$ is the probability of reaching B from s at some time t. For $s' \in B$, $\pi_{s'}(s)$ is given by $\pi_{s'}(s) = Prob(s, \Diamond B) \cdot \pi_B(s')$ where $\pi_B(s') = 1$ if $B = \{s'\}$, and otherwise

$$\pi_B(s') = \frac{\pi'_B(s')/\mathbf{E}(s')}{\sum_{s \in B} \pi'_B(s)/\mathbf{E}(s)}$$

for which $\pi'_B(s'')$ satisfies the linear equation system

$$\pi'_B(s'') = \sum_{s \in B} \mathbf{P}(s, s'') \cdot \pi'_B(s) \text{ such that } \sum_{s \in B} \pi'_B(s) = 1.$$

Example 2. Consider $\mathcal{S}_{>0.5}(\Phi)$ where $\Phi = (a \wedge b) \vee \mathcal{P}_{\leqslant 0.8}(a\,\mathcal{U}^{\leqslant 2}\, b)$, for the CTMC of Example 1. Note that the CTMC is not strongly connected, since e.g. s_3 cannot be reached from s_2 (and vice versa). Assume that Φ is valid in states s_2 and s_3, and invalid otherwise (as we will see later on). Then we have $s_0 \models \mathcal{S}_{>0.5}(\Phi)$, since from s_0 both the bscc $B_1 = \{\, s_3 \,\}$ and the bscc $B_2 = \{\, s_1, s_2 \,\}$ can be reached with probability $\mathbf{P}(s_0, s_3) = \mathbf{P}(s_0, s_2) = 1/2$, and s_2 has a non-zero steady-state probability in B_2. Thus, the steady-state probability for Φ exceeds 0.5. Formally: $\pi_{Sat(\Phi)}(s_0) = \pi_{\{\, s_2, s_3 \,\}}(s_0) = \pi_{s_2}(s_0) + \pi_{s_3}(s_0)$ where $\pi_{s_2}(s_0) = Prob(s_0, \Diamond B_2) \cdot \pi_{B_2}(s_2)$ and $\pi_{s_3}(s_0) = Prob(s_0, \Diamond B_1) \cdot \pi_{B_1}(s_3)$. We have $Prob(s_0, \Diamond B_1) = Prob(s_0, \Diamond B_2) = 1/2$, $\pi_{B_1}(s_3) = 1$, and obtain $\pi'_{B_2}(s_2) = 1/2$ by solving the equation system $\pi'_{B_2}(s_2) = \pi'_{B_2}(s_1)$, $\pi'_{B_2}(s_1) = \pi'_{B_2}(s_2)$, $\pi'_{B_2}(s_1) + \pi'_{B_2}(s_2) = 1$. Subsequently, calculation of $\pi_{B_2}(s_2)$ yields 2/3. Thus, $\pi_{\{\, s_2, s_3 \,\}}(s_0) = 1/2 \cdot 2/3 + 1/2 \cdot 1 = 5/6$ which indeed exceeds 0.5.

Computing $Prob(s, \varphi)$. The basis for calculating the probabilities $Prob(s, \varphi)$ is the following result.

Theorem 1. *For $s \in S$, $t \in \mathbb{R}_{\geqslant 0}$ and Φ, Φ_1, Φ_2 state-formulas in* **CSL**:

1. $Prob(s, X\Phi) = \sum_{s' \in Sat(\Phi)} \mathbf{P}(s, s')$.
2. *The function $S \to [0,1]$, $s \mapsto Prob(s, \Phi_1 \,\mathcal{U}\, \Phi_2)$ is the least fixed point of the higher-order operator $\Theta : (S \to [0,1]) \to (S \to [0,1])$ where*[2]

$$\Theta(F)(s) = \begin{cases} 1 & \text{if } s \models \Phi_2 \\ \sum_{s' \in S} \mathbf{P}(s, s') \cdot F(s') & \text{if } s \models \Phi_1 \wedge \neg\Phi_2 \\ 0 & \text{otherwise.} \end{cases}$$

3. *The function $S \times \mathbb{R}_{\geqslant 0} \to [0,1]$, $(s,t) \mapsto Prob(s, \Phi_1 \,\mathcal{U}^{\leqslant t}\, \Phi_2)$ is the least fixed point of the higher-order operator $\Omega : (S \times \mathbb{R}_{\geqslant 0} \to [0,1]) \to (S \times \mathbb{R}_{\geqslant 0} \to [0,1])$ where*[3]

$$\Omega(F)(s,t) = \begin{cases} 1 & \text{if } s \models \Phi_2 \\ \sum_{s' \in S} \mathbf{Q}(s, s') \cdot \int_0^t e^{-\mathbf{E}(s) \cdot x} \cdot F(s', t-x)\, dx & \text{if } s \models \Phi_1 \wedge \neg\Phi_2 \\ 0 & \text{otherwise.} \end{cases}$$

The first two results of Theorem 1 are identical to the discrete-time probabilistic case, cf. [14, 19, 4]. This entails that model checking for these formulas can be carried out by well-known methods:

- $(Prob(s, X\Phi))_{s \in S}$ can be obtained by multiplying the transition probability matrix $\mathbf{P}$ with the (boolean) vector $\mathbf{i}_\Phi = (i_\Phi(s))_{s \in S}$ characterising $Sat(\Phi)$, i.e. $i_\Phi(s) = 1$ if $s \models \Phi$, and 0 otherwise.
- $(Prob(s, \Phi_1 \,\mathcal{U}\, \Phi_2))_{s \in S}$ can be obtained by solving a linear equation system of the form $\mathbf{x} = \overline{\mathbf{P}} \cdot \mathbf{x} + \mathbf{i}_{\Phi_2}$ where $\overline{\mathbf{P}}(s, s') = \mathbf{P}(s, s')$ if $s \models \Phi_1 \wedge \neg\Phi_2$ and 0 otherwise. $Prob(s, \Phi_1 \,\mathcal{U}\, \Phi_2)$ is the least solution of this set of equations. Note, however, that this system of equations can, in general, have more than one solution. The least solution can be obtained by applying an iterative approximative method or a graph analysis combined with standard methods (like Gaussian elimination) to solve regular linear equation systems. The worst case time complexity of this step is linear in the size of φ and polynomial in the number of states.

Example 3. Consider our running CTMC example, $\Phi = (a \wedge b) \vee \mathcal{P}_{\leqslant 0.8}(a \,\mathcal{U}^{\leqslant 2}\, b)$ and suppose we want to check $s_1 \models \Phi$. It follows from $\mathbf{Q}(s_1, s_2) = 1$ that the probability of reaching b-state s_2 from s_1 within two time-units equals $1 - e^{-1 \cdot 2} \approx 0.864664$. Formally, we have $s_2 \models \Phi$, since $s_2 \models b$, and $s_1 \not\models \Phi$, since $s_1 \not\models a \wedge b$ and using Theorem 1 we have that $Prob(s_1, a \,\mathcal{U}^{\leqslant 2}\, b)$ equals

$$\sum_{s' \in S} \mathbf{Q}(s_1, s') \cdot \int_0^2 e^{-\mathbf{E}(s_1) \cdot x} \cdot F(s', 2-x)\, dx = \int_0^2 e^{-x} dx = [-e^{-x}]_0^2 = 1 - e^{-2}$$

which exceeds 0.8.

[2] The underlying partial order on $S \to [0,1]$ is defined for F_1, $F_2 : S \to [0,1]$ by $F_1 \leqslant F_2$ iff $F_1(s) \leqslant F_2(s)$ for all s.

[3] The underlying partial order on $S \times \mathbb{R}_{\geqslant 0} \to [0,1]$ is defined for F_1, $F_2 : S \times \mathbb{R}_{\geqslant 0} \to [0,1]$ by $F_1 \leqslant F_2$ iff $F_1(s,t) \leqslant F_2(s,t)$ for all s, t.

The last result of Theorem 1 is due to the fact that the probability density function of the sojourn time in state s is given by $\mathbf{E}(s) \cdot e^{-\mathbf{E}(s) \cdot t}$. The resulting recursive integral formula can be reformulated into a *heterogeneous linear differential equation* of the form

$$\mathbf{y}'(t) \;=\; \overline{\mathbf{Q}} \cdot \mathbf{y}(t) \;+\; \mathbf{b}(t)$$

where $\mathbf{y}(t)$ denotes the vector $(Prob(s, \Phi_1 \,\mathcal{U}^{\leqslant t} \Phi_2))_{s \in S}$, and $\overline{\mathbf{Q}}$ is derived from $\mathbf{Q}$, by $\overline{\mathbf{Q}}(s, s') = \mathbf{Q}(s, s')$ if $s, s' \models \Phi_1 \wedge \neg \Phi_2$, and otherwise $\overline{\mathbf{Q}}(s, s') = 0$. The vector $\mathbf{b}(t) = (b_s(t))_{s \in S}$ is given by $b_s(t) = \sum_{s' \in Sat(\Phi_2)} \mathbf{Q}(s, s') \cdot e^{-\mathbf{E}(s)t}$, if $s \models \Phi_1 \wedge \neg \Phi_2$, and otherwise $b_s(t) = 0$. The vector $(Prob(s, \Phi_1 \,\mathcal{U}^{\leqslant t} \Phi_2))_{s \in S}$ agrees with the following solution of the above heterogeneous linear differential equation:

$$\mathbf{y}(t) \;=\; e^{\overline{\mathbf{Q}}t} \cdot \left(\mathbf{i}_{\Phi_2} + \int_0^t e^{-\overline{\mathbf{Q}}x} \cdot \mathbf{b}(x)\, dx \right), \text{ where } \; e^{\overline{\mathbf{Q}}x} \;=\; \sum_{k=0}^{\infty} \frac{(\overline{\mathbf{Q}}x)^k}{k!}.$$

Unfortunately, it is not clear (at least to the authors) how to obtain a closed solution for the above integral. Using a numerical approximation method instead is also not an accurate way out, essentially because known approximative methods for computing $e^{\mathbf{A}x}$ (for some square matrix $\mathbf{A}$) are instable, yet computationally expensive [25]. For that reasons, our algorithm to compute $Prob(s, \Phi_1 \,\mathcal{U}^{\leqslant t} \Phi_2)$ is directly based on the last result of Theorem 1. The result suggests the following *iterative* method to approximate $Prob(s, \Phi_1 \,\mathcal{U}^{\leqslant t} \Phi_2)$: let $F_0(s, t) = 0$ for all s, t and $F_{k+1} = \Omega(F_k)$. Then,

$$\lim_{k \to \infty} F_k(s, t) \;=\; Prob(s, \Phi_1 \,\mathcal{U}^{\leqslant t} \Phi_2).$$

(The general nested time-bounded until in [1] can be treated in a similar way.) Each step in the iteration amounts to solve an integral of the following form:

$$F_{k+1}(s, t) = \int_0^t \sum_{s' \in S} \mathbf{Q}(s, s') \cdot e^{-\mathbf{E}(s) \cdot x} \cdot F_k(s', t - x)\, dx,$$

if $s \models \Phi_1 \wedge \neg \Phi_2$. These integrals can be solved numerically based on quadrature formulas of the type

$$\int_a^b f(x)\, dx \;\approx\; \sum_{j=0}^{N} \alpha_j \cdot f(x_j)$$

with interval points $x_0, \ldots, x_N \in [a, b]$ and weights $\alpha_0, \ldots, \alpha_N$ that do not depend on f (but may be on N). In our model checking algorithm we focus on equally-spaced abscissas, i.e. $x_j = a + j \cdot h$ where $h = (b - a)/N$. Well-known methods applied in practice, like trapezoidal, Simpson, and Romberg integration schemes belong to this category [24]. For instance, for the trapezoidal method $\alpha_0 = \alpha_N = \frac{h}{2}$ and $\alpha_i = h$ for $0 < i < N$.

4 Multi-terminal decision diagrams

BDDs and MTBDDs. While (ordered) binary decision diagrams (BDDs) are data structures for representing boolean functions $f : \{0,1\}^n \rightarrow \{0,1\}$, multi-terminal BDDs (MTBDDs [12], also called algebraic decision diagrams [3]) allow terminals to be labelled with values of some domain D (usually $\mathbb{R}$ or $[0,1]$), i.e. they represent functions of the type $f : \{0,1\}^n \rightarrow D$. The main idea behind the MTBDD representation is the use of acyclic rooted directed graphs for a simplified (more compact) representation of the (binary) decision tree which results from the Shannon expansion: $f(b_1,\ldots,b_n) = (1-b_1)\cdot f(0,b_2,\ldots,b_n) + b_1 \cdot f(1,b_2,\ldots,b_n)$.

For model checking discrete-time Markov chains against PCTL-formulas [19] it has been shown that MTBDDs can be effectively used [5, 20]. These techniques can potentially be adapted to our continuous-time setting, but are not able to cope with numerical integration, a technique needed for the time-bounded until operator of **CSL** with the iterative method sketched above. Therefore, we introduce a variant of MTBDDs that is focussed on dealing with numerical integration.

MTDDs. *Multi-terminal decision diagrams* (MTDDs) are a variant of MTBDDs that yield a discrete representation of real-valued functions whose arguments are either boolean variables (called *state variables*, since they represent the encoding of states) or real variables (called *integral variables*, since they represent variables over which numerical integration takes place). For instance, MTDDs can represent functions of the type $\{0,1\}^n \times \mathbb{R} \rightarrow \mathbb{R}$. For the state variables, the aforementioned Shannon expansion is used. For an integral variable x, a finite set $\{x_0,\ldots,x_N\}$ is chosen from the range of x. The function $(\ldots,x,\ldots) \mapsto f(\ldots,x,\ldots)$ is represented by the function values $f(\ldots,x_j,\ldots)$, for $0 \leqslant j \leqslant N$. To accomplish this, we use a representation of f that is based on a discrete fragment of the decision tree where the branches for the integral variables represent the cases where $x \in \{x_0,\ldots,x_N\}$.

Formally, with each integral variable x (over interval $[0,t]$) the following components are associated: (i) a natural number $N(\mathsf{x})$ that denotes the number of abscissas of x, (ii) a set of abscissas where $abs_j(\mathsf{x})$ denotes the j-th abscissa, (iii) a range $rng(\mathsf{x}) = \{abs_0(\mathsf{x}),\ldots,abs_{N(\mathsf{x})}(\mathsf{x})\}$, and (iv) a number of weights $wt_j^J(\mathsf{x})$ for $J \leqslant N(\mathsf{x})$, and $0 \leqslant j \leqslant J$. The basic idea is that this representation facilitates numerical integration based on the quadrature formula:

$$\int_0^{\mathsf{x}_J} f(\mathsf{x})\, d\mathsf{x} \;\approx\; \sum_{j=0}^{J} wt_j^J(\mathsf{x}) \cdot f(abs_j(\mathsf{x}))$$

where $abs_j(\mathsf{x}) = x_j = j \cdot h$ for step-size $h = t/N(\mathsf{x})$. This corresponds to a quadrature formula in which the interval points $abs_j(\mathsf{x})$ are *equally-spaced* abscissas [24].

For state variable z, we define $rng(\mathsf{z}) = \{0,1\}$ and $N(\mathsf{z}) = 1$. Let $<$ be a fixed total order on Var, the set of state and integral variables, such that $\mathsf{z} < \mathsf{x}$ for all state variables z and integral variables x.

Definition 2. *A* multi-terminal decision diagram *(MTDD) over* $\langle \mathsf{Var}, < \rangle$ *is a rooted acyclic directed graph with vertex set* V *containing 3 types of vertices:*

- *each* state vertex v *is labelled by a state variable* $var(v)$ *and has two children* $child_0(v)$, $child_1(v) \in V$.
- *each* integral vertex v *is labelled by an integral variable* $var(v) = \mathsf{x}$ *and a natural number* $epnt(v) \leqslant N(\mathsf{x})$, *(endpoint) and has* $N(\mathsf{x})+1$ *children* $child_0(v)$, $\ldots, child_{N(\mathsf{x})}(v)$.
- *each* terminal vertex v *is labelled by a real number* $val(v)$,

such that $var(v) < var(w)$ *for each non-terminal vertex* v *and non-terminal child* w *of* v.

(For $\mathsf{Var} = \{\mathsf{v}_1, \ldots, \mathsf{v}_n\}$ with $\mathsf{v}_i < \mathsf{v}_{i+1}$ we refer to an MTDD over $\langle \mathsf{Var}, < \rangle$ as an MTDD over $(\mathsf{v}_1, \ldots, \mathsf{v}_n)$.) The constraint on the labelling of the non-terminal vertices is standard for (ordered) BDDs, and requires that on any path from the root to a terminal vertex, the variables respect the given ordering $<$. An MTDD M over $(\mathsf{v}_1, \ldots, \mathsf{v}_n)$ represents a partial function f_{M}, the values of which are obtained by traversing M starting at the root vertex as follows. For state vertex v, the edge from v to $child_0(v)$ represents the case $var(v)$ is false; the edge from v to $child_1(v)$ the case $var(v)$ is true. For integral vertex v, the edge from v to $child_j(v)$ stands for the case where the value of the real variable $var(v) = \mathsf{x}$ is $abs_j(\mathsf{x})$. The value $epnt(v)$ is needed to perform the operator INTEGRATE (defined below). If $epnt(v) = J$ then in vertex v the range of integration is $[0, x_J]$ where $x_J = abs_J(\mathsf{x})$.[4] For efficiency reasons, an implementation will internally represent MTDDs in a *reduced* form [9], a compact and canonical representation.

The relationship between BDDs, MTBDDs and MTDDs is as follows. An MTBDD is an MTDD without integral vertices; a BDD is an MTBDD with $val(v) \in \{0, 1\}$ for all terminal vertices v.

Remark 1. Note that an MTDD over $\langle \mathsf{Var}, < \rangle$ is also an MTDD over $\langle \mathsf{Var}', <' \rangle$ for any superset Var' of Var and total order $<'$ on Var' such that $\mathsf{v}_1 < \mathsf{v}_2$ iff $\mathsf{v}_1 <' \mathsf{v}_2$ for all $\mathsf{v}_1, \mathsf{v}_2 \in \mathsf{Var}$.

Encoding CTMCs by MT(B)DDs. In BDD-approaches transition systems are symbolically represented by encoding states by bit vectors, and encoding the transition relation by its characteristic function. To represent the generator matrix of a CTMC by a MTDD we abstract from the names of the states, and instead, similar to [13], use binary tuples of atomic propositions that are true in that state. Using this scheme, CTMCs are encoded as MTDDs as follows. Let $\mathcal{M} = (S, \mathbf{Q}, L)$ be a labelled CTMC. We assume that $|S| = 2^n$ and that the labelling function L is injective. (Any labelled CTMCs may be transformed into one satisfying these conditions by adding dummy states and new propositions.)

[4] In our model checking procedure, for any integral vertex v with $var(v) = \mathsf{x}$ and $epnt(v) = J < N(\mathsf{x})$, the branches representing the cases where $\mathsf{x} = x_j$ and $j > epnt(v)$ (i.e. the edges to the children $child_j(v)$) are not of importance. Accordingly, we may assume that any integral vertex v with $epnt(v) = J$ has exactly $J+1$ children $child_0(v), \ldots, child_J(v)$.

We fix an enumeration $a_1, \ldots, a_n$ of atomic propositions and identify each state s with the boolean n-tuple $(b_1, \ldots, b_n)$ where $b_i = 1$ iff $a_i \in L(s)$. In what follows, we assume that $S = \{0,1\}^n$ where we identify each state s with its encoding and the generator matrix $\mathbf{Q}$ with the function $F : \{0,1\}^{2n} \to \mathbb{R}$ where $F(\mathsf{z}_1, \mathsf{z}'_1, \ldots, \mathsf{z}_n, \mathsf{z}'_n) = \mathbf{Q}((\mathsf{z}_1, \ldots, \mathsf{z}_n), (\mathsf{z}'_1, \ldots, \mathsf{z}'_n))$. We represent $\mathcal{M}$ by the MTBDD Q for $\mathbf{Q}$ over $(\mathsf{z}_1, \mathsf{z}'_1, \ldots, \mathsf{z}_n, \mathsf{z}'_n)$, in other words $f_{\mathsf{Q}} = F$. Note that Q does not contain integral variables and hence is a MTBDD.

Example 4. Consider the CTMC of Example 1. According to the above scheme we encode the states by $s_0 \mapsto 00$, $s_1 \mapsto 01$, $s_2 \mapsto 10$ and $s_3 \mapsto 11$. The function $F = f_{\mathsf{Q}}$ and the MTDD Q are given by:

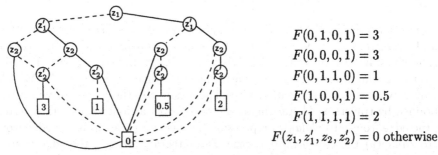

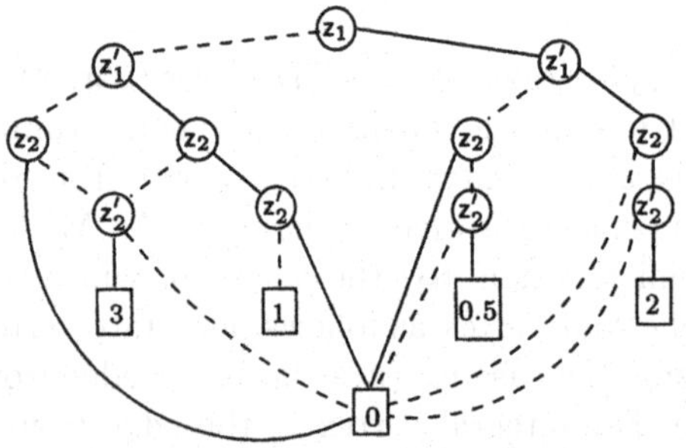

$$F(0,1,0,1) = 3$$
$$F(0,0,0,1) = 3$$
$$F(0,1,1,0) = 1$$
$$F(1,0,0,1) = 0.5$$
$$F(1,1,1,1) = 2$$
$$F(z_1, z'_1, z_2, z'_2) = 0 \text{ otherwise}$$

where dotted lines denote zero-edges and solid lines one-edges.

Operators on MTDDs. The symbolic model checking algorithm in this paper uses several operators on MTDDs that are slight modifications of equivalent operators on BDDs [9] and MTBDDs [12, 3]. For space reasons we only briefly describe these operators and focus on the new operators, in particular substitution and computing integrals. As it is standard in the BDD setting, hash tables can be used to generate a reduced MTDD during its construction.

- *Combining MTDDs via binary operators.* Operator APPLY allows a pointwise application of the binary operator *op* (like summation or multiplication) to two MTDDs. For MTDDs M_1 and M_2 over $(\mathsf{v}_1, \ldots, \mathsf{v}_n)$, APPLY$(\mathsf{M}_1, \mathsf{M}_2, op)$ yields an MTDD M over $(\mathsf{v}_1, \ldots, \mathsf{v}_n)$ for the function $f_{\mathsf{M}} = f_{\mathsf{M}_1} \; op \; f_{\mathsf{M}_2}$.
- *Variable renaming.* Operator RENAME changes the variable labelling of any v_i-labelled vertex of MTDD M over $(\mathsf{v}_1, \ldots, \mathsf{v}_n)$ into w, for $\mathsf{w} \neq \mathsf{v}_j$, $0 < j \leqslant n$. RENAME$(\mathsf{M}, \mathsf{v}_i, \mathsf{w})$ yields a MTDD over $(\mathsf{v}_1, \ldots, \mathsf{v}_{i-1}, \mathsf{w}, \mathsf{v}_{i+1}, \ldots, \mathsf{v}_n)$.
- *Restriction.* For state variable $\mathsf{v}_i = \mathsf{z}$ and boolean b, RESTRICT$(\mathsf{M}, \mathsf{z}, b)$ denotes the MTDD over $(\mathsf{v}_1, \ldots, \mathsf{v}_{i-1}, \mathsf{v}_{i+1}, \ldots, \mathsf{v}_n)$ that is obtained from M by replacing any edge from a vertex v to an z-labelled vertex w by an edge from v to $child_b(w)$, followed by removing all z-labelled vertices. In a similar way, RESTRICT$(\mathsf{M}, \mathsf{x}, x_j)$ is defined for $\mathsf{v}_i = \mathsf{x}$ an integral variable, $0 \leqslant j \leqslant N(\mathsf{x})$ and $x_j = abs_j(\mathsf{x})$.
- *Comparison operators.* Given MTDD M without integral vertices and over n state variables, and interval I, COMPARE(M, I) is the BDD representing the function that equals 1 if $f_{\mathsf{M}}(b_1, \ldots, b_n) \in I$ and 0, otherwise.
- *Matrix/vector multiplication.* Let MTBDDs Q and B without integral vertices over $2n$ and n state variables, respectively, represent the matrix $\mathbf{Q}$

and vector **b**. Then MULTI(Q, B) denotes the MTBDD over n variables that represents the vector $\mathbf{Q} \cdot \mathbf{b}$. This operator can easily be modified for MTDDs. E.g. if Q is a MTDD over $(\mathsf{v}_1, \ldots, \mathsf{v}_n, \mathsf{w}_1, \ldots, \mathsf{w}_m)$ and B a MTDD over $(\mathsf{w}_1, \ldots, \mathsf{w}_m)$ then MULTI(Q, B) represents the function

$$(\mathsf{v}_1, \ldots, \mathsf{v}_n) \mapsto \sum_{\mathsf{w}_1, \ldots, \mathsf{w}_m} f_{\mathsf{Q}}(\mathsf{v}_1, \ldots, \mathsf{v}_n, \mathsf{w}_1, \ldots, \mathsf{w}_m) \cdot f_{\mathsf{B}}(\mathsf{w}_1, \ldots, \mathsf{w}_m).$$

- *Substitution.* Let M be a MTDD over $(\mathsf{v}_1, \ldots, \mathsf{v}_n)$ where $\mathsf{v}_n = \mathsf{x}$ is an integral variable with $N(\mathsf{x}) = N$ and assume $\mathsf{y} \neq \mathsf{v}_i$ for all i. Assume that for any x-labelled vertex v in M we have $epnt(v) = N$. Then SUBST(M, y, x) denotes the MTDD over $(\mathsf{v}_1, \ldots, \mathsf{v}_{n-1}, \mathsf{y}, \mathsf{x})$ which represents the partial function that equals $f_{\mathsf{M}}(\ldots, \mathsf{y}-\mathsf{x})$ for $0 \leqslant \mathsf{x} \leqslant \mathsf{y}$ and is undefined otherwise. SUBST(M, y, x) results from M by replacing any x-labelled vertex v by the subgraph depicted as:

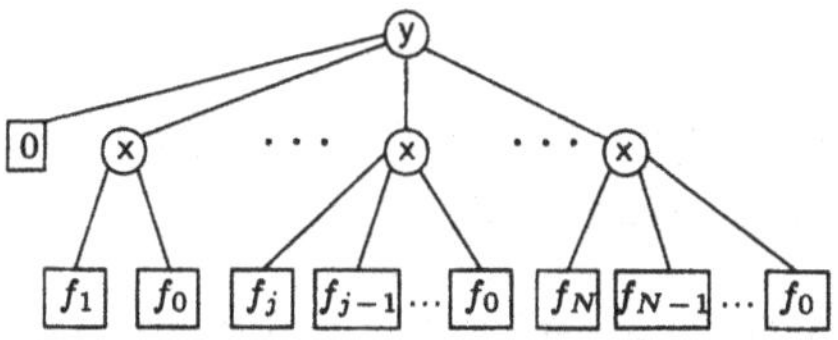

  In the figure (visualising the decision tree instead of the reduced MTDD), children are depicted from left to right and vertices $child_i(v'_j)$ where $i > j$ are omitted. Here $f_j = val(child_j(v))$ for $0 \leqslant j \leqslant N$. More precisely, new vertices $v', v'_1, \ldots, v'_N$ are introduced with $var(v') = \mathsf{y}$, $var(v'_j) = \mathsf{x}$, $child_j(v') = v'_j$, $epnt(v') = N$ and $epnt(v'_j) = j$.
- *Computing integrals.* Let $\mathsf{v}_n = \mathsf{x}$ be an integral variable with $N(\mathsf{x}) = N$, $rng(\mathsf{x}) = \{x_0, \ldots, x_N\}$ and $wt^J_j(\mathsf{x}) = \alpha^J_j$. Then, INTEGRATE(M, x) denotes the MTDD over $(\mathsf{v}_1, \ldots, \mathsf{v}_{n-1})$ that results from M by replacing any x-labelled vertex v with $epnt(v) = J$ by the terminal vertex labelled by

$$\sum_{j=0}^{J} \alpha^J_j \cdot val(child_j(v))$$

  where $\alpha^0_0 = 0$. [5] Thus, if $epnt(v) = N(\mathsf{x})$ for each x-labelled vertex v in M then INTEGRATE(M, x) represents an approximation of the function $(\ldots) \mapsto \int_0^t f_{\mathsf{M}}(\ldots, \mathsf{x})d\mathsf{x}$.

Besides the described operators on MTDDs, our model checking algorithm uses methods for boolean combinators and for a BDD-based graph analysis, e.g. to obtain the bottom strongly connected components of the graph underlying a CTMC, and MTBDD-based methods for solving linear equation systems, e.g. to compute the probabilities $\pi_{s'}(s)$. For these algorithms we refer to [9, 12, 3].

[5] Note that we assume that $\mathsf{x} = \mathsf{v}_n$; hence v_n is the largest in the variable ordering, and the children of a x-labelled vertex are terminal vertices.

5 Symbolically model checking CSL

Our symbolic model checking algorithm for **CSL** works as follows. Let $\mathcal{M} = (S, \mathbf{Q}, L)$ be a CTMC which is represented by a MT(B)DD Q over $2n$ variables as explained in the previous section. For each **CSL**-state-formula Φ we define a BDD $\mathsf{Sat}[\![\Phi]\!]$ over $(\mathsf{z}_1, \ldots, \mathsf{z}_n)$ that represents the characteristic function of the set $Sat(\Phi)$; for each **CSL**-path formula φ we define a MTBDD $\mathsf{PR}[\![\varphi]\!]$ representing the function $s \mapsto Prob(s, \varphi)$. By applying standard operators on MTBDDs we determine the MTBDDs P, representing the transition probability matrix $\mathbf{P}$ of $\mathcal{M}$, and SP representing the steady-state probabilities $\pi_{s'}(s)$ for $s, s' \in S$. $\mathsf{Sat}[\![\Phi]\!]$ is defined as follows:

$$
\begin{aligned}
\mathsf{Sat}[\![\mathrm{tt}]\!] &= \mathbf{1}\\
\mathsf{Sat}[\![a_i]\!] &= \text{the BDD for the boolean function } (\mathsf{z}_1, \ldots, \mathsf{z}_n) \mapsto \mathsf{z}_i\\
\mathsf{Sat}[\![\neg\Phi]\!] &= \neg\mathsf{Sat}[\![\Phi]\!]\\
\mathsf{Sat}[\![\Phi_1 \wedge \Phi_2]\!] &= \mathsf{Sat}[\![\Phi_1]\!] \wedge \mathsf{Sat}[\![\Phi_2]\!]\\
\mathsf{Sat}[\![\mathcal{S}_{\bowtie p}(\Phi)]\!] &= \textsc{Compare}(\textsc{Multi}(\mathsf{SP}, \mathsf{Sat}[\![\Phi]\!]'), I_{\bowtie p})\\
\mathsf{Sat}[\![\mathcal{P}_{\bowtie p}(\varphi)]\!] &= \textsc{Compare}(\mathsf{PR}[\![\varphi]\!], I_{\bowtie p}).
\end{aligned}
$$

Here, **1** denotes the BDD consisting of a single, terminal vertex labelled by 1. $\mathsf{Sat}[\![a_i]\!]$ is a BDD consisting of a single state-vertex v labelled with z_i such that $child_0(v)$ and $child_1(v)$ are labelled with 0 and 1, respectively. $\mathsf{Sat}[\![\Phi]\!]'$ denotes $\mathsf{Sat}[\![\Phi]\!]$ where z_i is renamed into z_i' (using nested applications of Rename). The definition of $\mathsf{Sat}[\![\mathcal{S}_{\bowtie p}(\Phi)]\!]$ is justified by the characterisation of $\pi_{Sat(\Phi)}(s)$ in Section 3.

MTBDD $\mathsf{PR}[\![\varphi]\!]$ is defined by induction over the structure of φ. For $\varphi = X\Phi$ and $\varphi = \Phi_1\,\mathcal{U}\,\Phi_2$, the MTBDD $\mathsf{PR}[\![\varphi]\!]$ can be obtained in the same way as for the discrete-time probabilistic case [5]. This follows directly from the first two clauses of Theorem 1. For the time-bounded until-operator we define:

$$
\mathsf{PR}[\![\Phi_1\,\mathcal{U}^{\leqslant t}\,\Phi_2]\!] = \textsc{BoundedUntil}(\mathsf{Q}, \mathsf{Sat}[\![\Phi_1]\!], \mathsf{Sat}[\![\Phi_2]\!], t, k_{max}, \epsilon)
$$

where k_{max} indicates the maximum number of iterations and ϵ is the maximum desired tolerance of the approximation. The algorithm for BoundedUntil is listed in Table 1. Here, F_0 represents the first approximation $F_0(s,t) = 0$. First the MT(B)DD H for the function $H(s, s', x) = \mathbf{Q}(s, s') \cdot e^{-\mathbf{E}(s)\cdot x}$ is constructed. This requires as input the MTBDD-representations of $\mathbf{E}$ that can easily be obtained from the MTDD Q representing the generator matrix (cf. Section 2). X consists of a single, integral vertex labelled by x with $N{+}1$ terminal vertices labelled with the values $x_0, \ldots, x_N$. Here we assume that N is sufficiently large. In the first five steps of the iteration, the MTDD-representation of F_{k+1} is constructed systematically. More precisely, F_k represents (approximations) for the values $F_k(s, x_j)$ $(0 \leqslant j \leqslant N)$ where $x_j = j \cdot h$ and $h = t/N$. I' represents the function $f_{\mathsf{I}'}(s', y, x) = F_k(s', y{-}x)$. MTDD J represents the function

$$
f_{\mathsf{J}}(s, s', y, x) = \mathbf{Q}(s, s') \cdot e^{-\mathbf{E}(s)\cdot x} \cdot F_k(s', y{-}x).
$$

algorithm BOUNDEDUNTIL $(\mathsf{Q}, \mathsf{B}_1, \mathsf{B}_2, t, k_{max}, \epsilon)$:
begin $\mathsf{F}_0 := \mathbf{0}; k := 0;$
    $\mathsf{H} :=$ APPLY(Q, APPLY(E, X, $(q_1, q_2) \mapsto e^{-q_1 \cdot q_2}$), $\cdot$);
    repeat
        $\mathsf{I} :=$ SUBST(F_k, x, y);
        $\mathsf{J} :=$ APPLY(H, I$'$, $\cdot$);
        $\mathsf{K} :=$ MULTI(J, **1**);
        $\mathsf{L} :=$ RENAME(INTEGRATE(K, x), y, x);
        $\mathsf{F}_{k+1} :=$ APPLY(APPLY(L, B_1, *min*), B_2, *max*);
        $\mathsf{D}_{k+1} :=$ APPLY(F_{k+1}, F_k, $-$);
        $\Delta_{k+1} := \max_{s,j} \left| f_{\mathsf{D}_{k+1}}(s, x_j) \right|$;
        $k := k + 1$;
    until ($k = k_{max}$ or $\Delta_k \leqslant \epsilon$);
    if $\Delta_k \leqslant \epsilon$ **then return** RESTRICT(F_k, x, t) **else return** *error*;
end.

Table 1. Algorithm for BOUNDEDUNTIL

For this, we consider J as the MTDD representation of a matrix whose rows are indexed by triples (s, y, x), and whose columns are indexed by s'. Matrix-vector multiplication with $\mathbf{1}$, the MTDD over $(\mathsf{z}'_1, \ldots, \mathsf{z}'_n)$ that represents the constant function $s' \mapsto 1$, yields MTDD K over $(\mathsf{z}_1, \ldots, \mathsf{z}_n, \mathsf{y}, \mathsf{x})$ representing

$$f_{\mathsf{K}}(s, y, x) = \sum_{s' \in S} f_{\mathsf{J}}(s, s', y, x).$$

By INTEGRATE(K, x) the integrals $\int_0^{x_J} f_{\mathsf{K}}(s, x_J, x)dx$ are approximated by $\sum_j \alpha_j^J \cdot f_{\mathsf{K}}(s, x_J, x_j)$. For generating the MTDD F_{k+1} that represents function $(s, x) \mapsto 1$ if $s \models \Phi_2$, $(s, x) \mapsto f_{\mathsf{L}}(s, x)$ if $s \models \Phi_1 \wedge \neg\Phi_2$ and $(s, x) \mapsto 0$ otherwise, we use the fact that $F_{k+1}(s, x) = \max\{ \min\{ f_{\mathsf{Sat}[\![\Phi_1]\!]}(s), f_{\mathsf{L}}(s, x) \}, f_{\mathsf{Sat}[\![\Phi_2]\!]}(s) \}$.

Finally, after the calculation of F_{k+1}, the result is compared with the result of the previous iteration, by an inspection of the terminal nodes of D_{k+1} which represents the difference between F_k and F_{k+1}. The iteration is finished if either the indicated maximum number of iterations is reached, or the tolerance of an "acceptable" approximation results.

Example 5. Consider our running example and check $s_1 \models \mathcal{P}_{\leqslant 0.8}(a\,\mathcal{U}^{\leqslant 2}\, b)$. We assume that N equals 4 and adopt the trapezoidal method for numerical integration. The MTBDD Q is the same as in Example 4. In the first iteration we obtain for F_1 a single state vertex v labelled z_1 with $child_1(v) = \mathbf{1}$, the terminal vertex labelled 1. In the second iteration we obtain the MTDD F_2 depicted on the rigth. Here $n_1 = \frac{1}{2} \cdot e^{-0} + \frac{1}{2} \cdot e^{-0.5}$, $n_2 = \frac{1}{4} \cdot e^{-0} + \frac{1}{2} \cdot e^{-0.5} + \frac{1}{4} \cdot e^{-1}$, $n_3 = \frac{1}{4} \cdot e^{-0} + \frac{1}{2} \cdot (e^{-0.5} + e^{-1}) + \frac{1}{4} \cdot e^{-1.5}$, and

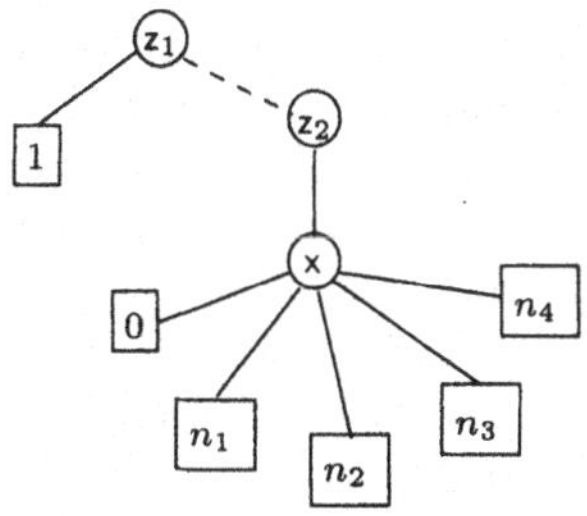

$n_4 = \frac{1}{4} \cdot e^{-0} + \frac{1}{2} \cdot (e^{-0.5} + e^{-1} + e^{-1.5}) + \frac{1}{4} \cdot e^{-2}$. The third iteration reveals that $\mathsf{F}_3 = \mathsf{F}_2$, so the algorithm finishes and returns $\mathsf{R} = \textsc{Restrict}(\mathsf{F}_3, \mathsf{x}, 2)$, obtained from F_3 by replacing the subgraph starting in vertex x by the terminal vertex with label $n_4 \approx 0.882604$. Finally, $\textsc{Compare}(\mathsf{R}, I_{\leqslant 0.8})$ reveals that $\mathcal{P}_{\leqslant 0.8}(a\,\mathcal{U}^{\leqslant 2}\,b)$ is indeed violated in s_1. Increasing the number of abscissas increases the accuracy: e.g. $N = 64$ leads to 0.8647350 as an approximation for $1 - e^{-2}$.

6 Concluding remarks

We have presented a symbolic model checking algorithm for verifying properties stated in **CSL** over continuous-time Markov chains. The basis of this model checking procedure is a characterisation of time-bounded until in terms of a Volterra integral equation system that can be solved by iteration. To solve the integrals in a symbolic way we generalised MTBDDs into multi-terminal decision diagrams (MTDDs) and presented suitable operators on these structures that facilitate a numerical integration using quadrature formulas based on equally-spaced abscissas. Due to their suitability for numerical integration, the potential application of MTDDs is much wider than model checking CTMCs.

An important direction for future research is the implementation of the proposed algorithm that should provide evidence about the adequacy of our approach. Amongst others, the size of intermediate MTDDs is unclear yet, and we want to compare our technique with standard methods to extract performance measures from Markov chains [26]. In fact, **CSL** and the algorithm can be generalised such that transient and steady state measures are expressible and can be approximated. We also plan to consider CTMCs that may contain non-determinism, like stochastic transition systems [17] or interactive Markov chains [21, 22]. We believe that by extending our approach with schedulers [27] in a similar way as for the discrete-time probabilistic case this is feasible.

Acknowledgement. We thank Markus Siegle for discussion about our initial ideas concerning MTDDs. Ed Brinksma provided valuable comments on an earlier version.

References

1. A. Aziz, K. Sanwal, V. Singhal and R. Brayton. Verifying continuous time Markov chains. In *CAV*, LNCS 1102, pp. 269–276, 1996.
2. A. Aziz, V. Singhal, F. Balarin, R. Brayton and A. Sangiovanni-Vincentelli. It usually works: the temporal logic of stochastic systems. In *CAV*, LNCS 939, pp. 155–165, 1995.
3. I. Bahar, E. Frohm, C. Gaona, G. Hachtel, E. Macii, A. Padro and F. Somenzi. Algebraic decision diagrams and their applications. *Formal Methods in System Design*, **10**(2/3): 171–206, 1997.
4. C. Baier. On algorithmic verification methods for probabilistic systems. Habilitation thesis (submitted), Univ. Mannheim, 1998.
5. C. Baier, E. Clarke, V. Hartonas-Garmhausen, M. Kwiatkowska, and M. Ryan. Symbolic model checking for probabilistic processes. In *ICALP*, LNCS 1256, pp. 430-440, 1997.

6. C. Baier and M. Kwiatkowska. Model checking for a probabilistic branching-time logic with fairness. *Distr. Comp.*, **11**(3): 125–155, 1998.
7. D. Beauquier and A. Slissenko. Polytime model checking for timed probabilistic computation tree logic. *Acta Inf.*, **35**: 645–664, 1998.
8. A. Bianco and L. de Alfaro. Model checking of probabilistic and nondeterministic systems. In *FSTTCS*, LNCS 1026, pp. 499–513, 1995.
9. R. Bryant. Graph-based algorithms for boolean function manipulation. *IEEE Trans. on Comp.*, **C-35**(8): 677-691, 1986.
10. W. Chan, R. Anderson, P. Beame, S. Burns, F. Modugno, D. Notkin and J.D. Reese. Model checking large software specifications. *IEEE Trans. on Softw. Eng.*, **24**(7): 498–519, 1998.
11. I. Christoff and L. Christoff. Reasoning about safety and liveness properties for probabilistic systems. In *FSTTCS*, LNCS 652, pp 342-355, 1992.
12. E. Clarke, M. Fujita, P. McGeer, J. Yang and X. Zhao. Multi-terminal binary decision diagrams: an efficient data structure for matrix representation. In *Proc. IEEE Int. Workshop on Logic Synthesis*, pp. 1–15, 1993.
13. E. Clarke, O. Grumberg and D. Long. Verification tools for finite-state concurrent programs. In *A Decade of Concurrency*, LNCS 803, pp. 124–175, 1993.
14. C. Courcoubetis and M. Yannakakis. Verifying temporal properties of finite-state probabilistic programs. In *FOCS*, pp. 338–345, 1988.
15. C. Courcoubetis and M. Yannakakis. The complexity of probabilistic verification. *J. ACM*, **42**(4): 857–907, 1995.
16. L. de Alfaro. How to specify and verify the long-run average behavior of probabilistic systems. In *LICS*, 1998.
17. L. de Alfaro. Stochastic transition systems. In *CONCUR*, LNCS 1466, pp. 423–438, 1998.
18. G. Hachtel, E. Macii, A. Padro and F. Somenzi. Markovian analysis of large finite-state machines. *IEEE Trans. on Comp. Aided Design of Integr. Circ. and Sys.*, **15**(12): 1479–1493, 1996.
19. H. Hansson and B. Jonsson. A logic for reasoning about time and probability. *Form. Asp. of Comp.*, **6**: 512–535, 1994.
20. V. Hartonas-Garmhausen, S. Campos and E.M. Clarke. ProbVerus: probabilistic symbolic model checking. In *Formal Methods for Real-Time and Probabilistic Systems (ARTS)*, LNCS 1601, pp. 96–110, 1999.
21. H. Hermanns. *Interactive Markov Chains.* Ph.D thesis, U. Erlangen-Nürnberg, 1998.
22. H. Hermanns and J.-P. Katoen. Automated compositional Markov chain generation for a plain-old telephone system. *Sci. of Comp. Programming*, 1999.
23. A. Pnueli, L. Zuck. Probabilistic verification. *Inf. and Comp.*, **103**,: 1–29, 1993.
24. W. Press, B. Flannery, S. Teukolsky and W. Vetterling. *Numerical Recipes in C: The Art of Scientific Computing.* Cambridge Univ. Press, 1989.
25. W. Stewart. *Introduction to the Numerical Solution of Markov Chains.* Princeton Univ. Press, 1994.
26. K.S. Trivedi, J.K. Muppala, S.P. Woolet, and B.R. Haverkort. Composite performance and dependability analysis. *Performance Evaluation*, **14**: 197–215, 1992.
27. M.Y. Vardi. Automatic verification of probabilistic concurrent finite state programs. In *FOCS*, pp 327–338, 1985.

From Synchrony to Asynchrony *

Albert Benveniste, Benoît Caillaud, and Paul Le Guernic

Irisa/Inria**, Campus de Beaulieu, 35042 Rennes cedex, France

Abstract. We present an in-depth discussion of the relationships between synchrony and asynchrony. Simple models of both paradigms are presented, and we state theorems which guarantee *correct desynchronization*, meaning that the original synchronous semantics can be reconstructed from the result of this desynchronization. Theorems are given for both the desynchronization of single synchronous programs, and for networks of synchronous programs to be implemented using asynchronous communication. Assumptions for these theorems correspond to proof obligations that can be checked on the original synchronous designs. If the corresponding conditions are not satisfied, suitable synchronous mini-programs which will ensure correct desynchronization can be composed with the original ones. This can be seen as a systematic way to generate "correct protocols" for the asynchronous distribution of synchronous designs. The whole approach has been implemented, in the framework of the SACRES project, within the SILDEX tool marketed by TNI, as well as in the SIGNAL compiler.

1 Introduction

Synchronous programming [5, 10, 14] has been proposed as an efficient approach for the design of reactive and real-time systems. It has been widely publicized, using the idealized picture of "zero time" computation and instantaneous broadcast communication [9]. Efficient techniques for code generation and verification have resulted [10, 20, 5, 15].

Criticisms have been addressed to this approach. It has been argued that, very frequently, real-life architectures do not obey the ideal model of perfect synchrony. Counter-examples are numerous: operating systems with multi-threading or multitasking, distributed architectures, asynchronous hardware, etc.

However, similarities and formal links between synchrony and asynchrony have already been discussed in the literature, thus questioning the oversimplified vision of "zero time" computation and instantaneous broadcast communication. Early paper [6] informally discussed the link between perfect synchrony and token-based asynchronous data-flow networks, see in particular section V therein. The first formal and deep study can be found in [13]. It establishes a precise

* This work is or has been supported in part by the following projects: Esprit R&D-SACRES (Esprit project EP 20897), Esprit LTR-SYRF (Esprit project EP 22703).

** Email: firstname.lastname@irisa.fr, Web: http://www.irisa.fr/

relation between so-called well-clocked synchronous functional programs and the subset of Kahn networks amenable to "buffer-less" evaluation.

Distributed code generation from synchronous programs requires to address the issue of the relationship between synchrony and asynchrony in a different way. Mapping synchronous programs to a network of automata, communicating asynchronously via unbounded FIFOs, has been implemented for the Lustre language and formalized in [12]. Mapping SIGNAL programs to distributed architectures was proposed in [19, 4], based on an early version of the theory we present in this paper. The SYNDEX tool [22, 21] also implements a similar approach. Recent work [11] on the POLIS system proposes to reuse the "constructive semantics" approach for the ESTEREL synchronous language, with CFSM (Codesign Finite State Machines) as a model for synchronous machines which can be desynchronized; this can be seen as a refinement of [13], although the referred model of asynchrony is not fully stated.

Independently, another approach relating synchrony and asynchrony has been followed. In [7, 18] it is shown how *nondeterministic* SIGNAL programs can be used to model asynchronous communication media such as queues and buffers. *Reactive Modules* [1] were proposed as a synchronous language for hardware modeling, in which asynchrony is emulated by the way of nondeterminism. Although this is of interest, we believe this approach is not suited to the analysis of true asynchrony, in which no notion of global synchronization state is available, unlike for synchrony.

In this paper we provide an extensive, in depth, analysis of the links between synchrony and asynchrony. Our vision of asynchrony encompasses distributed systems, in which no global synchronization state is available, and communications/actions are not instantaneous. This extension allows us to handle incomplete designs, specifications, properties, architectures, and executable programs, in a unified framework, for both synchronous and asynchronous semantics.

In section 2 we informally discuss the essentials of synchrony and asynchrony. Synchronous Transition Systems are defined in section 3, and their asynchronous counterpart is defined in section 4, where also desynchronization is formally defined. The rest of the paper is devoted to the analysis of desynchronization and its inverse, namely resynchronization.

2 The Essentials of the Synchronous Paradigm

There have been several attempts to characterize the essentials of the synchronous paradigm [5, 14]. With some experience and after many attempts to address the issue of moving from synchrony to asynchrony (and back), we feel the following features are indeed essential for characterizing this paradigm:

1. Programs progress via an infinite sequence of *reactions*: $P = R^\omega$, where R denotes the family of possible reactions[1].

[1] In fact, "reaction" is a slightly restrictive term, as we shall see in the sequel that "reacting to the environment" is not the only possible kind of interaction a synchronous system may have with its environment.

2. Within a reaction, decisions can be taken on the basis of the *absence* of some events, as exemplified by the following typical statements, taken from ESTEREL, LUSTRE, and SIGNAL respectively:
```
present S else 'stat'
y = current x
y := u default v
```
The first statement is self-explanatory. The "`current`" operator delivers the most recent value of `x` at the clock of the considered node, it thus has to test for absence of `x` before producing `y`. The "`default`" operator delivers its first argument when it is present, and otherwise its second argument.
3. Parallel composition is given by taking the pairwise conjunction of associated reactions, whenever they are composable: $P_1 \| P_2 = (R_1 \wedge R_2)^\omega$. Typically, if specifying is the intention, then the above formula is a perfect definition of parallel composition. In contrast, if programming is the intention, then the need for this definition to be compatible with an operational semantics complicates very much the "when it is defined" prerequisite[2].

Of course, such a characterization of the synchronous paradigm makes the class of synchronous formalisms much larger than usually considered. But it has been our experience that these were the key features for the techniques we have developed so far. Clearly, this calls for a simplest possible formalism with the above features, on which fundamental questions should be investigated: The design of the STS formalism[3] described in the next section has been guided by these objectives.

Keeping in mind the essentials of the synchronous paradigm, we are now ready to discuss informally how asynchrony relates to synchrony. Referring to points 1, 2, and 3 above, the following can be stated about asynchrony:

1. Reactions cannot be observed any more: since no global clock exists, global synchronization barriers which indicate the transition from one reaction to the next one are no more observable. Instead, a reliable communication medium is assumed, in which messages are not lost, and for each individual channel, messages are sent and received in the same order. We call a *flow* the totally ordered sequence of values sent or received on a given communication channel.
2. Absence cannot be detected, and thus cannot be used to exercise control.
3. Composition occurs by means of unifying each flow shared between two processes. This models in particular the communications via asynchronous unbounded FIFOs, such as those in Kahn networks. Rendez-vous type of communication can also be abstracted in this way.

Synchrony and asynchrony are formalized in sections 3 and 4, respectively. Section 7 details how these results can be put into practice.

[2] For instance, most of the effort related to the semantics of ESTEREL has been directed toward solving this issue satisfactorily [10].

[3] We thank Amir Pnueli for having proposed this formalism, in the course of the SACRES research project, as a minimal framework capturing the paradigm of perfect synchrony.

3 Synchronous Transition Systems (STS)

*Synchronous Transition Systems (*STS*).* We assume a vocabulary $\mathcal{V}$ which is a set of typed variables. All types are implicitly extended with a special element $\perp$, interpreted as ***absence***. Among the types we consider, there are the type of ***pure signals*** with domain $\{\mathrm{T}\}$, and the ***boolean*** type with domain $\{\mathrm{T}, \mathrm{F}\}$ (recall both types are extended with the distinguished element $\perp$). We define a ***state*** s to be a type-consistent interpretation of $\mathcal{V}$, assigning a value to each variable. We denote by S the set of all states. For a subset of variables $V \subseteq \mathcal{V}$, a V-state is a type-consistent interpretation of V. Thus a V-state s assigns a value $s[v]$ to each variable v in set V; the tuple of values assigned to the set of variables V is denoted by $s[V]$.

We define a *Synchronous Transition System* (STS) to be a tuple $\Phi = \langle V, \Theta, \rho \rangle$ consisting of the following components: V is a finite set of typed ***variables***, Θ is an assertion on V-states characterizing the set of ***initial states*** $\{s | s \models \Theta\}$ and ρ is the ***transition relation*** relating past and current V-states, s^- and s, by referring to both past[4] and current values of variables in V. For example the assertion $x = x^- + 1$ states that the value of x in s is greater by 1 than its value in s^-. If $(s^-, s) \models \rho$ then we say that state s^- is a ***ρ-predecessor*** of state s.

Runs. A *run* $\sigma : s_0, s_1, s_2, \ldots$ is a sequence of states such that $s_0 \models \Theta \;\bigwedge\; \forall i > 0\;,\; (s_{i-1}, s_i) \models \rho$.

Composition. The *composition* of two STS $\Phi = \Phi_1 \;||\; \Phi_2$ is defined as follows: $\Phi = \langle V = V_1 \cup V_2,\ \Theta = \Theta_1 \wedge \Theta_2,\ \rho = \rho_1 \wedge \rho_2 \rangle$. The composition is thus the pairwise conjunction of initial and transition relations. It should be noticed that, in STS composition, interaction occurs through common variables only.

Notations for STS. For the convenience specification, STS will have a set of ***reactive*** variables written $V_{\mathbf{r}}$, implicitly augmented with associated ***auxiliary*** variables: the whole constitutes the set V of variables. We shall use the following generic notations in the sequel:

- $b, c, v, w, \ldots$ denote reactive variables, and b, c are used to refer to variables of boolean type.
- for v a variable, $h_v \in \{\mathrm{T}, \perp\}$ denotes its *clock*: $[h_v \neq \perp] \Leftrightarrow [v \neq \perp]$
- for v a reactive variable, ξ_v denotes its associated ***state*** variable, defined by:

$$\textbf{if } h_v \textbf{ then } \xi_v = v \textbf{ else } \xi_v = \xi_v^-$$

Values can be given to $s_0[\xi_v]$ as part of the initial condition. Then, ξ_v is always present after the first occurrence of v. Finally, $\xi_{\xi_v} = \xi_v$, therefore "state variables of state variables" need not be considered.

[4] Usually, variables and *primed* variables are used to refer to current and *next* states. This is equivalent to our present notation. We have preferred to consider s^- and s, just because the formulas we shall write mostly involve current variables, rather than past ones. Using the standard notation would have resulted in a burden of primed variables in the formulas.

As modularity is desirable, every STS should be permitted to do nothing while its environment is possibly working. This feature has been yet identified in the literature and is known as *stuttering* invariance or robustness [16, 17]. For a STS Φ, stuttering invariance is defined as follows: If $\sigma = s_0, s_1, s_2, \ldots$ is a run of Φ, so is

$$\sigma' = s_0, \underbrace{\perp_{s_0}, \ldots, \perp_{s_0}}_{0 \leq \#\{\perp_{s_0}\} < \infty}, s_1, \perp_{s_1}, \ldots, \perp_{s_1}, s_2, \perp_{s_2}, \ldots, \perp_{s_2}, \ldots$$

where, for s an arbitrary state, symbol $\perp_s$ denotes the *silent state* associated with s, defined by

$$\forall v \in V_{\mathbf{r}} : \begin{cases} \perp_s[v] = \perp \\ \perp_s[\xi_v] = s[\xi_v] \end{cases}$$

meaning that state variables are kept unchanged whenever their associated reactive variables are absent. It should be noticed that stuttering invariance allows for runs possessing only a finite number of present states. We shall require in the sequel that all STS we consider are stuttering invariant. They should indeed satisfy: $(s^-, s) \models \rho \Rightarrow (s^-, \perp_{s^-}) \models \rho$ and $(\perp_{s^-}, s) \models \rho$. When this condition is not satisfied, we extend ρ minimally so that stuttering invariance is satisfied. By convention, we shall simply write $\perp$ instead of $\perp_s$ when mentioning a particular state s is not required.

4 Desynchronizing STS, and Two Fundamental Problems

From the definition of a run of a STS, we can say that a run is a sequence of tuples of values in domains extended with the extra symbol $\perp$. Desynchronizing a run amounts to discarding the synchronization barriers defining the successive reactions. Hence, for each variable $v \in V$, we only know the ordered sequence of *present* values. Thus desynchronizing a run amounts to mapping a *sequence of tuples* of values in domains extended with the extra symbol $\perp$, into a *tuple of sequences* of present values, one sequence per variable. This is formalized below.

For $\sigma = s_0, s_1, s_2, \ldots$ a run of Φ, we decompose state s_k as $s_k = (s_k[v])_{v \in V}$. Thus we can rewrite run σ as follows: $\sigma = (\sigma[v])_{v \in V}$, where $\sigma[v] = s_0[v]$, $s_1[v]$, $\ldots$, $s_k[v]$, $\ldots$. Now, each $\sigma[v]$ is compressed by deleting those $s_k[v]$ that are equal to $\perp$. Formally, let $k_0, k_1, k_2, \ldots$ be the subsequence of $k = 0, 1, 2, \ldots$ such that $s_k[v] \neq \perp$. Then we set: $\sigma^a = (\sigma^a[v])_{v \in V}$, where $\sigma^a[v] = s_{k_0}[v]$, $s_{k_1}[v]$, $s_{k_2}[v]$, $\ldots$. This defines our *desynchronization mapping* $\sigma \longmapsto \sigma^a$, and each $\sigma^a[v] = s_{k_0}[v], s_{k_1}[v], s_{k_2}[v], \ldots$ is called a *flow* in the sequel.

The asynchronous abstraction of a STS $\Phi = \langle V, \Theta, \rho \rangle$, is defined as follows:

$$\Phi^a =_{\text{def}} \langle V, \Sigma^a \rangle \,, \tag{1}$$

where Σ^a is the family of all (asynchronous) runs σ^a, with σ ranging over the set of (synchronous) runs of Φ. For $\Phi_i = \langle V_i, \Theta_i, \rho_i \rangle, i = 1, 2$, we define:

$$\Phi_1^a \parallel_a \Phi_2^a =_{\text{def}} \langle V, \Sigma^a \rangle \text{ , where } \begin{cases} V = V_1 \cup V_2 \\ \Sigma^a = \Sigma_1^a \ \wedge^a \ \Sigma_2^a \end{cases} \tag{2}$$

and $\wedge^a$ denotes conjunction of sets of asynchronous runs, which we define now. For $\sigma_i^a \in \Sigma_i^a, i = 1, 2$, we say that σ_1^a and σ_2^a are *unifiable*, written $\sigma_1^a \bowtie^a \sigma_2^a$, if the following condition holds: $\forall v \in V_1 \cap V_2 : \sigma_1^a[v] = \sigma_2^a[v]$. If σ_1^a and σ_2^a are unifiable, then we define $\sigma^a =_{\text{def}} \sigma_1^a \wedge^a \sigma_2^a$ as:

$$\begin{aligned}
&\forall v \in V_1 \cap V_2 : \sigma^a[v] = \sigma_1^a[v] = \sigma_2^a[v] \\
&\forall v \in V_1 \setminus V_2 : \sigma^a[v] = \sigma_1^a[v] \\
&\forall v \in V_2 \setminus V_1 : \sigma^a[v] = \sigma_2^a[v]
\end{aligned}$$

Finally, Σ^a is the set of the so defined σ^a. Thus asynchronous composition proceeds via unification of shared flows.

Synchrony vs. Asynchrony? At this point two natural questions arise, namely:

Question 1 (desynchronizing a single STS*).* **Is resynchronization feasible and uniquely defined?** More precisely, is it possible to reconstruct uniquely a synchronous run σ of our STS from a desynchronized run σ^a?

Question 2 (desynchronizing a communication). **Does communication behave equivalently for both the synchronous and asynchronous compositions?** More precisely, does the following property hold:

$$\Phi_1^a \parallel_a \Phi_2^a = (\Phi_1 \parallel \Phi_2)^a \ ? \tag{3}$$

If question 1 had a positive answer, then we could desynchronize a run of the considered STS, and then still recover the original synchronous run. Thus a positive answer to question 1 would guarantee that the synchronous semantics is preserved when desynchronization is performed on a single STS.

On the other hand, if question 2 had a positive answer, then we could interpret our STS composition equivalently as synchronous or asynchronous.

Unfortunately, neither 1 nor 2 have positive answers in general, due to the possibility of exercising control by the way of absence in synchronous composition $\parallel$. In the following section, we show that questions 1 and 2 have positive answers under certain sufficient conditions, in which the two notions of *endochrony* (for point 1) and *isochrony* (for point 2) play a central role.

5 Endochrony and Re-synchronization

5.1 Formal Results

In this section, we use notations from section 3. For an STS $\Phi = \langle V, \Theta, \rho \rangle$, and s a reachable state of Φ, the clock-abstraction of s (denoted by s^h) is defined as follows:

$$\forall v \in V : \ s^h[v] \in \{\perp, \top\}, \text{ and } s^h[v] = \perp \Leftrightarrow s[v] = \perp \tag{4}$$

For a STS $\Phi = \langle V, \Theta, \rho \rangle$, s^- a reachable state for Φ, and $W' \subseteq W \subseteq V$, we say that W' is a *clock inference of* W *given* s^-, written $W' \hookrightarrow_{s^-} W$, if for each state s of Φ, reachable from s^-, knowing the presence/absence and actual value carried by each variable belonging to W', allows us to determine exactly the presence/absence of each variable belonging to W. In other words $s[W']$ uniquely determines $s^h[W]$.

If both $W' \hookrightarrow_{s^-} W_1$ and $W' \hookrightarrow_{s^-} W_2$ hold, then $W' \hookrightarrow_{s^-} (W_1 \cup W_2)$ follows, thus there exists a greatest W such that $W' \hookrightarrow_{s^-} W$ holds. Hence we can consider the unique maximal increasing sequence of subsets of V, for a given s^-,

$$\emptyset = V(0) \hookrightarrow_{s^-} V(1) \hookrightarrow_{s^-} V(2) \hookrightarrow_{s^-} \dots \tag{5}$$

in which, for each $k > 0$, $V(k)$ is the greatest set of variables such that $V(k-1) \hookrightarrow_{s^-} V(k)$ holds. As $\emptyset = V(0)$, $V(1)$ consists in the subset of variables that are present as soon as the considered STS gets activated or which are always absent in successor states of s^-. Of course sequence (5) must become stationary at some finite $k_{\max}$: $V(k_{\max}+1) = V(k_{\max})$. In general, we only know that $V(k_{\max}) \subseteq V$. Sequence (5) is called the *synchronization sequence* of Φ in state s^-.

Definition 1 (Endochrony). *A* STS Φ *is said to be* endochronous *if, for each reachable state* s^- *of* Φ, $V(k_{\max}) = V$, *i.e., if the synchronization sequence:*

$$\emptyset = V(0) \hookrightarrow_{s^-} V(1) \hookrightarrow_{s^-} V(2) \hookrightarrow_{s^-} \dots \text{ converges to } V \tag{6}$$

Condition (6) expresses that presence/absence of all variables can be inferred *incrementally* from already known values carried by present variables and state variables of the STS in consideration. Hence no test for presence/absence on the environment is needed. The following theorem justifies our approach:

Theorem 1. *Consider a* STS $\Phi = \langle V, \Theta, \rho \rangle$.

1. *Conditions (a) and (b) given below are equivalent:*
   (a) Φ *is endochronous.*
   (b) For each $\delta \in \Sigma^a$, *we can reconstruct the corresponding synchronous run* σ *such that* $\sigma^a = \delta$, *in a unique way up to silent reactions.*
2. *Let us assume* Φ *is endochronous and stuttering invariant. If* $\Phi' = \langle V, \Theta, \rho' \rangle$ *is another endochronous and stuttering invariant* STS *then*

$$(\Phi')^a = \Phi^a \Rightarrow \Phi' = \Phi \tag{7}$$

Proof. We prove successively points 1 and 2.

1. We consider a previous state s^- and prove the result by induction. We pick out a $\delta \in \Sigma^a$, and assume for the moment that it can be decomposed in:

$$\underbrace{s_1\,,\, s_2\,,\, \dots\,,\, s_n}_{\text{initial segment of } \sigma \text{ of length } n} \qquad \delta_n \tag{8}$$

   i.e., into a sequence of length n, made of non-silent states s_i (the head of the synchronous run σ we wish to reconstruct), followed by the tail of the asynchronous run δ, which we denote by δ_n, and we assume that such a decomposition is unique. Then we claim that

$$\text{(8) is also valid with } n \text{ substituted by } n+1. \tag{9}$$

To prove (9), we note that, whenever STS Φ is activated in the considered state, the presence/absence of each variable belonging to $V(1)$ is known. By assumption, the state $s^h_{n+1}[V(1)]$ resulting from clock-abstraction, having $V(1)$ as variables, is uniquely determined. In the sequel we write $s^h_{n+1}(1)$ for short instead of $s^h_{n+1}[V(1)]$. Thus, presence/absence of variables for state $s_{n+1}(1)$ is known, the values carried by present variables still have to be determined.
For any $v \in V_1$, we simply take the value carried by the minimal element of the sequence associated with variable v in δ_n. Values carried by corresponding state variables are updated accordingly. Thus we know the presence or absence and the value of each individual variable in state $s_{n+1}(1)$.
Next we move on constructing $s_{n+1}(2)$. From $s_{n+1}(1)$ we know $s^h_{n+1}(2)$. Thus we know how to split V_2 into present and absent variables for the considered state. We pick up the present ones, and repeat the same argument as before to get $s_{n+1}(2)$.
Repeating this argument until $V(k) = V$ for some finite k (by endochrony assumption), proves claim (9).
Given the initial condition for δ, we get from (9), by induction, the desired proof that *(a)* $\Rightarrow$ *(b)*.
We shall now prove *(b)* $\Rightarrow$ *(a)*. We assume that Φ is not endochronous, and show that condition (b) cannot be satisfied. If Φ is not endochronous, there must be some reachable state s^- for which sequence (6) does not converge to V. Thus, again, we pick out a $\delta \in \Sigma^a$, decomposed in the same way as in formula (8):

$$\underbrace{s_1\,,\, s_2\,,\,\ldots\,,\, s_n}_{n-\text{initial segment of } \sigma}\ \delta_n$$

and we assume in addition that $s_n = s^-$, the given state for which endochrony is violated. We now show that (9) is not satisfied. Let $k_* \geq 0$ be the smallest index such that $V(k) = V(k+1)$, we know $V_{k_*} \neq V$. Thus we can apply the algorithm of case 1 for reconstructing the reaction, until variables of V_{k_*}. Then presence/absence for variables belonging to $V \setminus V_{k_*}$ cannot be determined based on the knowledge of variables belonging to V_{k_*}. This means that there exist several possible extensions of $s^h_{n+1}(k_* + 1)$ and the $(n+1)$-th reaction is not determined in a unique way. Hence condition (b) does not hold.

2. Let us assume Φ is endochronous, and consider Φ' as in point 2 of the theorem. As both Φ and Φ' are stuttering invariant, point 2 is an immediate consequence of point 1. $\diamond$

COMMENTS.

1. Endochrony is not decidable in general. However, it is decidable for STS only involving variables with finite domains of values, and model checking can be used for that. For general STS, model checking can be used, in combination with abstraction techniques. The case of interest is when the chain $V(0), V(1), \ldots$ does not depend upon the particular state s^-, and we write simply $V(k) \hookrightarrow V(k+1)$ in this case. This abstraction yields to a sufficient condition of endochrony.
2. The proof of this theorem in fact provides an effective algorithm for the on-the-fly reconstruction of the successive reactions from a desynchronized run of an endochronous program.

(COUNTER-)EXAMPLES.

Examples:

- a single-clocked STS.
- STS "**if** b = T **then** *get* u", where b, u are the two inputs, and b is boolean. The clock of b coincides with the activation clock for this STS, and thus $V(1) = \{b\}$. Then, knowing the value for b indicates whether or not u is present, thus $V(2) = \{b, u\} = V$.

Counter-example: STS "**if** ([**present** a] $\|$ [**present** b]) **then** ..." is not endochronous, as the environment is free to offer any combination of presence/absence for the two inputs a, b. Thus $\emptyset = V(0) = V(1) = V(2) = \ldots \subsetneq V$, and endochrony does not hold.

5.2 Practical Consequences

A first use of endochrony is shown in the following figure:

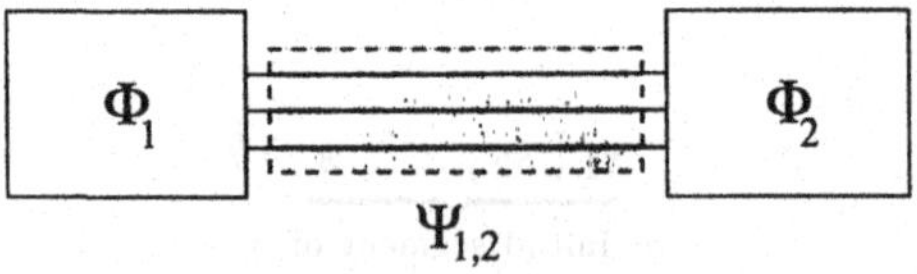

In this figure, a pair (Φ_1, Φ_2) of STS is depicted, with W as the set of shared variables. Their composition is rewritten as follows: $\Phi_1 \| \Phi_2 = \Phi_1 \| \Psi_{1,2} \| \Phi_2$, where $\Psi_{1,2}$ is the restriction of $\Phi_1 \| \Phi_2$ to W, hence $\Psi_{1,2}$ models a synchronous communication medium. We obtain by using property $\Phi \| \Phi = \Phi$ for every STS Φ:

$$\Phi_1 \| \Phi_2 = \underbrace{(\Phi_1 \| \Psi_{1,2})}_{\widetilde{\Phi}_1} \| \underbrace{(\Psi_{1,2} \| \Phi_2)}_{\widetilde{\Phi}_2} = \widetilde{\Phi}_1 \| \widetilde{\Phi}_2 \tag{10}$$

This model of communication medium $\Psi_{1,2}$ is endochronous, and composition $\Phi_1 \| \Phi_2$ is implemented by the (equivalent) composition $\widetilde{\Phi}_1 \| \widetilde{\Phi}_2$. Since all runs of $\Psi_{1,2}$ are also runs of $\widetilde{\Phi}_1$ and the former is endochronous, then communication can be equivalently implemented according to perfect synchrony or full asynchrony.

This answers question 2, however it does not extend to networks of STS involving more than two nodes. The following figure shows a counter-example:

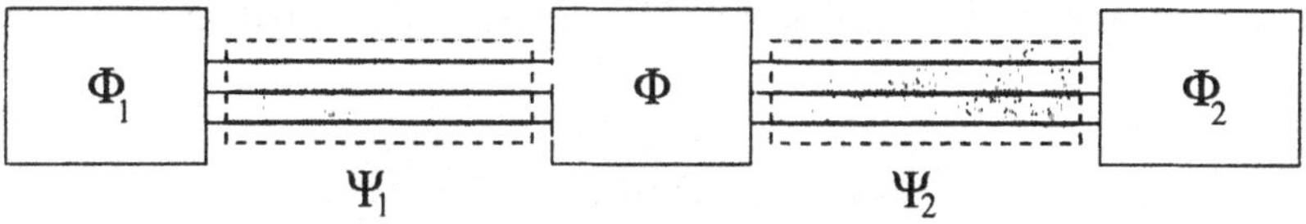

Transition systems Ψ_1 and Ψ_2 are assumed to be endochronous. Then communication between Φ_1 and Φ on the one hand, and Φ and Φ_2 on the other hand, can be desynchronized. Unfortunately, communication between Φ_1 and Φ_2 via Φ cannot, as it is not true in general that $\Psi_1 \parallel \Phi \parallel \Psi_2$ is endochronous. The problem is that endochrony is not compositional, hence even ensuring in addition that Φ itself is endochronous does not work out. Thus we would need to ensure that Ψ_1, Ψ_2 as well as $\Psi_1 \parallel \Phi \parallel \Psi_2$ are all endochronous. This cannot be considered as an adequate solution when networks of processes are considered. Therefore we move on introducing the alternative notion of *isochrony*, which focusses on communication, and is compositional.

6 Isochrony, and Synchronous/Asynchronous Compositions

The next result addresses the question of when property (3) holds. We are given two STS $\Phi_i = \langle V_i, \Theta_i, \rho_i \rangle, i = 1, 2$. Let $W = V_1 \cap V_2$ be the set of their common variables, and $\Phi = \Phi_1 \parallel \Phi_2$ their synchronous composition. For each reachable state s of Φ, we denote by $s_1 =_{\text{def}} s[V_1]$ and $s_2 =_{\text{def}} s[V_2]$ the restrictions of state s respectively to Φ_1 and Φ_2. It should be reminded that, for $i = 1, 2$, s_i is a reachable state of Φ_i. Corresponding notations s^-, s_1^-, s_2^- for past states are used accordingly.

Definition 2 (Isochrony). *Let* (Φ_1, Φ_2) *be a pair of* STS *and* $\Phi = \Phi_1 \parallel \Phi_2$ *be their parallel composition. Transitions of* $\Phi_i, i = 1, 2$, *are written* (s_i^-, s_i). *The following conditions* **(i)** *and* **(ii)** *are defined on pairs* $((s_1^-, s_1), (s_2^-, s_2))$ *of transitions of* (Φ_1, Φ_2)*:*

(i) *1.* $s_1^- = s^-[V_1]$ *and* $s_2^- = s^-[V_2]$ *holds for some reachable state* s^- *of* Φ, *in particular* s_1^- *and* s_2^- *are unifiable;*
2. none of the states $s_i, i = 1, 2$ *are silent on the common variables, i.e., it is not the case that, for some* $i = 1, 2$*:* $s_i[v] = \perp$ *holds for all* $v \in W$ *;*
3. s_1 *and* s_2 *coincide over the set of* present *common variables*[5], *i.e.:*

$$\forall v \in W \ : \ (s_1[v] \neq \perp \text{ and } s_2[v] \neq \perp) \Rightarrow s_1[v] = s_2[v] \ ;$$

(ii) *States* s_1 *and* s_2 *coincide over the* whole set *of common variables, i.e., states* s_1 *and* s_2 *are unifiable, i.e.,*

$$s_1 = s[V_1] \text{ and } s_2 = s[V_2] \text{ holds for some reachable state } s \text{ for } \Phi \ .$$

The pair (Φ_1, Φ_2) *is said to be* isochronous *if and only if for each pair* $((s_1^-, s_1), (s_2^-, s_2))$ *of transitions of* (Φ_1, Φ_2), *condition* **(i)** *implies condition* **(ii)**.

[5] By convention this is satisfied if the set of present common variables is empty.

COMMENT. Roughly speaking, condition of isochrony expresses that unifying over *present* common variables is enough to guarantee the unification of the two considered states s_1 and s_2. Condition of isochrony is illustrated on the following figure:

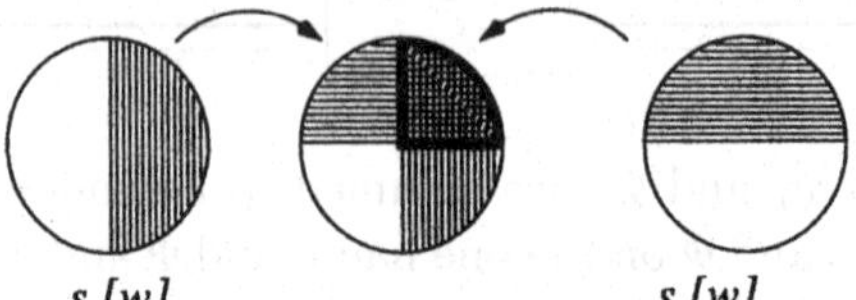

The figure depicts, for unifiable previous states s_1^-, s_2^-, the corresponding states s_1, s_2 where (s_i^-, s_i) is a valid transition for Φ_i. The figure depicts the interpretation of s_1 (circle on the left) and s_2 (circle on the right) over shared variables W. White and dashed areas represent absent and present values, respectively. The two left and right circles are superimposed in the mid circle. In general, vertically and horizontally dashed areas do not coincide, even if s_1 and s_2 unify over the subset of shared variables that are present for both transitions (double-dashed area). Pictorially, unification over double-dashed area does not imply in general that dashed areas coincide. Isochrony indeed requires that unification over double-dashed area does imply that dashed areas coincide, hence unification of s_1 and s_2 follows.

The following theorem justifies introducing this notion of isochrony.

Theorem 2.

1. *If the pair (Φ_1, Φ_2) is isochronous, then it satisfies property (3).*
2. *Conversely, we assume in addition that Φ_1 and Φ_2 are both endochronous. If the pair (Φ_1, Φ_2) satisfies property (3), then it is isochronous.*

Thus, isochrony is a sufficient condition of property (3), and it is also in fact necessary when the components are endochronous.

COMMENTS:

1. We have already discussed the importance of enforcing property (3). Now, why is this theorem interesting? Mainly because it replaces condition (3), which involves infinite runs, by condition **(i)** $\Rightarrow$ **(ii)** of isochrony, which only involves pairs of reactions of the considered pair of STS.
2. Comment 1 about endochrony also applies for isochrony.

Proof. We successively prove points 1 and 2.

1. Isochrony Implies Property (3). The proof proceeds from two steps:

1. Let Φ^a be the desynchronization of Φ, defined in equation (1), and $\delta \in \Sigma^a$ be an asynchronous run of Φ^a. There is at least one corresponding synchronous run σ of Φ such that $\delta = \sigma^a$. Any such σ is clearly the synchronous composition of two unifiable runs σ_1 and σ_2 for Φ_1 and Φ_2, respectively. Hence

associated asynchronous runs σ_1^a and σ_2^a are also unifiable, and their asynchronous composition $\sigma_1^a \wedge^a \sigma_2^a$ belongs to $\Sigma_1^a \wedge^a \Sigma_2^a$. Thus we always have the inclusion:

$$\Phi_1^a \parallel_a \Phi_2^a \supseteq (\Phi_1 \parallel \Phi_2)^a \tag{11}$$

Proving (3) now amounts to the proof of the converse inclusion. So far we have only used the definition of desynchronization and asynchronous composition, isochrony has not yet been used.

2. Proving the opposite inclusion, requires to prove that, when moving from asynchronous composition to the synchronous one, the additional constraints resulting from a reaction-per-reaction matching of unifiable runs will not result in rejecting pairs of runs that otherwise would be unifiable in the asynchronous sense. This is where isochrony is used.
A pair (δ_1, δ_2) of asynchronous runs is picked out such that $\delta_1 \bowtie^a \delta_2$: they can be combined with the asynchronous composition to form some run $\delta = \delta_1 \wedge^a \delta_2$ (cf. (2)). By definition of desynchronization (cf. section 4), there exist a (synchronous) run σ_1 of Φ_1, and a (synchronous) run σ_2 of Φ_2, such that δ_i is obtained by desynchronizing σ_i, $i = 1, 2$ (as we do not assume endochrony at this point, run σ_i is not uniquely determined). Thus each run σ_i is a succession of states. Clearly, inserting finitely many silent states between successive states of σ_i would also provide valid candidates for recovering δ_i after desynchronization. We shall show, by induction over the set of runs, that:

properly inserting such silent states in the appropriate component produces two runs which are *unifiable* in the synchronous sense. (12)

This means that, from a pair (δ_1, δ_2) such that $\delta_1 \bowtie^a \delta_2$, we can reconstruct (at least) one pair (σ_1, σ_2) of runs of Φ_1 and Φ_2 that are unifiable in the synchronous sense, and thus it proves the converse inclusion:

$$\Phi_1^a \parallel_a \Phi_2^a \subseteq (\Phi_1 \parallel \Phi_2)^a \ . \tag{13}$$

From (11) and (13) we then deduce property (3). Thus we move on proving (12) by induction over pairs of runs.
Let (σ_1, σ_2) be a pair of runs of Φ_1 and Φ_2. the induction hypothesis is:

$$\sigma_1^a \bowtie^a \sigma_2^a \Rightarrow \exists(\rho_1, \rho_2) \, \text{runs of} \, \Phi_1 \, \text{and} \, \Phi_2, \, \text{s.t.} \, \sigma_i^a = \rho_i^a \, \text{and} \, \rho_1 \bowtie \rho_2 \tag{14}$$

Let us assume that (14) holds for every pair of runs of ordinal strictly less than that of (σ_1, σ_2) and that σ_1^a and σ_2^a are asynchronously composable. These two runs may start with infinitely or finitely many silent states over the common variables W, therefore three cases may occur:

CASE 1 : Both runs contain some non silent state over W, therefore they can be decomposed as follows: $\sigma_1 = s_{1,1}, \dots s_{1,k_1}, s_{1,k_1+1}, \sigma_1'$ and $\sigma_2 = s_{2,1}, \dots s_{2,k_2}, s_{2,k_2+1}, \sigma_2'$, where the first k_1 states of σ_1 and the first k_2

states of σ_2 are all silent over W and s_{1,k_1+1}, s_{2,k_2+1} are both non-silent over W. We concentrate on those variables $v \in W$ that are present in both states s_{1,k_1+1} and s_{2,k_2+1}. As $\sigma_1^a \bowtie^a \sigma_2^a$ holds, then we must have $s_{1,k_1+1}[v] = s_{2,k_2+1}[v]$ for any such v. Thus points 1,2 and 3 of condition **(i)** of isochrony are satisfied. Hence, by isochrony, s_{1,k_1+1} and s_{2,k_2+1} are indeed unifiable in this case. Moreover $\sigma_1'^a \bowtie^a \sigma_2'^a$ and since the ordinal of (σ_1', σ_2') are strictly less than that of (σ_1, σ_2), induction hypothesis (14) holds, and there exists (ρ_1', ρ_2') a pair of composable runs such that $\sigma_i'^a = \rho_i'^a, i = 1, 2$. We now define two runs by inserting silent states in σ_1 and σ_2:

$$\rho_1 = s_{1,1}, \ldots s_{1,k_1}, \underbrace{\bot, \ldots \bot}_{h_1 \text{ silent states}}, s_{1,k_1+1}, \rho_1'$$

$$\rho_2 = s_{2,1}, \ldots s_{2,k_2}, \underbrace{\bot, \ldots \bot}_{h_2 \text{ silent states}}, s_{2,k_2+1}, \rho_2'$$

Where $h_1 = \max(0, k_2 - k_1)$, $h_2 = \max(0, k_1 - k_2)$. The first $\max(k_1, k_2)$ states of ρ_1 and ρ_2 are composable because they are silent over W. Recall that s_{1,k_1+1} and s_{2,k_2+1} are composable states and that $\rho_1' \bowtie \rho_2'$. Therefore ρ_1 and ρ_2 are composable and $\rho_i^a = \sigma_i^a$.

CASE 2 : Both runs $\sigma_1 = s_{1,1}, \ldots s_{1,i}, \ldots$ and $\sigma_2 = s_{2,1}, \ldots s_{2,i}, \ldots$ are silent over W. Therefore they are synchronously composable.

CASE 3 : One of the two runs σ_1, σ_2 is silent over W, while the other contains a non-silent state. This violates the left-hand part of the implication in the induction hypothesis (14): $\sigma_1^a \bowtie^a \sigma_2^a$ does not hold.

This proves that induction hypothesis (14) holds for runs (σ_1, σ_2). By induction principle it also holds for every pair of runs.

2. Under Endochrony of the Components, Property (3) Implies Isochrony. From Theorem 1 we know that, in our proof of point 1 of theorem 2, the synchronous runs σ_i are uniquely defined, up to silent states, from their desynchronized counterparts σ_i^a. If isochrony is not satisfied, then, for some pair (σ_1^a, σ_2^a) of unifiable asynchronous runs, and their decompositions $\sigma_i = (s_{i,j})_{j>0}, i = 1, 2$, of them, it follows that points 1,2,3 of condition **(i)** of isochrony are satisfied, and there exists $n > 0$ such that states $s_{1,n}$ and $s_{2,n}$ are *not* unifiable. As our only possibility is to try to insert silent states in the two components our process of incremental unification on a per reaction basis fails. Thus (13) is violated, and so is property (3). This finishes the proof of the theorem. ◇

An interesting immediate byproduct is the extension of these results on desynchronization to networks of communicating synchronous components:

Corollary 1 (desynchronizing a network of components). *Let $(\Phi_k)_{k=1,\ldots,K}$ be a family of* STS. *Let us assume that each pair $(\Phi_k, \Phi_{k'})$ is isochronous, then:*

1. *For each disjoint subsets I and J of set $\{1, \ldots, K\}$, the pair*

$$\left(\ \|_{k \in I} \ \Phi_k \ , \ \|_{k' \in J} \ \Phi_{k'} \ \right) \tag{15}$$

is isochronous.

2. Also, desynchronization extends to the network:

$$(\Phi_1 \parallel \dots \parallel \Phi_K)^a = \Phi_1^a \parallel_a \dots \parallel_a \Phi_K^a . \tag{16}$$

Proof. 1. It is sufficient to prove the following restricted case of (15):

$$(\Psi, \Phi_1) \text{ and } (\Psi, \Phi_2) \text{ are isochronous} \Rightarrow (\Psi, \Phi_1 \parallel \Phi_2) \text{ is isochronous} \tag{17}$$

as (15) follows via obvious induction on the cardinality of sets I and J. Thus we focus on proving (17). Let (s^-, s) and (t^-, t) be two pairs of successive states of Ψ and $\Phi_1 \parallel \Phi_2$ respectively, which satisfy condition **(i)** of isochrony, in definition 2. Let t be the composition (unification) of the two states s_1 and s_2 of Φ_1 and Φ_2, respectively. By point 2 of **(i)**, at least one of these two states is not silent, and we assume s_1 is not silent. From point 3 of **(i)**, s and s_1 coincide over the set of *present* common variables, and thus, since pair (Ψ, Φ_1) is isochronous, states s and s_1 coincide over the *whole* set of common variables of Ψ and Φ_1. Thus s and s_1 are unifiable. But, on the other hand, s_1 and s_2 are also unifiable since they are just restrictions of the same global state t of $\Phi_1 \parallel \Phi_2$. Thus states s and t are unifiable, and pair $(\Psi, \Phi_1 \parallel \Phi_2)$ is isochronous. This proves (17).

2. The second statement is proved via induction on the number of components: $(\Phi_1 \parallel \dots \parallel \Phi_K)^a = ((\Phi_1 \parallel \dots \parallel \Phi_{K-1}) \parallel \Phi_K)^a = (\Phi_1 \parallel \dots \parallel \Phi_{K-1})^a \parallel_a \Phi_K^a$, and the induction step follows from (15). ◇

(COUNTER-)EXAMPLES.

Examples:
- a single-clocked communication between two STS.
- the pair $(\widetilde{\Phi}_1, \widetilde{\Phi}_2)$ of formula (10).

Counter-example: two STS communicating with one another through two unconstrained reactive variables x and y. Both STS exhibit the following reactions: x present and y absent, or alternatively x absent and y present.

7 Getting GALS Architectures

In practice, only partial desynchronization of networks of communicating STS may be considered. This means that system designers may aim at generating *Globally Asynchronous* programs made of *Locally Synchronous* components communicating with one another via asynchronous communication media — this is referred to as GALS architectures.

In fact, theorems 1 and 2 provide the adequate solution to this problem. Let us assume that we have a finite collection Φ_i of STS such that

1. *each Φ_i is endochronous, and*
2. *each pair (Φ_i, Φ_j) is isochronous.*

Then, from corollary 1 and theorem 1, we know that

$$(\Phi_1 \parallel \dots \parallel \Phi_K)^a = \Phi_1^a \parallel_a \dots \parallel_a \Phi_K^a$$

and each Φ_k^a is in one-to-one correspondence with its synchronous counterpart Φ_k. Here is the resulting execution scheme for this GALS architecture:

- For communications involving a pair (Φ_i, Φ_j) of STS, each flow is preserved individually, but global synchronization is loosened.
- Each STS Φ_i reconstructs its own successive reactions by just observing its (desynchronized) environment, and then locally behaves as a synchronous STS.
- Finally, each Φ_i is allowed to have an internal activation clock which is *faster* than communication clocks. Resulting local activation clocks evolve asynchronously from one another.

8 Conclusion

We have presented an in depth study of the relationship between synchrony and asynchrony. The overall approach consists in characterizing those networks of STS which can be safely desynchronized, without semantic loss. Actual implementation of the communications only requests that 1/ message shall not be lost, and 2/ messages on each individual channel are sent and delivered in the same order. This type of communication can be implemented either by FIFOs or by rendez-vous.

The next questions are: 1/ how to test for endo/isochrony? and, 2/ if such properties are not satisfied, how to modify the given network of STS in order to guarantee them? It turns out that both points are easily handled on abstractions of synchronous programs, using the so-called *clock calculus* which is part of the SIGNAL compiler. We refer the reader to [2, 3, 8] for additional details. Enforcing endo/isochrony amounts to equipping each STS with a suitable additional STS which can be regarded as a kind of "synchronization protocol". When this is done, desynchronization can be performed safely.

This method has been implemented in particular in the SILDEX tool for the SIGNAL language, marketed by TNI, Brest, France. It is also implemented in the SIGNAL compiler developed at Inria, Rennes.

References

[1] R. Alur and T. A. Henzinger. Reactive modules. In *Proceedings of the 11th IEEE Symposium on Logic in Computer Science, LICS'96*, pages 207–218, 1996. extended version submitted for publication.

[2] T. P. Amagbegnon, L. Besnard, and P. Le Guernic. Arborescent canonical form of boolean expressions. Technical Report 2290, Inria Research Report, June 1994.

[3] T. P. Amagbegnon, L. Besnard, and P. Le Guernic. Implementation of the dataflow language SIGNAL. In *Proceedings of the ACM SIGPLAN Conference on Programming Languages Design and Implementation, PLDI'95*, pages 163–173, 1995.

[4] P. Aubry. *Mises en œuvre distribuées de programmes synchrones.* PhD thesis, université de Rennes 1, 1997.
[5] A. Benveniste and G. Berry, editors. *Proceedings of the IEEE*, volume 79, chapter The special section on another look at real-time programming, pages 1268–1336. IEEE, September 1991.
[6] A. Benveniste and G. Berry. Real-time systems design and programming. *Another look at real-time programming, special section of Proc. of the IEEE*, 79(9):1270–1282, September 1991.
[7] A. Benveniste and P. Le Guernic. Hybrid dynamical systems theory and the signal language. *IEEE Transactions on Autom. Control*, 35(5):535–546, May 1990.
[8] A. Benveniste, P. Le Guernic, and P. Aubry. Compositionality in dataflow synchronous languages: specification & code generation. research report RR–3310, INRIA, November 1997. `http://www.inria.fr/RRRT/publications-eng.html`, see also a revised version co-authored with B. Caillaud, March 1998.
[9] G. Berry. Real time programming: Special purpose or general purpose languages. In *Proceedings of the IFIP World Computer Congress*, San Francisco, 1989.
[10] G. Berry. The constructive semantics of esterel. Draft book, `http://www.inria.fr/meije/esterel`, December 1995.
[11] G. Berry and E. M. Sentovich. An implementation of construtive synchronous programs in POLIS. manuscript, November 1998.
[12] B. Caillaud, P. Caspi, A. Girault, and C. Jard. Distributing automata for asynchronous networks of processors. *European Journal on Automated Systems*, 31(3):503–524, 1997.
[13] P. Caspi. Clocks in dataflow languages. *Theoretical Computer Science*, 94:125–140, 1992.
[14] N. Halbwachs. *Synchronous programming of reactive systems.* Kluwer Academic Pub., 1993.
[15] N. Halbwachs, F. Lagnier, and Ratel C. Programming and verifying real-time systems by means of the synchronous dataflow language lustre. *IEEE Trans. on Software Engineering*, 18(9):785–793, September 1992.
[16] L. Lamport. Specifying concurrent program modules. *ACM Transactions on Programming Languages and Systems*, 5(2):190–222, 1983.
[17] L. Lamport. What good is temporal logic? In R. E. A. Mason, editor, *Proc. IFIP 9th World Congress*, pages 657–668. North Holland, 1983.
[18] P. Le Guernic, T. Gautier, M. Le Borgne, and Le Maire C. Programming real-time applications with SIGNAL. *Another look at real-time programming, special section of Proc. of the IEEE*, 79(9):1321–1336, September 1991.
[19] O. Maffeis and P. Le Guernic. Distributed implementation of Signal : scheduling and graph clustering. In *3rd International School and Symposium on Formal Techniques in Real-Time and Fault-Tolerant Systems*, volume 863 of *Lecture Notes in Computer Science*, pages 149–169. Springer Verlag, September 1994.
[20] EP-ATR Project. Signal : a formal design environment for real time systems. In *6th International Joint Conference on Theory and Practice of Software Development, TAPSOFT '95*, volume 915 of *Lecture Notes in Computer Science*, Aarhus, Denmark, 1995. Springer-Verlag.
[21] Y. Sorel. Sorel: Real-time embedded image processing applications using the a3 methodology. In *Proc. IEEE International Conf. on Image Processing*, Lausanne, September 1996.
[22] Y. Sorel and C. Lavarenne. Syndex v4.2 user guide. `http://www-rocq.inria.fr/syndex/.articles/doc/doc/SynDEx42.html`

Reachability Analysis of (Timed) Petri Nets Using Real Arithmetic

Béatrice Bérard and Laurent Fribourg

LSV - ENS de Cachan/CNRS - 61 av. Pdt. Wilson - 94235 Cachan - France
email: {berard,fribourg}@lsv.ens-cachan.fr

Abstract. In this paper, we address the issue of reachability analysis for Petri nets, viewed as automata with counters. We show that exact reachability analysis can be achieved by treating Petri nets integer variables (counters) as real-valued variables, and using Fourier-Motzkin procedure instead of Presburger elimination procedure. As a consequence, one can safely analyse Petri nets with performant tools, e.g. HyTech, originally designed for analysing automata with real-valued variables (clocks). We also investigate the use of *meta-transitions* (iterative application of a transition in a single step) and give sufficient conditions ensuring an exact computation in this case. Experimental results with HyTech show an impressive speed-up with respect to previous experiences performed with a Presburger arithmetic solver. The method extends for analysing Petri nets with inhibitors and with timing constraints, but difficulties arise for the treatment of meta-transitions in the latter case.

1 Introduction

Reachability analysis with Presburger arithmetic. As surveyed by Esparza [18], there are several classes of Petri nets whose set of reachable markings can be described as a formula of Presburger arithmetic (i.e. a linear arithmetic formula over integers). In practice, it turns out that iterative symbolic computation of reachable markings, when it converges, often yields a set *post** expressible in Presburger arithmetic, even for Petri nets which do not belong to the classes mentioned above [20, 29]. Such a practical result was also observed in [9, 32] in the case of infinite-state systems with integers, called "extended finite state automata (EFSM)". These observations have led independently several researchers to implement procedures, based on Presburger arithmetic solvers, for trying and compute the reachability sets of extended forms of automata [9, 20, 32]. Note that there is generally no guarantee that the computation terminates. Moreover, the complexity of Presburger arithmetic is extremely high, a major obstacle for dealing with more than small-size examples. In this paper we will try to alleviate this complexity problem.

Can real arithmetic be used instead? It is well known that real arithmetic is more tractable than integer arithmetic: for instance, checking the existence of a solution over $\mathbb{R}$ for a system of linear inequations is polynomial while it is NP-hard over $\mathbb{N}$ [31]. In general resolution over $\mathbb{R}$ leads to an ***upper approximation*** of the set of integer solutions. However Shostack [33] noticed that, for a particular class of formulas (made of constraints of the form $x + c \leq y$, $x \leq c$ or $c \leq x$), elimination of variables over $\mathbb{R}$ using Fourier-Motzkin procedure is *exact*, i.e. produces the same result as elimination of variables over $\mathbb{N}$. For this class, it is safe to focus on "polyhedric" constraints (i.e. sets of linear inequalities), and discard the cumbersome divisibility constraints that are *a priori* needed when dealing with integers.

Contribution of the paper. Following Shostack, we show here that reachability analysis of Petri nets can be done by treating integer variables as real-valued variables, and eliminating them with Fourier-Motzkin. As a consequence, given a Petri net and an initial marking, iterative computation of the successor markings with Fourier-Motzkin yields the exact reachability set $post^*$, when it terminates. The same is true for backward computation of predecessors pre^*.
The idea of solving over reals rather than integers was recently used by Delzanno and Podelski in [17]. They deal with EFSM rather than Petri nets, and verify not only safety properties but also liveness properties. On the other hand, in contrast with us, they do not obtain in general an exact result for the reachability sets but only an overapproximation. Besides, we treat here the case of "(cycle) meta-transitions" (see [7, 35]). A meta-transition simulates the iteration of the same transition applied an arbitrary number μ of times, often entailing a better convergence for $post^*$ and pre^*. Fourier-Motzkin elimination (of μ) does not give any longer an exact result with meta-transitions, but only an upper approximation. We still give a sufficient condition on the intermediate formulas generated by the procedure, that guarantees an exact computation.
Our method extends to Petri nets with inhibitors and Petri nets with timing constraints although problems arise with meta-transitions in the latter case. We experimented the procedure on examples of various sizes, including a part of a communication protocol (PNCSA, [22]) in which a deadlock was discovered.

Plan of the paper. After recalling some basic definitions in §2, we give results about ordinary Petri nets in §3, about Petri nets with meta-transitions in §4, and about timed nets in §5; we conclude in §6.

2 Preliminaries

2.1 Fourier-Motzkin variable elimination

This presentation is adapted from [25]. An *atomic constraint* is a linear inequality of the form $\Sigma_{i=1}^{n} a_i x_i \geq c$ or equality $\Sigma_{i=1}^{n} a_i x_i = c$ with coefficients in $\mathbb{Z}$. A *constraint* is a conjunction of atomic constraints. We assume every atomic constraint is written in a simplified form so that no variable appears twice in the same atomic constraint. Let $vars(\phi)$ be the set of variables appearing in constraint ϕ. For each variable $x \in vars(\phi)$ we can partition ϕ in the following way. Let ϕ_x^0 be the set of atomic constraints of ϕ which do not contain any occurrence of x. Let ϕ_x^+ (resp ϕ_x^-) be the set of inequalities of ϕ equivalent to $ax \leq e$ (resp. $e \leq bx$), where e is a linear expression not involving x. The projection algorithm shown below eliminates a variable x from constraint ϕ and returns constraint $\psi \leftrightarrow \exists x \phi$ using Fourier-Motzkin elimination (and replacement of equals).

```
project (φ, x)
if φ ↔ φ' ∧ x = e then
% Replacement of equals
   ψ := φ' with every occurrence of x replaced by e
else
% Fourier-Motzkin elimination
   ψ := φ⁰ₓ
   foreach A ∈ φ⁺ₓ with A ↔ ax ≤ e⁺
            foreach B ∈ φ⁻ₓ with B ↔ e⁻ ≤ bx
                     ψ := ψ ∧ ae⁻ ≤ be⁺
            endforeach
   endforeach
endif
return ψ;
```

Formula ψ is only an upper approximation of the set of *integer* solutions because the algorithm assumes that there is always a solution for x to $ae^- \leq abx \leq be^+$, as soon as $ae^- \leq be^+$. This is true over $\mathbb{R}$ but not over $\mathbb{N}$. Note however that if $a = 1$ or $b = 1$, such an integer solution always exists. Therefore, in the particular case where, in ϕ, all the coefficients of x in lower bounds on x are 1, or all the coefficients of x in upper bounds on x are 1, Fourier-Motzkin is *exact* over $\mathbb{N}$: formula ψ characterizes the set of integer solutions (see [30]). This observation will be used in §4.
In the following, elimination will be done implicitly via a projection algorithm over $\mathbb{R}$, such as Fourier-Motzkin. Experimental results given throughout the paper were obtained using HyTech[1] [26], which has a built-in procedure of projection over reals (taken from Halbwachs' polyhedral manipulation library [24]).

2.2 Petri Nets

A *net* N is a tuple $\langle X, R, \alpha, \beta \rangle$, where
- $X = \{x_1, \ldots, x_m\}$ is a finite set of counters[2],
- R is a finite set of transitions,
- $\alpha : X \times R \mapsto \mathbb{N}$ and $\beta : R \times X \mapsto \mathbb{N}$ are valuation functions.

The value $\alpha(x_i, r)$ represents the weight of an ***input-arc*** from ***input-counter*** x_i to r, and $\beta(r, x_i)$ the weight of an ***output-arc*** from r to ***output-counter*** x_i. In the

[1] on a SUN station ULTRA-1 with 64 Megabytes of RAM memory
[2] The usual terminology is *places*, but it is convenient here to refer to *counters*.

figures, counters are drawn as circles, transitions as bars. An arc is represented as a weighted arrow, only if it has a non zero weight. The weights of the drawn arcs are implicitly 1. With each transition $r \in R$ in the net is associated a couple of vectors $(\overline{c}_r, \overline{a}_r)$, where the i-th component c_r^i of $\overline{c}_r$ is $\beta(r, x_i) - \alpha(x_i, r)$, and the i-th component a_r^i of $\overline{a}_r$ is $\alpha(x_i, r)$. A *marking* is a mapping from X to $\mathbb{N}$. A set of markings is represented as a formula $\varphi(\overline{x})$ interpreted over $\mathbb{N}$, where $\overline{x}$ is the vector of variables $x_1, \cdots, x_m$. A *Petri net* is a pair (N, φ_{init}) where N is a net, and φ_{init} a formula, called "initial formula", characterizing the set of initial markings. Formula φ_{init} is generally of the form $\overline{x} = \overline{x}_{init}$, with $\overline{x}_{init} \in \mathbb{N}^m$. We also consider cases where the initial marking is *parametric*: some counters are assigned parameters instead of constant values. Henceforth we will assume given a set $P = \{p_1, p_2, \dots\}$ of parameters which are additional variables, the value of which are left unchanged by transitions. A parametric initial formula $\varphi_{init}(\overline{x}, \overline{p})$ is of the form $\overline{x} = \overline{x}_{init}$ with $\overline{x}_{init} \in (\mathbb{N} \cup P)^m$ (and possibly some additional constraints over $\overline{p}$). For a transition $r \in R$ corresponding to $(\overline{c}_r, \overline{a}_r)$, the *reachability relation via* r, written $\xrightarrow{r}$, is defined by:

$$\overline{x} \xrightarrow{r} \overline{x}' \quad \Leftrightarrow \quad \overline{x}' = \overline{x} + \overline{c}_r \wedge \overline{x} \geq \overline{a}_r.$$

Given a net $N : \langle X, R, \alpha, \beta \rangle$, the *reachability relation via* N, written $\xrightarrow{N}$, is $\bigcup_{r \in R} \xrightarrow{r}$. (This means: $\overline{x} \xrightarrow{N} \overline{x}'$ iff $\exists r \in R \;\; \overline{x} \xrightarrow{r} \overline{x}'$.) Without loss of understanding in the following, we will use R and $\xrightarrow{R}$ in place of N and $\xrightarrow{N}$ respectively. For lack of space we do not explain how to treat Petri nets with *inhibitors* ("0-tests") in this paper, but all our definitions and results easily extend to them.

Example 1. The Petri net in figure 1 corresponds to the "swimming-pool" example from M. Latteux (see [8, 16]). All weights are equal to 1. Transition r_1 (incoming guest) is enabled if $x_6 \geq 1$, when there is at least one free cabine. When r_1 is fired, one token is removed from counter x_6 and added into counter x_1, meaning the beginning of undressing. The initial marking $\overline{x}_{init}$ involves two parameters p_1 and p_2 (assumed to be greater than or equal to 1), which represent the number of available cabins and baskets respectively. We have $\varphi_{init} : \; x_1 = x_2 = x_3 = x_4 = x_5 = 0 \; \wedge \; x_6 = p_1 \wedge x_7 = p_2 \wedge p_1 \geq 1 \wedge p_2 \geq 1$. The reachability relation via r_1 is: $\quad \overline{x} \xrightarrow{r_1} \overline{x}'$ iff
$x_1' = x_1 + 1 \wedge x_6' = x_6 - 1 \wedge x_6 \geq 1 \wedge x_2' = x_2 \wedge x_3' = x_3 \wedge x_4' = x_4 \wedge x_5' = x_5$.
Relations for transitions r_2 to r_6 are similar. For the sake of brevity, all the equations of the form $x_i' = x_i$ (counters unchanged), will be omitted in the next examples. Also recall that inequations of the form $x_i \geq 0$ are always implicit.

3 Reachability analysis in Petri nets

Given a formula φ and a transition r, the classical notion of successor (resp. predecessor), characterizing the markings reachable via r (resp. r^{-1}) from a marking satisfying φ, is defined as follows.

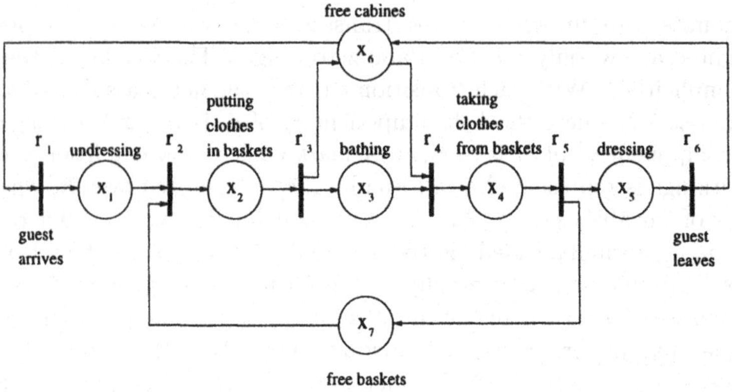

Fig. 1. A Petri net for the swimming pool example

The *successor of* φ *via* r, written $post_r(\varphi)$, is the result of eliminating $\overline{x}'$ over $\mathbb{N}$ from: $\exists \overline{x}' \; \varphi(\overline{x}') \wedge \overline{x}' \xrightarrow{r} \overline{x}$.
The *predecessor of* φ *via* r, written $pre_r(\varphi)$, is the result of eliminating $\overline{x}'$ over $\mathbb{N}$ from: $\exists \overline{x}' \; \varphi(\overline{x}') \wedge \overline{x} \xrightarrow{r} \overline{x}'$.
Since $\overline{x} \xrightarrow{r} \overline{x}'$ means $\overline{x}' = \overline{x} + \overline{c}_r \wedge \overline{x} \geq \overline{a}_r$, we have (by replacement of equals):

$$post_r(\varphi) \Leftrightarrow \varphi(\overline{x} - \overline{c}_r) \wedge \overline{x} \geq \overline{a}_r + \overline{c}_r.$$
$$pre_r(\varphi) \Leftrightarrow \varphi(\overline{x} + \overline{c}_r) \wedge \overline{x} \geq \overline{a}_r.$$

As usual, given a set R of transitions, $post_R(\varphi)$ is $\bigcup_{r \in R} post_r(\varphi)$ (the subscript R is sometimes omitted when it stands for the set of all transitions of the Petri net) and the iterated operator $post^j$ for a formula φ is defined recursively by: $post^0(\varphi) = \varphi$, $post^{j+1}(\varphi) = post(post^j(\varphi))$.
The infinite union $\bigcup_{j \geq 0} post^j(\varphi)$, written $post^*(\varphi)$, characterizes the set of all the markings reachable from φ through a finite sequence of transitions. We say that the iterative computation of $post^*$ *stabilizes* or *terminates at step* k when k is the first index such that $post^{k+1}(\varphi) \subseteq \bigcup_{0 \leq j \leq k} post^j(\varphi)$. In this case $post^*(\varphi) = \bigcup_{0 \leq j \leq k} post^j(\varphi)$. One defines similarly $pre^j(\varphi)$ and $pre^*(\varphi)$.

We are interested in (dis)proving that all the reachable markings of a Petri net satisfy a certain "safety" property, e.g. deadlock-freeness. This will be done by considering a formula φ_{safe}, which characterizes the set of all the safe markings, and checking: $post^*(\varphi_{init}) \cap \neg\varphi_{safe} = \emptyset$ or $pre^*(\neg\varphi_{safe}) \cap \varphi_{init} = \emptyset$.
In the examples hereafter, we will (dis)prove the absence of deadlock. The property $\varphi_{safe}(\overline{x})$, expressing that $\overline{x}$ enables at least one transition, is: $\bigvee_{r \in R} \overline{x} \geq \overline{a}_r$. Its negation $\neg\varphi_{safe}$, characterizing the set of all the deadlock markings, is of the form $\bigvee_{k \in K} \varphi_{dead_k}$ where K is a finite set of indices and φ_{dead_k} characterizes a subset of deadlock markings. There is a deadlock iff :
$post^*(\varphi_{init}) \cap \varphi_{dead_k} \neq \emptyset$ or $pre^*(\varphi_{dead_k}) \cap \varphi_{init} \neq \emptyset$, for some $k \in K$.

Definition 1. *A formula* $\varphi(\overline{x})$ *is a* counter region *iff it is a conjunction of inequalities of the form:*

$$x_i + d \geq x_j, \;\; x_i \geq x_j + d, \;\; x_i \geq d, \;\; d \geq x_i, \qquad \textit{with } x_i, x_j \in X \cup P, \, d \in \mathbb{N}.$$

We say that a formula $\varphi(\overline{x})$ belongs to counter class **C** *iff* φ *is a finite disjunction of counter regions.*

Proposition 1. *Let* φ *be a formula of* **C**. *Then:*

1. *Variable elimination with Fourier-Motzkin is exact for computing* $post_R(\varphi)$.
2. $post_R(\varphi)$ *belongs to* **C**.
3. *When it terminates, the iterative computation of* $post^j_R(\varphi)$ *(using Fourier-Motzkin) yields a formula* $post^*_R(\varphi)$ *belonging to* **C**.

Proof. Statement 1 follows from the fact that elimination of $\overline{x}'$ in $\exists \overline{x}' \varphi(\overline{x}') \wedge \overline{x}' \xrightarrow{r} \overline{x}$ is merely an operation of replacement of equals, and does not use the integer assumption for $\overline{x}'$. Besides the resulting formula $\varphi(\overline{x} - \overline{c}_r) \wedge \overline{x} \geq \overline{a}_r + \overline{c}_r$ belongs to class **C** since $\varphi \in$ **C**: this is because replacing x_i (resp. x_j) with $x_i - c^i_r$ (resp. $x_j - c^j_r$) in a constraint of the form $x_i \geq x_j + d$, $x_i + d \geq x_j$, $x_i \geq d$, or $d \geq x_i$ yields a constraint of the same form. Therefore statement 2 holds. Statement 3 follows from statements 1 and 2.

The proposition also holds with *pre* instead of *post*. Since φ_{init} belongs to **C**, it follows from the proposition that computation of *post** yields a formula of **C** when it terminates.

Example 2. This example is taken from [10] (cf. [36]) and illustrated by the Petri net of figure 2. It models an automated manufacturing system with four

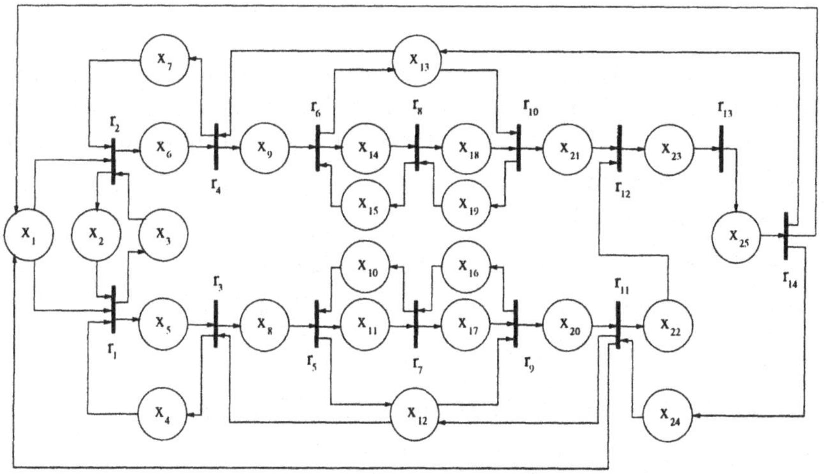

Fig. 2. A manufacturing system

machines, two robots, two buffers (x_{10} and x_{15}) and an assembly cell. The initial marking $\overline{x}_{init}$ is such that: $x_1 = p$ for some nonnegative parameter p, $x_2 = x_4 = x_7 = x_{12} = x_{13} = x_{16} = x_{19} = x_{24} = 1$, $x_{10} = x_{15} = 3$ (buffers have capacity 3). All other counters are empty (that is, all other variables are 0). The task is to discover for which values of parameter p the system may end up in a deadlock.
Previous works: In [36], deadlock-freeness is shown only for $1 \le p \le 4$. In [10], deadlock-freeness is proved using some mixed integer programming techniques for $1 \le p \le 8$. A path leading to a deadlock is then generated for $p = 9$. In [20] the exact reachability set has been automatically computed using variable elimination over $\mathbb{N}$, showing that a deadlock is reached from $\overline{x}_{init}$ if and only if $p \ge 9$, but the computation takes several hours (2.5 hours on a SUN station ULTRA-1).
New results: The computation of $post^*(\varphi_{init})$ with HYTECH stabilizes after 81 iterations, and takes only a couple of minutes (200 seconds). The output set $post^*$ is a disjunction of 560 counter regions. More precisely, each region is of the form $(p = x_1 + d_0 \wedge x_1 \ge d_1 \wedge \bigwedge_{i=2}^{25} x_i = d_i)$ or $(p = d_0 \wedge x_1 = d_1 \wedge \bigwedge_{i=2}^{25} x_i = d_i)$, with $d_i \in \{0, 1, \cdots, 12\}$ for all $0 \le i \le 25$. The intersection of $post^*$ with the formula characterizing deadlock markings gives $p = x_1 + 9 \ \wedge \ x_1 \ge 0 \wedge x_{11} = 3 \wedge x_{15} = 2 \wedge x_3 = x_4 = x_6 = x_8 = x_{14} = x_{17} = x_{18} = x_{21} = x_{24} = 1$, all the other variables being null. We thus retrieve the necessary and sufficient condition of deadlock-freeness ($p \le 8$). Note the impressive computation time speed-up gained here w.r.t. [20].

4 Petri Nets with Meta-Transitions

4.1 Meta-transitions

The computation of $post^*$ (or pre^*), performed in a brute manner as explained above, often converges slowly or does not terminate at all. A classical way to speed-up the convergence and improve the termination of $post^*$ is to use "(cycle) meta-transitions" ([7, 35]), i.e. repeated application of a transition r an arbitrary number μ of times in a single step. Let us define formally this notion in our context. Given $\mu \in \mathbb{N} \setminus \{0\}$, the *$\mu$-reachability* relation via transition $r : (\overline{c}_r, \overline{a}_r)$, written $\xrightarrow{r^\mu}$, is defined by:

$$\overline{x} \xrightarrow{r^\mu} \overline{x}' \ \Leftrightarrow \ \overline{x} \xrightarrow{r} \overline{x}' \qquad \text{if } \mu = 1,$$

$$\overline{x} \xrightarrow{r^\mu} \overline{x}' \ \Leftrightarrow \ \exists \overline{x}_1 \ \ \overline{x} \xrightarrow{r} \overline{x}_1 \xrightarrow{r^{\mu-1}} \overline{x}' \qquad \text{if } \mu \ge 2.$$

The *transitive closure* of $\xrightarrow{r}$, written $\xrightarrow{r^+}$, is defined by:

$$\overline{x} \xrightarrow{r^+} \overline{x}' \ \Leftrightarrow \ \exists \mu \in \mathbb{N} \setminus \{0\} \ \ \overline{x} \xrightarrow{r^\mu} \overline{x}'$$

Proposition 2. *[20] Given a transition $r : (\overline{c}_r, \overline{a}_r)$, we have:*

$$\overline{x} \xrightarrow{r^\mu} \overline{x}' \ \textit{ iff } \ \overline{x}' = \overline{x} + \mu.\overline{c}_r \ \wedge \overline{x} + (\mu - 1).\overline{c}'_r \ge \overline{a}_r,$$

where $\overline{c}'_r$ is obtained from $\overline{c}_r$ by replacing all the positive components with 0.

Proof. $\quad \overline{x} \xrightarrow{r^\mu} \overline{x}'$

$\Leftrightarrow \quad \exists \overline{x}_1, \overline{x}_2, \ldots, \overline{x}_{\mu-1} \quad \overline{x} \xrightarrow{r} \overline{x}_1 \wedge \overline{x}_1 \xrightarrow{r} \overline{x}_2 \wedge \cdots \wedge \overline{x}_{\mu-1} \xrightarrow{r} \overline{x}'$
$\Leftrightarrow \quad \overline{x}' = \overline{x} + \mu.\overline{c}_r \ \wedge \ (\overline{x} \geq \overline{a}_r \wedge \overline{x} + 1.\overline{c}_r \geq \overline{a}_r \wedge \cdots \wedge \overline{x} + (\mu - 1).\overline{c}_r \geq \overline{a}_r)$.
Conjunction $(\overline{x} \geq \overline{a}_r \wedge \overline{x} + 1.\overline{c}_r \geq \overline{a}_r \wedge \cdots \wedge \overline{x} + (\mu - 1).\overline{c}_r \geq \overline{a}_r)$ simplifies to $\overline{x} + (\mu - 1).\overline{c}'_r \geq \overline{a}_r$, with $\overline{c}'_r$ as defined above (see [20] for details).

Example 3. Consider transition r_1 of the swimming-pool example:
$\overline{x} \xrightarrow{r_1} \overline{x}'$ iff $x'_1 = x_1 + 1 \wedge x'_6 = x_6 - 1 \wedge x_6 \geq 1$.
From $\overline{c}_r = (1,0,0,0,0,-1,0)$, we obtain $\overline{c}'_r = (0,0,0,0,0,-1,0)$. Besides $\overline{a}_r$ is $(0,0,0,0,0,1,0)$, hence: $\overline{x} \xrightarrow{r_1^\mu} \overline{x}'$ iff $x'_1 = x_1 + \mu \wedge x'_6 = x_6 - \mu \wedge x_6 - (\mu - 1) \geq 1$, i.e. $x'_1 = x_1 + \mu \wedge x'_6 = x_6 - \mu \wedge x_6 \geq \mu$.

Definition 2. *The* μ-successor of φ via r, *denoted* $post_{r^\mu}(\varphi)$, *is such that:*
$$post_{r^\mu}(\varphi) \quad \Leftrightarrow \quad \exists \overline{x}' \ \varphi(\overline{x}') \wedge \overline{x}' \xrightarrow{r^\mu} \overline{x}.$$
The meta-successor of φ via r, *written* $Mpost_r(\varphi)$, *is such that*
$$Mpost_r(\varphi) \quad \Leftrightarrow \quad \exists \overline{x}' \ \varphi(\overline{x}') \wedge \overline{x}' \xrightarrow{r^+} \overline{x} \quad \Leftrightarrow \quad \exists \mu \in \mathbb{N} \setminus \{0\} \ \ post_{r^\mu}(\varphi).$$

It is straightforward to show (using proposition 2):

$$post_{r^\mu}(\varphi) \quad \Leftrightarrow \quad \varphi(\overline{x} - \mu.\overline{c}_r) \wedge \overline{x} + \mu(\overline{c}'_r - \overline{c}_r) \geq \overline{a}_r + \overline{c}'_r.$$

We define also: $post_{R^\mu}(\varphi) = \bigcup_{r \in R} post_{r^\mu}(\varphi)$, and $Mpost_R(\varphi) = \bigcup_{r \in R} Mpost_r(\varphi)$. Since only transitions of R are used in $Mpost_R$, we have: $\quad Mpost^*_R = post^*_R$.

If one regards μ not as a natural number but as a *real*, and eliminate μ via Fourier-Motzkin from $\exists \mu \geq 1 \ post_{r^\mu}(\varphi)$, one does not generally get an exact result, but only an upper approximation of $Mpost_r(\varphi)$. However, in the special case where all the coefficients of μ in $post_{r^\mu}(\varphi)$ for lower bounds of μ are 1, Fourier-Motzkin elimination is guaranteed to be exact (see §2), and yields a formula equivalent to $Mpost_r(\varphi)$. The same result holds if all the nonull coefficients of μ in upperbounds of μ are 1. This suggests to compute an upper approximation, denoted $Upost_R(\varphi)$, of $Mpost_R(\varphi)$ as follows.

Proposition 3. *Given a formula* φ, *let* $Upost_R(\varphi)$ *be the result of eliminating* μ *via Fourier-Motzkin from* $\exists \mu \geq 1 \ post_{R^\mu}(\varphi)$. *We have:*

$$Upost^*_R(\varphi) \supseteq Mpost^*_R(\varphi) = post^*_R(\varphi).$$

Besides, if iterative computation of $Upost^j_R(\varphi)$ *stabilizes at step* k, *and all the coefficients of* μ *on lower bounds (resp. upper bounds) of* μ *in* $post_{r^\mu}(Upost^j_R(\varphi))$ *are 1, for every* $0 \leq j \leq k$ *and every* $r \in R$, *then:*

$$Upost^*_R(\varphi) = Mpost^*_R(\varphi) = post^*_R(\varphi).$$

One defines similarly the notions $Mpre_r, Upre_r, Mpre_R, Upre_R$. The counterpart of the above proposition holds. The examples given at the end of the section show that iterative computation of $Upre_R$ is a practical and efficient way of computing an approximation of pre^*_R. Inspection of μ coefficients within intermediate formulas $pre_{r^\mu}(Upre^j_R(\varphi))$ will guarantee additionally that $Upre^*_R = pre^*_R$.

4.2 Fused transitions

In the rest of this section we consider, instead of the original set $R = \{r_1, \cdots, r_n\}$ of transitions of the given Petri net, a new set R' obtained from R by adding sequences of transitions, called *fused transitions*, of the form $r_{i_1} \cdots r_{i_k}$. Fused transitions can be derived in practice through a process of "decomposition" of R (see [21]), which uses a generalized form of fusion rules originally proposed by Berthelot [5] for transforming Petri nets.

The reachability relation via $r_{i_1} \cdots r_{i_k}$, written $\xrightarrow{r_{i_1} \cdots r_{i_k}}$, is defined by:

$$\overline{x} \xrightarrow{r_{i_1} \cdots r_{i_k}} \overline{x}' \text{ iff } \exists \overline{x}_1, \cdots, \overline{x}_{k-1} \quad \overline{x} \xrightarrow{r_{i_1}} \overline{x}_1 \xrightarrow{r_{i_2}} \cdots \xrightarrow{r_{i_{k-1}}} \overline{x}_{k-1} \xrightarrow{r_{i_k}} \overline{x}'.$$

Alternatively, we can consider a fused transition $\sigma = r_{i_1} \cdots r_{i_k}$ as a transition associated with a couple $(\overline{c}_\sigma, \overline{a}_\sigma)$. Writing ε for the empty sequence, we have:

Definition 3. *Given a fused transition* σ, *the* reachability relation via σ, *written* $\xrightarrow{\sigma}$, *is defined by* $\quad \overline{x} \xrightarrow{\sigma} \overline{x}' \;\Leftrightarrow\; \overline{x}' = \overline{x} + \overline{c}_\sigma \wedge \overline{x} \geq \overline{a}_\sigma$,
with $\overline{c}_\sigma$ *and* $\overline{a}_\sigma$ *recursively defined as:*

$$\begin{array}{llll} \overline{c}_\varepsilon = \overline{0} & \quad & \overline{c}_{r.\sigma} = \overline{c}_r + \overline{c}_\sigma \\ \overline{a}_\varepsilon = \overline{0} & & \overline{a}_{r.\sigma} = max(\overline{a}_r, \overline{a}_\sigma - \overline{c}_r) \end{array}$$

Since R' is obtained from R by adding sequences of transitions of R, we have: $post^*_{R'} = post^*_R$ and $pre^*_{R'} = pre^*_R$.

Example 4. Consider the transitions r_1, r_2 of the swimming-pool example:
$\overline{x} \xrightarrow{r_1} \overline{x}'$ iff $x'_1 = x_1 + 1 \wedge x'_6 = x_6 - 1 \wedge x_6 \geq 1$.
$\overline{x} \xrightarrow{r_2} \overline{x}'$ iff $x'_1 = x_1 - 1 \wedge x'_2 = x_2 + 1 \wedge x'_7 = x_7 - 1 \wedge x_1 \geq 1 \wedge x_7 \geq 1$.
The fused transition $r_1 r_2$ is characterized by $(\overline{c}_{r_1 r_2}, \overline{a}_{r_1 r_2})$ where $\overline{c}_{r_1 r_2}$ is $(0, 1, 0, 0, 0, -1, -1)$ and $\overline{a}_{r_1 r_2}$ is $(0, 0, 0, 0, 0, 1, 1)$. We have:
$\overline{x} \xrightarrow{r_1 r_2} \overline{x}'$ iff $x'_2 = x_2 + 1 \wedge x'_6 = x_6 - 1 \wedge x'_7 = x_7 - 1 \wedge x_6 \geq 1 \wedge x_7 \geq 1$.

4.3 Examples

In the following examples we consider an extended set R' obtained from R by adding fused transitions. We compute $Upre^*_{R'}$ and show that: $Upre^*_{R'} = Mpre^*_{R'}$ $(= pre^*_{R'} = pre^*_R)$. The set R' has been obtained by a preliminary decomposition of R (see [21]) up to level k, i.e. until k distinct fused transitions have been generated. The computation time of the decomposition process is neglectable w.r.t. computation time of $post^*$ (or pre^*). Note that the choice of the appropriate value k is the result of a compromise: the bigger k is, the better the convergence of $post^*$ is likely to be, but the space search may explode for too large values of k. In each of the examples below, we first had to try a couple of values for k before obtaining a convenient one. One can also improve the convergence by dropping certain "troublesome" fused transitions, but this issue is beyond the scope of the paper.[3]

[3] The decomposition process also infers an *ordering* of application among the set of ordinary and fused transitions, but this information is not exploited here.

Example 5. Consider the set of transitions $R = \{r_1, r_2, \cdots, r_6\}$ for the swimming pool Petri net. The Petri net derived by decomposition has a set R' with the following 6 fused transitions $\{r_2r_3,\ r_1r_2r_3,\ r_6r_4,\ r_4r_5,\ r_5r_6r_4,\ r_1r_2r_3r_4r_5\}$ added to R. The deadlock formula is of the form $\varphi_{dead_1} \vee \varphi_{dead_2}$ with:

$\varphi_{dead_1} : x_2 = x_4 = x_5 = x_6 = x_7 = 0, \quad \varphi_{dead_2} : x_1 = x_2 = x_4 = x_5 = x_6 = 0.$

The iterated computation of $Upre_{R'}$ applied to φ_{dead_1} terminates after 22 steps (and takes 50 seconds). Besides for all $j \leq 22$, one can check that all the nonnull coefficients of μ in $pre_{R'^{\mu}}(Upre^{j}_{R'}(\varphi_{dead_1}))$ are equal to 1. Therefore there is no upper approximation: $Upre^{*}_{R'}(\varphi_{dead_1}) = pre^{*}_{R}(\varphi_{dead_1})$. It is set of 10 constraints:

$\{x_2 = x_4 = x_5 = x_6 = x_7 = 0, \quad x_2 = x_4 = x_5 = x_7 = 0 \wedge x_6 \geq 1,$
$x_2 = x_4 = x_7 = 0 \wedge x_5 \geq 1, \quad x_2 \geq 1, \quad x_1 \geq 1 \wedge x_7 \geq 1, \quad x_6 \geq 1 \wedge x_7 \geq 1,$
$x_4 \geq 1, \quad x_3 \geq 1 \wedge x_6 \geq 1, \quad x_5 \geq 1 \wedge x_7 \geq 1, \quad x_3 \geq 1 \wedge x_5 \geq 1\}$

One of these constraints, $x_6 \geq 1 \wedge x_7 \geq 1$, has ϕ_{init} as an instance. It follows that the system always has a deadlock, for any values of parameters p_1 and p_2 (greater than or equal to 1). It is often interesting to try and keep track of a path linking the initial marking to a deadlock, but this requires to store additional information during the construction of $Upre^*$ at a cost generally prohibitive. In this simple example however, it is possible to retrieve such a path (see [4]). Note that, in [20], we were not able to treat this example (but only a simplified version) due to memory space saturation.

Example 6. This example is a modelization of a part of communication protocol PNCSA (Standard Protocol for Connection to the Authorization System [22]), borrowed from [19]. The Petri net has 38 transitions and 31 counters. The initial marking $\overline{x}_{init}$ has two components equal to 1, and all the other ones null. The deadlock formula is of the form: $\varphi_{dead_1} \vee \cdots \vee \varphi_{dead_4}$.

Previous works. In [19] the system was analyzed through the *minimal coverability graph*, an abstract form of the reachability set, which gives some information, e.g. about boundedness, but cannot certify the absence of deadlocks. We also tried to generate the reachability set using the Presburger based procedure of [20, 29], but the system ran out of memory.

New results. The extended Petri net R' is obtained by decomposing R up to level 20. The iterated computation of $Upre_{R'}$ applied to φ_{dead_1} terminates after 53 steps, and takes 15 minutes. $Upre^{*}_{R'}$ contains 376 constraints. Besides, for all $j \leq 53$, one can here again check that all the nonnull coefficients of μ in $pre_{R'^{\mu}}(Upre^{j}_{R'})$ appearing in lower bounds on μ are equal to 1. Therefore $Upre^{*}_{R'} = pre^{*}_{R'} = pre^{*}_{R}$. One constraint of $Upre^{*}_{R'}$ has ϕ_{init} as an instance, which proves that the system has a deadlock. It is here also possible to exhibit a path linking $\overline{x}_{init}$ to a deadlock marking (see [4]).

Note that the existence of this deadlock was unknown so far. There are several possible explanations to this deadlock ("local" but non "global" deadlock, flaws in the modelling, ...) but such an issue is beyond the scope of the paper.

5 Timed Nets

In this section, we consider "timed nets", obtained by adding to timed automata a finite set X of places (or counters) [23]. This allows us to combine known results on timed automata with those of §3 on Petri nets. Note that many other models have been proposed to build timed extensions of Petri nets (see e.g. [28]), but we do not consider them here. We first recall the definitions of timed automata, then treat the case of timed automata with counters (timed nets), with a short discussion on the issue of meta-transitions.

5.1 Timed automata

We consider here Alur-Dill's timed automata [2, 1]. As finite state automata, timed automata have a finite set of locations. In addition, they use a finite set Y of variables $y_1, \cdots, y_n$, called *clocks*, which evolve continuously at the same rate. A transition is enabled only if some relation, called *guard*, is satisfied by the current values of the clocks. When a transition is fired, some clocks are reset to fixed values. Moreover, inside each location, the clock values are required to satisfy a relation called *invariant.* Guards and invariants are conjunctions of constraints of the form $y_i \ll a_i$ or $a_i \ll y_i$ with $\ll \in \{<, \leq\}$. Formally, a timed automaton T is a tuple $\langle L, \ell_{init}, Y, I, E\rangle$, where

- L is a finite set of locations,
- $\ell_{init} \in L$ is the *initial location,*
- $Y = \{y_1, \cdots, y_n\}$ is a finite set of clocks ,
- I is a mapping that labels each location ℓ in L with some location invariant, simply written I_ℓ instead of $I(\ell)$ in the following,
- E is a set of *action transitions* of the form e: $\langle(\ell, \ell'), \psi, \rho\rangle$ where ℓ and ℓ' belong to L, ψ is a guard and ρ is a mapping over Y such that, for all $y \in Y$, either $\rho(y) = y$ or $\rho(y) = 0$.

A *state* is a mapping from $\{s\} \times Y$ to $L \times \mathbb{R}^+$ where s denotes a location variable. A set of states will be represented by a formula $\phi(s, \overline{y})$ where $\overline{y}$ is vector $y_1 \cdots y_n$ interpreted as (nonnegative) real-valued variables. The semantics of timed automata are defined by initial formula and reachability relations:
- the *initial formula* $\phi_{init}(s, \overline{y})$ is: $s = \ell_{init} \wedge \overline{y} = 0$.
- the *action reachability relation* via e : $\langle(\ell, \ell'), \psi, \rho\rangle$, written $\xrightarrow{e}$, is such that:

$$(s, \overline{y}) \xrightarrow{e} (s', \overline{y}') \quad \text{iff} \quad s = \ell \wedge s' = \ell' \wedge\ I_s(\overline{y}) \wedge I_{s'}(\overline{y}') \wedge\ \psi(\overline{y})\ \wedge \overline{y}' = \rho(\overline{y}).$$

- the *delay reachability relation,* written $\xrightarrow{\delta}$, is such that:

$$(s, \overline{y}) \xrightarrow{\delta} (s, \overline{y}') \quad \text{iff} \quad \exists \varepsilon \in \mathbb{R}^+ \ \ (I_s(\overline{y}) \wedge I_s(\overline{y}') \wedge \overline{y}' = \overline{y} + \varepsilon).$$

- the *reachability relation via automaton* T, written $\xrightarrow{T}$, is $\bigcup_{e \in E}(\xrightarrow{\delta} \ . \ \xrightarrow{e} \ . \ \xrightarrow{\delta})$

We also recall the notion of "clock region" [1].

Definition 4. *A formula $\phi(\overline{y})$ is a* clock region *iff it is a conjunction of inequalities of the form:* $\quad y_i + d \ll y_j, \quad y_i \ll y_j + d, \quad d \ll y_i, \quad y_i \ll d$ *with $y_i, y_j \in Y$, $d \in \mathbb{N}$ and $\ll \in \{<, \leq\}$.*

5.2 Timed automata with counters

We now show that proposition 3 also holds for timed automata with counters. For the sake of conciseness we give only results concerning the set *pre* of predecessors, but similar results also hold for *post*. An (extended) action transition f does not go any longer from s to s' over L, but from a pair $(\overline{x}, s)$ to a pair $(\overline{x}', s')$ over $\mathbb{N}^m \times L$ such that: $\overline{x}' = \overline{x} + \overline{c} \wedge \overline{x} \geq \overline{a}$, for a vector $\overline{c}$ of integers and a vector $\overline{a}$ of natural numbers associated with f. A *state* is now a mapping from $X \times \{s\} \times Y$ to $\mathbb{N} \times L \times \mathbb{R}^+$ and a set of states will be represented by a formula $\Phi(\overline{x}, s, \overline{y})$ where $\overline{x}$ is interpreted over $\mathbb{N}$ and $\overline{y}$ over $\mathbb{R}^+$.
A *timed net* U is a tuple $\langle X, \varphi_{init}, L, \ell_{init}, Y, I, F\rangle$, where X is the set of counters, φ_{init} the initial formula for counters, L the set of locations, ℓ_{init} the initial location, I the invariant mapping, Y the set of clocks and F the set of transitions.
We define:
- the *initial formula* $\Phi_{init}(\overline{x}, s, \overline{y})$: $\varphi_{init}(\overline{x}) \wedge s = \ell_{init} \wedge \overline{y} = 0$
- the *extended action reachability relation* via f: $\langle(\overline{c}, \overline{a}), (\ell, \ell'), \psi, \rho\rangle$, written $\xrightarrow{f}$:
$$(\overline{x}, s, \overline{y}) \xrightarrow{f} (\overline{x}', s', \overline{y}') \text{ iff}$$
$$\overline{x}' = \overline{x} + \overline{c} \wedge \overline{x} \geq \overline{a} \;\; \wedge \;\; s = \ell \wedge s' = \ell' \wedge I_s(\overline{y}) \wedge I_{s'}(\overline{y}') \wedge \psi(\overline{y}) \wedge \overline{y}' = \rho(\overline{y})$$
- the *delay reachability relation* $\xrightarrow{\delta}$:
$$(\overline{x}, s, \overline{y}) \xrightarrow{\delta} (\overline{x}, s, \overline{y}') \text{ iff } \exists \varepsilon \in \mathbb{R}^+ \; (I_s(\overline{y}) \wedge I_s(\overline{y}') \wedge \overline{y}' = \overline{y} + \varepsilon).$$
- the *reachability relation via* U, written $\xrightarrow{U}$: $\bigcup_{f \in F}(\xrightarrow{\delta} . \xrightarrow{f} . \xrightarrow{\delta})$.

Given a formula Φ and a timed net U, we define $pre_U(\Phi)$ as the result of eliminating s' over L, $\overline{x}'$ over $\mathbb{N}$ and $\overline{y}'$ over $\mathbb{R}$ from
$$\exists \overline{x}', s', \overline{y}' \;\; ((\overline{x}, s, \overline{y}) \xrightarrow{U} (\overline{x}', s', \overline{y}') \;\; \wedge \;\; \Phi(\overline{x}', s', \overline{y}')).$$

We now extend the notions of counter regions and counter class **C** as follows:

Definition 5. *A formula* $\Phi(\overline{x}, \overline{y})$ *is a* mixed region *iff it is a conjunction of inequalities of the form:*
$$x_i + d \geq x_j, \quad x_i \geq x_j + d, \quad x_i \geq d, \quad d \geq x_i,$$
$$y_i + d \ll y_j, \quad y_i \ll y_j + d, \quad d \ll y_i, \quad y_i \ll d$$
with $x_i, x_j \in X \cup P$, $y_i, y_j \in Y$, $d \in \mathbb{N}$ *and* $\ll \in \{<, \leq\}$.

Definition 6. *A formula* $\Phi(\overline{x}, s, \overline{y})$ belongs to mixed class **D** *iff it is a disjunction of the form* $\bigvee_{k \in K} s = \ell_k \wedge \Phi_k(\overline{x}, \overline{y})$, *where* K *denotes a finite set of indices,* ℓ_k *is in* L, *and* $\Phi_k(\overline{x}, \overline{y})$ *is a mixed region.*

Proposition 4. *Let* Φ *be a formula of* **D** *and* U *a timed net. Then:*

1. *Variable elimination with Fourier-Motzkin is exact for computing* $pre_U(\Phi)$.
2. $pre_U(\Phi)$ *belongs to* **D**.
3. *When it terminates, the iterative computation of* $pre_U^j(\Phi)$ *(using Fourier-Motzkin) yields a formula* $pre_U^*(\Phi)$ *belonging to* **D**.

Proof. (sketch) By proposition 3, statements 1 and 2 hold in the special case of an ordinary Petri net ($Y = \emptyset$). It easy to see that they hold also in the special case of an ordinary timed automata ($X = \emptyset$), using the closure property of "clock regions" [1].[4] The truth of statements 1 and 2 for a general timed net U, follows from the combination of these special cases, using the fact that, when computing pre_U, there is no interaction between variables of X and variables of Y in Φ. Statement 3 then follows from statements 1 and 2.

The proposition above allows us to consider all the variables as if they were real ones (including variables of X) and to use Fourier-Motzkin elimination for computing predecessors in class **D**.

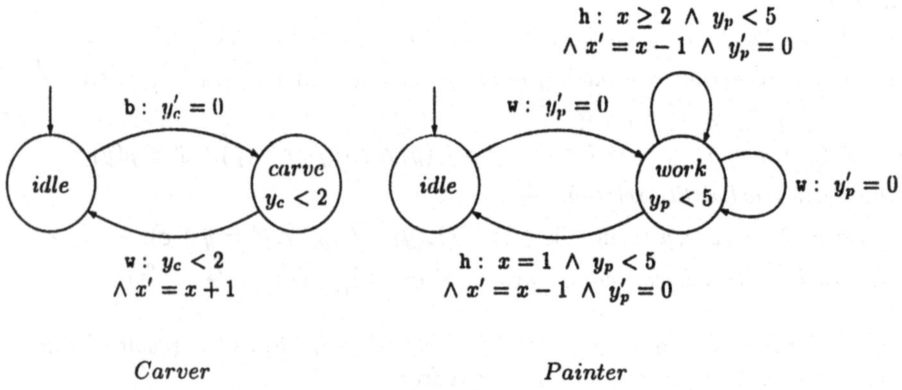

Fig. 3. Horse Manufactory

Example 7. This example, taken from [23], is a simple model of a "Horse Manufactory", where wooden black horses are produced by a carver and a painter. The system is represented in figure 3 by two timed nets, for the carver and the painter respectively. The carver has two locations *idle*, *carve* and one clock y_c, while the painter has two locations *idle* and *work* and the clock y_p. The processes communicate through the transition labeled w, which corresponds to the production of a wooden horse by the carver and its reception by the painter. This transition operates on the place x, which counts the number of carved horses that are not yet painted. The carver begins his work upon arrival of a block of wood, corresponding to the transition from *idle* to *carve*, with label b, which resets the carver clock y_c. From this point on, carving must be finished in at most 2 time units: invariant $y_c < 2$ is associated with the location *carve* and the transition labeled w increments the counter x and returns to location *idle*. With the same transition w, the painter first stores the piece. Then, transition h represents the production by the painter of one black horse, which decrements x

[4] Statement 1 is trivial in this case because all the variables belong to Y, and are real.

and must occur in at most 5 time units, unless a new wooden horse is produced. Thus, both transitions w and h reset the painter clock y_p.

We consider properties relative to the maximal delay between arrival of a block of wood and production of a painted horse. The strong property (S) simply requires that this delay be equal to 7: each time an action b occurs, there is an action h within a delay strictly less than 7 time units. Property (S) is expressed by an "observer automaton" and proved *false* (in the system composed of *Carver*, *Painter* and *Observer*) in seven iterations. The weak version (W) of this property expresses as follows: if b occurs and is not followed by another action b for 7 time units, then an action h occurs within this delay. Property (W) is proved *true* after three iterations (see [4] for details).

5.3 Meta-transitions

The notions of fused and meta-transitions transitions, as defined above in the context of ordinary Petri nets, do not easily extend to timed nets. The problems already arise with timed automata (without counters). One difficulty stems from the fact that the application of two action transitions e_1 and e_2 interleaved with delay-transitions ($\xrightarrow{\delta} . \xrightarrow{e_1} . \xrightarrow{\delta} . \xrightarrow{e_2} . \xrightarrow{\delta}$) cannot be reduced to the application of some "fused" action-transition, say e_3, interleaved with delay-transitions ($\xrightarrow{\delta} . \xrightarrow{e_3} . \xrightarrow{\delta}$), so the situation is radically different from the one prevailing in the case of ordinary Petri nets.[5] Still the finite application of a sequence of action transitions between two states $\overline{y}$ and $\overline{y}'$ defines a linear arithmetic relation, and the transitive closure of such a relation (meta-transition) is still definable in the additive theory of $\mathbb{R}$, as shown by Comon and Jurski ([12], p.274; see also [13]). However the complexity of such a relation is untractable (double exponential in the number of clocks in the worst case). A possible manner to overcome this problem might be to approximate this transitive closure relation using, e.g., methods proposed in [27], but this would deserve further study.

6 Final Remarks

We have given a theoretical justification for using tools designed for the analysis of automata with real-valued variables in order to analyse automata with integer-valued variables such as Petri nets. The constraint solving procedure was very easy to implement in HyTech, as well as the encoding of meta-transitions. Such a task could also have been achieved with a constraint programming language such as CLP($\mathbb{R}$) (as done in [17] in the context of EFSM). Other real-time analysis tools like Uppaal [3] might also be used when no parameter is involved (although encoding meta-transitions may be a problem). To our knowledge, the idea of using variable elimination over $\mathbb{R}$ instead of $\mathbb{N}$ for computing the set *post**

[5] In [34] there is an attempt to extend fusion rules à la Berthelot [5] within another model of time Petri nets [6], but with severe conditions for rule application.

of reachable markings for Petri nets, had not been exploited so far.[6] Experiences show an impressive gain of performance: all the examples from [29] have been reproduced with a speed-up from 10 to 50 times, a deadlock has been found for the first time in the PNCSA protocol (as modeled in [19]), ... The use of real arithmetic with meta-transitions may yield an upper approximation of the computed reachability set. Such an approximation is the price to pay for improving the convergence of *post**. Using real arithmetic can be seen here as an alternative to classical approximation methods as those used in Abstract Interpretation (e.g., "widening" or "convex hull") [14, 15, 24]. Finally our method extends to Petri nets with inhibitors and to timed nets. However difficulties arise when dealing with fused and meta-transitions for timed nets, and applicability of the approach in this context should be further studied on significant examples.

References

1. R. Alur, C. Courcoubetis and D. Dill. "Model-Checking in Dense Real-Time". *Information and Computation 104:1*, 1993, pp. 2–34.
2. R. Alur and D. Dill. "Automata for Modeling Real-Time Systems". *ICALP'90*, LNCS 443, Springer-Verlag, 1990, pp. 322–335.
3. J. Bengtsson, K.G. Larsen, F. Larsson, P. Pettersson and W. Yi. "UPPAAL – a Tool Suite for Automatic Verification of Real-Time Systems". *Hybrid Systems III*, LNCS 1066, Springer-Verlag, 1996, pp. 232-243.
4. B. Bérard and L. Fribourg. "Reachability Analysis of (Timed) Petri Nets Using Real Arithmetic". *Technical Report LSV-99-3*, CNRS & Ecole Normale Supérieure de Cachan, March 1999 (http://www.lsv.ens-cachan.fr/Publis/).
5. G. Berthelot. "Transformations and Decompositions of Nets". *Advances in Petri Nets*, LNCS 254, Springer-Verlag, 1986, pp. 359–376.
6. B. Berthomieu, M. Diaz. "Modeling and verification of time dependent systems using time Petri nets". *IEEE Trans. Softw. Eng.* 17:3, 1991, pp. 259–273.
7. B. Boigelot and P. Wolper. "Symbolic Verification with Periodic Sets". *CAV'94*, LNCS 818, Springer-Verlag, 1994, pp. 55–67.
8. G.W. Brams. *Réseaux de Petri: Théorie et Pratique.* Masson, Paris, 1983.
9. T. Bultan, R. Gerber and W. Pugh. "Symbolic Model Checking of Infinite State Systems Using Presburger Arithmetic". *CAV'97*, LNCS 1254, 1997, pp. 400–411.
10. F. Chu and X. Xie. "Deadlock Analysis of Petri Nets Using Siphons and Mathematical Programming". To appear in *IEEE Trans. on Robotics and Automation.*
11. J.M. Colom and M. Silva. "Convex geometry and semiflows in P/T nets. A comparative study of algorithms for computation of minimal p-semiflows". *Advances in Petri Nets*, LNCS 483, Springer-Verlag, 1990, pp.79–112.
12. H. Comon and Y. Jurski. "Multiple Counters Automata, Safety Analysis and Presburger Arithmetic". *CAV'98*, LNCS 1427, Springer-Verlag, 1998, pp. 268–279.
13. H. Comon and Y. Jurski. "Timed automata and the theory of real numbers". *CAV'99*, LNCS, Springer-Verlag, 1999. (This volume)
14. P. Cousot and R. Cousot. "Abstract Interpretation: a unified lattice model for static analysis of programs by construction or approximation of fixpoints". *POPL'77*, ACM Press, 1977, pp. 238–252.

[6] In [11], Fourier-Motzkin elimination is used for computing "p-semiflows" of Petri nets, but p-semiflows are only rough approximations of *post**.

15. P. Cousot and N. Halbwachs. "Automatic Discovery of Linear Restraints Among Variables of A Program". *POPL'78*, ACM Press, 1978, pp. 84–97.
16. R. David and H. Alla. *Du Grafcet aux Réseaux de Petri*, Hermès, Paris, 1989.
17. Delzanno and A. Podelski. "Model-Checking in CLP". *TACAS'99*, LNCS 1579, Springer-Verlag, 1999, pp. 223–239.
18. J. Esparza and M. Nielsen. "Decidability Issues for Petri Nets – a Survey". *EATCS Bulletin* 52, 1994.
19. A. Finkel. "The Minimal Coverability Graph for Petri Nets". *Advances in Petri Nets*. LNCS 674, Springer-Verlag, 1993, pp. 211–243.
20. L. Fribourg and H. Olsén. "Proving safety Properties of Infinite State Systems by Compilation into Presburger Arithmetic". *CONCUR'97*, LNCS 1243, Springer-Verlag, 1997, pp. 213–227.
21. L. Fribourg and H. Olsén. "A Decompositional Approach for Computing Least Fixed-Points of Datalog Programs with $\mathbb{Z}$-counters". *Constraints: An International Journal*, 2, 1997, pp. 305–335.
22. GIE CB. "Protocole normalisé de connexion au système d'autorisation: spécifications PNCSA Version 2". Document C *Communication OSI*, 1988.
23. R. Gorrieri and G. Silipandri. "Real-Time System Verification using P/T Nets". *CAV'94*, LNCS 818, Springer-Verlag, 1994, pp. 14–26.
24. N. Halbwachs. "Delay Analysis in Synchronous Programs". *CAV'93*, LNCS 697, Springer-Verlag, 1993, pp.333–346.
25. W. Harvey, P.J. Stuckey and A. Borning. "Compiling Constraint Solving Using Projection". *CP'97*, LNCS 1330, Springer-Verlag, 1997, pp. 491–505.
26. T. Henzinger, P.-H. Ho and H. Wong-Toi. "A user Guide to HyTech". *TACAS'95*, LNCS 1019, Springer-Verlag, 1995, pp. 41–71.
27. W. Kelly, W. Pugh, E. Rosser and T. Shpeisman. *Transitive closure of infinite graphs and its applications*. Technical Report CS-TR-3457, UMIACS-TR-95-48, University of Maryland, 1994.
28. P. Merlin and D.J. Farber. "Recoverability of communication protocols". *IEEE Trans. on Communications* 24:9, Sept. 1976.
29. H. Olsén. *Automatic Verification of Petri Nets in a CLP framework*. Ph.D Thesis, University of Linköping, Sweden, 1997.
30. W. Pugh. "A practical integer algorithm for exact array dependence analysis". *C.ACM* 35:8, 1992, pp. 102–114.
31. A. Schrijver, *Theory of Linear and Integer Programming*, Wiley, 1986.
32. T.R. Shiple, J.H. Kukula and R.K. Ranjan. "A Comparison of Presburger Engines for EFSM Reachability". *CAV'98*, LNCS 1427, Springer-Verlag, 1998, pp. 280–292.
33. R. Shostack. "Deciding linear inequalities by computing loop residues". *J. ACM* 28:4, 1981, pp. 769–779.
34. R. Sloan and U. Buy. "Reduction rules for time Petri nets". *Acta Informatica* 33, 1996, pp. 687–706.
35. P. Wolper and B. Boigelot. "Verifying Systems with Infinite but Regular State Spaces". *CAV'98*, LNCS 1427, Springer-Verlag, 1998, pp. 88–97.
36. M.C. Zhou, F. Dicesare and A.A. Desrochers. "A Hybrid Methodology for Synthesis of Petri Net Models for Manufacturing Systems". *IEEE Trans. on Robotics and Automation* 8:3, 1993, pp. 350–361.

Weak and Strong Composition of High-Level Petri Nets

Eike Best and Alexander Lavrov

Carl von Ossietzky Universität, Fachbereich 10, D-26111 Oldenburg
{eike.best, alexander.lavrov}@informatik.uni-oldenburg.de

Abstract. We propose generic schemes for basic composition operations (sequential composition, choice, iteration, and refinement) for high-level Petri nets. They tolerate liberal combinations of place types (equal, disjoint, intersecting) and allow for weak and strong versions of compositions (owing to a parameterised scheme of type construction). Important properties, including associativity, commutativity, and coherence w.r.t. unfolding, are preserved.

1 Introduction

The lack of compositionality, attributed to Petri net models, has inspired intensive research aimed at overcoming this disadvantage. Part of the efforts is concerned with introducing algebraic features into net formalisms [2, 10]. The operators defined on place-transition nets usually share intuitively well-founded, uniform schemes of node connection (as a rule, based on cross-products of participating frontier node sets) [2], while the diversity of high-level models induces an even more noticeable multitude of corresponding operator schemes. Usually these schemes either involve complex type formation constructs [6], or are accompanied by substantial restrictions on structural and/or notational characteristics of the interface nodes [5, 7], or are not concerned with algebraic treatment [9]. Such schemes usually represent polar views of the dichotomy 'strong vs. weak (layered [8]) composition' (the most frequent being the 'strong' case).

The work presented in this paper aims at systematically developing a set of place-based composition operations for high level Petri nets, which:

(1) impose as few as possible restrictions on the types of places (allowing, in particular, for equal, disjoint, and intersecting types);
(2) allow a wider interpretation of the interface place assignment (permitting, in particular, side conditions next to refined transitions, as well as combined entry/exit-places of a net);
(3) possess a well justified and uniform intuition regarding the causality aspects;
(4) are able to represent weak and strong composition aspects simultaneously;
(5) are still tractable in algebraic contexts, preserve reasonable properties such as commutativity, associativity, and are coherent w.r.t. net unfolding.

The above goals are reached in the following way. A liberal place typing (goal (1)) and an integrated treatment of weak and strong compositionality (goal (4)) become possible thanks to the proposed universal parameterised type construction scheme (section 3). This scheme is incorporated in auxiliary net-based operations (section 4) that serve as a repertoire of the elementary building blocks of which the main operators are composed (sections 6-8). These operations are defined in such a manner (section 5) that a number of desirable algebraic (goal (5)), causality (goal (3)), and interface-related (goal (2)) properties are obtained. As an auxiliary uniform and consistent means for expressing and proving properties of the operations, an 'M-nets with a basis' kit is developed (section 4).

Proofs of the results claimed in this paper can be found in [4].

2 Basic notions

$\mathbb{N}$ is the set of natural numbers. A multiset μ over a set U is a function $\mu: U \to \mathbb{N}$; $u \in \mu$ iff $\mu(u) > 0$, and μ is a set iff $\forall u \in U: \mu(u) \leq 1$. The support of a multiset μ is defined as the set $sp(\mu) = \{u \mid u \in \mu\}$, and μ is finite if its support is finite. The set of all finite multisets over U is denoted by $\mathcal{M}_f(U)$. Moreover:
$\mu_1 \leq \mu_2$ iff $\forall u \in U: \mu_1(u) \leq \mu_2(u)$ (this is $\subseteq$ if both μ_1 and μ_2 are sets),
$\mu_1 + \mu_2$ is a multiset such that $\forall u \in U: (\mu_1 + \mu_2)(u) = \mu_1(u) + \mu_2(u)$,
$\mu_1 - \mu_2$ is a multiset such that $\forall u \in U: (\mu_1 - \mu_2)(u) = \max\{0, \mu_1(u) - \mu_2(u)\}$,
$\mu_1 \setminus \mu_2$ is a multiset such that $\forall u \in U: (\mu_1 \setminus \mu_2)(u) = \begin{cases} \mu_1(u) & \text{if } \mu_2(u)=0 \\ 0 & \text{otherwise,} \end{cases}$
$\mu_1 \cup \mu_2$ is a multiset such that $\forall u \in U: (\mu_1 \cup \mu_2)(u) = \max\{\mu_1(u), \mu_2(u)\}$,
$\mu_1 \cap \mu_2$ is a multiset such that $\forall u \in U: (\mu_1 \cap \mu_2)(u) = \min\{\mu_1(u), \mu_2(u)\}$.

When using the generalised union $\bigcup$ (and similarly $\bigcap$), we consider this as a unary operator on a set of arguments. Thus, $\bigcup\{\mu_i \mid i \in \mathbb{N}\}$ is tantamount to $\mu_0 \cup \mu_1 \cup \mu_2 \cup \ldots$, and $\bigcap\{\mu_1, \mu_2\}$ equals $\mu_1 \cap \mu_2$. For the sum, we use the traditional notation; thus, for instance, $\sum_{i \in I} \mu_i$ equals (in the same convention as for union and intersection) $\sum\{\mu_i \mid i \in I\}$. Restriction of a function f to the domain U is denoted by $f|_U$.

The high-level net model we use is close to that from the standard [1] and represents a relation-oriented version of M-nets [3]. Its definition (given next) is parameterised by pairwise disjoint sets of values (VAL), modes (MOD), and actions ($\mathcal{A}$). However, MOD and $\mathcal{A}$ play very minor roles in the subsequent parts of the paper, and their involvement in the definition is mainly for the sake of completeness. These sets are, therefore, often omitted in examples.

Definition 1. **(M-net)** *An M-net is a triple (S, T, ι), where S is a set of places, T is a set of transitions, with $S \cap T = \emptyset$, and ι is an inscription function with domain $S \cup (S \times T) \cup (T \times S) \cup T$ such that:*

- $\forall s \in S$: $\iota(s) = (\lambda(s) \mid \alpha(s))$, *where* $\lambda(s) \in \{\{e\}, \{x\}, \{e, x\}, \{i\}\}$ *is called the status of* s, *and* $\alpha(s) \subseteq VAL$ *is the type of* s;
- $\forall t \in T$: $\iota(t) = (\lambda(t) \mid \alpha(t))$, *where* $\lambda(t) \subseteq \mathcal{A}$ *is called the label of* t, *and* $\alpha(t) \subseteq MOD$ *is the set of modes (or type) of* t;
- $\forall (s,t) \in S \times T$: $\iota(s,t) \in \mathcal{M}_f(\{(v,m) \mid v \in \alpha(s) \wedge m \in \alpha(t)\})$;
- $\forall (t,s) \in T \times S$: $\iota(t,s) \in \mathcal{M}_f(\{(m,v) \mid v \in \alpha(s) \wedge m \in \alpha(t)\})$. ■ 1

Nets $N_1 = (S_1, T_1, \iota_1)$ and $N_2 = (S_2, T_2, \iota_2)$ are *disjoint* if S_1, S_2, T_1, T_2 (and hence, ι_1 and ι_2) are mutually disjoint. The notion of a type $\alpha(.)$ is extended to the sets of nodes: if $Y \subseteq S$ or $Y \subseteq T$ then $\alpha(Y) = \bigcup\{\alpha(y) \mid y \in Y\}$.

Given a net N, the following interface-related notations will be used: entry places ${}^\bullet N = \{s \mid s \in S \wedge e \in \lambda(s)\}$, exit places $N^\bullet = \{s \mid s \in S \wedge x \in \lambda(s)\}$, combined entry-exit places $S_N^{\{e,x\}} = \{s \mid s \in S \wedge \lambda(s) = \{e, x\}\}$ ($= {}^\bullet N \cap N^\bullet$), pure entry places ${}^\circ N = {}^\bullet N \setminus N^\bullet$, pure exit places $N^\circ = N^\bullet \setminus {}^\bullet N$, interface places ${}^\bullet N \cup N^\bullet$. The other places with status $\{i\}$ are called internal. In contrast with [2], we allow input arcs into entry places, output arcs from exit places, and places that are both entry and exit (the latter make sense for unguarded loop operators).

Given a net N and a node $y \in S \cup T$, we denote by ${}^\bullet y$ and $y^\bullet$ the sets of entry and exit nodes of y, respectively: ${}^\bullet y = \{y' \mid y' \in S \cup T \wedge \iota(y', y) \neq \emptyset\}$, $y^\bullet = \{y' \mid y' \in S \cup T \wedge \iota(y, y') \neq \emptyset\}$, also, ${}^\circ y = {}^\bullet y \setminus y^\bullet$ and $y^\circ = y^\bullet \setminus {}^\bullet y$. In addition, for $t \in T$ we denote by $S_t^{(side)}$ the set of side places of t ($S_t^{(side)} = {}^\bullet t \cap t^\bullet$).

We will assume that *VAL* is the powerset of a plain set Ω_{BAS} of symbols, i.e., every value is a set of basic symbols. The rationale for this will by explained later. For simplicity, we will identify Ω_{BAS} with $\mathbb{N}$. We will use v, w as generic names for values from *VAL*. Examples of values: $v = \{1\}$, $w = \{0, 7, 8\}$, ... As generic names for value sets (i.e. for types) we will use ξ, β, φ.

The unfolding of an M-net N is obtained by disassembling N in such a way that every place in the resulting net gets a singleton type, and every transition exactly one mode. The behaviour of N ought to be the same as the behaviour of its unfolding; hence the importance of the preservation of unfolding-related equivalences in composition operations. The places and the transitions of the unfolding are given names s^v and t^m, to remind of the values and modes they arise from. The unfolding is unique up to the choice of these names (definition 3).

Definition 2. (M-net unfolding) *Let* $N = (S, T, \iota)$ *be an M-net.*
$\mathcal{U}(N) = (\mathcal{U}(S), \mathcal{U}(T), \iota_{\mathcal{U}})$, *where*
- $\mathcal{U}(S) = \{s^v \mid s \in S,\ v \in \alpha(s)\}$, $\mathcal{U}(T) = \{t^m \mid t \in T,\ m \in \alpha(t)\}$,
- $\forall s^v \in \mathcal{U}(S)$: $\iota_{\mathcal{U}}(s^v) = (\lambda_{\mathcal{U}}(s^v) \mid \alpha_{\mathcal{U}}(s^v))$, *and* $\lambda_{\mathcal{U}}(s^v) = \lambda(s)$, $\alpha_{\mathcal{U}}(s^v) = \{v\}$,
- $\forall t^m \in \mathcal{U}(T)$: $\iota_{\mathcal{U}}(t^m) = (\lambda_{\mathcal{U}}(t^m) \mid \alpha_{\mathcal{U}}(t^m))$, *and* $\lambda_{\mathcal{U}}(t^m) = \lambda(t)$, $\alpha_{\mathcal{U}}(t^m) = \{m\}$,
- $\forall (s^v, t^m) \in \mathcal{U}(S) \times \mathcal{U}(T)$: $\iota_{\mathcal{U}}(s^v, t^m) = \iota(s, t)(v, m) \cdot \{(v, m)\}$,
- $\forall (t^m, s^v) \in \mathcal{U}(T) \times \mathcal{U}(S)$: $\iota_{\mathcal{U}}(t^m, s^v) = \iota(t, s)(m, v) \cdot \{(m, v)\}$. ■ 2

We will also speak of the unfolding of a (sub)set of nodes of a net (e.g., $\mathcal{U}({}^\bullet N)$), considering this as the set of the corresponding nodes in the net unfolding (i.e., for instance, $\mathcal{U}({}^\bullet N) = {}^\bullet \mathcal{U}(N)$).

Definition 3. (Net isomorphism up to node renaming) *Let* $N = (S, T, \iota)$ *and* $N' = (S', T', \iota')$ *be M-nets.* N *and* N' *are called isomorphic up to node renaming, denoted by* $N \cong N'$, *if there is a sort-preserving bijection* $\rho: S \cup T \to S' \cup T'$ *such that* $\forall y \in S \cup (S \times T) \cup (T \times S) \cup T$: $\iota(y) = \iota'(\rho(y))$. ■ 3

3 Parameterised type construction

The composition operations presented in this paper are based on merging interface places of the participating nets. The difference between the weak and strong compositions comes into effect by the discipline of token propagation through the 'border' resulting from merging these interface places. A ***weak discipline*** implies that different values can proceed independently: tokens which correspond to a certain value and which enter a net component are allowed to go further without waiting for those tokens which are consumed by the same component but correspond to other values. A ***strong discipline*** assumes joint treatment of the values entering or leaving a participating net fragment (and considered as belonging to an initial or final marking).

This distinction is illustrated in fig. 1, where only places types, place names and (simplified) transition labels are shown. Assuming that places s_1 and s_2 are initially marked by values $\{1\}$ and $\{2\}$, respectively, then under strong sequence, interleaving *abcd* (but not *acbd*) is possible; under weak sequence, *abcd* and *acbd* (but not *adbc*) are possible. We will formalise this by using a parameter value set ψ, whose members specify the weak type of composition. In this case, $\psi = \emptyset$ would correspond to the strong discipline, and $\psi = \{\{1\},\{2\}\}$ would correspond to the weak discipline.

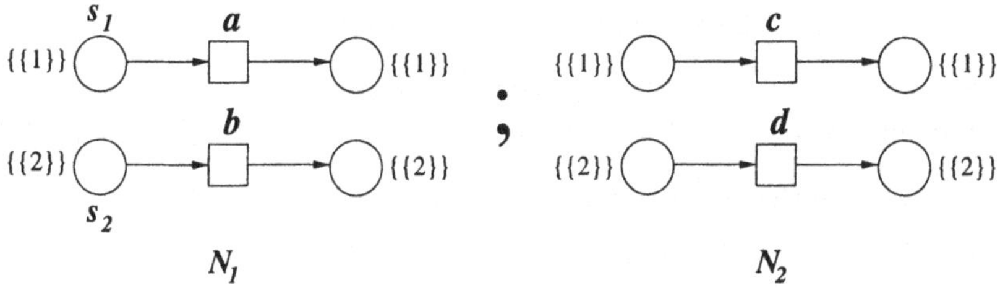

Fig. 1. Illustrating the difference between weak and strong sequential composition.

Since we deal with high-level nets, the creation of new places (from old ones by merging) must be accompanied by a *type construction* for these places. For the weak discipline, the values remain untouched; this mechanism is implemented by the intersection part of our type creation operator below. Under the strong composition, however, the new values must be bound to others, which implies that values reflect the corresponding dependency information.

The choice of sets as the form of value representation (i.e. that each value is a subset of Ω_{BAS}) is based primarily on the intention to obtain commutativity and associativity of composition operators, and, in particular, of the refinement, as the most complicated case. If two transitions, which share some adjacent place(s), are to be refined by different nets, commutativity means that the two different sequences of refinements must result in the same (including names and types) sets of new places. Therefore, type creation must be commutative: each new value must be able to absorb information gradually from the (types of the) contributing places, while the final outcome should not depend on the order of the contributions. Sets are used, since they are simple structured objects, yet – when used in the particular way we propose – are able to meet these wishes.

In general, let $\psi \subseteq VAL$ be a value set with respect to which the following notions will be parameterised. For example, in fig. 2, $\psi = \{\{3\}, \{4\}, \{5\}, \{6\}, \{7\}\}$. Values in ψ are called *weak*, others are called *strong*.

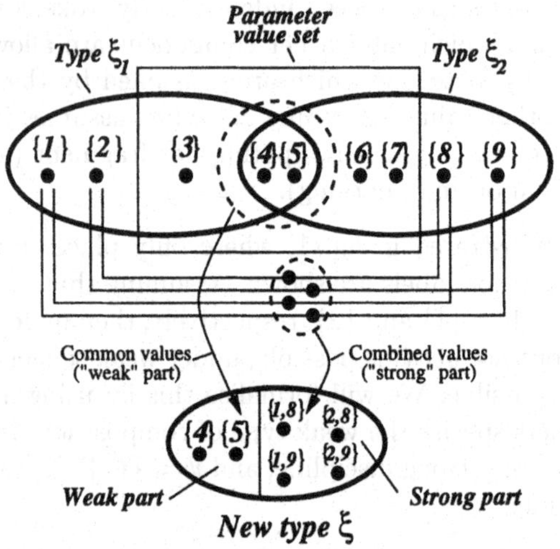

Fig. 2. Illustrating the origin of values in a new type.

In def. 6 below, we will describe a type construction facility which, given a parameter set ψ and a multiset of types $\mu \in \mathcal{M}_f(2^{VAL})$, creates a new type $\bowtie_{\psi} \mu$. In fig. 2, for example, we wish to create $\xi = \bowtie_{\psi} \{\xi_1, \xi_2\}$ with $\psi = \{\{3\}, \{4\}, \{5\}, \{6\}, \{7\}\}$, $\xi_1 = \{\{1\}, \{2\}, \{3\}, \{4\}, \{5\}\}$ and $\xi_2 = \{\{4\}, \{5\}, \{6\}, \{7\}, \{8\}, \{9\}\}$. The idea is to let the weak part of ξ arise as the intersection of the original weak types (after which, in fig. 2, values $\{4\}$ and $\{5\}$ survive), and the strong part of ξ as all pairwise combinations of the original strong types (which yields, in fig. 2, the new values $\{1,8\}, \{1,9\}, \{2,8\}$, and $\{2,9\}$). We have to ensure that the new values can again be characterised as weak or strong w.r.t. the same ψ. Definition 4 and lemma 1 serve this purpose.

Definition 4. (Conformable multiset of types) *Let* $\psi \subseteq VAL$, $\emptyset \neq \mu \in \mathcal{M}_f(2^{VAL})$. μ *is* ψ*-conformable iff* $\forall \xi \in \mu\ \forall v \in \xi: (v \in \psi) \vee (\forall w \in \psi \cup (\bigcup_{\xi' \in \mu - \{\xi\}} \xi'): v \cap w = \emptyset)$. ■ 4

Definition 4 (illustrated in table 1) means that each value v contained in one of the original types ξ is contained in ψ (then it is and remains weak) or, if it is strong, does not intersect with any value w from ψ or any other set (including ξ itself, if ξ is contained twice or more in μ) in μ. The last condition implies, by the use of multiset minus ($-$) rather than $\setminus$, that the types with membership in μ exceeding 1 must be completely weak. This is a technical subtlety which guarantees coherence in certain cases; details are in [4].

From def. 4 it follows that if μ is ψ-conformable then so is each multiset $\emptyset \neq \mu' \leq \mu$. In particular, the set $\{\xi \mid \xi \in \mu\}$ is itself ψ-conformable.

Value sets ξ and φ	$\psi = \{\{1\}, \{2\}\}$	**Comments**
$\xi = \{\{1\}, \{2\}\}$, $\varphi = \{\{1\}, \{4\}\}$	$\{\xi\}, \{\xi, \xi\}, \{\varphi\}, \{\xi, \varphi\}$, $\{\xi, \xi, \varphi\}$ are all conformable	
	$\{\varphi, \varphi\}, \{\xi, \varphi, \varphi\}$ not conformable	$\{4\} \notin \psi$ while $\mu(\varphi) > 1$
$\xi = \{\{1\}, \{3\}\}$, $\varphi = \{\{1\}, \{4\}\}$	$\{\xi, \varphi\}$ is conformable	
$\xi = \{\{1\}, \{3\}\}, \varphi = \{\{1\}, \{3, 4\}\}$	–"– not conformable	$\{3\}, \{3,4\} \notin \psi$
$\xi = \{\{1, 3\}, \{2\}\}$, $\varphi = \{\{2\}\}$	–"– not conformable	$\{1,3\} \notin \psi$

Table 1. Examples of conformable and not conformable multisets.

The following auxiliary operator yields a value combination from values of a set of value sets, taking each of the participating value sets once.

Definition 5. (Value combination)
Let $\nu \subseteq 2^{VAL}$. *Then* $\biguplus \nu \stackrel{\text{def}}{=} \{v \mid v = \bigcup\{v_\xi \mid \xi \in \nu\}$ *with* $\forall \xi \in \nu: v_\xi \in \xi\}$. ■ 5

Definition 6. (Type construction) *Let* $\psi \subseteq VAL$. *Let* $\mu \in \mathcal{M}_f(2^{VAL})$ *be* ψ-*conformable. Then* $\underset{\psi}{\bowtie} \mu \stackrel{\text{def}}{=} \bigcap\{\xi \cap \psi \mid \xi \in \mu\} \cup \biguplus\{\xi \setminus \psi \mid \xi \in \mu\}$. ■ 6

The first disjunct is the new weak part, the second is the new strong part. Fig. 2 and table 2 give examples of type construction. Moreover,

Lemma 1. (Consistency of $\bowtie$) *If* μ *is* ψ-*conformable, then so is* $\{\underset{\psi}{\bowtie} \mu\}$.

Proposition 1. (Properties of type construction)
Let $\psi \subseteq VAL$, $\emptyset \neq \mu \in \mathcal{M}_f(2^{VAL})$ *be* ψ-*conformable. Then:*

(1) Associativity: *If* $\emptyset \neq \kappa \leq \mu$, *then* $\underset{\psi}{\bowtie} \mu = \underset{\psi}{\bowtie}((\mu \setminus \kappa) \cup \{\underset{\psi}{\bowtie} \kappa\})$;
(2) Distributivity w.r.t. set union:
If $\forall \xi \in \mu$: $\xi = \bigcup\{\alpha \mid \alpha \in \mu_\xi\}$ *(i.e.* μ_ξ *is a collection of types whose union is* ξ*), then* $\underset{\psi}{\bowtie} \mu = \bigcup\{\beta \mid \beta = \underset{\psi}{\bowtie}\{\alpha_\xi \mid \xi \in \mu\}$ *with* $\forall \xi \in \mu: \alpha_\xi \in \mu_\xi\}$.

From the remark after def. 4, we infer that for any μ', μ, if $sp(\mu) \leq \mu' \leq \mu$ and μ is ψ-conformable, then $\underset{\psi}{\bowtie} \mu' = \underset{\psi}{\bowtie} \mu$. Thus we can use $\underset{\psi}{\bowtie}\{\xi \mid \xi \in \mu\}$ instead of $\underset{\psi}{\bowtie} \mu$.

Definitions 4 and 6 form the technical groundwork on which our results are based. They have been chosen rather judiciously, and [4] contains various counterexamples to some attempted simplifications or modifications of them.

4 Basic operations on nets

In this section, we define some auxiliary operations: (1) net union (a step which allows to treat two nets as a single starting point), (2) place multiplication

Participating value sets	Parameter set ψ	$\bowtie_{\psi} \mu$
$\xi = \{\{1\}\}$ $(\mu = \{\xi\})$	$\emptyset$	$\{\{1\}\}$
$\xi = \{\{1\}\}$ $(\mu = \{\xi\})$	$\{\{1\},\{2\}\}$	$\{\{1\}\}$
$\xi = \{\{1\},\{2\}\}, \varphi = \{\{1\},\{3\}\}$ $(\mu = \{\xi, \varphi\})$	$\{\{1\},\{2\}\}$	$\{\{1\}\}$
$\xi = \{\{1\}\}, \varphi = \{\{2\}\}$ $(\mu = \{\xi, \varphi\})$	$\{\{1\},\{2\}\}$	$\emptyset$
$\xi = \{\{1,2\}\}, \varphi = \{\{2,3\}\}$ $(\mu = \{\xi, \varphi\})$	$\{\{1,2\},\{2,3\}\}$	$\emptyset$
$\xi = \{\{1\},\{2,3\}\}, \varphi = \{\{2,3\},\{4\}\},$ $\beta = \{\{2,3\},\{5\},\{6\}\}$ $(\mu = \{\xi, \varphi, \beta\})$	$\{\{2,3\}\}$	$\{\{2,3\},\{1,4,5\},$ $\{1,4,6\}\}$

Table 2. Examples of type construction.

(to create 'border' places), (3) place addition (to insert new places into a net), (4) place or transition restriction (to eliminate old nodes). We will define the auxiliary operations and the main composition operations in the framework of *M-nets with basis* (MNB). Such a net is an M-net which is associated with another M-net, called basis. Elements (nodes, types, values) from this basis play the role of an 'alphabet', in terms of which the corresponding elements of the MNB are expressed. The benefit of such a mechanism – which is embodied in def. 8 below – is that the new (i.e. resulting from some composition) incidence relation can be constructed directly from (new) node names.

The next definition transfers the notion of type set conformability to the types of given place sets. The net in question will be a basis, whence the index $_B$.

Definition 7. (Compatible set of place sets) *Let* (S_B, T_B, ι_B) *be an M-net. Let* $I \neq \emptyset$. *Then* $\{S_i \subseteq \mathcal{M}_f(S_B) \mid i \in I\}$ *is called* ψ*-compatible if for any* $s = \sum_{i \in I} s_i$ *with* $\forall i \in I: s_i \in S_i$, *the multiset* $\mu_s = \sum_{s_B \in s} s(s_B) \cdot \{\alpha(s_B)\}$ *is* ψ*-conformable.* ■ 7

Let, for example, S_B be the place set of net N_B in fig. 3, i.e. $S_B = \{s_1, s_2\}$, and let $S' = \{\{s_1\}\}$ and $S'' = \{\{s_2\}\}$. Then $\{S', S''\}$ is $\{\{1\},\{2\}\}$-compatible because the multiset $\mu = \{\alpha(s_1), \alpha(s_2)\}$ $(= \{\{\{1\},\{3\}\}, \{\{1\},\{2\},\{4\}\}\}$, which is, in this example, a set) is $\{\{1\},\{2\}\}$-conformable.

Definition 8. (M-net with a basis) *Let* $N = (S, T, \iota)$, $N_B = (S_B, T_B, \iota_B)$ *be M-nets and let* ψ *be a parameter set. Then* N *is called an MNB (M-net with basis) over* N_B *and* ψ *if the following conditions are satisfied:*
$\forall s_B \in S_B \forall v \in \alpha(s_B): |v| = 1$;
$T \subseteq T_B$; $\forall t \in T: \iota(t) = \iota_B(t)$;
$S \subseteq \mathcal{M}_f(S_B)$; $\{S\}$ *is* ψ*-compatible; and* $\forall s \in S: \alpha(s) = \bowtie_{\psi} \{\alpha(s_B) \mid s_B \in s\}$;
$\forall (s,t) \in S \times T\ \forall v \in VAL\ \forall m \in MOD: \iota(s,t)(v,m) = \sum_{s_B \in s} \sum_{\varrho \in v} s(s_B) \cdot \iota_B(s_B, t)(\{\varrho\}, m)$;
$\forall (t,s) \in T \times S\ \forall v \in VAL\ \forall m \in MOD: \iota(t,s)(m,v) = \sum_{s_B \in s} \sum_{\varrho \in v} s(s_B) \cdot \iota_B(t, s_B)(m, \{\varrho\})$.
■ 8

Fig. 3 gives an example of an MNB N with basis N_B. Note how the type $\{\{1\},\{3,4\}\}$ of the 'combined' place $\{s_1, s_2\}$ arises from the types of s_1,s_2 via

ψ, and how its connections to t arise through the connections of s_1 and s_2 to t in N_B: weak value $\{1\}$, contained in the types of both s_1 and s_2, is preserved in the type of $\{s_1, s_2\}$, while weak value $\{2\}$, contained only in $\alpha(s_2)$, is eliminated via intersection and is not inherited by $\alpha(\{s_1, s_2\})$; strong values $\{3\}$ and $\{4\}$ are combined into a new strong value $\{3,4\}$; the coefficients 3 and 1 next to $(\{1\}, m1)$ and $(m1, \{1\})$, respectively, are equal to the corresponding coefficients in N_B; and similarly with coefficients 2 next to $(\{3,4\}, m2)$ and 4 next to $(m2, \{3,4\})$.

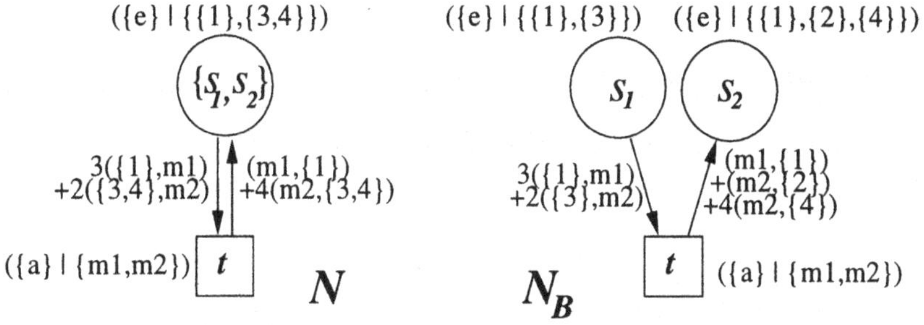

Fig. 3. Example of an MNB N over N_B and $\psi = \{\{1\}, \{2\}\}$.

Given an MNB N over N_B and ψ we sometimes do not give N_B and ψ explicitly, if they are clear from the context. If N is an MNB over N_B and ψ, then, up to place renaming, $\mathcal{U}(N)$ is an MNB over $\mathcal{U}(N_B)$ and ψ.

Definition 9. (Basis-disjoint MNB's) *Let N and N' be MNB's over the same ψ and the bases, respectively, N_B and N'_B. N and N' are called basis-disjoint if N_B and N'_B are disjoint.* ■ 9

Note that if N and N' are basis-disjoint then they are disjoint.

Definition 10. (MNB union) *Let $N{=}(S, T, \iota)$ and $N'{=}(S', T', \iota')$ be basis-disjoint MNB's over the same ψ and bases $N_B = (S_B, T_B, \iota_B)$ and $N'_B = (S'_B, T'_B, \iota'_B)$, respectively. Then,*
$N_B{\cup}N'_B{=}(S_B{\cup}S'_B, T_B{\cup}T'_B, \iota_B{\cup}\iota'_B)$ *and* $N{\cup}N'{=}(S{\cup}S', T{\cup}T', \iota{\cup}\iota')$. ■ 10

Note that $N{\cup}N'$ is an MNB over $N_B{\cup}N'_B$. Moreover, MNB union is commutative, associative, and commutes with unfolding, i.e.: if $\mathcal{U}(N{\cup}N')$ is defined then so is $\mathcal{U}(N) \cup \mathcal{U}(N')$, and they are equal.

The place multiplication operator $\otimes$ (def. 11 and table 3) takes an MNB and a collection of subsets of its places, and creates a new set of places through a cartesian product based scheme, by taking all possible combinations of places from the participating sets (one from each of the sets), but leaving only those which obtain non-empty types from the type construction operator $\bowtie$. The non-emptiness condition for the resulting types is necessary in order to preserve the coherence of this operator w.r.t. unfolding [4], and it just excludes useless places.

Definition 11. (Place multiplication) *Let $N = (S, T, \iota)$ be an MNB, and $I{\neq}\emptyset$. Let $\{S_i \mid i{\in}I\}$ be a ψ-compatible set of subsets of S, and $\tau \in \{\{e\}, \{x\}, \{e,x\}, \{i\}\}$.*

Then,
$\underset{\psi}{\overset{\tau}{\otimes}}\{S_i \mid i\in I\} \overset{\text{def}}{=} \{\, s \mid s=\sum_{i\in I} s_i,$ *with* $(\forall i\in I\colon s_i\in S_i)$ *and* $(\underset{\psi}{\bowtie}\{\alpha(s_B)|s_B\in s\}\neq\emptyset)\,\}$.
Moreover, for $s\in\underset{\psi}{\overset{\tau}{\otimes}}\{S_i|i\in I\}$ *we put* $\lambda(s)=\tau$ *and* $\alpha(s)=\underset{\psi}{\bowtie}\{\alpha(s_B)|s_B\in s\}$. ■ 11

Proposition 2. (Properties of place multiplication) *Let* (S,T,ι) *be an MNB,* $I\neq\emptyset$*, and* $\forall i\in I\colon S_i\subseteq S$*. Let* $\tau,\tau_1\in\{\{e\},\{x\},\{e,x\},\{i\}\}$*. Then,*

(1) Associativity: *If* $\emptyset\neq K\subseteq I$*, then* $\overset{\tau}{\otimes}\{S_i|i\in I\}=\overset{\tau}{\otimes}\left(\{S_j \mid j\in I\backslash K\}\cup\overset{\tau_1}{\otimes}\{S_k|k\in K\}\right)$*;*

(2) Distributivity w.r.t. set union: *If* $\forall i\in I\colon S_i=\bigcup\{S_i^{(j_i)} \mid j_i\in J_i\}$*, where* $J_i\neq\emptyset$ *is an index set, then* $\overset{\tau}{\otimes}\{S_i|i\in I\}=\bigcup\{S' \mid S'=\overset{\tau}{\otimes}\{S_i^{(j_i)}|i\in I\}$ *with* $\forall i\in I\colon j_i\in J_i\}$*;*

(3) Coherence w.r.t. unfolding: $\mathcal{U}(\overset{\tau}{\otimes}\{S_i \mid i\in I\})\cong\overset{\tau}{\otimes}\{\mathcal{U}(S_i) \mid i\in I\}$.

$S_i, i\in I$	**Parameter set** ψ	$\underset{\psi}{\overset{\{e\}}{\otimes}}\{S_i \mid i\in I\}$
$S_1=\left\{\overset{(\{e\}\mid\{\{1\}\})}{\{s_1\}\bigcirc}\right\}$ $(I=\{1\})$	$\{\{1\}\}$	$\left\{\overset{(\{e\}\mid\{\{1\}\})}{\{s_1\}\bigcirc}\right\}$ $(=S_1)$
$S_1=\left\{\overset{(\{e\}\mid\{\{1\},\{2\}\})}{\{s_1\}\bigcirc},\overset{(\{x\}\mid\{\{2\}\})}{\{s_2\}\bigcirc}\right\}$, $S_2=\left\{\overset{(\{e\}\mid\{\{1\},\{3\}\})}{\{s_3\}\bigcirc}\right\}$ $(I=\{1,2\})$	$\{\{1\}\}$	$\left\{\overset{(\{e\}\mid\{\{1\},\{2,3\}\})}{\{s_1,s_3\}\bigcirc},\overset{(\{e\}\mid\{\{2,3\}\})}{\{s_2,s_3\}\bigcirc}\right\}$
$S_1=\left\{\overset{(\{e\}\mid\{\{1\},\{2\}\})}{\{s_1\}\bigcirc},\overset{(\{x\}\mid\{\{2\}\})}{\{s_2\}\bigcirc}\right\}$, $S_2=\left\{\overset{(\{e\}\mid\{\{3\},\{4\}\})}{\{s_3\}\bigcirc}\right\}$ $(I=\{1,2\})$	$\{\{2\}\}$	$\left\{\overset{(\{e\}\mid\{\{1,3\},\{1,4\}\})}{\{s_1,s_3\}\bigcirc}\right\}$

Table 3. Examples of place multiplication.

Place addition $\oplus$ (def. 12 and fig. 4) serves for adding new places to an MNB: given an MNB and a set of places (which, in our context, will usually be outcomes of the previously introduced operator $\otimes$), it creates new incidences, depending on the types of the new places, and correspondingly modifies the MNB.

Definition 12. (Place addition) *Let* $N=(S,T,\iota)$ *be an MNB over basis* $N_B=(S_B,T_B,\iota_B)$ *and* ψ*. Let* $S'\subseteq\mathcal{M}_f(S_B)$ *s.t. for all* $s\in S'$, $\mu_s=\sum_{s_B\in s}s(s_B)\cdot\{\alpha(s_B)\}$ *is* ψ*-conformable. Then* $N\oplus S'\overset{\text{def}}{=}(S\cup S',T,\iota')$. ■ 12

Note that ι' arises from def. 8, since $(S\cup S',T,\iota')$ has the same basis as N. In fig. 4, adding to MNB N (over N_B,ψ) a place $\{s_1,s_1,s_2\}$ results in MNB N_1, whereas adding to the same MNB two places, $\{s_1,s_2\}$ and $\{s_2,s_3,s_3\}$, yields MNB N_2. New places are highlighted by bold lines.

Proposition 3. (Properties of place addition) *Let* N *be an MNB and* $S',S''\subseteq\mathcal{M}_f(S_B)$*. Then the following hold:* **(1)** Empty set as a zero element: $N\oplus\emptyset=N$,

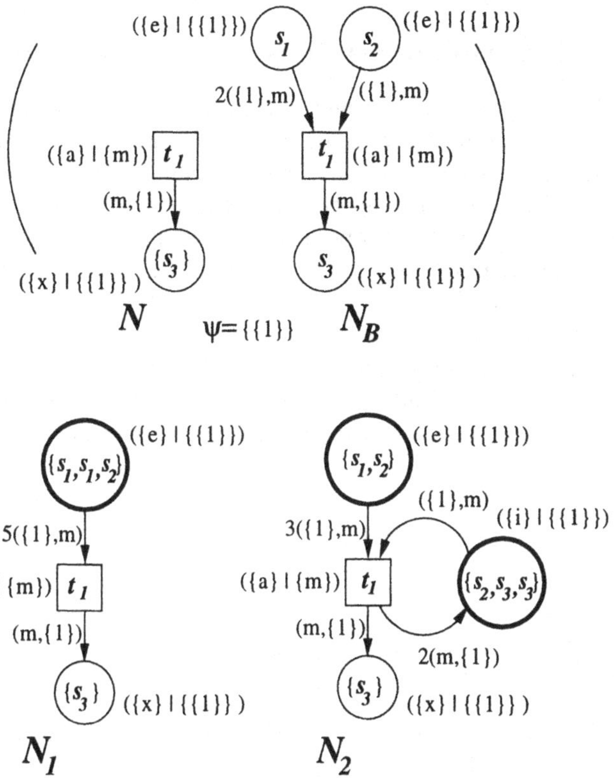

Fig. 4. Illustrating place addition.

(2) Duplication prevention: $N \oplus S' = N \oplus (S' \backslash S)$, **(3)** Commutativity: $(N \oplus S') \oplus S'' = (N \oplus S'') \oplus S'$, **(4)** Distributivity w.r.t. place set union: $N \oplus (S' \cup S'') = (N \oplus S') \oplus S''$, **(5)** Context sensitivity: *If N_1 is an MNB which is basis-disjoint with N, and $S' \subseteq \mathcal{M}_f(S_B)$, then $(N \cup N_1) \oplus S' = (N \oplus S') \cup N_1$*, **(6)** Coherence w.r.t. unfolding: $\mathcal{U}(N \oplus S') \cong \mathcal{U}(N) \oplus \mathcal{U}(S')$.

Restriction (next definition) drops elements from a net.

Definition 13. (Restriction) *Let $N = (S, T, \iota)$ be an MNB over (S_B, T_B, ι_B), and $Y \subseteq \mathcal{M}_f(S_B) \cup T_B$. Then N **rs** $Y \stackrel{\text{def}}{=} (S \backslash Y, T \backslash Y, \iota|_{S' \cup (S' \times T') \cup (T' \times S') \cup T'})$.* ■ 13

Proposition 4. (Properties of restriction) *Let $N = (S, T, \iota)$ be an MNB over $N_B = (S_B, T_B, \iota_B), \psi$. If $S' \subseteq \mathcal{M}_f(S_B)$, $Y, Y_1, Y_2 \subseteq S \cup T$, then:* **(1)** Empty set as a zero element: N **rs** $\emptyset = N$, **(2)** Idempotence: $(N$ **rs** $Y)$ **rs** $Y = N$ **rs** Y, **(3)** Commutativity: $(N$ **rs** $Y_1)$ **rs** $Y_2 = (N$ **rs** $Y_2)$ **rs** Y_1, **(4)** Context sensitivity: *If $Y \cap (S \cup T) = \emptyset$ then N* **rs** *$Y = N$*, **(5)** Distributivity w.r.t. net union: *If N_1 is MNB which is basis-disjoint with N, then $(N \cup N_1)$* **rs** *$Y = (N$* **rs** *$Y) \cup (N_1$* **rs** *$Y)$*, **(6)** Suppression of place addition: $(N \oplus S')$ **rs** $S' = N$ **rs** S', **(7)** Interchangeability with place addition: *If $Y \cap S' = \emptyset$ then $(N \oplus S')$* **rs** *$Y = (N$* **rs** *$Y) \oplus S'$*, **(8)** Coherence w.r.t. unfolding: $\mathcal{U}(N$ **rs** $Y) \cong \mathcal{U}(N)$ **rs** $\mathcal{U}(Y)$.

5 Net composition operations: general scheme

Using the auxiliary operators defined in section 4, we now construct our main composition operators. In [4], we consider the most important and widespread – in net theoretic and process algebraic contexts – ones: parallel composition ($\|$), sequence (;), choice ($\Box$), iteration (two versions), and transition refinement. The application of each of the above operators, except $\|$, can be considered as split into four subsequent phases, which correspond to the application of the auxiliary operators introduced earlier: $\cup$, $\otimes$, $\oplus$, **rs** . Which interface sets are involved and how exactly they are impacted under different operations, is determined by the underlying causality semantics of the intended composition, and is illustrated in the interface set correspondence diagrams in fig. 5; pairs and triples of place sets which participate in each place multiplication operator are connected via the symbol $\otimes$ in this figure.

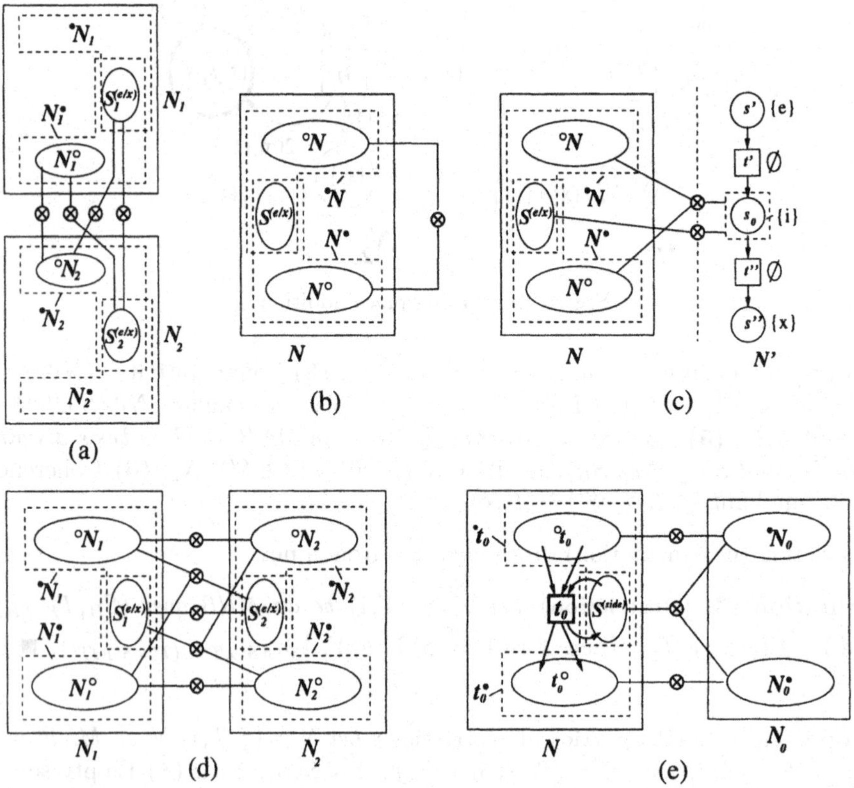

Fig. 5. Multiplication of interface place sets under different operators: (a) sequence, (b) iteration-I, (c) iteration-II, (d) choice, (e) transition refinement.

Suppose that two disjoint nets N_1 and N_2 have $\{e\}$-places, $\{x\}$-places, but no $\{e, x\}$-places. Then $N_1; N_2$ needs to be constructed by first taking the union of the two nets, and then $\otimes$-multiplying the $\{x\}$-places of N_1 with the $\{e\}$-places of N_2, parameterised by ψ. If N_1 or N_2 or both have $\{e, x\}$-places, then the logic behind our definition is as follows (fig. 5(a)). The $\{e,x\}$-places of the first net, while being combined with the pure entry places of the second one, 'exhaust' their exit capability, but retain the entry capability; the resulting places should thus be treated as *entry* for the resulting net. Analogously, the $\{e,x\}$-places of the second net retain their x-capability. When $\{e,x\}$-places of both nets are combined together, these new places are consequently treated as $\{e,x\}$-places.

Under iteration, the pure entry and exit sets of a net are merged either directly (fig. 5(b)), or via a place of some auxiliary net (fig. 5(c)), when an outcome without combined $\{e,x\}$-places is preferred. One of the distinctions between versions I and II of iteration will be exemplified in section 7.

Under choice composition (fig. 5(d)), the symmetry of the treatment of the nets implies that, as soon as a token has entered an $\{e,x\}$-place of one net, it can at once be considered as contained in an exit place of this net and consequently as being at the level of the exit places of the other net. Therefore, systematically, under choice composition, the $\{e,x\}$-places of each net have to be combined with both pure entry and pure output places of the other net. (Thus, for instance, 

in choice with 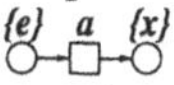

is the same as an iteration, a^*, of a.)

Under refinement composition (fig. 5(e)), the intuition is similar to that corresponding to the choice, a difference being, however, that the (pure) entry, exit and side places of the refined transition play now the role of the (pure) entry, exit, and $\{e,x\}$-places of a net in the choice composition, respectively.

For reasons of space, we consider in this paper only three operations: parallel composition, sequence and refinement. The ideas, formal definitions, and properties of the others are exposed in [4]. The following definitions of sequential composition and transition refinement are in their most general representation; they reduce to a noticeably simpler form if, for example, participating net(s) do not have combined $\{e,x\}$-places, or if the to-be-refined transition does not have side places.

6 Parallel composition

Parallel composition of two basis-disjoint nets is synonymous to the union of MNB's; the former serves as an independent operator, the latter is used as an auxiliary step in the definitions of operators.

Definition 14. (Parallel composition) *Let N_1 and N_2 be two basis-disjoint MNB's over the same ψ. Then $N_1 \parallel N_2 \stackrel{\text{def}}{=} N_1 \cup N_2$.* ■ 14

Parallel composition is commutative, associative, and coherent w.r.t. unfolding.

7 Sequential composition

Definition 15. (Sequential composition) *Let N_1 and N_2 be two basis-disjoint MNB's over the same ψ. Then,*

$$N_1; N_2 \stackrel{\text{def}}{=} \Big((N_1 \cup N_2) \oplus \big(\overset{\{i\}}{\otimes}\{N_1^\circ, {}^\circ N_2\} \cup \overset{\{e\}}{\otimes}\{S_1^{\{e,x\}}, {}^\circ N_2\} \cup \overset{\{x\}}{\otimes}\{N_1^\circ, S_2^{\{e,x\}}\} \cup \overset{\{e,x\}}{\otimes}\{S_1^{\{e,x\}}, S_2^{\{e,x\}}\} \big) \Big) \ \mathbf{rs}\ (N_1^\bullet \cup {}^\bullet N_2). \quad \blacksquare\ 15$$

If there are no $\{e, x\}$-places in either net, then only the first of the four disjuncts remains in this definition: $N_1; N_2 = \Big((N_1 \cup N_2) \oplus \overset{\{i\}}{\otimes}\{N_1^\circ, {}^\circ N_2\} \Big)\ \mathbf{rs}\ (N_1^\bullet \cup {}^\bullet N_2)$. Consider the example in fig. 6, where only place statuses and transition labels are shown; the types of all places are assumed to be the same singleton set, e.g. $\{\{0\}\}$ (this corresponds to ordinary place/transition nets).

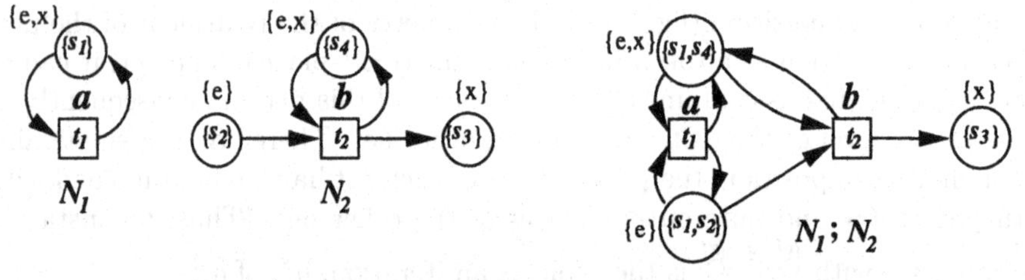

Fig. 6. Motivating place connections under sequential composition.

The rationale behind the above definition in its $\{e, x\}$-places-related part, in addition to being systematically motivated in section 5, can be supported by behavioural reasons. Since the initial marking of N_1 is also final, and thus can instantaneously be re-interpreted as the initial marking of N_2 (tokens on $\{s_2\}$ and $\{s_4\}$), it can be expected that, under the $\{0\}$–marking of the entry places (i.e. those having status $\{e\}$ or $\{e, x\}$), the firing sequences in the sequential composition of nets N_1 and N_2 are a^*b. In the right-hand side of the figure, net $N_1; N_2$ is shown, obtained corresponding to def. 15. It can be checked that $N_1; N_2$, with the initial marking consisting of tokens $\{0\}$ on $\{s_1, s_2\}$ and $\{s_1, s_4\}$, produces the same firing sequences.

The case of a sequential composition of two nets corresponding to iterations a^* and b^*, i.e. the sequence $a^*; b^*$, is worth being pointed out. Suppose that a^* is modelled as N_1 if fig. 6, and similarly b^* by another such net. This corresponds to using iteration-I in both cases. Then our operator for sequential composition yields a net with two transitions and a single place, with status $\{e, x\}$, which is a side place for both of transitions, i.e. a net with behaviour $(a|b)^*$. Suppose, on the other hand, that a^* is modelled using iteration-II. This yields a net with an a-labelled transition and a side place s to it, as before. However, s now has status $\{i\}$, and a silent (i.e., $\emptyset$-labelled) transition leads from an $\{e\}$-labelled place to s, and another silent transition leads from s to an $\{x\}$-labelled place. Suppose that b^* is modelled similarly. Then the sequential composition yields a net having behaviour $\emptyset a^* \emptyset \emptyset b^* \emptyset$ (i.e., a^*b^* if silent transitions are neglected).

This example can hardly be interpreted as pointing to a deficiency in the definition of sequential composition. Rather, the effect seems to be inherent in the M-net framework, since it can be shown that there is no M-net without silent transitions whose behaviour is a^*b^*. It is also one of the reasons for distinguishing two versions of iteration.

Theorem 1. **(Properties of sequential composition)** *Let N_1, N_2, and N_3 be mutually basis-disjoint MNB's over the same ψ. Then the following hold:*
(1) *(Associativity)* $(N_1;N_2);N_3 = N_1;(N_2;N_3)$;
(2) *(Coherence w.r.t. unfolding)* $\mathcal{U}(N_1;N_2) \cong \mathcal{U}(N_1);\mathcal{U}(N_2)$.

8 Transition refinement

In the definition of transition refinement, we permit side places for the to-be-refined transition. However we do not allow $\{e,x\}$-places in the refining net, because otherwise the commutativity of the refinement in the general case could be destroyed [4].[1] Under refinement, the statuses of new places are determined by the statuses of the contributing places from the main net. Under this convention, we do not put any τ over $\otimes$ in the following definition.

Definition 16. **(Single-mode transition refinement)** *Let N and N_0 be two basis-disjoint MNB's over the same ψ, $t_0 \in T$ with $|\alpha(t_0)| = 1$, ${}^\bullet N_0 \cap N_0^\bullet = \emptyset$, and let $\{{}^\bullet N_0, N_0^\bullet\}$ be ψ-compatible. The net obtained by refining transition t_0 in N with N_0 is defined as:*

$$N[t_0 \leftarrow N_0] \overset{\text{def}}{=} \Big(((N \textbf{ rs } t_0) \cup N_0) \oplus \big(\otimes\{{}^\circ t_0, {}^\bullet N_0\} \cup \otimes\{t_0^\circ, N_0^\bullet\} \cup \otimes\{S_{t_0}^{(side)}, {}^\bullet N_0, N_0^\bullet\}\big)\Big) \textbf{ rs } ({}^\bullet t_0 \cup t_0^\bullet \cup {}^\bullet N_0 \cup N_0^\bullet). \qquad \blacksquare\ 16$$

If there are no side places next to transition t_0, then only two disjuncts, namely $\otimes\{{}^\circ t_0, {}^\bullet N_0\}$ and $\otimes\{t_0^\circ, N_0^\bullet\}$, remain in the above definition.

Two examples of transition refinement are shown in fig. 7. They are of mixed - weak and strong - composition type and represent two different cases from the structural point of view: in the first one, t_0 does not have side places; in the second one, t_0 has a side place, and, in addition, an entry place of N_0 is at the same time a side place of some transition in N_0.

Theorem 2. **(Properties of transition refinement)** *Let N, N_1, and N_2 be mutually disjoint MNB's. Then the following hold:*
(1) *(Commutativity-I) If $t_1, t_2 \in T$ then*
$(N[t_1 \leftarrow N_1])[t_2 \leftarrow N_2] = (N[t_2 \leftarrow N_2])[t_1 \leftarrow N_1]$;
(2) *(Commutativity-II) If $t_1 \in T$ and $t_2 \in T_1$ and* $\left(S_{t_1}^{(side)} = \emptyset\right) \vee \left(({}^\bullet t_2 \cup t_2^\bullet) \cap ({}^\circ N_1 \cup N_1^\circ) = \emptyset\right)$, *then* $(N[t_1 \leftarrow N_1])[t_2 \leftarrow N_2] = N[t_1 \leftarrow (N_1[t_2 \leftarrow N_2])]$;
(3) *(Coherence w.r.t. unfolding)* $\mathcal{U}(N[t_1 \leftarrow N_1]) \cong \mathcal{U}(N)[t_1 \leftarrow \mathcal{U}(N_1)]$.

Because of this theorem and the possibility of splitting a transition with many modes into a series of transitions with single modes, def. 16 can be generalised directly to multi-mode transitions.

[1] Actually, we have a hunch that this restriction is non-essential. We are presently working to overcome this limitation.

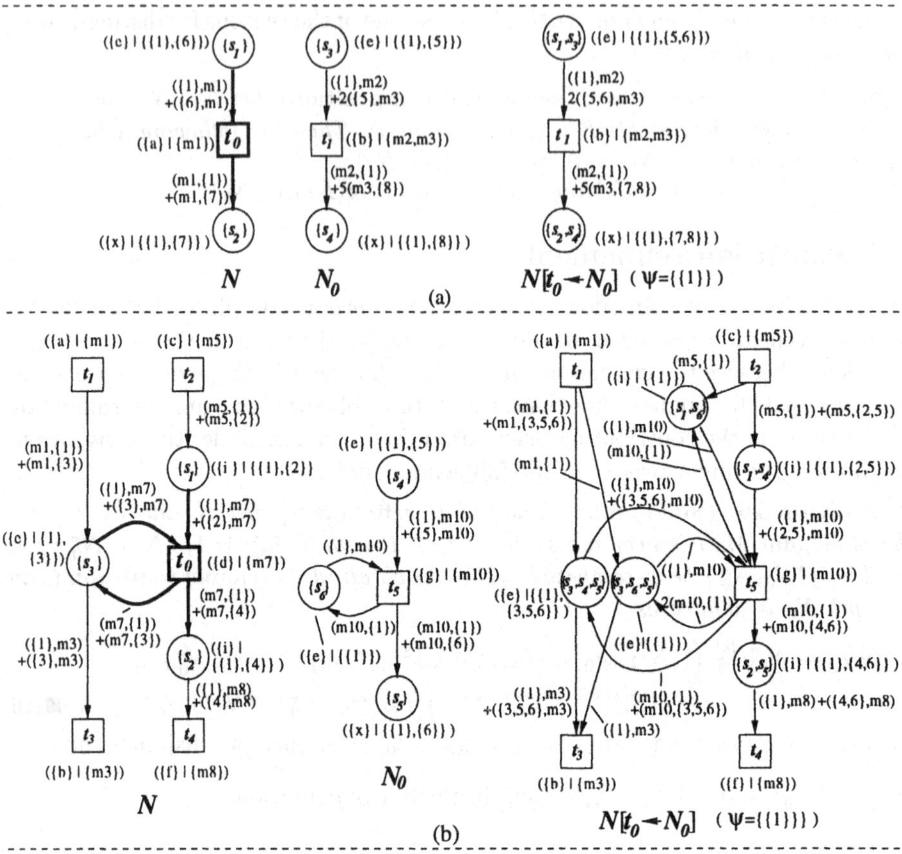

Fig. 7. $N[t_0 \leftarrow N_0]$: (a) t_0 without side places, (b) t_0 with side places.

9 Conclusions

We have introduced a set of generic schemes for high-level net composition operations. A uniform approach has allowed us to obtain operations which preserve nice algebraic properties while remaining coherent with respect to net unfolding. Equipped with a parameter value set, the proposed schemes enable us to implement different - strong, weak, and mixed - composition disciplines. These definitions have been justified by examples, by proving the desired algebraic (structural) laws (and also by the fact that in the purely strong case, and neglecting $\{e, x\}$ interface, they reduce to better known constructions).

Our approach differs from other research in similar direction, e.g. [5, 6], in particular by the following: (a) combined entry/exit places are allowed, (b) values are sets rather than more complicated labelled trees, (c) associativity and commutativity of the operations hold as equalities, and the coherence w.r.t. unfolding holds as isomorphism up to place renaming.

A formal behavioural justification of our operations (along the line hinted at in discussing fig. 6) has not been the objective of this paper, but this is a task that is planned to be done in future research. In [11], Thielke provides such a formal justification for a subtheory (without weak values, without intersecting types and without $\{e, x\}$ places).

Other future research includes elaborating place-oriented refinement operations, developing a formal definition of a universal, place- and transition-oriented, refinement scheme, extending it to a 'net-for-subnet substitution' case, and possibly embedding linear algebraic elements as inherent parts into composition operations.

Acknowledgments Comments from the anonymous referees are greatly appreciated and taken into account. The authors would also like to thank Hans Fleischhack and Burkhard Graves for useful comments. The work was done during a research fellowship granted to the second author by the Alexander-von-Humboldt-Stiftung, whose support is appreciated.

References

1. *High-level Petri Nets - Concepts, Definitions and Graphical Notation.* Committee Draft ISO/IEC 15909, October 2, 1997, Version 3.4 (1997).
2. E. Best, R. Devillers, and J. G. Hall: The Petri Box Calculus: a New Causal Algebra with Multilabel Communication. *Advances in Petri Nets 1992*, G. Rozenberg (Ed.). Springer-Verlag, LNCS Vol. 609, 21–69 (1992).
3. E. Best, H. Fleischhack, W. Fraczak, R. P. Hopkins, H. Klaudel, and E. Pelz: A Class of Composable High Level Petri Nets. *Application and Theory of Petri Nets 1995*, Springer-Verlag, LNCS Vol. 935, 103-120 (1995).
4. E. Best, A. Lavrov: Generalised Composition Operations for High-Level Petri Nets. Longer version of the present paper (submitted); available from the authors (1999).
5. E. Best, Th. Thielke: Refinement of Coloured Petri Nets. B. S. Chlebus, L. Czaja (Eds.): *Fundamentals of Computation Theory. Proceedings of 11th International Symposium, FCT'97*, Poland, Springer-Verlag, LNCS Vol. 1279, 105-116 (1997).
6. R. Devillers, H. Klaudel, and R.-C. Riemann: General Refinement for High Level Petri Nets. *Foundation of Software Technology and Theoretical Computer Science'97*, Springer-Verlag, LNCS Vol. 1346, 297 - 311 (1997).
7. H. Fleischhack and B. Grahlmann: A Petri Net Semantics for $B(PN)^2$ with Procedures. *Proc. of PDSE'97*, Boston, IEEE Computer Society Press (1997).
8. M. Fokkinga, M. Poel, and J. Zwiers. Modular Completeness for Communication Closed Layers. *CONCUR '93: 4th International Conference on Concurrency Theory*, Hildesheim, Germany, Springer-Verlag, LNCS Vol. 715, 50-65 (1993).
9. K. Jensen: Coloured Petri Nets. Basic Concepts, Analysis Methods and Practical Use. *EATCS Monographs on Theoretical Computer Science*, Springer-Verlag (1992).
10. E. R. Olderog: Nets, Terms and Formulas. *Cambridge Tracts in Theoretical Computer Science*, Vol. 23 (1991).
11. Th. Thielke: Multirelationenkalkül für höhere Petrinetze. Dissertation, Universität Hildesheim (1999).

Model Checking of Time Petri Nets Based on Partial Order Semantics

Burkhard Bieber and Hans Fleischhack

Burkhard Bieber and Hans Fleischhack
Fachbereich Informatik
Carl-von-Ossietzky – Universität Oldenburg
D-26111 Oldenburg
{bieber,fleischhack}@informatik.uni-oldenburg.de

Abstract. Model checking of place/transition-nets based on partial order semantics has been applied successfully to the analysis of causal behaviour of distributed systems. Here, this approach is extended to the causal behaviour of time Petri nets. Expansion of a time Petri net to an equivalent P/T-net is defined, and it is shown that (an abstraction of) the McMillan unfolding of the expanded net is sufficient for model checking w.r.t. formulae of a simple branching time temporal logic $\mathcal{L}$.

1 Introduction

Model checking is a widely accepted method for proving properties of distributed systems but faces the problem of 'state explosion'. To tackle this problem, besides partial order reductions [13] or BDD-based techniques [7, 16] also methods based on partial order semantics have been applied. The latter have proven especially successful with systems with a high degree of asynchronous parallelism [5, 9].

Safety critical applications often require verification of real time constraints, in addition to functional or qualitative temporal properties. For this task, model checking algorithms have been developed based on interleaving semantics (cf. e.g. [1, 19]), but much less work has been done starting from partial order semantics (cf. section 5 for some exceptions). By extending McMillan's [17] technique of unfolding safe Petri nets to the class of safe time Petri nets [18] this paper takes a step in this direction.

In [9], a branching time temporal logic $\mathcal{L}$ for safe nets is introduced. A model checking algorithm is given based on the finite prefix of the maximal branching process of a net N, the McMillan-unfolding $McM(N)$ of N (cf. also [10]). One might naively assume that in order to extend this approach to safe time Petri nets, it is sufficient to take the McMillan-unfolding of the underlying P/T-net and reduce it to that part which is not prohibited by the time restrictions. The following example of Figure 1 shows that this is not true.

First, consider the net TN_1 (Figure 1 (i)) without time restrictions. Its possible (concurrent) behaviour is described in Figure 1 (ii). All behaviour happens to remain realizable under the time requirements in the sense that for each concurrent run of the system there is a timing schedule respecting the requirements.

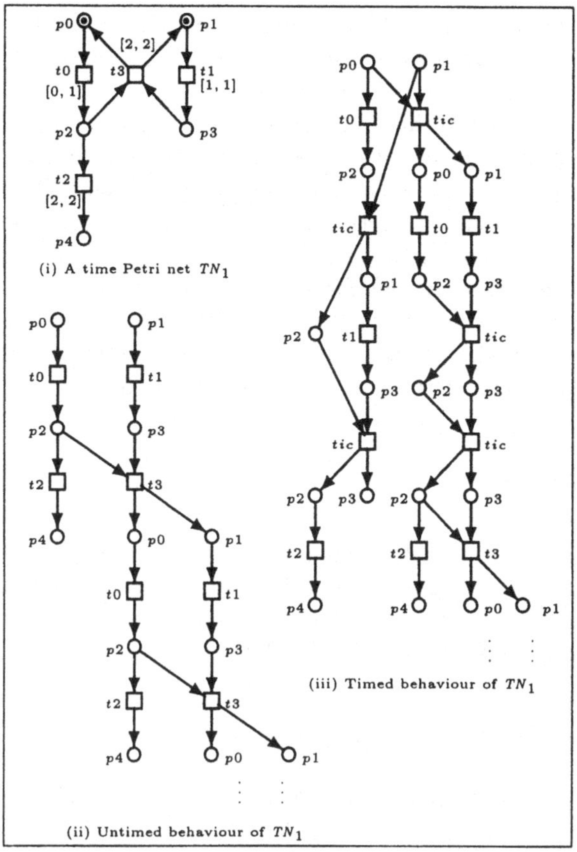

(i) A time Petri net TN_1

(ii) Untimed behaviour of TN_1

(iii) Timed behaviour of TN_1

Fig. 1. A time Petri net TN_1 and its behaviour

The untimed behaviour of TN_1 (cf. Figure 1(ii)) satisfies the following property Q: It always holds that, if p_2 and p_3 are marked, then possibly p_0 becomes marked.

But, if t_0 occurs in TN_1 at time 0, then t_2 becomes enabled at time 0 and has to occur at time 2. On the other hand, t_1 is forced to occur at time 1, so t_3 is enabled at time 1. However, t_3 cannot occur before time 3, and, since p_2 is not marked after time 2, t_3 can never occur at all. Hence, this property is not satisfied by TN_1. This is reflected by the net's behaviour if we represent lapse of time by inserting special events called *tic-events* (cf. Figure 1 (iii)): If t_0 occurs at time 0 (i.e. before any tic-event), then t_3 never becomes enabled.

The key idea of the approach presented here consists in transforming time restrictions into net structure, i.e. representing them by additional places, transitions, and arcs. Following this approach, for a given time Petri net TN the time-expansion $X(TN)$ is constructed as an ordinary P/T-net. Moreover, it is shown that an $\mathcal{L}$-formula ϕ is satisfied by TN iff it is satisfied by the time-expansion $X(TN)$. Subsequently, the McMillan-unfolding $McM(TN)$ of TN is defined by (an abstraction of) $McM(X(TN))$. As will be shown in [11], for the actual computation of $McM(TN)$ the explicit construction of the (complex) net $X(TN)$ is not necessary.

An additional benefit of this approach is given by the fact that corresponding tools for P/T-nets may be reused for the causal analysis of time Petri nets, which applies especially to the model checking component of the PEP tool [5].

We proceed as follows: In section 2 the notions used throughout the paper are introduced. Section 3 contains the definition of time expansion of time Petri nets. The new method for model checking of time Petri nets is given in section 4, based on the 'finite prefix' of the time expansion. We conclude with discussing some related work (section 5) and with remarks on further research (section 6).

An extended version of this paper is presented in [6].

2 Basic Notions

In this section, a partial order semantics as well as temporal logics for safe Place/Transition-nets (P/T-nets) is defined and a notion of time is added to P/T-nets. Moreover, the temporal logics is extended to time Petri nets. Note that most notions carry over to bounded P/T-nets.

2.1 Processes of Place/Transition-nets

Let $N = (P, T, F, M_0)$ be a P/T-net. N is called *safe* iff $M(p) \in \{0, 1\}$ for all $p \in P$ holds for any reachable marking. For the rest of the paper, only safe nets are considered. Nodes $x_1, x_2 \in (P \cup T)$ are *in conflict* ($x_1 \# x_2$) iff there exist distinct transitions $t_1, t_2 \in T$ such that $^\bullet t_1 \cap {}^\bullet t_2 \neq \emptyset$ and $(t_1, x_1), (t_2, x_2) \in F^*$, the reflexive and transitive closure of F. $x \in (P \cup T)$ is in *self-conflict* iff $x \# x$.

Often, the behaviour of a P/T-net N is described via the *reachability graph*. Its vertices consist of the set of all reachable markings of N and there is an edge leading from M to M', which is labeled by t iff $M \rightarrow^t M'$. However, using the reachability graph the concurrent behaviour of N cannot be retrieved easily and all interleavings of concurrent behaviour have to be represented explicitly, which often leads to an unnecessary exponential blow up.

In part, these problems are avoided by the maximal branching process, which associates a partial order semantics to each safe net N [9]. A *causal process* of a safe P/T-net N describes a possible run of N, displaying the causal dependencies of the events that take place during the run. A *branching process* may represent different alternative runs of N in one structure and hence may seen as the union of some causal processes. It consists of an occurrence net and a homomorphism.

An *occurrence net* $CN = (B, E, G)$ is an acyclic net such that $| {}^\bullet b | \leq 1$ for all $b \in B$, no event $e \in E$ is in self conflict, and for all $x \in (B \cup E)$, the set of elements $y \in (B \cup E)$ such that $(y, x) \in G^*$ is finite. The elements of B (E, respectively) are called *conditions* (*events*, respectively). $\leq$ denotes the partial order induced by G on $B \cup E$; $<$ denotes the corresponding strict partial order;

$Min(CN)$ denotes the set of minimal elements of CN w.r.t. $\leq$. A *causal net* is an occurrence net which also satisfies $\mid b^\bullet \mid \leq 1$ for every condition b.

A *homomorphism* from an occurrence net CN to a P/T-net N is a mapping $\pi : (B \cup E) \to (P \cup T)$ such that $\pi(B) \subseteq P$ and $\pi(E) \subseteq T$, for all $e \in E$, the restriction of π to ${}^\bullet e$ is a bijection between ${}^\bullet e$ and ${}^\bullet \pi(e)$ and the restriction of π to $e^\bullet$ is a bijection between $e^\bullet$ and $\pi(e)^\bullet$, the restriction of π to $Min(CN)$ is a bijection between $Min(CN)$ and M_0, and for all $e_1, e_2 \in E$ it holds that if ${}^\bullet e_1 = {}^\bullet e_2$ and $p(e_1) = p(e_2)$ then $e_1 = e_2$.

A *branching process* of a P/T-net N is a pair $\beta = (CN, \pi)$, consisting of an occurrence net CN, and a homomorphism π from CN to N. β is a *causal process* of N iff CN is a causal net. For each P/T-net N, there exists a unique (up to renaming of conditions and events) maximal branching process β_m, where 'maximal' is related to the prefix ordering (cf. [8]). An initial part of a causal process may be represented uniquely by the set of events contained in that part: A *configuration* C of a process β is a downward closed conflict free set of events, i.e., a set $C \subseteq E$ such that $e \in C$ implies $e' \in C$ for all $e' \leq e$, and, for all $e, e' \in C$, $e\#e'$ does not hold. For each event e, $[e] = \{e' \in E \mid e' \leq e\}$ denotes the configuration generated by e. Note, that each configuration C of a branching process $\beta = (CN, \pi)$ uniquely determines a causal process, containing C as set of events and all conditions connected to elements of C in CN as set of conditions. Arcs and labels are also inherited from CN. Hence, notions defined for (causal) processes may also be applied to configurations and vice versa.

A configuration C defines a unique marking, consisting of exactly all conditions of CN which are marked after occurrence of all events of C: A *co-set* is a set B' of conditions of an occurrence net CN such that, for all $b \neq b' \in B'$, neither $b < b'$ nor $b' < b$ nor $b\#b'$. A *cut* is a maximal co-set B' (w.r.t. to set inclusion). The cut of a finite configuration C is defined by $Cut(C) = (Min(C) \cup C^\bullet) \setminus {}^\bullet C$. A configuration C of a process $\beta = (CN, \pi)$ defines a marking of CN by $Mark(C) = p(Cut(C))$. Note that a marking M is reachable in a P/T-net N iff the maximal branching process β_m of N contains some configuration C such that $M = Mark(C)$.

Figure 1(ii) shows a causal branching process of the P/T-net underlying the time Petri net TN_1. Names of conditions and events are omitted; the image of a vertex x under π (also called *label of* x) is written beside it.

2.2 A Temporal Logic for Safe P/T-Nets

In this section we introduce the temporal logic $\mathcal{L}$ for safe P/T-nets defined in [9]. Later on we will extend this logic to time Petri nets and show how the model checking algorithm of [9] can be applied.

Let N be a P/T-net. The syntax of $\mathcal{L}$ over N is presented in Figure 2.

Properties of the current marking of a P/T-net N are expressed using place assertions. E. g., a formula '$s_1 \wedge s_2$' expresses that places s_1 and s_2 are marked. '$\Diamond\phi$' means that a marking satisfying 'ϕ' is reachable. The derived operator '$\Box\phi$' signifies that 'ϕ' is satisfied at all reachable markings.

ϕ ::=	**true**	(Truth)
	$\mid s \in P_N$	(Place Assertion)
	$\mid \neg\phi$	(Negation)
	$\mid \phi_1 \wedge \phi_2$	(Conjunction)
	$\mid \Diamond\phi$	(Possibly ϕ)
false	$= \neg$**true**	(Falsehood)
$(\phi_1 \vee \phi_2)$	$= \neg(\neg\phi_1 \wedge \neg\phi_2)$	(Disjunction)
$(\phi_1 \Rightarrow \phi_2)$	$= \neg\phi_1 \vee \phi_2$	(Implication)
$\Box\phi$	$= \neg\Diamond\neg\phi$	(Always ϕ)

Fig. 2. Esparza's Temporal Logic

Satisfaction of a formula ϕ w.r.t. a P/T-net N is defined inductively as follows. Let C be a finite configuration of a branching process β of N. The superscript N is dropped if it is clear from the context.

$(\beta, C) \models^N$ **true**
$(\beta, C) \models^N p$ iff $p \in \mathit{Mark}(C)$
$(\beta, C) \models^N \neg\phi$ iff not $(\beta, C) \models^N \phi$
$(\beta, C) \models^N \phi_1 \wedge \phi_2$ iff $(\beta, C) \models^N \phi_1$ and $(\beta, C) \models^N \phi_2$
$(\beta, C) \models^N \Diamond\phi$ iff $(\beta, C') \models^N \phi$ for some finite configuration $C' \supseteq C$
$\beta \models^N \phi$ iff $(\beta, \emptyset) \models^N \phi$
$N \models^N \phi$ iff $\beta_m \models^N \phi$.

The McMillan-unfolding of a P/T-net N was defined in [16] as a finite prefix of the maximal branching process β_m of N such that each reachable marking of N occurs as an image of some cut of this prefix. Let N be a P/T-net and β_m its maximal branching process. Let $\bot$ be a new event ('pseudo-event') and define $[\bot] = \emptyset$. An event $e \in E_m \cup \{\bot\}$ is a *cut-off event* iff there exists an event e' such that $\mid [e'] \mid < \mid [e] \mid$ and $\mathit{Mark}([e]) = \mathit{Mark}([e'])$. Let E_f be the set of events of β_m given by: $e \in E_f$ iff no event $e' < e$ is a cut-off event. $\beta_f = (B_f, E_f, F_f, \pi_f)$, the prefix of β_m having E_f as set of events, is called *McMillan-unfolding* of N and denoted by $\mathit{McM}(N)$. It can be shown to be unique and finite.

Let $\mathit{Sat}(\phi)_N$ and $\mathit{Sat}_f(\phi)_N$ denote, respectively, the set of finite configurations of the maximal branching process and the set of configurations of the McMillan-unfolding of the P/T-net N that satisfy ϕ. Since $\mathit{Sat}(\phi)_N = \emptyset$ implies (not $N \models \phi$) which is equivalent to $N \models \neg\phi$, the model checking problem reduces to checking emptiness of $\mathit{Sat}(\phi)_N$. Esparza's theorem [9] states that for the latter it is sufficient to inspect $\mathit{Sat}_f(\phi)_N$.

To simplify notation, a different but equivalent definition of the satisfaction relation will be applied in this paper. Let N be a P/T-net, ϕ an $\mathcal{L}$-formula, and M a marking of N. Then:

$M \models' \texttt{true}$
$M \models' p$ iff $p \in M$
$M \models' \neg\phi$ iff not $M \models' \phi$
$M \models' \phi_1 \wedge \phi_2$ iff $M \models' \phi_1$ and $M \models' \phi_2$
$M \models' \Diamond\phi$ iff $M' \models' \phi$ for some marking M' reachable from M
$N \models' \phi$ iff $M_0 \models' \phi$.

Proposition 1. *Let N be a P/T-net and ϕ a formula over N. Then:*

$$N \models \phi \textit{ iff } N \models' \phi.$$

2.3 Time Petri Nets

Time Petri nets were introduced in [18]. Following the reasoning of [20, 22], we consider time Petri nets with discrete time (also cf. [6]).

A *safe time Petri net TN* net consists of a safe P/T-net (P_{TN}, T_{TN}, F_{TN}) and a transition inscription $\chi_{TN} : T \to \mathcal{T}$ (where $\mathcal{T} = \mathbf{N} \times \mathbf{N}$) by closed time intervals with nonnegative integer bounds. The subscripts are omitted if clear from the context.

For $\chi(t) = (\mathit{eft}(t), \mathit{lft}(t))$ we call $\mathit{eft}(t)$ ($\mathit{lft}(t)$, respectively) the *earliest firing time* (*latest firing time*, respectively) of t. The intended meaning is that $\mathit{eft}(t)$ ($\mathit{lft}(t)$, respectively) denotes the minimal (maximal, respectively) number of time units which may pass between enabling and occurrence of t. $\chi(t)$ is denoted by $(\mathit{eft}, \mathit{lft})$ if t is clear from the context. A time Petri net *TN* is called safe if its underlying P/T-net N is safe.

A *state* (M, I) of a time Petri net is given by a marking M of the underlying P/T-net together with a clock vector $I : T \to (\mathbf{N} \cup \{\$\})$ such that $I(t) = \$$ or $I(t) \le \mathit{lft}(t)$ for all $t \in T$. The clock vector associates a clock to each transition, showing the number of time units that have elapsed since enabling of the transition, the \$-symbol indicating that the corresponding transition is not enabled at all. State (M, I) is called *consistent* iff, for all $t \in T$, $I(t) \ne \$ \Leftrightarrow t \in \mathit{Enabled}(M)$. Only consistent states will be considered.

For a clock vector I and a time delay $\theta \in \mathbf{N}$ such that $I(t) + \theta \le \mathit{lft}(t)$ for all $t \in \mathit{Enabled}(M)$, $(I + \theta)$ is defined by

$$(I+\theta)(t) = \begin{cases} I(t) + \theta \text{ if } t \in \mathit{Enabled}(M) \\ \$ \qquad\qquad \text{otherwise.} \end{cases}$$

The *initial state* (M_0, I_0) is given by the initial marking M_0 of the underlying P/T-net and the *initial clock vector* I_0 defined by

$$I_0(t) = \begin{cases} 0 \text{ if } t \in \mathit{Enabled}(M) \\ \$ \text{ otherwise.} \end{cases}$$

Two types of events are considered for a time Petri net, namely events during which time passes (called *tic-events*) and events during which a transition occurs (called *occur-events*):

1. *Tic-event*: The tic-event *tic* is *fireable at state* (M, I) iff no transition is forced to occur in between, i.e. iff for all $t \in \mathit{Enabled}(M)$, $I(t) < \mathit{lft}(t)$. In this case, the successor state (M', I') is given by $M' = M$ and $I' = (I + 1)$. The tic-event is denoted by $(M, I) \rightarrow^{tic} (M', I')$.
2. *Occur-events*: An occur-event is *fireable at state* (M, I) iff some transition t may occur, i.e. if $t \in \mathit{Enabled}(M)$ and $\mathit{eft}(t) \le I(t) \le \mathit{lft}(t)$. In this case, the successor state (M', I') is given by $M' = (M \setminus {}^\bullet t) \cup t^\bullet$ and

$$I'(t') = \begin{cases} \$ & \text{if } t' \notin \mathit{Enabled}(M') \\ 0 & \text{if } t' \notin \mathit{Enabled}(M \setminus {}^\bullet t) \wedge t' \in \mathit{Enabled}(M') \\ I(t') & \text{otherwise.} \end{cases}$$

An occur-event is denoted by $(M, I) \rightarrow^{t} (M', I')$.

A set $\{t_1, ...t_n\}$ of transitions is *concurrently fireable* from state (M, I) iff, for $1 \le i \le n$, t_i is fireable from state (M, I) and, for all $1 \le i < j \le n$, ${}^\bullet t_i \cap {}^\bullet t_j = \emptyset$. A *firing schedule* of a time Petri net TN is a finite or infinite sequence $\sigma = (\overline{t_0}\ \overline{t_1}\ \overline{t_2} \ ...)$ such that, for all i, $\overline{t_i} \in T$ or $\overline{t_i} = tic$. σ is *fireable at state* $S = S'_0$ iff there exist states $S'_1, S'_2, S'_3, ...$ such that $\overline{t_i}$ is fireable at S'_i. State S is called *reachable* from S_0 via σ iff some fireable firing schedule σ leads from S_0 to S. A marking M is *reachable via* σ iff, for some clock vector I, there is a state (M, I) reachable via σ.

Consider the net TN_1 of Figure 1. The initial clock vector is given by $I_0(t_0) = I_0(t_1) = 0$ and $I_0(t_2) = I_0(t_3) = \$$. For example, the firing schedule $\sigma = (tic\ t_0\ t_1\ tic\ tic\ t_3\ t_0\ tic\ t_1)$ is fireable at (M_0, I_0).

The behaviour of a time Petri net TN can be described by the reachability graph [20], which – in the case of P/T-nets – faces the same problems of not reflecting the concurrent behaviour, but instead representing all interleavings of concurrent events.

The logic $\mathcal{L}$ may be extended to time Petri nets. Let TN be a time Petri net, ϕ a formula, and (M, I) a marking of TN. The superscript TN is dropped if it is clear from the context.

$$\begin{array}{lll} (M,I) \models_t^{TN} \mathtt{true} & & \\ (M,I) \models_t^{TN} p & \text{iff} & p \in M \\ (M,I) \models_t^{TN} \neg\phi & \text{iff} & \text{not } (M,I) \models_t^{TN} \phi \\ (M,I) \models_t^{TN} \phi_1 \wedge \phi_2 & \text{iff} & (M,I) \models_t^{TN} \phi_1 \text{ and } (M,I) \models_t^{TN} \phi_2 \\ (M,I) \models_t^{TN} \Diamond\phi & \text{iff} & (M',I') \models_t^{TN} \phi \text{ for some state} \\ & & (M',I') \text{ reachable from } (M,I) \\ TN \models_t^{TN} \phi & \text{iff} & (M_0,I_0) \models_t^{TN} \phi. \end{array}$$

E.g. property Q of the introductory example is expressed by the formula $\phi = \Box((p_2 \wedge p_3) \Rightarrow (\Diamond p_0))$, which is not satisfied by TN_1.

3 Expanding Time Restrictions

In this section, the notion of time-expansion of a time Petri net TN into a P/T-net $N = X(TN)$ is introduced. It is shown that N captures the behaviour of TN

in the sense that satisfaction w.r.t. the temporal logic $\mathcal{L}$ is preserved. In general, the size of $X(TN)$ may be exponential in the size of TN, but the unfolding of TN may be generated without explicitly constructing $X(TN)$ (cf. [11]). Still, the definition of $X(TN)$ is needed to obtain a proper definition of the finite prefix of TN.

For the construction of N we will start with the P/T-net underlying TN and then add parts representing the components of TN related to time. First, clock vectors have to be represented in N. This is done by adding new places for the possible clock positions for each transition t.

Next, new transitions are introduced modelling tic-events. Since we would like to represent time dependence of conditions and events in TN by causal dependence in the time-expansion, these have to take into account also TN's 'original' places. The obvious solution is to introduce, for each state (M, I) of TN which enables a tic-event, a new transition, having M and the places representing I as preset and M and the places representing $(I+1)$ as postset. But, this solution causes a problem: If $M' \subseteq M$, then the tic-transition for (M', I) instead of the tic-transition for (M, I) may fire, possibly leaving some places untouched. To avoid this problem, for each place p of TN, we introduce a new 'complementary' place p^c. Consequently, the marking part of (M, I) is represented by $M \cup M^c$ where $M^c = \{p^c \mid p \in P \wedge p \notin M\}$ in N.

At last, we have to re-adjust the transitions of TN, since these have to respect the complementary places and the clock places (fire-transitions). The first task is fulfilled by having ${}^\bullet t \cup \{p^c \mid p \in t^\bullet\}$ as part of the preset and $t^\bullet \cup \{p^c \mid p \in {}^\bullet t\}$ as part of the postset (except for those places which are contained in both preset and postset of t). The second task raises new problems (see figure 3).

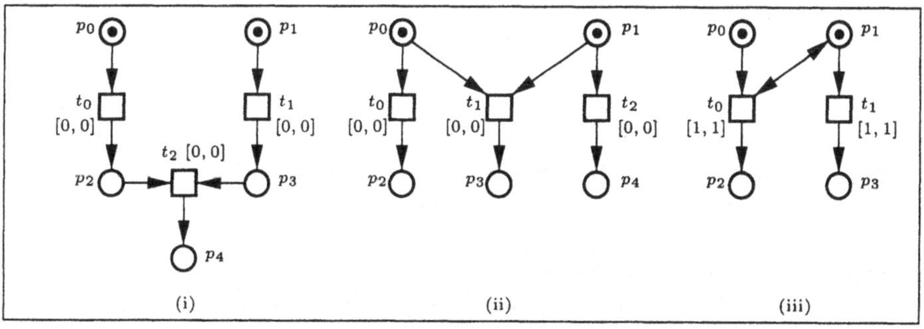

Fig. 3. Problems with updating of clocks

First, consider the time Petri net in part (i) of figure 3. If both t_0 and t_1 fire, then the clock for t_2 has to be started. But this can be decided neither by t_0 nor by t_1 alone. We will solve this problem by allowing fire-transitions to produce an intermediate marking representing an inconsistent state and by introducing additional repair-transitions called update-transitions. Subsequently, also the tic-transitions have to be re-adjusted to act properly on inconsistent states. Otherwise, faulty time distributions would be generated.

Second, consider (ii) of figure 3. Note that t_0 and t_2 are concurrently enabled. Upon firing, any of t_0 and t_2 has to switch off the clock for t_1. Hence, transitions modelling t_0 and t_2 in N have to synchronize on the clock-places for t_1. This introduces a conflict between t_0 and t_2, i.e. concurrency gets lost. This problem may also be fixed by leaving switching off the clock for t_1 to some update- or tic-transition.

Update- and tic-transitions are not sufficient to treat correctly the net of part (iii) of figure 3. Consider e.g. the state $(\{p_0, p_1\}, 11)$. Here, t_0 may occur, leading to $(\{p_1, p_2\}, \$0)$. But in the approach sketched here, the result would be $(\{p_1, p_2\}, \$1)$, and there is no inconsistency between the marking part and the clock part of this state. Thus a repair-transition cannot help. The problem relies on the fact that t_2 becomes disabled and newly enabled at the same instance of time or equivalently, that a token is removed from p_1 and a token is put on p_1 at the same instance of time.

In this special case the problem could be solved by introducing a private set of clock places w.r.t. t_1 for both t_0 and t_1, and by allowing firing of t_1 only if both its clocks are running. This solution also works in the general case, but leads to a much more involved definition of time-expansion and to a bigger time-expanded net $X(TN)$. Instead, we will restrict the class of nets under consideration by introducing a requirement of (local) time-divergence: If a place p loses its token at time θ, then no token will arrive at p before time $\theta + 1$[1].

Places of Petri nets denote (local) state. Hence, the divergent-time condition demands that if a local state is left it may not be entered again at the same instance of time. The latter property is fulfilled in many practical applications. Consider e.g. a resource with exclusive access: Here a local state (i.e. a place) may serve as a guard for the resource, a token in the place indicating availability of the resource. In this case, the time elapsing between the token leaving the place and a new token entering the place corresponds to the time for one allocation cycle of the resource. If time requirements are at all taken into account it is very reasonable to assume that this needs some non-zero amount of time.

These considerations lead to the following definitions:

Definition 1. *[Divergent-Time Property]*

Let $TN = (P, T, F, \chi, M_0)$ be a time Petri net. TN satisfies the *divergent-time property* (and is called *DT-net*) iff the following holds: If $(t_1 ... t_n)$ is enabled at (M, I) in TN such that $\bigcup_{1 \leq i \leq n} {}^\bullet t_i \cap \bigcup_{1 \leq i \leq n} t_i{}^\bullet \neq \emptyset$, then, for some $i \in \{1, ..., n\}$, $t_i = tic$. ◇

Definition 2. *[Time-Expansion]*

Let $TN = (P, T, F, \chi, M_0)$ be a DT-net. The *time-expansion* of TN is given by the P/T-net $X(TN) = (P_X, T_X, F_X, M_{X_0})$ defined as follows:

[1] Note, that the requirement of locally divergent time is slightly stronger than that given in [3], but may be fully released if the 'multiple clock solution' is chosen.

1. For each place $p \in P$, a new (complementary) place p^c is introduced. The set of these places is denoted by P^c.
2. For each transition $t \in T$ new places $p_t^{\$}, p_t^0, ..., p_t^{lft(t)}$ are introduced. The intended meaning of these places is to describe all possible clock positions for the corresponding transition t. The set of the new places is denoted as *Clock*, and the places are referred to as ***clock-places***.
3. For a clock function I of TN, let $PL(I) = \{p_t^{I(t)} \mid t \in T\}$. The time-expansion of a state (M, I) of TN is defined by $X(M, I) = (M \cup M^c \cup PL(I))$, where $M^c = \{p^c \mid p \in (P \setminus M)\}$.
4. For each state (M, I) of TN such that a time-event may occur a new transition $tic(M, I)$ is introduced, having $M \cup M^c \cup PL(I)$ as its preset, and $M \cup M^c \cup PL(I')$ as its postset, where, for all $v \in T$,

$$I'(v) = \begin{cases} I(v) + 1 & I(v) \neq \$ \wedge v \in Enabled(M) \\ 1 & I(v) = \$ \wedge v \in Enabled(M) \\ \$ & \text{otherwise.} \end{cases}$$

   These transitions are intended to model time-events and, in addition, updating clock vectors. The set of these transitions is denoted as *Tic*, and the transitions are referred to as ***tic-transitions***.
5. For each transition t in TN and for $eft(t) \leq i \leq lft(t)$, a new transition $t(i)$ is introduced. The preset of $t(i)$ is given by $^\bullet t \cup \{p^c \mid p \in (t^\bullet \setminus {}^\bullet t)\} \cup \{p_t^i\}$ and the postset by $t^\bullet \cup \{p^c \mid p \in ({}^\bullet t \setminus t^\bullet)\} \cup \{p_t^{\$}\}$.
   These transitions are intended to model changing of the marking. Moreover, the clock of the firing transition is switched off. The set of these transitions is denoted as *Fire*, and the transitions are referred to as ***fire-transitions***.
6. For each inconsistent state (M, I) of TN a new transition $update(M, I)$ is generated. The preset of the transition $update(M, I)$ is given by $M \cup M^c \cup PL(I)$ and the postset by $M \cup M^c \cup PL(I')$, where, for all $v \in T$,

$$I'(v) = \begin{cases} I(v) & I(v) \neq \$ \wedge v \in Enabled(M) \\ 0 & I(v) = \$ \wedge v \in Enabled(M) \\ \$ & \text{otherwise.} \end{cases}$$

   These transitions are intended to model updating the clocks of newly activated or deactivated transitions. The set of these transitions is denoted as *Update*, and the transitions are referred to as ***update-transitions***.

Hence, altogether we have $P_X = P \cup P^c \cup Clock$, $T_X = Tic \cup Fire \cup Update$, F_X as described above, and $M_{X_0} = M_0 \cup M_0^c \cup PL(I_0)$. ◇

Definition 3. Let TN be a DT-net and $X(TN)$ its time-expansion. A marking $\overline{M}$ of $X(TN)$ is called *consistent* iff $\mid \overline{M} \cap \{p, p^c\} \mid = 1$ for each $p \in P$ and $\mid \overline{M} \cap \{p_t^{\$}, p_t^0, ... p_t^{lft(t)}\} \mid = 1$ and $(p_t^{\$} \in \overline{M} \Leftrightarrow t \notin Enabled(\overline{M} \cap P))$ for all $t \in T$. ◇

Theorem 1. *Let $TN = (P, T, F, \chi, M_0)$ be a DT-net. Then the following hold:*

(i) $(M, I) \rightarrow^{\{t_1,\ldots,t_n\}} (M', I')$ *in TN iff either*
$M \cup M^c \cup PL(I) \rightarrow^{\{t_1(I(t_1)),\ldots,t_n(I(t_n))\}} M' \cup M'^c \cup PL(I')$ *or*
$M \cup M^c \cup PL(I) \rightarrow^{\{t_1(I(t_1)),\ldots,t_n(I(t_n))\}} \overline{M} \cup \overline{M}^c \cup PL(\overline{I})$ *and*
$\overline{M} \cup \overline{M}^c \cup PL(\overline{I}) \rightarrow^{update(\overline{M},\overline{I})} M' \cup M'^c \cup PL(I')$ *in $X(TN)$.*

(ii) The consistent state (M, I) is reachable in TN iff the consistent marking $(M \cup M^c \cup PL(I))$ is reachable in $X(TN)$. In particular, M is a reachable marking of TN iff $M = M' \cap P$ for some reachable marking M' of $X(TN)$.

Figure 4 (i) shows a simple time Petri net TN_2. The time expansion $X(TN_2)$ (see figure 4 (ii)) shows up the strong connection between the 'clock-net' (on the right) and the 'normal-net' (on the left). Some update-transitions are omitted, c.f. *update*$(\{p_0\}, 1\$\$)$ because the state $(M, I) = (\{p_0\}, 1\$\$)$ is not consistent.

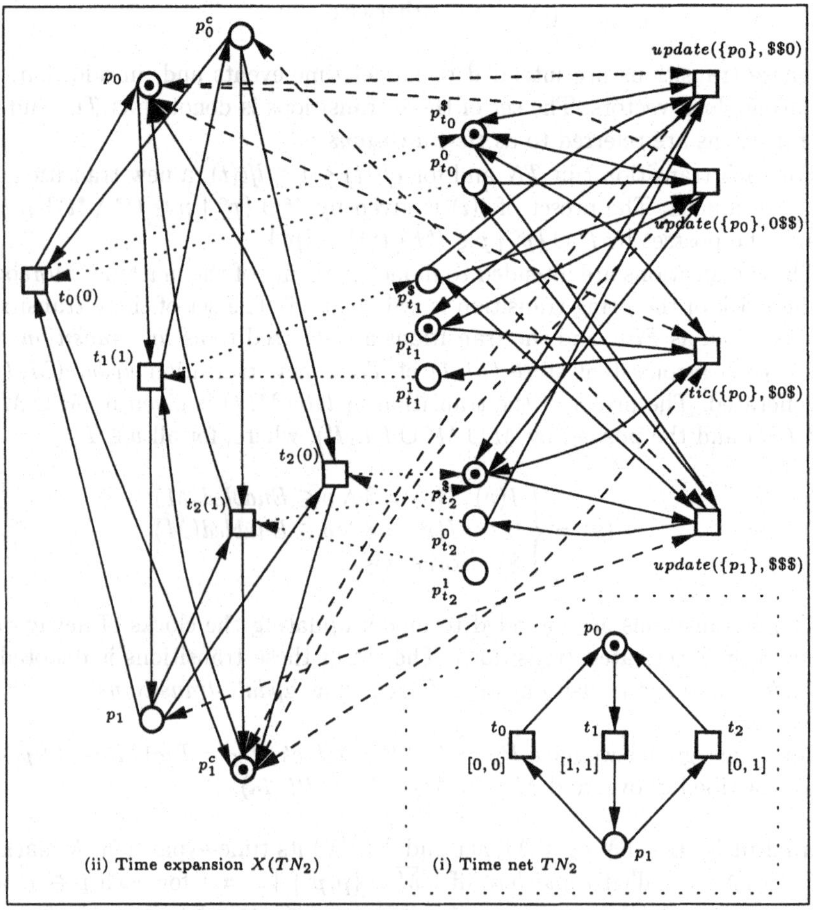

Fig. 4. Time expansion example

4 Partial Order Model Checking of Time Petri Nets

In section 2.3 we have extended the temporal logic $\mathcal{L}$ to time Petri nets. In this section we will show that, for a time Petri net *TN*, the model checking problem w.r.t. $\mathcal{L}$ can be solved by applying Esparza's model checking algorithm to the time expansion $X(TN)$ of *TN*.

Let e be a cut off event of a P/T-net N, and e^0 an event such that $| [e^0] | < | [e] |$ and $Mark([e]) = Mark([e^0])$. As pointed out in [9], in this case the possible future after $Cut([e])$ is isomorphic to the possible future after $Cut([e^0])$. This property is crucial for the proof of Esparza's theorem.

Figure 1 (iii) shows, that for a time Petri net *TN* the possible future after some finite configuration C also depends on the timing associated to C. It follows, that for a time Petri net *TN* that part of the McMillan-unfolding $McM(N)$ of the underlying P/T-net N, which is realizable under the time restrictions, does not contain sufficient information to serve as a basis for model checking (cf. Figure 1 (ii)).

The time-expansion $X(TN)$ of a time Petri net *TN*, on the other hand, has all necessary information about future behaviour, cf. theorem 1. The connection of *TN* and $X(TN)$ w.r.t. satisfaction of formulae is given by the following lemma, which states that a temporal logic formula ϕ is time-satisfied by a state of a time Petri net *TN* iff it is satisfied by the corresponding marking of the time-expansion $X(TN)$ of *TN*.

Lemma 1. *Let TN be a DT-net, (M, I) a reachable state of TN, and ϕ a formula over TN. Let $\overline{M} = M \cup M^c \cup PL(I)$ be the time-expansion of (M, I). Then*

$$(M, I) \models_t^{TN} \phi \text{ iff } \overline{M} \models'^{X(TN)} \phi.$$

Corollary 1. *Let TN be a DT-net and ϕ a formula over TN. Then*

$$TN \models_t^{TN} \phi \text{ iff } X(TN) \models^{X(TN)} \phi.$$

Hence, by Esparza's theorem, in order to check if *TN* satisfies a given formula ϕ, it is sufficient to consider the McMillan-unfolding of the time expansion $X(TN)$. On the other hand, $X(TN)$ – and therefore also its McMillan-unfolding – contains also some unnecessary information; e.g., conditions labeled by clock-places or complementary places which do not occur in formulas. If these are removed, also events labeled by update-transitions become meaningless, because then these just reproduce their presets. Events labeled by tic-transitions, on the other hand, will still be relevant because these are representing timed ordering of conditions. According to these considerations, we will simplify $McM(X(TN))$.

Definition 4 (Reduced unfolding).
Let *TN* be a DT-net and let $\beta = (B, E, F, \pi)$ be any branching process of its time-expansion. The *reduced* branching process $R(\beta)$ of β is constructed by performing the following steps:

1. Remove from β all conditions labeled by clock-places or complementary places and all incidental arcs, resulting in a process β_1.

2. Remove from β_1 all parts which are in conflict with an update-transition resulting in a process β_2.
3. Remove from β_2 all events labeled by update-transitions and all incidental arcs. Each time such an event e is taken out, $^\bullet e$ and $e^\bullet$ are fused. This results in a process β_3.
4. Rename the labels of all events: Label $t(i)$ of event e is replaced by t. This results in a process β_4.
5. For a configuration C of β the reduction $R(C)$ is defined by $R(C) = C \cap E_{R(\beta)}$.

◇

It is easy to see that reduction of a process β of $X(TN)$ is well defined, and that reduction preserves inclusion of configurations i.e.,

$$C \subseteq C' \text{ in } \beta \text{ iff } R(C) \subseteq R(C') \text{ in } R(\beta)$$

for all configurations C, C' in β.

Theorem 2. *Let TN be a DT-net, β a process of $X(TN)$, and ϕ a $\mathcal{L}$-formula over TN. Then it holds that*

$$\beta \models^{X(TN)} \phi \textit{ iff } R(\beta) \models^{X(TN)} \phi.$$

Theorem 2 motivates the following definition:

Definition 5 (McMillan-unfolding). Let *TN* be a DT-net. The McMillan-unfolding of *TN* is defined by $McM(TN) = R(McM(X(TN)))$. ◇

The McMillan-unfolding $McM(TN_1)$ of the time Petri net TN_1 is shown in Figure 1 (iii).

Next we will prove that, for a time Petri net *TN*, $McM(TN)$ satisfies exactly the same formulae as the McMillan-unfolding of $X(TN)$. Let $Sat_{rf}(\phi)_N$ denote the set of configurations of the reduced McMillan-unfolding of $X(TN)$ that satisfy the $\mathcal{L}$-formula ϕ over *TN*. The following proposition is an easy consequence of theorem 2:

Proposition 2. *Let TN be a time Petri net and ϕ be a $\mathcal{L}$-formula over TN. Then*

$$Sat_{rf}(\phi)_{X(TN)} = \emptyset \textit{ iff } Sat_f(\phi)_{X(TN)} = \emptyset.$$

Finally, putting together the puzzle, we can establish the key result of this paper, which states that a formula ϕ is time-satisfied by a time Petri net *TN* iff it is satisfied by the McMillan-unfolding of the time-expansion of *TN*.

Theorem 3. *Let TN be a time Petri net, $X(TN)$ its time expansion, and ϕ a formula over TN. Then*

$$TN \models_t \phi \textit{ iff } Sat_{rf}(\neg\phi)_{X(TN)} = \emptyset.$$

Consequently, to check of an $\mathcal{L}$-formula ϕ against a time Petri net *TN* one has to compute the McMillan-unfolding $McM(TN)$ and to check ϕ against $McM(TN)$.

Note, that neither the explicit construction of $X(TN)$ nor of $McM(X(TN))$ is needed for the construction of $McM(TN)$. In [11] we present an algorithm, which is a refinement of the algorithm for the finite prefix given in [10]. We prove that with this algorithm, $McM(TN)$ may be computed in $O(n \cdot log_n)$ time, where n denotes the size (number of nodes) of $McM(TN)$. It is shown that the number of events in $McM(TN)$ is bounded by the number of reachable states in *TN*, i.e. that the size of $McM(TN)$ is at most linear in the size of the reachability graph of *TN*. Moreover, some examples of time Petri nets are presented whose finite prefix is substantially smaller than the reachability graph.

5 Related Work

Approaches to model checking of time Petri nets w.r.t. (timed) temporal logics are either based on the timed state graph (TSG) [23, 24], or on the timed branching process [3, 15, 23].

In [24], model checking of safe time Petri nets with rational firing bounds w.r.t. a real time extension of linear time temporal logic is considered, which is based on differences between timing variables rather than absolute time points. A finite graph representing the state space is computed whose nodes consist of reachable markings augmented by additional sets of inequalities describing timing conditions. Since each reachable marking is represented by at least one node, this appraoch faces the state explosion problem.

The time state graph construction of [24] is improved in [15] in several ways. First, a coarser semantics which does not take into account implicit ordering of events is considered. Second, the finite prefix of the underlying net is used to speed-up searching the time state graph.

Model checking against TCTL is explored in [23]. Here, a region graph construction is used to compute a finite representation of the timed state space, cf. the construction for timed automata in [1].

A notion of timed process of a time Petri net, denoting a causal process which is realizable under the time constraints of the net, is introduced in [3]. A method is developed to compute all valid timings for a causal process of a time Petri net with divergent time.

In [21], the class of safe time Petri nets is restricted to *time independent choice* nets and analysed w.r.t. reachability. The reachability checking algorithm is based on a finite prefix of the underlying net. Like in [24], timing conditions are represented explicitely by sets of inequalities as additional inscriptions of events. Using the prefix construction, an asynchronous circuit with bounded delay is verified w.r.t. hazard-freedom.

The approach presented in this paper may also be compared to timed automata (cf. e.g. [1]). E.g. checking reachability in a time Petri net *TN* can be reduced to reachability analysis in a timed automaton *TA* in the following way: Given a safe time Petri net *TN*, first compute its reachability graph $G = (V, E)$, giving the states and the transitions of *TA*. Introduce a clock c_t for each transition t of *TN*. The insription of a transition $M \rightarrow^t M' \in E$ consists of the guard $\mathit{eft}(t) \leq c_1 \leq \mathit{lft}(t)$ and the set of clocks to be reset is given by $\{t' \mid t' \in \mathit{Enabled}(M - {}^\bullet t)\}$. Moreover, for each state M, the invariant is given by `true`. Unfortunately, this construction is very expensive: First, the size of G may be exponential in the size of *TN*; second, in order to check reachability in *TA*, the region graph construction is needed, bringing along an additional exponential blow up.

On the other hand, simulation of general timed automata by time Petri nets seems to be unfeasible.

6 Conclusion

We have defined the finite prefix of a time Petri net *TN*, based on the time-expansion of *TN*. By showing that this is sufficient for applying Esparza's model checking algorithm for safe P/T-nets, we have established a first – but still substantial – step towards quantitative analysis of distributed real-time systems based on partial order semantics. The next steps will consist in extending our approach to ***real-time temporal logic***.

In [12], the language SDL/R was introduced which extends SDL (Specification and Description Language) by adding real-time requirements. For SDL/R, a denotational semantics was defined in terms of time M-nets, which form a class of compositional timed high level Petri nets. As pointed out in [12], time M-nets may be unfolded into safe time Petri nets. Hence, the result of this paper also suggests a new method for proving functional and qualitative temporal properties of SDL/R-specifications. Currently, the new method is being integrated into the MOBY-tool (cf. [2]) and the PEP-tool (cf. [5]).

Acknowledgements

The authors would like to thank Javier Esparza, Eike Best, Alexander Lavrov, and an anonymous referee for helpful comments on preliminary versions of the paper.

References

1. R. Alur, C. Courcoubetis, and D. Dill: Model Checking for Real-time Systems. *Proc. 5th Symp. LICS* (1990).
2. P. Amthor, H. Fleischhack, and J. Tapken: *MOBY – more than a Tool for the Verification of SDL-Specifications.* Technical Report, University of Oldenburg (1996).
3. T. Aura, J. Lilius: Time Processes for Time Petri Nets. *Proc. ICATPN*, Toulouse (1997).

4. B. Berthomieu, M. Diaz: Modelling and Verification of Time Dependent Systems Using Time Petri Nets. *IEEE Transactions on Software Engineering*, Volume 17/3, 259–273 (1991).
5. E. Best: Partial Order Verification with PEP. Proc. *POMIV'96, Partial Order Methods in Verification*, G. Holzmann, D. Peled, and V. Pratt (eds.), American Mathematical Society (1996).
6. B. Bieber, H. Fleischhack: *Model Checking of Time Petri Nets Based on Partial Order Semantics.* Techn. Rep., CvO University of Oldenburg (1999). URL: http://theoretica.informatik.uni-oldenburg.de/~max/paper/
7. E.M. Clarke, E. A. Emerson, and A. P. Sistla: Automatic Verification of Finite-state Concurrent Systems Using Temporal Logic Specifications. *ACM TOPLAS* 8, 244–263 (1986).
8. J. Engelfriet: Branching Processes of Petri Nets, *Acta Informatica*, Volume 28, pages 575–591 (1991).
9. J. Esparza: Model Checking Using Net Unfoldings. *Science of Computer Programming*, Volume 23, 151–195, Elsevier (1994).
10. J. Esparza, S. Römer, and W. Vogler: An Improvement of McMillan's Unfolding Algorithm. *Proc. of TACAS'96* (1996).
11. H. Fleischhack: *Efficient computation of processes for time Petri nets.* University of Oldenburg (1999). (in preparation)
12. H. Fleischhack, and J. Tapken: An M-Net Semantics for a Real-Time Extension of μSDL, *Proc. FME'97*, Graz (1997).
13. P. Godefroid: Using partial orders to improve automatic verification methods. in *Proc. Workshop on Computer Aided Verification* (1990).
14. B. Graves: Identification of Specific Processes Contained in McMillan's Finite Unfolding. in *Proceedings of FSTTCS17*, Kharagpur, India (1997).
15. J. Lilius: Efficient state space search for time Petri nets. *MFCS Workshop on Concurrency* (1998).
16. K. MacMillan: Symbolic Model Checking – an Approach to the State Explosion Problem. PhD Thesis, SCS, Carnegie Mellon University (1992).
17. K.L. McMillan: Using unfoldings to avoid the state explosion problem in the verification of asynchronous circuits. *Proc. CAV '92, Fourth Workshop on Computer-Aided Verification*, Vol. 663 of Lecture Notes in Computer Science, 164–174 (1992).
18. P. Merlin, D. Farber: Recoverability of Communication Protocols – Implication of a Theoretical Study. *IEEE Transactions on Software Communications*, Vol. 24, 1036–1043 (1976).
19. X. Nicollin, J. Sifakis, and S. Yovine: From ATP to Timed Graphs and Hybrid Systems. Lecture Notes in Computer Science, *Real Time: Theory in Practice* (1992).
20. L. Popova: On Time Petri Nets. *Journal of Information Processing and Cybernetics*, Volume 1055 of *Lecture Notes in Computer Science*, Springer (1991).
21. A. Semenov, A. Yakovlev, and A. Koelmans: Time Petri net unfoldings and hardware verification. University of Newcastle (1998) (Draft).
22. P. Starke: Analyse von Petri-Netz-Modellen. Teubner Verlag, Stuttgart (1990) (in German).
23. I. Virbitskaite, E. Pokozy: On Partial Order Verification of Time Petri Nets. University of Novosibirsk (1998) (Draft).
24. T. Yoneda, B.-H. Schlingloff: Efficient Verification of Parallel Real-Time Systems. *Journal of Formal Methods in System Design*, Vol. 11-2, 187–215 (1997).

Generic Process Algebras for Asynchronous Communication

F.S. de Boer[1] and G. Zavattaro[2]

[1] Utrecht University, Department of Computer Science, The Netherlands.
Email: frankb@cs.ruu.nl
[2] Dipartimento di Scienze dell'Informazione, Università di Bologna, Italy.
Email: zavattar@cs.unibo.it

Abstract. We study at different levels of abstraction general semantic and algebraic properties of languages which are based on asynchronous communication. These different levels of abstraction concern the specific nature of the communication mechanism. At the highest level we introduce a process algebra which characterizes asynchronous communication in general, that is, when abstracting from the specific nature of the communication mechanism. This generic process algebra we further instantiate to algebras for different classes of languages. Considered are classes of languages which are based on a general monoid structure of the actions and classes of languages which are based on read/write operations.

1 Introduction

In this paper we investigate the semantics and equational theories, or process algebras as we call them, for the general paradigm of asynchronously communicating processes which has been introduced in [5]. This paradigm encompasses such diverse systems as described by concurrent logic languages [23], concurrent constraint languages [22], imperative languages in which processes communicate by means of shared variables [13], or asynchronous channels [18], dataflow languages [19], and coordination languages [15].

These systems have in common that processes communicate via some shared data structure. The asynchronous nature of the communication lies in the way access to this shared data structure is modelled: the data structure is updated by means of write primitives that have free access whereas the read primitives may suspend in case the data structure does not contain the information required by it. The execution of the read and write primitives are independent in the sense that they can take place at different times. This marks an essential difference with synchronously communicating processes, like CSP [17], where reading from and writing to a channel has to take place at the same time.

In [5] a generic concurrent language $\mathcal{L}$ is introduced which assumes given a set of basic (or atomic) actions. Statements are constructed from these actions by means of sequential composition, the plus operator for non-deterministic choice, the parallel operator, and recursion. The basic actions are interpreted by means of an interpretation function I as partially defined *state* transformations. When the interpretation $I(a)$ of a pure read action a is undefined in a state this means that the execution of a is suspended. A suspended process is forced to wait until actions of other processes produce a state in which it is enabled. A pure write action a is characterized by the fact that $I(a)$ is a totally defined function. It can always proceed autonomously. In general, an action can embody both a read and a write component.

Many languages for asynchronously communicating processes can be obtained as instances of $\mathcal{L}$ by choosing the appropriate set of actions, the set of states and the interpretation function for the basic actions. For example, the imperative language described in [13], based on shared variables, can be modelled by taking as states functions from variables to values, as actions the set of assignments, and then the usual interpretation of an assignment as a state transformation. Languages based on the blackboard model [14], like Linda [15] and Shared Prolog [8] can be modelled analogously, by taking as states the configurations of a centralized data structure (the blackboard) and as actions checks and updates of the blackboard. Another example is the class of concurrent constraint languages [22]. These are modelled by interpreting the abstract set of states as a constraint system and the actions as ask/tell primitives. Concurrent logic languages, like Flat Concurrent Prolog [23], can be obtained by interpreting the states as the bindings established on the logical variables, and the actions as the unification steps. An asynchronous variant of CCS [20, 21] is modelled by considering the state as a set (or a multi-set) of actions. Performing an action then corresponds to adding it to the set, while performing the complementary action corresponds to testing whether the action is already in the set. Finally, a variant of CSP [17], based on asynchronous channels (see also [18]), can be obtained by taking as states the configurations of the channels and as actions the input-output primitives on these channels.

The basic computation model of $\mathcal{L}$ is described in [5] by means of a labelled transition system which is defined parametric with respect to the interpretation function I. It specifies for every statement what (initial) steps it can take. Each step results in a state transformation, which is registered in the label: as labels we use pairs of states. Based on this transition system, we define the initial/final state semantics of statements which distinguishes between the results of successfully terminating and deadlocking computations.

In this paper we study the congruence relation induced by the initial/final state semantics at different levels of abstraction. Our first result is a denotational characterization of this congruence relation at the highest level of abstraction, that is, when abstracting from the actual interpretation function I. This denotational model is based on *reactive sequences* as introduced in [5]. It is also defined parametric with respect to the interpretation function. In this model a computation is described as a sequence of pairs of states, a so-called reactive sequence. A pair of states encodes the state transformation which occurred during a single (atomic) transition step. These sequences are not necessarily connected, i.e. the final state of a pair can be different from the initial state of the following pair.

These "gaps" represent the possible steps made by the environment. We show that this model based on reactive sequences is fully abstract with respect to the initial/final state semantics in the following *generic* sense: two processes have in *every* interpretation the same set of reactive sequences if and only if in *every* interpretation they are observationally congruent.

In general the standard full abstraction statement for a given interpretation I does not hold. For example, the interpretation I which models shared variable concurrency requires some further abstractions (notably, abstraction from *finite stuttering* and the *granularity of interleaving*). In general these abstractions concern the specific nature of the underlying computational model as described by the given interpretation.

Next, we show that two processes have in *every* interpretation the same set of reactive sequences if and only if they have the same *failure* semantics. This failure semantics as described in [3] is defined directly in terms of the uninterpreted actions of $\mathcal{L}$. As such we have obtained a correspondence between the *state-based* interpretation of the language $\mathcal{L}$ and its *action-based* interpretation as studied in process algebras like PA or ACP [1]. That is, the intersection of the congruence relations induced by the initial/final state semantics of any instantiation of $\mathcal{L}$ coincides with the failure semantics of $\mathcal{L}$. It is worthwhile to observe here that the (action-based) failure semantics of $\mathcal{L}$ does *not* coincide with the *maximal traces* (of actions) of $\mathcal{L}$ because $\mathcal{L}$ does not include the *encapsulation* operator of ACP which allows one to hide actions. As such the maximal traces semantics of $\mathcal{L}$ already constitutes a congruence relation. However this action-based semantics of $\mathcal{L}$ does not capture the deadlock behavior as described by the initial/final state semantics of $\mathcal{L}$. This deadlock behavior instead is captured by the failure semantics. Consequently the standard failure axioms of [3] provide a general complete axiomatization of asynchronous communication, that is, these axioms characterize what holds in *any* parallel programming language based on an asynchronous communication mechanism.

We proceed our investigation by considering the specific nature of the basic actions of certain classes of instantiations of $\mathcal{L}$. First we consider instantiations which are based on a monoid structure of the basic actions. This monoid describes the internal structure of actions in terms of a state-based composition operation. That is, actions as state-transformations now can be viewed as being composed of other state-transformations. In this view, the 'silent step' τ of process algebras like ACP, denotes the (total) identity transformation, and the 'inaction' δ denotes the transformation which is undefined in every state. Our second main result is that a generic fully abstract state-based model for languages based on the (functional) composition of actions requires abstraction from *finite stuttering* and the *granularity of interleaving*. A complete and generic algebraic characterization is given of these abstractions.

We conclude with an axiomatization of read and write actions. In general a read action is a partially defined identity function (on the set of states) whereas a write action is defined in every state, that is, it never blocks. Moreover we further refine this axiomatization to the class of so-called monotonic languages. These languages are based on a monotonic interpretation of the actions with respect to a given information-ordering on states. Characteristic instances of this class are the concurrent logic (constraint) languages.

1.1 Comparison with related work

Most existing works on process algebras of asynchronous communication concern particular programming languages [4, 6, 7, 10–12, 18]. One of the main contributions of this paper is to provide a general framework for studying algebraic properties of *classes* of languages, instead of one particular language. These classes represent different asynchronous communication mechanisms which are defined in terms of some general semantic characteristics of the basic actions.

Our approach allows for interesting generalizations of various existing semantic models and algebras. For example, in [9] a fully abstract model is given for a concurrent imperative language based on shared-variables. We generalize this model in two dimensions. First of all we give a general semantic treatment of deadlock in languages based on asynchronous communication (in [9] only the results of successfully terminating computations are considered). Moreover, we show that the abstraction from finite stuttering and granularity of interleaving defined in [9] and [5] apply to all instantiations of $\mathcal{L}$ which are based on a monoid structure of actions.

Other generalizations concern the axiomatization of read and write actions and the axiomatization of certain commutativity properties of actions as described in [7, 4, 6]. We axiomatize these properties in terms of a general class of monotonic interpretations.

2 The language $\mathcal{L}$ and its operational semantics

In this section we introduce the language $\mathcal{L}$ and define its operational semantics by means of a transition system.

Definition 1. *Let $(a \in)A$ be an infinite but countable set of actions and let $(x \in)$ Var be a set of (statement) variables. We define the set $(s \in)\mathcal{L}$ of statements as follows:*

$$s ::= a \mid x \mid s;t \mid s+t \mid s \parallel t \mid \mu x.s$$

Moreover, $\mathcal{L}$ contains a special E, the terminated statement.

The symbols ';', '+' and '||' represent the sequential, the choice and the parallel operator respectively. In the statement $\mu x.s$, the free ocurrences of x in s are interpreted recursively as the statement $\mu x.s$ itself. By $s[\mu x.s/x]$ we denote the statement obtained by substituting $\mu x.s$ for each free occurrence of x in s. In the following we consider only closed statements (without free variables).

Definition 2. *An* interpretation *I consists of a set $(\sigma \in)\Sigma$ of states and an assignment of a state-transformation $I(a) \in \Sigma \rightarrow \Sigma_\perp$ to each action a, where $\perp \notin \Sigma$ and $\Sigma_\perp = \Sigma \cup \{\perp\}$.*

Actions are thus interpreted as partially defined functions on the set of states: $I(a)(\sigma) = \perp$ indicates that $I(a)$ is not defined in σ.

Given an interpretation I the computational model of $\mathcal{L}$ is described by a *labelled transition-system* $(\mathcal{L}, Label, \rightarrow)$. The set $(\lambda \in) Label$ of labels is defined by $Label = \Sigma \times \Sigma$. A label represents the state tranformation caused by the action performed during the transition step.

Definition 3. *The transition relation* $\rightarrow \subseteq \mathcal{L} \times Label \times \mathcal{L}$ *is defined as the smallest relation satisfying the following axiom and rules:*

- *For* $\sigma' = I(a)(\sigma)$ *we have* $a \xrightarrow{\langle\sigma,\sigma'\rangle} E$.
- *If* $s \xrightarrow{\lambda} s'$ *then*

  $$s;t \xrightarrow{\lambda} s';t \quad s+t \xrightarrow{\lambda} s' \quad t+s \xrightarrow{\lambda} s' \quad s \parallel t \xrightarrow{\lambda} s' \parallel t \quad t \parallel s \xrightarrow{\lambda} t \parallel s'$$

  If $s' = E$ *then read* t *for* $s';t$, $s' \parallel t$ *and* $t \parallel s'$ *in the clauses above.*
- *If* $s[\mu x.s/x] \xrightarrow{\lambda} s'$ *then* $\mu x.s \xrightarrow{\lambda} s'$.

For $s \neq E$ *we introduce* $s \xrightarrow{\langle\sigma,\Delta\rangle} E$ *to indicate that there does not exist a state* σ' *and a statement* s' *such that* $s \xrightarrow{\langle\sigma,\sigma'\rangle} s'$.

Note that the transition system depends on a given I (this dependence we leave implicit). On the basis of the transition system we introduce the following notion of *observables*.

Definition 4. *Let* $\Sigma^+ = \Sigma \times \{\Delta, E\}$. *Given an interpretation* I, *we define* $I(s) \in \Sigma \rightarrow \mathcal{P}(\Sigma^+)$ *recursively by*

$$I(s)(\sigma) = \bigcup \{I(s')(\sigma') \mid s \xrightarrow{\langle\sigma,\sigma'\rangle} s'\} \cup \{\langle\sigma,\Delta\rangle \mid s \xrightarrow{\langle\sigma,\Delta\rangle} E\}$$

We define $I(E)(\sigma) = \{\langle\sigma, E\rangle\}$, *for every* $\sigma \in \Sigma$. *This recursive definition of* $I : \mathcal{L} \rightarrow (\Sigma \rightarrow \mathcal{P}(\Sigma \times \{\Delta, E\}))$ *can be justified formally as the least fixed point (defined with respect pointwise extended set-inclusion) of a corresponding operator.*

The observables $I(s)$ records for each input state σ both the final results of successfully terminating and deadlocking computations of s in σ. Successfull termination is denoted by E and deadlock is indicated by Δ.

3 A generic process algebra

In this section we introduce a generic process algebra for asynchronous communication. Its genericity lies in the fact that we abstract from the specific nature of the asynchronous communication mechanism. This abstraction is captured by the following congruence relation.

Definition 5. *We define $s \cong t$ if and only if for every interpretation I, $I(c[s]) = I(c[t])$, for every context $c[x]$. A context $c[x]$ is simply a statement with a free statement-variable x. The result of replacing every occurrence of x in $c[x]$ by a statement s is denoted by $c[s]$.*

The relation $s \cong t$ thus describes the identities between statements which hold in every interpretation; this is reflected by the presence of universal quantification on both contexts and interpretations. The following definition (which already appeared in [5]) gives a denotational characterization of this relation which assigns to each statement a set of *reactive sequences.*

Definition 6. *Let I be an interpretation. Let $D = (\Sigma \times \Sigma_{\perp})^*$, with typical element ω. Elements of D are also called reactive sequences. For $R \subseteq D$ and label λ we define $\lambda \cdot R = \{\lambda \cdot \omega \mid \omega \in R\}$ (here $\cdot$ denotes the prefix operation). We define $\tilde{I}(s) \subseteq D$ recursively by*

$$\tilde{I}(s) = \bigcup \{\lambda \cdot \tilde{I}(t) \mid s \xrightarrow{\lambda} t\} \cup \{\langle \sigma, \Delta \rangle \mid s \xrightarrow{\langle \sigma, \Delta \rangle} E\}$$

We define $\tilde{I}(E)(s) = \{\epsilon\}$ (the empty sequence is denoted by ϵ). The formal justification of this recursive definition of $\tilde{I}$ can be given as the least fixed point (defined with respect pointwise extended set-inclusion) of a corresponding operator.

The proof of the following theorem can be found in [5].

Theorem 1. *The relation $s \approx t$ defined by $\tilde{I}(s) = \tilde{I}(t)$, for all I, is a congruence relation.*

We are now able to state the following *generic* full abstractness result.

Theorem 2. *For any statement s and t, $s \cong t$* if and only if *$s \approx t$.*

Proof The *if* part directly follows from the compositionality result of Theorem 1 and by the fact that for every s (distinct from E) and σ the set $I(s)(\sigma)$ can be obtained from $\tilde{I}(s)$ in the following way

$$\begin{aligned} I(s)(\sigma_1) = \{\langle \sigma_{n+1}, E \rangle \mid \langle \sigma_1, \sigma_2 \rangle \langle \sigma_2, \sigma_3 \rangle \ldots \langle \sigma_n, \sigma_{n+1} \rangle \in \tilde{I}(s)\} \cup \\ \{\langle \sigma_n, \Delta \rangle \mid \langle \sigma_1, \sigma_2 \rangle \langle \sigma_2, \sigma_3 \rangle \ldots \langle \sigma_n, \Delta \rangle \in \tilde{I}(s)\} \end{aligned}$$

Note that we consider sequences of connected pairs of states, namely, the second state of a pair is the first one of the next pair. Such sequences represent computations of s viewed as a *closed* system.

In order to prove the *only if* part we proceed by contraposition. We assume $s \not\approx t$; this ensures the existence of an interpretation I such that $\tilde{I}(s) \neq \tilde{I}(t)$. Without loss of generality we may assume the existence of a reactive sequence ω such that $\omega \in \tilde{I}(s) \setminus \tilde{I}(t)$. There are two cases, $\omega = \langle \sigma_1, \sigma_1' \rangle \langle \sigma_2, \sigma_2' \rangle \ldots \langle \sigma_n, \sigma_n' \rangle$ or $\omega = \langle \sigma_1, \sigma_1' \rangle \langle \sigma_2, \sigma_2' \rangle \ldots \langle \sigma_n, \Delta \rangle$. We present only the proof for the first case; the other one is treated similarly.

Let Σ be the set of states of I. We define a new interpretation I' with the state space $(w \in)\Sigma^*$ consisting of all (finite) sequences of states in Σ and the following interpretation of the basic actions:

$$I'(a) \;=\; \lambda w.\text{if } w \neq \epsilon \text{ then } w \cdot I(a)(last(w)) \text{ else } \perp$$

where $\cdot$ denotes the append operation, as a special case we define $w \cdot \perp = \perp$; the last element of the non-empty sequence w is denoted by $last(w)$. In this way we have encoded the history of the computation in the actual state.

We now consider a statement $r = a_1; \ldots; a_{n-1}$ such that each action a_i is fresh in the sense that it does not occur in s or in t. Such an action a_i is used to fill the gap between the states $\sigma_1 \ldots \sigma'_i$ and $\sigma_1 \ldots \sigma'_i \sigma_{i+1}$:

$$I'(a_i) = \lambda\omega.\text{if } \omega = \sigma_1 \ldots \sigma'_i \text{ then } \omega \cdot \sigma_{i+1} \text{ else } \perp$$

It is now easy to see that $I'(s \parallel r)(\sigma_1) \neq I'(t \parallel r)(\sigma_1)$ (hence also $s \not\approx t$). In fact $\langle \sigma_1 \ldots \sigma'_{n-1}\sigma_n\sigma'_n, E\rangle \in I(s \parallel r)(\sigma_1) \setminus I(t \parallel r)(\sigma_1)$, otherwise we would have $\omega \in \tilde{I}(t)$ which contradicts the above assumption. □

However, given an interpretation I, in general the *usual* full abstraction statement

$$\tilde{I}(s) = \tilde{I}(t) \text{ if and only if for every context } c[\,], I(c[s]) = I(c[t])$$

does not hold. For example, the interpretation I which models shared variable concurrency requires some further abstractions (notably, abstraction from *finite stuttering* and the *granularity of interleaving*). These abstractions in general however concern the specific nature of the underlying computational model as described by the given interpretation.

We now show that the relation $\approx$ coincides with the *failure* semantics of the (uninterpreted) language $\mathcal{L}$.

Definition 7. *Consider the usual labelled transition system $s \xrightarrow{a} t$ indicating that the statement s can reach t performing an action a (e.g. $a; s \xrightarrow{a} s$). With $s \xrightarrow{w} t$ we mean that t can be reached after having performed the (finite) sequence of actions w (i.e. if w is the empty sequence then $s = t$, otherwise, if w is the sequence $a_1 \ldots a_n$ then $s \xrightarrow{a_1} \ldots \xrightarrow{a_n} t$). Let $Init(s)$ be the set of labels a such that $s \xrightarrow{a} t$ for some t. We define*

$$\begin{aligned} F(s) = \{&(w, E) \mid s \xrightarrow{w} E\} \cup \\ &\{(w, \mathcal{F}) \mid s \xrightarrow{w} t,\; t \neq E, \text{ and } \mathcal{F} \subseteq A \text{ s.t. } \mathcal{F} \cap Init(t) = \emptyset\} \end{aligned}$$

Given a statement s, $F(s)$ is the set of the pairs (w, E), where the sequence of actions w (performed by s) leads to the termination statement E, and $(w, \mathcal{F})$, where the sequence of actions w (performed by s) leads to a statement different from E that is not able to perform the actions in $\mathcal{F}$.

Theorem 3. *For any statement s and t, $s \approx t$ if and only if $F(s) = F(t)$.*

Proof The *if* part follows from the fact that for every s, $\tilde{I}(s)$ can be obtained from the set $F(s)$ (for example, a deadlock state $\langle \sigma, \Delta \rangle$ is derived from a failure set $\mathcal{F}$ by the condition that $I(a)(\sigma) = \perp$, for $a \notin \mathcal{F}$).

In order to prove the *only if* part we proceed by contraposition. Assume that $F(s) \neq F(t)$; without loss of generality we may assume the existence of a pair $(w, \mathcal{F})$ (or (w, E)) contained in $F(s) \setminus F(t)$. We present only the case of $(w, \mathcal{F})$, as the other one is treated similarly.

Consider the interpretation I whose actions are defined on (finite) sequences $A_1 \ldots A_n$ of *sets* of actions (i.e. $A_i \subseteq A$) in the following manner:

$$I(a) = \lambda\, A_1 \ldots A_n.\text{if } a \in A_1 \text{ then } A_2 \ldots A_n \text{ else } \perp.$$

Let $w = a_1 \ldots a_n$; we have that $I(s)(\{a_1\} \ldots \{a_n\}\mathcal{F}) \neq I(t)(\{a_1\} \ldots \{a_n\}\mathcal{F})$. In fact $\langle \mathcal{F}, \Delta \rangle \in I(s)(\{a_1\} \ldots \{a_n\}\mathcal{F}) \setminus I(t)(\{a_1\} \ldots \{a_n\}\mathcal{F})$, otherwise we have $(w, \mathcal{F}) \in F(t)$ which contradicts the above assumption. This ensures that $s \not\cong t$, and so we have $s \not\approx t$ by Theorem 2. □

By Theorem 3 the well-known algebraic laws which characterize the failure equivalence for PA processes (that is, ACP processes without the synchronous communication function, see [1]) in Table 1 give a complete axiomatization of $s \approx t$ for finite statements.

A1	$s + t \;=\; t + s$
A2	$(s + t) + u \;=\; s + (t + u)$
A3	$s + s \;=\; s$
A4	$(s + t); u \;=\; s; u + t; u$
A5	$(s; t); u \;=\; s; (t; u)$
M1	$s \parallel t \;=\; s \mathbin{\lfloor\!\lfloor} t + t \mathbin{\lfloor\!\lfloor} s$
M2	$a \mathbin{\lfloor\!\lfloor} s \;=\; a; s$
M3	$a; s \mathbin{\lfloor\!\lfloor} t \;=\; a; (s \parallel t)$
M4	$(s + t) \mathbin{\lfloor\!\lfloor} u \;=\; s \mathbin{\lfloor\!\lfloor} u + t \mathbin{\lfloor\!\lfloor} u$
F1	$a; (b; s + t) + a; (b; s' + t') \;=\; a; (b; s + b; s' + t) + a; (b; s + b; s' + t')$
F2	$a; (b + t) + a; (b; s' + t') \;=\; a; (b + b; s' + t) + a; (b + b; s' + t')$
F3	$a; s + a; (t + u) \;=\; a; s + a; (s + u) + a; (t + u)$

Table 1. Axioms for failure equivalence.

The failure axioms thus provide a general characterization of asynchronous communication when abstracting from the particular nature of the actions.

4 The monoid of actions

As the set of actions we now consider not just a generic set A but a monoid $(A, \circ)$, with neutral element τ and zero element δ (i.e. $\tau \circ a = a \circ \tau = a$ and $\delta \circ a = a \circ \delta = \delta$). The operation $\circ$ is interpreted as functional composition of state-transformations. We restrict our analysis to the class of interpretation functions which are *monoidal*.

Definition 8. *A monoidal interpretation I consists of a set $(\sigma \in)\Sigma$ of states and an assignment of a state-transformation $I(a) \in \Sigma \to \Sigma_\perp$ to each action a, which satisfies the monoid structure:*

- $I(\tau) = \lambda\sigma.\ \sigma,$
- $I(\delta) = \lambda\sigma.\ \perp,$
- $I(a \circ b) = \lambda\sigma.$ *if* $I(a)(\sigma) \neq \perp$ *then* $I(b)(I(a)(\sigma))$ *else* $\perp$.

The action τ denotes the identity function and δ denotes the nowhere defined function. The composite action $a \circ b$ is interpreted as the composition of the interpretations of a and b.

Given the additional monoid structure on actions the congruence $s \approx t$ no longer coincides with $s \cong t$. For example we have that $a \cong a + a;\tau$ (in fact, $a \cong a;\tau$) and $a;b \cong a;b + a \circ b$ but we do not have $a \approx a;\tau$ or $a;b \approx a;b + a \circ b$: For every monoidal interpretation I we have

$$\tilde{I}(a) = \{\langle\sigma, \sigma'\rangle \mid I(a)(\sigma) = \sigma'\} \cup \{\langle\sigma, \Delta\rangle \mid I(a)(\sigma) = \perp\}$$

whereas

$$\tilde{I}(a;\tau) = \{\langle\sigma, \sigma'\rangle\langle\sigma'', \sigma''\rangle \mid I(a)(\sigma) = \sigma'\} \cup \{\langle\sigma, \Delta\rangle \mid I(a)(\sigma) = \perp\}.$$

Moreover,

$$\begin{array}{l} \tilde{I}(a;b) = \\ \{\langle\sigma_1, \sigma_2\rangle\langle\sigma_3, \sigma_4\rangle \mid I(a)(\sigma_1) = \sigma_2,\ I(b)(\sigma_3) = \sigma_4\} \cup \\ \{\langle\sigma_1, \sigma_2\rangle\langle\sigma_3, \Delta\rangle \mid I(a)(\sigma_1) = \sigma_2,\ I(b)(\sigma_3) = \perp\} \cup \\ \{\langle\sigma, \Delta\rangle \mid I(a)(\sigma) = \perp\} \end{array}$$

whereas

$$\tilde{I}(a;b + a \circ b) = \tilde{I}(a;b) \cup \tilde{I}(a \circ b).$$

Therefore we introduce the following abstraction operation.

Definition 9. *Let $\alpha(R)$, for $R \subseteq D$, be the smallest set containing R which is closed under the following conditions:*

$$\begin{array}{lll} \textbf{C1}\ w_1 w_2 \in R & \Rightarrow w_1\langle\sigma, \sigma\rangle w_2 \in R & (w_1 \neq \epsilon) \\ \textbf{C2}\ w \in R & \Rightarrow \langle\sigma, \sigma\rangle w \in R & (\langle\sigma, \Delta\rangle \notin R) \\ \textbf{C3}\ w_1\langle\sigma, \sigma'\rangle\langle\sigma'\sigma''\rangle w_2 \in R & \Rightarrow w_1\langle\sigma, \sigma''\rangle w_2 \in R & \end{array}$$

The first two closure conditions introduce *finite stuttering.* The third condition 'internalizes' interleaving points. The restrictions on the conditions **C1** and **C2** ensure correctness (without these restrictions we would identify, for example, δ and $\tau;\delta$).

Definition 10. *We introduce the semantics $\tilde{I}_\alpha$ defined by $\tilde{I}_\alpha(s) = \alpha(\tilde{I}(s))$. Moreover, we define $s \approx_\alpha t$ if and only if $\tilde{I}_\alpha(s) = \tilde{I}_\alpha(t)$, for every I.*

We have the following generic full abstraction result for monoidal interpretations.

Theorem 4. *For any statement s and t, $s \approx_\alpha t$ if and only if $s \cong t$.*

Proof The implication $s \approx_\alpha t \Rightarrow s \cong t$ follows from the compostionality of I_α and its correctness with respect to the initial/final state semantics.

With respect to the proof of the implication $s \not\approx_\alpha t \Rightarrow s \not\cong t$ we first observe that the construction of a distinguishing context in the proof of Theorem 2 which is based on sequences of states (of the state-space of the given interpretation) does not apply here because the resulting interpretation does not satisfy the monoid structure: the interpretation of τ does not correspond with the identity function and the interpretation of a composed action $a \circ b$ does not correspond with the (functional) composition of the interpretations of a and b.

Instead we define now the distinguishing context in terms of the state-space of the interpretation I for which $\tilde{I}_\alpha(s) \neq \tilde{I}_\alpha(t)$ holds: Let $\omega = \langle \sigma_1, \sigma_1' \rangle \dots \langle \sigma_n, \sigma_n' \rangle \in \tilde{I}_\alpha(s) \setminus \tilde{I}_\alpha(t)$ (the case of ω ending in deadlock is treated similarly). We define a new interpretation J with the *same* state space Σ as that of the given interpretation I such that for every action a occurring in s or t, $J(a) = I(a)$. Furthermore let a_i, $1 \leq i < n$, be some new actions not occurring in s or t. Without loss of generality we may assume that these actions are *atomic* in the following sense: If $a = b \circ c$ then either $a = b$ and $c = \tau$ or $a = c$ and $b = \tau$. We define $J(a_i) = \{\langle \sigma_i', \sigma_{i+1} \rangle\}$ (so a_i transforms σ_i' into σ_{i+1} and blocks in every other state). Let $r = a_1; \dots; a_{n-1}$. It follows that $\sigma_n' \in J(s \parallel r)(\sigma_1)$. Moreover, for any statement u, with $a_1, \dots, a_{n-1}$ not occurring in u, we have that $\sigma_n' \in J(u \parallel r)(\sigma_1)$ implies $\omega \in \tilde{J}_\alpha(u)$. We sketch the basic reasoning pattern underlying this implication: Let ω' be a succesfully terminating computation (more precisely, a *connected* reactive sequence) of $u \parallel r$ starting in σ_1 and resulting in σ_n'. It follows that for every action a_i there exists a computation step $\langle \sigma_i', \sigma_{i+1} \rangle$ in ω' such that the execution of u has transformed σ_i by some finite number of computation steps into σ_i'. By the closure condition **C3** these consecutive computation steps of u are represented in $\tilde{J}_\alpha(u)$ also by the one computation step $\langle \sigma_i, \sigma_{i+1}' \rangle$. In case $\sigma_i = \sigma_{i+1}'$ we can also apply the closure condition **C1** or **C2** (it is not difficult to see that $\sigma_n' \in J(u \parallel r)(\sigma_1)$ implies $\langle \sigma_1, \Delta \rangle \notin \tilde{J}_\alpha(u)$). □

Also here we have that the usual full abstractness result for a given $\tilde{I}_\alpha$ does not hold. For example, the monoidal interpretation based on concurrent constraint programming requires some further abstractions which stem from the monotonicity of the basic actions (see [6] and below).

With the introduction of the monoid structure on actions the failure equivalence defined above must be changed in order to cope with the fixed interpretation of the τ and δ actions and the composition operator $\circ$.

Definition 11. *Let a be an action different from τ and δ: by $s \xrightarrow{a}_\alpha t$ we mean that there exist $a_1 \dots a_n$ and $s_1 \dots s_{n-1}$ such that $s \xrightarrow{a_1} s_1 \xrightarrow{a_2} \dots \xrightarrow{a_n} t$ and $a = a_1 \circ a_2 \circ \dots \circ a_n$. Let w be a sequence of actions different from τ and δ; by $s \xrightarrow{w}_\alpha t$ we mean the following: if w is the empty sequence then t can be reached from s via a non-empty sequence of τ actions, otherwise, if w is the sequence $a_1 \dots a_n$ then $s \xrightarrow{a_1}_\alpha \dots \xrightarrow{a_n}_\alpha t$. We define*

$$\begin{aligned} F_\alpha(s) = \{&(w, E) \mid s \xrightarrow{w}_\alpha E\} \cup \\ &\{(w, \mathcal{F}) \mid s \xrightarrow{w}_\alpha t,\ \tau \notin Init(t),\ and\ \mathcal{F} \subseteq A\ s.t.\ \mathcal{F} \cap Init(t) = \emptyset\} \end{aligned}$$

where Init has been defined in Definition 3.5 on the initial transition system $\xrightarrow{a}$.

In the definition of F_α we consider the condition $\tau \notin Init(t)$, where $Init(t)$ denotes the set of initial actions of t according to the original transition relation $t \xrightarrow{a} t'$ $(t \neq E)$, because only failures of *stable* statements (as they are called in [20, 21]) must be taken into account. A statement s which is able to perform an initial τ action is not stable, in the sense that it is always able to perform this kind of action whatever is the actual state of the computation.

Theorem 5. *For any statement s and t, $s \approx_\alpha t$ if and only if $F_\alpha(s) = F_\alpha(t)$.*

Proof We have to resort to a different proof than that of Theorem 3 because the interpretation defined in that proof does *not* satisfy the monoidal structure.
Instead, we show in the full paper that for every two statements s and t there exist statements s' and t' such that $I_\alpha(s) = I_\alpha(t)$ if and only if $I(s') = I(t')$, for every monoidal interpretation I, and $F_\alpha(s) = F_\alpha(t)$ if and only if $F(s') = F(t')$. And so we can apply Theorem 3. In fact, the statements s' and t' can be obtained by the rewrite rules $s;\tau \Rightarrow s$ and $a;(b;s+t) \Rightarrow a;(b;s+t) + a \circ b;s$ (or $a;(b+t) \Rightarrow a;(b+t) + a \circ b$). □

In order to give a complete axiomatization (for finite statements) of this new failure semantics we have to add to the axioms in Table 1 new axioms dealing with the special δ and τ actions, and the composition operator $\circ$. In Table 2 we present the deadlock laws D for the δ action, the axioms T for τ, and the contraction laws C for the composition operator $\circ$.

Theorem 6. *For any statement s and t,*

$$s \approx_\alpha t \text{ if and only if } \mathrm{A} + \mathrm{M} + \mathrm{F} + \mathrm{D} + \mathrm{T} + \mathrm{C} \vdash s = t$$

It is of interest to observe that we do not consider the standard τ-law $s = s;\tau$ but a weaker version $s = s + s;\tau$. In fact, we will see in the following that the standard law is now derivable by its weaker version due to the presence of the contraction laws. Moreover, we introduced this kind of axiomatization because it

D1	$s + \delta = s$	
D2	$\delta; s = \delta$	
T1	$s = s + s; \tau$	
T2	$a; (\tau; s + t) = a; (\tau; s + t) + a; \tau; s$	
T3	$a; (\tau + t) = a; (\tau + t) + a; \tau$	
C1	$a; (b; s + t) = a; (b; s + t) + a \circ b; s$	
C2	$a; (b + t) = a; (b + t) + a \circ b$	

Table 2. The Laws for δ, τ and $\circ$.

is possible to prove that there is a strong connection between, on the one hand, abstraction of granularity and the contraction laws C1 and C2, and, on the other hand, finite stuttering and the τ-laws T1, T2, and T3.

Theorem 7. *Let α_1 denote the abstraction operation corresponding to the closure conditions* **C1** *and* **C2**. *We define $\tilde{I}_{\alpha_1}(s) \subseteq D$ by $\alpha_1(\tilde{I}(s))$. Here $\alpha_1(R)$, for $R \subseteq D$, is defined as the smallest set containing R which is closed under the closure conditions* **C1** *and* **C2**. *We have*

$$s \approx_{\alpha_1} t \text{ if and only if } \mathrm{A} + \mathrm{M} + \mathrm{F} + \mathrm{D} + \mathrm{T} \vdash s = t$$

where $s \approx_{\alpha_1} t$ if and only if $\tilde{I}_{\alpha_1}(s) = \tilde{I}_{\alpha_1}(t)$, for every I.

Let α_2 denote the abstraction operation corresponding to the closure condition **C3**. *We define $\tilde{I}_{\alpha_2}(s) \subseteq D$ by $\alpha_2(\tilde{I}(s))$. Here $\alpha_2(R)$, for $R \subseteq D$ is defined as the smallest set containing R which is closed under the closure condition* **C3**. *We have*

$$s \approx_{\alpha_2} t \text{ if and only if } \mathrm{A} + \mathrm{M} + \mathrm{F} + \mathrm{D} + \mathrm{C} \vdash s = t$$

where $s \approx_{\alpha_2} t$ if and only if $\tilde{I}_{\alpha_2}(s) = \tilde{I}_{\alpha_2}(t)$, for every I.

An alternative axiomatization of the contraction laws consists of the introduction of the $\circ$ operator to statements.

Definition 12. *We have the following transition rules for $\circ$:*

- *If $s \xrightarrow{\lambda} s'$ then $s \circ t \xrightarrow{\lambda} s' \circ t$ where s' is assumed to be distinct from E.*
- *If $s \xrightarrow{\langle\sigma,\sigma'\rangle} E$ and $t \xrightarrow{\langle\sigma',\sigma''\rangle} t'$ then $s \circ t \xrightarrow{\langle\sigma,\sigma''\rangle} t'$*

The operator $\circ$ is associative and is distributive wrt $+$ and $;$ (see Table 3). The 'contraction' law can be now described by the axiom in Table 4.

B1	$(s \circ t) \circ u$	$= s \circ (t \circ u)$
B2	$s \circ (t; u)$	$= (s \circ t); u$
B3	$(s; t) \circ u$	$= s; (t \circ u)$
B4	$s \circ (t + u)$	$= (s \circ t) + (s \circ u)$
B5	$(s + t) \circ u$	$= (s \circ u) + (t \circ u)$

Table 3. Associativity and Distributivity of $\circ$.

B6	$s; t = s; t + s \circ t$

Table 4. The Contraction Law.

Theorem 8. *For any statement s and t,*

$$s \approx_\alpha t \text{ if and only if } \mathrm{A} + \mathrm{M} + \mathrm{F} + \mathrm{D} + \mathrm{T} + \mathrm{B} \vdash s = t$$

Under this alternative characterization it is clear that the standard τ-law $s = s; \tau$ is derivable. Indeed, we have that $s = s + s; \tau$ by $T1$ and $s; \tau = s; \tau + s \circ \tau$; but it is easy to prove that $s \circ \tau = s$ hence also $s; \tau = s; \tau + s$.

Of interest is that a similar operator $\circ$ has been introduced already in [2] for modeling mutual exclusion. The main difference with our axiomatization is that we use the operator of mutual exclusion to model the abstraction from internal interleaving points. This abstraction is captured by the Contraction Law.

5 Read and write actions

In this section we briefly discuss a generic process algebra for read and write actions (in the context of a monoid of actions). We assume a partitioning of the set of actions into the sets *write* and *read* of write and read actions. Write actions are autonomous actions which never block, i.e. for every interpretation I and write action a we have that $I(a)(\sigma) \neq \perp$. For $a \in$ *read* we require that $I(a)(\sigma) = \sigma$ if $I(a)(\sigma)$ is defined. Note that thus τ and δ are special kind of read actions. Table 5 gives an algebraic characterization of the read and write actions. The axioms W1 and W2 describe the autonomous nature of the write action with respect to the plus operator. The axiom R captures the interplay between write and read actions. Note that we do *not* have $a = b; a$ because the read action b in general introduces deadlock. For example, the initial/final state semantics of $a + b$ and $b; a + b$ in general differ: Let I be an interpretation such that $I(b)(\sigma)$ is undefined. The execution of $b; a + b$ in σ then will deadlock whereas $a + b$ will result in $I(a)(\sigma)$.

Next we consider monotonic read and write actions.

W1	$a;(b;s+t)$	$= a;(b;s+t) + a;b;s$	$b \in \mathit{write}$
W2	$a;(b+t)$	$= a;(b+t) + a;b$	$b \in \mathit{write}$
R	a	$= a + b;a$	$a \in \mathit{write},\ b \in \mathit{read}$

Table 5. Read and write actions.

Definition 13. *Given an* information *ordering* $\sqsubseteq$ *on the set of states a monotonic interpretation assigns to each action a monotonic function.*

Characteristic instances of this class are the concurrent constraint languages. Table 6 gives an algebraic characterization of these monotonic interpretations. The axioms MR1 and MR2 describe a characteristic commutativity property of the read actions in monotonic interpretations.

MR1	$a;(b;s+t)$	$= a;(b;s+t) + b;a;s$	$b \in \mathit{read}$
MR2	$a;(b+t)$	$= a;(b+t) + b;a$	$b \in \mathit{read}$

Table 6. Monotonic interpretations.

We refer to the full paper for the completeness proof of the above axiomatizations.

6 Conclusion and future work

We have introduced a general framework $\mathcal{L}$ for the study of process algebras of concurrent languages based on asynchronous communication. We have argued that in this framework the failure semantics as developed for action-based process algebras like ACP coincides with the initial/final state semantics of $\mathcal{L}$. Furthermore we have given a generic process algebra for instantiations of $\mathcal{L}$ which are based on a state-based composition operation of the actions. Finally, we concluded with a generic characterization of read and write actions.

Currently we are investigating instantiations of $\mathcal{L}$ which involve an information ordering on the set of states. An interesting class of such instantiations are those which interpret the actions as *closure* operators (concurrent constraint programming is such an instantiation), i.e.

- $I(a)$ is extensive ($\sigma \sqsubseteq I(a)(\sigma)$)
- $I(a)$ is idempotent ($I(a)(I(a)(\sigma)) = I(a)(\sigma)$).

Of interest also is a generic process algebra of so-called *get* actions which remove information, i.e. $I(a)(\sigma) \sqsubseteq \sigma$.

Another line of research involves the introduction of a general state-based *hiding* operator in the language $\mathcal{L}$. To this end we envisage the introduction of instantiations of $\mathcal{L}$ which are based on *cylindric algebras* [16]. The general notion of existential quantification as given by these algebras then will provide a basis for the introduction of a corresponding programming construct in the language $\mathcal{L}$.

References

1. J.C.M. Baeten and W.P. Weijland. Process algebra. Cambridge Tracts in Theoretical Computer Science 18, Cambridge University Press 1990.
2. J.A. Bergstra and J.W. Klop. Process algebra for communication and mutual exclusion. Technical report CS-R8409, CWI, 1984.
3. J.A. Bergstra, J.W. Klop, and E.-R. Olderog. Readies and failures in the algebra of communicating systems. *SIAM J. Comp.*, 17(6):1134–1177, 1988.
4. F.S. de Boer, J.W. Klop and C. Palamidessi. Asynchronous communication in process algebra. Proc. of the seventh annual IEEE symposium on Logics in Computer Science (LICS), pages 137–147. IEEE Computer Society Press, Los Alamitos, California, 1992.
5. F.S. de Boer, J.N. Kok, C. Palamidessi, and J.J.M.M. Rutten. The failure of failures: Towards a paradigm for asynchronous communication. Proc. of Concur '91, Lecture Notes in Computer Science, Vol. 527, pages 111-126, Amsterdam, The Netherlands, 1991.
6. F.S. de Boer and C. Palamidessi. A process algebra of concurrent constraint programming. Proc. of the Joint International Conference and Symposium on Logic Programming, JICSLP'92, pages 463–477, The MIT press, 1992.
7. M. Boreale, R. De Nicola, R. Pugliese. Asynchronous observations of processes. Proc. of FoSSaCS'98, Lecture Notes in Computer Science, 1998.
8. A. Brogi and P. Ciancarini. The concurrent language Shared Prolog. ACM Transaction on Programming Languages and Systems, 13(1):99–123, 1991.
9. S. D. Brookes. Full abstraction for a shared-variable parallel language. Information and Computation, 127(2):145-163, 15 June 1996.
10. N. Busi, R. Gorrieri, and G. Zavattaro. A process algebraic view of Linda coordination primitives. Theoretical Computer Science, 192(2): 167–199, 1998.
11. P. Ciancarini, R. Gorrieri, and G. Zavattaro. An alternative semantics for the parallel operator of the calculus of gamma programs. Coordination Programming: Mechanism, Models and Semantics, pages 232–248, Imperial College Press, 1996.
12. R. De Nicola and R. Pugliese. A process algebra based on Linda. Proc. of Coordination'96, Lecture Notes of Computer Science, Vol. 1061, pages 160–178, 1996.
13. E.W. Dijkstra. Co-operating sequential processes. In Academic Press, New York, 1968.
14. D. Erman, F. HayesRoth, V. Lesser, and D. Reddy. The Hearsay2 speech understanding system: Integrating knowledge to resolve uncertainty. ACM Computing Surveys, 12:213–253, 1980.
15. D. Gelenter. Generative communication in Linda. ACM TOPLAS, 7(1):80–112, 1986.

16. L. Henkin, J.D. Monk, and A. Tarski. *Cylindric Algebras (Part I)*. North-Holland, 1971.
17. C.A.R. Hoare. Communicating sequential processes. Communications of the ACM, 21(8):666–677, 1978.
18. He Jifeng, M.B. Josephs, and C.A.R. Hoare. A theory of synchrony and asynchrony. Proc. of IFIP Working Conference on Programming Concepts and Methods, pages 459–478, 1990.
19. G. Kahn. The semantics of a simple language for parallel programming. Information Processing 74: Proc. of IFIP Congress, pages 471–475, New York, 1974. North-Holland.
20. R. Milner. A Calculus of Communicating Systems, Lecture Notes in Computer Science, Vol. 92, New York, 1980.
21. R. Milner. Calculi for synchrony and asynchrony. Theoretical Computer Science, 25:267–310, 1983.
22. V.A. Saraswat. Concurrent Constraint Programming Languages. PhD thesis, january 1989. The MIT Press.
23. E.Y. Shapiro. The family of concurrent logic programming languages. ACM Computing Surveys, 21(3):412–510, 1989.

Timed Automata and the Theory of Real Numbers

Hubert Comon and Yan Jurski

LSV, ENS Cachan, 94235 Cachan cedex, France
{comon,jurski}@lsv.ens-cachan.fr
http://www.lsv.ens-cachan.fr/~comon
Tel: 01 47 40 24 30. Fax: 01 47 40 24 64

Abstract. A configuration of a timed automaton is given by a control state and finitely many clock (real) values. We show here that the binary reachability relation between configurations of a timed automaton is definable in an additive theory of real numbers, which is decidable. This result implies the decidability of model checking for some properties which cannot be expressed in timed temporal logics and provide with alternative proofs of some known decidable properties. Our proof relies on two intermediate results: 1. Every timed automaton can be effectively emulated by a timed automaton which does not contain nested loops. 2. The binary reachability relation for counter automata without nested loops (called here *flat automata*) is expressible in the additive theory of integers (resp. real numbers). The second result can be derived from [10].

1 Introduction

Timed automata have been introduced in [4] to model real time systems and became quickly a standard. They roughly consist in adding to finite state automata a finite number of clocks which grow at the same speed. Each transition comes together with some clock resets and an enabling condition, whose satisfaction depends on the current clock values. Temporal properties of real time systems have been expressed and studied through temporal logics such as TPTL [7], MITL [6], TCTL [2,15,19],timed μ-calculi [15,16]. These logics are in general undecidable, with the notable exception of MITL. On the other hand, the model-checking is decidable for the (real-time) branching time logics, though hard in general.

Timed models are harder than untimed ones since they can be seen as infinite state systems in which every configuration consists of a pair of a control state (out of a finite set) and a vector of real clock values. Reasoning about possible clocks values in each state is the core of the difficulty. In this paper, we adopt the following point of view: infinite sets of configurations can be finitely described using *constraints*. For instance, "$(q, x \geq y + z)$" is the set of configurations in control state q and such that the clock x is larger than the sum of clocks y and z. This point of view is not new, as the *regions* of [2], which are used in a crucial way in the verification algorithm, are indeed a representation of sets of configurations indeed. Here, we go one step further: we express not only sets of configurations, but also *relations* between configurations in a (decidable) constraint system. Then temporal properties of the model are described through the binary reachability relation $\xrightarrow{*}$ relating clock values, which is expressible in the constraint system. Since we may always assume that there is a clock τ which is never reset by the automaton (and hence is a witness of the total elapsed time), we may express for instance some delay conditions such as "d is a delay between q and q'" as a constraint: $\exists \boldsymbol{x}, \boldsymbol{x}', \tau.(q, \boldsymbol{x}, \tau) \xrightarrow{*} (q', \boldsymbol{x}', \tau+d)$. Now it is possible to analyse delays between some events such as finding minimal or maximal delays. There are already algorithms which find such extremal delays [12], but we may also decide properties such as: "the delay between event a and event b is never larger than twice the delay between event a' and event b'" (which is, up to our knowledge, a new decidability result). More generally, our main result is that the binary reachability relation between clocks values, which is defined by a timed automaton, is effectively expressible in the additive theory of real numbers. Since the additive theory of real numbers is decidable, any property which can be expressed in this theory using the reachability relation, can be decided. In particular, we can compute reachable configurations from a definable set of configurations as well as the set of configurations from which we can reach a definable set. Hence we have forward and backward model-checking algorithms of safety properties as simple instances of our result. But we may also check properties which express relations between the original and final clock values. Also, some parametric verification is possible as we may keep free variables in the description of original and final configurations: for safety properties, the results of [18] can be derived from our main result. Finally, we can handle more general models than timed automata: transitions may be guarded by arbitrary first-order formulas over clocks, provided that such transitions can only be fired a bounded number of times.

On the negative side, not all timed temporal properties can be expressed in the first-order theory of $\xrightarrow{*}$. Typically, unavoidability is not expressible. This is not surprising since our logic is decidable, whereas the timed temporal logics are not in general.

Our main result is proved in two steps: first we show that any timed automaton can be emulated by an automaton without nested loops, hereafter called *flat automaton*. The notion of emulation will be precised, but keep only in mind that it preserves the reachability relation. Hence, in some sense, timed automata with a star height n are not more expressive than timed automata with star height 1. (This is not true, of course, if we consider the accepted language instead of the reachability relation as an equivalence on automata). The second step consists of applying one of our former results, which shows that the reachability relation is effectively expressible in the additive theory of real (resp. integer) numbers for flat counter automata [10]. We go from timed automata to automata with counters using an encoding due to L. Fribourg [13].

The emulation result itself is proved in three steps: first we define an equivalence relation on transition sequences, which we show to be right compatible and of finite index. This is similar to a region construction, though the equivalence is rather on pairs of configurations than on configurations. Second, we show some commutation properties of equivalent transition sequences: roughly, equivalent transition sequences can be performed in any order, without affecting the reachability relation. The third (and last) step consists in using combinatorial arguments on words and proving that there is a flat automaton whose language contains a set of representatives for the congruence generated by the commutation properties. (This result can be stated as a formal language property which is independent from the rest of the paper). This provides with a mechanical way to choose which transition sequence to commute with another: the representatives of the congruence classes are the normal forms w.r.t. a regular string rewriting system.

From this proof, we can also derive some other decidability results. For instance, we can decide whether a sequence of transitions can be iterated.

In section 2, we recall the basic definitions of timed automata and we introduce our constraint system. Next, we sketch the proof of the emulation result in section 3 and derive in section 4 the definability of the reachability relation. In section 5, we show some examples of temporal properties which can be expressed in the theory of real numbers. In particular we give examples showing the expressiveness of the binary reachability relation. We conclude in section 6. Many constructions are only sketched in this abstract paper. More details can be found in [11].

2 Timed Automata

We start with a classical notion of timed automaton, which includes invariants in the states and guarded transitions. The syntax and semantics of timed transition systems we use here is not important: we can switch from the following definitions to others (such as [5]) without changing our main result. The events and the accepted language are also irrelevant here, as we are interested in reachability.

2.1 Syntax and semantics

Let B be a finite set of real numbers (we will assume later that these constants are in $\mathbb{Z}$) and C a finite number of variables called *clocks*. $\Phi(B, C)$ is the set of conjunctions[1] of atomic formulas of the form $x \leq c, x \geq c, x < c, x > c, x = c$ where $x \in C$ and $c \in B$.

Definition 1 ([1]). *A* timed automaton *is a tuple* $< \Sigma, Q, Q_0, C, I, E >$ *where*

- *Σ is a finite alphabet*
- *Q is a finite set of states (and $Q_0 \subseteq Q$ is a set of start states, irrelevant here)*

[1] Having arbitrary Boolean combination does not increase the expressive power. We choose to consider conjunctions only since they guarantee a convexity property for the invariant constraints.

- *C is a finite set of clocks*
- *I is a mapping from Q to* $\Phi(B, C)$ *(the* invariant *associated with each state).*
- $E \subseteq Q \times Q \times \Sigma \times 2^C \times \Phi(B, C)$ *gives the set of transitions. In each transition* $< q, q', a, \lambda, \phi >$, λ *is a set of* clock resets *and* ϕ *is a* clock constraint.

A *configuration* of the automaton is a pair $(q, \boldsymbol{V})$ where $q \in Q$ and $\boldsymbol{V} \in \mathbb{R}_+^{|C|}$ is a clock value. There is a *move* of a timed automaton $\mathcal{A}$ from a configuration $(q, \boldsymbol{V})$ to $(q', \boldsymbol{V'})$, which we write $(q, \boldsymbol{V}) \underset{\mathcal{A}}{\longrightarrow} (q', \boldsymbol{V'})$, iff

- Either $q = q'$ and $\boldsymbol{V} \models I(q), \boldsymbol{V'} \models I(q)$ and there is a positive real number t such that, for every component i, $v'_i = v_i + t$.
- Or else there is a transition $< q, q', a, \lambda, \phi >$ and a positive real number t such that $\boldsymbol{V} \models \phi$ and for every component i, either $v_i \in \lambda$ and then $v'_i = t$ or else $v_i \notin \lambda$ and $v'_i = v_i + t$. Moreover, $\boldsymbol{V'} \models I(q')$.

$\overset{*}{\underset{\mathcal{A}}{\longrightarrow}}$ is the reflexive transitive closure of $\underset{\mathcal{A}}{\longrightarrow}$. We also write $\overset{*}{\underset{q_1, q_2}{\longrightarrow}} \subseteq \mathbb{R}^{|C|} \times \mathbb{R}^{|C|}$ the relation on clocks vectors defined by $\boldsymbol{V} \overset{*}{\underset{q_1, q_2}{\longrightarrow}} \boldsymbol{V'}$ iff $(q_1, \boldsymbol{V}) \overset{*}{\underset{\mathcal{A}}{\longrightarrow}} (q_2, \boldsymbol{V'})$. We will always assume without loss of generality that there is a clock τ which is never reset.

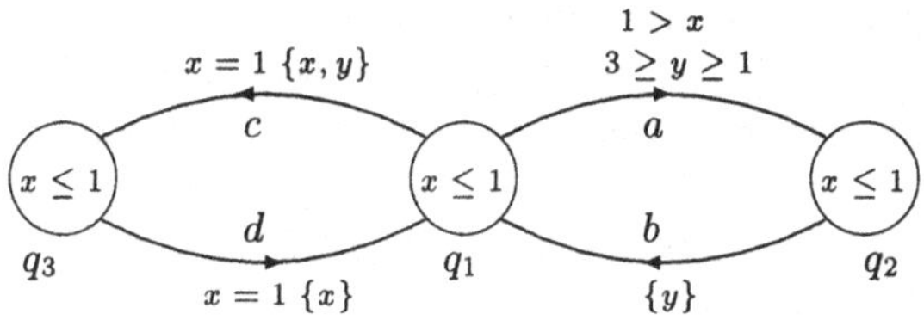

Fig. 1. A timed automaton

Example 1. An example of a timed automaton is displayed on figure 1. As usual, invariants are written in the states and enabling conditions label the edges. The variables which are reset by a transition are written inside brackets. We assume that there are three clocks x, y, τ.

If we consider for instance transitions c, d only, we can express $\overset{*}{\underset{q_1, q_1}{\longrightarrow}}$ using the formula:

$$\begin{aligned} \exists \tau_1, x_1, y_1, t_1, t_2.\ & \tau_1 = \tau + t_1 \wedge x_1 = x + t_1 \wedge y_1 = y + t_1 \wedge x_1 \leq 1 \\ & \wedge \exists n. \tau' = \tau_1 + 2n - 1 + t_2 \wedge x' = t_2 \wedge x' \leq 1 \wedge y' = 1 + t_2 \end{aligned}$$

t_1 is the time spent before the first transition c is fired and is specified on the first line. τ_1, x_1, y_1 are the values of the clocks at that date. Then n is the number of times the loop cd is executed and t_2 is the time spent in q_1 after the last transition d has been fired.

This is typically what we will get from our formula computation.

A *flat automaton* is a timed automaton which does not contain nested loops: for every state q there is at most one non-empty path from q to itself.

Example 2. The automaton of figure 1 is not flat. If we remove any of the four transitions, we have a flat automaton.

2.2 Emulation

Definition 2. *A timed automaton $\mathcal{A}'$ emulates $\mathcal{A}$ if there is a mapping ϕ from the set of states of $\mathcal{A}'$ into the set of states of $\mathcal{A}$ such that, for every states q, q' of $\mathcal{A}$ and every clock vectors $\boldsymbol{V}, \boldsymbol{V}'$, $(q, \boldsymbol{V}) \xrightarrow[\mathcal{A}]{*} (q', \boldsymbol{V}')$ iff there are states $q_1 \in \phi^{-1}(q)$, $q_1' \in \phi^{-1}(q')$ such that $(q_1, \boldsymbol{V}) \xrightarrow[\mathcal{A}']{*} (q_1', \boldsymbol{V}')$.*

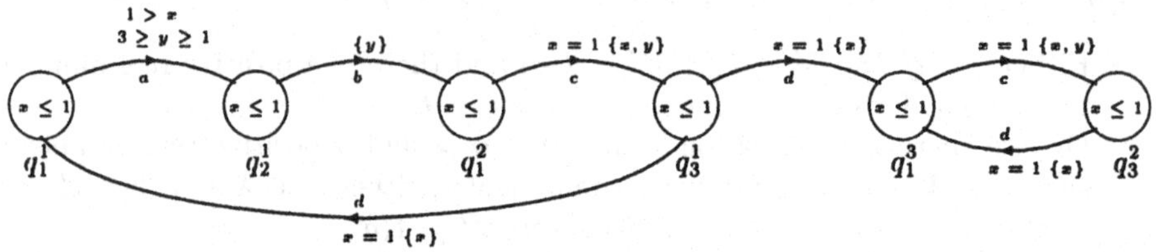

Fig. 2. An emulation automaton

Example 3. The automaton of figure 2 emulates the automaton of figure 1. The states q_i^j are mapped to q_i. It is a flat automaton. It is not straightforward that this automaton indeed emulates the original one. Note for instance that the possible event sequences are different as *abcdcdabcd* is a possible sequence in the automaton of figure 1 and is not a possible sequence in the automaton of figure 2. However, this sequence yields the same binary relation between configurations as the sequence *abcdabcdcd* which is possible in the automaton of figure 2.

The automaton of figure 2 is typically what we want to compute from the automaton of figure 1.

Lemma 1. *If $\mathcal{A}'$ emulates $\mathcal{A}$ then* $\xrightarrow[q_1, q_2]{*} \;=\; \bigcup_{q_1' \in \phi^{-1}(q_1), q_2' \in \phi^{-1}(q_2)} \xrightarrow[q_1', q_2']{*}$

Hence, as far as $\xrightarrow{*}$ is concerned, we may consider any automaton emulating $\mathcal{A}$ instead of $\mathcal{A}$ itself.

2.3 The additive theory of real numbers

The theory $\mathcal{T}$ we consider here is defined as follows. Terms are built from variables, the constants $0, 1$ and the function symbol $+$. Formulas are built using first-order quantifiers and the usual logical connectives on atomic formulas which are either equations $u = v$ between terms or predicates $Int(u)$ where u is a term.

The domain of interpretation of such formulas is the set of non-negative real numbers, with the usual interpretation of function symbols. Int is the set of natural numbers.

This theory can be encoded in S1S using (infinite) binary representation of real numbers. Hence it is a decidable theory.[2]

Example 4. The formula of example 1 is a formula of this theory.

[2] We do not know the complexity of this theory, nor of the fragment of $\mathcal{T}$ which is used here.

3 Every timed automaton can be emulated by a flat timed automaton

The automaton of figure 1 is not flat because there are two loops *ab* and *cd* on state q_1. If the order of the two loops was irrelevant to the reachability relation, we could switch them and assume that all sequences *ab* are performed before sequences *cd*. Then we would get a flat automaton, first considering the loop *ab* and then the loop *cd*. However, in this example, we cannot switch the two sequences because, for instance, *abcd* and *cdab* do not induce the same relations on the initial values of the clocks. Then, the question is: when can we switch two sequences of transitions w and w' without altering the reachability relation? Let us look first at some necessary conditions.

If w and w' do not induce the same relations on initial clock values, then their order is relevant since, for instance w may occur after some other transition sequence, whereas w' cannot. This is the case in our example: *ab* and *cd* do not induce the same relations on initial clock values and *ab* cannot occur after another *ab*, whereas *cd* can occur after *ab*. *ababcd* is impossible and *abcdab* is possible. Hence a first necessary condition is that w and w' induce the same constraints on initial clock values.

Similarly, w and w' should enable the same transitions: whereas w or w' has been executed last should not be relevant for further transitions. This means that w and w' should induce the same constraints on final clock values, or at least constraints that can be met by the same enabling conditions of further transitions.

There are further necessary conditions for two transition sequences to commute.

Example 5. For instance consider the automaton displayed on figure 3. On this

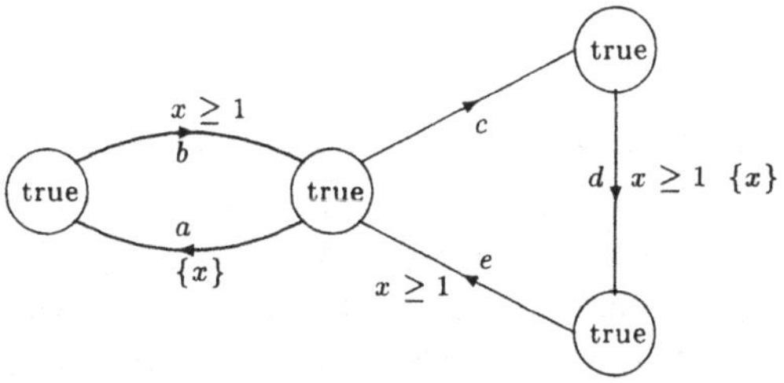

Fig. 3. Another example of non-commuting loops

example, let $w = ab$ and $w' = cde$. Executing ww' yields a constraint $x' - x \geq \tau' - \tau - 1$ on the final and initial x's values whereas $w'w$ yields a strictly weaker constraint $x' - x \geq \tau' - \tau - 2$. (To see this, start with $\tau = 1$ and $x = 0$. $w'w$ may yield a configuration in which $\tau' = 4$ and $x' = 1$, which is not reachable using ww'). On the other hand both w and w' induce the same constraints on initial clock values and on final clock values. Hence another necessary condition for switching w and w' is that they induce the same discrepancy between $x - x'$ and $\tau - \tau'$.

We will see that, very roughly, these three necessary conditions are also sufficient.

We are going to define a right compatible equivalence $\sim$ on transition sequences such that, in particular, the above situations cannot occur when $w \sim w'$. $\sim$ will be of finite index. Hence we will be able to split the states according to its equivalence classes, ensuring that two sequences starting from the initial state and which have the same final extended state can be switched, without changing the reachability relation.

3.1 A right compatible equivalence on transition sequences

This is the analog of *regions*, considering pairs of configurations instead of single configurations. Roughly, in the regions construction, two configurations $(q, \boldsymbol{v})$ and $(q, \boldsymbol{v}')$ are considered as equivalent, if they satisfy the same constraints $x \geq y+c$, $x > y+c$ where x, y are clocks and c is a constant (which is bounded by the largest constant of the model). Here, we define a right compatible equivalence on pairs of configurations $(q_1, \boldsymbol{v_1}), (q_2, \boldsymbol{v_2})$ and $(q_1, \boldsymbol{v}_1'), (q_2, \boldsymbol{v}_2')$. Two such pairs are equivalent, roughly, if they satisfy the same constraints $x \geq y' + c, x \geq x + c \ldots$, i.e. not only constraints relating clock values at a given time, but also constraints relating clock values before and after a sequence of transitions. The situation is not as simple as in the region case, however. Indeed, the relevant constants c now range over an a priori infinite set.

Let E be the set of transitions and w, w' be transition words. Let moreover ϕ_w be the formula with free variables $\boldsymbol{X}, \boldsymbol{X}'$ which expresses the relationship between clock values before and after w:

$$(q, \boldsymbol{V}) \xrightarrow[\mathcal{A}]{w} (q', \boldsymbol{V}') \quad \text{iff} \quad \boldsymbol{V}, \boldsymbol{V}' \models \phi_w$$

Note that this formula is independent of the states q, q' since we gave a different name for each transition of $\mathcal{A}$. Hence, given w there is only one starting and one target state for w. (Actually, q, q' are implicitly "encoded" into w).

As we want to commute sequences of transitions of the same class, we will need to keep a control property in the equivalence relation. $\sim_0$ ensures that two equivalent words are computations between the same states : $w \sim_0 w' \stackrel{\text{def}}{=}$ *the source and target states of w, w' are identical.*

The first property we want to keep is the relation between the initial clock values: let $\sim_1$ be defined by ($\#$ is either $\leq$ or $<$; in the following we will only consider $\leq$ for sake of simplicity):

$$w \sim_1 w' \stackrel{\text{def}}{=} \forall x, y \in \boldsymbol{X}, \forall c \in \mathbb{R}, \;\; \phi_w \models x \# y + c \; \text{ iff } \; \phi_{w'} \models x \# y + c$$

Example 6. Let us consider again the automaton of figure 1 and let S be the transition sequence ab and T be the transition sequence cd. Then $S \not\sim_1 T$. However, we have $ST^n \sim_1 ST^m$ and $T^{n+1} \sim_1 T^{n+k}$ for every n, m, k.

The finite index property of $\sim_1$ will be guaranteed if constants belong to $\mathbb{Z}$ since, if K is the maximal constant occurring the enabling conditions or invariants of the automaton:

Lemma 2. *For every w, x, y, $min\{c \in \mathbb{R} \mid \phi_w \models x \leq y + c\} \in [-K; K] \cup \{+\infty, -\infty\}$.*

However, $\sim_1$ is not right compatible as we may have $w \sim_1 w'$ and, for some transition t, $w \cdot t \not\sim_1 w' \cdot t$. That is because the transition t may introduce some constraint which is backward propagated to the initial values of the clocks. And this propagation does not depend on the initial values of the clocks only. In addition to this first difficulty, if we define a similar relation on the final values of the clocks: $w \sim_2 w' \stackrel{\text{def}}{=} \phi_w \models x' \leq y' + c$ iff $\phi_{w'} \models x' \leq y' + c$, we also loose the finite index property since the analog of lemma 2 does not hold for final values of the clocks: the minimal difference imposed by a sequence of transitions on the final values of the clocks can be arbitrarily large. (Think of a clock which is reset at each transition of w and another clock which is never reset). Hence we have to define $\sim_2$ as an approximation w.r.t the above definition:

$$w \sim_2 w' \stackrel{\text{def}}{=} \forall c \in [-K; K] \cup \{-\infty, +\infty\},\ \phi_w \models x' \leq y' + c \ \text{ iff } \ \phi_{w'} \models x' \leq y' + c$$

Example 7. In our example, there are only two classes of sequences in $(S + T)^*$ w.r.t. $\sim_1$: $ST(S+T)^*$ and $T(S+T)^*$. Now, w.r.t. $\sim_2$, sequences are also distinguished according to their last transition: constraints on final clock values depend on whether the last transition is S or T.

It is out of the scope of this extended abstract to show in details how we solve the above mentioned difficulties (see [11] for more details, in particular all definitions and proofs concerning $\sim_i$). The main idea is to define other relations $\sim_i$ which, altogether, will give a right compatible equivalence of finite index. The idea behind $\sim_3$ is to restore the compatibility of $\sim_1$ w.r.t. the right composition. Hence $\sim_3$ anticipates possible relations between initial clock values. More precisely, assume that $w \sim_1 w'$. $w \cdot t$ may induce a new constraint on the initial clock values in the following situation: $\phi_w \models x \leq y' + c_1 \wedge z' \leq u + c_2$ and $t \models y \leq z + c_3$. Then composing w and t, we get $\phi_{w \cdot t} \models \exists y_1, z_1 . x \leq y_1 + c_1 \wedge z_1 \leq u + c_2 \wedge y_1 \leq z_1 + c_3$, which implies $x \leq u + c_1 + c_2 + c_3$, possibly a new relation between initial clock values.

Roughly $w \sim_3 w'$ if the sums $c_1 + c_2$ such that we have such relations are the same for w and for w'. $c_1 + c_2$ may take arbitrary real values. However, we know that $c_1 + c_2 + c_3 \in [-K, K]$, thanks to lemma 2, otherwise $x \leq u + c_1 + c_2 + c_3$ would already be a consequence of ϕ_w. Hence, we only need to consider the sums $c_1 + c_2$ which belong to $[-2K, 2K]$, and we keep a finite index relation.

$\sim_4$ is more complicated. In the spirit, it is the same as $\sim_3$, guaranteeing the compatibility of $\sim_2$. Now, one could think that we have to define $\sim_5, \sim_6$ in order to guarantee the compatibility of the new relations $\sim_2, \sim_3$. Fortunately, this is not the case and, altogether, the union of $\sim_i$ for $i = 0, ..., 4$ is already a right compatible equivalence of finite index.

Still, we have to define more equivalences which take care of e.g. example 5. $\sim_5$ and $\sim_6$ express that the discrepancies between constraints on $x' - u$ and $y' - z$ are the same, up to the constant K. Finally $\sim_7$ is the symmetric of $\sim_3$, roughly reversing the ordering on time. This relation is also necessary for the commutation property (See [11] for more details).

Then we have the expected results, defining $\sim$ as $\bigcup_{i=0}^{7} \sim_i$:

Lemma 3. *$\sim$ is a right compatible equivalence on E^* and $E^*/\sim$ can be effectively computed in time $O(K^{n^2} \times q^2)$ where n is the number of clocks and q the number of states of $\mathcal{A}$.*

Example 8. Continuing example 1, we display on figure 4 the automaton for $\sim$: $u \sim w$ iff u and v are accepted in the same state of this automaton. Let us recall that $S = ab$ and $T = cd$, hence figure 4 is actually an abstraction of the automaton; each transition should be split into two transitions. In addition there should be a trash state for every impossible transition sequence. The complete automaton contains 16 states.

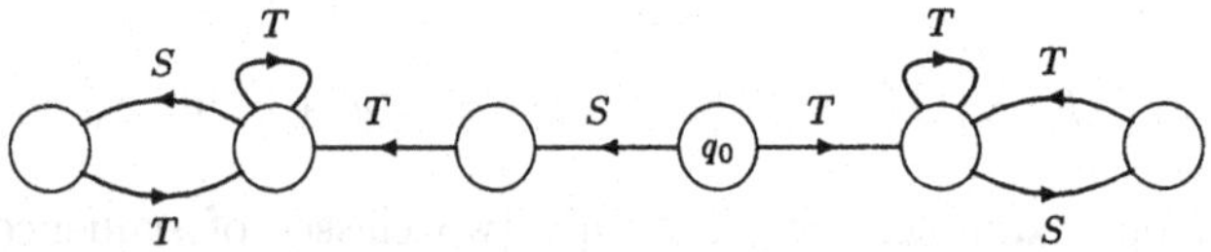

Fig. 4.

3.2 Commutation properties

The first result is that, if we have equivalent sequences of transitions then we can (almost) perform them in any order, without changing the constraint they induce on the clocks values:

Lemma 4. *If $w \sim w' \sim w''$ then $\phi_{w \cdot w' \cdot w''} \models\!\!\!\dashv \phi_{w' \cdot w \cdot w''}$.*

This lemma shows that, if we have two sequences of transitions w and w' from q to itself and such that $w \sim w'$, then the iteration of both loops i.e. the set of transitions $(w + w')^*$ has the same effect on clocks values as $w^* w'^* (w + \epsilon)$. This shows a flattening operation on regular expressions: $(w + w')^*$ is not flat whereas $w^* w'^* (w + \epsilon)$ is flat.

However, we cannot conclude yet since it is not always possible to compute an automaton emulating $\mathcal{A}$ and such that any two loops on the same state are equivalent for $\sim$. We need a more complex construction which proves that the automaton can be flattened when we have such commutation properties. This is the subject of the next section.

3.3 A formal language property

In this section, we sketch a formal language property which relates commutation properties and flat automata:

Theorem 1. *Let $\sim$ be a congruence of finite index on A^* and let $\approx$ be the least congruence on A^* such that $w \sim w' \sim w''$ implies $w \cdot w' \cdot w'' \approx w' \cdot w \cdot w''$. Then there is a flat automaton (whose all states are final states) which accepts a language containing a set of representatives for $\approx$.*

It is a consequence of the following series of lemmas:

Lemma 5. *Let k be the index of $\sim$. There is a constant $n(k)$ such that any word w of length larger than $n(k)$ can be factorised into $w = w_0 \cdot w_1 \cdot w_2 \cdot w_3 \cdot w_4$ with $w_1 \sim w_2 \sim w_3$.*

This states that any long enough word contains a factor to which the commutation property can be applied. Now, consider the following word rewrite system:

$$R \stackrel{\text{def}}{=} \{uxvyz \rightarrow vyuxz \mid u >_{lex} v; ux \sim vy \sim z; u, v \in A^{k_1}\}$$

This rewrite system compares lexicographically prefixes of length k_1 of equivalent words and commutes them according to the ordering.

Lemma 6. *R is a terminating rewrite system. The set of irreducible words $NF(R)$ w.r.t R is recognisable. Moreover, if $u \xrightarrow[R]{*} v$ then $u \approx v$.*

Now, we claim that $NF(R)$ is accepted by a flat automaton for some well-chosen k_1. This relies on the following lemma:

Lemma 7. *Let $\mathcal{A}$ be an automaton accepting $NF(R)$.There is a constant k_2 such that, if w is a transition sequence from a state q to itself, then there is a word u whose length is smaller than k_2 and an integer m satisfying $w = u^m$.*

Using this lemma together with standard combinatorial arguments [17], we get:

Lemma 8. *Let $\mathcal{A}$ be an automaton accepting $NF(R)$.There is a constant k_2 such that, if $w_1, \ldots, w_n$ are transition sequences, of length greater than k_2, from a state q to itself, then there is a word u whose length is smaller than k_2 and integers $m_1, \ldots, m_n$ satisfying $w_i = u^{m_i}$ for all i.*

Finally, thanks to arithmetical properties, we can flatten such automata:

Lemma 9. *$(u^{m_1} + \ldots + u^{m_n})^*$ is accepted by a flat automaton.*

Which, altogether, allows to prove theorem 1.

Example 9. Let us continue our running example. If we apply the algorithm sketched in this section and if we let $\alpha = ST(= abcd)$ and $\beta = T(= cd)$, then the rewrite system is

$$R = \{\alpha\beta x\alpha\alpha y\alpha \rightarrow \alpha\alpha y\alpha\beta x\alpha;\ \beta\beta x\beta\alpha y\beta \rightarrow \beta\alpha y\beta\beta x\beta \mid x, y \in (\alpha + \beta)^*\}$$

and the flat automaton (or rather its abstraction using the letters α, β instead of a, b, c, d: the whole flat automaton for $NF(R)$ would contain 76 states) is displayed on figure 5. All states are final.

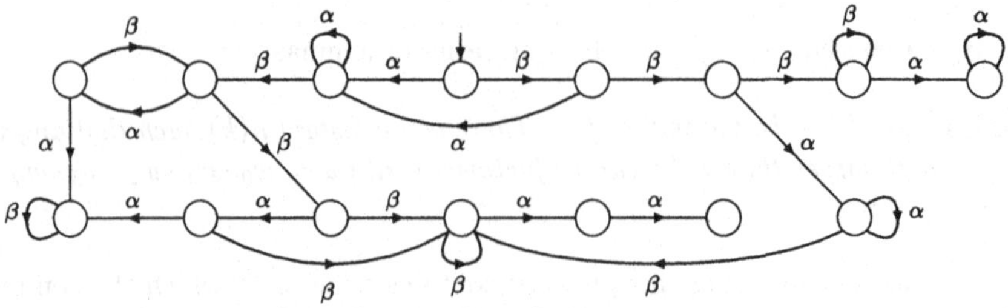

Fig. 5. The resulting flat automaton

3.4 The main theorem

Summing up what we have so far, thanks to lemma 3 and theorem 1, we can compute a flat automaton $\mathcal{A}_1$ which accepts at least one sequence of transitions for each equivalence class of $\approx$. We may moreover restrict the transition sequences accepted by $\mathcal{A}_1$ to transition sequences which are possible in the original automaton $\mathcal{A}$, thanks to a closure property for flat automata:

Lemma 10. *Let $\mathcal{A}$ be any finite automaton and $\mathcal{A}_1$ be a flat automaton. Then there is a flat automaton $\mathcal{A}'$ which accepts $L(\mathcal{A}) \cap L(\mathcal{A}_1)$.*

Moreover, there are two mappings f_1, f_2 from states of $\mathcal{A}'$ into respectively the states of $\mathcal{A}$ and the states of $\mathcal{A}_1$ such that, for every w and every states

$$q, q', q_1, q_1', \quad \left.\begin{array}{r} q \xrightarrow[\mathcal{A}]{w} q' \\ q_1 \xrightarrow[\mathcal{A}_1]{w} q_1' \end{array}\right\} \Leftrightarrow \exists q_2, q_2'. \left\{\begin{array}{l} q_2 \xrightarrow[\mathcal{A}']{w} q_2' \\ f_1(q_2) = q,\, f_2(q_2) = q_1 \\ f_1(q_2') = q',\, f_2(q_2') = q_1' \end{array}\right.$$

The proof of this lemma is similar to that of lemma 9: we construct $\mathcal{A}'$ from the product automaton $\mathcal{A} \times \mathcal{A}_1$, then all loops on a state are power of a same word, thanks to the flatness of $\mathcal{A}_1$. Now, we are able to prove our main theorem:

Theorem 2. *Every timed automaton can be emulated by a flat timed automaton.*

Proof. Let $\mathcal{A}$ be any timed automaton and let $\mathcal{A}_1$ be a flat automaton which accepts at least one transition sequence for each class modulo $\approx$. Such an automaton does exist thanks to lemmas 3, 4 and theorem 1 ($w \approx w'$ implies $\phi_w \models\!\mid \phi_{w'}$ as $\approx$ is the least congruence relation which satisfies the commutation properties of lemma 4). $\mathcal{A}$ can also be seen as a finite automaton on transition sequences and we construct $\mathcal{A}'$ as in lemma 10. $\mathcal{A}'$ can be seen as a timed automaton.

Now $\mathcal{A}'$ emulates $\mathcal{A}$: let f_1, f_2 be as in lemma 10. Then if $(q, \boldsymbol{V}) \xrightarrow[\mathcal{A}]{w} (q', \boldsymbol{V}')$, let $w' \approx w$. $(q, \boldsymbol{V}) \xrightarrow[\mathcal{A}]{w'} (q', \boldsymbol{V}')$, thanks to lemma 4. In particular if w' is the representative of the class of w w.r.t. $\approx$ which is accepted by $\mathcal{A}_1$ in state q_1', then $q_1 \xrightarrow[\mathcal{A}_1]{w'} q_1'$ where q_1 is the initial state of $\mathcal{A}_1$. It follows, by lemma 10, that there are states q_2, q_2' of $\mathcal{A}'$ such that $q_2 \xrightarrow[\mathcal{A}']{w'} q_2'$. Then $(q_2, \boldsymbol{V}) \xrightarrow[\mathcal{A}']{w'} (q_2', \boldsymbol{V}')$. Conversely, if $(q_2, \boldsymbol{V}) \xrightarrow[\mathcal{A}']{w} (q_2', \boldsymbol{V}')$, then $(f_1(q_2), \boldsymbol{V}) \xrightarrow[\mathcal{A}]{w} (f_1(q_2'), \boldsymbol{V}')$. Then it suffices to choose f_1 for the emulation function ϕ of definition 2.

Example 10. Considering the automaton of figure 1, our automatic construction will not yield the automaton of figure 2 (unfortunately). We actually obtain an automaton which is isomorphic to the automaton of figure 5.

4 Expressibility of the reachability relation

We use here a transformation of timed automata into automata with counters and use a result on flat automata with counters.

Definition 3 ([10]). *An* automaton with (real valued) counters *is a tuple* $(Q, q_i, C, \delta \subseteq Q \times G(C, C') \times Q)$ *where*

- Q *is a finite set of* states *and* $q_i \in Q$ *is an* initial state
- C *is a finite set of* counter names*;* C' *is the set of primed counter names.*
- $G(C, C')$ *is the set of* guards *built on the alphabets* C, C'*. A member of* $G(C, C')$ *is a conjunction of atomic formulas of one of the forms* $x\#y + c$, $x\#c$ *where* $x, y \in C \cup C'$, $\# \in \{\geq, \leq, =, >, <\}$ *and* $c \in \mathbb{R}$.

The automaton may *move* from a configuration $(q, \boldsymbol{v})$ to a configuration $(q', \boldsymbol{v}')$, which we write $(q, \boldsymbol{v}) \rightarrow (q', \boldsymbol{v}')$ if there is a triple $(q, g, q') \in \delta$ such that $\boldsymbol{v}, \boldsymbol{v}' \models g$, with the standard interpretation of relational symbols.

Example 11. Consider the automaton of figure 6.

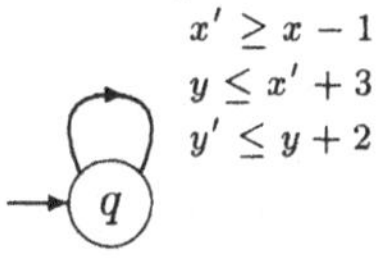

Fig. 6. A flat counter automaton

Possible moves are for instance: $(q, \binom{1}{1}) \rightarrow (q, \binom{0}{2})$ or $(q, \binom{1}{1}) \rightarrow (q, \binom{3}{3})$.

Following [13], timed automata can be seen as a particular class of automata with counters: we add a clock τ which is never reset and never used in the constraints (it measures the total elapsed time) and we use the variable transformation $x \mapsto \tau - x$. This yields a transformation on clocks valuations from $\boldsymbol{V}$ to $\boldsymbol{V_c}$. Then, if $< q, q', a, \phi, \lambda > \in E$, we translate it into a transition $\delta = < q, q', g >$ where g is the translation of ϕ together with the constraints $c' = \tau$ for each $c \in \lambda$ and $c' = c$ for each $c \notin \lambda$, plus the constraints on time positiveness: $\tau' \geq \tau$ and $c \leq \tau$ for every c. In this way each timed automaton $\mathcal{A}$ can be translated into an automaton with counters $\mathcal{A}_c$:

Theorem 3 ([13]). $(q, \boldsymbol{V}) \xrightarrow[\mathcal{A}]{*} (q', \boldsymbol{V'})$ *iff there is a vector* $\boldsymbol{V^1}$ *such that* $\exists t \geq 0$, $(q, \boldsymbol{V_c^1}) \xrightarrow[\mathcal{A}_c]{*} (q', \boldsymbol{V_c'})$ *and* $\boldsymbol{V^1} \models I(q), \boldsymbol{V} \models I(q), \boldsymbol{V^1} = \boldsymbol{V} + t\,\boldsymbol{1}$.

In other words, if we start a computation of $\mathcal{A}$ by firing a transition, then $\xrightarrow[\mathcal{A}]{*}$ is identical to $\xrightarrow[\mathcal{A}_c]{*}$, modulo the variable change.

On the other hand, we have the following result on flat automata:

Theorem 4 ([10]). *For every flat real (resp. integer) counter automata, the relations* $\xrightarrow[q,q']{*}$ *are effectively definable in* $\mathcal{T}$ *(resp. Presburger arithmetic).*

Now, from theorem 2, every timed automaton $\mathcal{A}$ can be emulated by a flat timed automaton $\mathcal{A}'$ and, by lemma 1, the reachability relations in $\mathcal{A}$ can be expressed as finite unions of the reachability relations of $\mathcal{A}'$. By theorem 4, the reachability relations of the flat automaton $\mathcal{A}'$ are expressible in the additive theory of real numbers. Hence it follows that:

Theorem 5. *For every timed automata, the binary reachability relations* $\xrightarrow[q,q']{*}$ *are effectively definable in the additive theory of real numbers.*

5 Examples of properties which can be decided on timed automata

Using the reachability relation, it is possible to express that some "good" (resp. "bad") thing may (resp. may not) happen. This is typical for safety properties. However it is not (at least in an obvious way) possible to express that something *must* happen. Typically, we cannot express inevitability. Hence, only a fragment of timed temporal logics can be expressed in the first-order theory of the reachability relation. However, our formalism offers other possibilities.

Example 12. On the simple automaton of figure 7, using the binary reachability relation, we can check properties of configurations, that are not properties of regions. This points out a difference in nature between our result and [18].

Consider the automaton of figure 7 wich contains only one clock x, and no constant. Assume that we are interested in the following property: " After firing a in configuration c, what is the minimal delay spent before reaching c again ?" For example, from state q_0 with $x = 0$ we can fire a and then b without waiting in q_1, then we are again in q_0 with $x = 0$. In this case the minimal delay is 0.

If the initial value x_0 of x is a parameter, then the minimal delay d is a function of x_0. In our example, we have $d(x_0) = x_0$. This result can be obtained using the binary reachability relation, since the set of possible delays is $\mathcal{D}(x_0) = \{\tau' - \tau \mid (q_0, x_0, \tau) \xrightarrow{ab} (q_0, x_0, \tau')\}$ and $d(x_0)$ is such that $d(x_0) \in \mathcal{D}(x_0)$ and $\forall d' \in \mathcal{D}(x_0), d(x_0) \leq d'$.

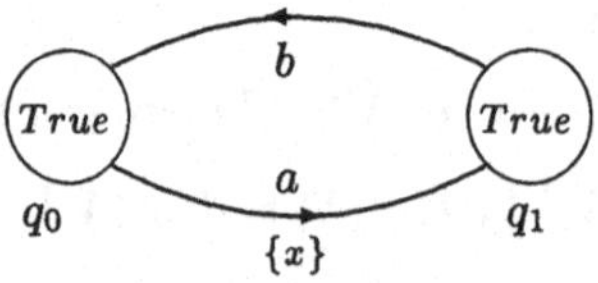

Fig. 7. A very simple timed automaton

Such a minimal delay property cannot be obtained using classical methods since, usually, the computed delays cannot depend on the initial configuration.

For instance, for the automaton of figure 7, there are two clock regions (depending on $x = 0$ or not). The minimal delay between two configurations of the region automaton is always 0.

More generally, it is possible to express sets of configurations which are not necessary unions of regions; any first-order definable set of clock values can be used in the logic. For instance we could express "each time the clock x is the double of clock y, we can reach a state in which y is a third of z, within a delay which is the half of x":

$$\forall \boldsymbol{X}, \exists \boldsymbol{X}', (x = 2y) \Rightarrow (\boldsymbol{X} \xrightarrow[\mathcal{A}]{*} \boldsymbol{X}' \wedge z' = 3y' \wedge 2(\tau' - \tau) = x)$$

Not only such sets of configurations are expressible, but also relations between configurations. For instance we can express that "each time we are in state q, we can reach a configuration in which the clock x has doubled". This corresponds to a relation $x' = 2x$.

Using free variables, it is also possible to define values of clocks (or delays). For instance, the minimal delay between configurations c_1, c_2 can be defined by:

$$\phi(x) \stackrel{\text{def}}{=} \exists \tau, (c_1, \tau) \xrightarrow[\mathcal{A}]{*} (c_2, \tau + x) \wedge \forall y < x, \forall \tau, \neg (c_1, \tau) \xrightarrow[\mathcal{A}]{*} (c_2, \tau + x)$$

and hence can be computed or used in further verifications.

We can take advantage of the *binary* relation: it is as easy to express properties about the past as about the future. For instance: "each time we reach a state q, then the clock x was never larger than $y + z$ in the past".

We conclude this section with an example which looks more relevant: assume we have a server which receives requests from several users. Assume that the server receives requests from two users at time t and $t + \epsilon$ and that these requests are granted at time $t + \delta_1$ and $t + \delta_2$. We may want to check that the server is "fair" and that the delay δ_2 is always smaller than $2 \times \delta_1 + \epsilon$, for instance. This is again a typical property which can be expressed and checked in our theory.

6 Conclusion and perspectives

We have shown that, for timed automata, the binary reachability relation is definable in a decidable theory. The formula which results from our computation may be, in principle, quite huge since for instance the number of equivalence classes for $\sim$ is exponential. Hence our result is, so far, more of theoretical nature. One possible future research direction is then to have more precise informations on the complexity of the method (both theoretically and practically). However, beyond an hypothetical practical verification technique for timed automata, we believe that our result shows another possible direction of research. It may suggest other (more tractable) real-time computation models, starting from the logical side (the theory of real numbers). For instance, we could start from flat timed automata (without loosing expressiveness in the clock valuations sense). It also separates the expressiveness of the properties to be checked (in which it is possible to express much more relationships between clocks) from the expressiveness of the model. There are also several side effects of our proof. For instance we can decide whether or not a loop (or several loops) can be iterated.

The ability to express the reachability relation as a constraint between initial and final clock values allows to replace a whole automaton (or a piece of it corresponding to a timed automaton) with a single *meta-transition* [9], hence faithfully abstracting complex models. This can be used in verifying complex systems.

Conversely, we can mechanically check properties of models which are more expressive than timed automata.

Example 13. We consider two timed automata which are connected by a single transition whose enabling condition is an arbitrary first-order formula ψ over clock values (see figure 8). Properties of such a network can be verified as easily (or as hardly) as properties of a single timed automaton: it suffices to compute the binary reachability relation for each individual automaton and then to connect the two formulas with the enabling condition ψ.

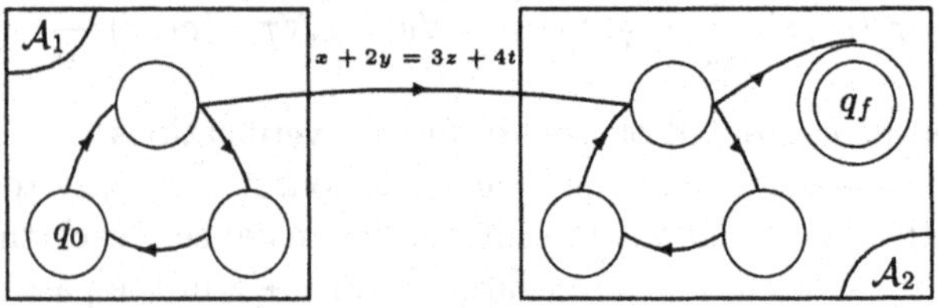

Fig. 8. An extended timed automaton

This can be extended of course to any network of timed automata, provided that there is no cycle through such general transitions.

Adding a stop watch to the model yields undecidability of the reachability [14]. However, as shown in [3], this does not imply that we cannot check properties involving accumulated delays. It is not possible, at least in an obvious way, to express accumulated delay constraints using the first-order theory of the reachability relation. We believe that it is still worth to study more deeply what can (and what cannot) be automatically checked using a similar approach.

Another interesting possible investigation consists in considering *parametrised* timed automata, as in [8]. Though the authors show that emptiness of such automata is undecidable as soon as there are at least three clocks, our method seems to be well-suited for parametric reasoning and, for instance, we may derive conditions on the control instead of on the number of clocks, which yield decision techniques for such parametrised automata.

Finally, our method seems to be well-suited for models which combine timed automata with additional global variables: assume we add registers to the model, with simple operations on them such as in definition 3, then, for flat timed automata with counters, we expect to get again a decision procedure.

References

1. R. Alur. Timed automata. In *Verification of Digital and Hybrid Systems, Proc. NATO-ASI Summer School, Antalya, Turkey*, 1997. To appear.
2. R. Alur, C. Courcoubetis, and D. Dill. Model-checking in dense real time. *Information and Computation*, 104(1):2–24, 1993.
3. R. Alur, C. Courcoubetis, and T. Henzinger. Computing accumulated delays in real-time systems. In *Proc. 5th Conf. on Computer Aided Verification*, volume 818 of *Lecture Notes in Computer Science*, pages 181–193. Springer-Verlag, 1993.
4. R. Alur and D. Dill. Automata for modeling real-time systems. In *Proc. 17th Int. Coll. on Automata, Languages and Programming, Warwick, LNCS 443*, pages 322–335. Springer-Verlag, 1990.
5. R. Alur and D. Dill. A theory of timed automata. *Theoretical Computer Science*, 126:183–235, 1994.
6. R. Alur, T. Feder, and T. Henzinger. The benefits of relaxing punctuallity. *J. ACM*, 43:116–146, 1996.
7. R. Alur and T. Henzinger. A really temporal logic. *J. ACM*, 41:181–204, 1994.
8. R. alur, T. Henzinger, and M. Vardi. Parametric real-time reasoning. In *Proc. 25th Annual ACM Symposium on Theory of Computing*, 1993.
9. B. Boigelot and P. Wolper. Symbolic verification with periodic sets. In *Computer Aided Verification, Proc. 6th Int. Conerence*, LNCS, Stanford, June 1994. Springer-Verlag.
10. H. Comon and Y. Jurski. Multiple counters automata, safety analysis and presburger arithmetic. In A. Hu and M. Vardi, editors, *Proc. Computer Aided Verification*, volume 1427 of *LNCS*, pages 268–279, Vancouver, 1998. Springer-Verlag.
11. H. Comon and Y. Jurski. Timed automata and the theory of real numbers. Technical report, LSV Research Report, June 1999. avalaible at http://www.lsv.ens-cachan.fr/Publis/publis-lsv-index.html.
12. C. Courcoubetis and M. Yannakakis. Minimal and maximal delay problems in real time systems. In K. Larsen and A. Skou, editors, *Proc. CAV 91: Computer Aided Verification*, volume 575 of *Lecture Notes in Computer Science*, pages 399–409. Springer-Verlag, 1991.
13. L. Fribourg. A closed form evaluation for extending timed automata. Technical Report 1998-02, Laboratoire Spécification et Vérification, ENS Cachan, Mar. 1998.
14. T. Henzinger, P. Kopke, A. Puri, and P. Varaiya. What is decidable about hybrid automata ? In *Proc. 27th Symposium on Theory of Computing*, pages 373–382. ACM Press, 1995.
15. T. Henzinger, X. Nicollin, J. Sifakis, and S. Yovine. Symbolic model checking for real time systems. *Information and Computation*, 111(2):193–244, 1994.
16. F. Laroussinie, K. Larsen, and C. Weise. From timed automata to logic – and back. In *Proc. 20th Conf. on Foundations of Computer Science*, volume 969 of *Lecture Notes in Computer Science*, Prag, 1995. Springer-Verlag.
17. M. Lothaire. *Combinatorics on words*, volume 17 of *Encyclopedia of mathematics and its applications*. Cambridge University Press, 1982.
18. F. Wang. Timing behaviour analysis for real time systems. In *Tenth Annual IEEE Symposium on Logic in Computer Science*, San Diego, CA, June 1995. IEEE Comp. Soc. Press.
19. T. Wilke and M. Dickhfer. The automata-theoretic method works for tctl model checking. Technical Report 9811, Inst. f. Informatik u. Prakt. Math., CAU Kiel, 1998.

Metrics for Labeled Markov Systems

Josée Desharnais[1], Vineet Gupta[2*], Radha Jagadeesan[3], and Prakash Panangaden[1]

[1] School of Computer Science, McGill University, Montreal, Quebec, Canada
[2] Autonomy and Robotics Area, NASA Ames Research Center, Moffett Field CA 94035, USA
[3] Dept. of Math. and Computer Sciences, Loyola University-Lake Shore Campus, Chicago IL 60626, USA

Abstract. Partial Labeled Markov Chains (`plMc`) generalize process algebra and traditional Markov chains. They provide a foundation for interacting discrete probabilistic systems. Existing notions of process equivalence are too sensitive to the exact probabilities of transitions in `plMc`s. This paper studies more robust notions of "approximate" equivalence between `plMc`s .

1 Introduction

Probability, like nondeterminism, is an abstraction mechanism used to hide inessential or unknown details. Statistical mechanics — originated by Boltzmann, Gibbs, Maxwell and others — is the fundamental successful example of the use of the probabilistic abstraction. Our investigations are concerned with the development of contextual reasoning principles for concurrent interacting probabilistic systems. Consider the following paradigmatic examples.

Example 1. [1] analyzes a component (say c) of the Lucent Technologies' 5ESS® telephone switching system that is responsible for detecting malfunctions on the hardware connections between switches. This component responds to alarms generated by another complicated system that is only available as a black-box. A natural model to consider for the black-box is a stochastic one, representing the timing and duration of the alarm by random variables with a given probability distribution. [1] shows that the desired properties hold with high probability, showing that the component being analyzed approximates the idealized behavior (say i) with sufficient accuracy.

Example 2. Consider model-based diagnosis settings. Often information about *failure models* and their associated probabilities is obtained from field studies and studies of manufacturing practices. Failure models can be incorporated by assigning a variable, called the *mode* of the component, to represent the physical state of the component, and associating a failure model with each value of the mode variable. Probabilistic information can be incorporated by letting the mode vary according to the given probability distribution [13]. The diagnostic engine computes the most probable diagnostic hypothesis, given observations about the current state of the system.

* Caelum Research Corporation

These examples illustrate the modes of contextual reasoning that interest us. In the first example, we are interested in exploring whether c can substitute for i in arbitrary program contexts; i.e. for some context $C[]$, does $C[c]$ continue to approximate $C[i]$. Similarly, in the second example, we are looking to see the extent to which systems with similar failure behaviors are intersubstitutable. Such a question perforce generalizes the study of congruences elaborated by the theory of concurrency. The theory of concurrency performs a study of "exactly intersubstitutable" processes with temporal behavior. In the probabilistic context, the extant notions of bisimulation (or any process equivalence for that matter) are too sensitive to the probabilities; a slight perturbation of the probabilities would make two systems non-bisimilar. The examples motivate a shift to the study of the more robust notion of "approximately intersubstitutable".

The next example illustrates a deeper interaction of the temporal and probabilistic behavior of processes.

Example 3. Consider a producer and a consumer process connected by a buffer, where the producer is say a model of a network. Examples of this kind are studied extensively in the performance modeling of systems. In a model of such a system, probability serves to abstract the details of the producer (resp. consumer) process by considering rates of production (resp. consumption) of data based on empirical information. This model can be analyzed to calculate the number of packets lost as a function of the probabilities and the buffer size. The analysis aids in tuning system parameters, e.g. to optimize the buffer size. These studies are often couched in terms of asymptotic/stationary behavior to abstract over the transient behavior associated with system initialization (such as large bursts of communication) evident when the system begins execution.

Such examples motivate the study of equality notions based on "eventually approximately intersubstitutable" processes.

1.1 Our results

Partial labeled Markov chains (`plMcs`) are the discrete probabilistic analogs of labeled transition systems. In this model "internal choice" is modeled probabilistically and the so-called "external choice" is modeled by the indeterminate actions of the environment. The starting point of our investigation is the study of strong bisimulation for `plMcs`. This study was initiated by [26] for `plMcs` in a style similar to the queuing theory notion of "lumpability". This theory has been extended to continuous state spaces and continuous distributions [5, 11]. These papers provided a characterization of bisimulation using a negation-free logic $\mathcal{L}$.

In the context of the earlier discussion, we note that probabilistic bisimulation is too "exact" for our purposes — intuitively, two states are bisimilar only if the probabilities of outgoing transitions match exactly, motivating the search for a relaxation of the notion of equivalence of probabilistic processes. Jou and Smolka [22] note that the idea of saying that processes that are close should have probabilities that are close does not yield a transitive relation, as illustrated by an example of van Breugel [8]. This leads them to propose that the correct formulation of the "nearness" notion is via a metric.

A metric d is a function that yields a real number distance for each pair of processes. It should satisfy the usual metric conditions: $d(P, Q) = 0$ implies P is bisimilar to Q,

$d(P,Q) = d(Q,P)$ and $d(P,R) \leq d(P,Q) + d(Q,R)$. Inspired by the Hutchinson metric on probability measures [19], we demand that d be "Lipschitz" with respect to probability numbers, an idea best conveyed via a concrete example.

Example 4. Consider the family of `plMcs` $\{P_\epsilon \mid 0 \leq \epsilon < r\}$ where $P_\epsilon = a_{r-\epsilon}.Q$, i.e. P_ϵ is the `plMc` that does an a with probability $r - \epsilon$ and then behaves like Q. We demand that: $d(P_{\epsilon_1}, P_{\epsilon_2}) \leq |\epsilon_1 - \epsilon_2|$. This implies that P_ϵ converges to P_0 as ϵ tends to 0.

Metrics on `plMcs`*.* Our technical development of these intuitions is based on the key idea expounded by Kozen [24] to generalize logic to handle probabilistic phenomena.

Classical logic	Generalization
Truth values $\{0,1\}$	Interval $[0,1]$
Propositional function	Measurable function
State	Measure
Evaluation of prop. functions	Integration

Following these intuitions, we consider a class $\mathcal{F}$ of functions that assign a value in the interval $[0,1]$ to states of a `plMc`. These functions are inspired by the formulas of $\mathcal{L}$ — the result of evaluating these functions at a state corresponds to a quantitative measure of the extent to which the state satisfies a formula of $\mathcal{L}$. The identification of this class of functions is a key contribution of this paper, and motivates a metric d:

$$d(P,Q) = \sup\{|f(s_P) - f(s_Q)| \mid f \in \mathcal{F}\}.$$

In section 4, we formalize the above intuitions to define a family of metrics $\{d^c \mid c \in (0,1]\}$. These metrics support the spectrum of possibilities of relative weighting of the two factors that contribute to the distance between processes: the complexity of the functions distinguishing them versus the amount by which each function distinguishes them. d^1 captures only the differences in the probability numbers; probability differences at the first transition are treated on par with probability differences that arise very deep in the evolution of the process. In contrast, d^c for $c < 1$ give more weight to the probability differences that arise earlier in the evolution of the process, i.e. differences identified by simpler functions. As c approaches 0, the future gets discounted more.

As is usual with metrics, the actual numerical values of the metric are less important than properties like the significance of zero distance, relative distance of processes, contractivity and the notion of convergence.

Example 5. Consider the `plMc` P with two states, and a transition going from the start state to the other state with probability p. Let Q be a similar process, with the probability q. Then in section 4, we show that $d^c(P,Q) = c|p-q|$. Now if we consider P' with a new start state, which makes a b transition to P with probability 1, and similarly Q' whose start state transitions to Q on b with probability 1, then $d^c(P',Q') = c^2|p-q|$, showing that the next step is discounted by c.

Each of these metrics agree with bisimulation:

$$d^c(P,Q) = 0, \text{ iff } P \text{ and } Q \text{ are bisimilar.}$$

For $c < 1$, we show how to compute $d^c(P,Q)$ to within ϵ.

An "asymptotic" metric on `plMcs`*.* The d^c metric (for $c < 1$) is heavily influenced by the initial transitions of a process — processes which can be differentiated early are far apart. For each $c \in (0, 1]$, we define a dual metric d^c_∞ (Section 6) on `plMcs` to capture the idea that processes are close if they have the same behavior "eventually", thus disregarding their initial behavior. Informally, we proceed as follows. Let P **after** s stand for the `plMc` P after exhibiting a trace s. Then, the j'th distance d^c_j between P, Q after exhibiting traces of length j is given by $\sup\{d^c(P \textbf{ after } s, Q \textbf{ after } s) \mid length(s) = j\}$. The asymptotic distance between P, Q is given by the appropriate limit of the d^c_j's:

$$d^c_\infty(P, Q) = \lim_{i \to \infty} \sup_{j > i} d^c_j(P, Q).$$

A process algebra of probabilistically determinate processes. In order to illustrate the properties of the metrics via concrete examples, we use an algebra of probabilistically determinate processes and a (bounded) buffer example coded in the algebra (Section 5). This process algebra has input and output prefixing, parallel composition and a probabilistic choice combinator. We do not consider hiding since this paper focuses on strong (as opposed to weak) probabilistic bisimulation.

We show that bisimulation is a congruence for all these operations. Furthermore, we generalize the result that bisimulation is a congruence, by showing that process combinators do not increase distance in any of the d^c metrics. Formally, let $d^c(P_i, Q_i) = \epsilon_i$. For every n-ary process combinator $C[X_1, \ldots, X_n]$, we have

$$d^c(C(P_1, \ldots, P_n), C(Q_1, \ldots, Q_n)) \leq \sum_i \epsilon_i.$$

Prefixing and parallel composition combinators do not increase d^c_∞. However, the probabilistic choice combinator is not contractive for d^c_∞.

Continuous systems. While this paper focuses on systems with a countable number of states, all the results extend to systems with continuous state spaces. The technical development of continuous systems requires measure theory apparatus and will be reported in a separate paper.

Related and future work. In this paper, we deal with probabilistic nondeterminism. In a probabilistic analysis, quantitative information is recorded and used in the reasoning. In contrast, a purely qualitative nondeterministic analysis does not require and does not yield quantitative information. In particular when one has no quantitative information at all, one has to work with indeterminacy — using a uniform probability distribution is not the same as expressing complete ignorance about the possible outcomes.

The study of the interaction of probability and nondeterminism, largely in the context of exact equivalence of probabilistic processes, has been explored extensively in the context of different models of concurrency. Probabilistic process algebras add a notion of randomness to the process algebra model and have been studied extensively in the traditional framework of (different) semantic theories of (different) process algebras (to name but a few, see [16, 23, 26, 18, 2, 32, 9]) e.g. bisimulation, theories of (probabilistic) testing, relationship with (probabilistic) modal logics etc. Probabilistic

Petri nets [28, 33] add Markov chains to the underlying Petri net model. This area has a well developed suite of algorithms for performance evaluation. Probabilistic studies have also been carried out in the context of IO Automata [30, 34].

In contrast to the above body of research the primary theme of this paper is the the study of intersubstitutivity of (eventually) (approximately) equivalent processes. The ideas of approximate substitutivity in this paper are inspired by the work of Jou and Smoka [22] referred to earlier and the ideas in the area of performance modeling as exemplified in on the work on process algebras for compositional performance modeling (see for example [15]). The extension of the methods of this paper to systems which have both probability and traditional nondeterminism remains open and will be the object of future study.

The verification community has been active in developing model checking tools for probabilistic systems, for example [6, 4, 3, 10, 17]. Approximation techniques in the spirit of those of this paper have been explored for hybrid systems [14]. In future work, we will explore efficient algorithms and complexity results for our metrics.

Our work on the asymptotic metric is closely related to, at least in spirit, the work of Lincoln, Mitchell, Mitchell and Scedrov [25] in the context of security protocols. Both [25] and this paper consider the asymptotic behavior of a single process, rather than the limiting behavior of a probabilistically described family of processes as is performed in some analysis performed in Markov theory.

Organization of this paper The rest of this paper is organized as follows. First, in section 2, we review the notions of plMc and probabilistic bisimulation and associated results to make the paper self-contained. We next present (section 3) an alternate way to study processes using real-valued functions and show that this view presents an alternate characterization of probabilistic bisimulation. In section 4, we define a family of metrics, illustrate with various examples and describe a decision procedure to evaluate the metric. The following section 5 describes a process algebra of probabilistically determinate processes. We conclude with a section 6 on the asymptotic metric.

2 Background

This section on background briefly recalls definitions from previous work on partial labeled Markov processes [5, 11, 26] and sets up the basic notations and framework for the rest of the paper.

Definition 1. *A* partial labeled Markov chain (`plMc`) *with a label set L is a structure $(S, \{k_l \mid l \in L\}, s)$, where S is a countable set of states, s is the start state, and $\forall l \in L. k_l : S \times S \rightarrow [0,1]$ is a transition function such that $\forall s \in S. \sum_t k_l(s,t) \leq 1$.*

A `plMc` is finite if S is finite. There is no finite branching restriction on a `plMc`; $k_l(s,t)$ can be non-zero for countably many t's. k_l is extended to a function $S \times \mathcal{P}(S) \rightarrow [0,1]$ by defining: $k_l(s,A) = \sum_{t \in A} k_l(s,t)$. Given a `plMc` $P = (S, \{k_l \mid l \in L\}, s)$, we shall refer to its state set, transition probability and initial state as S_P, k_l^P and s_P respectively, when necessary. We will assume a fixed finite set of labels L, and that k_l is a rational function, this does not restrict the theory.

We could have alternatively presented a plMc as a structure $(S, \{k_l \mid l \in L\}, \mu)$ where μ is an initial distribution on S. Given a plMc with initial distribution P, one can construct an equivalent plMc with initial state Q as follows. $S_Q = S_P \cup \{u\}$ where u is a new state not in S_P. u will be the start state of Q. $k_l^Q(s,t) = k_l^P(s,t)$ if $s, t \in S_P$; $k_l^Q(s,u) = 0$, and $k_l^Q(u,t) = \sum k_l^P(s,t)\mu^P(s)$. We will freely move between the notions of initial state and initial distribution. For example, when a transition on label l occurs in a plMc P, there is a new initial distribution given by $\mu'(t) = \sum k_l(s,t) \times \mu(s)$.

We recall the definition of bisimulation on plMc from [26].

Definition 2. *An equivalence relation, R, on the set of states of a* plMc *P is a* ***bisimulation*** *if whenever two states s_1 and s_2 are R-related, then for any label l and any R-equivalence class of states T, $k_l(s_1, T) = k_l(s_2, T)$.*

Two plMcs *P, Q are bisimilar if there is a bisimulation R on the disjoint union of P, Q such that $s_P \; R \; s_Q$.*

In [11] it is shown that bisimulation can be characterized using a *negation free* logic $\mathcal{L}$: $\mathsf{T} | \phi_1 \wedge \phi_2 | \langle a \rangle_q \phi$, where a is an label from the set of labels L and $q \in [0,1)$ is a rational number. Given a plMc $P = (S, \Sigma, k_a, s)$ we write $t \models_P \phi$ to mean that the state t satisfies the formula ϕ. The definition of the relation $\models$ is given by induction on formulas.

$$\begin{array}{ll} t \models_P \mathsf{T} & \\ t \models_P \phi_1 \wedge \phi_2 \Leftrightarrow t \models_P \phi_1, \quad t \models_P \phi_2 & \\ t \models_P \langle a \rangle_q \phi & \Leftrightarrow \exists A \subseteq S. (\forall t' \in A. t' \models_P \phi) \wedge (q < k_a(t, A)). \end{array}$$

In words, $t \models_P \langle a \rangle_q \phi$ if the system P in state t can make an a-move to a set of states that satisfy ϕ with probability strictly greater than q. We write $[\![\phi]\!]_P$ for the set $\{s \in S_P | s \models \phi\}$. We often omit the P subscript when no confusion can arise. The result of [11] that is relevant to the current paper is that two plMcs are bisimilar if and only if their start states satisfy the same formulas.

Definition 3. *P is a sub-*plMc *of Q if $S_P \subseteq S_Q$ and $(\forall l) \; [k_l^P(s,t) \leq k_l^Q(s,t)]$*

Thus, a sub plMc of a plMc has fewer states and lower probabilities. The logic $\mathcal{L}$, since it does not have negation, satisfies a basic monotonicity property with respect to substructures: If P is a sub-plMc of Q, then $(\forall s \in S_P) \; [s \models_P \phi \; \Rightarrow \; s \models_Q, \phi]$. Every formula satisfied in a state of a plMc is witnessed by a finite sub-plMc.

Lemma 1. *Let P be a* plMc, *$s \in S_P$, such that $s \models_P \phi$. Then there exists a finite sub-*plMc *of P, Q_ϕ^s, such that $s \in S_Q$, and $s \models_Q \phi$.*

3 An alternate characterization of probabilistic bisimulation

In this section, following Kozen [24], we present an alternate characterization of probabilistic bisimulation using functions into the reals instead of the logic $\mathcal{L}$. We define a set of functions which are sufficient to characterize bisimulation. It is worth clarifying our terminology here. We define a set of *functional expressions* by giving an explicit syntax. A functional expression becomes a function when we interpret it in a system. Thus we may loosely say "the same function" when we move from one system to another. What we really mean is the "same functional expression"; obviously it cannot be the same function when the domains are different. This is no different from having syntactically defined formulas of some logic which become boolean-valued functions when they are interpreted on a structure.

We now give the class of functional expressions. First, some notation. Let $\lfloor r \rfloor_q = r - q$ if $r > q$, and 0 otherwise. $\lceil r \rceil^q = q$ if $r > q$, and r otherwise. Note that $\lfloor r \rfloor_q + \lceil r \rceil^q = r$.

For each $c \in (0, 1]$, we consider a family $\mathcal{F}^c$ of functional expressions generated by the following grammar. Here q is a rational in $[0, 1]$.

$$\begin{array}{lll} f^c ::= & \lambda s.1 & \text{Constant schema} \\ \mid & \lambda s.1 - f^c(s) & \text{Negation schema} \\ \mid & \lambda s.\min(f_1^c(s), f_2^c(s)) & \text{Min schema} \\ \mid & \lambda s.\sup_{i\in\mathbb{N}}\{f_i^c(s)\} & \text{Sup schema} \\ \mid & \lambda s.c\int_{t\in S}\tau_a(s,t) f^c(t) & \text{Prefix schema} \\ \mid & \lambda s.\lfloor f^c(s) \rfloor_q & \\ \mid & \lambda s.\lceil f^c(s) \rceil^q & \text{Conditional schemas} \end{array}$$

$\mathcal{F}_+^c$ is the sub-collection of $\mathcal{F}^c$ that does not use the negation schema.

The functional expressions generated by these schemas will be written as $1, 1 - f$, $\min(f_1, f_2)$, $\sup_i f_i$, $\langle a \rangle.f$, $\lfloor f \rfloor_q$ and $\lceil f \rceil^q$ respectively. We will use $\langle a \rangle^n.f$ to represent $\underbrace{\langle a \rangle.\ldots.\langle a \rangle}_{n}.f$.

One can informally associate functional expressions with every connective of the logic $\mathcal{L}$ in the following way. T is represented by $\lambda s.1$ and conjunction by min. The contents of the connective $\langle a \rangle_q$ is split up into two expression schemas: the $\langle a \rangle.f$ schema that intuitively corresponds to prefixing and the conditional schema $\lfloor f \rfloor_q$ that captures the "greater than q" idea.

Lemma 2. *The functions* $1, 1 - f, \min(f_1, f_2), \lfloor f \rfloor_q, \lceil f \rceil^q$ *can be used to approximate any continuous Lipschitz function from* $[0, 1]$ *to* $[0, 1]$.

This shows that we can replace the constant schema, negation schema and conditional schemas with one schema: $\lambda s.g(f(s))$, where g is any continuous Lipschitz function. To get positive functions $\mathcal{F}_+^c$, we can restrict g to monotone continuous Lipschitz functions.

Example 6. Consider the `plMcs` A_1 and A_2 of figure 1. All transitions are labeled a. The functional expression $(\langle a \rangle.1)^c$ evaluates to c at states s_0, s_2 of both A_1 and A_2; it evaluates to 0 at states s_1, s_3 of A_1 and s_3, s_4 of A_2, and it evaluates to $c/2$ at state s_1 of A_2. The functional expression $(\langle a \rangle.\langle a \rangle.1)^c$ evaluates to $3c^2/4$ at states s_0 of A_1, A_2 and to 0 elsewhere. The functional expression $(\langle a \rangle.\lfloor \langle a \rangle.1 \rfloor_{\frac{1}{2}})^c$ evaluates to $3c^2/8$ at state s_0 of A_1 and to $c^2/4$ at state s_0 of A_2.

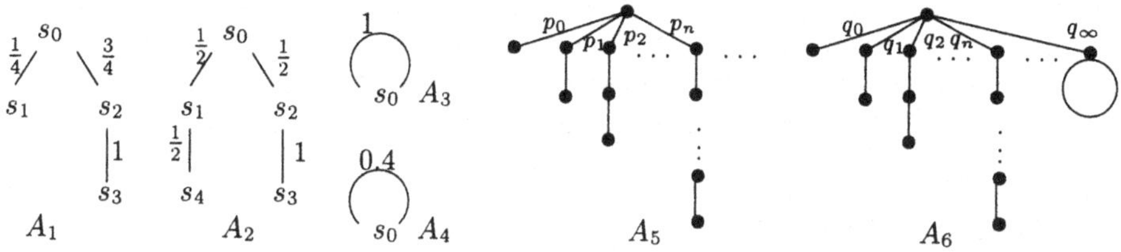

Fig. 1. Examples of plMcs

Example 7. Consider the plMc A_3 of figure 1. All transitions are labeled a. A functional expression of the form $(\underbrace{\langle a\rangle.\dots.\langle a\rangle}_{n}.1)^c$ evaluates to c^n at state s_0. On state s_0 of plMc A_4 the same functional expression evaluates to $(c \times 0.4)^n$.

A routine induction on the structure of the functional expression $f^c \in \mathcal{F}_+^c$, shows:

Lemma 3. *If P is a sub-plMc of Q, then $(\forall f \in \mathcal{F}_+^c)$ $(\forall s \in S_P)$ $[f_P^c(s) \leq f_Q^c(s)]$.*

For any state in a finite plMc that satisfies a formula, there is a partial witness from $\mathcal{F}_+^c$.

Lemma 4. *Given any $\phi \in \mathcal{L}$ and a finite plMc P, and any $c \in (0,1]$, there is a functional expression $f^c \in \mathcal{F}_+^c$ such that*

1. *$\forall s \in S_P . f_P^c(s) > 0$ iff $s \models_P \phi$.*
2. *for any plMc Q, $\forall s \in S_Q . s \not\models_P \phi \Rightarrow f_Q^c(s) = 0$.*

Proof. The proof is by induction on the structure of ϕ. The key case is $\phi = \langle a\rangle_q.\psi$, let g^c be the functional expression corresponding to ψ yielded by induction. Let S_ψ be the set of states in P satisfying ψ, and let $x = \min\{g(s) \mid s \in S_\psi\}$. By induction hypothesis, $x > 0$. Consider the functional expression f^c given by $\lfloor\langle a\rangle.\lceil g\rceil^x\rfloor_{cxq}$. For all $t \in [\![\psi]\!]$, $(\lceil g\rceil^x)(t) = x$. Now for any state $s \in S_P$,

$$(\langle a\rangle.\lceil g\rceil^x)^c(s) = cx \sum_{t \in [\![\psi]\!]} k_a(s,t) = cxk_a(s, [\![\psi]\!]).$$

Now for each state $s \in [\![\phi]\!]$, $k_a(s, [\![\psi]\!]) > q$. Thus f^c satisfies the first condition.

The second condition holds because for any $s \in s_Q$, $(\langle a\rangle.\lceil g\rceil^x)(s) \leq cxk_a(s, [\![\psi]\!]_Q)$, so if $k_a(s, [\![\psi]\!]_Q) \leq q$ then $(\lfloor\langle a\rangle.\lceil g\rceil^x\rfloor_{cxq})(s) = 0$.

Lemma 3 permits us to show that $\mathcal{F}_+^c$ is complete for reasoning about logical satisfiability. In fact, logical satisfiability is sound for reasoning about all functions in $\mathcal{F}^c$. Thus, we get:

Theorem 1. *For any plMc P, $(\forall c \in (0,1])$, $\forall s, s' \in S_P$*

$$[(\forall \phi \in \mathcal{L})\ s \models_P \phi \Leftrightarrow s' \models_P \phi] \Leftrightarrow (\forall f \in \mathcal{F}^c)\ [f_P^c(s) = f_P^c(s')].$$

Example 8. Consider the plMcs A_1, A_2 of figure 1. The calculations of example 6 show that the s_0 states of A_1, A_2 are distinguishable. Furthermore, the states are indistinguishable if we use only the function schemas Constant, Min and Prefixing. Thus, example 6 shows that the conditional functional expressions are necessary.

4 A Metric on Processes

Each collection of functional expression $\mathcal{F}^c$ be the set of all such expressions induces a distance function as follows:

$$d^c(P,Q) = \sup_{f^c \in \mathcal{F}^c} |f^c_P(s_P) - f^c_Q(s_Q)|.$$

Theorem 2. *For all* $c \in (0,1]$, d^c *is a metric.*

Proof. The transitivity and symmetry of d^c is immediate. $d^c(P,Q) = 0$ iff P and Q are bisimilar follows from theorem 1.

Example 9. The analysis of example 8 yields $d^c(A_1, A_2) = c^2/8$.

Example 10. Example 7 shows the fundamental difference between the metrics $d^c, c < 1$ and d^1, explaining why we can get an algorithm for $d^c, c < 1$. For $c < 1$, $d^c(A_3, A_4)$ is witnessed by some $(\langle a \rangle^n.1)^c$ and is given by $d^c(A_3, A_4) = c^n(1 - (0.4)^n)$ for that n. In contrast, $d^1(A_3, A_4) = \sup\{1 - (0.4)^n \mid n = 0, 1, \ldots\} = 1$; no finitary function witnesses this.

Example 11. (Analysis of Example 4) Consider the family of plMcs $\{P_\epsilon \mid 0 \leq \epsilon < r\}$ where $P_\epsilon = a_{r-\epsilon}.Q$, i.e. P_ϵ is the plMc that does an a with probability $r - \epsilon$ and then behaves like Q. The function expression $(\langle a \rangle.1)^c$ evaluates to $(r - \epsilon)c$ at P_ϵ. This functional expression witnesses the distance between any two P's (other functions will give smaller distances). Thus, we get $d(P_{\epsilon_1}, P_{\epsilon_2}) = c|\epsilon_1 - \epsilon_2|$. This furthermore ensures that P_ϵ converges to P_0 as ϵ tends to 0.

Example 12. (from [11]) Consider the plMcs A_5 and A_6 of figure 1. A_6 is just like A_5 except that there is an additional transition to a state which then has an a-labeled transition back to itself. The probability numbers are as shown. If both plMcs have the same values on all functional expressions we will show that $q_\infty = 0$, i.e. it really cannot be present. The functional expression $(\langle a \rangle.1)^c$ yields $c(\sum_{i \geq 0} p_i)$ on A_5 and $c(q_\infty + \sum_{i \geq 0} q_i)$ on A_6. The functional expression $(\langle a \rangle.\langle a \rangle.1)^c$ yields $c^2(\sum_{i \geq 1} p_i)$ on A_5 and $c^2(q_\infty + \sum_{i \geq 2} q_i)$ on A_6. Thus, we deduce that $p_0 = q_0$. Similarly, considering functional expressions $(\langle a \rangle.\langle a \rangle.\langle a \rangle.1)^c$ etc, we deduce that $p_n = q_n$. Thus, $q_\infty = 0$.

A decision procedure for $d^c, c < 1$. Given finite plMcs P, Q, we now provide a decision procedure for computing $d^c(P,Q)$ for $c < 1$ to any desired accuracy c^n, where n is a natural number. We do this by computing $\sup_F |f^c(s_P) - f^c(s_Q)|$ for a finite set of functions F, and then show that for this F, $d^c(P,Q) - \sup_F |f^c(s_P) - f^c(s_Q)| \leq c^n$.

Define the depth of a finitary functional expression (i.e. without sup) inductively as follows: $depth(\lambda s.1) = 0$, $depth(\min(f^c_1, f^c_2)) = max(depth(f^c_1), depth(f^c_2))$

and $depth(1-f) = depth(\lfloor f^c \rfloor_q) = depth(\lceil f^c \rceil^q) = depth(f^c)$, $depth(\langle a \rangle.f^c) = depth(f^c) + 1$. Then it is clear that $|f^c(s_P) - f^c(s_Q)| \leq c^{depth(f)}$. Now if we include in F all functions of depth $\leq n$, then $d^c(P,Q) - \sup_F |f^c(s_P) - f^c(s_Q)| \leq c^n$.

However there are infinitely many functional expressions of depth $\leq n$. We now construct a finite subset of these, such that the above inequality still holds. Let $A^i = \{\frac{k}{3^{m+1+n-i}} \mid k = 0, \ldots 3^{m+1+n-i}\}$, where $1/3^m < c^n$. We construct the set of functions inductively as follows. We use $\overline{F}$ to stand for $F \cup \{1 - f \mid f \in F\}$. Let F^i be the set of all functions of depth $\leq i$. Define:

$$\begin{aligned}
F_1^{i+1} &= \overline{\{\langle a \rangle.f \mid f \in F^i\}} & F_2^{i+1} &= \overline{\{\lfloor f \rfloor_q \mid f \in F_1^{i+1}, q \in A^{i+1}\}} \\
F_3^{i+1} &= \overline{\{\lceil f \rceil^q \mid f \in F_2^{i+1}, q \in A^{i+1}\}} & F_4^{i+1} &= \overline{\{\lceil f \rceil^q \mid f \in F_3^{i+1}, q \in A^{i+1}\}} \\
F_5^{i+1} &= \{\min(f_1, \ldots, f_n) \mid f_i \in F_4^{i+1} \cup F^i\} & F^{i+1} &= \{\max(f_1, \ldots, f_n) \mid f_i \in F_5^{i+1}\}
\end{aligned}$$

We can prove that for any $f^c \in \mathcal{F}^c$ of depth $\leq n$, there is a function in F^n that approximates it closely enough.

Lemma 5. *Let $f^c \in \mathcal{F}^c$ and $\epsilon > 0$. Then $\exists g_f^c \in F^i$ such that:* (∀`plMc` P) $(\forall s \in S_P)$ $[|f^c(s) - g_f^c(s)| \leq \epsilon]$.

The proof relies on the following identities that show that repeating steps 2–5 on F^{i+1} does not get any new functions.

$$\begin{aligned}
&\lfloor \lfloor f \rfloor_q \rfloor_r = \lfloor f \rfloor_{q+r} & &\lceil \lceil f \rceil^q \rceil^r = \lceil f \rceil^{\min(q,r)} \\
&\lfloor \lceil f \rceil^q \rfloor_r = \lceil \lfloor f \rfloor_r \rceil^{q-r} & & \\
&\lfloor \min(f_1, f_2) \rfloor_r = \min(\lfloor f_1 \rfloor_r, \lfloor f_2 \rfloor_r) & &\lceil \min(f_1, f_2) \rceil^r = \min(\lceil f_1 \rceil^r, \lceil f_2 \rceil^r) \\
&1 - (1 - f) = f & & \\
&1 - \max(f_1, f_2) = \min(1 - f_1, 1 - f_2) & &1 - \min(f_1, f_2) = \max(1 - f_1, 1 - f_2) \\
&\lfloor 1 - \lfloor f \rfloor_q \rfloor_r = \lceil \lfloor 1 - f \rfloor_{r-q} \rceil^{1-r} & &\lceil 1 - \lceil f \rceil^r \rceil^q = 1 - \lceil 1 - \lceil 1 - f \rceil^q \rceil^r \\
&\lfloor 1 - \lceil f \rceil^q \rfloor_r = \lfloor 1 - f \rfloor_r, \text{if } q + r \geq 1 & &\lfloor 1 - \lceil f \rceil^q \rfloor_r = \lceil \lfloor 1 - f \rfloor_r \rceil^{1-q-r}, \text{if } q + r < 1 \\
&\min(\max(f_1, f_2), \max(f_3, f_4)) = \max(\min(f_1, f_3), \min(f_1, f_4), \ldots) & &
\end{aligned}$$

5 Examples of metric reasoning principles

In this section, we use a process algebra and an example coded in the process algebra to illustrate the type of reasoning provided by our study. We also show that small perturbations of a process results in a nearby process.

5.1 A process algebra

The process algebra describes probabilistically determinate processes. The processes are input-enabled [27, 12, 20] in a weak sense ($(\forall s \in S_P)$ $(\forall a \in L)$ $k_{a?}(s, S_P) > 0$) and communication is via CSP style broadcast. The process combinators that we consider are parallel composition, prefixing and probabilistic choice. We do not consider hiding since this paper focuses on strong probabilistic bisimulation. Though we do not enforce the fact that output actions do not block, this assumption can safely be added to the algebra to make it an IO calculus [31].

We assume an underlying set of labels $\mathcal{A}$. Let $L? = \{a? \mid a \in \mathcal{A}\}$ be the set of input labels, and $L! = \{a! \mid a \in \mathcal{A}\}$ the set of output labels. The set of labels are given by $L = L? \cup L!$. Every process P is associated with a subset of labels: $P_O \subseteq L!$, the set of relevant output labels. This signature is used to constrain parallel composition.

Prefixing. $P = a?_r.Q$ where r is a rational number, is the process that accepts input a and then performs as Q. The number r is the probability of accepting $a?$. With probability $(1-r)$ the process $P = a?_r.Q$ will *block* on an $a?$ label. S_P is given by adding a new state, q to S_Q. Add a transition labeled $a?$ from q to the start state of Q with probability r. For all other labels l, add a $l?$ labeled self-loop at q with probability 1. q is the start state of P.

Output prefixing, $P = a!_r.Q$, where r is a rational number, the process that performs output action $a!$ and then functions as Q, is defined analogously. In this case, $P_O = Q_O \cup \{a!\}$. For both input and output prefixing, we have: $d^c(a_r.P, a_s.P) \leq c \mid r - s \mid$.

Probabilistic choice. $P = Q +_r Q'$ is the probabilistic choice combinator [21] that chooses$Q$ is with probability r and Q' is chosen with probability $1-r$. $P_O = Q_O \cup Q'_O$. $S_P = S_Q \uplus S_{Q'}$. Now $k_l^P(q, A \uplus A') = k_l^Q(q, A)$ if $q \in S_Q$, and $k_l^P(q, A \uplus A') = k_l^{Q'}(q, A')$ if $q \in S_{Q'}$. We define an initial distribution μ: $\mu(\{s_Q\}) = r, \mu(\{s_{Q'}\}) = 1 - r$, referring the reader to section 2 for a way to convert to an initial state format.

We have: $d^c(P +_r Q, P +_s Q) \leq \mid r - s \mid d^c(P, Q)$; $d^c(P +_r Q, P' +_r Q) \leq r d^c(P, P')$.

Parallel composition. $P = Q \mid\mid Q'$ is permitted if the output actions of Q, Q' are disjoint, i.e. $Q_O \cap Q'_O = \emptyset$. The parallel composition synchronizes on all labels in $Q_L \cap Q'_L$. $P_O = Q_O \uplus Q'_O$. $S_P = S_Q \times S_{Q'}$. The k_l^P definition is motivated by the following idea. Let s (resp. s') be a state of Q (resp. Q'). We expect the following synchronized transitions from the product state (s, s').

$$\frac{s \xrightarrow{c?} t \quad s' \xrightarrow{c?} t'}{(s, s') \xrightarrow{c?} (t, t')} \qquad \frac{s \xrightarrow{c!} t \quad s' \xrightarrow{c?} t'}{(s, s') \xrightarrow{c!} (t, t')} \qquad \frac{s \xrightarrow{c?} t \quad s' \xrightarrow{c!} t'}{(s, s') \xrightarrow{c!} (t, t')}.$$

The disjointness of the output labels of Q, Q' ensures that there is no non-determinism. Formally, if $l = a! \in Q_O$, then $k_{a?}^P((s, s'), (t, t')) = k_{a!}^P((s, s'), (t, t')) = k_{a!}^Q(s, t) \times k_{a?}^{Q'}(s', t')$. The case when $a! \in Q'_O$ and $l = a?$ is similar.

Theorem 3. *(Contractivity of process combinators)*

- $d^c(l_r.P, l_r.Q) \leq c d^c(P, Q)$ *for any label* l
- $d^c(P +_r R, Q +_r R) \leq d^c(P, Q)$ *for any* R
- $d^c(P \mid\mid R, Q \mid\mid R) \leq d^c(P, Q)$ *for any* R *for which* $P \mid\mid R, Q \mid\mid R$ *are defined.*

Thus, theorem 1 allows us to conclude that bisimulation is a congruence with respect to these operations.

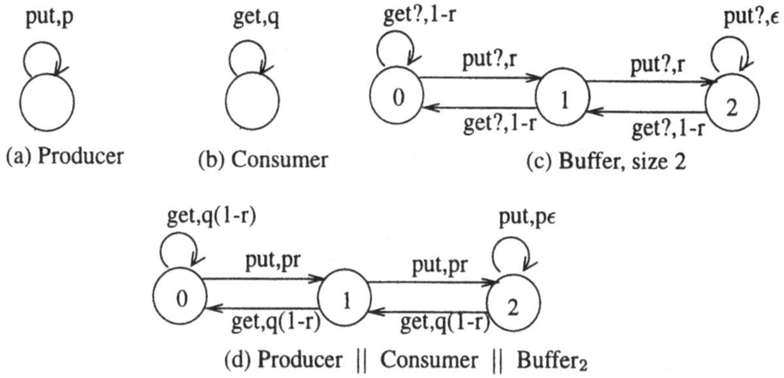

Fig. 2. The producer consumer example.

5.2 A bounded buffer example

We specify a producer consumer process with a bounded buffer (along the lines of [29]). The producer is specified by the 1 state finite automaton shown in Figure 2(a) — it outputs a *put*, corresponding to producing a packet, with probability p (we omit the ! in the labels). To keep the figure uncluttered, we also omit the input-enabling arcs, all of which have probability 1. The consumer (Figure 2(b)) is analogous — it outputs a *get* with probability q, corresponding to consuming a packet. The buffer is an n-state automaton, the states are merely used to count the number of packets in the buffer, while the probabilities code up the probability of scheduling either the producer or the consumer (thus the producer gets scheduled with probability r, and then produces a packet with probability p). Upon receiving a *put* in the last state, the buffer accepts it with a very small probability ϵ, modeling a blocked input. The parallel composition of the three processes is shown in Figure 2(d).

As the buffer size increases, the distance between the bounded buffer and the unbounded buffer decreases to 0. Let P_k = Producer || Consumer || Buffer$_k$, where Buffer$_k$ denotes the process Buffer with k states. Then by looking at the structure of the process, we can compute that $d(P_k, P_\infty) \propto (cpr)^k$. Thus we conclude the following:

- As the bounded buffer becomes larger, it approximates an infinite buffer more closely: if $m > k$ then $d^c(P_k, P_\infty) > d^c(P_m, P_\infty)$.
- As the probability of a put decreases, the bounded buffer approximates an infinite buffer more closely. Thus if $p < p'$, $d^c(P^p, P^p_\infty) < d^c(P^{p'}, P^{p'}_\infty)$, where the superscripts indicate the producer probability.
- Similarly, as the probability of scheduling the Producer process (r) decreases, the buffer approximates an infinite buffer more closely.

5.3 Perturbation results

Let P be a `plMc`. Define $Q = (S_P, k_l^Q, s_P)$ to be an ϵ-perturbation of P if for all labels a, $\forall s \in S_P$. $\forall A \subseteq S_P$ $|k_a^P(s, A) - k_a^Q(s, A)| < \epsilon$. Small perturbations of the probabilities in a `plMc` yields a `plMc` that is within a small distance.

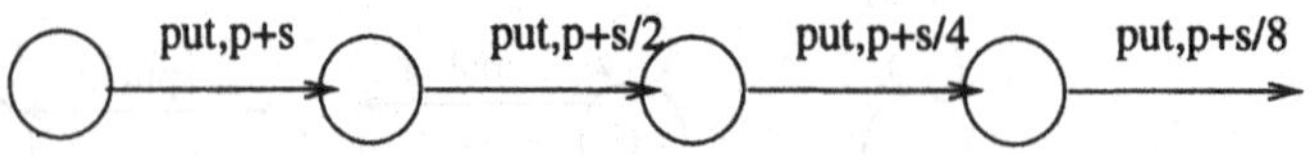

Fig. 3. A producer with transient behavior

Lemma 6. *If $c < 1$, and Q is an ϵ perturbation of P, then $d^c(P,Q) < k\epsilon$ where $k = \sup_n nc^n$.*

A note on $c = 1$. For $c = 1$, nc^n increases without limit, and example 7 shows that the above lemma does not hold for $c = 1$. However, suppose P is unfolded and has no loops. Let $\epsilon_i, i \in \mathbb{N}$ be non-negative rationals such that $\sum_i \epsilon_i = \epsilon < 1/3$. Now, let Q have the same state set S_P, and for each state s at depth n, $|k_a^P(s,A) - k_a^Q(s,A)| < \epsilon_n$ for each label a and each measurable set A. Then, $d^1(P,Q) < 1 - \exp(-2\epsilon)$, thus as $\epsilon \to 0$, $d^1(P,Q) \to 0$.

5.4 Extracting behavioral information

The definition of the metric d^c has a quantification over all functional expressions. However we show that for $c < 1$, there is a universal function that characterizes the ϵ balls around a given state.

Lemma 7. *Let s be a state in a labeled Markov process. Let $\epsilon \in (0, 0.5), c \in (0,1)$. Then, there is a function $f \in \mathcal{F}^c$ such that $f(s) = \epsilon$, and for any state t in any process, $f(t) = 0$ iff $d^c(s,t) \geq \epsilon$*

For finite processes, we can perform a "bisimulation" style matching of transition probabilities for nearby states. Let $\mathcal{P}$ and $\mathcal{P}'$ be finite processes with start states p_0 and p_0' such that $d^c(p_0, p_0') < \epsilon$. Let there be a transition from p_0 to p on label a with probability r,. Then, then there is a state p' of $\mathcal{P}'$ to which p_0' has an a-transition and $d(p,p') < \epsilon/rc$.

We can also use the metric distance to construct bounds on probabilities. Let $\mathcal{P}$ be a finite process with initial state p_0 and $\mathcal{S} = (S, i, \Sigma, \tau)$ be any process such that $d(\mathcal{P}, \mathcal{S}) < \epsilon$. Let f be any function, whose values on the states of P are $y_1, \ldots, y_n$, where $y_1 < y_2 < \ldots < y_n$. Let the states be numbered $1 \ldots n$, lumping together states with equal values. Let $p_1, \ldots, p_n$ be the probability of going to states $1 \ldots n$ from p_0 in $\mathcal{P}$ on label a. Then,

$$\begin{aligned}
\textstyle\sum_{k<i} p_k - \epsilon/(y_i - y_{i-1}) &< \tau_a(i, \{t \mid f(t) < y_i\}) \\
\textstyle\sum_{k\leq i} p_k + \epsilon/(y_{i+1} - y_i) &> \tau_a(i, \{t \mid f(t) \leq y_i\}) \\
\textstyle\sum_{k>i} p_k - \epsilon/(y_{i+1} - y_i) &< \tau_a(i, \{t \mid f(t) > y_i\}) \\
\textstyle\sum_{k\geq i} p_k + \epsilon/(y_i - y_{i-1}) &> \tau_a(i, \{t \mid f(t) \geq y_i\})
\end{aligned}$$

Thus, we have bounded $\tau_a(i, \{t \mid f(t) \geq y_i\})$ by two numbers. The gaps are caused by the fact that some states in $\mathcal{S}$ may have $f(t) = y_i$.

6 The asymptotic metric

Let P be a `plMc`. Then P **after** a is the same `plMc` but with start distribution given by $\nu(t) = k_a(s,t)$. We perform some normalization based on the total probability of the resulting initial configuration $\nu(S)$: If $\nu(S) > 0$, it is normalized to be 1; if $\nu(S) = 0$, it is left untouched. This definition extends inductively to P **after** s, where s is a finite sequence of labels $(a_0, a_1, a_2, \ldots, a_k)$. Note that P **after** s is identical to P except that its initial configuration may be different.

Define the j distance between P, Q, $d^c_j(P,Q) = \sup\{d^c(P \textbf{ after } s, Q \textbf{ after } s) \mid length(s) = j\}$. We define the asymptotic distance between processes P and Q, $d^c_\infty(P,Q)$ to be

$$d^c_\infty(P,Q) = \lim_{i\to\infty} \sup_{j>i} d^c_j(P,Q).$$

The fact that d^c_∞ satisfies the triangle inequality and is symmetric immediately follows from the same properties for d.

Example 13. For any `plMc` P, $d^c_\infty(a_r.P, a_s.P) = 0$, where $r, s > 0$. Consider A_3 from Figure 1. Without the normalization in the definition of A_3 **after** s, we would have got $d^c_\infty(a_r.A_3, a_s.A_3) = c|r - s|$

Example 14. Consider the producer process P_2 shown in Figure 3. This is similar to the producer P_1 in Figure 2, except that initially the probability of producing *put* is more than p, however as more *put*'s are produced, it asymptotically approaches p. If we consider the asymptotic distance between these two producers, we see that $d^c(P_2 \textbf{ after } put^n, P_1 \textbf{ after } put^n) \propto 2^{-(n+1)}$. Thus $d^c_\infty(P_1, P_2) = 0$. Now by using the compositionality of parallel composition (see below), we see that $d^c_\infty(P_1 \mid\mid \text{Consumer} \mid\mid \text{Buffer}_k, P_2 \mid\mid \text{Consumer} \mid\mid \text{Buffer}_k) = 0$, which is the intuitively expected result.

Asymptotic equivalence is preserved by parallel composition and prefixing.

Theorem 4. *1.* $d^c_\infty(l_r.P, l_r.Q) \leq d^c_\infty(P,Q)$ *for any label* l.
2. $d^c_\infty(P \mid\mid R, Q \mid\mid R) \leq d_\infty(P,Q)$.

Acknowledgements. We have benefited from discussions with Franck van Breugel about the Hutchinson metric.

References

1. R. Alur, L. J. Jagadeesan, J. J. Kott, and J. E. von Olnhausen. Model-checking of real-time systems: A telecommunications application. In *Proceedings of the 19th International conference on Software Engineering*, pages 514–524, 1997.
2. J.C.M. Baeten, J.A. Bergstra, and S.A. Smolka. Axiomatizing probabilistic processes: Acp with generative probabilities. *Information and Computation*, 121(2):234–255, 1995.
3. C. Baier, E. Clark, V. Hartonas-Garmhausen, M. Kwiatkowska, and M. Ryan. Symbolic model checking for probabilistic processes. In *Proceedings of the 24th International Colloquium On Automata Languages And Programming*, Springer Verlag LNCS vol 1256, pages 430–440, 1997.

4. A. Bianco and L. de Alfaro. Model checking of probabilistic and nondeterministic systems. In P. S. Thiagarajan, editor, *Proceedings of the 15th Annual Conference on Foundations of Software Technology and Theoretical Computer Science*, Springer Verlag LNCS vol 1026, pages 499–513, 1995.
5. R. Blute, J. Desharnais, A. Edalat, and P. Panangaden. Bisimulation for labelled Markov processes. In *Proceedings of the Twelfth IEEE Symposium On Logic In Computer Science, Warsaw, Poland.*, 1997.
6. J. Bengtsson, K. G. Larsen, F. Larsson, P. Pettersson, and W. Yi. Uppaal: A tool suite for automatic verification of real-time systems. In R. Alur, T. Henzinger, and E. Sontag, editors, *Hybrid Systems III*, Springer Verlag LNCS vol 1066, pages 232–243, 1996.
7. N. Bourbaki. *Elements of Mathematics: General Topology Chapters 1-4*. Springer-Verlag, 1989.
8. F.van Breugel. private communication.
9. R. Cleaveland, S. A. Smolka, and A. Zwarico. Testing preorders for probabilistic processes. *Lecture Notes in Computer Science*, 623, 1992.
10. C. Courcoubetis and M. Yannakakis. The complexity of probabilistic verification. *Journal of the ACM*, 42(4):857–907, 1995.
11. J. Desharnais, A. Edalat, and P. Panangaden. A logical characterization of bisimulation for labeled Markov processes. In *Proceedings of the 13th IEEE Symposium On Logic In Computer Science, Indianapolis*. IEEE Press, June 1998.
12. D. Dill. *Trace Theory for Automatic Hierarchical Verification of Speed-Independent Circuits*. ACM Distinguished Dissertations. MIT Press, 1988.
13. Johan de Kleer and B. C. Williams. Diagnosis with behavioral modes. In *Proceedings of the Eleventh International Joint Conference on Artificial Intelligence*, pages 1324–1330, August 1989.
14. V. Gupta, T. A. Henzinger, and R. Jagadeesan. Robust timed automata. In Oded Maler, editor, *Hybrid and Real-Time Systems*, LNCS Vol 1201, pages 331–345. Springer Verlag, March 1997.
15. Jane Hillston. *A Compositional Approach to Performance Modelling*. PhD thesis, University of Edinburgh, 1994. To be published as a Distinguished Dissertation by Cambridge University Press.
16. H. Hansson and B. Jonsson. A calculus for communicating systems with time and probabilities. In *Proceedings of the 11th IEEE Real-Time Systems Symposium*, pages 278–287. IEEE Computer Society Press, 1990.
17. M. Huth and M. Kwiatkowska. Quantitative analysis and model checking. In *proceedings of the 12 IEEE Symposium On Logic In Computer Science*, pages 111–122. IEEE Press, 1997.
18. S. Hart and M. Sharir. Probabilistic propositional temporal logics. *Information and Control*, 70:97–155, 1986.
19. J. Hutchinson. Fractals and self-similarity. *Indiana University Journal of Mathematics*, 30:713–747, 1981.
20. M. B. Josephs. Receptive process theory. *Acta Informatica*, 29(1):17–31, February 1992.
21. C. Jones and G. D. Plotkin. A probabilistic powerdomain of evaluations. In *Proceedings, Fourth Annual Symposium on Logic in Computer Science*, pages 186–195, Asilomar Conference Center, Pacific Grove, California, 1989.
22. C. Jou and S. Smolka. Equivalences, congruences and complete axiomatizations for probabilistic processes. In *CONCUR 90*, Springer Verlag LNCS vol 458, 1990.
23. B. Jonsson and W. Yi. Compositional testing preorders for probabilistic processes. In *Proceedings, Tenth Annual IEEE Symposium on Logic in Computer Science*, pages 431–441, San Diego, California, 1995.
24. D. Kozen. A probabilistic PDL. *Journal of Computer and Systems Sciences*, 30(2):162–178, 1985.
25. P. Lincoln, J. Mitchell, M. Mitchell, and A. Scedrov. A probabilistic poly-time framework for protocol analysis. In *ACM Computer and Communication Security (CCS-5)*, 1998.

26. Kim G. Larsen and Arne Skou. Bisimulation through probabilistic testing. *Information and Computation*, 94(1):1–28, September 1991.
27. N. A. Lynch and M. R. Tuttle. An introduction to input/output automata. *CWI Quarterly*, 2(3):219–246, 1989.
28. M. Ajmone Marsan. Stochastic petri nets: an elementary introduction. In *Advances in Petri Nets 1989*, pages 1–29. Springer, June 1989.
29. J L Peterson and A Silberschatz. *Operating System Concepts*. Addison-Wesley Inc., 1985.
30. R. Segala. *Modeling and Verification of Randomized Distributed Real-Time Systems*. PhD thesis, MIT, Dept. of Electrical Engineering and Computer Science, 1995. Also appears as technical report MIT/LCS/TR-676.
31. F. W. Vaandrager. On the relationship between process algebra and input/output automata. In *Proceedings, Sixth Annual IEEE Symposium on Logic in Computer Science*, pages 387–398, Amsterdam, The Netherlands, 15–18 July 1991. IEEE Computer Society Press.
32. R. van Glabbeek, S.A. Smolka, and B.U. Steffen. Reactive, generative, and stratified models of probabilistic processes. *Information and Computation*, 121(1):59–80, 1995.
33. N. Viswanadham and Y. Narahari. *Performance Modeling of Automated Manufacturing Systems*. Prentice-Hall Inc., 1992.
34. S.-H. Wu, S.A. Smolka, and E. Stark. Composition and behaviors for probabilistic i/o automata. *Theoretical Computer Science*, 176(1–2):1–36, April 1997.

The Expressive Power of Temporal Logic of Actions

(Extended Abstract)

Arkady Estrin[1] and Michael Kaminski[2]

[1] Avant! Corporation, 46781 Bayside Parkway, Fremont CA 94538, U.S.A.
[2] Department of Computer Science, Technion – Israel Institute of Technology, Haifa 32000, Israel

Abstract. It is shown that the satisfiability/validity questions in Temporal Logic of Actions can be translated into satisfiability/validity questions in Second Order Temporal Logic and vice versa. The translation from Second Order Temporal Logic into Temporal Logic of Actions is linear and the translation from Temporal Logic of Actions into Second Order Temporal Logic is quadratic.

1 Introduction

A concurrent algorithm is usually specified by a program and correctness of the algorithm means that the program satisfies some desired properties. A number of methods for reasoning about concurrent programs (and hardware devices) are based on proof systems for the Temporal Logic (TL). A few years ago Leslie Lamport in [6] introduced a simpler approach in which both the algorithm and the property were specified by formulas in a single logic — the *temporal logic of actions* (TLA). In TLA correctness of an algorithm means that the formula specifying the algorithm implies the formula specifying the property, where "implies" is an ordinary logical implication.

TLA combines two logics: a logic of actions and a standard temporal logic. An *action* is a formula without temporal operators containing variables, primed variables, and constant symbols. In general, an action represents a relation between the current state and the next state, where the unprimed variables refer to the current state and the primed ones refer to the next state. For example, $x' = y + 1$, where x and y are variables, is an action stating that the value of x in the next state equals the value of y in the current state plus 1. Elementary formulas of TLA are those not containing primed variables or temporal operators and formulas of the form $\Box[\mathcal{A}]_{\sqcup}$, where $\mathcal{A}$ is an *action*. The formula $[\mathcal{A}]_{\sqcup}$ states that either $\mathcal{A}$ holds between the current and the next states or the value of the term t does not change when passing to the next state. In this way "stuttering" steps which leave all variables unchanged are allowed. General TLA formulas are obtained from the elementary ones using, under certain restrictions, boolean connectives, quantification, and the unary temporal operator $\Box$ (*always*).

TLA was invented to reduce the expressive power of TL and to allow modularization and refinement. Lamport conjectured in [6] that TLA is less powerful then TL, because TLA can only express invariant under stuttering formulas, whereas TL can express non-invariant ones as well.

In this paper we show that TLA and Second Order Temporal Logic (SOTL) have the same expressive power, cf. [7, 8]. Namely, we present a syntactical satisfiability preserving translation of SOTL formulas (not necessarily invariant under stuttering) into TLA formulas and a syntactical satisfiability preserving translation of TLA formulas into SOTL formulas.[1] Since validity is dual to satisfiability we obtain the corresponding validity preserving translations as well. The translation from SOTL into TLA is linear and the translation from TLA into SOTL is quadratic. In addition, the proofs of correctness of the translations show the modifications to be made in temporal interpretations in order to preserve satisfiability/validity of the translations. It should be pointed out that, contrary Lamport's claim in [6], our translation from TLA into SOTL it relatively simple. Also, the existence of a translation from SOTL into TLA is quite surprising in view of the expressive power of SOTL.[2]

The above transformations, apart from being just of a theoretical interest, have a possible practical value. For example, to check whether a program (given by a temporal interpretation) satisfies a specification (given by a temporal formula) one can translate the specification formula into TLA, modify the temporal interpretation, and check, using the Temporal Logic Prover based on TLA – TLP ([4]) whether the TLA formula describing the modified temporal interpretation implies the translation of the temporal formula. Conversely, to check whether a program (given by a temporal interpretation) satisfies a specification (given by a TLA formula) one can translate the specification formula into SOTL, modify the temporal interpretation, and check, using temporal logic provers, e.g., Stanford Temporal Prover – STeP ([3]) whether the modified temporal interpretation satisfies the translation of the TLA formula.

The paper is organized as follows. In the next section we recall the notion of *temporal interpretation*. In Section 3 we introduce an "intermediate" logic ERTLA and define the syntax and semantics of TLA. In Section 4 we recall the syntax and semantics of SOTL and embed ERTLA into that logic. In Section 5 we introduce the *time variable* needed for simulation of the ordinary (ERTLA) satisfiability by the TLA satisfiability and vice versa, and for the embedding of SOTL into ERTLA. The corresponding results are presented in Sections 6,7, and 8, respectively. In order to use the time variable in the translations in the above three sections we must assume that the interpretation domain is infinite. Thus, in Section 9 we show how finite domains can be simulated my infinite ones. Finally, in Section 10 we combine all the results from the previous sections to obtain the desired intertranslations between SOTL and TLA.

---

[1] That is, a formula is SOTL-satisfiable if and only if its translation is TLA-satisfiable and vice versa.

[2] Note that the expressive power of SOTL is like that of Second Order Arithmetic.

2 Temporal Interpretations

All temporal logics considered in this paper have two types of variables: one – ordinary variables whose interpretation is rigid, i.e., does not change in time, and the other – variables whose interpretation is flexible, i.e., may have different values in different states. The latter will be referred to as *flexible* variables and will be denoted by bold-face letters $\boldsymbol{x}, \boldsymbol{y}, \boldsymbol{z}$, etc.

Semantics of all temporal logics considered in this paper is based on the notion of *temporal interpretations* defined below. First we recall the definition of interpretations of Predicate Calculus (PC).

A PC interpretation s consists of a non-empty domain D_s, an assignment to each n-place function symbol f of an n-place function $f^s : D_s^n \to D_s$,[3] an assignment to each n-place predicate symbol P of an n-place relation P^s on D_s, such that $=^s$ is the identity relation on D_s, and an assignment to each (rigid) variable x of an element x^s of D_s.

A *temporal interpretation* $\boldsymbol{M}$ is a PC interpretation together with an assignment to each flexible variable under consideration $\boldsymbol{x}$ of a function $\boldsymbol{x}^{\boldsymbol{M}} : \mathbb{N} \to D_{\boldsymbol{M}}$,[4] where $\mathbb{N}$ denotes the set of non-negative integers. For a non-negative integer i, $\boldsymbol{x}^{\boldsymbol{M}}(i)$ is the value of $\boldsymbol{x}$ at the ith state (or in the ith moment of time). That is, (interpretations of) flexible variables may change in time.

One can think of a temporal interpretation $\boldsymbol{M}$ as of a sequence of states $\boldsymbol{M} = s_0, s_1, \ldots$ which are PC interpretations over the same domain which differ only in (PC) assignments to flexible variables. Namely, for a flexible variable $\boldsymbol{x}$, $\boldsymbol{x}^{s_i}$ is $\boldsymbol{x}^{\boldsymbol{M}}(i)$.

Let $\boldsymbol{M}_1$ and $\boldsymbol{M}_2$ be temporal interpretations and let x ($\boldsymbol{x}$) be a rigid (flexible) variable. We write $\boldsymbol{M}_1 =_x \boldsymbol{M}_2$ ($\boldsymbol{M}_1 =_{\boldsymbol{x}} \boldsymbol{M}_2$) if $\boldsymbol{M}_1$ and $\boldsymbol{M}_2$ assign the same values to all variables other than x ($\boldsymbol{x}$).

Finally, let $\boldsymbol{M} = s_0, s_1, \ldots$ be a temporal interpretation and let i be a non-negative integer. We denote by $\boldsymbol{M}^{+i}$ a temporal interpretation obtained from $\boldsymbol{M}$ by removing its first i states, i.e., $\boldsymbol{M}^{+i} = s_i, s_{i+1}, \ldots$. In particular, for a flexible variable $\boldsymbol{x}$, $\boldsymbol{x}^{\boldsymbol{M}^{+i}}(j) = \boldsymbol{x}^{\boldsymbol{M}}(j+i)$, $j = 0, 1, \ldots$.

3 Temporal Logics of Actions

First we introduce a new temporal logic called Extended Raw Temporal Logic of Actions (ERTLA). ERTLA is a second order[5] temporal logic that contains as fragments both the Raw Temporal Logic of Actions (RTLA) and Simple Temporal Logic of Actions (STLA) introduced in [6] and used as preliminary steps in the definition of TLA. In addition, the syntax of ERTLA contains the syntax of TLA. The language of TLA formulas with the ordinary (ERTLA) semantics serves as an intermediate station between the intertranslation of SOTL

[3] We treat the constants as 0-place function symbols.

[4] For a temporal interpretation $\boldsymbol{M}$ based on a PC interpretation s we write $D_{\boldsymbol{M}}$, $f^{\boldsymbol{M}}$, $P^{\boldsymbol{M}}$, and $x^{\boldsymbol{M}}$ for D_s, f^s, P^s, and x^s, respectively.

[5] Because quantification over flexible variables is allowed.

and TLA. Namely we first translate SOTL and TLA into the intermediate logic and then we translate that logic into TLA and SOTL, respectively.

The language of ERTLA is obtained from the language of PC by

- dividing its variables in two sorts: one – ordinary variables, whose (temporal) interpretation is rigid, and the other – variables whose interpretation is flexible. The latter are referred to as *flexible* variables and are denoted by bold-face letters $\boldsymbol{x}, \boldsymbol{y}, \boldsymbol{z}$, etc.
- adding to the language of PC a unary function $'$ – *prime* and a unary temporal operator $\Box$ – *always*. As usual, $\Diamond$ – *eventually* is an abbreviation for $\neg\Box\neg$.

The formula formation rules of ERTLA are the ordinary ones with the exception that $'$ can be applied to unprimed flexible variables only: the meaning of $\boldsymbol{x}'$ is the value of $\boldsymbol{x}$ in the next state.

Next we define the semantics of ERTLA.

For a temporal interpretation $\boldsymbol{M}$, a non-negative integer i, and a term t we define the *value* of t in the ith state of $\boldsymbol{M}$, denoted $t^{\boldsymbol{M},i}$ by induction, as follows.

- If t is a rigid variable x, then $t^{\boldsymbol{M},i} = x^{\boldsymbol{M}}$;
- if t is a flexible variable $\boldsymbol{x}$, then $t^{\boldsymbol{M},i} = \boldsymbol{x}(i)$;
- if t is of the form $\boldsymbol{x}'$, then $t^{\boldsymbol{M},i} = \boldsymbol{x}(i+1)$; and
- if t is of the form $f(t_1, \ldots, t_k)$, where f is a k-place function symbol, then $t^{\boldsymbol{M},i} = f^{\boldsymbol{M}}(t_1^{\boldsymbol{M},i}, \ldots, t_k^{\boldsymbol{M},i})$. Recall that we treat constant symbols as 0-place function symbols.

We say that a temporal interpretation $\boldsymbol{M}$ *satisfies* an ERTLA formula φ, denoted $\boldsymbol{M} \models \varphi$, if the following holds.

- If φ is an atomic formula $P(t_1, \ldots, t_k)$, then $\boldsymbol{M} \models \varphi$ if and only if $(t_1^{\boldsymbol{M},0}, \ldots, t_k^{\boldsymbol{M},0}) \in P^{\boldsymbol{M}}$;
- $\boldsymbol{M} \models \neg\varphi$ if and only if $\boldsymbol{M} \not\models \varphi$;
- $\boldsymbol{M} \models \varphi \supset \psi$ if and only if $\boldsymbol{M} \models \varphi$ implies $\boldsymbol{M} \models \psi$;
- $\boldsymbol{M} \models \exists x \varphi$ if and only if there exists a temporal interpretation $\boldsymbol{M}_x$ such that $\boldsymbol{M}_x =_x \boldsymbol{M}$ and $\boldsymbol{M}_x \models \varphi$;
- $\boldsymbol{M} \models \exists \boldsymbol{x} \varphi$ if and only if there exists a temporal interpretation $\boldsymbol{M}_{\boldsymbol{x}}$ such that $\boldsymbol{M}_{\boldsymbol{x}} =_{\boldsymbol{x}} \boldsymbol{M}$ and $\boldsymbol{M}_{\boldsymbol{x}} \models \varphi$; and
- $\boldsymbol{M} \models \Box\varphi$ if and only if for each $i = 0, 1, \ldots$, $\boldsymbol{M}^{+i} \models \varphi$.

A formula is (ERTLA-)*satisfiable/valid*, if it is satisfied by some/any temporal interpretation.

Definition 1. *The Raw Temporal Logic of Actions (RTLA) is the fragment of ERTLA consisting of all ERTLA formulas without quantified flexible variables which do not contain $\Box$ in the scope of rigid variable quantifiers. RTLA formulas not containing $\Box$ are called* actions.[6]

[6] An action can be thought as a relation between two consecutive states of a temporal interpretation.

Next we define the Simple Temporal Logic of Actions (STLA) that is extended to TLA by allowing, under certain restrictions, quantification over rigid and flexible variables outside of $\Box$. We shall need the following notation. For an expression E not containing $'$, we denote by E' the expression that is obtained from E by replacing all occurrences of every flexible variable x in E with x'.

Let $\mathcal{A}$ be an action and t be a term not containing $'$. We shall denote the formulas $\mathcal{A} \vee \sqcup = \sqcup'$ and $\mathcal{A} \wedge \sqcup \neq \sqcup'$ by $[\mathcal{A}]_\sqcup$ and $\langle \mathcal{A} \rangle_\sqcup$, respectively.

Now we define STLA as the fragment of RTLA (see Definition 1) where each action $\mathcal{A}$ containing $'$ appears in the form $\Box[\mathcal{A}]_\sqcup$. Since $\neg\Box[\neg\mathcal{A}]_\sqcup$ is equivalent to $\Diamond\langle\mathcal{A}\rangle_\sqcup$, the latter is an STLA formula as well. We refer the reader to [6] for a comprehensive study of STLA.

Finally, for two tuples of terms $(u_1, u_2, \ldots, u_n)$ and $(v_1, v_2 \ldots, v_n)$, denoted $\bar{u}$ and $\bar{v}$, respectively, we abbreviate the formula $\bigwedge_{i=1}^{n} u_i = v_i$ by $\bar{u} = \bar{v}$. It can be easily verified that $[\mathcal{A}]_{\bar{\sqcup}}$ is equivalent to $\bigwedge_{i=1}^{n} [\mathcal{A}]_{\sqcap_i}$. Thus, both $\Box(\mathcal{A} \vee \bar{\sqcup} = \bar{\sqcup}')$ and $\Diamond(\mathcal{A} \wedge \bar{\sqcup} \neq \bar{\sqcup}')$ are STLA formulas.

As it has been pointed out in [6], satisfiability of an STLA formula by a temporal interpretation $\boldsymbol{M}$ is not affected if we add to (or remove from) $\boldsymbol{M}$ stuttering steps. To be more specific, we need the following notation.

Let $\boldsymbol{M} = s_0, s_1, \ldots$ be a temporal interpretation.

For a non-negative integer i we denote by $\mu(i)$ the maximal integer $j \geq i$ such that for all $k = i, i+1, \ldots, j$, $s_k = s_i$. That is, all the computation steps between i and $\mu(i)$ are stuttering. Note that if $s_k = s_i$ for all $k \geq i$, i.e., the computation is *halted*, $\mu(i)$ is undefined.

Next we define a function $\natural : \mathbb{N} \to \mathbb{N}$ by

$$\natural(0) = \begin{cases} \mu(0) & \text{if } \mu(0) \text{ is defined} \\ 0 & \text{otherwise} \end{cases}$$

and

$$\natural(i+1) = \begin{cases} \mu(\natural(i)+1) & \text{if } \mu(\natural(i)+1) \text{ is defined} \\ \natural(i)+1 & \text{otherwise} \end{cases}.$$

Finally we define a temporal interpretation $\natural\boldsymbol{M}$ by $\natural\boldsymbol{M} = s_{\natural(0)}, s_{\natural(1)}, \ldots$. By definition, $\natural\boldsymbol{M}$ is obtained from $\boldsymbol{M}$ by removing all stuttering steps, i.e., all states s_i such that no flexible variable changes its value when passing to the state s_{i+1}, or, in other words, by removing all states s_i such that $s_{i+1} = s_i$.

Now the precise statement of invariance under stuttering is as follows.

Definition 2. *We say that satisfiability of a formula* φ *is* invariant under stuttering *if for all temporal interpretations* $\boldsymbol{M}_1$ *and* $\boldsymbol{M}_2$ *such that* $\natural\boldsymbol{M}_1 = \natural\boldsymbol{M}_2$, $\boldsymbol{M}_1 \models \varphi$ *if and only if* $\boldsymbol{M}_2 \models \varphi$.

At last we have arrived to the definition of the syntax of TLA. The definition is by induction, as follows.

- Each STLA formula is an *atomic* TLA formula;

- if φ and ψ are TLA formulas then $\neg\varphi$ and $\varphi \supset \psi$ are TLA formulas; and
- if φ is a TLA formula, x is a rigid variable and $\boldsymbol{x}$ is a flexible variable, then $\exists x\varphi$ and $\exists \boldsymbol{x}\varphi$ are TLA formulas.

It was pointed out in [6] that (the ordinary ERTLA-) satisfiability of TLA formulas is not invariant under stuttering.

Example 1. Suppose, for the sake of argument, that there is an STLA formula $F(\boldsymbol{x}, \boldsymbol{y})$ stating that always $\boldsymbol{x}$ changes before $\boldsymbol{y}$ and $\boldsymbol{x}$ and $\boldsymbol{y}$ never change simultaneously. It follows that a temporal interpretation $\boldsymbol{M}$ with

$$\boldsymbol{y}^{\boldsymbol{M}}(i) = \begin{cases} a\ i = 0,1 \\ b\ i = 2,3,\ldots \end{cases} \quad a \neq b$$

satisfies $\exists \boldsymbol{x} F(\boldsymbol{x}, \boldsymbol{y})$, because the temporal interpretation $\boldsymbol{M}_{\boldsymbol{x}}$ obtained from $\boldsymbol{M}$ by interpreting $\boldsymbol{x}$ as

$$\boldsymbol{x}^{\boldsymbol{M}_{\boldsymbol{x}}}(i) = \begin{cases} a\ i = 0 \\ b\ i = 1,2,\ldots \end{cases}$$

satisfies $F(\boldsymbol{x}, \boldsymbol{y})$. Let $\boldsymbol{M}_1$ be a temporal interpretation where

$$\boldsymbol{y}^{\boldsymbol{M}_1}(i) = \begin{cases} a\ i = 0 \\ b\ i = 1,2,\ldots \end{cases}.$$

Obviously, $\natural\boldsymbol{M} = \natural\boldsymbol{M}_1$. However, $\boldsymbol{M}_1 \not\models \exists \boldsymbol{x} F(\boldsymbol{x}, \boldsymbol{y})$, because, in $\boldsymbol{M}_1$, $\boldsymbol{y}$ changes already in the initial state. Thus, satisfiability of TLA formulas is not invariant under stuttering.

To obtain invariance under stuttering Lamport in [6] introduced a new notion of satisfiability of TLA formulas that will be referred to as the TLA-satisfiability and is defined as follows.

Let φ be a TLA formula and let $\boldsymbol{M}$ be a temporal interpretation. We say that $\boldsymbol{M}$ *TLA-satisfies* φ, denoted $\boldsymbol{M} \models_{\text{TLA}} \varphi$, if the following holds.

- If φ is an STLA formula, then $\boldsymbol{M} \models_{\text{TLA}} \varphi$ if and only if $\boldsymbol{M} \models \varphi$;
- $\boldsymbol{M} \models_{\text{TLA}} \neg\varphi$ if and only if $\boldsymbol{M} \not\models_{\text{TLA}} \varphi$;
- $\boldsymbol{M} \models_{\text{TLA}} \varphi \supset \psi$ if and only if $\boldsymbol{M} \models_{\text{TLA}} \varphi$ implies $\boldsymbol{M} \models_{\text{TLA}} \psi$;
- $\boldsymbol{M} \models_{\text{TLA}} \exists x\varphi$ if and only if there exists a temporal interpretation $\boldsymbol{M}_1$ such that $\boldsymbol{M}_1 =_x \boldsymbol{M}$ and $\boldsymbol{M}_1 \models_{\text{TLA}} \varphi$; and
- $\boldsymbol{M} \models_{\text{TLA}} \exists \boldsymbol{x}\varphi$ if and only if there exist temporal interpretations $\boldsymbol{M}_1$ and $\boldsymbol{M}_2$ such that $\natural\boldsymbol{M}_1 = \natural\boldsymbol{M}$, $\boldsymbol{M}_2 =_{\boldsymbol{x}} \boldsymbol{M}_1$, and $\boldsymbol{M}_2 \models_{\text{TLA}} \varphi$.

Example 2. Let $F(\boldsymbol{x}, \boldsymbol{y})$, $\boldsymbol{M}$, $\boldsymbol{M}_{\boldsymbol{x}}$, and $\boldsymbol{M}_1$ be as in Example 1. Then $\boldsymbol{M}_1 \models_{\text{TLA}} \exists \boldsymbol{x} F(\boldsymbol{x}, \boldsymbol{y})$, because $\natural\boldsymbol{M} = \natural\boldsymbol{M}_1$, $\boldsymbol{M}_{\boldsymbol{x}} =_{\boldsymbol{x}} \boldsymbol{M}$, and $\boldsymbol{M}_{\boldsymbol{x}} \models_{\text{TLA}} F(\boldsymbol{x}, \boldsymbol{y})$.

In general, it was shown in [6] that TLA-satisfiability (of TLA formulas) is invariant under stuttering.

We conclude this section with an additional example which illustrates the difference between the ordinary satisfiability and the TLA-satisfiability.

Example 3. This example deals with the formula $\exists x \forall y \Box [x \neq x']_y$ stating that some x changes at least as fast as any other y.[7] This formula is ERTLA-valid. Indeed, let M be a temporal interpretation. If D_M consists of only one element, obviously, $M \models \exists x \forall y [x \neq x']_y$. If there are two distinct elements a and b in D_M, then for the temporal interpretation $M_\circ =_\circ M$ with

$$x^M(i) = \begin{cases} a & i \text{ is even} \\ b & i \text{ is odd} \end{cases}$$

we have $M_\circ \models \forall y[x \neq x']_y$.

On the other hand, the negation $\forall x \exists y \Diamond \langle x = x' \rangle_y$ of $\exists x \forall y \Box [x \neq x']_y$ is TLA-satisfied by all temporal interpretations whose domain contains two or more distinct elements. That is, $\exists x \forall y \Box [x \neq x']_y$ is "almost a TLA-contradiction." Indeed, let M be a temporal interpretation such that there are two distinct elements a and b in D_M. It suffices to show that for all "stretchings" M_1 of M and all assignments to x in M_1 there is a stretching M_2 of M_1 and an assignment to y in M_2 such that, in M_2, in some moment, y changes before x. Let $M_1 = s_{1,0}, s_{1,1}, \ldots$ be a stretching of M with any assignment to x. We have to show that there are two temporal interpretations $\tilde{M}_1$ and M_2 such that $\natural \tilde{M}_1 = \natural M_1$, $M_2 =_y \tilde{M}_1$, and $M_2 \models \Diamond \langle x = x' \rangle_y$. We define $\tilde{M}_1$ by "doubling" the first state $s_{1,0}$ of M_1, i.e., $\tilde{M}_1 = s_{1,0}, s_{1,0}, s_{1,1}, \ldots$, and obtain M_2 from $\tilde{M}_1$ assigning to y

$$y^{M_2}(i) = \begin{cases} a & i = 0 \\ b & i = 1, 2, \ldots \end{cases} .$$

Then $M_2 \models x = x' \wedge y \neq y'$, implying $M_2 \models \Diamond \langle x = x' \rangle_y$.

In Section 6 we show that the TLA satisfiability can be interpreted by the ordinary (ERTLA) satisfiability and in Section 7 we show that the ordinary satisfiability can be interpreted by the TLA satisfiability.

4 Second Order Temporal Logic

In this section we recall syntax and semantics of Second-Order Temporal Logic (SOTL) and embed ERTLA into that logic. The language of SOTL is obtained from the language of PC by

- dividing its variables in rigid and flexible and
- adding to the language of PC two unary temporal operators $\circ$ – *next* and $\Box$ – *always*, and one binary temporal operator $\mathcal{U}$ – *until*.

The satisfiability of a SOTL formula φ by a temporal interpretation M, also denoted $M \models \varphi$, is defined by the following induction.

---

[7] That is, in each computation step where y changes, x changes as well.

- The cases of atomic formulas, boolean connectives, quantifiers over rigid and flexible variables, and the temporal operator $\Box$ are exactly as those in ERTLA and are omitted;
- $\boldsymbol{M} \models \circ\varphi$ if and only if $\boldsymbol{M}^{+1} \models \varphi$;
- $\boldsymbol{M} \models \varphi \mathcal{U} \psi$ if and only if there is an $i = 0, 1, \ldots$, such that $\boldsymbol{M}^{+i} \models \psi$ and for each $j = 0, 1, \ldots, i-1$, $\boldsymbol{M}^{+j} \models \varphi$.

A (SOTL) formula is SOTL-*satisfiable/valid*, if it is satisfied by some/any temporal interpretation.

Next we present a satisfiability preserving translation of ERTLA formulas into SOTL formulas.

Let $\boldsymbol{y}_1, \ldots, \boldsymbol{y}_n$ be all primed flexible variables that appear in an ERTLA formula φ. For each $j = 1, \ldots, n$ we pick a "new" rigid variable y_j° and for a term t that appears in φ we denote by t° the result of simultaneous replacement of each $\boldsymbol{y}_j'$ with y_j°, $j = 1, \ldots, n$. Now we define the translation of φ into SOTL, denoted φ^{SOTL}, by induction as follows.

- If φ is an atomic formula $P(t_1, \ldots, t_k)$, then φ^{SOTL} is

$$\forall y_1^\circ \ldots \forall y_n^\circ ((\bigwedge_{j=1}^{n} \circ(\boldsymbol{y}_j = y_j^\circ)) \supset P(t_1^\circ, \ldots, t_k^\circ)).$$

- $(\neg\varphi)^{\text{SOTL}}$, $(\varphi \supset \psi)^{\text{SOTL}}$, $(\exists x\varphi)^{\text{SOTL}}$, $(\exists \boldsymbol{x}\varphi)^{\text{SOTL}}$, and $(\Box\varphi)^{\text{SOTL}}$ are $\neg\varphi^{\text{SOTL}}$, $\varphi^{\text{SOTL}} \supset \psi^{\text{SOTL}}$, $\exists x\varphi^{\text{SOTL}}$, $\exists \boldsymbol{x}\varphi^{\text{SOTL}}$, and $\Box\varphi^{\text{SOTL}}$, respectively.

Theorem 1. *An ERTLA formula φ is (ERTLA-) satisfiable if and only if the SOTL formula φ^{SOTL} is (SOTL-) satisfiable.*

5 The Time Variable

In this section we introduce the *time variable* $\boldsymbol{t}$ needed for translations of ERTLA into TLA, of TLA into ERTLA, and of SOTL into ERTLA. The definition is similar to that of the *time predicate* in [5]. The time variable is intended to simulate the external time. In particular, let φ be an ERTLA formula, $\boldsymbol{x}$ be a free flexible variable of φ, and let $\boldsymbol{M} = s_0, s_1, \ldots$ be a temporal interpretation. If for some $i = 0, 1, \ldots$, $\boldsymbol{M}^{+i} \models \boldsymbol{x} \neq \boldsymbol{x}'$, then (the assignment to) the time variable $\boldsymbol{t}$ advances to the next value in the state s_{i+1}. To define the time variable $\boldsymbol{t}$ we shall need the following notation.

Let $<$ be a two place predicate symbol that do not belong to the original language of ERTLA. We denote by *DORD* the sentence stating that $<$ is an infinite discrete partial order. That is, *DORD* is the conjunction of the partial order axioms with $\forall x \exists y NEXT(x, y)$, where $NEXT(x, y)$ denotes the formula

$$x < y \wedge \forall z(x < z \supset (y < z \vee y = z))$$

saying that y follows x in the order imposed by $<$.

Finally, let V_φ denote the set of all free flexible variables of φ.

For an ERTLA formula φ we define the "time variable axiom of φ," denoted T_φ, as

$$DORD \wedge \Box[NEXT(t,t')]_{\langle V_\varphi, t\rangle} \wedge \Box\Diamond\langle true\rangle_t.$$

Note that T_φ is an STLA formula where the conjunct $\Box[NEXT(t,t')]_{\langle V_\varphi, t\rangle}$ implies that t changes at least as fast as any of the free flexible variables in φ, and the conjunct $\Box\Diamond\langle true\rangle_t$ implies that the time t never stops, i.e., changes forever.

Obviously, the time axiom T_φ cannot be satisfied by temporal interpretations over finite domains. Thus, till Section 9, we assume that all temporal interpretations under consideration are over infinite domains. In Section 9 we show how finite domains can be simulated by infinite ones.

6 Simulating the ERTLA Satisfiability by the TLA Satisfiability

In this section, using the time variable, we present a translation of TLA formulas into TLA formulas such that a TLA formula is ERTLA-satisfiable (by a temporal interpretation with an infinite domain) if and only if its translation is TLA-satisfiable. The translation $\mathcal{F}(\varphi, t)$ of a formula φ depends on the time variable t and is defined by induction as follows.

- If φ is an STLA formula, then $\mathcal{F}(\varphi, t)$ is φ itself;
- $\mathcal{F}(\neg\varphi, t)$, $\mathcal{F}(\varphi \supset \psi, t)$, and $\mathcal{F}(\exists\S\varphi, t)$ are $\neg\mathcal{F}(\varphi, t)$, $\mathcal{F}(\varphi, t) \supset \mathcal{F}(\psi, t)$, and $\exists x\mathcal{F}(\varphi, t)$, respectively; and
- $\mathcal{F}(\exists \boldsymbol{x}\varphi, t)$ is $\exists \boldsymbol{x}(\Box[t \neq t']_{\boldsymbol{x}} \wedge \mathcal{F}(\varphi, t))$.

The relativization of the existential quantifier over the flexible variable $\boldsymbol{x}$ to $\Box[t \neq t']_{\boldsymbol{x}}$ in the last point is intended to neutralize the influence of adding/ removing stuttering steps before choosing an appropriate assignment to $\boldsymbol{x}$ when passing to the TLA-satisfiability. Namely, the assignment to $\boldsymbol{x}$ cannot change faster than the assignment to t that does not change in the stuttering steps.

The relationship between the formulas φ and $\mathcal{F}(\varphi, t)$ is as follows.

Theorem 2. *A TLA formula φ is ERTLA-satisfiable by a temporal interpretation over an infinite domain if and only if $T_\varphi(t) \wedge \mathcal{F}(\varphi, t)$ is TLA-satisfiable.*

Example 4 below illustrates the influence of the conjunct $\Box[t \neq t']_{\boldsymbol{x}}$ on the assignment to $\boldsymbol{x}$ in interpretations which satisfy the formula translation.

Example 4. Let $F(\boldsymbol{x}, \boldsymbol{y})$ and M_1 be as in Example 1. Then, as it has been shown in that example, $M_1 \not\models \exists\boldsymbol{x}F(\boldsymbol{x}, \boldsymbol{y})$, but $M_1 \models_{\mathrm{TLA}} \exists\boldsymbol{x}F(\boldsymbol{x}, \boldsymbol{y})$. We shall prove that there is no assignment to t (in M_1) such that M_1 TLA-satisfies the translation $T_{\exists\boldsymbol{x}F(\boldsymbol{x},\boldsymbol{y})}(t) \wedge \exists\boldsymbol{x}(\Box[t \neq t']_{\boldsymbol{x}} \wedge F(\boldsymbol{x}, \boldsymbol{y}))$ of $\exists\boldsymbol{x}F(\boldsymbol{x}, \boldsymbol{y})$. Assume to the contrary that for some $t^{M_1} : \mathbb{N} \to D_M$, $M_1 \models_{\mathrm{TLA}} T_{\exists\boldsymbol{x}F(\boldsymbol{x},\boldsymbol{y})}(t) \wedge \exists\boldsymbol{x}(\Box[t \neq t']_{\boldsymbol{x}} \wedge F(\boldsymbol{x}, \boldsymbol{y}))$. That is, there exist temporal interpretations $\tilde{M}_1$ and M_2 such that $\natural\tilde{M}_1 = \natural M_1$,

$M_2 =_{\circ} \tilde{M}_1$, and $M_2 \models T_{\exists_{\circ} F(\circ, y)}(t) \wedge \Box[t \neq t']_{\circ} \wedge F(x, y)$. However, by the conjunct $\Box[t \neq t']_{\circ}$, in M_2, the first change of x cannot occur before the first change of t. The latter occurs together with the first change of y, because, in M_1, y changes already in the first state. Therefore, in M_2, y changes before x, which is impossible, because $M_2 \models F(x, y)$. That is, we have arrived to contradiction, which completes the proof.

7 Simulating the TLA Satisfiability by the ERTLA Satisfiability

In this section, using the time variable, we present a translation of TLA formulas into TLA formulas such that a TLA formula is TLA-satisfiable (by a temporal interpretation with an infinite domain) if and only if its translation is ERTLA-satisfiable. The idea lying behind our translation is illustrated by Example 5 below.

Example 5. Again, let $F(x, y)$ and M_1 be as in Example 1. Then, as it has been shown in that example, $M_1 \models_{\mathrm{TLA}} \exists x F(x, y)$, but $M_1 \not\models \exists x F(x, y)$. To simulate the TLA satisfiability of $\exists x F(x, y)$ by the ERTLA satisfiability in M_1 we first replace the free flexible variable y with a new free flexible variable u_y that behaves like y after adding/removing stuttering steps, and then to find an assignment to x such that the resulting interpretation $\tilde{M}_1$ satisfies $F(x, u_y)$.

For example, we can put

$$u_y^{\tilde{M}_1}(i) = \begin{cases} a & i = 0, 1 \\ b & i = 2, 3 \ldots \end{cases}$$

and

$$x^{\tilde{M}_1}(i) = \begin{cases} a & i = 0 \\ b & i = 1, 2 \ldots \end{cases}.$$

The free variable u_y is introduced in three steps.

1. First, we "add to M_1" a new time variable t whose assignment changes in each computation step.
2. Then we add an additional time variable $t_{\circ}$ whose assignment simulates t after adding/removing stuttering steps to M_1.
3. Finally, we add u_y whose assignment respects $t_{\circ}$.

After all that we choose an appropriate assignment to x. For example, this can result in the following sequence of states (temporal interpretation) $\tilde{M}_1$ that satisfies $F(x, u_y)$.

$$\begin{gathered}
\langle x^{\tilde{M}_1} = a, y^{\tilde{M}_1} = a, t^{\tilde{M}_1} = \tau_0, t_{\circ}^{\tilde{M}_1} = \tau_0, u_y^{\tilde{M}_1} = a \rangle \\
\langle x^{\tilde{M}_1} = b, y^{\tilde{M}_1} = b, t^{\tilde{M}_1} = \tau_1, t_{\circ}^{\tilde{M}_1} = \tau_0, u_y^{\tilde{M}_1} = a \rangle \\
\langle x^{\tilde{M}_1} = b, y^{\tilde{M}_1} = b, t^{\tilde{M}_1} = \tau_2, t_{\circ}^{\tilde{M}_1} = \tau_1, u_y^{\tilde{M}_1} = b \rangle \\
\vdots
\end{gathered}$$

Our translation involves the (TLA) formula

$$T_{true}(t) \wedge \forall u(T_{true}(u) \supset \Box[t \neq t']_u)$$

denoted $FT(t)$. This formula states that t is the fastest time variable. That is, the assignment to t in temporal interpretations (ERTLA-) satisfying $FT(t)$ must change from state to state.[8]

We shall also need the (TLA) formula

$$T_{true}(t_2) \wedge (t_1 = t_2)$$

denoted $S(t_1, t_2)$, whose intended meaning is that t_2 simulates t_1 after adding/removing stuttering steps. Namely, $S(t_1, t_2)$ will appear in conjunction with $FT(t_1)$. In this case, it states that t_2 is a time variable whose initial value is equal to that of the fastest time variable t_1. Thus, the whole range of t_2 is equal to that of t_1, even though t_1 may change faster than t_2.

Now we define the desired translation of TLA formulas. For a TLA formula φ and a new flexible variable t (the fastest time variable) we define a TLA formula $\mathcal{G}(\varphi, t)$ by induction as follows.

- If φ is an STLA formula, then $\mathcal{G}(\varphi, t)$ is φ itself;
- $\mathcal{G}(\neg\varphi, t)$, $\mathcal{G}(\varphi \supset \psi, t)$, and $\mathcal{G}(\exists\S\varphi, t)$ are $\neg\mathcal{G}(\varphi, t)$, $\mathcal{G}(\varphi, t) \supset \mathcal{G}(\psi, t)$, and $\exists x\mathcal{G}(\varphi, t)$, respectively.
- Assume that all free flexible variables of φ are among $x, y_1, \ldots, y_m$. Then $\mathcal{G}(\exists x\varphi(x, y_\infty, \ldots, y_\Updownarrow), t)$ is

$$\begin{aligned}\exists t_\bullet \exists u_{y_1} \ldots \exists u_{y_m} \exists x(S(t, t_\bullet) \wedge \\ \forall t \forall y \bigwedge_{i=1}^{m} (\Diamond(t = t \wedge y_i = y) \equiv \Box(t_\bullet = t \supset u_{y_i} = y)) \wedge \\ \mathcal{G}(\varphi(x, u_{y_\infty}, \ldots, u_{y_\Updownarrow}), t)).\end{aligned}$$

In the last point of the definition of $\mathcal{G}$, the quantifier part $\exists t_\bullet \exists u_{y_1} \ldots \exists u_{y_m}$ reflects adding/removing stuttering steps to the original temporal interpretation and $\exists x$ states that there is an appropriate assignment to x in the temporal interpretation "after adding/removing stuttering steps;" the conjunct $S(t, t_\bullet)$ assures that (subject to satisfiability of $FT(t)$) $t_\bullet$ simulates t in the stretched temporal interpretation; the conjunct

$$\forall t \forall y \bigwedge_{i=1}^{m} (\Diamond(t = t \wedge y_i = y) \equiv \Box(t_\bullet = t \supset u_{y_i} = y))$$

assures that the new flexible variable u_{y_i} simulates y_i, $i = 1, \ldots, m$, in the stretched temporal interpretation; and the last conjunct $\mathcal{G}(\varphi(x, u_\infty, \ldots, u_\backslash), t)$ is the recursive call of the translation of $\varphi(x, u_{y_1}, \ldots, u_{y_m})$.

Theorem 3. *A (TLA) formula φ is TLA-satisfiable by a temporal interpretation over an infinite domain if and only if the formula $FT(t) \wedge \mathcal{G}(\varphi, t)$ is ERTLA-satisfiable.*

[8] In particular, $FT(t)$ implies $T_\varphi(t)$ for all formulas φ.

Example 6 below continues Example 5.

Example 6. Let M_1 be as in Example 5. It can be easily verified that $M_1 \models FT(\boldsymbol{t}) \wedge \mathcal{G}(\exists \boldsymbol{x} \mathcal{F}(\boldsymbol{x}, \boldsymbol{y}), \boldsymbol{t})$.

8 Embedding SOTL into ERTLA

In this section we embed SOTL into ERTLA. For a SOTL formula φ, a rigid variable u that does not appear in φ, and the time variable $\boldsymbol{t}$ we define an ERTLA formula $\mathcal{H}(\varphi, \boldsymbol{t}, \sqcap)$ whose intended meaning is that φ holds at the moment u of the time $\boldsymbol{t}$. The formula $\mathcal{H}(\varphi, \boldsymbol{t}, \sqcap)$ is defined by induction as follows.

- If φ is an atomic formula, then $\mathcal{H}(\varphi, \boldsymbol{t}, \sqcap)$ is $\Box(\boldsymbol{t} = u \supset \varphi)$;
- $\mathcal{H}(\neg\varphi, \boldsymbol{t}, \sqcap)$, $\mathcal{H}(\varphi \supset \psi, \boldsymbol{t}, \sqcap)$, and $\mathcal{H}(\exists \S \varphi, \boldsymbol{t}, \sqcap)$ are $\neg\mathcal{H}(\varphi, \boldsymbol{t}, \sqcap)$, $\mathcal{H}(\varphi, \boldsymbol{t}, \sqcap) \supset \mathcal{H}(\psi, \boldsymbol{t}, \sqcap)$, and $\exists x \mathcal{H}(\varphi, \boldsymbol{t}, \sqcap)$, respectively;
- $\mathcal{H}(\exists \boldsymbol{x} \varphi, \boldsymbol{t}, \sqcap)$ is $\exists \boldsymbol{x}(\Box[\boldsymbol{t} \neq \boldsymbol{t}']_{\infty} \wedge \mathcal{H}(\varphi, \boldsymbol{t}, \sqcap))$
- $\mathcal{H}(\circ\varphi, \boldsymbol{t}, \sqcap)$ is $\forall v(NEXT(u, v) \supset \mathcal{H}(\varphi, \boldsymbol{t}, \sqsubseteq))$;
- $\mathcal{H}(\Box\varphi, \boldsymbol{t}, \sqcap)$ is $\forall v((v \geq u \wedge \Diamond(\boldsymbol{t} = v)) \supset \mathcal{H}(\varphi, \boldsymbol{t}, \sqsubseteq))$; and
- $\mathcal{H}(\varphi \mathcal{U} \psi, \boldsymbol{t}, \sqcap)$ is

$$\exists v((v \geq u) \wedge \Diamond(\boldsymbol{t} = v) \wedge \mathcal{H}(\psi, \boldsymbol{t}, \sqsubseteq) \wedge \forall \sqsupseteq (\sqcap \leq \sqsupseteq < \sqsubseteq \supset \mathcal{H}(\varphi, \boldsymbol{t}, \sqsupseteq)).$$

Theorem 4. *A SOTL formula φ is (SOTL-) satisfiable by a temporal interpretation over an infinite domain if and only if the formula $T_\varphi(\boldsymbol{t}) \wedge (\boldsymbol{t} = u) \wedge \mathcal{H}(\varphi, \boldsymbol{t}, \sqcap)$ is (ERTLA-) satisfiable. Thus, combining Theorems 2 and 4, we obtain a satisfiability preserving translation from SOTL into TLA.*

Remark 1. Since $T_\varphi(\boldsymbol{t})$ is a TLA formula and $\mathcal{H}(\varphi, \boldsymbol{t}, \sqcap)$ does not contain primed flexible variables, $T_\varphi \wedge (\boldsymbol{t} = u) \wedge \mathcal{H}(\varphi, \boldsymbol{t}, \sqcap)$ is a TLA formula. Thus, combining Theorems 2 and 4, we obtain a satisfiability preserving translation from SOTL into TLA.

9 Simulating Finite Domains by Infinite Domains

In this section, using the "relativization technique" we show how satisfiability by temporal interpretations over an infinite domain can be simulated by the standard satisfiability. This simulation step is required because we need infinitely many domain elements to simulate time. It is combined with the translations from the previous sections, which results in an "unconditional" validity preserving translation.

We shall use the following notation. Let R be a new unary predicate symbol and let φ be a formula over the original language. We denote the formulas $\exists x(R(x) \wedge \varphi(x))$ and $\exists \boldsymbol{x}(\Box R(\boldsymbol{x}) \wedge \varphi(\boldsymbol{x}))$ by $\exists_R x \varphi(x)$ and $\exists_R \boldsymbol{x} \varphi(\boldsymbol{x})$, respectively.

The *R-relativization* of φ, denoted φ_R is the formula that is obtained from φ by replacing the quantifier $\exists$ with $\exists_R$. Also, we denote by F^φ the conjunction of the all formulas of the form $\forall x_1 \ldots \forall x_n \exists x(f(x_1, \ldots, x_n) = x)$, all formulas of the form $\exists y(y = x)$, and all formulas of the form $\exists \boldsymbol{y} \Box(\boldsymbol{y} = \boldsymbol{x})$ where f is a function symbol that appears in φ and $x/\boldsymbol{x}$ is a free rigid/flexible variable of φ.

Proposition 1. *Let φ be an SOTL/ERTLA/TLA formula. Then φ is SOTL/ ERTLA/TLA-satisfiable if and only if $\exists x R(x) \wedge F_R^{\varphi} \wedge \varphi_R$ is SOTL/ ERTLA/ TLA-satisfiable by a temporal interpretation over an infinite domain.*[9]

10 Summary

Combining Theorems 1,2,3 and 4 with Proposition 1 we obtain satisfiability preserving translations of TLA into SOTL and vice versa. Since validity is dual to satisfiability, the above translations can be easily transformed into validity preserving ones.

Our last remark is that the translation from SOTL into TLA is linear, whereas the converse translation is quadratic.

Acknowledgment

The work of Michael Kaminski was supported by the fund for the promotion of research in the Technion.

References

1. M. Abadi, The power of temporal proofs, *Theoretical Computer Science* **65** (1989), 35–83.
2. M. Abadi, An Axiomatization of Lamport's Temporal Logic of Actions, in: J.C.M. Baeten and J.W. Klop, eds., *Proceedings of CONCUR'90: Theories of concurrency: unification and extension*, Springer Verlag, 1990, pp. 57–69. (*Lecture Notes in Computer Science* **458**)
3. N. Bjørner, A. Browne, E. Chang, M. Colón, A. Kapur, Z. Manna, H.B. Simpa, and T.E. Uribe, STeP: Deductive-Algorithmic Verification of Reactive and Real-time Systems, in: R. Alur and T.A. Henzinger, eds., *Proceedings of the 8th International Conference on Computer-Aided Verification*, Springer Verlag, 1996, pp. 415–418. (*Lecture Notes in Computer Science* **1102**)
4. U. Engberg, *Reasoning in the Temporal Logic of Actions – The design and implementation of an interactive computer system*, Ph.D. thesis, Department of Computer Science, University of Aarhus, 1996.
5. M. Kaminski and C.K. Wong, The power of the "always" operator in first-order temporal logic, *Theoretical Computer Science B* **160** (1996), 271–281.
6. L. Lamport, The temporal logic of actions, *ACM Transactions on Programming Languages and Systems* **16** (1994), 872–923.
7. A. Rabinovich, Expressive Completeness of Temporal Logic of Actions, in: L. Brim, J. Gruska, and J. Zlatuška, eds., *Proceedings of the 23rd International Symposium on Mathematical Foundations of Computer Sciece*, Springer Verlag, 1998, pp. 229–238. (*Lecture Notes in Computer Science* **1450**)

[9] We need the conjunct $\exists x R(x)$ to ensure that R is not empty and the conjunct F_R^{φ} is required because for each (n-place) function symbol f (a rigid/flexible variable $x/\boldsymbol{x}$) that appears in φ we have an "implicit" axiom $\forall x_1 ... \forall x_n \exists x(f(x_1, \ldots, x_n) = x)$ $(\exists y(y = x)/\exists \boldsymbol{y}\Box(\boldsymbol{y} = \boldsymbol{x}))$ that also must be relativized.

8. A. Rabinovich, On Translation of Temporal Logic of Actions into Monadic Second Order Logic, *Theoretical Computer Science* **193** (1998), 197–214
9. M. Tiomkin and M. Kaminski, Semantical analysis of logic of actions, *Journal of Logic and Computation* **5** (1995), 203–212.

Object Types against Races

Cormac Flanagan and Martín Abadi

Systems Research Center, Compaq
[flanagan|ma]@pa.dec.com

Abstract. This paper investigates an approach for statically preventing race conditions in an object-oriented language. The setting of this work is a variant of Gordon and Hankin's concurrent object calculus. We enrich that calculus with a form of dependent object types that enables us to verify that threads invoke and update methods only after acquiring appropriate locks. We establish that well-typed programs do not have race conditions.

1 Introduction

Concurrent object-oriented programs suffer from many of the errors common in concurrent programs of other sorts. In particular, the use of objects does not diminish the importance of careful synchronization. With objects or without them, improper synchronization may lead to race conditions (that is, two processes accessing a shared resource simultaneously) and ultimately to incorrect behavior.

A standard approach for eliminating race conditions consists in protecting each shared resource with a lock, requiring that a process acquires the corresponding lock before using the resource [5]. Object-oriented programs often rely on this approach, but with some peculiar patterns. It is common to group related resources into an object, and to attach the lock that protects the resources to this object. Processes may acquire the lock before invoking the methods of the object; alternatively, the methods may acquire this lock at the start of their execution. With constructs such as Java's `synchronized` methods [2, 10], some object-oriented languages support these synchronization patterns. However, standard object-oriented languages do not enforce proper synchronization; it remains possible, even easy, to write programs with race conditions.

This paper investigates a static-analysis approach for preventing race conditions in an object-oriented language. The approach consists in associating locks with shared object components and in verifying that appropriate locks have been acquired before each operation. In the object-oriented language that we treat, the object components are methods; they can be both invoked and updated. (Fields are a special case of methods [1].) Thus, the approach consists in associating locks with methods and in verifying that appropriate locks have been acquired before each method invocation and update. Because a method invocation may trigger other operations, several locks may be required for it; only one lock is required for a method update.

The annotations and checks necessary in our static analysis are expressed in a type system. Like standard type systems, this type system assigns a type to each of the methods of an object. In addition, it gives the set of locks that must be held before invoking the method and the lock that must be held before updating the method (or an indication that the update is forbidden). Each of these locks may be external to the object, but it may also be a special lock associated with self, that is, with this object.

Thus, we are led to a type system with dependent types, in which the type of an object refers to values, namely to locks. However, the type system is restrictive enough to preserve the important phase distinction between compile-time and run-time [7, 11]. All our checking takes place at compile-time (without excluding the possibility of further run-time checking).

The checking guarantees the absence of race conditions: if a program is well-typed, then during its execution no two threads attempt to access an object component at the same time. In addition, the checking guarantees the absence of standard run-time type errors ("message-not-understood" errors). Our approach can handle some interesting, common examples, as we demonstrate. Although it is far from complete, we believe that it represents a sensible compromise between simplicity and generality, and a worthwhile step in the ongoing investigation of the use of types for safe locking.

Background

In a recent paper [8], we developed an analogous technique for a basic calculus with reference cells but no data structures. In that paper, singleton types (types with one single element) enable the tracking of locks; existential types permit the hiding of singleton types for locks. That paper also describes a technique for avoiding deadlocks, which it should be possible to adapt to the setting of the present paper. The substantial novelty of the present paper is the treatment of objects, object types, and subtyping. Here we avoid the use of singleton types and existential types, and resort to specialized dependent types instead. These dependent types require somewhat less conceptual machinery and support more flexible subtyping relations.

In addition to our own previous work, we rely on Gordon and Hankin's concurrent object calculus **conc**ς [9]. This calculus is a small but extremely expressive concurrent object-oriented language; it features a compact and elegant presentation of the concepts of expression, process, store, and configuration. We refer the reader to Gordon and Hankin's work for motivations for this calculus and additional examples and technical developments.

The calculus **conc**ς extends a sequential calculus of Abadi and Cardelli [1], adopting the basic type structure and subtyping relation of that sequential calculus. Here we extend the type system with our form of dependent types. Gordon and Hankin's type system does not attempt to guarantee the absence of race conditions; our type system provides that guarantee.

For simplicity, we omit the cloning construct from **conc**ς. We also replace the synchronization primitives that Gordon and Hankin presented as an extension to

concς. Those primitives are two separate operations for acquiring and releasing a lock. Instead, we use the expression *lock v in a*, which acquires the lock denoted by v, evaluates a, then releases the lock. Like Java's `synchronized` statement, the expression *lock v in a* automatically guarantees the proper nesting of lock operations, helping static checking. Moreover, our calculus associates locks with objects (unlike **concς**, but like Java).

There are some other languages that we might have used as a starting point for this work instead of **concς**, in particular Di Blasio and Fisher's concurrent object calculus [6]. However, we prefer to base our work on that of Gordon and Hankin, for two main reasons. First, Di Blasio and Fisher's calculus permits object extension but not subtyping, unlike **concς** and unlike most typed object-oriented languages; we wish to treat subtyping. Furthermore, Di Blasio and Fisher's calculus combines synchronization mechanisms with the primitive operations on objects. Like Gordon and Hankin, we prefer to keep synchronization separate from object operations, although our object types do mention locks. Di Blasio and Fisher's interesting study does not address race conditions, but shows that certain synchronization guards do not have side-effects.

Other pieces of related work are discussed in our recent paper. These rely on a variety of techniques, including program-verification methods and data-flow analyses, for example. One of the most relevant is the work of Kobayashi and Sumii [13, 16], which develops a type-based techniques for avoiding deadlocks (not necessarily race conditions) in the context of a process calculus. Another one is Warlock [15], a system for partial detection of race conditions and deadlocks in ANSI C programs. We are not aware of any work that specifically addresses race conditions in object-oriented programs. In another direction, there have been intriguing explorations of the combination of dependent types with objects and subtyping, with an emphasis on logical frameworks rather than programming languages [3, 12].

Outline

The next section presents the syntax and informal semantics of the concurrent object-oriented language that we treat; an appendix contains a formal semantics. Section 3 develops our type system for this calculus. Section 4 shows some example applications of our type system. Section 5 considers formal properties. Finally, section 6 concludes. Proofs are omitted.

2 A Concurrent Object Calculus

This section describes our variant of Gordon and Hankin's concurrent object calculus. It is largely a review.

2.1 Syntax

We define the sets of *results*, *denotations*, *lock states*, and *terms* by the grammar:

Syntax

$u, v ::=$	results
x	variable
p	name
$d ::=$	denotations
$[\ell_i = \varsigma(x_i)b_i\ ^{i \in 1..n}]^m$	object
$m ::=$	lock states
$\bullet$	locked
$\circ$	unlocked
$a, b, c ::=$	terms
u	result
$(\nu p)a$	restriction
$p \mapsto d$	denomination
$u.\ell$	method invocation
$u.\ell \Leftarrow \varsigma(x)b$	method update
$let\ x{=}a\ in\ b$	let
$a \upharpoonleft b$	parallel composition
$lock\ u\ in\ a$	lock acquisition
$locked\ p\ in\ a$	lock acquired

Results include both variables and names. Variables represent intermediate values, and are bound by methods $\varsigma(x)b$ and by let expressions $let\ x{=}a\ in\ b$; both of these constructs bind the variable x with scope b. Names represent the addresses of stored objects. They are introduced by a restriction $(\nu p)a$, which binds the name p with scope a. We let $fn(a)$ and $fv(a)$ denote the sets of free names and free variables in a, respectively. We write $a\{\!\{u \leftarrow b\}\!\}$ to denote the capture-free substitution of b for all free occurrences of u in a. We write $a = b$ to mean that a and b are equal up to the renaming of bound variables and bound names, and the reordering of object methods.

2.2 Informal Semantics

A denotation $[\ell_i = \varsigma(x_i)b_i\ ^{i \in 1..n}]^m$ describes an object containing a collection of methods. Each method has a label ℓ_i and consists of a self parameter x_i and a body b_i. In addition, the object has an associated lock whose state is described by m. If $m = \bullet$, the lock is held by some term in the program; if $m = \circ$, the lock is unlocked. (As a straightforward extension, each object could have several associated locks.)

A denotation may appear in a denomination $p \mapsto d$, which maps the name p to the denotation d. Intuitively, this term represents the portion of the store containing the object d, and p represents the address of that object. The term $(\nu p)a$ introduces a fresh name p and then evaluates a. This operation corresponds to allocating a fresh address p at which objects can be stored. Thus the language separates name introduction $(\nu p)a$ from name definition $p \mapsto d$; the type system forbids programs with multiple definitions of the same name.

A method invocation $u.\ell$ invokes the method ℓ of the object u. A method update $u.\ell \Leftarrow \varsigma(x)b$ replaces the method ℓ of u with $\varsigma(x)b$. The term *let* x=a *in* b first evaluates a to yield a result, binds x to this result, and then evaluates b. A parallel composition $a \upharpoonright b$ evaluates both a and b in parallel. The result of this parallel composition is the result of b; the subterm a is evaluated only for effect.

The lock operation *lock* u *in* a functions in a similar manner to Java's `synchronized` statement: the lock on the object u is acquired; then the subterm a is evaluated; and finally the lock is released. The implementation of this construct relies on an auxiliary construct *locked* p *in* a, which indicates that the lock p has been acquired and that the term a is being evaluated. Locks are not reentrant, so expressions like *lock* u *in lock* u *in* a will deadlock; similarly, calls to the recursive method ℓ of $[\ell = \varsigma(x)\mathit{lock}\ x\ \mathit{in}\ x.\ell]^{\circ}$ will deadlock (cf. Java).

The appendix contains a detailed formal semantics of the language. It is a chemical semantics in the style of Berry and Boudol [4]. It consists of a group of structural congruence rules, which permit the rearrangement of terms and imply for example that $\upharpoonright$ is associative, and a group of reduction rules, which model proper computation steps.

A typical structural congruence rule is:

$$a \upharpoonright \mathcal{E}[\,b\,] \equiv \mathcal{E}[\,a \upharpoonright b\,] \qquad\qquad (\text{if } \mathit{fn}(a) \cap \mathit{bn}(\mathcal{E}) = \varnothing)$$

where $\mathcal{E}$ denotes an evaluation context:

$$\mathcal{E} ::= [\,\cdot\,] \mid \mathit{let}\ x{=}\mathcal{E}\ \mathit{in}\ b \mid \mathcal{E} \upharpoonright b \mid a \upharpoonright \mathcal{E} \mid (\nu p)\mathcal{E} \mid \mathit{locked}\ p\ \mathit{in}\ \mathcal{E}$$

and the *binding names* $\mathit{bn}(\mathcal{E})$ of an evaluation context are the names p bound by a restriction $(\nu p)\mathcal{E}'$ that encloses the hole $[\,\cdot\,]$.

For our purposes, the two most interesting reduction rules are:

$$\begin{array}{rcll} (p \mapsto [\ldots]^{\circ}) \upharpoonright \mathit{lock}\ p\ \mathit{in}\ a & \to & (p \mapsto [\ldots]^{\bullet}) \upharpoonright \mathit{locked}\ p\ \mathit{in}\ a & \text{(Red Lock)} \\ (p \mapsto [\ldots]^{\bullet}) \upharpoonright \mathit{locked}\ p\ \mathit{in}\ u & \to & (p \mapsto [\ldots]^{\circ}) \upharpoonright u & \text{(Red Locked)} \end{array}$$

where $[\ldots]$ represents an object (excluding its lock state). The rule (Red Lock) evaluates a lock operation by acquiring the lock associated with p, and yielding the term *locked* p *in* a. Subsequent reduction steps may then evaluate a. Once a is reduced to some result u, the rule (Red Locked) releases the lock on p and returns u as the result of the locked expression.

2.3 An Example

For clarity, we present example programs in an extended language with integer constants and operations, and we abbreviate *let* x=a *in* b to $a; b$ when $x \notin \mathit{fv}(b)$.

A counter that has a read method and an increment method, and initially contains the integer n, can be defined as:

$$\begin{array}{rl} \mathit{count}_n \triangleq [\mathit{val} & = \varsigma(x)n, \\ \mathit{read} & = \varsigma(x)x.\mathit{val}, \\ \mathit{inc} & = \varsigma(x)\mathit{let}\ t{=}x.\mathit{val}+1\ \mathit{in}\ x.\mathit{val} \Leftarrow \varsigma(x)t] \end{array}$$

The following program allocates a counter (initially containing 0), increments the counter, and then reads the value of the counter.

$$(\nu p)(p \mapsto \mathit{count}_0 \upharpoonright^{\!\ast} (p.\mathit{inc}; p.\mathit{read}))$$

As expected, this program reduces to $(\nu p)(p \mapsto \mathit{count}_1 \upharpoonright^{\!\ast} 1)$, since the counter works correctly in a sequential setting. In the presence of concurrency, however, the counter may exhibit unexpected behavior. To illustrate this danger, we consider the following program, which creates a counter and then increments it twice, in parallel.

$$(\nu p)(p \mapsto \mathit{count}_0 \upharpoonright^{\!\ast} p.\mathit{inc} \upharpoonright^{\!\ast} p.\mathit{inc})$$

This program is non-deterministic. It may reduce to $(\nu p)(p \mapsto \mathit{count}_2 \upharpoonright^{\!\ast} p \upharpoonright^{\!\ast} p)$, as expected. Alternatively, if the evaluations of the two calls to *inc* are interleaved, the program may also reduce to $(\nu p)(p \mapsto \mathit{count}_1 \upharpoonright^{\!\ast} p \upharpoonright^{\!\ast} p)$, which is presumably not what the programmer intended. Thus the program has a race condition: two threads may attempt to update the method *val* simultaneously, with incorrect results.

We can fix this error by adding appropriate synchronization to the counter:

$$\begin{aligned} \mathit{sync\_count}_n \triangleq [&\mathit{val} = \varsigma(x)n, \\ &\mathit{read} = \varsigma(x)\mathit{lock}\ x\ \mathit{in}\ x.\mathit{val}, \\ &\mathit{inc} = \varsigma(x)\mathit{lock}\ x\ \mathit{in}\ \mathit{let}\ t{=}x.\mathit{val}+1\ \mathit{in}\ x.\mathit{val} \Leftarrow \varsigma(x)t] \end{aligned}$$

In this synchronized counter, the method *val* is protected by the lock of the counter. This lock should be held whenever the method *val* is invoked or updated, and thus the methods *read* and *inc* both acquire that lock. The modified counter implementation is race-free and will behave correctly even if used by multiple threads, provided those threads access *val* only through *read* and *inc*, or acquire the lock before accessing *val* directly. We revisit this example in later sections.

3 The Type System

Race conditions, such as that in count_n, are a common bug in concurrent object-oriented programs, just as they are in concurrent programs of other kinds. In practice, race conditions are often avoided by the same strategy that we employed in $\mathit{sync\_count}_n$; each mutable component is protected by a lock, and is only accessed when that lock is held. In this section, we describe a type system that supports this programming discipline.

3.1 The Type Language

The set of types in our system is described by the grammar:

Types

$A, B ::= [\ell_i : \varsigma(x_i)A_i \cdot r_i \cdot s_i \ ^{i \in 1..n}] \mid \mathit{Proc} \mid \mathit{Exp}$	types
$r ::= \{u_1, \ldots, u_n\}$	permissions
$s ::= u \mid +$	protection annotations

An object type $[\ell_i : \varsigma(x_i)A_i \cdot r_i \cdot s_i \ ^{i \in 1..n}]$ describes an object containing n methods labeled $l_1, \ldots, l_n$. Each method l_i has result type A_i, *permission* r_i, and *protection annotation* s_i. The permission r_i is a set of results describing the locks that must be held before invoking l_i. Because the method invocation may trigger other operations, we allow r_i to contain more than one lock. The protection annotation s_i is a result describing the lock that must be held before updating l_i; we refer to s_i as the lock that protects l_i (although additional locks may be required for invoking l_i). In the case where l_i is never updated, s_i may alternatively be the symbol '+'. Since methods are commonly protected by the *self lock* (that is, the lock of the object itself), the description of each method also binds the self variable x_i; this variable may occur free in A_i, r_i, and s_i.

An example type is $[l : \varsigma(x)A \cdot \varnothing \cdot +]$, which describes an object containing a single method l with result type A. The permission $\varnothing$ indicates that no locks need to be acquired before invoking this method; the protection annotation '+' indicates that the method cannot be updated. The type $[l : \varsigma(x)A \cdot \{x\} \cdot x]$ is similar, except that it describes an object whose method l can be updated. The self lock of the object must be acquired before invoking or updating that method. The type $[l : \varsigma(x)A \cdot \{x\} \cdot x, l' : \varsigma(x)A \cdot \{x, p\} \cdot p]$ mentions an additional method l' protected by an external lock p. Since the lock that protects l (the self lock x) must be acquired before invoking l', the code for l' may update or invoke l without any further locking.

As a slightly more complicated example, a suitable type for the synchronized counter $\mathit{sync\_count}_n$ described earlier is:

$$[\mathit{val} : \varsigma(x)\mathit{Int} \cdot \{x\} \cdot x, \quad \mathit{read} : \varsigma(x)\mathit{Int} \cdot \varnothing \cdot +, \quad \mathit{inc} : \varsigma(x)[\ ] \cdot \varnothing \cdot +]$$

This type states that the method *val* is protected by the self lock, which must be acquired before invoking or updating that method. The methods *read* and *inc* are read-only; they cannot be updated. Furthermore, since these methods perform the necessary synchronization internally, no locks need to be held when invoking these methods.

In addition to object types, the type language also includes the types *Exp* and *Proc*. The type *Exp* describes all results that may be returned by expressions; the type *Proc* is a supertype of *Exp* that also covers terms that never return results, such as a denomination $p \mapsto d$.

3.2 Clean and Defined Names

In addition to checking that the appropriate locks are held whenever a method is invoked or updated, the type system also verifies that each lock is held by at most one thread at any time. That is, for each name p, there is at most one term of the form *locked p in* ... in the program.

Verifying this mutual exclusion property is a little tricky, since any term that contains the denomination $p \mapsto [\ldots]^\circ$ can potentially acquire the lock on p via the reduction rule (Red Lock). Therefore, we introduce the notion of *clean names*, and we say that p is a clean name of a term if the term includes either *locked p in* ... or $p \mapsto [\ldots]^\circ$ in an evaluation context. (The restriction to evaluation contexts excludes some nonsensical programs.) The set of clean names of a term is preserved during evaluation, even though the set of locks held by the term may vary. The type system checks that for any parallel composition $a \upharpoonright b$, the clean names of the subterms a and b are distinct. This check ensures that a lock cannot be simultaneously held by two terms executing in parallel.

The type system also verifies that every name that is introduced is associated with a unique denotation. We say that a name is *defined* by a term if it is associated with a denotation in an evaluation context.

Clean and defined names

$p \in clean(a)$ if $a = \mathcal{E}[\, p \mapsto [\ldots]^\circ \,]$ or $a = \mathcal{E}[\, locked\ p\ in\ b \,]$ and $p \notin bn(\mathcal{E})$
$p \in defined(a)$ if $a = \mathcal{E}[\, p \mapsto d \,]$ and $p \notin bn(\mathcal{E})$

3.3 Type Rules

We define the type system using the following six judgments and associated rules. In these judgments, an *environment* E is a sequence of bindings of results to types, of the form $\varnothing, u_1 : A_1, \ldots, u_n : A_n$.

Judgments

$E \vdash \diamond$	E is a well-formed environment
$E \vdash A$	given E, type A is well-formed
$E \vdash r$	given E, permission r is well-formed
$E \vdash A <: B$	given E, A is a subtype of B
$E \vdash r <: r'$	given E, r is a subpermission of r'
$E; r \vdash a : A$	given E and r, term a has type A

Type rules

(Env $\varnothing$)
$$\frac{}{\varnothing \vdash \diamond}$$

(Env u)
$$\frac{E \vdash A \quad u \notin dom(E)}{E, u : A \vdash \diamond}$$

(Perm)
$$\frac{E \vdash \diamond \quad r \subseteq dom(E)}{E \vdash r}$$

(Type Proc)
$$\frac{E \vdash \diamond}{E \vdash Proc}$$

(Type Exp)
$$\frac{E \vdash \diamond}{E \vdash Exp}$$

(Type Object) (ℓ_i distinct)

$$\frac{E \vdash \diamond \quad E, x_i : [\,] \vdash B_i <: Exp \quad E, x_i : [\,] \vdash r_i \quad s_i \in r_i \cup \{+\} \quad \forall i \in 1..n}{E \vdash [\ell_i : \varsigma(x_i) B_i \cdot r_i \cdot s_i \ ^{i \in 1..n}]}$$

(Val Object) (where $A = [\ell_i : \varsigma(x_i) B_i \cdot r_i \cdot s_i \ ^{i \in 1..n}]$)

$$\frac{\begin{array}{c} E = E_1, p : A, E_2 \quad E \vdash \diamond \quad E; r_i \{\!\{x_i \leftarrow p\}\!\} \vdash b_i \{\!\{x_i \leftarrow p\}\!\} : B_i \{\!\{x_i \leftarrow p\}\!\} \\ \mathit{defined}(b_i) = \mathit{clean}(b_i) = \varnothing \quad \forall i \in 1..n \end{array}}{E; \varnothing \vdash p \mapsto [\ell_i = \varsigma(x_i) b_i \ ^{i \in 1..n}]^m : \mathit{Proc}}$$

(Val u)

$$\frac{E, u : A, E' \vdash \diamond}{E, u : A, E'; \varnothing \vdash u : A}$$

(Val Select)

$$\frac{E; \varnothing \vdash u : [\ell_i : \varsigma(x_i) B_i \cdot r_i \cdot s_i \ ^{i \in 1..n}] \quad j \in 1..n}{E; r_j \{\!\{x_j \leftarrow u\}\!\} \vdash u.\ell_j : B_j \{\!\{x_j \leftarrow u\}\!\}}$$

(Val Update) (where $A = [\ell_i : \varsigma(x_i) B_i \cdot r_i \cdot s_i \ ^{i \in 1..n}]$)

$$\frac{\begin{array}{c} E; \varnothing \vdash u : A \quad E; r_j \{\!\{x_j \leftarrow u\}\!\} \vdash b \{\!\{x_j \leftarrow u\}\!\} : B_j \{\!\{x_j \leftarrow u\}\!\} \quad s_j \neq + \quad j \in 1..n \\ \mathit{defined}(b) = \mathit{clean}(b) = \varnothing \end{array}}{E; \{s_j \{\!\{x_j \leftarrow u\}\!\}\} \vdash u.\ell_j \Leftarrow \varsigma(x_j) b : A}$$

(Val Let)

$$\frac{\begin{array}{c} E; r \vdash a : A \quad E, x : A; r \vdash b : B \quad E \vdash A <: Exp \quad E \vdash B <: Exp \\ \mathit{defined}(b) = \mathit{clean}(b) = \varnothing \end{array}}{E; r \vdash \mathit{let}\ x{=}a\ \mathit{in}\ b : B}$$

(Val Res)

$$\frac{\begin{array}{c} E, p : A; r \vdash a : B \quad E \vdash r \quad E \vdash B \\ p \in \mathit{defined}(a) \quad p \in \mathit{clean}(a) \end{array}}{E; r \vdash (\nu p) a : B}$$

(Val Par)

$$\frac{\begin{array}{c} E; \varnothing \vdash a : \mathit{Proc} \quad E; r \vdash b : B \\ \mathit{defined}(a) \cap \mathit{defined}(b) = \varnothing \quad \mathit{clean}(a) \cap \mathit{clean}(b) = \varnothing \end{array}}{E; r \vdash a \upharpoonright b : B}$$

(Val Lock)

$$\frac{\begin{array}{c} E; \varnothing \vdash u : [\,] \quad E; r \cup \{u\} \vdash a : A \\ \mathit{defined}(a) = \mathit{clean}(a) = \varnothing \end{array}}{E; r \vdash \mathit{lock}\ u\ \mathit{in}\ a : A}$$

(Val Locked)

$$\frac{\begin{array}{c} E; \varnothing \vdash p : [\,] \quad E; r \cup \{p\} \vdash a : A \\ p \notin \mathit{clean}(a) \end{array}}{E; r \vdash \mathit{locked}\ p\ \mathit{in}\ a : A}$$

(Val Subsumption)

$$\frac{E; r \vdash a : A \quad E \vdash A <: B \quad E \vdash r <: r'}{E; r' \vdash a : B}$$

(Subperm)

$$\frac{E \vdash r \quad E \vdash r' \quad r \subseteq r'}{E \vdash r <: r'}$$

(Sub Refl)

$$\frac{E \vdash A}{E \vdash A <: A}$$

(Sub Trans)

$$\frac{E \vdash A <: B \quad E \vdash B <: C}{E \vdash A <: C}$$

(Sub Exp)

$$\frac{E \vdash A \quad A \neq \mathit{Proc}}{E \vdash A <: Exp}$$

(Sub Proc)

$$\frac{E \vdash A}{E \vdash A <: \mathit{Proc}}$$

(Sub Object)

$$\frac{E \vdash [\ell_i : \varsigma(x_i)B_i \cdot r_i \cdot s_i \ ^{i \in 1..n+m}]}{E \vdash [\ell_i : \varsigma(x_i)B_i \cdot r_i \cdot s_i \ ^{i \in 1..n+m}] <: [\ell_i : \varsigma(x_i)B_i \cdot r_i \cdot s_i \ ^{i \in 1..n}]}$$

Many of the rules of the type system are based on corresponding rules in Gordon and Hankin's system, which is in turn based on Abadi and Cardelli's calculi. The novel aspects of our system mainly pertain to locking; they include the treatment of permissions and dependent types.

The core of the system is the set of rules for the judgment $E; r \vdash a : A$ (read "a is a well-typed expression of type A in typing environment E with permission r"). Our intent is that, if this judgment holds, then a is race-free and yields results of type A, provided the free variables of a are given bindings consistent with the typing environment E, and the current thread holds at least the locks in r.

The type system thus tracks the set of locks that are assumed to be held at each program point. The rule (Val Object) checks that each method body is race-free under the assumption that the locks described by the method's permission are held. The rule (Val Select) ensures that these locks are held whenever the method is invoked. The rule (Val Update) ensures that the lock protecting a method is held whenever that method is updated. The rule (Val Lock) for *lock* u *in* a typechecks a under the assumption that the lock u is held. The rule (Val Subsumption) allows for subsumption on both types and permissions: if $E \vdash r <: r'$, then any term that is race-free with permission r is also race-free with the superset r' of r.

The type system provides dependent types, that is, a type may contain a result that refers to an object. In some cases, an object can be the referent of several results, for example, its self variable and some external name for the object. The type rules contain a number of substitutions that support changing the result used to refer to a particular object. For example, the rule (Val Select) for a method invocation $u.\ell_j$ replaces occurrences of the self variable x_j in the type B_j with the result u, since x_j and u refer to the same object, and x_j is going out of scope. A similar substitution is performed on the permission r_j. The rules (Val Object) and (Val Update) rely on analogous substitutions.

In order to accommodate self-dependent types, where the description of an object's method may refer to the object itself, the rule (Type Object) checks that the result type and permission of each method is well-formed in an extended environment that contains a binding for the self variable. Because types may refer to results, the rules (Val Let) and (Val Res) ensure that a type that is lifted outside a result binding is still well-formed. The rule (Val Res) also has a similar requirement on permissions.

The type rules include conditions on the clean and defined names of subterms. The rule (Val Par) for $a \upharpoonleft b$ requires that the defined names of the subterms a and b be disjoint. Furthermore, the clean names of the subterms must also be disjoint. The latter condition implies that the two subterms cannot simultaneously hold the same lock. The rule (Val Res) for $(\nu p)a$ requires that the name p being introduced be defined in a, and that the lock associated with p is either unlocked or is held by a. The rule (Val Locked) disallows nested acquisitions of the same lock. In addition, in order to ensure that the clean and defined names of a term are invariant under evaluation, the type rules require that terms not in evaluation contexts do not have any clean or defined names.

The rule (Sub Object) defines the usual subtyping relation on object types (appropriately adapted to our type syntax). Since the protection annotation + can be considered a variance annotation [1], we could extend the type system with a more powerful subtyping rule. This rule would allow the result types and permissions of immutable components to behave covariantly. We conjecture that the extended system would still be race-free.

4 Examples

In this section we show a few applications of our type system in examples. For convenience, we use the abbreviation $b.\ell \triangleq \mathit{let}\ x{=}b\ \mathit{in}\ x.\ell$ when b is not a result.

4.1 Counters

The unsynchronized counter implementation count_n described earlier can be assigned the type:

$$[\mathit{val} : \varsigma(x)\mathit{Int}\cdot\{x\}\cdot x,\quad \mathit{read} : \varsigma(x)\mathit{Int}\cdot\{x\}\cdot +,\quad \mathit{inc} : \varsigma(x)[\,]\cdot\{x\}\cdot +]$$

This type states that the method *val* is protected by the self lock of the object, and this self lock must be acquired before invoking the methods *read* and *inc*.

The method *val* may be considered private to the implementation of the counter, and can be dropped via subtyping, yielding:

$$[\mathit{read} : \varsigma(x)\mathit{Int}\cdot\{x\}\cdot +,\quad \mathit{inc} : \varsigma(x)[\,]\cdot\{x\}\cdot +]$$

This type describes the public interface to the counter; it states that the self lock of the counter must be acquired before invoking the counter's methods. This interface expresses a synchronization protocol that is sufficient to ensure that the counter operates correctly. The type system requires that this protocol be obeyed by each client of the counter. Programs that do not obey this synchronization protocol, such as $(\nu p)(p \mapsto \mathit{count}_0 \upharpoonright p.\mathit{inc} \upharpoonright p.\mathit{inc})$, are forbidden.

4.2 Input Streams

In some cases, we may wish to provide similar synchronized and unsynchronized interfaces to the same object. For example, an input stream may provide both a synchronized method *read* for reading characters from the stream, and a faster but unsynchronized method *read'*. (The Modula-3 I/O package provides both of these methods [14].)

An outline implementation of such an input stream might be:

$$\begin{array}{ll} instream \triangleq [buffer = \varsigma(x)\ldots, & \text{internal data structure} \\ \quad read' = \varsigma(x)\cdots buffer\cdots, & \text{fast, unsynchronized read} \\ \quad read = \varsigma(x) lock\ x\ in\ x.read'] & \text{slower, synchronized read} \end{array}$$

The method *buffer* contains some internal data structures of the input stream and is protected by the self lock. The method *read'* assumes that the self lock is held, and returns the next input character after some manipulation of *buffer*. The method *read* does not assume that the self lock is held; it first acquires that lock and then dispatches to *read'*.

A suitable type for this input stream is:

$$[buffer : \varsigma(x)Buffer\cdot\{x\}\cdot x,\quad read' : \varsigma(x)Char\cdot\{x\}\cdot+,\quad read : \varsigma(x)Char\cdot\varnothing\cdot+]$$

Subtyping then allows us to view an input stream as having either the synchronized interface $[read : \varsigma(x)Char\cdot\varnothing\cdot+]$ or the faster but unsynchronized interface $[read' : \varsigma(x)Char\cdot\{x\}\cdot+]$.

4.3 Lines and Points

The examples above describe objects whose components are protected by the self lock of the object. In addition, object components can also be protected by a lock external to the object. To illustrate this possibility, we consider the following example consisting of point and line objects.

$$\begin{array}{l} point \triangleq [x = \varsigma(s)0, \\ \quad y = \varsigma(s)0, \\ \quad bmp = \varsigma(s) let\ t{=}s.x + 1\ in\ s.x \Leftarrow \varsigma(s)t] \end{array}$$

$$\begin{array}{l} line \triangleq [start = \varsigma(s)pt_1, \\ \quad end = \varsigma(s)pt_2, \\ \quad bmp = \varsigma(s) lock\ s\ in\ (s.start.bmp;\ s.end.bmp)] \end{array}$$

A point contains a method *bmp* that increments the x-coordinate of the point. (An analogous method for y is omitted for brevity.) Each line object includes two methods for its end points, *start* and *end*, and a method *bmp* that increments the x-coordinate of both end points of the line. This method first acquires the self lock of the line, then calls the method *bmp* of both end points. These points do not perform any synchronization internally; their mutable methods x and y are protected by the lock of the enclosing line object. Appropriate types for lines and points are:

$$\begin{array}{l} Point_z \triangleq [x : \varsigma(s)Int\cdot\{z\}\cdot z,\ y : \varsigma(s)Int\cdot\{z\}\cdot z,\ bmp : \varsigma(s)[\,]\cdot\{z\}\cdot+] \\ Line \triangleq [start : \varsigma(s)Point_s\cdot\varnothing\cdot+,\ end : \varsigma(s)Point_s\cdot\varnothing\cdot+,\ bmp : \varsigma(s)[\,]\cdot\varnothing\cdot+] \end{array}$$

where the type $Point_z$ describes a point whose mutable methods are protected by the lock z. The type *Line* states that the methods *start* and *end* yield points whose mutable components are protected by the lock of the enclosing line object.

Interestingly, the type *Line* also permits the following line object:

$$\begin{aligned} line' \triangleq [&start = \varsigma(s)pt_1, \\ &end = \varsigma(s)pt_2, \\ &bmp = \varsigma(s)(lock\ s\ in\ s.start.bmp);\ (lock\ s\ in\ s.end.bmp)] \end{aligned}$$

When a thread runs the method *bmp* of this object, it acquires and releases the self lock twice; it does not hold the lock continuously throughout the execution of the method. Therefore, another thread may observe *line'* in an intermediate state where *start* has been updated but *end* has not. Such an intermediate state might violate higher-level invariants, so the interleaving of the two threads might be regarded as a higher-level race condition. Our type system does not address such errors directly.

4.4 Encoding Functions as Race-Free Objects

We encode function abstraction and application in our calculus as follows, much as in other object calculi [1, 9]:

Encoding functions

$$\lambda(x)b \triangleq [new = \varsigma(s)[arg = \varsigma(s)s.arg,\ val = \varsigma(s)let\ x{=}s.arg\ in\ b]] \quad \text{for } s \notin fv(b)$$
$$b(a) \triangleq let\ y{=}b.new\ in\ lock\ y\ in\ (y.arg \Leftarrow \varsigma(z)a).val \quad \text{for } y, z \notin fv(a)$$

In the absence of cloning, we need to use a method *new* to create a fresh object with the usual methods *arg* and *val*. The self lock is acquired before accessing *arg* and *val*. Locking is needed for both methods because *arg* is mutable and *val* invokes *arg*.

This translation provides an encoding for the simply-typed call-by-value λ-calculus; a function of type $A \to B$ is mapped to an object of type:

$$[new : \varsigma(s)[arg : \varsigma(s)A\cdot\{s\}\cdot s,\ val : \varsigma(s)B\cdot\{s\}\cdot +]\cdot\varnothing\cdot +]$$

This translation cannot encode dependent function types, in which the result type depends on the argument value. Encoding dependent function types in our calculus seems to require an extension, for example allowing the use of terms (and not just results) as locks.

4.5 Other Encodings (Sketches)

We can translate programs of the imperative object calculus **impς** [1] into our calculus in a straightforward manner, much as Gordon and Hankin. (Since our calculus does not include cloning, this translation only works for clone-free programs.) Each translated program includes a single global lock, which protects all object components in the program. Since the program is single-threaded, this lock needs to be acquired only once, at the start of the program's execution;

it is then held throughout the execution, allowing unrestricted invocations and updates of object components.

Gordon and Hankin describe an encoding of the π-calculus into their concurrent object calculus. Their encoding is based on an implementation of channels. A similar approach works in our setting, but not as neatly. Because the locks of our calculus are not semaphores, our implementation of channels uses busy-waiting; for example, reading from a channel may involve looping until a value is written to that channel by some other thread.

5 Well-Typed Programs Don't Have Races

The fundamental property of the type system is that well-typed programs do not have race conditions. We formalize the notion of a race condition as follows. A term b *reads* $p.\ell$ if there exists some $\mathcal{E}$ such that $b = \mathcal{E}[\ p.\ell\ ]$ and $p \notin bn(\mathcal{E})$. Similarly, a term b *writes* $p.\ell$ if there exists some $\mathcal{E}$, x, and c such that $b = \mathcal{E}[\ p.\ell \Leftarrow \varsigma(x)c\ ]$ and $p \notin bn(\mathcal{E})$. A term *accesses* $p.\ell$ if it either reads or writes $p.\ell$. A term b has an *immediate race condition* if there exists some c_1, c_2, and $p.\ell$ such that $b \equiv \mathcal{E}[\ c_1 \upharpoonright c_2\ ]$, c_1 and c_2 both access $p.\ell$, and at least one of those accesses is a write. Finally, a term b has a *race condition* if its evaluation may yield a term with an immediate race condition, that is, if there exists a term c such that $b \rightarrow^* c$ and c has an immediate race condition.

The type system ensures that, in a well-typed program, a thread that accesses a method holds the appropriate locks. The following lemma is crucial in establishing that property.

Lemma 1. *If $E; r \vdash b : B$ and b accesses $p.\ell$ then $E; \varnothing \vdash p : [\ell : \varsigma(x)A \cdot r' \cdot s]$ and $s\{\!\{x \leftarrow p\}\!\} \in clean(b) \cup r \cup \{+\}$. Furthermore, if the access is a write, then $s \neq +$.*

Since each lock can be held by at most one term at any time, a well-typed program does not have an immediate race condition.

Lemma 2. *If $E; \varnothing \vdash b : B$ then b does not have an immediate race condition.*

Furthermore, typing is invariant under reduction.

Lemma 3. *If $E; r \vdash b : B$ and $b \rightarrow c$ then $E; r \vdash c : B$.*

Using the previous lemmas, we can prove that well-typed programs do not have race conditions.

Theorem 1. *If $E; \varnothing \vdash b : B$ then b does not have a race condition.*

6 Conclusion

As this paper shows, a simple type system can help detect and avoid some synchronization errors in concurrent object-oriented programs. Our type system builds on the underlying object constructs: it extends standard object types with locking information. Through operational arguments, we establish that well-typed programs do not have race conditions.

A static-analysis technique such as ours is necessarily incomplete. In practice, it probably should be complemented with mechanisms for escaping its requirements, that is, with means for asserting that program fragments do not have race conditions even when these fragments do not typecheck. Complementarily, we are currently investigating type systems more sophisticated and liberal than the one presented in this paper. In the context of Java, we are also considering implementations of our methods.

Acknowledgments

We thank Andrew Gordon and Paul Hankin for permission to adapt their macros and inference rules.

Appendix: Formal Semantics

The formal semantics of our calculus closely follows that given by Gordon and Hankin. It consists of a group of structural congruence rules ($\equiv$), which permit the rearrangement of terms, and a group of reduction rules ($\rightarrow$), which model proper computation steps. In addition to the rules listed here, we also use a set of rules that imply that $\equiv$ is a congruence relation.

Structural congruence rules

(Struct Res $\mathcal{E}$)

$$\frac{p \notin fn(\mathcal{E}) \cup bn(\mathcal{E})}{(\nu p)\mathcal{E}[\, a\,] \equiv \mathcal{E}[\,(\nu p)a\,]}$$

(Struct Par $\mathcal{E}$)

$$\frac{fn(a) \cap bn(\mathcal{E}) = \varnothing}{a \upharpoonright \mathcal{E}[\, b\,] \equiv \mathcal{E}[\, a \upharpoonright b\,]}$$

Reduction rules

(Red Let)

$$\frac{}{let\ x{=}p\ in\ b \ \rightarrow\ b\{\!\{x \leftarrow p\}\!\}}$$

(Red Select)

$$\frac{d = [\ell_i = \varsigma(x_i)b_i\ ^{i \in 1..n}]^m \quad j \in 1..n}{(p \mapsto d) \upharpoonright p.\ell_j \ \rightarrow\ (p \mapsto d) \upharpoonright b_j\{\!\{x_j \leftarrow p\}\!\}}$$

(Red Update)

$$\frac{d = [\ell_i = \varsigma(x_i)b_i\ ^{i \in 1..n}]^m \quad j \in 1..n \quad d' = [\ell_j = \varsigma(x)b, \ell_i = \varsigma(x_i)b_i\ ^{i \in (1..n)-\{j\}}]^m}{(p \mapsto d) \upharpoonright (p.\ell_j \Leftarrow \varsigma(x)b) \ \rightarrow\ (p \mapsto d') \upharpoonright p}$$

(Red Lock) (where $[\ldots] = [\ell_i = \varsigma(x_i)b_i\ ^{i \in 1..n}]$)

$$\frac{}{(p \mapsto [\ldots]^\circ) \upharpoonright lock\ p\ in\ a \ \rightarrow\ (p \mapsto [\ldots]^\bullet) \upharpoonright locked\ p\ in\ a}$$

(Red Locked) (where $[\ldots] = [\ell_i = \varsigma(x_i)b_i\ ^{i \in 1..n}]$)

$$\frac{}{(p \mapsto [\ldots]^\bullet) \upharpoonright locked\ p\ in\ u \ \rightarrow\ (p \mapsto [\ldots]^\circ) \upharpoonright u}$$

(Red $\mathcal{E}$)

$$\frac{a \rightarrow a'}{\mathcal{E}[\,a\,] \rightarrow \mathcal{E}[\,a'\,]}$$

(Red Struct)

$$\frac{a \equiv a' \quad a' \rightarrow b' \quad b' \equiv b}{a \rightarrow b}$$

References

1. Martín Abadi and Luca Cardelli. *A Theory of Objects.* Springer-Verlag, 1996.
2. Ken Arnold and James Gosling. *The Java Programming Language.* Addison-Wesley, 1996.
3. David Aspinall and Adriana Compagnoni. Subtyping dependent types. In *Proceedings of the 11th Annual IEEE Symposium on Logic in Computer Science*, pages 86–97, July 1996.
4. Gérard Berry and Gérard Boudol. The chemical abstract machine. *Theoretical Computer Science*, 96:217–248, 1992.
5. Andrew D. Birrell. An introduction to programming with threads. Research Report 35, Digital Equipment Corporation Systems Research Center, 1989.
6. Paolo Di Blasio and Kathleen Fisher. A calculus for concurrent objects. In *CONCUR'96: Concurrency Theory*, volume 1119 of *Lecture Notes in Computer Science*, pages 655–670. Springer Verlag, 1996.
7. Luca Cardelli. Phase distinctions in type theory. Unpublished, 1988.
8. Cormac Flanagan and Martín Abadi. Types for safe locking. In *Proceedings of the 8th European Symposium on Programming, ESOP '99*, pages 91–108. Springer-Verlag, March 1999.
9. Andrew D. Gordon and Paul D. Hankin. A concurrent object calculus: Reduction and typing. In Uwe Nestmann and Benjamin C. Pierce, editors, *HLCL '98: High-Level Concurrent Languages*, volume 16.3 of *Electronic Notes in Theoretical Computer Science.* Elsevier Science Publishers, 1998. An extended version appears as Technical Report No. 457 of the University of Cambridge Computer Laboratory, February 1999.
10. James Gosling, Bill Joy, and Guy L. Steele. *The Java Language Specification.* Addison-Wesley, 1996.
11. Robert Harper, John C. Mitchell, and Eugenio Moggi. Higher-order modules and the phase distinction. In *Conference Record of the Seventeenth Annual ACM Symposium on Principles of Programming Languages*, pages 341–354, January 1990.
12. Martin Hofmann, Wolfgang Naraschewski, Martin Steffen, and Terry Stroup. Inheritance of proofs. *Theory and Practice of Object Systems*, 4(1):51–69, 1998.
13. Naoki Kobayashi. A partially deadlock-free typed process calculus. In *Proceedings of the 12th Annual IEEE Symposium on Logic in Computer Science*, pages 128–139, 1997.
14. Greg Nelson, editor. *Systems Programming in Modula-3.* Prentice Hall, 1991.
15. Nicholas Sterling. Warlock: A static data race analysis tool. In *USENIX Winter Technical Conference*, pages 97–106, 1993.
16. Eijiro Sumii and Naoki Kobayashi. A generalized deadlock-free process calculus. In Uwe Nestmann and Benjamin C. Pierce, editors, *HLCL '98: High-Level Concurrent Languages*, volume 16.3 of *Electronic Notes in Theoretical Computer Science.* Elsevier Science Publishers, 1998.

Open Bisimulations on Chi Processes

Yuxi Fu*

Department of Computer Science
Shanghai Jiaotong University, Shanghai 200030, China

Abstract. The paper carries out a systematic investigation into the axiomatization problem of the asymmetric chi calculus. As a crucial step in attacking the problem, an open style bisimilarity is defined for each of the eighteen L-bisimilarities and the two are proved to be equal. On top of the open bisimilarities, explicit definitions of the eighteen L-congruences are given, which suggest immediately possible axioms for the congruence relations. In addition to the axioms for strong bisimilarity, the paper proposes altogether twenty one additional axioms, three of which being the well-known tau laws and the other eighteen being new. These axioms help to lift a complete system for the strong bisimilarity to complete systems for the eighteen L-congruences.

1 Introduction

The χ-calculus ([1, 2, 5]) is a recent addition to the family of calculi of mobile process ([8]). It is a process algebraic formalization of reaction graph ([4]). The latter is proposed to emphasize the graphical aspect of concurrent computational objects. The language is a further step towards a more abstract model of concurrent computation. One of its novel features is a uniform treatment of names. Uniformity supports the idea that there should be no difference between input and output prefixes. The followings are examples of communication in χ:

$$(x)(a[x].P|\overline{a}[y].Q|R) \xrightarrow{\tau} P[y/x]|Q[y/x]|R[y/x] \tag{1}$$

$$(x)(\overline{a}[x].P|a[y].Q|R) \xrightarrow{\tau} P[y/x]|Q[y/x]|R[y/x] \tag{2}$$

$$(x)a[x].P|\overline{a}[y].Q \xrightarrow{\tau} P[y/x]|Q \tag{3}$$

$$(x)\overline{a}[x].P|a[y].Q \xrightarrow{\tau} P[y/x]|Q \tag{4}$$

Here $a[x].P$ and $\overline{a}[y].Q$ are processes in prefix form, in which x and y are global. In $(x)(a[x].P|\overline{a}[y].Q|R)$ the name x is local as it is restricted by a localization operator (x). In (1) and (2) the interactions between $a[x].P$ and $\overline{a}[y].Q$ cause the local name x to be replaced by y throughout the term over which the localization operator (x) applies. In (3) and (4) the interactions do not affect Q as it is not restricted by (x). The four reductions should demonstrate the symmetry of communications in χ-calculus.

* Supported by NNSFC (69873032) and 863 Hi-Tech Project (863-306-ZT06-02-2).

If one insists that there should be a difference between positive prefix operation $a[x]$ and negative prefix operation $\overline{a}[x]$ then one obtains an asymmetric version of χ-calculus. In asymmetric χ-calculus, reductions (1) and (3) are admissible whereas reductions (2) and (4) are illegal. Asymmetric and polyadic versions of χ-calculus have been studied by Parrow and Victor in [10, 11].

The equational theory of mobile processes has attracted a lot of attention. Lin has axiomatized successfully some weak congruences on mobile processes ([7]). He concluded that Milner's three tau laws are enough to lift system for strong congruences to system for weak congruences in calculus of mobile processes. So far all complete systems for weak congruences on mobile processes are essentially of symbolic nature. An alternative is used by Sangiorgi in his study of open bisimulation ([12]). Compared to symbolic approach, the open approach has the virtue of simplicity. Strong open bisimilarity on finite mobile processes can be easily axiomatized. Axiomatization of weak open congruence however has not been seriously considered.

In this paper we answer some of the open problems in the theory of mobile processes. Our main contributions are as follows:

- The paper improves our understanding of the asymmetric χ-calculus by studying open bisimilarities. For each of the eighteen distinct L-bisimilarities, we introduce an open bisimilarity that coincides with the L-bisimilarity.
- Axiomatization for L-congruences on asymmetric χ-processes has not been considered before. We give in this paper complete systems for all the eighteen distinct L-congruences. Our result brings out the importance of the open counterparts of L-congruences.
- In [11, 3] attempts have been made to give complete systems for weak congruence on polyadic χ-processes, and respectively, four L-congruences on symmetric χ-processes. In this paper it is pointed out that all the proofs establishing the claimed completeness are wrong. A way to correct the mistake is proposed .
- As a byproduct, the paper provides a complete system for barbed congruence ([9]) on asymmetric χ-processes. It is demonstrated that bisimulation lattice is of great help in obtaining such a system.
- Axiomatization for weak open congruence on π-processes has not been paid enough attention. The approach used in this paper can be applied to give immediately a complete system for weak open congruence on π-processes.
- The paper refutes the general belief that Milner's three tau-laws are sufficient, in calculi of mobile processes, to lift a complete system for a strong congruence to a complete system for the corresponding weak congruence. This is related to the failure of Hennessy Lemma in such calculi.

Due to space restriction, most of the proofs are omitted in this extended abstract. The proofs given are sketchy. Some of the intermediate lemmas are excluded. A much more detailed account can be found in the full paper [6]. In the rest of the paper we will leave out the adjective in "asymmetric χ".

2 Background

Let $\mathcal{N}$ be the set of names ranged over by small case letters and $\overline{\mathcal{N}}$ the set $\{\overline{x} \mid x \in \mathcal{N}\}$ of conames. The Greek letter α ranges over $\mathcal{N} \cup \overline{\mathcal{N}}$. For $\alpha \in \mathcal{N} \cup \overline{\mathcal{N}}$, $\overline{\alpha}$ is defined as a if $\alpha = \overline{a}$ and as $\overline{a}$ if $\alpha = a$. The χ-processes are defined by the following abstract grammar: $P := \mathbf{0} \mid \alpha[x].P \mid P|P \mid (x)P \mid [x{=}y]P \mid P{+}P$. Most of the combinators have completely the same reading as those of the π-calculus. The name x in $(x)P$ is local. A name is global in P if it is not local in P. For instance the name x in both $a[x].P$ and $\overline{a}[x].P$ is global. We will write $gn(P)$ for the set of global names in P. As this paper is mainly concerned with axiomatization of finite χ-processes, we have omitted the replication operator. The set of χ-processes will be denoted by $\mathcal{C}$. The well-known α-convention will be adopted.

Let δ range over the set $\{\tau\} \cup \{m[x], \overline{m}[x], mx, \overline{m}(x), [y/x], (y/x] \mid m, x, y \in \mathcal{N}\}$ of transition labels and μ over $\{\tau\} \cup \{m[x], \overline{m}[x], mx, \overline{m}(x) \mid m, x \in \mathcal{N}\}$. In $(y/x]$, x and y must be different. A name in δ is local if it appears as x in $\overline{m}(x)$ or $(x/y]$; it is global otherwise. Let $ln(\delta)$, respectively $gn(\delta)$, denote the set of local, respectively global, names appearing in δ; and let $n(\delta)$ denote the set of names in δ. The sets $ln(\mu)$, $gn(\mu)$ and $n(\mu)$ are defined accordingly.

The following rules define the operational semantics of χ-calculus:

$$\frac{}{\alpha[x].P \xrightarrow{\alpha[x]} P} \qquad \frac{P \xrightarrow{\delta} P'}{[x{=}x].P \xrightarrow{\delta} P'} \qquad \frac{P \xrightarrow{\delta} P'}{P{+}Q \xrightarrow{\delta} P'}$$

$$\frac{P \xrightarrow{\mu} P' \quad ln(\mu) \cap gn(Q) = \emptyset}{P|Q \xrightarrow{\mu} P'|Q}$$

$$\frac{P \xrightarrow{[y/x]} P'}{P|Q \xrightarrow{[y/x]} P'|Q[y/x]} \qquad \frac{P \xrightarrow{(y/x]} P' \quad y \notin gn(Q)}{P|Q \xrightarrow{(y/x]} P'|Q[y/x]}$$

$$\frac{P \xrightarrow{mx} P' \quad Q \xrightarrow{\overline{m}[x]} Q'}{P|Q \xrightarrow{\tau} P'|Q'} \qquad \frac{P \xrightarrow{mx} P' \quad Q \xrightarrow{\overline{m}(x)} Q' \quad x \notin gn(P)}{P|Q \xrightarrow{\tau} (x)(P'|Q')}$$

$$\frac{P \xrightarrow{m[x]} P' \quad Q \xrightarrow{\overline{m}[y]} Q'}{P|Q \xrightarrow{[y/x]} P'[y/x]|Q'[y/x]} \qquad \frac{P \xrightarrow{m[x]} P' \quad Q \xrightarrow{\overline{m}(y)} Q' \quad y \notin gn(P)}{P|Q \xrightarrow{(y/x]} P'[y/x]|Q'[y/x]}$$

$$\frac{P \xrightarrow{\delta} P' \quad x \notin n(\delta)}{(x)P \xrightarrow{\delta} (x)P'} \qquad \frac{P \xrightarrow{m[x]} P' \quad x \neq m}{(x)P \xrightarrow{my} P'[y/x]} \qquad \frac{P \xrightarrow{\overline{m}[x]} P' \quad x \neq m}{(x)P \xrightarrow{\overline{m}(x)} P'}$$

$$\frac{P \xrightarrow{[y/x]} P' \quad x \neq y}{(x)P \xrightarrow{\tau} P'} \qquad \frac{P \xrightarrow{[y/x]} P' \quad x \neq y}{(y)P \xrightarrow{(y/x]} P'} \qquad \frac{P \xrightarrow{(y/x]} P'}{(x)P \xrightarrow{\tau} (y)P'} \qquad \frac{P \xrightarrow{[x/x]} P'}{(x)P \xrightarrow{\tau} (x)P'}$$

The semantics is different from the one in [10, 3]. Here $\overline{m}[x].P|m[x].Q \xrightarrow{[x/x]} P|Q$ but not $\overline{m}[x].P|m[x].Q \xrightarrow{\tau} P|Q$. In [10, 3], $[x/x]$ is identified with τ; here they are different.

Some notations need be fixed before we proceed to next section. Let $\Longrightarrow$ be the reflexive and transitive closure of $\stackrel{\tau}{\longrightarrow}$. We will write $\stackrel{\delta}{\Longrightarrow}$ for $\Longrightarrow\stackrel{\delta}{\longrightarrow}\Longrightarrow$. We will also write $\stackrel{\widehat{\delta}}{\Longrightarrow}$ for $\stackrel{\delta}{\Longrightarrow}$ if $\delta \neq \tau$ and for $\Longrightarrow$ otherwise. The notation $\Longrightarrow_x$ will stand for $\Longrightarrow\stackrel{[x/x]}{\longrightarrow}\Longrightarrow\stackrel{[x/x]}{\longrightarrow}\ldots\Longrightarrow\stackrel{[x/x]}{\longrightarrow}\Longrightarrow$, where the transition $\stackrel{[x/x]}{\longrightarrow}$ occurs zero or a finite number of times. It follows from definition that $\Longrightarrow\subseteq\Longrightarrow_x$. For simplification $\stackrel{\delta}{\Longrightarrow}\Longrightarrow_x$ will be abbreviated to $\stackrel{\delta}{\Longrightarrow}_x$. An atomic substitution of y for x is denoted by $[y/x]$. A general substitution σ is the composition of atomic substitutions, whose effect on a process P is defined by $P[y_1/x_1]\ldots[y_n/x_n] \stackrel{\text{def}}{=} (P[y_1/x_1]\ldots[y_{n-1}/x_{n-1}])[y_n/x_n]$. The composition of zero atomic substitution is an empty substitution $[]$ whose effect on a process is vacuous. A sequence of names $x_1, \ldots, x_n$ will be abbreviated as $\boldsymbol{x}$; and consequently $(x_1)\ldots(x_n)P$ will be abbreviated to $(\boldsymbol{x})P$. When the length of $\boldsymbol{x}$ is zero, $(\boldsymbol{x})P$ is just P.

In the rest of the paper M and N, and their indexed forms, denote finite lists of equalities $x{=}y$. Let M be $x_1{=}y_1, \ldots, x_n{=}y_n$. Then $[M]P$ denotes $[x_1{=}y_1]\ldots[x_n{=}y_n]P$. If M logically implies N, we write $M \Rightarrow N$; and if both $M \Rightarrow N$ and $N \Rightarrow M$ we write $M \Leftrightarrow N$. If M is an empty list, it plays the role of logical truth, in which case $[M]P$ is just P. Clearly a list M of match equalities defines an equivalence relation on the set $n(M)$ of names appearing in M. We use σ_M to denote an arbitrary substitution that sends all members of an equivalence class to a representative of that class and sends a name not in $n(M)$ to itself. For a finite number of processes P_i, $i \in I$, we write $\sum_{i \in I} P_i$ for $P_1 + \ldots + P_n$. We have leave out the parentheses in $P_1 + \ldots + P_n$ as $+$ is associative both semantically and proof theoretically.

In order to axiomatize the congruence relations of this paper, we need to internalize, as it were, the labels of the transition system. In the following definition a is fresh: $\alpha(x).P \stackrel{\text{def}}{=} (x)\alpha[x].P$, where $x \notin \{\alpha, \overline{\alpha}\}$; $\tau.P \stackrel{\text{def}}{=} (a)[a/a].P$; $[y/x].P \stackrel{\text{def}}{=} (a)(\overline{a}[y]|a[x].P)$; $(y/x].P \stackrel{\text{def}}{=} (y)[y/x].P$, where $x \neq y$. The prefix $[y/x]$, first introduced in [1, 10], is called an update. It is clear from definition that both x and y in $[y/x].P$ are global. On the other hand the y in the restricted update $(y/x].P$ is local.

We state below some technical results to be used in the rest of the paper. The proofs of which are simple inductions on derivation.

Lemma 1. *Let $n(\sigma)$ denote the names appearing in the substitution σ.*
(i) If $n(\sigma) \cap ln(\mu) = \emptyset$ and $P \stackrel{\mu}{\longrightarrow} P'$ then $P\sigma \stackrel{\mu\sigma}{\longrightarrow} P'\sigma$.
(ii) If $P \stackrel{[y/x]}{\longrightarrow} P'$ then $P\sigma \stackrel{[y\sigma/x\sigma]}{\longrightarrow} P'\sigma[y\sigma/x\sigma]$.
(iii) If $y \notin n(\sigma)$ and $P \stackrel{(y/x]}{\longrightarrow} P'$ then $P\sigma \stackrel{(y/x\sigma]}{\longrightarrow} P'\sigma[y/x\sigma]$.

Lemma 2. *If $(x)P \Longrightarrow (y)P'$, then either $P \Longrightarrow_x P'$ and $x = y$, or $P \Longrightarrow_x \stackrel{(y/x]}{\Longrightarrow}_y P'$, or, for $y_1, \ldots, y_n$, $n \geq 1$, $P \Longrightarrow_x \stackrel{(y_1/x]}{\Longrightarrow}_{y_1} \stackrel{(y_2/y_1]}{\Longrightarrow}_{y_2} \cdots \stackrel{(y_n/y_{n-1}]}{\Longrightarrow}_{y_n} \stackrel{(y/y_n]}{\Longrightarrow}_y P'$.*

We refer the reader to [2, 3] for more on the semantics of χ-calculus.

3 Bisimulation Lattice

Bisimulation equalities are the finest equivalence relation on processes. For a particular process calculus, there is not just one weak bisimulation equality but a whole range of them. These equalities differ in the extent actions are admitted. In practice one uses one bisimulation equality in preference to others because the processes one is interested in are capable of performing only certain kinds of actions.

We will define a class of bisimulation equalities on χ-processes induced by different sets of admissible actions. For that purpose, we introduce the following notations. Let fo denote the set $\{\overline{a}[x] \mid a, x \in \mathcal{N}\}$ of free outputs, fi the set $\{a[x] \mid x \in \mathcal{N}\}$ of free inputs, i the set $\{ax \mid x \in \mathcal{N}\}$ of inputs, ro the set $\{\overline{a}(x) \mid a, x \in \mathcal{N}\}$ of restricted outputs, u the set $\{[y/x] \mid x, y \in \mathcal{N}\}$ of updates and ru the set $\{(y/x) \mid x, y \in \mathcal{N}\}$ of restricted updates. Define $\mathcal{L}$ as $\{\cup S \mid S \subseteq \{fo, fi, i, ro, u, ru\} \wedge S \neq \emptyset\}$.

Contexts are certain processes with a hole. They are inductively defined as follows: (i) $[]$ is a context; (ii) if $C[]$ is a context then $C[]|P$, $P|C[]$, $(x)C[]$ and $\alpha[x].C[]$ are contexts. A binary relation $\mathcal{R}$ on $\mathcal{C}$ is closed under context if $P\mathcal{R}Q$ implies $C[P]\mathcal{R}C[Q]$ for every context $C[]$. It is closed under substitution if $P\mathcal{R}Q$ implies $P\sigma\mathcal{R}Q\sigma$ for every substitution σ.

Definition 3. *Let $\mathcal{R}$ be a binary symmetric relation on $\mathcal{C}$ and L be an element of $\mathcal{L}$. The relation $\mathcal{R}$ is an L-relation if whenever $P\mathcal{R}Q$ and $P \xrightarrow{\phi} P'$, for $\phi \in L \cup \{\tau\}$, then some Q' exists such that $Q \stackrel{\widehat{\phi}}{\Longrightarrow} Q'\mathcal{R}P'$. An open L-relation is an L-relation that is closed under substitution. An L-bisimulation is an L-relation that is closed under context. The L-bisimilarity, notation $\approx_L$, is the largest L-bisimulation.*

According to Definition 3, P is L-bisimilar to Q if an admissible action ϕ of P, that is $\phi \in L$, can be simulated by the same action from Q up to tau actions and vice versa. Closedness under context guarantees that L-bisimilarity is stable with respect to all but the summation operation.

Theorem 4. *If σ is a substitution, $L \in \mathcal{L}$, $P \approx_L Q$ and $O \in \mathcal{C}$ then (i) $\alpha[x].P \approx_L \alpha[x].Q$; (ii) $P|O \approx_L Q|O$; (iii) $(x)P \approx_L (x)Q$; (iv) $[x{=}y]P \approx_L [x{=}y]Q$; (v) $P\sigma \approx_L Q\sigma$.*

There are altogether 63 L-bisimilarities. Not all of them are distinct. The next theorem reveals the full picture of the order relationship among them.

Theorem 5. *Suppose $L, L_1, L_2 \in \mathcal{L}$. Then the following properties hold:*
(i) $\approx_L \subseteq \approx_i$ and if $L \neq i$ then the inclusion is strict.
(ii) $\approx_{L_1} \not\subseteq \approx_{L_2}$ if either $(fi \cap L_1 = \emptyset) \wedge (fi \subseteq L_2)$ or $(ru \cap L_1 = \emptyset) \wedge (ru \subseteq L_2)$ or $((ru \cup ro) \cap L_1 = \emptyset) \wedge (ro \subseteq L_2)$ or $(u \cap L_1 = \emptyset) \wedge (u \subseteq L_2)$ or $((u \cup fo) \cap L_1 = \emptyset) \wedge (fo \subseteq L_2)$.
(iii) $\approx_{ru} \subset \approx_{ro}$ and $\approx_u \subset \approx_{fo}$. Both inclusions are strict.

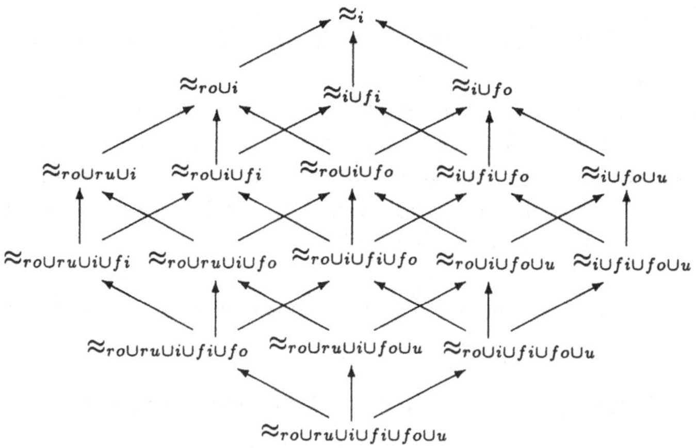

Fig. 1. The Bisimulation Lattice of Chi Calculus

It follows from Theorem 5 that there are altogether 18 distinct L-bisimilarities. These are described in Figure 1. In the diagram an arrow indicates a strict inclusion. Each labeled node is the principal representative of a number of L-bisimilarities that coincide. For instance, the node labeled by $\approx_{ro\cup ru\cup i\cup fi}$ is the principal representative of the equivalence class $\{\approx_L \mid ru \cup fi \subseteq L \subseteq ro \cup ru \cup i \cup fi$ and $L \in \mathcal{L}\}$. The order structure induced by the arrow relation is called the bisimulation lattice of χ-calculus. Due to space limitation we will be concentrating on $\approx_{ro\cup ru\cup i\cup fi\cup fo\cup u}$ and $\approx_{ro\cup ru\cup i\cup fi\cup fo}$ in this paper. Extensive studies of the other sixteen can be found in the full paper.

The proof of Theorem 5 is sketched in [3]. In this paper we give some examples to support the inequality claimed in the theorem. These examples are far more general than the ones in [3] in the sense that they are axiom generating.

Suppose $x \neq y$ and $L \cap fi = \emptyset$. Then

$$m(y).(P + [y/x].Q) \not\approx_{fi} m(y).(P + [y/x].Q) + m[x].Q[x/y] \tag{5}$$
$$m(y).(P + [y/x].Q) \approx_{L} m(y).(P + [y/x].Q) + m[x].Q[x/y] \tag{6}$$

Now (5) is obvious whereas (6) is slightly more subtle. None of the actions confined in L can tell the two processes apart. The reduction $(m(y).(P+[y/x].Q)+m[x].Q[x/y])|\overline{m}[z] \xrightarrow{[z/x]} Q[z/y][z/x]$, for instance, is matched up by $m(y).(P + [y/x].Q)|\overline{m}[z] \xrightarrow{\tau} [z/x].Q[z/y] \xrightarrow{[z/x]} Q[z/y][z/x]$.

Suppose Q cannot perform any restricted updates up to tau actions. Then

$$(y/x].Q + \tau.Q[x/y] \not\approx_{ru} \tau.Q[x/y] \tag{7}$$
$$(y/x].(P + (z/y].Q) \not\approx_{ru} (y/x].(P + (z/y].Q) + (z/x].Q[z/y] \tag{8}$$
$$(y/x].Q + \tau.Q[x/y] \approx_{L} \tau.Q[x/y] \tag{9}$$
$$(y/x].(P + (z/y].Q) \approx_{L} (y/x].(P + (z/y].Q) + (z/x].Q[z/y] \tag{10}$$

when $L \cap ru = \emptyset$. Both (7) and (8) are obvious. Intuitively (9) holds because if the admissible actions are confined in L then the first action of $(y/x].Q$ is not fireable. For it to be activated the global name x must be localized in a context. But then the action invoked by $(y/x]$ amounts to substituting the local y for the local x, the side effect being the same as that of applying the α-conversion. For similar reason (10) holds because replacing x by the local z in $Q[z/y]$ results in the same process as the one obtained by first substituting the local y for x in Q and then substituting the local z for y.

If Q can perform a restricted output action not matched up by P then

$$\overline{a}(x).(P + (y/x].Q) \not\approx_{ro} \overline{a}(x).(P + (y/x].Q) + \overline{a}(y).Q[y/x] \tag{11}$$
$$\overline{a}(x).(P + (y/x].Q) \approx_{L} \overline{a}(x).(P + (y/x].Q) + \overline{a}(y).Q[y/x] \tag{12}$$

when $L \cap (ru \cup ro) = \emptyset$. The inequality (11) holds because $\overline{a}(y).Q[y/x]$ can perform two consecutive restricted output actions not matchable by $\overline{a}(x).(P + (y/x].Q)$. The equality (12) holds as $\overline{a}(y).Q[y/x]$ can only be involved in a communication when restricted output and restricted update actions are banned.

Suppose $L \cap u = \emptyset$ and $x \neq y$. Let A be $[x/x].(P_1 + [y/x].Q)$ and B be $[y/y].(P_2 + [y/x].Q)$. Then

$$A + B \not\approx_{u} A + B + [y/x].Q \tag{13}$$
$$A + B \approx_{L} A + B + [y/x].Q \tag{14}$$

The inequality (13) is obvious. To understand (14) notice that if updates are banned then the component $[y/x].Q$ can be initiated when at least one of x and y is localized. If x is localized then $[x/x].(P_1 + [y/x].Q)$ can simulate $[y/x].Q$ and if y is localized then it is for $[y/y].(P_2 + [y/x].Q)$ to do the job.

Suppose C is $[x/x].(P_1 + \overline{a}[x].Q[x/y])$ and D is $[y/y].(P_2 + [x/y].Q)$. Then

$$C + \overline{a}(y).(P + D) \not\approx_{fo} C + \overline{a}(y).(P + D) + \overline{a}[x].Q[x/y] \tag{15}$$
$$C + \overline{a}(y).(P + D) \approx_{L} C + \overline{a}(y).(P + D) + \overline{a}[x].Q[x/y] \tag{16}$$

when $L \cap (u \cup fo) = \emptyset$. Now (15) is clear. Justification of (16) is as follows: If the component $\overline{a}[x].Q[x/y]$ induces a restricted output (update) action, then $[x/x].(P_1 + \overline{a}[x].Q[x/y])$ can simulate the action by performing first a tau action and then a restricted output (update). For example

$$a[z]|(x)(C + \overline{a}(y).(P + D) + \overline{a}[x].Q[x/y]) \xrightarrow{(x/z]} \mathbf{0}|Q[x/y][x/z]$$

is simulated by $a[z]|(x)(C + \overline{a}(y).(P + D)) \xrightarrow{\tau} a[z]|(x)(P_1 + \overline{a}[x].Q[x/y]) \xrightarrow{(x/z]} \mathbf{0}|Q[x/y][x/z]$. If the component $\overline{a}[x].Q[x/y]$ is involved in a communication as in $(z)a[z].R|(C + \overline{a}(y).(P + D) + \overline{a}[x].Q[x/y]) \xrightarrow{\tau} R[x/z]|Q[x/y]$. Then $\overline{a}(y).(P + [y/y].(P_2 + [x/y].Q))$ will put itself into action. The simulating sequence is:

$$\begin{aligned}(z)a[z].R|(C + \overline{a}(y).(P + D)) &\xrightarrow{\tau} (y)(R[y/z]|(P + [y/y].(P_2 + [x/y].Q))) \\ &\xrightarrow{\tau} (y)(R[y/z]|(P_2 + [x/y].Q)) \\ &\xrightarrow{\tau} R[x/z]|Q[x/y].\end{aligned}$$

The reader is advised to play with these examples before moving on.

4 Open Bisimilarities

The idea of open bisimilarity ([12]) is this: In order to show P and Q to be bisimilar, all one needs to consider are substitution instances of the pair. As a process contains only a finite number of names, it is usually enough to consider only a finite number of substitution instances. This is the basic reason for the effectiveness of open bisimilarity. The adjective "open" refers to the fact that in this approach the global names appearing in a process are treated very much like the free variables in, say, an open λ-term.

In this section we will define, for each $L \in \mathcal{L}$, an open bisimilarity that coincides with the L-bisimilarity. The proofs of coincidence not only support our definitions of open bisimulations for χ-processes but also reveals much deeper properties of L-bisimilarities from the technical point of view.

Definition 6. *(i) An open $ro \cup ru \cup i \cup fi \cup fo \cup u$-bisimulation is the same as an open $ro \cup ru \cup i \cup fi \cup fo \cup u$-relation.*

(ii) An open $ro \cup ru \cup i \cup fi \cup fo$-relation $\mathcal{R}$ is an open $ro \cup ru \cup i \cup fi \cup fo$-bisimulation if the following properties hold for P and Q whenever $P\mathcal{R}Q$:

- *If $P \xrightarrow{[x/x]} P'$ then Q' exists such that either $Q \Longrightarrow_x Q'\mathcal{R}P'$ or some $y_1, \ldots, y_n$, $n \geq 1$, exist such that $Q \Longrightarrow_x \overset{(y_1/x]}{\Longrightarrow}_{y_1} \overset{(y_2/y_1]}{\Longrightarrow}_{y_2} \cdots \overset{(y_n/y_{n-1}]}{\Longrightarrow}_{y_n} Q'$ and $Q'[x/y_n]\mathcal{R}P'$.*
- *If $P \xrightarrow{[y/x]} P'$, where $x \neq y$, then Q' exists such that either $Q \overset{[y/x]}{\Longrightarrow} Q'\mathcal{R}P'$ or both of the following properties hold:*
  - *either $Q \overset{[x/x]}{\Longrightarrow}_x \overset{[y/x]}{\Longrightarrow} Q'\mathcal{R}P'$ or some $y_1, \ldots, y_n$, for $n \geq 1$, exist such that $Q \Longrightarrow_x \overset{(y_1/x]}{\Longrightarrow}_{y_1} \overset{(y_2/y_1]}{\Longrightarrow}_{y_2} \cdots \overset{(y_n/y_{n-1}]}{\Longrightarrow}_{y_n} \overset{[y/y_n]}{\Longrightarrow} Q'\mathcal{R}P'$;*
  - *either $Q \overset{[y/y]}{\Longrightarrow}_y \overset{[y/x]}{\Longrightarrow} Q'\mathcal{R}P'$ or some Q'', $z_1, \ldots, z_m$, for $m \geq 1$, exist such that $Q''[y/z_m]\mathcal{R}P'$ and $Q \Longrightarrow_y \overset{(z_1/y]}{\Longrightarrow}_{z_1} \overset{(z_2/z_1]}{\Longrightarrow}_{z_2} \cdots \overset{(z_m/z_{m-1}]}{\Longrightarrow}_{z_m} \overset{[z_m/x]}{\Longrightarrow} Q''$.*

For each $L \in \mathcal{L}$, the open L-bisimilarity $\approx^L_{open}$ is the largest open L-bisimulation.

The above definition is not easy to digest. For motivations, the reader should check up the examples given in the previous section against the definition.

Theorem 7. *Suppose $L \in \mathcal{L}$. Then $\approx^L_{open}$ coincides with $\approx_L$.*

Proof. First we show that $\approx_L \subseteq \approx^L_{open}$. Let a be fresh throughout the proof.

(i) It follows from Definition 6 and Theorem 4 that $\approx_{ro \cup ru \cup i \cup fi \cup fo \cup u}$ is an open $ro \cup ru \cup i \cup fi \cup fo \cup u$-bisimulation.

(ii) Suppose $P \approx_{ro \cup ru \cup i \cup fi \cup fo} Q$ and $P \xrightarrow{[x/x]} P'$. Then $(x)(P|a[x]) \xrightarrow{\tau}\xrightarrow{ax} P'|\mathbf{0}$ must be matched up by $(x)(Q|a[x]) \overset{ax}{\Longrightarrow} Q'|\mathbf{0}$ for some Q' such that $P' \approx_{ro \cup ru \cup i \cup fi \cup fo} Q'$. Using Lemma 2, one derives that either $Q\sigma \Longrightarrow_x Q'\mathcal{R}P'$ or some $y_1, \ldots, y_n$, for $n \geq 1$, exist such that $Q\sigma \Longrightarrow_x \overset{(y_1/x]}{\Longrightarrow}_{y_1} \overset{(y_2/y_1]}{\Longrightarrow}_{y_2} \cdots \overset{(y_n/y_{n-1}]}{\Longrightarrow}_{y_n} Q'$ and $Q'[x/y_n]\mathcal{R}P'$.

If $P \xrightarrow{[y/x]} P'$ and $x \neq y$, then $(x)(P|a[x]) \xrightarrow{\tau} P'|a[y]$. To match up the reduction, there must exist some Q' such that $(x)(Q|a[x]) \overset{\tau}{\Longrightarrow} Q'|a[y]$. According to Lemma 2, there are only three possibilities: either $Q \stackrel{[y/x]}{\Longrightarrow} Q' \approx_{roUruUiUfiUfo} P'$, or $Q \stackrel{[x/x]}{\Longrightarrow}_x \stackrel{[y/x]}{\Longrightarrow} Q' \approx_{roUruUiUfiUfo} P'$, or $Q \Longrightarrow_x \stackrel{(y_1/x]}{\Longrightarrow}_{y_1} \stackrel{(y_2/y_1]}{\Longrightarrow}_{y_2} \cdots \stackrel{(y_n/y_{n-1}]}{\Longrightarrow}_{y_n} \stackrel{[y/y_n]}{\Longrightarrow} Q' \approx_{roUruUiUfiUfo} P'$ for some $y_1, \ldots, y_n$, $n \geq 1$. If the first case is not possible, there must also exist some Q'' such that $(y)(Q|a[x]) \stackrel{(y/x]}{\Longrightarrow} Q''|a[y]$ matches up $(y)(P|a[x]) \stackrel{(y/x]}{\longrightarrow} P'|a[y]$. Clearly $(y)(Q|a[x]) \stackrel{(y/x]}{\Longrightarrow} Q''|a[y]$ can be factorized as $(y)(Q|a[x]) \Longrightarrow (z)(Q_1|a[x]) \equiv (y)(Q_1[y/z]|a[x]) \stackrel{(y/x]}{\longrightarrow} Q_2|a[y] \Longrightarrow Q''|a[y]$. By Lemma 2, either $Q\sigma \stackrel{[y/y]}{\Longrightarrow}_y \stackrel{[y/x]}{\Longrightarrow} Q'\mathcal{R}P'$ or some Q'', $z_1, \ldots, z_m$, for $m \geq 1$, exist such that $Q''[y/z_m]\mathcal{R}P'$ and $Q\sigma \Longrightarrow_y \stackrel{(z_1/y]}{\Longrightarrow}_{z_1} \stackrel{(z_2/z_1]}{\Longrightarrow}_{z_2} \cdots \stackrel{(z_m/z_{m-1}]}{\Longrightarrow}_{z_m} \stackrel{[z_m/x]}{\Longrightarrow} Q''$. Notice that $Q\sigma \stackrel{[y/x]}{\Longrightarrow} Q'\mathcal{R}P'$ is impossible by assumption.

The inclusion $\approx^L_{open} \subseteq \approx_L$ amounts to showing that, for each $L \in \mathcal{L}$, $\approx^L_{open}$ is an L-bisimulation. □

With the help of the above theorem we can now define L-congruences by exploiting the explicit requirement in the definition of open L-bisimilarities.

Definition 8. *Two processes P and Q are L-congruent, written $P =_L Q$, if $P \approx^L_{open} Q$ and, for every substitution σ and every $\phi \in ru \cup u \cup \{\tau\}$, $P\sigma \stackrel{\phi}{\longrightarrow} P'$ must be matched up by a nonempty sequence of actions from Q and vice versa.*

So for instance, if $P =_{roUruUiUfiUfo} Q$ then it should not be the case that the only way Q can simulate $P \stackrel{[x/x]}{\longrightarrow} P'$ is by the vacuous action $Q \Longrightarrow Q$.

5 Prefix Laws

Let AS be the system given in Figure 2 plus the equivalence rules, the congruence rules and the following expansion law, in which π and γ range over $\{\tau\} \cup \{\alpha[x], [y/x] \mid x, y \in \mathcal{N}\}$:

$$P|Q = \sum_{i \in I} [M_i](\boldsymbol{x})\pi_i.(P_i|Q) + \sum_{\substack{\pi_i = a_i[x_i] \\ \gamma_j = \overline{b_j}[y_j]}} [M_i][N_j](\boldsymbol{x})(\boldsymbol{y})[a_i{=}b_j][y_j/x_i].(P_i|Q_j)$$

$$+ \sum_{j \in J} [N_j](\boldsymbol{y})\gamma_j.(P|Q_j) + \sum_{\substack{\pi_i = \overline{a_i}[x_i] \\ \gamma_j = b_j[y_j]}} [M_i][N_j](\boldsymbol{x})(\boldsymbol{y})[a_i{=}b_j][x_i/y_j].(P_i|Q_j)$$

where P is $\sum_{i \in I}[M_i](\boldsymbol{x})\pi_i.P_i$, Q is $\sum_{j \in J}[N_j](\boldsymbol{y})\gamma_j.Q_j$ and $\{\boldsymbol{x}\} \cap \{\boldsymbol{y}\} = \emptyset$. The second component in the right hand of the above equality captures the idea that whenever π_i is of the form $a_i[x_i]$ for some $i \in I$ and γ_j is of the form $\overline{b_j}[y_j]$ for some $j \in J$ then there is a summand $[M_i][N_j](\boldsymbol{x})(\boldsymbol{y})[a_i{=}b_j][y_j/x_i].(P_i|Q_j)$.

We will write $AS \cup \{R_1, \ldots, R_n\} \vdash P = Q$ to mean that the equality $P = Q$ is derivable from the axioms and rules of AS together with axioms and rules $R_1, \ldots, R_n$. When no confusion arises, we simply write $P = Q$. We will also write $P \stackrel{R}{=} Q$ to indicate that R is the major axiom applied to derive $P = Q$.

L1	$(x)\mathbf{0} = \mathbf{0}$	
L2	$(x)\alpha[y].P = \mathbf{0}$	$x \in \{\alpha, \overline{\alpha}\}$
L3	$(x)\alpha[y].P = \alpha[y].(x)P$	$x \notin \{y, \alpha, \overline{\alpha}\}$
L4	$(x)(y)P = (y)(x)P$	
L5	$(x)[y{=}z]P = [y{=}z](x)P$	$x \notin \{y, z\}$
L6	$(x)(P{+}Q) = (x)P{+}(x)Q$	
L7	$(x)[x{=}y]P = \mathbf{0}$	
L8	$(x)[y/x].P = \tau.P[y/x]$	
L9	$(x)[y/z].P = [y/z].(x)P$	$x \notin \{y, z\}$
L10	$(x)\tau.P = \tau.(x)P$	
M1	$[M]P = [N]P$	if $M \Leftrightarrow N$
M2	$[x{=}y]P = [x{=}y]P[y/x]$	
M3	$[x{=}y](P{+}Q) = [x{=}y]P{+}[x{=}y]Q$	
S1	$P{+}\mathbf{0} = P$	
S2	$P{+}Q = Q{+}P$	
S3	$P{+}(Q{+}R) = (P{+}Q){+}R$	
S4	$[x{=}y]P{+}P = P$	
U1	$[y/x].P = [y/x].[x{=}y]P$	
MD1	$[x{=}y].\mathbf{0} = \mathbf{0}$	derivable from S1 and S4
MD2	$[x{=}x].P = P$	derivable from M1
MD3	$[M]P = [M](P\sigma_M)$	derivable from M2
SD1	$P{+}P = P$	derivable from MD2 and S4
SD2	$[M]P{+}P = P$	derivable from S-rules
UD1	$[y/x].P = [y/x].P[y/x]$	derivable from U1 and M2

Fig. 2. Axioms for Strong Bisimilarity on Chi Processes

In AS every process P can be converted to a normal form process of the following shape: $\sum_{i\in I_1}[M_i]\alpha_i[x_i].P_i + \sum_{i\in I_2}[M_i]\alpha_i(x).P_i + \sum_{i\in I_3}[M_i][y_i/x_i].P_i + \sum_{i\in I_4}[M_i](y/x_i].P_i + \sum_{i\in I_5}[M_i]\tau.P_i$, in which neither x nor y appears global in P and P_i is in normal form for each $i \in I_1 \cup I_2 \cup I_3 \cup I_4 \cup I_5$. Here I_1, I_2, I_3, I_4 and I_5 are pairwise disjoint finite indexing sets.

AS is sound and complete for strong bisimilarity whose definition we omit, but see [10]. In order to lift AS to complete systems for L-congruences, we propose 17 axioms as given in Figure 3. We call them prefix laws as they are mainly dealing with prefix combinators. The first three are the well-known tau laws. We have seen axioms P4 and P14 in [3]. The other twelve axioms are new. In what follows, AS_τ denotes $AS \cup \{P1, P2, P3\}$.

6 Saturation Property

In the standard proof of completeness theorem for weak congruence on finite CCS processes, one verifies first that every normal form process is provably equal to a saturated normal form process using the three tau laws. Recall that a process P is saturated if , for every α, $P \xrightarrow{\alpha} P'$ whenever $P \overset{\alpha}{\Longrightarrow} P'$. Now if P and Q are weakly congruent saturated normal form processes and $P \xrightarrow{\alpha} P'$ then

P1	$\delta.\tau.P = \delta.P$
P2	$P + \tau.P = \tau.P$
P3	$\delta.(P + \tau.Q) = \delta.(P + \tau.Q) + \delta.Q$
P4	$a(y).(P + [y/x].Q) = a(y).(P + [y/x].Q) + a[x].Q[x/y]$
P5	$(y/x].P + \tau.P[x/y] = \tau.P[x/y]$
P6	$(y/x].(P + (z/y].Q) = (y/x].(P + (z/y].Q) + (z/x].Q[z/y]$
P7	$\overline{a}(x).(P + (y/x].Q) = \overline{a}(x).(P + (y/x].Q) + \overline{a}(y).Q[y/x]$
P8	$A + B = A + B + [y/x].Q$
P9	$C + \overline{a}(y).(P + D) = C + \overline{a}(y).(P + D) + \overline{a}[x].Q[x/y]$
P10	$E + a(y).(P + F) = E + a(y).(P + F) + a[x].Q[x/y]$
P11	$G + a(y).(P + F) = G + a(y).(P + F) + a[x].Q[x/y]$
P12	$(y/x].(P + [z/y].Q) = (y/x].(P + [z/y].Q) + [z/x].Q[z/y]$
P13	$(y/x].(P + a[y].Q) = (y/x].(P + a[y].Q) + a[x].Q[x/y]$
P14	$\overline{a}(x).(P + [y/x].Q) = \overline{a}(x).(P + [y/x].Q) + \overline{a}[y].Q[y/x]$
P15	$(y/x].P + [x/x].P[x/y] = [x/x].P[x/y]$
P16	$(y/x].P + [x/x].P[x/y] = (y/x].P$
P17	$[x/x].P + \tau.P = \tau.P$

In P8, $A \equiv [x/x].(P_1 + [y/x].Q)$ and $B \equiv [y/y].(P_2 + [y/x].Q)$.
In P9, $C \equiv [x/x].(P_1 + \overline{a}[x].Q[x/y])$ and $D \equiv [y/y].(P_2 + [x/y].Q)$.
In P10, $E \equiv [x/x].(P_1 + a[x].Q[x/y])$ and $F \equiv [y/y].(P_2 + [y/x].Q)$.
In P11, $G \equiv [x/x].(P_1 + a(y).(Q_1 + [x/x].(Q_2 + [y/x].Q)))$ and $F \equiv [y/y].(P_2 + [y/x].Q)$.

Fig. 3. The Prefix Laws

$Q \stackrel{\alpha}{\Longrightarrow} Q'$ for some Q' such that $Q' \approx P'$, where $\approx$ denotes weak bisimilarity. By saturation, $Q \stackrel{\alpha}{\longrightarrow} Q'$ and therefore $\alpha.Q'$ is a summand of Q. If, and this is a nontrivial if, we can deduce by induction hypothesis that $\alpha.P'$ is provably equal to $\alpha.Q'$, then we can conclude that every summand of P is provably equal to a summand of Q, and vice versa. This gives us the required completeness.

If one is focusing only on completeness proof, then the notion of saturated process is a distraction. All one really needs is the following saturation property:

> If $P \stackrel{\alpha}{\Longrightarrow} P'$ for normal form P, then P and $P + \alpha.P'$ are provably equal.

This is the first of the two crucial properties a completeness proof rests upon. Another one is to be discussed in next section. These properties suffice to establish the following absorption property:

> If two normal form processes P and Q are congruent then $P + Q$ is provably equal to P.

Of course, under the same assumption, $P + Q$ is also provably equal to Q. Hence the completeness.

In χ-calculus, a basic saturation lemma would say that P and $P + \delta.P'$ are provably equal whenever $P \stackrel{\delta}{\Longrightarrow} P'$ for normal form process P. But this is far from sufficient. Suppose $P =_{ro \cup ru \cup i \cup fi \cup fo} Q$ for normal form processes P and Q.

Suppose further that $[x/x].P'$ is a summand of P. Then some Q' must exist such that either $P' \approx_{ro \cup ru \cup iu \cup fi \cup fo} Q'$ and $Q \Longrightarrow_x Q'$ or $P' \approx_{ro \cup ru \cup iu \cup fi \cup fo} Q'[x/y_n]$ and some $y_1, \ldots, y_n$, for $n \geq 1$, exist such that $Q \Longrightarrow_x \overset{(y_1/x]}{\Longrightarrow}_{y_1} \overset{(y_2/y_1]}{\Longrightarrow}_{y_2} \cdots \overset{(y_n/y_{n-1}]}{\Longrightarrow}_{y_n} Q'$. In the former case we have, by the basic saturation lemma, that Q is provably equal to $Q + [x/x].Q'$. In the latter case we would also like to say the same. But it no longer follows from the basic saturation lemma. Extra axiom are necessary to derive the equality $Q = Q + [x/x].Q'$.

Lemma 9. *Suppose Q is in normal form. Then the following properties hold:*
(1) If $Q\sigma_M \overset{\tau}{\Longrightarrow} Q'$ then $AS_\tau \vdash Q = Q + [M]\tau.Q'$.
(2) If $Q\sigma_M \overset{\alpha[x]}{\Longrightarrow} Q'$ then $AS_\tau \vdash Q = Q + [M]\alpha[x].Q'$.
(3) If $z \notin gn(Q) \cup n(M)$ and $Q\sigma_M \overset{az}{\Longrightarrow} Q'$ then $AS_\tau \vdash Q = Q + [M]a(z).Q'$.
(4) If $z \notin gn(Q) \cup n(M)$ and $Q\sigma_M \overset{\overline{a}(z)}{\Longrightarrow} Q'$ then $AS_\tau \vdash Q = Q + [M]\overline{a}(z).Q'$.
(5) If $Q\sigma_M \overset{[y/x]}{\Longrightarrow} Q'$ then $AS_\tau \vdash Q = Q + [M][y/x].Q'$.
(6) If $y \notin gn(Q) \cup n(M)$ and $Q\sigma_M \overset{(y/x]}{\Longrightarrow} Q'$ then $AS_\tau \vdash Q = Q + [M](y/x].Q'$.
(7) If $Q\sigma_M \overset{\tau}{\Longrightarrow}_x Q'$ or $Q\sigma_M \overset{[x/x]}{\Longrightarrow}_x Q'$ then $AS_\tau \cup \{P8, P17\} \vdash Q = Q + [M][x/x].Q'$.
(8) If $Q\sigma_M \Longrightarrow_x \overset{(y_1/x]}{\Longrightarrow}_{y_1} \overset{(y_2/y_1]}{\Longrightarrow}_{y_2} \cdots \overset{(y_n/y_{n-1}]}{\Longrightarrow}_{y_n} Q'$ then $Q = Q + [M][x/x].Q'[x/y_n]$ is provable in the system $AS_\tau \cup \{P8, P16, P17\}$.

Proof. (8) Suppose $Q\sigma_M \Longrightarrow_x Q_1 \overset{(y_1/x]}{\Longrightarrow} Q_2 \cdots Q_{2n-1} \overset{(y_n/y_{n-1}]}{\Longrightarrow} Q_{2n} \Longrightarrow_{y_n} Q'$. Now $x \notin Q_2$, $x \notin Q_3$; $x, y_1 \notin Q_4$, $x, y_1 \notin Q_5$; …; $x, y_1, \ldots, y_{n-2} \notin Q_{2n-2}$, $x, y_1, \ldots, y_{n-2} \notin Q_{2n-1}$; $x, y_1, \ldots, y_{n-1} \notin Q_{2n}$, $x, y_1, \ldots, y_{n-1} \notin Q'$. Therefore $Q_{2i-2}[x/y_{i-1}] \Longrightarrow_x Q_{2i-1}[x/y_{i-1}] \overset{(y_i/x]}{\Longrightarrow} Q_{2i}$, for $2 \leq i \leq n$, and $Q_{2n}[x/y_n] \Longrightarrow_x Q'[x/y_n]$ by Lemma 1. With these observations one obtains the following inference, assuming $Q\sigma_M \Longrightarrow_x Q_1$ is not vacuous:

$$\begin{aligned}
Q &= Q + [M][x/x].(Q_1 + (y_1/x].Q_2) \\
&\overset{P16}{=} Q + [M][x/x].(Q_1 + (y_1/x].Q_2 + [x/x].Q_2[x/y_1]) \\
&= Q + [M]([x/x].(Q_1 + (y_1/x].Q_2 + [x/x].Q_2[x/y_1]) + [x/x].Q_2[x/y_1]) \\
&= Q + [M][x/x].Q_2[x/y_1] \\
&\ \ \vdots \\
&= Q + [M][x/x].Q'[x/y_n]
\end{aligned}$$

where the first equality holds by (6) and (7) of this lemma; the third equality is a consequence of P8. If $Q\sigma_M \Longrightarrow_x Q_1$ is vacuous then

$$\begin{aligned}
Q &= Q + [M](y_1/x].Q_2 \\
&\overset{P16}{=} Q + [M]((y_1/x].Q_2 + [x/x].Q_2[x/y_1]) \\
&= Q + [M][x/x].Q_2[x/y_1].
\end{aligned}$$

So the previous inference is valid anyway. □

Pr1	$\tau.P = \tau.(P + \sum_{i\in I}[M_i]\tau.P)$
Pr2	$\tau.P = \tau.(P + \sum_{i\in I_1}[M_i]\tau.P + \sum_{i\in I_2}[M_i](w/x_i].P)$
Pr3	$\tau.P = \tau.(P + \sum_{i\in I_1}[M_i]\tau.P + \sum_{i\in I_2}[M_i][x_i/x_i].P)$
Pr4	$\tau.P = \tau.(P + \sum_{i\in I_1}[M_i]\tau.P + \sum_{i\in I_2}[M_i](w/x_i].P + \sum_{i\in I_3}[M_i][x_i/x_i].P)$
In the above axioms, w is fresh, I, I_1, I_2, I_3 are finite indexing sets.	

Fig. 4. The Promotion Axioms

7 Promotion Property

In the proof of completeness theorem for weak congruence in CCS, the following result, due to Hennessy, plays a crucial role:

If $P \approx Q$ then either $\tau.P = Q$ or $P = Q$ or $P = \tau.Q$.

Here $\approx$ is the weak bisimilarity and $=$ is the congruence induced by $\approx$. In the proof of the completeness theorem by induction, Hennessy Lemma helps to lift $P \approx Q$ to either $\tau.P = Q$ or $P = Q$ or $P = \tau.Q$, thus allowing the induction hypothesis to apply. In π-calculus however Hennessy Lemma does not hold! For a counter example, consider the following three propositions

$$\tau.(\bar{a}x + [x{=}y]\tau.\bar{a}y) = \bar{a}x$$
$$\bar{a}x + [x{=}y]\tau.\bar{a}y = \bar{a}x$$
$$\bar{a}x + [x{=}y]\tau.\bar{a}y = \tau.\bar{a}x$$

None of them holds although $\bar{a}x + [x{=}y]\tau.\bar{a}y \approx \bar{a}x$ is true. This example explains the reason why nobody has given a proof that Sangiorgi's system ([12]) together with Milner's tau laws constitute a complete system for weak open congruence. We believe that the resulting system is not capable of establishing the equality $\tau.(\bar{a}x + [x{=}y]\tau.\bar{a}y) = \tau.\bar{a}x$.

The purpose of this section is to present our solution. The motivation comes from a careful examination of the role of Hennessy Lemma in CCS. What it really comes down to is the following promotion property:

If $P \approx Q$ for normal form processes P and Q then $\tau.P = \tau.Q$ is provable.

Motivated by this observation, we introduce four additional axioms as given in Figure 4. In the presence of a suitable set of prefix laws, these four axioms are capable of lifting AS to a complete system for an L-congruence. For this reason we call them promotion axioms. Clearly both Pr2 and Pr3 subsume Pr1 and are subsumed by Pr4.

Theorem 10. *Suppose P and Q are in normal form.*
(i) If $P \approx_{ro\cup ru\cup i\cup fi\cup fo\cup u} Q$ then $AS_\tau \cup \{Pr1\} \vdash \tau.P = \tau.Q$.
(ii) If $P \approx_{ro\cup ru\cup i\cup fi\cup fo} Q$ then $AS_\tau \cup \{P8, P16, P17, Pr3\} \vdash \tau.P = \tau.Q$.

Proof. We prove (ii) only. Suppose $P \approx_{ro\cup ru\cup i\cup fi\cup fo} Q$ for normal form processes P and Q. The proof is carried out by induction on the sum of the depths of P and Q. Let P be of the form $\sum_{i\in I_1}[M_i]\alpha_i[x_i].P_i + \sum_{i\in I_2}[M_i]\alpha_i(x).P_i + \sum_{i\in I_3}[M_i][y_i/x_i].P_i + \sum_{i\in I_4}[M_i](y/x_i).P_i + \sum_{i\in I_5}[M_i]\tau.P_i$ and Q be of the form $\sum_{j\in J_1}[N_j]\alpha_j[x_j].Q_j + \sum_{j\in J_2}[N_j]\alpha_j(x).Q_j + \sum_{j\in J_3}[N_j][y_j/x_j].Q_j + \sum_{j\in J_4}[N_j](y/x_j].Q_j + \sum_{j\in J_5}[N_j]\tau.Q_j$.

Suppose, for $i \in I_3$, $y_i\sigma_{M_i} = x_i\sigma_{M_i}$ and $([M_i][y_i/x_i].P_i)\sigma_{M_i} \stackrel{[y_i\sigma_{M_i}/x_i\sigma_{M_i}]}{\longrightarrow} P_i\sigma_{M_i}$ can only be matched up vacuously by $Q\sigma_{M_i}$. Then $AS_\tau \cup \{P8, P16, P17\} \vdash \tau.P_i\sigma_{M_i} = \tau.Q\sigma_{M_i}$ by induction hypothesis. It follows that

$$AS_\tau \cup \{P8, P16, P17\} \vdash [M_i][y_i/x_i].P_i = [M_i][x_i/x_i].Q.$$

Suppose, for some $i \in I_3$, $y_i\sigma_{M_i} \neq x_i\sigma_{M_i}$. Using Lemma 9 and axiom P8, one can show that $AS_\tau \cup \{P8, P16, P17\} \vdash [M_i][y_i/x_i].P_i + Q = Q$.

Similarly one shows that some $I' \subseteq I_5$ exists such that the equality $[M_i]\tau.P_i = [M_i]\tau.Q$ is provable in $AS_\tau \cup \{P8, P16, P17\}$ for each $i \in I'$ and that $AS_\tau \cup \{P8, P16, P17\} \vdash [M_i]\tau.P_i + Q = Q$ for each $i \in I_5 \setminus I'$.

It is also clear that $AS_\tau \cup \{P8, P16, P17\} \vdash [M_i]\alpha_i[x_i].P_i + Q = Q$, respectively $AS_\tau \cup \{P8, P16, P17\} \vdash [M_i]\alpha_i(x).P_i + Q = Q$, $AS_\tau \cup \{P8, P16, P17\} \vdash [M_i](y/x_i].P_i + Q = Q$, if $[M_i]\alpha_i[x_i].P_i$, respectively $[M_i]\alpha_i(x).P_i$, $[M_i](y/x_i].P_i$, is a summand of P.

We can now conclude that $P + Q = Q + \Sigma_{i\in I'}[M_i]\tau.Q + \Sigma_{i\in I}[M_i][x_i/x_i].Q$ is provable in the system $AS_\tau \cup \{P8, P16, P17\}$ for some $I \subseteq I_3$ and $I' \subseteq I_5$. Now $\tau.Q = \tau.(Q + \Sigma_{i\in I'}[M_i]\tau.Q + \Sigma_{i\in I}[M_i][x_i/x_i].Q)$ by Pr3. It follows that $AS_\tau \cup \{P8, P16, P17, Pr3\} \vdash \tau.(P + Q) = \tau.Q$.

Symmetrically $AS_\tau \cup \{P8, P16, P17, Pr3\} \vdash \tau.(P + Q) = \tau.P$. Therefore $AS_\tau \cup \{P8, P16, P17, Pr3\} \vdash \tau.P = \tau.Q$. □

8 Completeness Theorem

Having done all the preparations, we finally come to the completeness theorem. Its proof is so similar to the proof of the promotion lemma reported in the previous section as to render any reiteration redundant.

Theorem 11. *(i) $AS_\tau \cup \{Pr1\}$ is sound and complete for $=_{ro\cup ru\cup i\cup fi\cup fo\cup u}$. (ii) $AS_\tau \cup \{P8, P16, P17, Pr3\}$ is sound and complete for $=_{ro\cup ru\cup i\cup fi\cup fo}$.*

Proof. Let's see how (ii) is proved. The soundness is easy. Suppose both P and Q are in normal form and $P =_{ro\cup ru\cup i\cup fi\cup fo} Q$. Using (ii) of Theorem 10 and its proof, one concludes that $P + Q = Q + \Sigma_{i\in I'}[M_i]\tau.Q + \Sigma_{i\in I}[M_i][x_i/x_i].Q$ is provable in the system $AS_\tau \cup \{P8, P16, P17\}$ for some $I \subseteq I_3$ and $I' \subseteq I_5$. But Q must be able to simulate a first move of P by a nonempty sequence of moves. This implies that both I and I' are empty. It follows that $P + Q = Q$ is provable in $AS_\tau \cup \{P8, P16, P17, Pr3\}$. Similarly $P + Q = P$ is provable in $AS_\tau \cup \{P8, P16, P17, Pr3\}$. Hence $AS_\tau \cup \{P8, P16, P17, Pr3\} \vdash P = Q$. □

Congruence	Axioms in Addition to $AS \cup \{P1, P2, P3\}$	
$=_{ro \cup ru \cup i \cup fi \cup fo \cup u}$		Pr1
$=_{ro \cup ru \cup i \cup fo \cup u}$	P4	Pr1
$=_{ro \cup i \cup fi \cup fo \cup u}$	P5, P6, P15	Pr2
$=_{i \cup fi \cup fo \cup u}$	P5, P7, P15	Pr2
$=_{ro \cup i \cup fo \cup u}$	P4, P5, P6, P15	Pr2
$=_{i \cup fo \cup u}$	P4, P5, P7, P15	Pr2
$=_{ro \cup ru \cup i \cup fi \cup fo}$	P8, P16, P17	Pr3
$=_{ro \cup ru \cup i \cup fi}$	P8, P9, P16, P17	Pr3
$=_{ro \cup ru \cup i \cup fo}$	P4, P8, P10, P11, P16, P17	Pr3
$=_{ro \cup ru \cup i}$	P4, P8, P9, P10, P11, P16, P17	Pr3
$=_{ro \cup i \cup fi \cup fo}$	P5, P6, P12, P15, P16	Pr4
$=_{i \cup fi \cup fo}$	P5, P7, P12, P15, P16	Pr4
$=_{ro \cup i \cup fo}$	P4, P5, P6, P13, P15, P16	Pr4
$=_{i \cup fo}$	P4, P5, P7, P13, P15, P16	Pr4
$=_{ro \cup i \cup fi}$	P5, P6, P9, P12, P15, P16	Pr4
$=_{ro \cup i}$	P4, P5, P6, P9, P13, P15, P16	Pr4
$=_{i \cup fi}$	P5, P7, P9, P14, P15, P16	Pr4
$=_{i}$	P4, P5, P7, P9, P13, P14, P15, P16	Pr4

Fig. 5. Summary of the 18 Completeness Systems

9 Concluding Remarks

The work reported in this paper consists of two parts. The first part is a continuation of the study of L-bisimilarities on asymmetric χ-processes initiated in [3]. The result of this investigation is a finer description of L-bisimilarities in terms of open L-bisimilarities. This alternative view leads immediately to the explicit definition of the largest congruence relation, the L-congruence, contained in an L-bisimilarity. Building upon the first part, the second part explains a streamlined approach to derive complete systems for L-congruences. In addition to the axioms and rules for strong bisimilarity, 21 axioms are proposed. It is shown that these are enough to lift a complete system for the strong bisimilarity to complete systems for L-congruences. Due to space restriction, the paper discusses only two L-bisimilarities. The definitions of the other sixteen open L-bisimilarities fit into the pattern of Definition 6. Most of them are even more complex. It should be emphasized that these definitions are not simply a matter of putting things together. In Figure 5 all the 18 complete systems for L-congruences are given.

It can be easily shown that the top element of the bisimulation lattice coincides with the barbed bisimilarity on the χ-processes. So we have in fact given a complete system for the barbed congruence. It is surprising that axiomatization of the barbed congruence is almost as difficult as axiomatization of all the 18 L-congruences. This points out the importance of the bisimulation lattice. Even if we do not care much about most of the L-congruences, we are forced to pay attention to them. The author certainly could not have discovered the axioms for barbed congruence had he not discovered the bisimulation lattice of χ-calculus. As a digression, we remark that barbed bisimilarity, which we believe is a very sensible equality, is usually much weaker than the 'traditional bisimilarity' we have in mind. The coincidence, as might be the case in π-calculus with binary choice operator, is an exception rather than the rule.

The mistake made in [11, 3] is caused by the false assumption that Hennessy Lemma held in calculi of mobile processes. The result of this paper indicates that none of the systems given in [11, 3] is likely to be complete for the intended congruence. It is apparent from our work that Sangiorgi's system for strong open congruence on π-processes can be extended to a complete system for weak open congruence on π-processes by adding the tau laws and the promotion axiom Pr1.

Among the 17 prefix laws, P8, P9, P10 and P11 are most unusual. They share structural similarity that is quite different from the structures of the three tau laws shared by the rest of the laws. We leave for future study the question of if these laws can be simplified. The rest of the prefix laws are quite satisfactory.

The promotion axioms are very interesting. It is worth investigating the possibility of simplifying them.

Finally we remark that the prefix laws are not independent. For instance P6 is subsumed by P7 in the system AS. This however does not mean that P6 is redundant. It is used for example in the complete system of $=_{ro\cup i\cup fi\cup fo\cup u}$ for which P7 is too strong.

References

1. Fu, Y.: The χ-Calculus. Proceedings of the International Conference on Advances in Parallel and Distributed Computing, March 19-21, Shanghai, IEEE Computer Society Press (1997) 74-81
2. Fu, Y.: A Proof Theoretical Approach to Communications. ICALP'97, July 7-11, Bologna, Italy, Lecture Notes in Computer Science 1256 (1997) 325–335
3. Fu, Y.: Bisimulation Lattice of Chi Processes. ASIAN'98, December 8-10, Manila, The Philippines, Lecture Notes in Computer Science 1538 (1998) 245–262
4. Fu, Y.: Reaction Graphs. Journal of Computer Science and Technology **13** (1998) 510–530
5. Fu, Y.: Variations on Mobile Processes. To appear in Theoretical Computer Science, Elsevier Science Publisher
6. Fu, Y.: Algebraic Theory of Chi Calculus. Preprint
7. Lin, H.: Complete Inference Systems for Weak Bisimulation Equivalences in the π-Calculus. TAPSOFT'95, Lecture Notes in Computer Science 915 (1995) 187–201
8. Milner, R., Parrow, J., Walker, D.: A Calculus of Mobile Processes. Information and Computation **100** (1992) Part I:1–40, Part II:41–77
9. Milner, R., Sangiorgi, D.: Barbed Bisimulation. ICALP'92, Lecture Notes in Computer Science 623 (1992) 685–695
10. Parrow, J., Victor, B.: The Update Calculus. AMAST'97, December 13-17, Sydney, Australia, Lecture Notes in Computer Science 1119 (1997) 389–405
11. Parrow, J., Victor, B.: The Tau-Laws of Fusion. CONCUR'98, Lecture Notes in Computer Science 1466 (1998) 99–114
12. Sangiorgi, D.: A Theory of Bisimulation for π-Calculus. Acta Informatica **3** (1996) 69–97

Rectangular Hybrid Games*

Thomas A. Henzinger, Benjamin Horowitz, and Rupak Majumdar

Department of Electrical Engineering and Computer Sciences
University of California, Berkeley, CA 94720-1770, USA
and
Max-Planck Institute for Computer Science
Im Stadtwald, 66123 Saarbrücken, Germany

{tah,bhorowit,rupak}@eecs.berkeley.edu

Abstract. In order to study control problems for hybrid systems, we generalize hybrid automata to *hybrid games* —say, controller vs. plant. If we specify the continuous dynamics by constant lower and upper bounds, we obtain *rectangular games*. We show that for rectangular games with objectives expressed in LTL (linear temporal logic), the winning states for each player can be computed, and winning strategies can be synthesized. Our result is sharp, as already reachability is undecidable for generalizations of rectangular systems, and optimal —singly exponential in the size of the game structure and doubly exponential in the size of the LTL objective. Our proof systematically generalizes the theory of hybrid systems from automata (single-player structures) [9] to games (multi-player structures): we show that the successively more general infinite-state classes of timed, 2D rectangular, and rectangular games induce successively weaker, but still finite, quotient structures called game bisimilarity, game similarity, and game trace equivalence. These quotients can be used, in particular, to solve the LTL control problem.

1 Introduction

A *hybrid automaton* [1] is a mathematical model for a system with both discretely and continuously evolving variables, such as a digital computer that interacts with an analog environment. An important special case of a hybrid automaton is the *rectangular automaton* [14], where the enabling condition for each discrete state change is a rectangular region of continuous states, and the first derivative of each continuous variable x is bounded by constants from below and above; that is, $\dot{x} \in [a, b]$. Rectangular automata are important for several reasons. First, they generalize *timed automata* [2] (for which $a = b = 1$) and naturally model real-time systems whose clocks have bounded drift. Second, they can over-approximate with arbitrary precision the behavior of hybrid automata

* This research was supported in part by the NSF CAREER award CCR-9501708, by the NSF grant CCR-9504469, by the DARPA (NASA Ames) grant NAG2-1214, by the DARPA (Wright-Patterson AFB) grant F33615-98-C-3614, and by the ARO MURI grant DAAH-04-96-1-0341.

with general linear and nonlinear continuous dynamics, as long as all derivatives satisfy the Lipschitz condition [11,22]. Third, they form a most general class of hybrid automata for which the LTL *model-checking problem* can be decided: given a rectangular automaton $\mathcal{A}$ and a formula φ of linear temporal logic over the discrete states of $\mathcal{A}$, it can be decided in polynomial space if all possible behaviors of $\mathcal{A}$ satisfy φ [14].

Since hybrid automata are often used to model digital controllers for analog plants, an important problem for hybrid automata is the LTL control problem: given a hybrid automaton $\mathcal{A}$ and an LTL formula φ, can the behaviors of $\mathcal{A}$ be "controlled" so as to satisfy φ? However, the hybrid automaton per se is an inadequate model for studying this problem because it does not differentiate between the capabilities of its individual components —the controller and the plant, if you wish. Since the control problem is naturally formalized in terms of a two-player game, we define *hybrid games*.[1] Because our setup is intended to be as general as possible, following [3,19], we do not distinguish between a "discrete player" (which directs discrete state changes) and a "continuous player" (which advances time); rather, in a hybrid game, each of the two players can itself act like a hybrid automaton. The game proceeds in an infinite sequence of rounds and produces an ω-sequence of states. In each round, both players independently choose enabled moves; the pair of chosen moves either results in a discrete state change, or in a passage of time during which the continuous state evolves. In the special case of a *rectangular game*, the enabling condition of each move is a rectangular region of continuous states, and when time advances, then the derivative of each continuous variable is governed by a constant differential inclusion. Now, the LTL *control problem* for hybrid games asks: given a hybrid game $\mathcal{B}$ and an LTL formula φ over the discrete states of $\mathcal{B}$, is there a strategy for player-1 so that all possible outcomes of the game satisfy φ?

Our main result shows that the LTL control problem can be decided for rectangular games. Previously, beyond the finite-state case, control problems have been solved only for the special case of timed games (which corresponds to timed automata) [6,16,20], and for rectangular games under the assumption that the controller can move only at integer points in time (sampling control) [13]. Semi-algorithms for control have also been proposed for more general linear [27] and nonlinear [18,26] hybrid games, but in these cases termination is not guaranteed. The algorithms for timed games and sampling control are based on the fact that the underlying state spaces can be partitioned into finitely many bisimilarity classes, and the controller does not need to distinguish between bisimilar states. Our argument is novel, because rectangular games in general do not have finite bisimilarity quotients. Our result is sharp, because the control problem for a class of hybrid games is at least as hard as the reachability problem for the corresponding class of hybrid automata, and reachability has been proved undecidable for several minor extensions of rectangular automata [14]. The complexity of our algorithm, which requires singly exponential time in the size of

[1] For the sake of simplicity, in this paper we restrict ourselves to the two-player case. All results generalize immediately to more than two players.

the game $\mathcal{B}$ and doubly exponential time in the size of the formula φ, is optimal, because control is harder than model checking: reachability control for timed games is EXPTIME hard [13]; LTL control for finite-state games is 2EXPTIME hard [24].

Let us now take a more detailed preview of our approach. For the solution of infinite-state model-checking problems, such as those of hybrid automata, it is helpful if there exists a finite quotient space that preserves the properties under consideration [9]. Specifically, provided the duration of time steps is invisible, every timed automaton is bisimilar to a finite-state automaton [2]; every 2D rectangular automaton (with two continuous variables) is similar (simulation equivalent) to a finite-state automaton [10]; and every rectangular automaton is trace equivalent to a finite-state automaton [14]. Since LTL model checking can be reduced to model checking on the trace-equivalence quotient, the decidability of LTL model checking for rectangular automata follows. The three characterizations are sharp; for example, the similarity quotient of 3D rectangular automata can be infinite [12], and therefore the quotient approach does not lead to branching-time model-checking algorithms for rectangular automata.

By introducing an appropriate generalization of trace equivalence, which we call *game trace equivalence*, the argument for LTL model checking of rectangular automata (single-player structures) can be systematically carried over to LTL control of rectangular games (two-player structures). This is done in two steps. First, we show that given the game trace equivalence $\equiv$ on the (possibly infinite) state space of a two-player structure $\mathcal{B}$, an appropriately defined quotient game $\mathcal{B}/_{\equiv}$ can be used to answer the LTL control problem for $\mathcal{B}$, and to synthesize the corresponding control strategies (Proposition 2). Second, following the arguments of [14], we show that if $\mathcal{B}$ is a rectangular game, then $\equiv$ has only finitely many equivalence classes, and consequently $\mathcal{B}/_{\equiv}$ is a finite-state game (Theorem 6). Our main result follows (Corollary 7). Along the way, we also generalize bisimilarity and similarity to *game bisimilarity* and *game similarity*, which are finer than game trace equivalence, and we show that the special case of timed games has finite game bisimilarity relations (Theorem 3), and the special case of 2D rectangular games has finite game similarity relations (Theorem 4). This gives, on one hand, better bounds on the number of equivalence classes for the special cases, and on the other hand, cleanly generalizes the entire theory of rectangular automata to rectangular games.

2 Using Games For Modeling Control

In this section, we define a standard model of discrete-event control using games with simultaneous moves and LTL objectives [5, 23], review some known results [25], and introduce several equivalences on the state space of such a game.

2.1 Game Structures and the LTL Control Problem

One player. A *transition structure* (or single-player structure)

$$\mathcal{F} = (Q, \Pi, \langle\!\langle\cdot\rangle\!\rangle, \mathit{Moves}, \mathit{Enabled}, \delta)$$

consists of a set Q of states, a set Π of observations, an observation function $\langle\!\langle\cdot\rangle\!\rangle\colon Q \to 2^{\Pi}$ which maps each state to a set of observations, a set *Moves* of moves, an enabling function *Enabled*: *Moves* $\to 2^Q$ which maps each move to the set of states in which it is enabled, and a partial transition function δ: $Q \times Moves \to 2^Q$ which maps each move m and each state in *Enabled*(m) to a set of possible successor states. For each state $q \in Q$, we write $mov(q) = \{m \in Moves \mid q \in Enabled(m)\}$ for the set of moves that are enabled in q. We require that $mov(q) \neq \emptyset$ for all $q \in Q$. A *step of* $\mathcal{F}$ is a triple $q \xrightarrow{m} q'$ such that $m \in mov(q)$ and $q' \in \delta(q, m)$. A *run of* $\mathcal{F}$ is an infinite sequence $r = s_0 s_1 s_2 \ldots$ of steps $s_j = q_j \xrightarrow{m_j} q'_j$ such that $q_{j+1} = q'_j$ for all $j \geq 0$. The state q_0 is called the *source* of r. The run r induces a *trace*, denoted $\langle\!\langle r\rangle\!\rangle$, which is the infinite sequence $\langle\!\langle q_0\rangle\!\rangle m_0 \langle\!\langle q_1\rangle\!\rangle m_1 \langle\!\langle q_2\rangle\!\rangle m_2 \ldots$ of alternating observation sets and moves. For a state $q \in Q$, the *outcome* R^q *from* q is the set of all runs of $\mathcal{F}$ with source q. For a set R of runs, we write $\langle\!\langle R\rangle\!\rangle$ for the set $\{\langle\!\langle r\rangle\!\rangle \mid r \in R\}$ of corresponding traces.

Two players. A (two-player) *game structure*

$$\mathcal{G} = (Q, \Pi, \langle\!\langle\cdot\rangle\!\rangle, Moves_1, Moves_2, Enabled_1, Enabled_2, \delta)$$

consists of the same components as above, only that $Moves_1$ ($Moves_2$) is the set of moves of player-1 (player-2), $Enabled_1$ maps $Moves_1$ to 2^Q, $Enabled_2$ maps $Moves_2$ to 2^Q, and the partial transition function δ: $Q \times Moves_1 \times Moves_2 \to 2^Q$ maps each move m_1 of player-1, each move m_2 of player-2, and each state in $Enabled_1(m_1) \cap Enabled_2(m_2)$ to a set of possible successor states. For $i = 1, 2$, we define mov_i: $Q \to 2^{Moves_i}$ to yield for each state q the set $mov_i(q) = \{m \in Moves_i \mid q \in Enabled_i(m)\}$ of player-i moves that are enabled in q. We require that $mov_i(q) \neq \emptyset$ for all $q \in Q$ and $i = 1, 2$. At each step of the game, player-1 chooses a move $m_1 \in mov_1(q)$ that is enabled in the current state q, player-2 simultaneously and independently chooses a move $m_2 \in mov_2(q)$ that is enabled in q, and the game proceeds nondeterministically to a new state in $\delta(q, m_1, m_2)$. Formally, a *step* of $\mathcal{G}$ is a step of the underlying transition structure

$$\mathcal{F}_{\mathcal{G}} = (Q, \Pi, \langle\!\langle\cdot\rangle\!\rangle, Moves_1 \times Moves_2, Enabled, \delta'),$$

where $Enabled(m_1, m_2) = Enabled_1(m_1) \cap Enabled_2(m_2)$ and $\delta'(q, (m_1, m_2)) = \delta(q, m_1, m_2)$. We refer to the runs and traces of $\mathcal{F}_{\mathcal{G}}$ as *runs and traces of the game structure* $\mathcal{G}$.

A *strategy for player-i* is a function f_i: $Q^+ \to 2^{Moves_i}$ such that $\emptyset \subsetneq f_i(w \cdot q) \subseteq mov_i(q)$ for every state sequence $w \in Q^*$ and every state $q \in Q$. The strategy f_i is *memory-free* if $f_i(w \cdot q) = f_i(w' \cdot q)$ for all $w, w' \in Q^*$ and $q \in Q$. Let f_1 (f_2) be a strategy for player-1 (player-2). The *outcome* $R^q_{f_1,f_2}$ *from state* $q \in Q$ *for* f_1 *and* f_2 is a subset of the runs of $\mathcal{G}$ with source q: a run $s_0 s_1 s_2 \ldots$ is in $R^q_{f_1,f_2}$ if for all $j \geq 0$, if $s_j = q_j \xrightarrow{(m_{1,j}, m_{2,j})} q'_j$, then $m_{i,j} \in f_i(q_0 q_1 \cdots q_j)$ for $i = 1, 2$, and $q_0 = q$.

Linear temporal logic. The *formulas of linear temporal logic* (LTL) are generated inductively by the grammar

$$\varphi \ ::= \ \pi \mid \neg\varphi \mid \varphi_1 \vee \varphi_2 \mid \bigcirc\varphi \mid \varphi_1 \mathcal{U} \varphi_2,$$

where $\pi \in \Pi$ is an observation, $\bigcirc$ is the *next* operator, and $\mathcal{U}$ is the *until* operator. From these operators, additional operators such as $\Diamond\varphi \stackrel{\Delta}{=} (\mathit{true}\,\mathcal{U}\varphi)$ and $\Box\varphi \stackrel{\Delta}{=} \neg\Diamond\neg\varphi$ can be defined as usual. The LTL formulas are interpreted over traces in the standard way [7]. For example, the formula $\Box\pi$ is satisfied by the trace $\langle\!\langle q_0\rangle\!\rangle m_0 \langle\!\langle q_1\rangle\!\rangle m_1 \langle\!\langle q_2\rangle\!\rangle m_2 \ldots$ if $\pi \in \langle\!\langle q_j\rangle\!\rangle$ for all $j \geq 0$.

Player-1 can *control the state q of a game structure for the* LTL *formula φ* if there exists a strategy f_1 of player-1 such that for every strategy f_2 of player-2 and every run $r \in R^q_{f_1,f_2}$, the trace $\langle\!\langle r\rangle\!\rangle$ satisfies φ.[2] In this case, we say that the strategy f_1 *witnesses the player-1 controllability of q for φ*. The LTL *control problem* asks, given a game structure $\mathcal{G}$ and an LTL formula φ, which states of $\mathcal{G}$ can be controlled by player-1 for φ. The LTL *controller-synthesis problem* asks, in addition, for the construction of witnessing strategies. If the game structure $\mathcal{G}$ is finite, then the LTL control problem is PTIME-complete in the size of $\mathcal{G}$ [17] and 2EXPTIME-complete in the size of φ [24]. Whereas for simple LTL formulas such as safety (for example, $\Box\pi$ for an observation $\pi \in \Pi$), controllability ensures the existence of memory-free witnessing strategies, this is not the case for arbitrary LTL formulas [25].

2.2 State Equivalences and Quotients for Game Structures

One player. The following equivalences on the states of a transition structure will motivate our definitions for game structures. Consider a transition structure $\mathcal{F} = (Q, \Pi, \langle\!\langle\cdot\rangle\!\rangle, \mathit{Moves}, \mathit{Enabled}, \delta)$. A binary relation $\preceq^s \subseteq Q \times Q$ is a *(forward) simulation* if $p \preceq^s q$ implies the following three conditions:

1. $\langle\!\langle p\rangle\!\rangle = \langle\!\langle q\rangle\!\rangle$;
2. $\mathit{mov}(p) \subseteq \mathit{mov}(q)$;
3. $\forall m \in \mathit{mov}(p).\ \forall p' \in \delta(p,m).\ \exists q' \in \delta(q,m).\ p' \preceq^s q'$.

We say that *p is (forward) simulated by q*, in symbols $p \preceq^S q$, if there is a simulation $\preceq^s$ with $p \preceq^s q$. We write $p \cong^S q$ if both $p \preceq^S q$ and $q \preceq^S p$. The relation $\cong^S$ is called *similarity*. A binary relation $\cong^b$ on Q is a *bisimulation* if $\cong^b$ is a symmetric simulation. Define $p \cong^B q$ if there is a bisimulation $\cong^b$ with $p \cong^b q$. The relation $\cong^B$ is called *bisimilarity*. A binary relation $\preceq^{-s}$ on Q is a *backward simulation* if $p' \preceq^{-s} q'$ implies the following three conditions:

1. $\langle\!\langle p'\rangle\!\rangle = \langle\!\langle q'\rangle\!\rangle$;
2. $\mathit{mov}(p') \subseteq \mathit{mov}(q')$;
3. $\forall p \in Q.\ \exists q \in Q.\ \forall m \in \mathit{mov}(p).\ p' \in \delta(p,m) \ \Rightarrow\ q' \in \delta(q,m) \wedge p \preceq^{-s} q$.

Then, *p' is backward simulated by q'*, in symbols $p' \preceq^{-S} q'$, if there is a backward simulation $\preceq^{-s}$ with $p' \preceq^{-s} q'$. A binary relation $\preceq^l$ on Q is a *trace containment* if $p \preceq^l q$ implies $\langle\!\langle R^p\rangle\!\rangle \subseteq \langle\!\langle R^q\rangle\!\rangle$. Define $p \preceq^L q$ if there is a trace containment

[2] Our choice to control for LTL formulas rather than, say, ω-automata [25] is arbitrary. In the latter case, only the complexity results must be modified accordingly.

$\preceq^l$ with $p \preceq^l q$. We write $p \cong^L q$ if both $p \preceq^L q$ and $q \preceq^L p$. The relation $\cong^L$ is called *trace equivalence.*

Two players. The basic local requirement behind the preorders $\preceq^S$ and $\preceq^L$ on the states of a transition structure is that if $p \preceq q$, then the move and the observation set of each step from p can be matched by a step from q (the two preorders differ in how they globalize this local requirement). For the corresponding preorders $\preceq_g$ on the states of a game structure, we generalize this to requiring that if $p \preceq_g q$, and player-1 can enforce a certain observation set by a certain move from q in one step, then player-1 can enforce the same observation set by the same move also from p in one step. This gives rise to the following definitions. Consider a game structure $\mathcal{G} = (Q, \Pi, \langle\!\langle\cdot\rangle\!\rangle, Moves_1, Moves_2, Enabled_1, Enabled_2, \delta)$. A binary relation $\preceq_g^s \subseteq Q \times Q$ is a *(forward player-1) game simulation* if $p \preceq_g^s q$ implies the following three conditions:[3]

1. $\langle\!\langle p\rangle\!\rangle = \langle\!\langle q\rangle\!\rangle$;
2. $mov_1(q) \subseteq mov_1(p)$ and $mov_2(p) \subseteq mov_2(q)$;
3. $\forall m_1 \in mov_1(q), m_2 \in mov_2(p), p' \in \delta(p, m_1, m_2).\ \exists q' \in \delta(q, m_1, m_2).\ p' \preceq_g^s q'$.

A binary relation $\preceq_g^{-s}$ on Q is a *backward (player-1) game simulation* if $p' \preceq_g^{-s} q'$ implies the following three conditions:

1. $\langle\!\langle p'\rangle\!\rangle = \langle\!\langle q'\rangle\!\rangle$;
2. $mov_1(q') \subseteq mov_1(p')$ and $mov_2(p') \subseteq mov_2(q')$;
3. $\forall p \in Q.\ \exists q \in Q.\ \forall m_1 \in mov_1(q).\ \forall m_2 \in mov_2(p).$
   $p' \in \delta(p, m_1, m_2) \Rightarrow q' \in \delta(q, m_1, m_2) \wedge p \preceq_g^{-s} q$.

A binary relation $\preceq_g^l$ on Q is a *(player-1) game trace containment* if $p \preceq_g^l q$ implies that for all strategies f_1 of player-1, there exists a strategy f_1' of player-1 such that for all strategies f_2' of player-2 there exists a strategy f_2 of player-2 such that $\langle\!\langle R^q_{f_1', f_2'}\rangle\!\rangle \subseteq \langle\!\langle R^p_{f_1, f_2}\rangle\!\rangle$. From this, the maximal preorders $\preceq_g^S$, $\preceq_g^{-S}$, and $\preceq_g^L$, as well as the equivalence relations *game similarity* $\cong_g^S$, *game bisimilarity* $\cong_g^B$, and *game trace equivalence* $\cong_g^L$ are defined as in the single-player case.[4] The following proposition, which follows immediately from the definitions, characterizes the game equivalences in terms of the underlying transition structure.

Proposition 1. *Two states p and q of a game structure $\mathcal{G}$ are game bisimilar (game similar, game trace equivalent) if p and q are bisimilar (similar, trace equivalent) in the underlying transition structure $\mathcal{F}_\mathcal{G}$.*

It follows that $\cong_g^B$ refines $\cong_g^S$, that $\cong_g^S$ refines $\cong_g^L$, and that in general these refinements are proper.[5] It also follows that the standard partition-refinement

[3] There is also a dual, player-2 game simulation, which we do not need in this paper.

[4] Note that, being symmetric, the game equivalences $\cong_g^S$, $\cong_g^B$, and $\cong_g^L$ are not indexed by a player (unlike the game preorders $\preceq_g^S$, $\preceq_g^{-S}$, and $\preceq_g^L$). In particular, say, $p \cong_g^L q$ implies that $mov_1(p) = mov_1(q)$ and $mov_2(p) = mov_2(q)$.

[5] We say that the equivalence relation $\equiv_1$ *(properly) refines* the equivalence relation $\equiv_2$ if $p \equiv_1 q$ implies $p \equiv_2 q$ (but not vice versa).

algorithms for computing bisimilarity [21] and similarity [10] can be applied also to compute the game bisimilarity and the game similarity relations.

Game trace-equivalence quotient. Consider two states p and q of a game structure $\mathcal{G}$. By definition, if $p \preceq_g^L q$, then for every LTL formula φ, if player-1 can control p for φ, then player-1 can control also q for φ. The relations with this property are called *alternating trace containments* [5] and differ from the game trace containments defined here in that the names of the moves of both players are not observable.[6] We keep all moves observable, and include the names of moves in the definition of traces, so that $p \preceq_g^L q$ implies if the strategy f_1 witnesses the player-1 controllability of p for φ, then the same strategy f_1 also witnesses the player-1 controllability of q for φ. Consequently, the game trace equivalence on the game structure $\mathcal{G}$ suggests a quotient structure that can be used for controller synthesis. Let $\equiv$ be any equivalence relation on the states of $\mathcal{G}$ which refines the game trace equivalence $\cong_g^L$. The *quotient structure* $\mathcal{G}/_\equiv$ is the game structure $(Q/_\equiv, \Pi, \langle\!\langle\cdot\rangle\!\rangle/_\equiv, \mathit{Moves}_1, \mathit{Moves}_2, \mathit{Enabled}_1/_\equiv, \mathit{Enabled}_2/_\equiv, \delta/_\equiv)$ with

- $Q/_\equiv = \{[q]_\equiv \mid q \in Q\}$ is the set of equivalence classes of $\equiv$;
- $\langle\!\langle [q]_\equiv \rangle\!\rangle/_\equiv = \langle\!\langle q \rangle\!\rangle$ (note that $\langle\!\langle\cdot\rangle\!\rangle/_\equiv$ is well defined because $\equiv$ refines $\cong_g^L$, and hence $\langle\!\langle\cdot\rangle\!\rangle$ is uniform within each equivalence class);
- $[q]_\equiv \in \mathit{Enabled}_1/_\equiv(m)$ if $\exists p \in [q]_\equiv \,.\, p \in \mathit{Enabled}_1(m)$ (note that this is equivalent to $\forall p \in [q]_\equiv \,.\, p \in \mathit{Enabled}_1(m)$ because $\equiv$ refines $\cong_g^L$), and analogously for $\mathit{Enabled}_2/_\equiv(m)$;
- $[q']_\equiv \in \delta([q]_\equiv, m_1, m_2)/_\equiv$ if $\exists p' \in [q']_\equiv \,.\, \exists p \in [q]_\equiv \,.\, p' \in \delta(p, m_1, m_2)$.

The following proposition reduces control for an LTL formula φ in the game structure $\mathcal{G}$ to control for φ in the quotient structure $\mathcal{G}/_\equiv$.

Proposition 2. *Let $\mathcal{G}$ be a game structure, let q be a state of $\mathcal{G}$, and let $\equiv$ be an equivalence relation on the states of $\mathcal{G}$ which refines the game trace equivalence for $\mathcal{G}$. Player-1 can control q for φ in $\mathcal{G}$ if and only if player-1 can control $[q]_\equiv$ for φ in $\mathcal{G}/_\equiv$. Moreover, if the strategy f_1 witnesses the player-1 controllability of $[q]_\equiv$ for φ in $\mathcal{G}/_\equiv$, then the strategy f_1' defined by $f_1'(p_0 \ldots p_j) \triangleq f_1([p_0]_\equiv \ldots [p_j]_\equiv)$ witnesses the player-1 controllability of q for φ in $\mathcal{G}$.*

3 Control of Rectangular Games

In this section, we apply the framework developed in the previous section to a particular class of infinite-state game structures: rectangular hybrid games. We show that for every rectangular game, the game trace-equivalence quotient is finite. It follows from Proposition 2 that the LTL control and controller-synthesis problems are decidable for rectangular games.

[6] Similarly, our game (bi)similarity relations, which consider all moves to be observable, refine the *alternating (bi)similarity relations* of [5], where moves are not observable.

3.1 Rectangular Games

We generalize the rectangular automata of [14], which are single-player structures, to two-player structures called rectangular games. A *rectangle* $\mathfrak{r}$ *of dimension* n is a subset of $\mathbb{R}^n$ such that $\mathfrak{r}$ is the cartesian product of n closed intervals —bounded or unbounded— all of whose finite end-points are integers.[7] Let $\mathfrak{R}^n$ be the set of all rectangles of dimension n. Denote by $\mathfrak{r}_k$ the projection of $\mathfrak{r}$ on its kth coordinate, so that $\mathfrak{r} = \prod_{k=1}^{n} \mathfrak{r}_k$. A *rectangular game*

$$\mathcal{R} = (L, X, M_1, M_2, \mathit{enabled}_1, \mathit{enabled}_2, \mathit{flow}, E, \mathit{jump}, \mathit{post})$$

consists of the following components:

- A finite set L of locations which determine the discrete state of the game.
- A set $X = \{x_1, \ldots, x_n\}$ of real-valued variables which determine the continuous state of the game. The number n is called the *dimension* of $\mathcal{R}$.
- For $i = 1, 2$, a finite set M_i of moves of player-i. Let $M_i^{time} = M_i \uplus \{\mathit{time}\}$, where *time* is a special symbol not in M_1 or M_2 which denotes a move that permits the passage of time.
- For $i = 1, 2$, a function $\mathit{enabled}_i$: $M_i^{time} \times L \to \mathfrak{R}^n$ which specifies for each move m_i of player-i and each location ℓ, the rectangle in which m_i is enabled when the discrete state of the game is ℓ. Given a location $\ell \in L$, the rectangle $\mathit{enabled}_1(\mathit{time}, \ell) \cap \mathit{enabled}_2(\mathit{time}, \ell)$ is said to be the *invariant region* of ℓ, and is denoted $\mathit{inv}(\ell)$.
- A function *flow*: $L \to \mathfrak{R}^n$ which maps each location ℓ to a *bounded* rectangle that constrains the evolution of the continuous state of the game when the discrete state is ℓ.
- A set $E \subseteq (L \times M_1 \times M_2^{time} \times L) \cup (L \times M_1^{time} \times M_2 \times L)$ of edges which specifies how the discrete state may pass from one location to another.
- A function *jump*: $E \to 2^{\{1,\ldots,n\}}$ which maps each edge to the indices of those variables whose values may change when the discrete state proceeds along that edge.
- A function *post*: $E \to \mathfrak{R}^n$ which maps each edge to a *bounded* rectangle that constrains the new continuous state when the discrete state proceeds along that edge.

We require that for every edge $e = (\ell, \cdot, \cdot, \ell')$ and every coordinate $k = 1, \ldots, n$, if $\mathit{flow}(\ell)_k \neq \mathit{flow}(\ell')_k$, then $k \in \mathit{jump}(e)$. In [14], the corresponding requirement on rectangular automata is called *initialization* and is shown to be necessary for simple reachability questions to be decidable.

A *state* of the game $\mathcal{R}$ consists of a discrete part $\ell \in L$ and a continuous part $\boldsymbol{x} \in \mathbb{R}^n$ such that $\boldsymbol{x}$ lies in the invariant region of ℓ; that is, the state space of $\mathcal{R}$ is the infinite set $Q_{\mathcal{R}} = \{(\ell, \boldsymbol{x}) \in L \times \mathbb{R}^n \mid \boldsymbol{x} \in \mathit{inv}(\ell)\}$. Informally, when the game is in state $(\ell, \boldsymbol{x})$, time can progress as long as both players choose the move *time*

[7] It is straightforward to permit intervals with rational end-points. A generalization of our results to open and half-open intervals is technically involved, but possible along the lines of [14].

and the state, whose continuous part evolves over time obeying the differential inclusion $flow(\ell)$, remains in the invariant region $inv(\ell)$. The differential inclusion is obeyed by all differentiable trajectories whose first time derivative stays inside the rectangle $flow(\ell)$. Alternatively, a player may choose a move different from *time* which is enabled in the current state. In this case, the discrete part of the state changes along an edge $e \in E$, and the continuous part of the state changes as follows. For each $k \in jump(e)$, the value of x_k is nondeterministically assigned a new value in the interval $post(e)_k$. For each $k \notin jump(e)$, the value of x_k does not change. This semantics is captured formally by the following definition. With the game $\mathcal{R}$ we associate the game structure

$$\mathcal{G}_{\mathcal{R}} = (Q_{\mathcal{R}}, L, \langle\!\langle \cdot \rangle\!\rangle, M_1^{time}, M_2^{time}, Enabled_1, Enabled_2, \delta),$$

where $\langle\!\langle (\ell, x) \rangle\!\rangle = \{\ell\}$, $Enabled_i(m) = \{(\ell, x) \in Q_{\mathcal{R}} \mid x \in enabled_i(m, \ell)\}$, and $(\ell', x') \in \delta((\ell, x), m_1, m_2)$ if either of the following two conditions is met:

Time step of duration t and slope s. We have $m_1 = m_2 = time$ and $\ell' = \ell$, and $x' = x + t \cdot s$ for some real vector $s \in flow(\ell)$ and some real $t \geq 0$ such that $(x + t' \cdot s) \in inv(\ell)$ for all $0 \leq t' \leq t$.

Discrete step along edge e. There exists an edge $e = (\ell, m_1, m_2, \ell') \in E$ such that $(\ell, x) \in Enabled(m_i)$ for $i = 1, 2$, and $x'_k \in post(e)_k$ for all $k \in jump(e)$, and $x'_k = x_k$ for all $k \notin jump(e)$.

Runs and traces, as well as preorders and equivalences on the states of a rectangular game $\mathcal{R}$ are all inherited from the underlying game structure $\mathcal{G}_{\mathcal{R}}$.[8] In what follows, we shall relate also states of two different rectangular games $\mathcal{R}_1$ and $\mathcal{R}_2$, as long as they agree on the observation (location) and move sets, with the understanding that this refers to the disjoint union of the structures $\mathcal{G}_{\mathcal{R}_1}$ and $\mathcal{G}_{\mathcal{R}_2}$. The LTL *control problem for rectangular games* asks, given a rectangular game $\mathcal{R}$ and an LTL formula φ, which states of the underlying game structure $\mathcal{G}_{\mathcal{R}}$ can be controlled by player-1 for φ. As before, the controller-synthesis problem asks, in addition, for a witnessing strategy.

Example. Consider an assembly line scheduler that must assign each element from an incoming stream of parts to one of two assembly lines. The lines process jobs at different speeds: on the first line, each job travels between one and two meters per minute; on the second line, each job travels between two and three meters per minute. The first line is three meters long and the second is six meters long. Once an assembly line finishes a job, before the line can accept a new job there is a clean-up phase, which introduces a delay of two minutes for the first line and three minutes for the second. At least four minutes pass between the

[8] Along some runs of a rectangular game, the sum of durations of all time steps converges. We do not rule out such degenerate runs, because appropriate conditions on the divergence of time can be expressed in LTL once slight modifications are made to the given game. A typical condition may assert that player-1 achieves the control objective unless player-2 refuses to let time diverge by infinitely often resetting a clock from 1 to 0 [3].

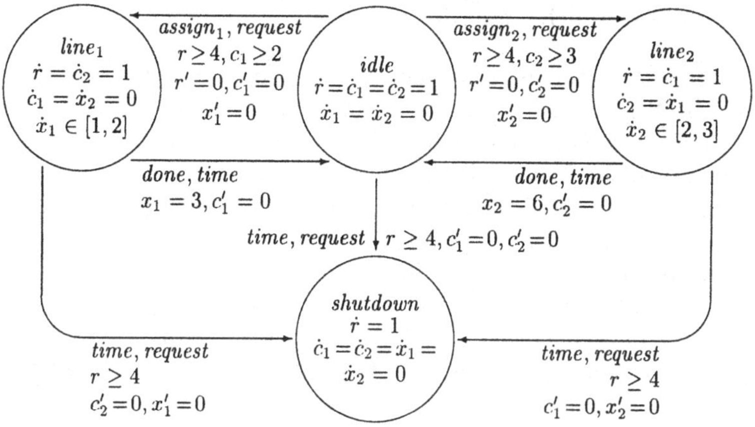

Fig. 1. Two assembly lines modeled as a rectangular game

arrival of two successive jobs. The system is able to accept a new job if neither line is processing a job and at most one line is cleaning up. If a job arrives when the system is unable to process it, the system shuts down.

We model the system as a rectangular game, pictured in Fig. 1. The discrete locations are *idle*, in which no job is being processed; $line_1$ ($line_2$), in which line 1 (line 2) is processing a job; and *shutdown*. The continuous variable r measures the time since the last job arrived. The variable c_1 (c_2) tracks the amount of time line 1 (line 2) has spent cleaning up its previous job. The variable x_1 (x_2) measures the distance a job has traveled along line 1 (line 2). Player-2 has a single move, *request*, which alerts player-1 to the arrival of a new job. The moves of player-1 are $assign_1$ ($assign_2$), which assigns a job to line 1 (line 2); and *done*, which signals the completion of a job. It can be seen that a strategy which assigns jobs first to one assembly line, then to the other, and so on, ensures that the system never shuts down if started from location *idle*, with $r \geq 4$, $c_1 \geq 2$, and $c_2 \geq 2$. However, a strategy that always chooses the same line does not ensure that the system never shuts down.

Special cases of rectangular games. Consider a variable x_k of a rectangular game. The variable x_k is a *finite-slope variable* if for each location ℓ of the game, the interval $flow(\ell)_k$ is a singleton. If $flow(\ell)_k = [1,1]$ for all locations ℓ, then x_k is called a *clock*. A rectangular game has *deterministic jumps* if for each edge e of the game, and each coordinate $k \in jump(e)$, the interval $post(e)_k$ is a singleton. A *singular game* is a rectangular game with deterministic jumps all of whose variables are finite-slope variables. Even more specific is the case of a *timed game*, which is a rectangular game with deterministic jumps all of whose variables are clocks. An essentially identical class of timed games has been defined and solved in [6], and closely related notions of timed games are studied in [3, 15, 16, 20].

3.2 Game Bisimilarity for Singular Games

Given an n-dimensional singular game $\mathcal{S}$, we define the region equivalence on the states of $\mathcal{S}$ following [1, 2]. For a real number u, let $\mathit{frac}(u)$ denote the fractional part of u. For a vector $\boldsymbol{u} \in \mathbb{R}^n$, let $\mathit{frac}(\boldsymbol{u})$ denote the vector whose kth coordinate is the fractional part of $\boldsymbol{u}_k$, for $k = 1, \ldots, n$. For an n-tuple $\boldsymbol{a}$ of integers, define the equivalence relation $\equiv_{\boldsymbol{a}}$ on $\mathbb{R}^n$ such that $\boldsymbol{x} \equiv_{\boldsymbol{a}} \boldsymbol{y}$ iff for $k, m = 1, \ldots, n$, (1) $\lfloor \boldsymbol{a}_k \boldsymbol{x}_k \rfloor = \lfloor \boldsymbol{a}_k \boldsymbol{y}_k \rfloor$, (2) $\mathit{frac}(\boldsymbol{a}_k \boldsymbol{x}_k) = 0$ iff $\mathit{frac}(\boldsymbol{a}_k \boldsymbol{y}_k) = 0$, and (3) $\mathit{frac}(\boldsymbol{a}_k \boldsymbol{x}_k) < \mathit{frac}(\boldsymbol{a}_m \boldsymbol{x}_m)$ iff $\mathit{frac}(\boldsymbol{a}_k \boldsymbol{y}_k) < \mathit{frac}(\boldsymbol{a}_m \boldsymbol{y}_m)$. Let c be the maximum, over all constants c' that appear in the definition of $\mathcal{S}$, of $|c'|$. For each location ℓ of $\mathcal{S}$, if $\mathit{flow}(\ell) = \prod_{k=1}^{n} [b_k^\ell, b_k^\ell]$, let $\boldsymbol{a}^\ell = (\boldsymbol{a}_1^\ell, \ldots, \boldsymbol{a}_n^\ell)$ such that $\boldsymbol{a}_k^\ell = b_k^\ell$ if $b_k^\ell \neq 0$, and $\boldsymbol{a}_k^\ell = 1$ if $b_k^\ell = 0$. In particular, if $\mathcal{S}$ is a timed game, then $\boldsymbol{a}^\ell = (1, 1, \ldots, 1)$ for all locations $\ell \in L$. Two states $(\ell, \boldsymbol{x})$ and $(\ell', \boldsymbol{y})$ of $\mathcal{S}$ are *region equivalent*, written $(\ell, \boldsymbol{x}) \cong^R (\ell', \boldsymbol{y})$, if (1) $\ell = \ell'$, (2) $\mathit{frac}(\boldsymbol{x}) \equiv_{\boldsymbol{a}^\ell} \mathit{frac}(\boldsymbol{y})$, and (3) for $k = 1, \ldots, n$, either $\lfloor \boldsymbol{x}_k \rfloor = \lfloor \boldsymbol{y}_k \rfloor$ or both $\lfloor \boldsymbol{x}_k \rfloor > c$ and $\lfloor \boldsymbol{y}_k \rfloor > c$. The arguments of [1, 2] show that the region equivalence $\cong^R$ is a bisimulation on the single-player structure $\mathcal{F}_{\mathcal{G}_\mathcal{S}}$ associated with $\mathcal{G}_\mathcal{S}$. Using Proposition 1, we conclude that $\cong^R$ is a game bisimulation for $\mathcal{S}$.

Theorem 3. *For every singular game, the region equivalence $\cong^R$ refines the game bisimilarity $\cong_g^B$.*

It follows that every singular game has a finite quotient structure with respect to game bisimilarity. Since game bisimilarity refines game trace equivalence, by Proposition 2, the finite quotient structure can be used for LTL controller synthesis. The game bisimilarity quotient of a singular game may have at most $|L| \cdot 2^{O(n \log nbc)}$ equivalence classes ("regions"), where b is the absolute value of the least common multiple of all nonzero, finite endpoints of flow intervals. We note that the singular games are a maximal class of hybrid games for which finite game bisimilarity quotients exist. In particular, there exists a 2D rectangular game $\mathcal{R}$ such that the equality relation on states is the only game bisimulation for $\mathcal{R}$ [8].

3.3 Game Similarity for 2D Rectangular Games

Given a 2D rectangular game $\mathcal{T}$, we define, following [10], the double-region equivalence on the states of $\mathcal{T}$ as the intersection of two region equivalences. For 2-tuples $\boldsymbol{a}$ and $\boldsymbol{b}$ of integers, define the equivalence relation $\equiv_{\boldsymbol{a},\boldsymbol{b}}$ on $\mathbb{R}^2$ as the intersection of $\equiv_{\boldsymbol{a}}$ and $\equiv_{\boldsymbol{b}}$. Let c be defined for $\mathcal{T}$ as it was for $\mathcal{S}$. For each location ℓ of $\mathcal{T}$, if $\mathit{flow}(\ell) = [g_1^\ell, h_1^\ell] \times [g_2^\ell, h_2^\ell]$, let $\boldsymbol{a}^\ell = (g_2^\ell, h_1^\ell)$ and $\boldsymbol{b}^\ell = (h_2^\ell, g_1^\ell)$. Two states $(\ell, \boldsymbol{x})$ and $(\ell', \boldsymbol{y})$ of $\mathcal{T}$ are *double-region equivalent*, written $(\ell, \boldsymbol{x}) \cong^{2R} (\ell', \boldsymbol{y})$, if (1) $\ell = \ell'$, (2) $\mathit{frac}(\boldsymbol{x}) \equiv_{\boldsymbol{a}^\ell, \boldsymbol{b}^\ell} \mathit{frac}(\boldsymbol{y})$, and (3) for $k = 1, 2$, either $\lfloor \boldsymbol{x}_k \rfloor = \lfloor \boldsymbol{y}_k \rfloor$ or both $\boldsymbol{x}_k > c$ and $\boldsymbol{y}_k > c$. The arguments of [10] show that the double-region equivalence $\cong^{2R}$ is a simulation on the single-player structure $\mathcal{F}_{\mathcal{G}_\mathcal{T}}$ associated with $\mathcal{G}_\mathcal{T}$. Using Proposition 1, we conclude that $\cong^{2R}$ is a game simulation for $\mathcal{T}$.

Theorem 4. *For every* 2D *rectangular game, the double-region equivalence* $\cong^{2R}$ *refines the game similarity* $\cong_g^S$.

This implies that every 2D rectangular game has a finite quotient structure with respect to game similarity. Since game similarity refines game trace equivalence, by Proposition 2, the finite quotient structure can be used for LTL controller synthesis. The game similarity quotient of a 2D rectangular game may have at most $O(|L| \cdot c^4)$ equivalence classes. We note that the 2D rectangular games are a maximal class of hybrid games for which finite game similarity quotients exist. In particular, there exists a 3D rectangular game $\mathcal{R}$ such that the equality relation on states is the only game simulation for $\mathcal{R}$ [12].

3.4 Game Trace Equivalence for Rectangular Games

Given an n-dimensional rectangular game $\mathcal{R}$, we define, following [14], a $2n$-dimensional singular game $\mathcal{S}_\mathcal{R}$ and a map *rect* between the states of $\mathcal{S}_\mathcal{R}$ and the states of $\mathcal{R}$ so that states that are related by *rect* are game trace equivalent. Since the singular game $\mathcal{S}_\mathcal{R}$ has a finite game trace-equivalence quotient (Theorem 3), it follows that the rectangular game $\mathcal{R}$ also has a finite game trace-equivalence quotient. The game $\mathcal{S}_\mathcal{R}$ has the same location and move sets as $\mathcal{R}$. We replace each variable x_k of $\mathcal{R}$ by two finite-slope variables $y_{l(k)}$ and $y_{u(k)}$ such that when $\mathit{flow}_\mathcal{R}(\ell)(x_k) = [a_k, b_k]$, then $\mathit{flow}_{\mathcal{S}_\mathcal{R}}(\ell)(y_{l(k)}) = [a_k, a_k]$ and $\mathit{flow}_{\mathcal{S}_\mathcal{R}}(\ell)(y_{u(k)}) = [b_k, b_k]$. Intuitively, the variable $y_{l(k)}$ tracks the least possible value of x_k, and the variable $y_{u(k)}$ tracks the greatest possible value of x_k. With each edge step of $\mathcal{S}_\mathcal{R}$, the values of the variables are appropriately updated so that the interval $[y_{l(k)}, y_{u(k)}]$ maintains the possible values of x_k; the details can be found in [14]. Call a state $(\ell, \boldsymbol{y})$ of $\mathcal{S}_\mathcal{R}$ an *upper-half state* if $\boldsymbol{y}_{l(k)} \leq \boldsymbol{y}_{u(k)}$ for all $k = 1, \ldots, n$. The function *rect*: $Q_{\mathcal{S}_\mathcal{R}} \to 2^{Q_\mathcal{R}}$, which maps each state of $\mathcal{S}_\mathcal{R}$ to a set of states of $\mathcal{R}$, is defined by $\mathit{rect}(\ell, \boldsymbol{y}) = \{\ell\} \times \prod_{k=1}^{n} [\boldsymbol{y}_{l(k)}, \boldsymbol{y}_{u(k)}]$ if $(\ell, \boldsymbol{y})$ is an upper-half state, and $\mathit{rect}(\ell, \boldsymbol{y}) = \emptyset$ otherwise. In [14] it is shown that a state q of the single-player structure $\mathcal{F}_{\mathcal{G}_{\mathcal{S}_\mathcal{R}}}$ forward simulates every state in $\mathit{rect}(q)$ of the single-player structure $\mathcal{F}_{\mathcal{G}_\mathcal{R}}$, and that every state $p \in \mathit{rect}(q)$ backward simulates q. In analogy to Proposition 1, these arguments carry over to the two-player structures $\mathcal{G}_{\mathcal{S}_\mathcal{R}}$ and $\mathcal{G}_\mathcal{R}$.

Lemma 5. *Let* $\mathcal{R}$ *be a rectangular game, let* q *be a state of the singular game* $\mathcal{S}_\mathcal{R}$, *and let* $p \in \mathit{rect}(q)$ *be a state of* $\mathcal{R}$. *Then* p *is forward game simulated by* q, *and* q *is backward game simulated by* p.

Lemma 5 holds even if the durations of time steps are observable. It ensures the game trace equivalence of p and q for *finite* traces. Since the rectangles used in the definition of rectangular games are closed, it follows, as in [14], that the trace set of $\mathcal{R}$ is limit-closed.[9] Hence, Lemma 5 is sufficient to show the game trace equivalence of p and q also for infinite traces.

[9] A set L of infinite sequences is *limit-closed* if for every infinite sequence w, when every finite prefix of w is a prefix of some sequence in L, then w itself is in L.

Theorem 6. *For every rectangular game $\mathcal{R}$, every state q of the singular game $\mathcal{S}_{\mathcal{R}}$, and every state $p \in rect(q)$ of $\mathcal{R}$, the states p and q are game trace equivalent.*

It follows that the game trace-equivalence quotient of every rectangular game $\mathcal{R}$ is finite, which can be used for LTL controller synthesis (Proposition 2). It may have at most $2^{O(\log|L|+n\log nc)}$ equivalence classes (corresponding to the regions of $\mathcal{S}_{\mathcal{R}}$), where b and c are defined as for singular games. (The constant factors hidden by the big-O notation may make the number of game trace-equivalence classes much larger than for a singular game with the same number of locations and continuous variables. Therefore, for the special cases that $\mathcal{R}$ is singular or 2-dimensional, the constructions of Sections 3.2 and 3.3 are superior in that they provide better bounds.) The EXPTIME-hardness part of the ensuing corollary follows from the fact that the structure complexity of LTL control is EXPTIME-hard already in the special case of timed games [13].

Corollary 7. *The control problem for a rectangular game $\mathcal{R}$ and an* LTL *formula φ is* EXPTIME*-complete in the size of $\mathcal{R}$ and* 2EXPTIME*-complete in the size of φ.*

We note that the rectangular games are a maximal class of hybrid games for which finite game trace-equivalence quotients are known to exist. A *triangle of dimension* n is a subset of $\mathbb{R}^n$ that can be defined by a conjunction of inequalities of the form $x_k \sim x_m + c$ and $x_k \sim c$, where c is an integer constant and $\sim \in \{\leq, \geq\}$. Let $\mathfrak{T}^n$ be the set of all triangles of dimension n. All results about timed automata and timed games still apply if triangular enabling conditions (that is, $enabled_i$: $M_i^{time} \times L \to \mathfrak{T}^n$) are permitted, because for a timed game $\mathcal{T}$, every triangle is a union of equivalence classes of the region equivalence $\cong^R$ of $\mathcal{T}$ [2]. This, however, in general is not the case for singular games. Indeed, the reachability problem for singular automata with triangular enabling conditions is undecidable [1], and therefore, so is the LTL control problem for singular games with triangular enabling conditions.

We also note that unlike for timed games with triangular enabling conditions, a witnessing strategy for the LTL control of a rectangular game may not be implementable as a rectangular controller automaton. This is because already for timed games without triangular enabling conditions, a winning strategy may have to be triangular, in the following sense. A memory-free strategy f for an n-dimensional rectangular game is *rectangular* (*triangular*) if there exists a finite set Γ of rectangles (triangles) such that (1) $\bigcup \Gamma = \mathbb{R}^n$ and (2) for every location ℓ of the game, and all $\boldsymbol{x}, \boldsymbol{y} \in \mathbb{R}^n$, if $\boldsymbol{x}$ and $\boldsymbol{y}$ belong to exactly the same sets in Γ, then f agrees on both $(\ell, \boldsymbol{x})$ and $(\ell, \boldsymbol{y})$. The following example illustrates a simple timed game for which no rectangular winning strategy exists. Consider Fig. 2 and the LTL objective $\Diamond \ell_3$. In the timed game $\mathcal{T}$ on the left, at the states whose discrete part is ℓ_0, the only moves enabled for player-1 are m_1 and m_2; in particular, *time* is not enabled there for player-1. Let $\mathfrak{t}_1 \stackrel{\Delta}{=} \{\boldsymbol{x} \in \mathbb{R}^2 \mid 0 < x_2 < x_1 < 1\}$, and $\mathfrak{t}_2 \stackrel{\Delta}{=} \{\boldsymbol{x} \in \mathbb{R}^2 \mid 0 < x_1 < x_2 < 1\}$. The right-hand side illustrates a portion of the finite quotient game structure $\mathcal{G}_{\mathcal{T}}/_{\cong^R}$. At the states in $\{\ell_0\} \times \mathfrak{t}_1$, if player-1 chooses the move m_2, then from location ℓ_2 player-1 will be unable to

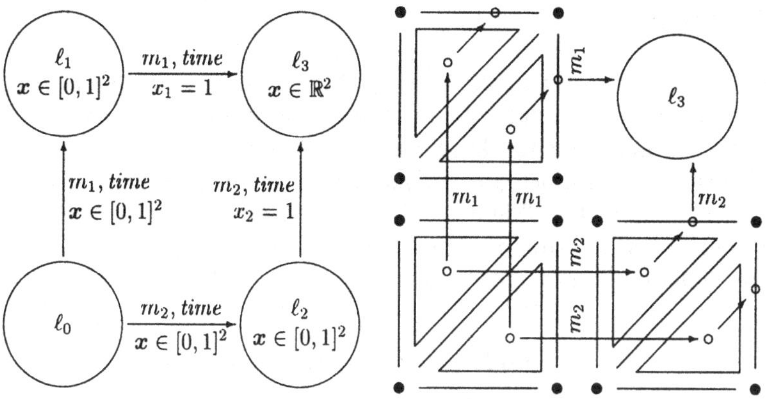

Fig. 2. A triangular strategy is necessary for winning $\Diamond \ell_3$

force a transition to location ℓ_3. On the other hand, if at the states in $\{\ell_0\} \times \mathfrak{t}_1$ player-1 chooses the move m_1, then player-1 will be able to force a transition to ℓ_3 (by first letting time progress until $x_2 = 1$, and then playing again the move m_1). Similarly, at the states in $\{\ell_0\} \times \mathfrak{t}_2$, player-1 must choose the move m_2 in order to eventually force entry into location ℓ_3.

Safety control. We conclude with an observation that is important for making the control of rectangular games practical. The most important special case of LTL control is safety control. The *safety control problem* asks, given a game structure $\mathcal{G} = (Q, \Pi, \langle\!\langle \cdot \rangle\!\rangle, Moves_1, Moves_2, Enabled_1, Enabled_2, \delta)$ and a subset $\Psi \subseteq \Pi$ of the observations, which states of $\mathcal{G}$ can be controlled by player-1 for the LTL formula $\Box \bigvee \Psi$. Define $R_\Psi = \{q \in Q \mid \Psi \cap \langle\!\langle q \rangle\!\rangle \neq \emptyset\}$. For every set $R \subseteq Q$ of states, define the *uncontrollable (player-1) predecessors* of R to be the set

$$upre(R) = \{p \in Q \mid \forall m_1 \in mov_1(p).\ \exists m_2 \in mov_2(p).\ \delta(p, m_1, m_2) \cap R \neq \emptyset\}.$$

Then the set of states that can be controlled by player-1 for the LTL formula $\Box \bigvee \Psi$ may be computed by iterating the *upre* operator: the answer to the safety control problem is $Q \setminus \bigcup_{i=0}^{\infty} upre^i(Q \setminus R_\Psi)$. This method is called the *fixpoint iteration* for safety control.

For every rectangular game $\mathcal{R}$, the *upre* operator can be computed effectively [27]. We say that a region R of $\mathcal{R}$ *corresponds* to the region S of the singular game $\mathcal{S}_\mathcal{R}$ if $R = \bigcup_{q \in S} rect(q)$. Notice that for every set Ψ of observations, the region R_Ψ of $\mathcal{R}$ corresponds to a union of game bisimilarity classes, and by Lemma 5, if R corresponds to such a union, then so does $upre(R)$. Since the number of game bisimilarity classes of $\mathcal{S}_\mathcal{R}$ is finite (Theorem 3), the fixpoint iteration for safety control terminates.

Corollary 8. *The safety control problem for rectangular games can be decided by fixpoint iteration.*

4 Conclusion

Our results for two-player hybrid games, which extend also to multiple players, are summarized in the right column of the table below. They can be seen to generalize systematically the known results for hybrid automata (i.e., single-player hybrid games), which are summarized in the center column. The number of equivalence classes for all finite equivalences in the table is exponential in the size of the given automaton or game. The infinitary results in the right column follow immediately from the corresponding results in the center column.

Table 1. Summary of results

	Hybrid automata (single-player)	Hybrid games (multi-player)
Timed, singular	finite bisimilarity [1, 2]	finite game bisimilarity
2D rectangular	infinite bisimilarity [8], finite similarity [10]	infinite game bisimilarity, finite game similarity
Rectangular	infinite similarity [12], finite trace equivalence [14]	infinite game similarity, finite game trace equivalence
Triangular	infinite trace equivalence [14]	infinite game trace equivalence

References

1. Alur, R., Courcoubetis, C., Halbwachs, N., Henzinger, T.A., Ho, P.-H., Nicollin, X., Olivero, A., Sifakis, J., Yovine, S.: The algorithmic analysis of hybrid systems. Theoretical Computer Science **138** (1995) 3–34
2. Alur, R., Dill, D.L.: A theory of timed automata. Theoretical Computer Science **126** (1994) 183–235
3. Alur, R., Henzinger, T.A.: Modularity for timed and hybrid systems. In: CONCUR 97: Concurrency Theory. Lecture Notes in Computer Science, Vol. 1243. Springer-Verlag (1997) 74–88
4. Alur, R., Henzinger, T.A., Kupferman, O.: Alternating-time temporal logic. In: Proceedings of the 38th Annual Symposium on Foundations of Computer Science. IEEE Computer Society Press (1997) 100–109
5. Alur, R., Henzinger, T.A., Kupferman, O., Vardi, M.Y.: Alternating refinement relations. In: CONCUR 98: Concurrency Theory. Lecture Notes in Computer Science, Vol. 1466. Springer-Verlag (1998) 163–178
6. Asarin, E., Maler, O., Pnueli, A., Sifakis, J.: Controller synthesis for timed automata. In: Proceedings of the IFAC Symposium on System Structure and Control. Elsevier Science Publishers (1998) 469–474
7. Emerson, E.A.: Temporal and modal logic. In: van Leeuwen, J. (ed.): Handbook of Theoretical Computer Science, Vol. B. Elsevier Science Publishers (1990) 995–1072
8. Henzinger, T.A.: Hybrid automata with finite bisimulations. In: ICALP 95: Automata, Languages, and Programming. Lecture Notes in Computer Science, Vol. 944. Springer-Verlag (1995) 324–335

9. Henzinger, T.A.: The theory of hybrid automata. In: Proceedings of the 11th Annual Symposium on Logic in Computer Science. IEEE Computer Society Press (1996) 278–292
10. Henzinger, M.R., Henzinger, T.A., and Kopke, P.W.: Computing simulations on finite and infinite graphs. In: Proceedings of the 36th Annual Symposium on Foundations of Computer Science. IEEE Computer Society Press (1995) 453–462
11. Henzinger, T.A., Ho, P.-H., Wong-Toi, H.: Algorithmic analysis of nonlinear hybrid systems. IEEE Transactions on Automatic Control **43** (1998) 540–554
12. Henzinger, T.A., Kopke, P.W.: State equivalences for rectangular hybrid automata. In: CONCUR 96: Concurrency Theory. Lecture Notes in Computer Science, Vol. 1119. Springer-Verlag (1996) 530–545
13. Henzinger, T.A., Kopke, P.W.: Discrete-time control for rectangular hybrid automata. In: ICALP 97: Automata, Languages, and Programming. Lecture Notes in Computer, Vol. 1256. Springer-Verlag (1997) 582–593
14. Henzinger, T.A., Kopke, P.W., Puri, A., Varaiya, P.: What's decidable about hybrid automata? Journal of Computer and System Sciences **57** (1998) 94–124
15. Heymann, M., Lin, F., Meyer, G.: Control synthesis for a class of hybrid systems subject to configuration-based safety constraints. In: HART 97: Hybrid and Real-Time Systems. Lecture Notes in Computer Science, Vol. 1201. Springer-Verlag (1997) 376–390
16. Hoffmann, G., Wong-Toi, H.: The input-output control of real-time discrete-event systems. In: Proceedings of the 13th Annual Real-time Systems Symposium. IEEE Computer Society Press (1992) 256–265
17. Immerman, N.: Number of quantifiers is better than number of tape cells. Journal of Computer and System Sciences **22** (1981) 384–406
18. Lygeros, J., Tomlin, C., Sastry, S.: Controllers for reachability specifications for hybrid systems. Automatica **35** (1999) 349–370
19. Lynch, N.A., Segala, R., Vaandrager, F., Weinberg, H.B.: Hybrid I/O Automata. In: Hybrid Systems III. Lecture Notes in Computer Science, Vol. 1066 Springer-Verlag (1996), 496–510
20. Maler, O., Pnueli, A., Sifakis, J.: On the synthesis of discrete controllers for timed systems. In: STACS 95: Theoretical Aspects of Computer Science. Lecture Notes in Computer Science, Vol. 900. Springer-Verlag (1995) 229–242
21. Paige, R., Tarjan, R.E.: Three partition-refinement algorithms. SIAM Journal of Computing **16** (1987) 973–989
22. Puri, A., Borkar, V., Varaiya, P.: ε-approximation of differential inclusions. In: Hybrid Systems III. Lecture Notes in Computer Science, Vol. 1066 Springer-Verlag (1996), 362–376
23. Ramadge, P.J., Wonham, W.M.: Supervisory control of a class of discrete-event processes. SIAM Journal of Control and Optimization **25** (1987) 206–230
24. Rosner, R.: Modular Synthesis of Reactive Systems. Ph.D. thesis, Weizmann Institute of Science (1992)
25. Thomas, W.: On the synthesis of strategies in infinite games. In: STACS 95: Theoretical Aspects of Computer Science. Lecture Notes in Computer Science, Vol. 900. Springer-Verlag (1995) 1–13
26. Tomlin, C.: Hybrid Control of Air Traffic Management Systems. Ph.D. thesis, University of California, Berkeley (1998)
27. Wong-Toi, H.: The synthesis of controllers for linear hybrid automata. In: Proceedings of the 36th Conference on Decision and Control. IEEE Computer Society Press (1997) 4607–4612

Localizability of Fairness Constraints and Their Distributed Implementations*
(Extended Abstract)

Yuh-Jzer Joung

joung@ccms.ntu.edu.tw
Department of Information Management
National Taiwan University
Taipei, Taiwan

Abstract. We propose a *localizability* criterion that allows local computations to be composed into a valid global one. We show that, in the presence of *equivalence-robustness*, most fairness notions proposed in the literature satisfy the localizability criterion. Moreover, we also present a general and efficient distributed algorithm to implement equivalence-robust fairness notions satisfying the localizability criterion. Our results therefore offer an appealing solution to the implementation problem for existing fairness notions for distributed programming languages and algebraic models of concurrency.

1 Introduction

The concept of *interactions* has been widely used in distributed programming languages (e.g., CSP [10], Ada, Script [9], Action Systems [2], IP [8], and DisCo [13, 12]) and algebraic models of concurrency (e.g., CCS [20], SCCS [19], LOTOS [3], π-calculus [21]) to model *synchronization* and *nondeterminism* among processes. An interaction is a set of actions (usually communications) to be executed jointly by a set of processes, which must synchronize in order to commence the actions.[1] Synchronization prevents a process from committing to a joint action too early before other participants are ready. Nondeterminism allows a process to choose one ready interaction to execute, from a set of potential interactions it has specified.

For example, the producers and consumers problem can be easily expressed in CSP as follows:

$Producer_i, i = 1, 2$ ::
  integer *data*;
  generate(*data*);
  $*$ [$Consumer_1$! *data* $\longrightarrow$ *generate*(*data*);
    $\Box$ $Consumer_2$! *data* $\longrightarrow$ *generate*(*data*);]

* Research supported by the National Science Council, Taipei, Taiwan, under Grants NSC 85-2213-E-002-059, NSC 86-2213-E-002-053, and NSC 88-2213-E-002-009.

[1] Although some languages like CSP and Ada permit interactions only between two processes, more recent proposals have relaxed this biparty limitation to multipartiness of *arbitrary* arity to offer a higher level of abstraction. See [16] for a taxonomy of programming languages offering linguistic support for multiparty interaction along with a detailed complexity analysis.

```
Consumer_i, i = 1,2 ::
    integer x;
    *[  Producer_1? x ⟶ process(x);
     □ Producer_2? x ⟶ process(x); ]
```

The program describes that when a producer has prepared its data, it is willing to establish an interaction with either consumer to process the data. Similarly, when a consumer is ready, it is willing to accept new data from any producer.

Note that conflicting interactions involving a common process cannot be executed simultaneously. So if they are enabled at the same time, then only one of them can be chosen for execution. (An interaction is *enabled* if and only if all of its participants are ready for the interaction.) However, since nondeterminism allows an arbitrary enabled interaction to be chosen, the resulting computation may be improper to the system if the underlying scheduling scheme in the implementation is prejudicial to some process or interaction. For example, the above CSP program may have a computation in which the two consumers alternately interact with one producer over and over again while blocking the other producer from sending out its data indefinitely.

Therefore, some semantic constraint (often called a *fairness notion*) is typically imposed on a program to exclude undesirable computations that would otherwise be valid. For example, the notion of *strong process fairness (SPF)*, which requires that a process infinitely often ready to participate in an enabled interaction be able to execute some interaction infinitely often, can be used in the above program to prevent a process from being forever locked out from interaction. An implementation of the interaction scheduling (usually offered by the language compiler) then satisfies a given semantic constraint if it guarantees that all possible computations are *valid*, i.e., satisfying the constraint.

We remark here that the use of semantic constraints in programming language allows a program to take advantage of nondeterminism so that the program can be naturally specified and easily proved, while hiding detailed run-time dependent scheduling activities into the implementation level. The problem could also be solved in the programming level by completely re-designing the program so that all computations meet the constraint. But such approaches usually require more sophisticated programming techniques, and often make a program hard to understand.

A number of semantic constraints have been proposed in the literature (see [6, 1]), but only few of them have been successfully implemented. In [14] we propose an implementability criterion to determine whether or not a given semantic constraint is implementable. In particular, we show that a semantic constraint satisfies the criterion if it is *strongly feasible* and *equivalence-robust*. Intuitively, strong feasibility means that when some interaction is enabled, there should be a continuation allowing some interaction to be executed regardless of whether non-ready processes will become ready or not. Clearly, given that each process decides on its own when it will be ready for interaction, no implementation can rely on non-ready processes to become ready whenever some interaction can be scheduled for execution; otherwise deadlock would occur. So strong feasibility is necessary for a semantic constraint to be implementable.

Equivalence-robustness requires that equivalent computations be either all valid or all invalid [1]. Computations are *equivalent* if they differ only in the interleavings of independent actions. Here we assume that the underlying semantics induces a *dependency* relation on actions of the system, which is usually a partial order reflecting Lamport's causality relation [17]. Equivalence-robustness ensures that different observations of the same computation obtain the same property of the system [18, 23]. It thus serves as a natural bridge over the gap between *interleaving semantics* and *partial order semantics*, which is highly desirable in distributed languages.

Although several important semantic constraints turn out to be non-equivalence-robust, Francez et al. [7] have proposed a notion of *completion* to convert them into equivalence-robust ones. The idea of completion is to treat the computations in a mixed equivalence class as all valid or all invalid. An equivalence class (as induced by the equivalence relation considered above) is *mixed* if it contains both valid and invalid computations. As such, a hierarchy of completions can be obtained, with the minimum (i.e., the strongest) being that all computations in a mixed class are treated as invalid, and the maximum (i.e., the weakest) being that all computations in a mixed class are treated as valid. More importantly, completions also shed some light on many existing unimplementable semantic constraints: so long as strong feasibility can be preserved, a completion suffices to warrant an implementation [15].

In [14] we propose a general algorithm to implement strongly feasible and equivalence-robust semantic constraints. The algorithm, however, is centralized because it employs a central coordinator in charge of the interaction scheduling. Our ultimate goal, of course, is to provide a distributed implementation. Note that the term "distributedness" is somewhat vague and sometimes subjective. Our interpretation here is that a distributed implementation should allow non-conflicting interactions to be established *concurrently* and *independently*. Under this interpretation, using a central coordinator in charge of the scheduling obviously cannot qualify for a distributed implementation because interactions are always established sequentially. Independency means that the decision for a process to choose either x or y to participate should not depend on another process's state if that process is not involved in x and y. Thus an implementation that replicates several copies of the above centralized coordinator for scheduling cannot be said distributed, even if nonconflicting interactions may be established concurrently.

From the semantic level, however, some semantic constraint may inherently preclude a distributed implementation. For example, let x_{12} and x_{34} be two interactions, where x_{12} involves p_1 and p_2, and x_{34} involves p_3 and p_4. The two interactions do not conflict as they involve mutually disjoint sets of processes. However, when all four processes are ready for interaction, if the semantic constraint requires x_{12} be established *before* x_{34}, then no distributed implementation (in the sense defined above) is possible.

One might then presume that, in the light of strong feasibility (which is necessary for every implementable semantic constraint), equivalence-robustness is sufficient to guarantee a distributed implementation. Unfortunaltely, this is not true as can be illustrated by the following example. Let $\mathbb{S}$ be a system consisting of four interactions x_1, x_2, y_1, and y_2, where x_1 and x_2 involve only process p, while y_1 and y_2 involve only q. Suppose that each process behaves as follows: It is initially idle. From time to time it becomes ready for interaction, where p is always ready for only one interaction, x_1 or x_2 nondeterministically, while q is always ready for both y_1 and y_2 (but of course it can choose only one of them to execute). After interaction a process returns to an idle state. Consider the following semantic constraint $\mathbb{C}$:

> If p does not execute x_2 then q must execute y_1 and y_2 alternately; otherwise, q eventually executes only y_1.

It can be seen that the semantic constraint is strongly feasible and equivalence-robust. However, the semantic constraint implies that when q is ready, the choice of y_1 or y_2 cannot be locally made by q without consulting p's state. This is because otherwise q would either execute y_1 and y_2 alternately, or always execute y_1 regardless of whether x_2 has been executed or not, and so the overall computation would not be valid.

Although semantic constraints like above are bizarre and rare, they do preclude us from obtaining a general distributed implementation for *all* possible strongly feasible and equivalence-robust semantic constraints. In this paper we propose a *localizability* criterion that requires a semantic constraint not only be strongly feasible, but also permit the decision to establish an interaction to be made locally by its participants, subject to the condition that local decisions do not infinitely often block other processes that have been continuously waiting for an interaction from establishing the interaction. Based on this criterion, we then present a general distributed algorithm to implement equivalence-robust semantic constraints.

Note that localizability only provides a sufficient condition for distributed implementation. We do not intend to provide a criterion that is both sufficient and necessary for distributed implementation, as doing so would require a formal distributed implementation model that can well justify the meaning of "distributedness." Instead, we appraise our localizability criterion by showing that, in the presence of equivalence-robustness, the notion of ***strong interaction fairness (SIF)***, which requires that an interaction that is infinitely often enabled be executed infinitely often, is localizable. Since SIF is considerably stronger than most semantic constraints proposed in the literature, our results imply that most equivalence-robust semantic constraints can be distributedly implemented as well.

The rest of the paper is organized as follows: Section 2 presents an abstract model for process interactions. Section 3 presents our localizability criterion. Section 4 sets up some useful lemmas to highlight the possibility of a distributed implementation for localizable and equivalence-robust semantic constraints. Section 5 then presents a real implementation. Section 6 concludes.

2 Preliminaries

An *interaction system* is a triple $\mathbb{IS} = (\mathsf{P}, \mathsf{I}, \mathsf{M})$, where P is a finite set of processes, I is a finite set of interactions, and $\mathsf{M} : \mathsf{P} \times \mathsf{I}^* \longrightarrow 2^{\mathsf{I}}$ is a function called the *program* of the system. Each interaction x involves a fixed set $P_x \subseteq \mathsf{P}$ of participant processes, and can be executed by the participants (and only the participants) only if they are all ready for the interaction. A process is either in an *idle* state or in a *ready* state. Initially, every process is idle. An idle process p may autonomously become ready, where it is ready for a set $p.aim$ of potential interactions of which it is a member. After executing one interaction in $p.aim$, p returns to an idle state. The set $p.aim$ is determined by $\mathsf{M}(p, \mathcal{I})$, where $\mathcal{I}$ is the sequence of interactions p has executed. As only the participants of an interaction can execute the interaction, we shall assume that $\mathsf{M}(p, \mathcal{I}) = x$ only if $p \in P_x$.

A state s of $\mathbb{IS}$ is a tuple consisting of the history of interactions executed so far, and for each $p \in \mathsf{P}$, the state (i.e., idle or ready) of p and the set of potential interactions p is ready to execute. We use $[s]_{hist}$ to denote the history, $[s]_p$ the state of p in s, and $[s]_{p.aim}$ the set of potential interactions p is ready to execute.

Moreover, $[s]_{hist,p}$ denotes the sequence of interactions in $[s]_{hist}$ that involve p, i.e., the history of interactions executed by p. An interaction x is *enabled* in s iff (if and only if) every process $p \in P_x$ is ready for x, i.e., $[s]_p = \mathit{ready}$ and $x \in [s]_{p.aim}$. Let S be the set of all possible states of $\mathbb{IS}$. State transitions of the system are written as $s \xrightarrow{a} s'$, where $s, s' \in \mathsf{S}$, and a is the action whose execution results in the transition. State transitions are of the following two forms:

ready: $s \xrightarrow{p.I} s'$ iff $[s]_{hist} = [s']_{hist}$, $[s]_p = \mathit{idle}$, $[s']_p = \mathit{ready}$, $\mathsf{M}(p, [s]_{hist,p}) = [s']_{p.aim} = I$, and $\forall q \in \mathsf{P} - \{p\} : [s]_q = [s']_q \wedge [s]_{q.aim} = [s']_{q.aim}$.
That is, the action $p.I$ transits process p from idle to a state ready for the set I of interactions.

interaction: $s \xrightarrow{x} s'$ iff $[s']_{hist} = [s]_{hist} \cdot x$, $\forall p \in P_x : [s]_p = \mathit{ready} \wedge x \in [s]_{p.aim} \wedge [s']_p = \mathit{idle} \wedge [s']_{p.aim} = \emptyset$, and $\forall q \in \mathsf{P} - P_x : [s]_q = [s']_q \wedge [s]_{q.aim} = [s']_{q.aim}$.
That is, the execution of interaction x transits all participants of x from state ready to idle.

A *run* π is a sequence of the form

$$s_0 \xrightarrow{a_1} s_1 \xrightarrow{a_2} s_2 \dots$$

where s_0 is the initial state (that is, $[s_0]_{hist} = \epsilon$ and $\forall p \in \mathsf{P} : [s_0]_p = \mathit{idle} \wedge [s_0]_{p.aim} = \emptyset$), and each $s_i \xrightarrow{a_{i+1}} s_{i+1}$ is a state transition of the system. In particular, π is *complete* if it is infinite or it ends up in a state in which all processes are ready but no interaction is enabled; otherwise, π is *partial.* We use $\mathrm{run}^*(\mathbb{IS})$ to denote the set of all finite runs of $\mathbb{IS}$, and $\mathrm{run}(\mathbb{IS})$ denotes the set of complete runs. Thus, $\mathrm{run}^*(\mathbb{IS}) \cap \mathrm{run}(\mathbb{IS})$ is the set of finite complete runs.

It should be noted that, like most systems, we have made a *minimal progress assumption* [22]—if some action can be executed in $\mathbb{IS}$, then some action will eventually be executed. So if no interaction is enabled and not all processes are ready for interaction, then some idle process will eventually become ready. However, the minimal progress assumption does not exclude the possibility that a process remains idle forever in an infinite run. Indeed, so long as some executable action will eventually be executed, we do not assume any bound on the time a process stays in a ready state. The reason for this is that making such a bounded transition assumption could result in an implementation where the scheduling of an interaction depends not only on the processes involved in the interaction, but also on other processes not involved in the interaction. For example, an implementation can simply wait until all processes are ready, and then chooses an interaction fulfilling the semantic constraint for execution. As pointed out by Buckley and Silberschatz [4], such implementation is impractical and inefficient. Besides, it could result in a deadlock for systems where local actions may not terminate.

Since each run $s_0 \xrightarrow{a_1} s_1 \xrightarrow{a_2} s_2 \dots$ is uniquely determined by the sequence of actions executed in the run, we often write the run as $a_1 a_2 \dots$. Conversely, we call a sequence of actions $a_1 a_2 \dots$ a run if there exist states $s_0, s_1, s_2, \dots$ of $\mathbb{IS}$, with s_0 being the initial state, such that $s_0 \xrightarrow{a_1} s_1 \xrightarrow{a_2} s_2 \dots$ is a run of $\mathbb{IS}$.

Definition 1. A *semantic constraint* $\mathbb{C}$ is a function which, given an interaction system $\mathbb{IS}$, returns a set of runs $\mathbb{C}[\![\mathbb{IS}]\!] \subseteq \mathrm{run}(\mathbb{IS})$. We say that π is $\mathbb{C}$-*valid* (or simply *valid* when the context is clear) if $\pi \in \mathbb{C}[\![\mathbb{IS}]\!]$.

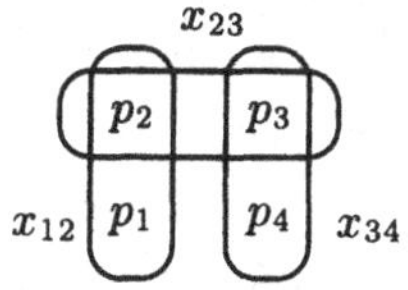

Fig. 1. A system $\mathbb{IS} = (\{p_1, p_2, p_3, p_4\}, \{x_{12}, x_{23}, x_{34}\}, \mathsf{M}^{\forall})$.

We assume that actions involving a common process in a run are totally ordered by the order in which the process executes them. These total orderings then induce a typical partial order dependency relation on the actions of a run such that $a \prec b$ iff some process executes a before b, or there exists c such that $a \prec c$ and $c \prec b$ [17]. Two runs π and ρ are *equivalent*, denoted by $\pi \equiv \rho$, iff for every process p, the sequence of actions involving p in π is the same as that in ρ. As can be seen, if $\pi \equiv \rho$, then one of them can be obtained from the other by transpositions of independent actions.

For example, consider the following run of the system shown in Fig. 1, where $P_{x_{12}} = \{p_1, p_2\}$, $P_{x_{23}} = \{p_2, p_3\}$, and $P_{x_{34}} = \{p_3, p_4\}$:

$$\pi = (p_1 p_2 x_{12} p_3 p_4 x_{34})^{\omega}$$

We use $\mathsf{M}^{\forall}$ to denote a program that allows a process to be ready for all interactions of which it is a member every time when the process is ready. For notational simplicity we overload the notation p_i to abbreviate the corresponding ready action $p_i.I$ where $I = \{x \in \mathsf{I} \,|\, p_i \in P_x\}$. This abbreviation will be adopted throughout the paper. Also, a^{ω} denotes the infinite sequence $aaaa\ldots$. Observe that every instance of x_{12} in π is independent of the following action p_3. So π is equivalent to $(p_1 p_2 p_3 x_{12} p_4 x_{34})^{\omega}$.

Definition 2. A semantic constraint $\mathbb{C}$ is *equivalence-robust* for $\mathbb{IS}$ iff

$$\forall \pi, \rho \in \text{run}(\mathbb{IS}) : \pi \in \mathbb{C}[\![\mathbb{IS}]\!] \wedge \rho \equiv \pi \Rightarrow \rho \in \mathbb{C}[\![\mathbb{IS}]\!]$$

The notion of strong feasibility is realized by a two-player game between an explicit scheduler S which copes with interaction scheduling, and an adversary A which captures the processes' autonomy in making their ready transitions.

Definition 3.

1. An *adversary* A for $\mathbb{IS}$ is a function which given a run $\pi \in \text{run}^*(\mathbb{IS})$ returns either an empty sequence ϵ or a sequence of actions $p_1.I_1 \ldots p_k.I_k$ as the continuation of π such that $\pi \cdot p_1.I_1 \ldots p_k.I_k$ represents a run of $\mathbb{IS}$. Moreover, $A(\pi) = \epsilon$ only if π is complete or some interaction is enabled in π (i.e., enabled in the last state of π).
2. A *nonblocking scheduler* S for $\mathbb{IS}$ is a function which given a run $\pi \in \text{run}^*(\mathbb{IS})$ returns either ϵ or an interaction x enabled in π as the continuation of π. Moreover, $S(\pi) = \epsilon$ only if no interaction is enabled in π.[2]

[2] For simplicity, we allow S to schedule only one interaction at a time even if there is more than one nonconflicting interaction enabled. This does not lose any generality because the game allows the adversary in response to suspend idle processes from becoming ready until all enabled interactions have been disabled.

3. The result of the S versus A game up to round i is defined by $r^i(S,A)$, where

$$r^i(S,A) = \begin{cases} \epsilon & : \ i = 0 \\ r^{i-1}(S,A) \cdot A(r^{i-1}(S,A)) & : \ i = 2n-1, n \in N \\ r^{i-1}(S,A) \cdot S(r^{i-1}(S,A)) & : \ i = 2n, n \in N \end{cases}$$

The run generated by S versus A, denoted by $r(S,A)$, is the result of the game proceeding in the maximum number of rounds.

Definition 4. A semantic constraint $\mathbb{C}$ is *strongly feasible* for $\mathbb{IS}$ iff there exists a nonblocking scheduler S for $\mathbb{IS}$ satisfying $\mathbb{C}$; that is, for every adversary A, $r(S,A) \in \mathbb{C}[\![\mathbb{IS}]\!]$.

We say that a semantic constraint $\mathbb{C}_1$ is *stronger* than $\mathbb{C}_2$ (or $\mathbb{C}_2$ is *weaker* than $\mathbb{C}_1$) if $\mathbb{C}_1[\![\mathbb{IS}]\!] \subseteq \mathbb{C}_2[\![\mathbb{IS}]\!]$. It follows that if $\mathbb{C}_1$ is strongly feasible and $\mathbb{C}_2$ is weaker than $\mathbb{C}_1$, then $\mathbb{C}_2$ is strongly feasible as well. Note that a nonblocking scheduler S satisfying $\mathbb{C}$ needs not be "faithful" [1] to $\mathbb{C}$ in the sense that not every run in $\mathbb{C}[\![\mathbb{IS}]\!]$ needs be generated by S.

Moreover, it is worth noting that a nonblocking scheduler S represents only an abstract scheduling policy for $\mathbb{C}$. It does not directly correspond to an implementation. This is because in a real system any scheduling decision must be made by a process p, either an existing one in the system or an auxiliary coordinator that is added to assist the scheduling. For p to make a scheduling decision, it must obtain processes' states, and only through communications. Since a communication takes time and since processes determine autonomously when they will become ready, it is not possible for p to obtain a global view of the system that is consistent with the view observed externally. For example, when p observes that only x is enabled and decides to schedule x, it is possible that some other conflicting interaction y is also enabled and the semantics insists that y shall be established. In contrast, a nonblocking scheduler S implicitly assumes an external global view of the system.

3 The Localizability Criterion

In this section we present our localizability criterion, for which we need the following definitions. Let $\mathbb{IS} = (\mathsf{P}, \mathsf{I}, \mathsf{M})$ be an interaction system, $Q \subseteq \mathsf{P}$ be a set of processes, and $\varphi \in \text{run}^*(\mathbb{IS}) \cup \text{run}(\mathbb{IS})$ be a run. The *projection of Q in φ*, denoted by $[\varphi]_Q$, is the result by extracting from φ every action a involving a process in Q and the actions b satisfying $b \prec a$; the relative order of the extracted actions in φ is preserved in $[\varphi]_Q$. Note that $[\varphi]_Q$ also represents a run. For notational simplicity, we often write $[\varphi]_Q$ as $[\varphi]_q$ if Q is a singleton consisting solely of q. For example, consider the interaction system shown in Fig. 1. Let $\pi = p_4(p_1p_2x_{12}p_2p_3x_{23})^\omega$ be a run of the system. Then, $[\pi]_{p_1} = [\pi]_{p_2} = [\pi]_{p_3} = (p_1p_2x_{12}p_2p_3x_{23})^\omega$, and $[\pi]_{p_4} = p_4$. Moreover, $[\pi]_{\{p_1,p_2\}} = (p_1p_2x_{12}p_2p_3x_{23})^\omega$, and $[\pi]_{\{p_2,p_4\}} = [\pi]_{\{p_1,p_2,p_4\}} = \pi$.

Moreover, let S be a nonblocking scheduler for $\mathbb{IS}$. An interaction x in π is *S-admissible* if $S([\pi|_x]_{P_x}) = x$, where $\pi|_x$ is the prefix of π up to but not including x. A nonempty set of processes Q is *S-neglected* in π if every process in Q remains ready forever from some point onward, but there exist two interactions y and z (where z could be an empty interaction ϵ with no participant) and infinitely many prefixes ρ of π such that (a) $P_y - P_z = Q$ and (b) $S([\rho]_{P_y}) = y$ but ρ is followed by z.

The localizability criterion is defined operationally by a "localizable" scheduler S as follows.

Definition 5. A semantic constraint $\mathbb{C}$ is *localizable* for $\mathbb{IS} = (\mathsf{P}, \mathsf{I}, \mathsf{M})$ if there exists a nonblocking scheduler S such that:

1. For every adversary A of $\mathbb{IS}$, $r(S, A) \in \mathbb{C}[\![\mathbb{IS}]\!]$.
2. For every π in run($\mathbb{IS}$), if every x in π is S-admissible and no nonempty subset of P is S-neglected in π, then $\pi \in \mathbb{C}[\![\mathbb{IS}]\!]$.

In this case, S is called a *localizable* scheduler for $\mathbb{IS}$ satisfying $\mathbb{C}$.

To help understand the definition, consider again the interaction system shown in Fig. 1. Assume that S is such that if no interaction is enabled in π, then $S(\pi)$ returns ϵ, otherwise it returns the enabled interaction that is executed the least often in π; tie is broken arbitrarily. Let $\pi = p_4(p_1p_2x_{12}p_2p_3x_{23})^\omega$ be a run of the system. Then every interaction x in π is S-admissible because only x is enabled in $[\pi|_x]_{P_x}$. However, the set $\{p_4\}$ is S-neglected because $\{p_4\} = P_{x_{34}} - P_{x_{23}}$, and there are infinitely many prefixes $\rho_i = p_4(p_1p_2x_{12}p_2p_3x_{23})^ip_1p_2x_{12}p_2p_3$ of π such that $S([\rho_i]_{P_{x_{34}}}) = x_{34}$, but ρ_i is followed by x_{23}.

By definition, localizability implies strong feasibility. Intuitively, admissibility means that the participants of an enabled interaction x can decide on their own whether or not to establish x. Localizability of $\mathbb{C}$ therefore allows interactions to be established locally by their participants, so long as no nonempty set Q of processes is forever neglected in the sense that infinitely often, the processes in Q together with the processes in some other set R can establish an interaction y, but the processes in R neglect Q by always choosing another interaction z. The fact that R could be empty means that if the processes of Q alone can establish an interaction, then they should be able to do so regardless of other process's progress.

The following lemma shows that the maximum completion of SIF, denoted as SIF$^+$, is localizable. It can be proved by presenting a nonblocking scheduler that schedules interactions by giving the priority to the one that is executed the least often, and showing that the scheduler satisfies SIF and is localizable.

Lemma 1. SIF$^+$ *is localizable for every* $\mathbb{IS} = (\mathsf{P}, \mathsf{I}, \mathsf{M})$.

The following lemma shows that localizability is respected by the weaker-than relation between semantic constraints. It follows from the definition of localizability.

Lemma 2. *If* $\mathbb{C}_1$ *is localizable and* $\mathbb{C}_2$ *is weaker than* $\mathbb{C}_1$*, then* $\mathbb{C}_2$ *is also localizable.*

Since SIF is considerably stronger than most semantic constraints proposed in the literature, Lemmas 1 and 2 imply that in the presence of equivalence-robustness most semantic constraints are localizable as well. (See [6, 1] for a survey of existing semantic constraints.)

For a counterexample of localizability, consider the semantic constraint $\mathbb{C}$ discussed in Section 1 for the system $\mathbb{IS} = (\{p, q\}, \{x_1, x_2, y_1, y_2\}, \mathsf{M})$ where q must execute y_1 and y_2 alternately if p does not execute x_2, or otherwise q must eventually execute only y_1. To see that $\mathbb{C}$ is not localizable, suppose otherwise S is a localizable scheduler satisfying $\mathbb{C}$. Then, since S is also a nonblocking scheduler

S satisfying $\mathbb{C}$, the following must hold: if q is ready for interaction in π and π does not contain any action involving p, then $S(\pi)$ must return y_1 if the last interaction executed by q (if any) is y_2, and returns y_2 if the last interaction is y_1. Since for any partial run ρ of $\mathbb{IS}$, $[\rho]_q$ cannot contain any action involving p, every instance of y_1 and y_2 in ρ is S-admissible if q executes y_1 and y_2 alternately in ρ. Moreover, since by the definition of M at most one interaction can be enabled in $[\rho]_p$, every instance of x_1 and x_2 in ρ (if any) is S-admissible. So, for any complete run π, if q executes y_1 and y_2 alternately, then every interaction in π must be S-admissible regardless of whether p has executed x_2 or not. As a result, if no process remains ready forever (e.g., $\pi = (p\,x_1\,p\,x_2\,q\,y_1 q\,y_2)^\omega$), then by the second condition of Definition 5, π must belong to $\mathbb{C}[\![\mathbb{IS}]\!]$. However, if p does execute x_2 in π, then by the definition of $\mathbb{C}$ π should not be in $\mathbb{C}[\![\mathbb{IS}]\!]$; contradiction. So $\mathbb{C}$ cannot be localizable because there is no localizable scheduler satisfying $\mathbb{C}$. To summarize, a semantic constraint that is strongly feasible and equivalent-robust may not necessarily be localizable.

4 An Abstract Distributed Scheduler

Recall that the definition of localizability involves a nonblocking scheduler S. By Definition 3, a nonblocking scheduler essentially assumes centralized scheduling because it takes as input the global state of the system and schedules all interactions among the processes. The centralized nature of the scheduler helps us determine whether or not a given semantic constraint is localizable (and strongly feasible), but it does not reveal too much about how the semantic constraint can be distributedly implemented. Therefore, the purpose of this section is to set up some lemmas to highlight the possibility of a distributed realization of a localizable scheduler at an abstract level. Based on these lemmas the next section then presents a real implementation for the scheduler.

We begin by distributing the task of a localizable scheduler S to each process. The game between S and a given adversary A in Definition 3 will be played by the local schedulers and A. A local scheduler for process p can be obtained from S by simply restricting it to the portion of input relevant to p, as defined below.

Definition 6. Let S be a nonblocking scheduler for $\mathbb{IS} = (\mathsf{P}, \mathsf{I}, \mathsf{M})$. A function $S{\downarrow_p}$: run$^*(\mathbb{IS}) \to \mathsf{I} \cup \{\epsilon\}$ is an S's *restriction to p* if $S{\downarrow_p}(\pi) = x \neq \epsilon$ implies $S([\pi]_{P_x}) = x$ and $p \in P_x$.

The game between the local schedulers and an adversary proceeds by letting the adversary schedule some processes to be ready for interaction, and then activating some local schedulers to schedule an interaction for their processes, and so on, as defined below:

Definition 7. Let S be a nonblocking scheduler for $\mathbb{IS} = (\mathsf{P}, \mathsf{I}, \mathsf{M})$, and let $\mathcal{S} = \{S{\downarrow_p} \mid p \in \mathsf{P}\}$ be a set of restrictions. Define a *trace* t to be a sequence $t = p_{1,1}p_{1,2}\cdots p_{1,n_1}, \cdots, p_{i,1}p_{i,2}\cdots p_{i,n_i}, \cdots$ such that each $p_{i,k} \in \mathsf{P}$. Then, given an adversary A, the result of the $\mathcal{S}$ versus A game generated by this trace up to round i, step j, where $0 \le j \le n_i$ and $n_0 = 0$, is defined by

$$d_t^{i,j}(\mathcal{S}, A) = \begin{cases} \epsilon & : \ i = j = 0 \\ d_t^{i-1,n_{i-1}}(\mathcal{S}, A) \cdot A(d_t^{i-1,n_{i-1}}(\mathcal{S}, A)) & : \ i > 0,\ j = 0 \\ d_t^{i,j-1}(\mathcal{S}, A) \cdot S{\downarrow_{p_{i,j}}}(d_t^{i,j-1}(\mathcal{S}, A)) & : \ i > 0,\ 1 \le j \le n_i \end{cases} \tag{1}$$

The run generated by $\mathcal{S}$ versus A, denoted by $d_t(\mathcal{S}, A)$, is the result of the game proceeding in the maximum number of rounds and steps. When no confusion is possible, we drop the subscript t from $d_t^{i,j}(\mathcal{S}, A)$ and $d_t(\mathcal{S}, A)$.

In the above definition, the sequence $p_{i,1}p_{i,2}\ldots p_{i,n_i}$ is referred to as the *ith segment of* t, and the local scheduler $S{\downarrow_{p_{i,j}}}$ is said to be *activated* in round i, step j of the game. Note that by the definition it is easy to see that the sequence $d^{i,j}(\mathcal{S}, A)$ represents a run, and every interaction in $d^{i,j}(\mathcal{S}, A)$ is S-admissible.

To illustrate the game, consider the interaction system shown in Fig. 1. Let S be a nonblocking scheduler that schedules the least-executed-interaction for execution; tie is broken arbitrarily. Furthermore, let each $S{\downarrow_{p_i}}$, $1 \leq i \leq 4$, be a restriction of S such that if there is some x satisfying $S([\pi]_x) = x$ and $p_i \in P_x$, then $S{\downarrow_{p_i}}(\pi) = x$; and if there is more than one interaction satisfying the condition, then $S{\downarrow_{p_i}}(\pi)$ returns the least executed one. Again, tie is broken arbitrarily. Suppose $d^{1,0}(\mathcal{S}, A) = p_1p_3p_2p_4$. Assume all four local schedulers are activated in the order of $S{\downarrow_{p_1}}, S{\downarrow_{p_3}}, S{\downarrow_{p_2}}$, and $S{\downarrow_{p_4}}$. Consider first $d^{1,1}(\mathcal{S}, A)$, which by definition is $d^{1,0}(\mathcal{S}, A) \cdot S{\downarrow_{p_1}}(p_1p_3p_2p_4)$. Since $S{\downarrow_{p_1}}(p_1p_3p_2p_4) = x_{12}$, $d^{1,1}(\mathcal{S}, A) = p_1p_3p_2p_4x_{12}$. Then, since $S{\downarrow_{p_3}}(p_1p_3p_2p_4x_{12}) = x_{34}$, $d^{1,2}(\mathcal{S}, A) = p_1p_3p_2p_4x_{12}x_{34}$. Since no more interaction is enabled, $d^{1,3}(\mathcal{S}, A) = d^{1,4}(\mathcal{S}, A) = d^{1,2}(\mathcal{S}, A)$.

One may have observed a "centralization" nuance in Definition 7 that the input to each local scheduler is the global run. As we shall see shortly in this section, the restriction imposed on local schedulers makes it no difference whether the input is local or global.

Definition 8. Let S be a nonblocking scheduler for $\mathbb{IS} = (\mathsf{P}, \mathsf{I}, \mathsf{M})$, and let $\mathcal{S} = \{S{\downarrow_p} \mid p \in \mathsf{P}\}$ be a set of restrictions. Given $d_t(\mathcal{S}, A)$, the trace t is *advertent* if $d_t(\mathcal{S}, A)$ is complete and no nonempty subset of P is S-neglected in $d_t(\mathcal{S}, A)$. The set of *advertent runs* of $\mathcal{S}$ versus A, denoted by $\mathbb{AR}(\mathcal{S}, A)$, is the set of complete runs

$$\mathbb{AR}(\mathcal{S}, A) = \left\{ d_t(\mathcal{S}, A) \;\middle|\; t \text{ is advertent.} \right\}$$

Since every interaction in $d_t(\mathcal{S}, A)$ is S-admissible, the following lemma follows immediately.

Lemma 3. *Assume that* $\mathbb{C}$ *is localizable for* $\mathbb{IS} = (\mathsf{P}, \mathsf{I}, \mathsf{M})$. *Let* S *be a localizable scheduler for* $\mathbb{IS}$ *satisfying* $\mathbb{C}$, *and let* $\mathcal{S} = \{S{\downarrow_p} \mid p \in \mathsf{P}\}$ *be an arbitrary set of restrictions. Then, for every adversary* A, $\mathbb{AR}(\mathcal{S}, A) \subseteq \mathbb{C}[\![\mathbb{IS}]\!]$.

In the following we present a scheme—our main lemma of the paper—to guarantee advertent traces. Based on this scheme a real distributed implementation for any given localizable semantic constraint will then be presented in the next section. The lemma statement is somewhat complex, and so some comments have been placed inside the lemma to help understand the lemma. The proof will be given in the full paper.

Lemma 4. *Let* S *be a nonblocking scheduler for* $\mathbb{IS} = (\mathsf{P}, \mathsf{I}, \mathsf{M})$, $\mathcal{S} = \{S{\downarrow_p} \mid p \in \mathsf{P}\}$ *be a set of restrictions,* A *be an adversary,* t *be a trace, and* $C = \{c(p, i) \mid p \in \mathsf{P}, i \in N\}$ *be a set of integers. In the game that generates* $d_t(\mathcal{S}, A)$, *associates with each* $p \in \mathsf{P}$ *in round* i *the integer* $c(p, i)$ *such that* $c(p, i)$ *is referred to as the* age *of* p *in round* i; *the larger the value, the younger the process. The ages of the processes satisfy the following two conditions:*

- *$c(p,i) \neq c(q,i)$ if $p \neq q$; that is, processes' ages are distinct in each round.*
- *$c(p,i) = c(p,i+1)$ if p is ready in round i (immediately after step 0 of the round), and remains ready in round $i+1$; that is, p's age remains the same in a ready state.*

Let $p_{i,1}p_{i,2}\cdots p_{i,n_i}$ be the ith segment of t. Then, t is advertent if the following three conditions are satisfied during the game:

1. *A process p's local scheduler is activated at most once from the time p enters a ready state until the time it leaves the state. Moreover, if x becomes enabled in round i and p is the youngest process among P_x in round i, then p's local scheduler will be activated (in round i or some round afterwards).*
Remark. *The condition implies that if x is disabled because some participant p has executed a conflicting interaction, then when p is ready again to enable x, if all the other participants of x remain ready for x and their local schedulers have been activated (since last time x is enabled), then p's age must be the youngest among P_x.*
2. *For each pair (i,j), $1 \leq j \leq n_i$, if there exists an x satisfying the following condition*
$$p_{i,j} \in P_x,\ S([d^{i,j-1}(S,A)]_{P_x}) = x, \text{ and } c(p_{i,j},i) = \max\{c(p,i) \mid p \in P_x\},$$
then $S{\downarrow}_{p_{i,j}}(d^{i,j-1}(S,A)) = x$. Otherwise $S{\downarrow}_{p_{i,j}}(d^{i,j-1}(S,A)) = \epsilon$. If more than one interaction satisfy the above condition, then $S{\downarrow}_{p_{i,j}}$ can return any one of them, except that if infinitely often $S{\downarrow}_{p_{i,j}}$ can return some x, then x must be executed infinitely often in the run.
Remark. *The condition implies that if x is enabled, then only the scheduler of the youngest process in P_x can establish x.*
3. *For each pair (i,j), $1 \leq j \leq n_i$, if $S{\downarrow}_{p_{i,j}}(d^{i,j-1}(S,A)) = x$ for some $x \neq \epsilon$, then for all $y \in \mathsf{I}$ such that $P_x \cap P_y \neq \emptyset$ and $S([d^{i,j-1}(S,A)]_{P_y}) = y$, if $c(p,i) = \max\{c(q,i) \mid q \in P_y\}$, then either $p = p_{i,j}$ or $p \notin P_x$.*
Remark. *The condition implies that while an interaction y remains enabled and p is the youngest process in P_y, no other scheduler except the scheduler of p can schedule p to execute an interaction.*

It should be noted that some trace t may cause conditions 2 and 3 of the lemma to conflict. For example, assume a system of two interactions x_{12} and x_{23}, where $P_{x_{12}} = \{p_1,p_2\}$, and $P_{x_{23}} = \{p_2,p_3\}$. Suppose for some given π, $S([\pi]_{P_{x_{12}}}) = x_{12}$ and $S([\pi]_{P_{x_{23}}}) = x_{23}$, and p_1 and p_2 are the youngest processes in $P_{x_{12}}$ and $P_{x_{23}}$, respectively, up to the end of π. If p_1's scheduler is activated before p_2, then, by condition 2 p_1's scheduler must choose x_{12}, while by condition 3 x_{12} cannot be chosen. On the other hand, no conflict occurs if p_2's scheduler is activated before p_1. In this case, p_2's scheduler must establish x_{23} because p_1 is younger than p_2 and so by condition 2 p_2's scheduler cannot establish x_{12}. Once x_{23} is established, all interactions are disabled, leaving p_1's scheduler no choice but to establish an empty interaction.

We have described how interactions can be scheduled locally in an abstract level so that the overall run rendered by the processes is valid. Still, the game defined in Definition 7 between local schedulers and a given adversary is in some sense "centralized." This is because the input given to each $S{\downarrow}_{p_{i,j}}$ is the entire global run proceeds so far. To make the game really "distributed", each local scheduler should be concerned with only the portion of the global run relevant to its scheduling. Fortunately, the restriction imposed on local schedulers makes it no difference whether the input is the entire global run or only the relevant portion. What needs to be taken care of is to ensure that the combined effect of independent local scheduling yields an advertent trace, and this is exactly why Lemma 4 is conceived.

Finally, before presenting a real implementation for the local schedulers, we need a way to combine projections. Let $V_\varphi = \{[\varphi]_{p_1}, \dots, [\varphi]_{p_m}\}$ be a set of projections of φ. The *join* ρ of V_φ (with respect to φ) is the projection $[\varphi]_{\{p_1,\dots,p_m\}}$. It follows directly that $\forall 1 \leq i \leq m : [\rho]_{p_i} = [\varphi]_{p_i}$. For example, consider Fig. 1, and let $\pi_2 = p_1p_2x_{12}p_2$ and $\pi_3 = p_3p_4x_{34}p_3$ be the projections of p_2 and p_3, respectively, in $\varphi = p_1p_2x_{12}p_3p_4x_{34}p_2p_3p_4$. Then the join of $\{\pi_2, \pi_3\}$ is $p_1p_2x_{12}p_3p_4x_{34}p_2p_3$. Note that, in general, different runs may yield the same projection. So given a set of projections $V_\varphi = \{[\varphi]_{p_1}, \dots, [\varphi]_{p_m}\}$, it is virtually impossible to compute their join without the knowledge of φ. For example, π_2 and π_3 are also the projections of p_2 and p_3, respectively, in $\varphi' = p_1p_3p_2x_{12}p_2p_4x_{34}p_3$. So $p_1p_3p_2x_{12}p_2p_4x_{34}p_3$ is the join of $\{\pi_2, \pi_3\}$ with respect to φ'. However, the join ρ of $\{[\varphi]_{p_1}, \dots, [\varphi]_{p_m}\}$ with respect to φ cannot include any action not in the projections $[\varphi]_{p_1}, \dots, [\varphi]_{p_m}$. So, if we know the relative order of the actions in the projections (e.g., by timestamps), then we can still compute their join without the complete knowledge of φ.

5 A Real Implementation

Based on the results in the previous section we now present a general distributed algorithm to implement semantic constraints that satisfy the localizability and equivalence-robustness criteria. By this we mean augmenting each process in an interaction system with variables and actions, and possibly introducing auxiliary processes, so that the resulting computations satisfy a given semantic constraint. Due to the space limitation, we shall only highlight the main ideas. The complete code and its analysis will be given in the full paper.

Let $\mathbb{C}$ be a localizable semantic constraint on $\mathbb{IS} = (\mathsf{P}, \mathsf{I}, \mathsf{M})$, and let S be a localizable scheduler satisfying $\mathbb{C}$. In the implementation we pair each process $p_i \in \mathsf{P}$ with a coordinator process $Coord_i$, which p_i "activates" upon entering a ready state. $Coord_i$ acts as a local scheduler for p_i by simulating the scheduling of some S's restriction to p_i. Processes communicate exclusively with coordinators and vice versa. We assume that communication is by reliable FIFO message-passing.

In the light of Lemma 3 we must ensure that the overall run rendered by $\mathbb{IS}$ is equivalent to some $d_t(\mathcal{S}, A)$, where $\mathcal{S}$ is the set of restrictions simulated by the coordinators, A is an adversary of $\mathbb{IS}$, and t is an advertent trace. To do so, when $Coord_i$ is activated, it chooses an interaction x from $p_i.aim$, and attempts to establish x by *capturing* the processes in P_x. When all processes are captured, $Coord_i$ computes the global run *viewed* by the participants of x. By this "global" run we mean the computation in which events (concerning only processes' readiness and their interactions) are ordered by their logical timestamps. Note that this run may not be the same as the one that is observed outside the system (i.e., the run where events are ordered by the universal time at which they take place.) However, the logical clocks maintained by the processes guarantee that the two runs are equivalent. Since $\mathbb{C}$ is equivalence-robust, it suffices to ensure that the global run envisaged by the processes is valid. We shall use α to denote this global run.

We do not need each p_j to maintain a copy of α. Instead, p_j maintains only its local view of α, which, as shall be clear, corresponds to p_j's projection in α. The local view is built incrementally by a variable β_j as follows. Initially, $\beta_j = \epsilon$. When p_j becomes ready, it appends the ready transition $p_j.aim$ to β_j. The timestamp of the ready transition is taken from p_j's logical clock at which the action occurs. Logical clocks are maintained through timestamped message-passing as described in [17]. When a coordinator $Coord_i$ attempting x has captured all processes in P_x, it computes the join π of $\{\beta_j \,|\, p_j \in P_x\}$, and checks if $S(\pi) = x$.[3] (Note that $\pi = [\alpha]_{P_x}$.) If so, $Coord_i$ has successfully established x for p_i. It then releases all processes it has captured and directs them to execute x by informing each p_j to update its β_j to $\pi \cdot x$, where x's timestamp is given by $Coord_i$'s logical clock at the time it establishes x. If $S(\pi) \neq x$, $Coord_i$ aborts its attempt on x by releasing all processes it has captured except p_i, which will still remain captured by $Coord_i$. $Coord_i$ then attempts another interaction in $p_i.aim$ for p_i. If all interactions have been attempted, then $Coord_i$ stops its coordination activity and releases p_i, which can now be captured by other coordinators. Note that in the above $Coord_i$ will also abort its attempt on x if it fails to capture a process in P_x.

The way that each β_j is maintained guarantees that at any time instance $\beta_j = [\alpha]_{p_j}$. As time proceeds ad infinitum, so long as we can guarantee that α corresponds to some $d_t(S, A)$ defined in Definition 7 such that the trace t is advertent, then by Lemma 3 α is valid. To relate α with $d_t(S, A)$, we assume that an activated coordinator $Coord_i$ establishes an empty interaction if it fails to establish any interaction for p_i. The timestamp of the empty interaction is given by the time $Coord_i$ stops its coordination activity. So the relative order of the coordinators' scheduling in $d_t(S, A)$ is distinguished by the time they finish their coordination activities (either because they have established a nonempty interaction or because they have failed to do so).

Recall that Lemma 4 provides a scheme to guarantee an advertent trace. To apply the lemma, we use the timestamp of p_i's ready transition to implement p_i's age in the ready state p_i enters. (Timestamps are made unique by additionally considering process id's.) The age of a coordinator $Coord_i$ is taken to be the same as the age of p_i. Before describing our techniques for coping with the three conditions of the lemma, we must note that coordinators may be activated concurrently and therefore attempt to establish interactions at the same time. So a necessary condition to apply Lemma 4 is to guarantee that conflicting interactions cannot be established simultaneously. The main techniques used in the implementation are summarized below.

1. A process can be captured by only one coordinator at a time.
2. Upon activating $Coord_i$, p_i is immediately captured by $Coord_i$ and remains captured until $Coord_i$ stops its coordination activity.
3. An activated coordinator $Coord_i$ will not stop its coordination activity until either it has established an interaction, or for each $x \in p_i.aim$ $Coord_i$ has failed to capture a process in an attempt to establish x.
4. A coordinator $Coord_i$ is only allowed to capture older processes (except p_i, which has the same age as $Coord_i$).
5. A coordinator $Coord_i$ attempting to capture p_j to establish x can fail only if (a) p_j is not ready for x, or (b) $Coord_i$ is older than p_j.

[3] Because each β_j is the projection of the global run α where actions are ordered by their timestamps, the join of $\{\beta_j \,|\, p_j \in P_x\}$ can be computed without the complete knowledge of α; see the comment at the end of Section 4.

6. Two or more coordinators seeking to establish interactions involving a common process are said to *conflict*. Specifically, a conflict arises when a coordinator attempts to capture a process that already belongs to another coordinator. Conflicts are resolved in favor of older coordinators to prevent a coordinator from being locked out from capturing processes. That is, we let the older of the two coordinators capture/retain the process p under contention. The younger coordinator has then to wait until p is released by the older coordinator.
7. A coordinator $Coord_i$ attempts the interactions in $p_i.aim$ in a round-robin style: if $Coord_i$ has successfully established x, then x will be placed into the end of the attempt list (which is initialized arbitrarily) so that when next time $Coord_i$ is activated, it will not attempt x until all other attempts have failed.

In the following we informally argue why these techniques are useful to guarantee the three conditions of Lemma 4, thereby ensuring that all computations of the system satisfy $\mathbb{C}$. A more formal proof will be given in the full paper. First of all, technique 1 guarantees that conflicting interactions are not established simultaneously. Next, consider the first condition of Lemma 4. There are two clauses in the condition. Observe that techniques 4 and 6 guarantee that no other coordinator $Coord_k$ can successfully capture p_i while p_i is being captured by $Coord_i$: If $Coord_k$ is older than p_i then it is not allowed to capture p_i. If $Coord_k$ is younger, then it is also younger than $Coord_i$ because $Coord_i$ has the same age as p_i. So $Coord_k$ must wait until $Coord_i$ has finished its coordination activity and released p_i. By technique 2, then, immediately after p_i activates $Coord_i$, no other coordinator can capture p_i until $Coord_i$ has stopped its coordination activity. The fact that a process activates its coordinator only once in a ready state, and that a coordinator establishes at most one interaction for its process per activation then ensure that a process's local scheduler is activated exactly once in a ready state. So the first clause of the condition is satisfied.

Based on the first clause, the second clause can be derived by showing that if x is enabled immediately after some process $p \in P_x$ enters a ready state, then it cannot be the case that p is not the youngest process in P_x and the coordinator of the youngest process in P_x has stopped its coordination activity. To see this, technique 3 guarantees that if a coordinator $Coord_i$ fails to establish x, then some $p_j \in P_x$ must have rejected $Coord_i$'s capture request. By the logical clock adjustment, when p_j receives $Coord_i$'s capture request, p_j's clock must be advanced to a value greater than $Coord_i$'s (and thus p_i's) age. By technique 5, for p_j to reject $Coord_i$'s capture request, either p_j is not yet ready for x, or p_j (and thus $Coord_j$) is younger than $Coord_i$. Note that here it suffices to consider the former case, because in the latter case p_j is in a ready state and by the implementation p_j must activate $Coord_j$ regardless of whether it is the youngest process in P_x or not. For the former case, it is clear that p_j must obtain an age younger than p_i when it enters a ready state. So the second clause is satisfied.

For condition 2, techniques 3, 4, and 5 and the fact that coordinators will not cyclicly block one another from capturing processes (see technique 6) ensure that at any given point if $S([\alpha]_{P_y}) = y$, then the youngest coordinator, say $Coord_l$, among $\{Coord_k : p_k \in P_y\}$ will establish y, unless some other coordinator $Coord_j$ has concurrently established a conflicting interaction z. In the later case, let p be a process in $P_y \cap P_z$. By technique 5 $Coord_l$ will not give up capturing p until p has rejected $Coord_l$'s capture request (because z has been established by $Coord_j$). So $Coord_j$ will finish its coordination activity before $Coord_l$ does.

By our ordering of coordinators' scheduling, z is added to α before $Coord_l$'s scheduling result is added (which may be an empty interaction). Note that by technique 4 $Coord_j$ must be the youngest coordinator among $\{Coord_k : p_k \in P_z\}$. This means that, when viewing α as some $d_t(S, A)$ in Lemma 4, if $p_{i,j} \in P_x$, $S([d^{i,j-1}(S,A)]_{P_x}) = x$, and $p_{i,j}$ is the youngest process among P_x, then $S{\downarrow_{p_{i,j}}}(d^{i,j-1}(S,A))$ must not return an empty interaction. Moreover, technique 7 guarantees that if there is another interaction x' satisfying the same condition (that $p_{i,j} \in P_{x'}$, $S([d^{i,j-1}(S,A)]_{P_{x'}}) = x'$, and $p_{i,j}$ is the youngest process among $P_{x'}$) but the associated coordinator of $p_{i,j}$ chooses x, then when next time the same situation occurs the coordinator will favor x'. So condition 2 of Lemma 4 is satisfied.

Finally, recall that techniques 2, 4, and 6 guarantee that immediately after p_i has entered a ready state and activated $Coord_i$, no other coordinator $Coord_k$ can successfully capture p_i until $Coord_i$ has finished its coordination activity. That is, $Coord_k$'s scheduling result is timestamped after $Coord_i$'s. So if $S([\alpha]_{P_x}) = x$ and p_i is the youngest process in P_x, then no other coordinator can capture p_i until $Coord_i$ has attempted x. So condition 3 of Lemma 4 is satisfied.

6 Conclusions

We have presented a localizability criterion that allows local runs to be composed into a valid global run. Based on this criterion, we have presented a general distributed algorithm to implement equivalence-robust semantic constraints. As we have shown, SIF$^+$ is localizable. Since SIF is stronger than most semantic constraints in the literature, most equivalence-robust semantic constraints are localizable as well. That is, they can all be distributedly implemented. (Note that SIF is not equivalence-robust, and cannot even be implemented by any centralized algorithm [14].) In particular, when interactions cannot contain interactions (y *contains* x if $x \neq y$ and $P_x \subseteq P_y$), we have shown in [15] that SIF$^+$ is the strongest implementable equivalence-robust semantic constraint one can impose on any system where processes interact by engaging in interactions. Our results in this paper then indicate that under such circumstances SIF$^+$ is also the strongest semantic constraint that can be distributedly implemented.

Semantic constraints are often defined over an abstract model (like the one presented in Section 2) where the execution of interactions is assumed to be "atomic." However, processes in the underlying system of execution are usually asynchronous and so must synchronize in order to execute an interaction. So the paper is concerned with whether or not a semantic constraint can be (distributedly) implemented in the underlying system. Other issues related to global vs. local views of fairness have also been studied in the literature. Howell, et al. [11] investigate how the global and local views of state fairness will affect the complexity and decidability of the *fair nontermination problem* (which concerns whether there exists an infinite fair computation). They show that the problem related to local fairness is, in some cases, harder to analyze than that related to global fairness. Burkhard [5] studies the effects of combination and of distribution for controls (i.e., an abstract scheduling) for multi-agent systems. In general, the effects of different individual controls cannot regard arbitrary global requirements, neither can the recombination of a distributed global control. Note that both papers study their problems in an abstract model as described above.

Acknowledgments. The work of the paper is stimulated from some earlier discussions with Reino Kurki-Suonio on the implementability of fairness notions. The author would also like to thank Nissim Francez for introducing him to the area of fairness semantics and for providing him with many valuable comments and suggestions in the area. Last, but not the least, the anonymous referees of CONCUR99 also help improve the content and the presentation of the paper.

References

1. K. R. Apt, N. Francez, and S. Katz. Appraising fairness in languages for distributed programming. *Distributed Computing*, 2(4):226–241, 1988.
2. R. J. R. Back and R. Kurki-Suonio. Distributed cooperation with action systems. *ACM TOPLAS*, 10(4):513–554, 1988.
3. T. Bolognesi and E. Brinksma. Introduction to the ISO specification language LOTOS. *Computer Networks and ISDN Systems*, 14:25–59, 1987.
4. G. N. Buckley and A. Silberschatz. An effective implementation for the generalized input-output construct of CSP. *ACM TOPLAS*, 5(2):223–235, 1983.
5. H.-D. Burkhard. Fairness and control in multi-agent systems. *TCS*, 189(1–2):109–127, 1997.
6. N. Francez. *Fairness*. Springer-Verlag, 1986.
7. N. Francez, R. J. R. Back, and R. Kurki-Suonio. On equivalence-completions of fairness assumptions. *Formal Aspects of Computing*, 4:582–591, 1992.
8. N. Francez and I. R. Forman. *Interacting Processes: A Multiparty Approach to Coordinated Distributed Programming*. Addison Wesley, 1996.
9. N. Francez, B. Hailpern, and G. Taubenfeld. Script: A communication abstraction mechanism. *Science of Computer Programming*, 6(1):35–88, 1986.
10. C. A. R. Hoare. Communicating sequential processes. *CACM*, 21(8):666–677, 1978.
11. R. R. Howell, L. E. Rosier, and H.-C. Yen. Global and local views of state fairness. *TCS*, 80(1):77–104, 1991.
12. H.-M. Järvinen and R. Kurki-Suonio. DisCo specification language: Marriage of actions and objects. In *Proc. 11th ICDCS*, pages 142–151, 1991.
13. H.-M. Järvinen, R. Kurki-Suonio, M. Sakkinen, and K. Systä. Object-oriented specification of reactive systems. In *Proc. 12th Int'l Conf. on Software Engineering*, pages 63–71, 1990.
14. Y.-J. Joung. Characterizing fairness implementability for multiparty interaction. In *Proc. 23rd ICALP*, pages 110–121, 1996. *LNCS 1099*.
15. Y.-J. Joung. On strong feasibilities of equivalence-completions. In *Proc. 15th ACM PODC*, pages 156–165, 1996.
16. Y.-J. Joung and S. A. Smolka. A comprehensive study of the complexity of multiparty interaction. *JACM*, 43(1):75–115, 1996.
17. L. Lamport. Time, clocks and the ordering of events in a distributed system. *CACM*, 21(7):558–565, 1978.
18. A. W. Mazurkiewicz. Concurrent program schemes and their interpretations. TR DAIMI PB-78, Aarhus University, Aarhus, Denmark, 1977.
19. R. Milner. Calculi for synchrony and asynchrony. *TCS*, 25:267–310, 1983.
20. R. Milner. *Communication and Concurrency*. Prentice Hall, 1989.
21. R. Milner, J. Parrow, and D. Walker. A calculus of mobile processes, I. *Information and Computation*, 100(1):1–40, 1992.
22. S. Owicki and L. Lamport. Proving liveness properties of concurrent programs. *ACM TOPLAS*, 4(3):455–495, 1982.
23. W. Reisig. Partial order semantics vs. interleaving semantics for CSP-like languages and its impact on fairness. In J. Paredaens, editor, *LNCS 172*, 1984.

Generating Type Systems for Process Graphs

Barbara König

Fakultät für Informatik, Technische Universität München
koenigb@in.tum.de

Abstract. We introduce a hypergraph-based process calculus with a generic type system. That is, a type system checking an invariant property of processes can be generated by instantiating the original type system. We demonstrate the key ideas behind the type system, namely that there exists a hypergraph morphism from each process graph into its type, and show how it can be used for the analysis of processes. Our examples are input/output-capabilities, secrecy conditions and avoiding vicious circles occurring in deadlocks.
In order to specify the syntax and semantics of the process calculus and the type system, we introduce a method of hypergraph construction using concepts from category theory.

1 Introduction

In this work we propose a framework for the generation of type systems checking invariant properties of processes. We introduce a graph-based, asynchronous process calculus, similar to the polyadic π-calculus [13, 15], and give a generic type system for this calculus. Specialized type systems can then be generated by instantiating the original system.

Type systems are a valuable tool for the static analysis of parallel processes. Applications range from checking the use of channels [19, 11], confirming confluence [17], avoiding deadlocks [10] and ascertaining security properties [1, 4]. In all these cases, types are considered as partial descriptions of process behaviour, staying invariant during reduction. Furthermore a method for inferring properties of a process out of its behaviour description is required. Generally types are computable and in some type systems there is a most general or principal type for every typable process, from which all other types of the process can be derived. The flip side to type systems is the fact that some correct processes may not be typable.

Examining the type systems mentioned above, one can observe that they share similarities concerning the structure of types and typing rules. Our idea is to present a framework making a first step towards the integration of different type systems. The generic type system presented in this paper satisfies the subject reduction property, and we can guarantee absence of runtime errors for well-typed processes, the existence of principal types and of a type inference algorithm.

In order to type communicating processes, recursive types are essential. They can be represented in several ways: as expressions with a recursion operator μ (e.g. in [21]), as infinite trees [19] or as graphs [20,23] (for different representations of recursive types for the λ-calculus see [22]). We chose graph representation for types as well as for processes. This enables us to establish a close correspondence between processes and types: there is a graph morphism from each process into its type. Thus, if a type graph satisfies a property which is closed under inverse graph morphisms (e.g. absence of circles, necessary for deadlock prevention), it is also valid for the process and—because of the subject reduction property—for all its successors.

Since in a general type system like ours the properties of a process which are to be analyzed are not fixed a priori, a close relationship between processes and their types is essential. Describing both processes and types by graphs seems a convenient method for allowing easy inference of process properties. This allows a systematic approach to obtaining correctness proofs for generated type systems. It is not clear to us how the same effect could be achieved by a string representation of processes.

There are several papers describing various ways of representing processes by graphs [14,16,9,18]. Our method is closest to [9], but differs in several aspects, the most prominent being that we employ hierarchical hypergraphs.

Pure graph structure (or hypergraph structure in our case) is ordinarily not sufficient to capture relevant properties of a process. We therefore enrich our types by annotating them with lattice elements, e.g. describing input/output-capabilities of ports or channels. Every set of mappings assigning lattice elements to nodes or edges of a graph is a lattice itself, and thus it is sufficient to assign only one lattice element to every graph. It is necessary to define, how these lattice elements behave under morphisms. This is described by a functor mapping graph morphisms to join-morphisms in lattices. It is not the only case where we make use of category theory. It also allows us to give an elegant definition of graph construction (related to [6]) in terms of co-limits.

The annotation of graphs with lattice elements is another argument in favour of the use of graphs. It is more convenient to add additional labels or structures to a type represented as a graph than to a type represented by a term. This point will become clearer in section 5 where we will assign lattice elements to pairs of nodes.

2 Categorical Hypergraph Construction

We work with a variant of graphs: so-called hypergraphs [7,2], where each edge has several (ordered) source nodes. There are two kinds of labels: edge sorts and and edge labels.

Definition 1. (Hypergraph) *Let Z be a fixed set of edge sorts and let L be a fixed set of labels. A* simple hypergraph *G is a tuple $G = (V, E, s, z, l)$ where V is a set of* nodes, *E is a set of* edges *disjoint from V, $s: E \to V^*$ maps each edge* *to a string of* source nodes, $z: E \to Z$ *assigns a sort to each edge and* $l: E \to L$ *assigns a label to each edge.*

A hypergraph *or* multi-pointed hypergraph $H = G[\chi]$ *is composed of a simple hypergraph* $G = (V, E, s, z, l)$ *and a string* $\chi \in V^*$. χ *is called the string of* external nodes. EXT_H *is the set of all external nodes in* H.

The components of a hypergraph H are denoted by $V_H, E_H, s_H, z_H, l_H, \chi_H$, while the components of a simple hypergraph G are denoted by V_G, E_G, s_G, z_G, l_G. Furthermore we define the arity of edges and hypergraphs as follows: $ar(e) := |s_H(e)|$, if $e \in E_H$, and $ar(G[\chi]) := |\chi|$.

External nodes are the interface of a hypergraph with its environment and are used to attach hypergraphs. In the process calculus, which will be presented in section 3, we have two edge sorts, dividing the edge set into processes and messages, while the label of a process specifies its behaviour. In the rest of this paper we use both terms graph and hypergraph interchangeably

The following definition of hypergraph morphism is quite straightforward. A morphism is expected to preserve graph structure, as well as edge sorts and labels:

Definition 2. (Hypergraph Morphism) *Let* G, G' *be two simple hypergraphs.* *A* hypergraph morphism $\phi: G \to G'$ *consists of two mappings* $\phi_E: E_G \to E_{G'}$, $\phi_V: V_G \to V_{G'}$ *satisfying for all* $e \in E_G$:

$$\phi_V(s_G(e)) = s_{G'}(\phi_E(e)) \quad z_G(e) = z_{G'}(\phi_E(e)) \quad l_G(e) = l_{G'}(\phi_E(e))$$

We write $\phi : G[\chi] \to G'[\chi']$ *if* $\phi : G \to G'$ *is a hypergraph morphism. If*[1] $\phi_V(\chi) = \chi'$, ϕ *is called a* strong morphism *and we write* $\phi : G[\chi] \twoheadrightarrow G'[\chi']$

$G[\chi]$ *and* $G'[\chi']$ *are called* strongly isomorphic *(*$G[\chi] \cong G'[\chi']$*) if there exists a bijective strong morphism from one graph into the other.*

Notation:

We call a hypergraph *discrete*, if its edge set is empty. **m** denotes a discrete graph of arity $m \in \mathbb{N}$ with m nodes where every node is external (see (a), external nodes are labelled (1), (2), ... in their respective order).

$H := z_n(l)$ is the hypergraph with exactly one edge e with sort z and label l where $s_H(e) = \chi_H$, $|\chi_H| = n$, $V_H = EXT_H$ (see (b), nodes are ordered from left to right).

(a)

(b)

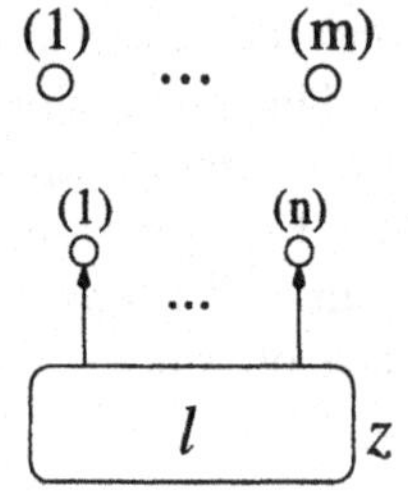

For the definition of the process calculus and its type system we need some basic concepts from category theory, namely categories, functors and co-limits. Detailed definitions can be found in [5].

Since we want to associate hypergraphs with lattice elements, we need a functor between the following two categories:

---

[1] Each morphism can be extended to strings of nodes in a canonical way, i.e. $\phi_V(v_1 \ldots v_n) = \phi_V(v_1) \ldots \phi_V(v_n)$

The category of hypergraphs (with hypergraph morphisms): The class of all simple hypergraphs (multi-pointed hypergraphs) forms a category together with the (strong) hypergraph morphisms.

The category of lattices (with join-morphisms): Let $(I_1, \leq_1)$, $(I_2, \leq_2)$ be two lattices with bottom elements $\bot_1$ respectively $\bot_2$. For two elements $a_1, b_1 \in I_1$ $(a_2, b_2 \in I_2)$ let $a_1 \vee_1 b_1$ $(a_2 \vee_2 b_2)$ be the *least upper bound* or *join* of the two elements.
A mapping $t : I_1 \to I_2$ is called a *join-morphism* iff $t(a_1 \vee_1 b_1) = t(a_1) \vee_2 t(b_1)$ and $t(\bot_1) = \bot_2$

Type graphs are hypergraphs $G[\chi]$ which are associated with a lattice element. A type functor F maps every type graph to a lattice $F(G)$ from which this associated lattice element can be taken. The concept of hypergraph morphisms can be extended to type graph morphisms, from which we demand, that they not only preserve graph structure but also the order in the corresponding lattices.

Definition 3. (Type Functors and Type Graphs) *A functor F from the category of simple hypergraphs into the category of lattices is called a* type functor.

$T = G[\chi, a]$ *where $G[\chi]$ is a hypergraph and $a \in F(G)$ is called a* type graph *wrt. F. The class of all type graphs wrt. F is denoted by $\mathcal{T}_F$.*

We write $\phi : G[\chi, a] \overset{F}{\to} G'[\chi', a']$ if $\phi : G \to G'$ is a hypergraph morphism and $F(\phi)(a) \leq a'$. ϕ is called type graph morphism.

We say ϕ is a strong type graph morphism *if additionally $\phi_V(\chi) = \chi'$ and it is denoted by $\phi : G[\chi, a] \overset{F}{\twoheadrightarrow} G'[\chi', a']$.*

Two type graphs $G[\chi, a]$, $G'[\chi', a']$ are called isomorphic (wrt. F) *if there exists a strong isomorphism $\phi : G[\chi] \twoheadrightarrow G'[\chi']$ such that $F(\phi)(a) = a'$. In this case we write $G[\chi, a] \cong_F G'[\chi', a']$.*

Note: If $H = G[\chi]$ we define $H[a] := G[\chi, a]$.

Example: We consider the following type functor F: let $(I, \leq)$ be an arbitrary lattice and let $k \in \mathbb{N}$. For any simple hypergraph G we define $F(G)$ as the set of all mappings from $(V_G)^k$ (cartesian product) into I (which yields a lattice with pointwise order).

Let $a : V_G^k \to I$, $\phi : G \to G'$, $s' \in V_{T'}^k$. We define:

$$F(\phi)(a) := a' \text{ where } a'(s') := \bigvee_{\phi_V(s)=s'} a(s)$$

It is straightforward to check that F is indeed a type functor.

We now introduce a mechanism for the construction of hypergraphs. Compared to string concatenation it is not so obvious how to build larger graphs out of smaller ones. We describe a construction plan with morphisms mapping discrete graphs into discrete graphs. This construction plan is then applied to hypergraphs by a co-limit construction. Our method is related to the double-pushout approach for graph rewriting described in [6].

If we define how to transform and sum up lattice elements, we can assemble type graphs in the same way.

Definition 4. (Construction of Hypergraphs and Type Graphs)
Let $H_1, \ldots, H_n$ be hypergraphs and let $\zeta_i : \mathbf{m_i} \to D$, $i \in \{1, \ldots, n\}$ be hypergraph morphisms where $ar(H_i) = m_i \in \mathbb{N}$ and D is a discrete graph. There is always a unique strong morphism $\phi_i : \mathbf{m_i} \twoheadrightarrow H_i$ for every $i \in \{1, \ldots, n\}$.

Let H (with morphisms $\phi : D \to H$, $\zeta_i' : H_i \to H$) be the co-limit of $\zeta_1, \ldots, \zeta_n, \phi_1, \ldots, \phi_n$ such that ϕ is a strong morphism. We define:

$$\bigotimes_{i=1}^{n}(H_i, \zeta_i) := H$$

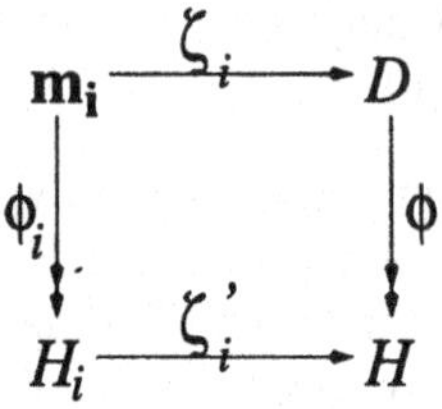

Let $T_i = H_i[a_i]$, $i \in \{1, \ldots, n\}$ be type graphs and let F be a fixed type functor. The construction of type graphs wrt. F is defined in the following way:

$$\bigotimes_{i=1}^{n}(T_i, \zeta_i) := \left(\bigotimes_{i=1}^{n}(H_i, \zeta_i)\right)[a] \text{ where } a := \bigvee_{i=1}^{n} F(\zeta_i')(a_i)$$

Generally, co-limits do not necessarily exist, but they always exist in our case. The co-limit is unique up to isomorphism (i.e. bijective morphisms), but *not* unique up to strong isomorphism. Therefore we demand above that the morphism from D into the co-limit is a strong morphism and thereby determine the string of external nodes of the result.

In order to clarify the intuition behind graph construction we give the following two examples:

Example 1: As stated above, the morphisms ζ_i can be regarded as a construction plan for assembling hypergraphs. The example in figure 1 will illustrate this: we describe how to construct H below out of smaller hypergraphs H_1, H_2, H_3. In this case $H \cong \bigotimes_{i=1}^{3}(H_i, \zeta_i)$ (see the graphical description of $\zeta_1, \zeta_2, \zeta_3$ below).
Example 2: Let H_1, H_2 be hypergraphs with $ar(H_1) = ar(H_2) = n$. Then $H_1 \Box H_2 := \bigotimes_{i=1}^{2}(H_i, \zeta_i)$ where $\zeta_1, \zeta_2 : \mathbf{n} \twoheadrightarrow \mathbf{n}$ are the unique strong morphisms from $\mathbf{n}$ into $\mathbf{n}$. That is $H_1 \Box H_2$ is constructed out of H_1, H_2 by fusing corresponding external nodes.

Every hypergraph can be decomposed into hyperedges and has the following normal form:

Proposition 1. (Graph construction out of hyperedges) *Let H be a hypergraph. Then there exists a natural number n, sorts z_i, labels l_i and morphisms $\zeta_i : \mathbf{m_i} \to D$ (where $i \in \{1, \ldots, n\}$ and D is a discrete hypergraph) such that*

$$H \cong \bigotimes_{i=1}^{n}((z_i)_{m_i}(l_i), \zeta_i)$$

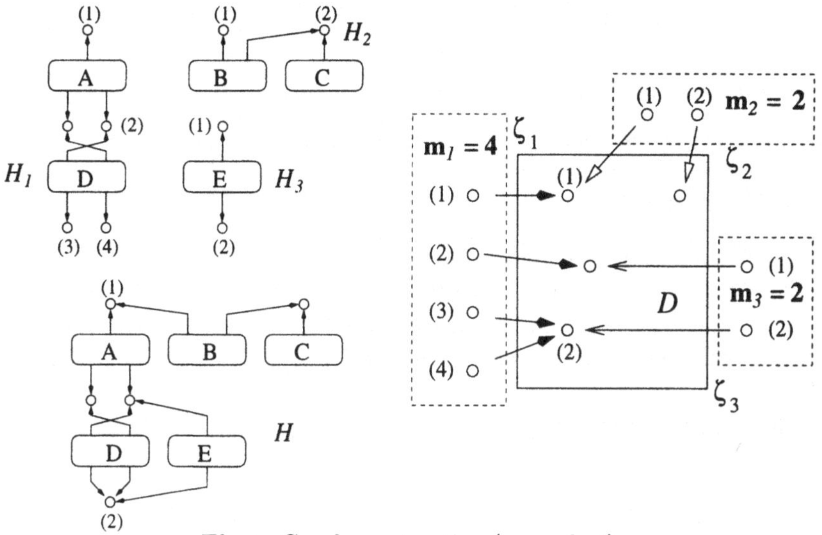

Fig. 1. Graph construction (example 1)

3 Process Graphs

We are now ready to introduce the calculus: an expression in our process calculus is a hierarchical hypergraph. Edges are representing either processes or messages (since we present an asynchronous calculus we distinguish processes and messages) and nodes are representing ports. In the rest of this paper we use the names "port" and "node" interchangeably.

Definition 5. (Process Graph) *A process graph P is inductively defined as follows: P is a hypergraph with edge sorts $Z = \{proc, mess\}$. Edges with $z_P(e) = proc$ are called processes and edges with $z_P(e) = mess$ are called messages.*

Processes are either labelled $!Q$ (Replication) or $\lambda_k^{(n)}.Q$ (the process receives a message with $n+1$ ports—one of it the send-port—on its k-th port and then behaves like Q) where Q is again a process graph. Messages have at least arity 1 *and remain unlabelled (or are labelled with dummies).*

By definition, a message is sent to its last port[2]: $send_P(e) := \lfloor s_P(e) \rfloor_{ar(e)}$ if $z_P(e) = mess$. Process graphs have an intuitive graphical representation which will be introduced step by step.

The most important form of reduction in our calculus is the reception of a message by a process, which means the replacement of a redex, consisting of process and message, by the hypergraph inside the process.

Redex: Let $P_1 := proc_m(l)$, $P_2 := mess_{n+1}$ and let $1 \le k \le m$. Furthermore let

$$\zeta_1 : \mathbf{m} \to \mathbf{m} + \mathbf{n} \text{ with } \zeta_1(\chi_\mathbf{m}) := \lfloor \chi_{\mathbf{m}+\mathbf{n}} \rfloor_{1\ldots m}$$

[2] We define the following operator on strings: if $s = a_1 \ldots a_n$ is a string, we define $\lfloor s \rfloor_{i_1 \ldots i_n} := a_{i_1} \ldots a_{i_n}$.

$$\zeta_2 : \mathbf{n+1} \to \mathbf{m+n} \text{ with } \zeta_2(\chi_{\mathbf{n+1}}) := \lfloor \chi_{\mathbf{m+n}} \rfloor_{m+1\ldots m+n\, k}$$

We define $Red_{k,m,n}(l) := \bigotimes_{i=1}^{2}(P_i, \zeta_i)$.

Graphical representation: We draw messages with dashed lines, thereby distinguishing them from processes.

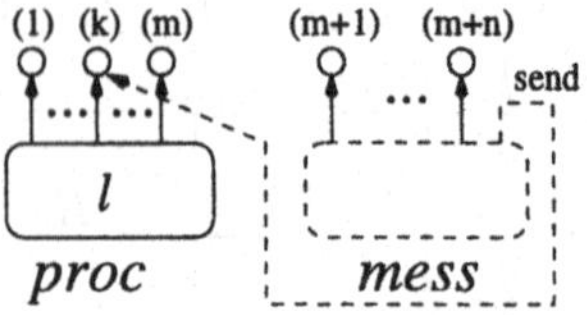

We favour reduction semantics in the spirit of the Chemical Abstract Machine [3]. Our calculus obeys the rules of structural congruence and reduction in table 1. $\equiv$ is the smallest equivalence which contains hypergraph isomorphism and which satisfies the rules (C-ABSTR), (C-REPL$_1$), (C-CON) and (C-REPL$_2$).

Note that runtime errors may occur if $ar(P) \neq m+n$ in (R-MR) or $ar(P) \neq n$

$$\frac{P_1 \equiv P_2}{proc_n(\lambda_k.P_1) \equiv proc_n(\lambda_k.P_2)} \text{ (C-ABSTR)} \qquad \frac{P_1 \equiv P_2}{proc_n(!P_1) \equiv proc_n(!P_2)} \text{ (C-REPL}_1\text{)}$$

$$\frac{P_i \equiv Q_i,\ i \in \{1,\ldots,n\}}{\bigotimes_{i=1}^{n}(P_i,\zeta_i) \equiv \bigotimes_{i=1}^{n}(Q_i,\zeta_i)} \text{ (C-CON)}$$

$$proc_n(!P) \equiv P \Box proc_n(!P) \text{ (C-REPL}_2\text{)}$$

$$Red_{k,m,n}(\lambda_k^{(n)}.P) \to P \text{ (R-MR)}$$

$$\frac{Q \equiv P,\ P \to P',\ P' \equiv Q'}{Q \to Q'} \text{ (R-EQU)} \qquad \frac{P_i \to P_i',\ (i \neq j \Rightarrow P_j \equiv P_j')}{\bigotimes_{i=1}^{n}(P_i,\zeta_i) \to \bigotimes_{i=1}^{n}(P_i',\zeta_i)} \text{ (R-CON)}$$

Table 1. Operational semantics of process graphs

in (C-REPL$_2$), i.e. if the left hand and the right hand side of a rule do not have the same arity, or if there is a mismatch in arities for the construction operator $\Box$.

Mobility of port addresses is inherent in rule (R-MR): For a process of the form $proc_m(\lambda_k^{(n)}.P)$ the arity of P should be $m+n$ in order not to cause runtime errors. If such a process receives a message with n ports attached to it, the rules cause the first m external ports of P to fuse with the ports of the process, while the rest of the ports fuses with the ports of the message. In this way a process can gain access to new ports which means dynamic restructuring of the entire process graph. This feature is often called mobility [13].

Our calculus, as presented here, is closely related to the asynchronous, polyadic π-calculus without summation (see appendix A). Asynchronous means, in this case, that the continuation of an output prefix is always 0, the nil process.

Example: In figure 2 we give a small example, illustrating message reception in the calculus. Note that messages are drawn with dashed lines, all other edges are representing processes. The dashed arrow leading away from a message indicates the send-port of a message. The arrows leading to the source nodes of an edge are ordered from left to right. Furthermore the external ports inside a process abstraction which are going to be fused with the ports of a message are filled with grey. The corresponding expression in the π-calculus would be

$$(\nu a)(\nu b)(a(a_1 a_2).\overline{a_2}\langle a_1\rangle.0 \mid b(b_1).\overline{a}\langle e_1 b_1\rangle.0 \mid \overline{b}\langle e_2\rangle.0)$$

where e_1, e_2 are the names representing external ports (the only free names). (See also appendix A.)

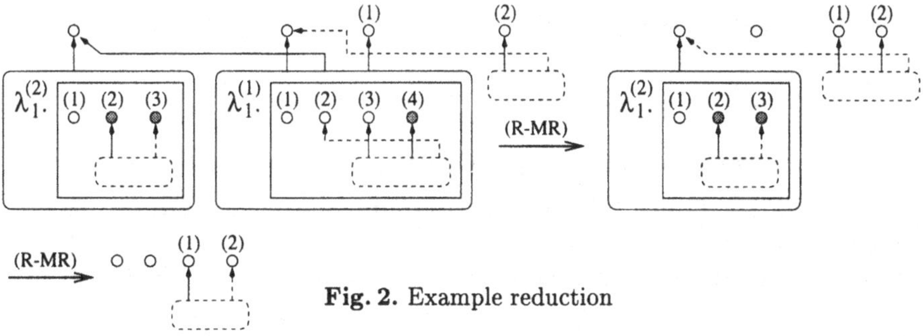

Fig. 2. Example reduction

4 The Type System

We assume that F is a fixed type functor. It is one of two parameters of the type system, we will now specify the second: we need a method for mapping a process graph to a corresponding type graph. It is only necessary to describe this mapping for graphs consisting of one edge only. The extension to arbitrary graphs is straightforward.

Definition 6. (Linear Mapping) *Let L be a function which maps graphs of the form $z_n(l)$ to type graphs in $\mathcal{T}_F$, satisfying $ar(L(z_n(l))) = n$. Furthermore we demand that*

$$P_1 \equiv P_2 \Rightarrow L(P_1) \cong_F L(P_2) \tag{1}$$

$$\exists \phi : mess_n[\bot] \overset{F}{\twoheadrightarrow} L(mess_n) \tag{2}$$

Since proposition 1 implies that all hypergraphs can be constructed out of graphs of the form $z_n(l)$ we can expand L to arbitrary hypergraphs in the following way:

$$L(\bigotimes_{i=1}^{n}(H_i, \zeta_i)) := \bigotimes_{i=1}^{n}(L(H_i), \zeta_i)$$

L is well-defined and is called a linear mapping.

Condition (2) in the definition of the linear mapping may seem somewhat out of place. It is however (together with condition (3) below and rule (T-ABSTR)) essential to the proof of the subject reduction property (see proof sketch below). Both conditions ensure that nodes that might get fused during reduction are already fused in the type graph.

The type system works with arbitrary linear mappings as long as they satisfy conditions (1) and (2). In practice, however, the structure of the graphs created by the linear mappings does not vary much (see also section 5) and the important part in defining the linear mappings is to choose sensible lattice elements.

We have now described the two parameters of the type system: the type functor F and the linear mapping L and the conditions imposed on them. The typing rules in table 2 describe how a type can be assigned to an expression, $P \triangleright G[\chi, a]$ meaning that the process graph P has type $G[\chi, a]$. We demand that in G every port is the send-port of at most one message, i.e. G satisfies:

$$e, e' \in E_G,\ z_G(e) = z_G(e') = \mathit{mess}_G,\ \mathit{send}_G(e) = \mathit{send}_G(e') \ \Rightarrow\ e = e' \qquad (3)$$

The main motivation behind the typing rules is to ensure that there exists a morphism $L(P) \overset{F}{\twoheadrightarrow} T$, if $P \triangleright T$, and that the subject reduction property holds. The former is ensured by the morphisms in rules (T-PROC), (T-MESS) and (T-CON) and the latter is mainly ensured by typing rule (T-ABSTR). In (T-ABSTR) we demand the existence of a message in the type graph which implies, with conditions (2) and (3), that, in the type graph, the images of ports attached to any message arriving at the k-th port are already fused with the images of the last n ports of P. (T-CON) checks that all parts of a hypergraph are typed and

$$\frac{l \triangleright G[\chi, a],\ \exists \phi : L(\mathit{proc}_n(l)) \overset{F}{\twoheadrightarrow} G[\chi, a]}{\mathit{proc}_n(l) \triangleright G[\chi, a]} \ \text{(T-PROC)}$$

$$\frac{\exists \phi : L(\mathit{mess}_n) \overset{F}{\twoheadrightarrow} G[\chi, a]}{\mathit{mess}_n \triangleright G[\chi, a]} \ \text{(T-MESS)} \qquad \frac{P \triangleright G[\chi, a]}{!P \triangleright G[\chi, a]} \ \text{(T-REPL)}$$

$$\frac{P \triangleright G[\chi, a],\ \exists \phi : \mathit{mess}_{n+1} \twoheadrightarrow G[\lfloor \chi \rfloor_{m+1 \ldots m+n\, k}],\ |\chi| = m + n}{\lambda_k^{(n)}.P \triangleright G[\lfloor \chi \rfloor_{1 \ldots m}, a]} \ \text{(T-ABSTR)}$$

$$\frac{\exists \phi : D \twoheadrightarrow G[\chi],\ \zeta_i : \mathbf{m_i} \to D,\ P_i \triangleright G[\phi(\zeta_i(\chi_{\mathbf{m_i}})), a],\ i \in \{1, \ldots, n\}}{\bigotimes_{i=1}^{n} (P_i, \zeta_i) \triangleright G[\chi, a]} \ \text{(T-CON)}$$

Table 2. Typing Rules

that their types overlap at least in the corresponding external ports (they may overlap in other places as well). (T-REPL) states that we can produce copies of a process without changing its type since all copies are represented by the same part of the type graph, and the join operation in lattices is idempotent.

Condition (3) can not be satisfied if there are messages with a different number of ports sent to the same port. It therefore ensures that the arities of expected and received message match and thus avoids runtime errors.

The type system satisfies the properties listed below.

Proposition 2. (Properties of the Type System)

Subject Reduction Property: $P \triangleright G[\chi, a],\ P \to^* Q \Rightarrow Q \triangleright G[\chi, a]$

Runtime Errors: $P \triangleright G[\chi, a]$ *implies that* P *will never cause a runtime error.*

Morphisms: $P \triangleright G[\chi, a] \Rightarrow \exists \phi : L(P) \overset{F}{\twoheadrightarrow} G[\chi, a]$

Principal Types: *If* P *is typable then there exists a* principal *type graph* $P \triangleright G[\chi, a]$ *with*

- $\phi : G[\chi, a] \overset{F}{\twoheadrightarrow} G'[\chi', a']$ *implies* $P \triangleright G'[\chi', a']$
- $P \triangleright G'[\chi', a']$ *implies the existence of* $\phi : G[\chi, a] \overset{F}{\twoheadrightarrow} G'[\chi', a']$, *where* ϕ *is a strong morphism*

And there exists a type inference algorithm constructing the principal type graph for every process graph, if it exists.

In order to expose the intuition behind the type system we show that reduction of a process does not change its type. In order to unravel the typing and to be able to trace it backwards, we need the following non-trivial lemma:

Lemma 1. *Let* $\bigotimes_{i=1}^{n}(P_i, \zeta_i) \triangleright G[\chi, a]$ *with* $\zeta_i : \mathbf{m_i} \to D$. *Then there is a strong morphism* $\phi : D \twoheadrightarrow G[\chi]$ *such that for all* $i \in \{1, \ldots, n\}$: $P_i \triangleright G[\phi(\zeta_i(\chi_{\mathbf{m_i}})), a]$

We now demonstrate how to prove the subject reduction property in the case of (R-MR):

Proof Sketch (Subject Reduction Property): *Let* $Red_{k,m,n}(\lambda_k.Q) \triangleright G[\chi, a]$. *It follows with lemma 1 that*

$$proc_m(\lambda_k.Q) \ \triangleright\ G[\lfloor\chi\rfloor_{1\ldots m}, a], \quad mess_{n+1} \ \triangleright\ G[\lfloor\chi\rfloor_{m+1\ldots m+n\,k}, a]$$

Since $proc_m(\lambda_k.Q)$ *was typed with (T-PROC) and (T-ABSTR) it follows that there exists a* $\chi' \in V_G^*$ *such that*

$$Q \ \triangleright\ G[\lfloor\chi\rfloor_{1\ldots m} \circ \chi', a], \quad mess_{n+1} \ \twoheadrightarrow\ G[\chi' \circ \lfloor\chi\rfloor_k]$$

And since (2) and (T-MESS) imply that

$$mess_{n+1}[\bot] \overset{F}{\twoheadrightarrow} L(mess_{n+1}) \overset{F}{\twoheadrightarrow} G[\lfloor\chi\rfloor_{m+1\ldots m+n\,k}, a]$$

it follows with condition (3) that $\chi' = \lfloor\chi\rfloor_{m+1\ldots m+n}$ *and therefore* $Q \triangleright G[\chi, a]$.

We now describe how the type system can be used for verification purposes: we introduce two predicates X and Y where X is a predicate on type graphs and Y is a predicate on process graphs. We want to show that Y is an invariant with the help of X.

Proposition 3. (Process Analysis with the Type System) *Let Y be a predicate on process graphs and let X be a predicate on type graphs of the form $G[\chi, a]$. We assume that X, Y satisfy*

$$X(L(P)) \Rightarrow Y(P) \tag{4}$$

$$\phi : G[\chi, a] \overset{F}{\twoheadrightarrow} G'[\chi', a'],\ X(G'[\chi', a']) \Rightarrow X(G[\chi, a]) \tag{5}$$

i.e. X is closed under inverse hypergraph morphisms.
Then $P \triangleright G[\chi, a]$ and $X(G[\chi, a])$ imply $Y(Q)$ for all $P \rightarrow^* Q$.

A full type system is determined by four components: the type functor F, a linear mapping L and the two predicates X, Y.

5 Examples

In the following examples we use a rather restricted linear mapping L with

$$L(proc_n(l)) \cong_F \mathbf{n}[a_p] \quad L(mess_n) \cong_F mess_n[a_m]$$

for lattice elements a_p, a_m, yet to be determined. That is, in this case, type graphs consist of messages only. Furthermore we use the type functor F introduced in the example after definition 3. The simplest version of this type system, where every lattice consists of one element only, corresponds to standard type systems for the π-calculus with recursive types, but without let-polymorphism [21].

In all of the following examples the predicate X is preserved by inverse hypergraph morphisms.

5.1 Input/Output-Capabilities

We want to ensure that some external ports are only used as input ports (i.e. for receiving messages) and that some are only used as output ports (i.e. for sending messages). For F we choose $k = 1$, $I = \{\bot, in, out, both\}$ where $\bot < in < both$ and $\bot < out < both$ and $in \vee out = both$. The linear mapping L is defined in the following way:

$$L(proc_m(l)) = \mathbf{n}[a_p] \text{ where } a_p(\lfloor\chi\rfloor_i) = \begin{cases} in \text{ if } l = \lambda_i^{(n)}.P,\ n \in \mathbb{N} \\ \bot \text{ otherwise} \end{cases}$$

$$L(mess_n) = mess_n[a_m] \text{ where } a_m(\lfloor\chi\rfloor_i) = \begin{cases} out \text{ if } i = n \\ \bot \quad \text{otherwise} \end{cases}$$

Since the only nodes of $proc_n$ and $mess_n$ are external, it is sufficient to define a_p and a_m on the respective string χ of external nodes.

We want to ensure that P will never reduce to a process graph P' where a message is sent to $\lfloor\chi_P\rfloor_i$. The corresponding predicate X is:

$$X(G[\chi, a]) := (a(\lfloor\chi\rfloor_i) \leq in)$$

If we replace *in* by *out* in X we can conclude that P will never reduce to a process graph P' where a process listens at $\lfloor \chi_P \rfloor_i$.

A similar version of this type system is presented in [19].

Typing the example process graph from section 3 (figure 2) yields the principal type in figure 3 (left). This implies that the first external part is not used for any I/O-operations at all, while the second external port is only an output port.

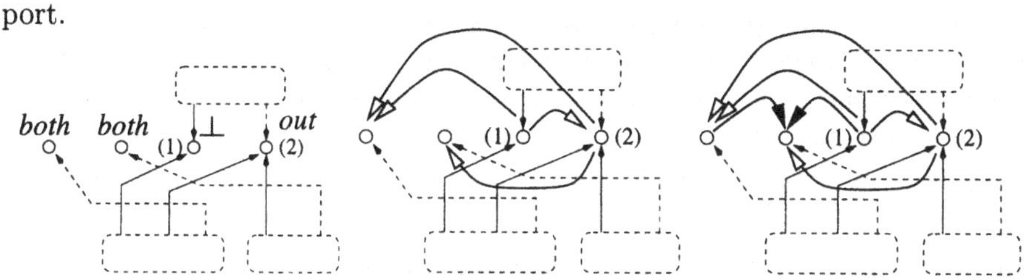

Fig. 3. Types for process graphs (input/output, secrecy, deadlocks)

5.2 Secrecy of External Ports

We assume that the external ports of a process graph can have different levels of secrecy. They might either be public or secret. Both sorts of ports can be used to send or receive message, but it is not allowed to forward a secret port to a receiver listening at a public port.

We choose $k = 2$, $I = \{\mathit{false}, \mathit{true}\}$ where $\{\mathit{false}, \mathit{true}\}$ is the boolean lattice with $\mathit{false} < \mathit{true}$. If a tuple (v_1, v_2) is associated with *true*, this means that the port v_2 is sent to v_1. The mapping L has the following form:

$$L(\mathit{proc}_m(l)) = \mathbf{n}[a_p] \text{ where } a_p(\lfloor \chi \rfloor_i, \lfloor \chi \rfloor_j) = \mathit{false}$$

$$L(\mathit{mess}_n) = \mathit{mess}_n[a_m] \text{ where } a_m(\lfloor \chi \rfloor_i, \lfloor \chi \rfloor_j) = \begin{cases} \mathit{true} \text{ if } i = n, j \neq n \\ \mathit{false} \text{ otherwise} \end{cases}$$

Let P be a process graph and we assume that the sets SEC and PUB form a partition of $\{1, \ldots, ar(P)\}$. If the type of P satisfies

$$X(G[\chi, a]) := (\forall i \in \mathit{PUB}, j \in \mathit{SEC} \colon a(\lfloor \chi \rfloor_i, \lfloor \chi \rfloor_j) = \mathit{false})$$

it follows that no message, with secret ports attached to it, is ever sent to a public port. In this case typing the example process graph yields the principal type in figure 3 (middle), where an arrow from port v_2 to v_1 indicates that $a(v_1, v_2) = \mathit{true}$. The predicate X is not satisfied only in the case where the first external port is secret and the second external port is public. In all other cases, the process graph is well-typed.

[4] presents a related method for checking the secrecy of ports.

5.3 Avoiding Deadlocks

We attempt to avoid vicious circles of processes and messages waiting for one another and causing a deadlock. Let P be a process graph with a non-empty edge set such that there is no process graph P' with $P \to P'$ and (C-REPL$_2$) is not applicable, i.e. P is stuck. Then at least one of the following conditions is satisfied:

(1) There is a message waiting or a process listening at an external port. This case is good-natured, since P is only waiting to perform an I/O-operation.
(2) There is an internal port where all edges connected to it are either messages, sent to this port, or processes listening at this port.
(3) There is a vicious circle, i.e. a sequence $v_0, \dots, v_n = v_0 \in V_P$ such that for every pair v_i, v_{i+1} there is *either* a message q with $send_P(q) = v_i$ and $\lfloor s_P(q) \rfloor_j = v_{i+1}$ for some $j \in \{1, \dots, ar(q)-1\}$ *or* a process p with $l_P(p) = \lambda_k^{(n)}.Q$, $\lfloor s_P(p) \rfloor_k = v_i$ and $\lfloor s_P(p) \rfloor_j = v_{i+1}$ for some $j \in \{1, \dots, ar(p)\}$, $j \neq k$.

Our aim is to avoid circles as described in **(3)**. We set $k := 2$ and I is again the boolean lattice with *false* and *true*. We define:

$$L(proc_m(l)) = \mathbf{n}[a_p] \text{ where } a_p(\lfloor \chi \rfloor_i, \lfloor \chi \rfloor_j) = \begin{cases} \mathit{true} \text{ if } l = \lambda_i^{(n)}.P, j \neq i,\ n \in \mathbb{N} \\ \mathit{false} \text{ otherwise} \end{cases}$$

$$L(mess_n) = mess_n[a_m] \text{ where } a_m(\lfloor \chi \rfloor_i, \lfloor \chi \rfloor_j) = \begin{cases} \mathit{true} \text{ if } i = n, j \neq n \\ \mathit{false} \text{ otherwise} \end{cases}$$

$$X(G[\chi, a]) := (\nexists v_0, \dots, v_n = v_0 \in V_G : a(v_i, v_{i+1}) = \mathit{true},\ 0 \leq i < n)$$

In this case we can retrieve the principal type of our example process graph rather easily from the type in section 5.2. All we have to do is add the arrows (with filled arrow heads) produced by a_p (see figure 3 (right)).

Since there is no circle of arrows we can conclude that the process graph will (during its reduction) never contain a vicious circle as described in **(3)**.

Similar methods for avoiding deadlocks are presented in [12, 10].

5.4 Composing Type Systems

We asumme that we have two type systems, with functors F_1, F_2, linear mappings L_1, L_2, predicates X_1, X_2 on type graphs and predicates Y_1, Y_2 on process graphs.

We define a functor F with $F(G) := (I_1 \times I_2, \leq)$ if $F_i(G) = (I_i, \leq_i)$, $i = 1, 2$ and $(a_1, a_2) \leq (a_1', a_2') \iff a_1 \leq_1 a_1'$ and $a_2 \leq_2 a_2'$.

Furthermore $F(\phi)((a_1, a_2)) := (F_1(\phi)(a_1), F_2(\phi)(a_2))$

Furthermore let $L(P) := G[\chi, (a_1, a_2)]$ if $L_1(P) \cong_{F_1} G[\chi, a_1]$ and $L_2(P) \cong_{F_2} G[\chi, a_2]$. This requires, of course, that L_1 and L_2 map process graphs to type graphs of the same structure. This is actually not a severe restriction since all practical examples can be defined in such a way that they satisfy this condition (see the examples in this section).

$$\text{We define } \quad X_\wedge(G[\chi, (a_1, a_2)]) := X_1(G[\chi, a_1]) \wedge X_2(G[\chi, a_2])$$
$$X_\vee(G[\chi, (a_1, a_2)]) := X_1(G[\chi, a_1]) \vee X_2(G[\chi, a_2])$$

Then $F, L, X_\wedge, Y_1 \wedge Y_2$ respectively $F, L, X_\vee, Y_1 \vee Y_2$ denote type systems checking the conjunction respectively disjunction of Y_1 and Y_2. A type system checking a negated property can only be constructed in very special cases.

6 Conclusion and Comparison to Related Work

Several type systems for process calculi have been proposed, each checking different properties of processes, e.g. I/O-capabilities [19], confluence [17], secrecy in security protocols [1] or deadlock-freedom [10]. This paper is an attempt to integrate these approaches and to propose one single generic type system which can be instantiated in order to verify invariant properties of processes. Our type system seems to be especially well-suited for properties related to the geometry of processes and messages represented in the process graph.

In [8] Kohei Honda proposes a general framework for type systems, satisfying the condition of *strict additivity*, i.e. two process connected to each other via several ports can be typed if and only if connections via one port only can be typed. As Honda remarks in his paper, strict additivity is sometimes too strong, e.g. in the case of deadlock-freedom. The type system presented in this paper is *semi-additive*, i.e. the "if and only if" is replaced by "only if". In contrast to [8] our method of instantiation is restricted, but this enables us to prove the subject reduction property for all possible type systems.

There is, of course, a trade-off between generality and the percentage of processes which can be typed: e.g. in the case of I/O-capabilities our type system is less powerful than the type system introduced in [19], which can partly be explained by the very general nature of our type system and partly by the fact that the type system in [19] does not have principal types.

We believe, however, that this type system can serve as a starting point for further research.

We will finish by dicussing two existing extensions of this type system which we were not able to present here due to limited space:

- Sometimes labelling processes with lattice elements does not seem to be sufficient. E.g. if we want to type confluent processes [17], typing involves the counting of processes and messages adjacent to a certain port. (In this case we have to demand that there is at most one process listening and at most one message waiting at a certain port.) Counting is also necessary if we want to design a type system checking **(2)** in the conditions for deadlocks or if we attempt to introduce linear types as in [11].
  An extension of our type system is based on latttice-ordered monoids rather than on lattices.
- It is not very difficult to extend the process calculus to a calculus where higher-order communication is possible, i.e. where entire processes can be sent as the content of a message. The corresponding extension of the type system is not very hard.
  We have to add environments, describing the type of the variables in a process graph. Furthermore it is necessary to slightly change the linear mapping L and rule (T-ABSTR).

Another future area of research is the use of the results of the type system in order to establish bisimilarity of processes (as in [1]).

References

1. Martín Abadi. Secrecy by typing in security protocols. In *Theoretical Aspects of Computer Software*, pages 611–638. Springer-Verlag, 1997.
2. Michel Bauderon and Bruno Courcelle. Graph expressions and graph rewritings. *Mathematical Systems Theory*, 20:83–127, 1987.
3. Gérard Berry and Gérard Boudol. The chemical abstract machine. *Theoretical Computer Science*, 96:217–248, 1992.
4. Chiara Bodei, Pierpaolo Degano, Flemming Nielson, and Hanne Riis Nielson. Control flow analysis for the pi-calculus. In *CONCUR '98*, pages 84–98. Springer-Verlag, 1998. LNCS 1466.
5. Roy L. Crole. *Categories for Types.* Cambridge University Press, 1993.
6. H. Ehrig. Introduction to the algebraic theory of graphs. In *Proc. 1st International Workshop on Graph Grammars*, pages 1–69. Springer-Verlag, 1979. LNCS 73.
7. Annegret Habel. *Hyperedge Replacement: Grammars and Languages.* Springer-Verlag, 1992. LNCS 643.
8. Kohei Honda. Composing processes. In *Proc. of POPL'96*, pages 344–357. ACM Press, 1996.
9. Kohei Honda and Nobuko Yoshida. Combinatory representation of mobile processes. In *POPL '94*, pages 348–360. ACM Press, 1994.
10. Naoki Kobayashi. A partially deadlock-free typed process calculus. In *LICS '97*, pages 128–139. IEEE, Computer Society Press, 1997.
11. Naoki Kobayashi, Benjamin C. Pierce, and David N. Turner. Linearity and the pi-calculus. In *POPL '96*, pages 358–371. ACM SIGACT/SIGPLAN, 1996.
12. Yves Lafont. Interaction nets. In *POPL '90*, pages 95–108. ACM Press, 1990.
13. R. Milner, J. Parrow, and D. Walker. A calculus of mobile processes, Part I. Tech. Rep. ECS-LFCS-89-85, University of Edinburgh, Laboratory for Foundations of Computer Science, 1989.
14. Robin Milner. Operational and algebraic semantics of concurrent processes. In Jan van Leeuwen, editor, *Formal Models and Semantics, Handbook of Theoretical Computer Science*, volume B, pages 1201–1242. Elsevier, 1990.
15. Robin Milner. The polyadic π-calculus: a tutorial. Tech. Rep. ECS-LFCS-91-180, University of Edinburgh, Laboratory for Foundations of Computer Science, 1991.
16. Robin Milner. Pi-nets: a graphical form of pi-calculus. In *European Symposium on Programming*, pages 26–42. Springer-Verlag, 1994. LNCS 788.
17. Uwe Nestmann and Martin Steffen. Typing confluence. In *ERCIM Workshop on Formal Methods in Industrial Critical Systems*, pages 77–101, 1997.
18. Joachim Parrow. Interaction diagrams. *Nordic Journal of Computing*, 2:407–443, 1995.
19. Benjamin Pierce and Davide Sangiorgi. Typing and subtyping for mobile processes. In *LICS '93*, pages 376–385, 1993.
20. Antonio Ravara and Vasco T. Vasconcelos. Behavioural types for a calculus of concurrent objects. In *Euro-Par '97*. Springer-Verlag, 1997.
21. David Turner. *The Polymorphic Pi-Calculus: Theory and Implementation.* PhD thesis, University of Edinburgh, 1995. ECS-LFCS-96-345.
22. P. Urzyczyn. Positive recursive type assignment. In J. Wiedermann and P. Hájek, editors, *Mathematical Foundations of Computer Science 1995.* Springer-Verlag, 1995. LNCS 969.
23. Nobuko Yoshida. Graph types for monadic mobile processes. In *FST/TCS '96*, pages 371–386. Springer-Verlag, 1996.

A Process graphs and the asynchronous polyadic π-calculus

In order to show how process graphs are related to the π-calculus we give an encoding, transforming a subset of all process graphs to expressions in the asynchronous polyadic π-calculus.

Definition 7. (Encoding) *Let P be a process graph such that χ_P is duplicate-free, i.e. $\lfloor\chi_P\rfloor_i = \lfloor\chi_P\rfloor_j$ implies $i = j$. And we assume that the same condition is satisfied for all process graphs occurring inside of P. Let $\mathcal{N}$ be the name set of the π-calculus and let $t \in \mathcal{N}^*$ such that $|t| = ar(P)$. We define $\Theta_t(P)$ inductively as follows:*

Message: $\Theta_{a_1\ldots a_n}(mess_n) := \overline{a_n}\langle a_1 \ldots a_{n-1}\rangle.0$
Replication: $\Theta_t(proc_m(!P)) :=!\,\Theta_t(P)$
Process Abstraction: $\Theta_{a_1\ldots a_m}(proc_m(\lambda_k^{(n)}(P))) := a_k(x_1 \ldots x_n).\Theta_{a_1\ldots a_m x_1\ldots x_n}(P)$
where $x_1, \ldots, x_n \in \mathcal{N}$ are fresh names.
Process Graph Construction:
$\Theta_t(\bigotimes_{i=1}^{n}(P_i, \zeta_i)) := (\nu\, \mu(V_D \backslash EXT_D))(\Theta_{\mu(\zeta_1(\chi_{\mathbf{m_1}}))}(P_1) \mid \ldots \mid \Theta_{\mu(\zeta_n(\chi_{\mathbf{m_n}}))}(P_n))$
where $\zeta_i : \mathbf{m_i} \to D$, $i \in \{1, \ldots, n\}$ and $\mu : V_D \to \mathcal{N}$ is an arbitrary mapping such that μ restricted to $V_D \backslash EXT_D$ is injective and $\mu(\chi_D) = t$.
If $n = 0$ (i.e. if the process graph is identical to D) we set $\Theta_t(\bigotimes_{i=1}^{0}(P_i, \zeta_i)) := 0$.

The set of all process graphs satisfying the condition in the definition above is closed under reduction and corresponds exactly to the asynchronous part of the polyadic π-calculus without summation. (We rely on the syntax and semantics given for its synchronous version in [19], omitting sort annotations.)

Proposition 4. *Let p be an arbitrary expression in the asynchronous polyadic π-calculus without summation. Then there exists a process graph P (satisfying the condition in definition 7) and a duplicate-free string $t \in \mathcal{N}^*$ such that $\Theta_t(P) \equiv p$.*

Furthermore for process graphs P, P' satisfying the condition in definition 7 and for every duplicate-free string $t \in \mathcal{N}^*$ with $|t| = ar(P) = ar(P')$ it is true that:

- $P \equiv P'$ *implies* $\Theta_t(P) \equiv \Theta_t(P')$
- $P \to P'$ *implies* $\Theta_t(P) \to \Theta_t(P)$
- $\Theta_t(P) \to p$ *with* $p \not\equiv wrong$ *implies that there exists a process graph Q with $P \to Q$ and $\Theta_t(Q) \equiv p$*

The proposition implies that one calculus can match the reductions of the other step by step. The main difference of the calculi lies in their interface towards the environment. How these interfaces are converted into one another is described by the string t.

Weak Bisimilarity with Infinite-State Systems Can Be Decided in Polynomial Time

Antonín Kučera*[1] and Richard Mayr**[2]

[1] Faculty of Informatics, Masaryk University, Botanická 68a, 60200 Brno, Czech Republic, tony@fi.muni.cz

[2] LFCS, Dept. of Computer Science, Univ. of Edinburgh, King's Buildings, Mayfield Road, Edinburgh EH9 3JZ, UK, mayrri@dcs.ed.ac.uk

Abstract. We prove that weak bisimilarity is decidable in polynomial time between BPA and finite-state processes, and between normed BPP and finite-state processes. To the best of our knowledge, these are the first polynomial algorithms for weak bisimilarity with infinite-state systems.

1 Introduction

Recently, a lot of attention has been devoted to the study of decidability and complexity of verification problems for infinite-state systems [27, 11, 5]. We consider the problem of weak bisimilarity between certain infinite-state processes and finite-state ones. The motivation is that the intended behavior of a process is often easy to specify (by a finite-state system), but a 'real' implementation can contain components which are essentially infinite-state (e.g. counters, buffers). The aim is to check if the finite-state specification and the infinite-state implementation are semantically equivalent, i.e. weakly bisimilar.

We concentrate on the classes of infinite-state processes definable by the syntax of BPA (Basic Process Algebra) and normed BPP (Basic Parallel Processes) systems. BPA processes can be seen as simple sequential programs (due to the binary operator of sequential composition). They have recently been used to solve problems of dataflow analysis in optimizing compilers [12]. BPP model simple parallel systems (due to the binary operator of parallel composition). A process is *normed* iff at every reachable state it can terminate via a finite sequence of computational steps.

The state of the art. Baeten, Bergstra, and Klop [1] proved that *strong bisimilarity* [29] is decidable for normed BPA processes. Simpler proofs have been given later in [18, 13], and there is even a polynomial-time algorithm [15]. The decidability result has later been extended to the class of all (not necessarily normed) BPA processes in [9], but the best known algorithm is doubly exponential [4]. Decidability of strong bisimilarity for BPP processes has been established in [8], but the algorithm has non-elementary complexity. However, there is a polynomial-time algorithm for the subclass of normed BPP [16]. Strong bisimilarity between normed BPA and normed BPP is also decidable [7]. This result even holds for parallel compositions of normed BPA and normed BPP processes [20].

* Supported by a Research Fellowship granted by the Alexander von Humboldt Foundation and by a Post-Doc grant GA ČR No. 201/98/P046.

** This work was partly supported by a DAAD Post-Doc grant.

For weak bisimilarity, much less is known. Semidecidability of weak bisimilarity for BPP is due to [10]. In [14] it is shown that weak bisimilarity is decidable for those BPA and BPP processes which are 'totally normed' (a process is totally normed if it can terminate at any moment via a finite sequence of computational steps, but at least one of those steps must be 'visible', i.e. non-internal). Decidability of weak bisimilarity for general BPA and BPP is open; those problems might be decidable, but they are surely intractable (assuming $\mathcal{P} \neq \mathcal{NP}$)—for BPP we have $\mathcal{NP}$-hardness, and for BPA even $PSPACE$-hardness [30].

The situation is dramatically different if we consider weak bisimilarity between certain infinite-state processes and finite-state ones. In [24] it is shown that weak bisimilarity between BPP and finite-state processes is decidable. A more general result has recently been obtained in [19], where it is shown that many bisimulation-like equivalences (including the strong and weak ones) are decidable between PAD and finite-state processes. The class PAD strictly subsumes not only BPA and BPP, but also PA [2] and pushdown processes. This result is obtained by a general reduction to the model-checking problem for the simple branching-time temporal logic *EF*. As the model-checking problem for *EF* is hard (for example, it is known to be $PSPACE$-complete for BPP [24] and $PSPACE$-hard for BPA [25]), this does not yield an efficient algorithm.

Our contribution. We show that weak (and hence also strong) bisimilarity is decidable in polynomial time between BPA and finite-state processes, and between normed BPP and finite-state processes. Due to the aforementioned hardness results for the 'symmetric case' (when we compare two BPA or two (normed) BPP processes) we know that our results cannot be extended in this direction. To the best of our knowledge, these are the first polynomial algorithms for weak bisimilarity with infinite-state systems. Moreover, the algorithm for BPA is the first example of an efficient decision procedure for a class of *unnormed* infinite-state systems (the polynomial algorithms for strong bisimilarity of [15, 16] only work for normed subclasses of BPA and BPP, respectively). It should also be noted that *simulation equivalence* between BPA/BPP and finite-state systems is co-$\mathcal{NP}$-hard [22].

The basic scheme of our constructions for BPA and normed BPP processes is the same. The main idea is that weak bisimilarity between BPA (or normed BPP) processes and finite-state ones can be generated from a finite base and that certain infinite subsets of BPA and BPP state-space can be 'symbolically' described by finite automata and context-free grammars, respectively. A more detailed intuition is given in Section 3. As weak bisimilarity is not a congruence w.r.t. sequencing (see Section 3), we propose its natural refinement called termination-sensitive bisimilarity which *is* a congruence and which is also decidable between BPA and finite-state processes in polynomial time.

2 Definitions

We use process rewrite systems [23] as a formal model for processes. Let $Act = \{a, b, c, \ldots\}$ and $Const = \{X, Y, Z, \ldots\}$ be disjoint countably infinite sets of *actions* and *process constants*, respectively. The class of *process expressions* $\mathcal{E}$ is defined by $E ::= \varepsilon \mid X \mid E \| E \mid E.E$, where $X \in Const$ and ε is a special constant that denotes the empty expression. Intuitively, '.' is sequential composition and '$\|$' is parallel composition. We do not distinguish between expressions related by *structural congruence* which is given by the following laws: '.' and '$\|$' are associative, '$\|$' is commutative, and 'ε' is a unit for '.' and '$\|$'.

A *process rewrite system* [23] is specified by a finite set of *rules* Δ which have the form $E \stackrel{a}{\rightarrow} F$, where $E, F \in \mathcal{E}$ and $a \in Act$. $Const(\Delta)$ and $Act(\Delta)$ denote the sets of process constants and actions which are used in the rules of Δ, respectively (note that these sets are finite). Each process rewrite system Δ defines a unique transition system where states are process expressions over $Const(\Delta)$, $Act(\Delta)$ is the set of labels, and transitions are determined by Δ and the following inference rules (remember that '$\|$' is commutative):

$$\frac{(E \stackrel{a}{\rightarrow} F) \in \Delta}{E \stackrel{a}{\rightarrow} F} \qquad \frac{E \stackrel{a}{\rightarrow} E'}{E.F \stackrel{a}{\rightarrow} E'.F} \qquad \frac{E \stackrel{a}{\rightarrow} E'}{E\|F \stackrel{a}{\rightarrow} E'\|F}$$

We extend the notation $E \stackrel{a}{\rightarrow} F$ to elements of Act^* in the standard way. F is *reachable* from E if $E \stackrel{w}{\rightarrow} F$ for some $w \in Act^*$.

Sequential and *parallel* expressions are those process expressions which do not contain the '$\|$' and the '.' operator, respectively. Finite-state, BPA, and BPP systems are subclasses of process rewrite systems obtained by putting certain restrictions on the form of the rules. Finite-state, BPA, and BPP allow only a single constant on the left-hand side of rules, and a single constant, sequential expression, and parallel expression on the right-hand side, respectively. The set of states of a transition system which is generated by a finite-state, BPA, or BPP process Δ is restricted to $Const(\Delta)$, the set of all sequential expressions over $Const(\Delta)$, or the set of all parallel expressions over $Const(\Delta)$, respectively. A constant $X \in Const(\Delta)$ is *normed* iff $X \stackrel{w}{\rightarrow} \varepsilon$ for some $w \in Act^*$. A process is normed, iff all constants of its underlying system Δ are normed.

The semantical equivalence we are interested in here is *weak bisimilarity* [26]. This relation distinguishes between 'observable' and 'internal' moves (computational steps); the internal moves are modeled by a special action which is denoted 'τ' by convention. In what follows we consider process expressions over $Const(\Delta)$ where Δ is some fixed process rewrite system.

Definition 1. *The* extended transition relation *'$\stackrel{a}{\Rightarrow}$' is defined by $E \stackrel{a}{\Rightarrow} F$ iff either $E = F$ and $a = \tau$, or $E \stackrel{\tau^i}{\rightarrow} E' \stackrel{a}{\rightarrow} E'' \stackrel{\tau^j}{\rightarrow} F$ for some $i, j \in \mathbb{N}_0$, $E', E'' \in \mathcal{E}$. A binary relation R over process expressions is a* weak bisimulation *iff whenever $(E, F) \in R$ then for every $a \in Act$: if $E \stackrel{a}{\rightarrow} E'$ then there is $F \stackrel{a}{\Rightarrow} F'$ s.t. $(E', F') \in R$ and if $F \stackrel{a}{\rightarrow} F'$ then there is $E \stackrel{a}{\Rightarrow} E'$ s.t. $(E', F') \in R$. Processes E, F are* weakly bisimilar, *written $E \approx F$, iff there is a weak bisimulation relating them.*

Let Γ be a finite-state system with n states, $f, g \in Const(\Gamma)$. It is easy to show that the problem whether $f \approx g$ is decidable in $\mathcal{O}(n^3)$ time. For example, we can compute the '$\stackrel{a}{\Rightarrow}$' relation of Γ and then start to refine $Const(\Gamma) \times Const(\Gamma)$ in a number of steps until it 'stabilizes' w.r.t. $\stackrel{a}{\Rightarrow}$. We note that the use of some advanced techniques (see e.g. [28]) could probably decrease the mentioned upper bound; however, the complexity of the algorithms which are designed in this paper is a bit worse (even if we could decide the problem $f \approx g$ in constant time), hence we do not try to improve this bound.

Sometimes we also consider weak bisimilarity between processes of *different* process rewrite systems, say Δ and Γ. Formally, Δ and Γ can be considered as a *single* system by taking their disjoint union.

3 BPA Processes

Let E be a BPA process with the underlying system Δ, F a finite-state process with the underlying system Γ s.t. $Const(\Delta) \cap Const(\Gamma) = \emptyset$. We assume (w.l.o.g.) that $E \in Const(\Delta)$; moreover, we also assume that for all $f, g \in Const(\Gamma)$, $a \in Act$ s.t. $f \neq g$ or $a \neq \tau$ we have that $f \stackrel{a}{\Rightarrow} g$ implies $f \stackrel{a}{\rightarrow} g \in \Gamma$. If those '$\stackrel{a}{\rightarrow}$' transitions are missing in Γ, we can add them safely—it does not influence our complexity estimations, as we always consider the worst case when Γ has all possible transitions (we do not want to add new transitions of the form $f \stackrel{\tau}{\rightarrow} f$, because then our proof for weak bisimilarity would not immediately work for termination-sensitive bisimilarity which is designed at the end of this section).

In this section, we use upper-case letters $X, Y, \ldots$ to denote elements of $Const(\Delta)$, and lower-case letters $f, g, \ldots$ to denote elements of $Const(\Gamma)$. Greek letters $\alpha, \beta, \ldots$ are used to denote elements of $Const(\Delta)^*$. The size of Δ is denoted by n, and the size of Γ by m (we measure the complexity of our algorithm in (n, m)).

The set $Const(\Delta)$ can be divided into two disjoint subsets of *normed* and *unnormed* constants (remember that $X \in Const(\Delta)$ is normed iff $X \stackrel{w}{\rightarrow} \varepsilon$ for some $w \in Act^*$). The set of all normed constants of Δ is denoted *Normed*(Δ). In our constructions we also use processes of the form αf; they should be seen as BPA processes with the underlying system $\Delta \cup \Gamma$.

Intuition: Our proof can be divided into two parts: first we show that the greatest weak bisimulation between processes of Δ and Γ is finitely representable. There is a finite relation $\mathcal{B}$ of size $\mathcal{O}(n\,m^2)$ (called *bisimulation base*) such that each pair of weakly bisimilar processes can be generated from that base (a technique first used by Caucal [6]). Then we show that the bisimulation base can be computed in polynomial time. To do that, we take a sufficiently large relation $\mathcal{G}$ which surely subsumes the base and 'refine' it (this refinement technique has been used in [15, 16]). The size of $\mathcal{G}$ is still $\mathcal{O}(n\,m^2)$, and each step of the refinement procedure possibly deletes some of the elements of $\mathcal{G}$. If nothing is deleted, we have found the base (hence we need at most $\mathcal{O}(n\,m^2)$ steps). The refinement step is formally introduced in Definition 4 (we compute the *expansion* of the currently computed approximation of the base). Intuitively, a pair of processes belongs to the expansion iff for each $\stackrel{a}{\rightarrow}$ move of one component there is a $\stackrel{a}{\Rightarrow}$ move of the other component s.t. the resulting pair of processes can be generated from the current approximation of $\mathcal{B}$. We have to overcome two fundamental problems:

1. The set of pairs which can be generated from $\mathcal{B}$ (and its approximations) is infinite.
2. The set of states which are reachable from a given BPA state in one '$\stackrel{a}{\Rightarrow}$' move is infinite.

We employ a 'symbolic' technique to represent those infinite sets (similar to the one used in [3]), taking advantage of the fact that they have a simple (regular) structure which can be encoded by finite-state automata (see Theorem 1 and 4). This allows to compute the expansion in polynomial time.

Definition 2. *A relation K is* fundamental *iff it is a subset of*

$$((Normed(\Delta) \cdot Const(\Gamma)) \times Const(\Gamma)) \cup (Const(\Delta) \times Const(\Gamma)) \cup ((\{\varepsilon\} \cup Const(\Gamma)) \times Const(\Gamma))$$

Note that the size of any fundamental relation is $\mathcal{O}(n\,m^2)$. The greatest fundamental relation is denoted by $\mathcal{G}$. The bisimulation base *for Δ and Γ, denoted $\mathcal{B}$, is defined as follows: $\mathcal{B} = \{(Yf, g) \mid Yf \approx g, Y \in Normed(\Delta)\} \cup \{(X, g) \mid X \approx g\} \cup \{(f, g) \mid f \approx g\} \cup \{(\varepsilon, g) \mid \varepsilon \approx g\}$.*

As weak bisimilarity is a left congruence w.r.t. sequential composition, we can 'generate' from $\mathcal{B}$ new pairs of weakly bisimilar processes by substitution (it is worth noting that weak bisimilarity is *not* a right congruence w.r.t. sequencing—to see this, it suffices to define $X \stackrel{\tau}{\to} X$, $Y \stackrel{\tau}{\to} \varepsilon$, $Z \stackrel{a}{\to} Z$. Now $X \approx Y$, but $XZ \not\approx YZ$). This generation procedure can be defined for any fundamental relation as follows:

Definition 3. *Let K be a fundamental relation. The* closure *of K, denoted $Cl(K)$, is the least relation M which satisfies the following conditions:*

1. *$K \subseteq M$*
2. *if $(f, g) \in K$ and $(\alpha, f) \in M$, then $(\alpha, g) \in M$*
3. *if $(f, g) \in K$ and $(\alpha h, f) \in M$, then $(\alpha h, g) \in M$*
4. *if $(Yf, g) \in K$ and $(\alpha, f) \in M$, then $(Y\alpha, g) \in M$*
5. *if $(Yf, g) \in K$ and $(\alpha h, f) \in M$, then $(Y\alpha h, g) \in M$*
6. *if $(\alpha, g) \in M$ and α contains an unnormed constant, then $(\alpha\beta, g), (\alpha\beta h, g) \in M$ for every $\beta \in Const(\Delta)^*$ and $h \in Const(\Gamma)$.*

Note that $Cl(K)$ contains elements of just two forms – (α, g) and $(\alpha f, g)$. Clearly $Cl(K) = \bigcup_{i=0}^{\infty} Cl(K)^i$ where $Cl(K)^0 = K$ and $Cl(K)^{i+1}$ consists of $Cl(K)^i$ and the pairs which can be immediately derived from $Cl(K)^i$ by the rules 2–6 of Definition 3.

Although the closure of a fundamental relation can be infinite, its structure is in some sense regular. This fact is precisely formulated in the following theorem:

Theorem 1. *Let K be a fundamental relation. For each $g \in Const(\Gamma)$ there is a finite-state automaton A_g of size $\mathcal{O}(n\,m^2)$ constructible in $\mathcal{O}(n\,m^2)$ time s.t. $L(A_g) = \{\alpha \mid (\alpha, g) \in Cl(K)\} \cup \{\alpha f \mid (\alpha f, g) \in Cl(K)\}$*

Proof. We construct a regular grammar of size $\mathcal{O}(n\,m^2)$ which generates the mentioned language. Let $G_g = (N, \Sigma, \delta, \overline{g})$ where

- $N = \{\overline{f} \mid f \in Const(\Gamma)\} \cup \{U\}$
- $\Sigma = Const(\Delta) \cup Const(\Gamma)$
- δ is defined as follows:
  - for each $(\varepsilon, h) \in K$ we add the rule $\overline{h} \to \varepsilon$.
  - for each $(f, h) \in K$ we add the rules $\overline{h} \to \overline{f}$, $\overline{h} \to f$.
  - for each $(Yf, h) \in K$ we add the rules $\overline{h} \to Yf$, $\overline{h} \to Y\overline{f}$.

- for each $(X, h) \in K$ we add the rule $\overline{h} \rightarrow X$ and if X is unnormed, then we also add the rule $\overline{h} \rightarrow XU$.
- for each $X \in Const(\Delta)$, $f \in Const(\Gamma)$ we add the rules $U \rightarrow XU, U \rightarrow X$, $U \rightarrow f$.

A proof that G_g indeed generates the mentioned language is routine. Now we translate G_g to $\mathcal{A}_g$ (see e.g. [17]). Note that the size of $\mathcal{A}_g$ is essentially the same as the size of G_g; $\mathcal{A}_g$ is non-deterministic and can contain ε-rules. □

As an immediate consequence of the previous theorem we obtain that the membership to $Cl(K)$ for any fundamental relation K is easily decidable in polynomial time. Another property of $Cl(K)$ is specified in the lemma below.

Lemma 1. *Let* $(\alpha f, g) \in Cl(K)$. *If* $(\beta h, f) \in Cl(K)$, *then also* $(\alpha\beta h, g) \in Cl(K)$. *Similarly, if* $(\beta, f) \in Cl(K)$, *then also* $(\alpha\beta, g) \in Cl(K)$.

The importance of the bisimulation base is clarified by the following theorem. It says that $Cl(\mathcal{B})$ subsumes the greatest weak bisimulation between processes of Δ and Γ.

Theorem 2. *For all* α, f, g *we have* $\alpha \approx g$ *iff* $(\alpha, g) \in Cl(\mathcal{B})$, *and* $\alpha f \approx g$ *iff* $(\alpha f, g) \in Cl(\mathcal{B})$.

Proof. The 'if' part is obvious in both cases, as $\mathcal{B}$ contains only weakly bisimilar pairs and all the rules of Definition 3 produce pairs which are again weakly bisimilar. The 'only if' part can, in both cases, be easily proved by induction on the length of α (we just show the first proof; the second one is similar).

- $\alpha = \varepsilon$. Then $(\varepsilon, g) \in \mathcal{B}$, hence $(\varepsilon, g) \in Cl(\mathcal{B})$.
- $\alpha = Y\beta$. If Y is unnormed, then $Y \approx g$ and $(Y, g) \in \mathcal{B}$. By the rule 6 of Definition 3 we obtain $(Y\beta, g) \in Cl(\mathcal{B})$. If Y is normed, then $Y\beta \stackrel{w}{\rightarrow} \beta$ for some $w \in Act^*$ and g must be able to match the sequence w by some $g \stackrel{w}{\Rightarrow} g'$ s.t. $\beta \approx g'$. By substitution we now obtain that $Yg' \approx g$. Clearly $(Yg', g) \in \mathcal{B}$, and $(\beta, g') \in Cl(\mathcal{B})$ by induction hypothesis. Hence $(\alpha, g) \in Cl(\mathcal{B})$ due to the rule 4 of Definition 3. □

The next definition formalizes one step of the 'refinement procedure' which is applied to $\mathcal{G}$ to compute $\mathcal{B}$.

Definition 4. *Let* K *be a fundamental relation. We say that a pair* (X, g) *of* K expands *in* K *iff the following two conditions hold:*

- *for each* $X \stackrel{a}{\rightarrow} \alpha$ *there is some* $g \stackrel{a}{\Rightarrow} g'$ *s.t.* $(\alpha, g') \in Cl(K)$
- *for each* $g \stackrel{a}{\rightarrow} g'$ *there is some* $X \stackrel{a}{\Rightarrow} \alpha$ *s.t.* $(\alpha, g') \in Cl(K)$

The expansion of a pair of the form (Yf, g), (f, g), (ε, g) *in* K *is defined in the same way—for each '*$\stackrel{a}{\rightarrow}$*' move of the left component there must be some '*$\stackrel{a}{\Rightarrow}$*' move of the right component such that the resulting pair of processes belongs to* $Cl(K)$, *and vice versa (note that* $\varepsilon \stackrel{\tau}{\Rightarrow} \varepsilon$*). The set of all pairs of* K *which expand in* K *is denoted by* $Exp(K)$.

The notion of expansion is in some sense 'compatible' with the definition of weak bisimulation. This intuition is formalised in the following lemma.

Lemma 2. *Let K be a fundamental relation s.t. $Exp(K) = K$. Then $Cl(K)$ is a weak bisimulation.*

Proof. We prove that every pair (α, g), $(\alpha f, g)$ of $Cl(K)^i$ has the property that for each '$\stackrel{a}{\rightarrow}$' move of one component there is a '$\stackrel{a}{\Rightarrow}$' move of the other component s.t. the resulting pair of processes belongs to $Cl(K)$ (we consider just pairs of the form $(\alpha f, g)$; the other case is similar). By induction on i.

- $i = 0$. Then $(\alpha f, g) \in K$; as $K = Exp(K)$, the claim follows directly from the definitions.
- **Induction step.** Let $(\alpha f, g) \in Cl(K)^{i+1}$. There are three possibilities:
  I. There is an h s.t. $(\alpha f, h) \in Cl(K)^i$, $(h, g) \in K$.
  Let $\alpha f \stackrel{a}{\rightarrow} \gamma f$ (note that α can be empty; in this case we have to consider moves of the form $f \stackrel{a}{\rightarrow} f'$. It is done in a similar way as below). As $(\alpha f, h) \in Cl(K)^i$, we can use the induction hypothesis and conclude that there is $h \stackrel{a}{\Rightarrow} h'$ s.t. $(\gamma f, h') \in Cl(K)$. We distinguish two cases:
  1) $a = \tau$ and $h' = h$. Then $(\gamma f, h) \in Cl(K)$ and as $(h, g) \in K$, we obtain $(\gamma f, g) \in Cl(K)$ due to Lemma 1. Hence g can use the move $g \stackrel{\tau}{\Rightarrow} g$.
  2) $a \neq \tau$ or $h \neq h'$. Then there is a transition $h \stackrel{a}{\rightarrow} h'$ (see the beginning of this section) and as $(h, g) \in K$, by induction hypothesis we know that there is some $g \stackrel{a}{\Rightarrow} g'$ s.t. $(h', g') \in Cl(K)$. Hence, $(\gamma f, g') \in Cl(K)$ due to Lemma 1.
  Now let $g \stackrel{a}{\rightarrow} g'$. As $(h, g) \in K$, there is $h \stackrel{a}{\Rightarrow} h'$ s.t. $(h', g') \in Cl(K)$. We distinguish two possibilities again:
  1) $a = \tau$ and $h' = h$. Then αf can use the move $\alpha f \stackrel{\tau}{\Rightarrow} \alpha f$; we have $(h, g') \in Cl(K)$ and $(\alpha f, h) \in Cl(K)$, hence also $(\alpha f, g') \in Cl(K)$.
  2) $a \neq \tau$ or $h \neq h'$. Then $h \stackrel{a}{\rightarrow} h'$ and as $(\alpha f, h) \in Cl(K)^i$, there is $\alpha f \stackrel{a}{\Rightarrow} \gamma f$ (or $\alpha f \stackrel{a}{\Rightarrow} f'$; it is handled in the same way) s.t. $(\gamma f, h') \in Cl(K)$. Hence also $(\gamma f, g') \in Cl(K)$ by Lemma 1.
  II. $\alpha = Y\beta$ and there is h s.t. $(Yh, g) \in K$, $(\beta f, h) \in Cl(K)^i$.
  Let $Y\beta f \stackrel{a}{\rightarrow} \gamma\beta f$. As $(Yh, g) \in K$, we can use induction hypothesis and conclude that there is $g \stackrel{a}{\Rightarrow} g'$ s.t. $(\gamma h, g') \in Cl(K)$. As $(\beta f, h) \in Cl(K)$, we obtain $(\gamma\beta f, g') \in Cl(K)$ by Lemma 1.
  Let $g \stackrel{a}{\rightarrow} g'$. As $(Yh, g) \in K$, by induction hypothesis we know that Yh can match the move $g \stackrel{a}{\rightarrow} g'$; there are two possibilities:
  1) $Yh \stackrel{a}{\Rightarrow} \gamma h$ s.t. $(\gamma h, g') \in Cl(K)$. Then also $Y\beta f \stackrel{a}{\Rightarrow} \gamma\beta f$. As $(\beta f, h) \in Cl(K)$, we immediately have $(\gamma\beta f, g') \in Cl(K)$ as required.
  2) $Yh \stackrel{a}{\Rightarrow} h'$ s.t. $(h', g') \in Cl(K)$. The transition $Yh \stackrel{a}{\Rightarrow} h'$ can be 'decomposed' into $Yh \stackrel{x}{\Rightarrow} h$, $h \stackrel{y}{\Rightarrow} h'$ where $x = a \wedge y = \tau$ or $x = \tau \wedge y = a$. If $y = \tau$ and $h' = h$, we are done immediately because then $Y\beta \stackrel{a}{\Rightarrow} \beta$ and as (h, g'), $(\beta, h) \in Cl(K)$, we also have $(\beta, g') \in Cl(K)$ as needed. If $y \neq \tau$ or $h' \neq h$, there is a transition $h \stackrel{y}{\rightarrow} h'$. As $(\beta f, h) \in Cl(K)^i$, due to induction hypothesis we know that there is some $\beta f \stackrel{y}{\Rightarrow} \gamma f$ (or $\beta f \stackrel{y}{\Rightarrow} f'$; this is

handled in the same way) with $(\gamma f, h') \in Cl(K)$. Clearly $Y\beta f \stackrel{a}{\Rightarrow} \gamma f$. As $(h', g'), (\gamma f, h') \in Cl(K)$, we also have $(\gamma f, g') \in Cl(K)$.

III. $\alpha = \beta\gamma$ where β contains an unnormed constant and $(\beta, g) \in Cl(K)^i$.
Let $\alpha \stackrel{a}{\rightarrow} \alpha'$. Then $\alpha' = \delta\gamma$ and $\beta \stackrel{a}{\rightarrow} \delta$. As $(\beta, g) \in Cl(K)^i$, there is $g \stackrel{a}{\Rightarrow} g'$ s.t. $(\delta, g') \in Cl(K)$ due to the induction hypothesis. Clearly δ contains an unnormed constant, hence $(\delta\gamma, g') \in Cl(K)$ by the last rule of Definition 3.
Let $g \stackrel{a}{\rightarrow} g'$. As $(\beta, g) \in Cl(K)^i$, there is $\beta \stackrel{a}{\Rightarrow} \delta$ s.t. $(\delta, g') \in Cl(K)$ and δ contains an unnormed constant. Hence $\alpha \stackrel{a}{\Rightarrow} \delta\gamma$ and $(\delta\gamma, g') \in Cl(K)$ due to the last rule of Definition 3. □

The notion of expansion allows to approximate $\mathcal{B}$ in the following way: $\mathcal{B}^0 = \mathcal{G}$, $\mathcal{B}^{i+1} = Exp(\mathcal{B}^i)$. A proof of the next theorem is now easy to complete.

Theorem 3. *There is a $j \in \mathbb{N}$, bounded by $\mathcal{O}(n\,m^2)$, such that $\mathcal{B}^j = \mathcal{B}^{j+1}$. Moreover, $\mathcal{B}^j = \mathcal{B}$.*

In other words, $\mathcal{B}$ can be obtained from $\mathcal{G}$ in $\mathcal{O}(n\,m^2)$ refinement steps which correspond to the construction of the expansion. The only thing which remains to be shown is that $Exp(K)$ is effectively constructible in polynomial time. To do that, we employ a 'symbolic' technique which allows to represent infinite subsets of BPA state-space in an elegant and succinct way.

Theorem 4. *For all $X \in Const(\Delta)$, $a \in Act(\Delta)$ there is a finite-state automaton $\mathcal{A}_{(X,a)}$ of size $\mathcal{O}(n^2)$ constructible in $\mathcal{O}(n^2)$ time s.t. $L(\mathcal{A}_{(X,a)}) = \{\alpha \mid X \stackrel{a}{\Rightarrow} \alpha\}$*

Proof. We define a left-linear grammar $G_{(X,a)}$ of size $\mathcal{O}(n^2)$ which generates the mentioned language. This grammar can be converted to $\mathcal{A}_{(X,a)}$ by a standard algorithm known from automata theory (see e.g. [17]). Note that the size of $\mathcal{A}_{(X,a)}$ is essentially the same as the size of $G_{(X,a)}$. First, let us realize that we can compute in $\mathcal{O}(n^2)$ time the sets M_τ and M_a consisting of all $Y \in Const(\Delta)$ s.t. $Y \stackrel{\tau}{\Rightarrow} \varepsilon$ and $Y \stackrel{a}{\Rightarrow} \varepsilon$, respectively. Let $G_{(X,a)} = (N, \Sigma, \delta, S)$ where

- $N = \{Y^a, Y^\tau \mid Y \in Const(\Delta)\} \cup \{S\}$. Intuitively, the index indicates whether the action 'a' has already been emitted.
- $\Sigma = Const(\Delta)$
- δ is defined as follows:
  - we add the rule $S \rightarrow X^a$ to δ, and if $X \stackrel{a}{\Rightarrow} \varepsilon$ then we also add the rule $S \rightarrow \varepsilon$.
  - for every transition $Y \stackrel{a}{\rightarrow} Z_1. \cdots .Z_k$ of Δ and every i s.t. $1 \leq i \leq k$ we test whether $Z_j \stackrel{\tau}{\Rightarrow} \varepsilon$ for every $0 \leq j < i$. If this is the case, we add to δ the rules
  $$Y^a \rightarrow Z_i \cdots Z_k,\; Y^a \rightarrow Z_i^\tau Z_{i+1} \cdots Z_k$$
  - for every transition $Y \stackrel{\tau}{\rightarrow} Z_1. \cdots .Z_k$ of Δ and every i s.t. $1 \leq i \leq k$ we do the following:
    * we test whether $Z_j \stackrel{\tau}{\Rightarrow} \varepsilon$ for every $0 \leq j < i$. If this is the case, we add to δ the rules
    $$Y^a \rightarrow Z_i^a Z_{i+1} \cdots Z_k,\; Y^\tau \rightarrow Z_i^\tau Z_{i+1} \cdots Z_k,\; Y^\tau \rightarrow Z_i \cdots Z_k$$

- $*$ we test whether there is a $t < i$ such that $Z_t \stackrel{a}{\Rightarrow} \varepsilon$ and $Z_j \stackrel{\tau}{\Rightarrow} \varepsilon$ for every $0 \leq j < i, j \neq t$. If this is the case, we add to δ the rules

$$Y^a \rightarrow Z_i^\tau Z_{i+1} \cdots Z_k,\ Y^a \rightarrow Z_i \cdots Z_k$$

The fact that $G_{(X,a)}$ generates the mentioned language is intuitively clear and a formal proof of that is easy. The size of $G_{(X,a)}$ is $\mathcal{O}(n^2)$, as Δ contains $\mathcal{O}(n)$ basic transitions of length $\mathcal{O}(n)$. $\square$

The crucial part of our algorithm (the 'refinement step') is presented in the proof of the next theorem. Our complexity analysis is based on the following facts: Let $\mathcal{A} = (Q, \Sigma, \delta, q_0, F)$ be a non-deterministic automaton with ε-rules, and let t be the total number of states and transitions of $\mathcal{A}$.

- The problem whether a given $w \in \Sigma^*$ belongs to $L(\mathcal{A})$ is decidable in $\mathcal{O}(|w| \cdot t)$ time.
- The problem whether $L(\mathcal{A}) = \emptyset$ is decidable in $\mathcal{O}(t)$ time.

Theorem 5. *Let K be a fundamental relation. The relation $Exp(K)$ can be effectively constructed in $\mathcal{O}(n^4\, m^5)$ time.*

Proof. First we construct the automata $\mathcal{A}_g$ of Theorem 1 for every $g \in Const(\Gamma)$. This takes $\mathcal{O}(n\, m^3)$ time. Then we construct the automata $\mathcal{A}_{(X,a)}$ of Theorem 4 for all X, a. This takes $\mathcal{O}(n^4)$ time. Furthermore, we also compute the set of all pairs of the form $(f,g), (\varepsilon, g)$ which belong to $Cl(K)$. It can be done in $\mathcal{O}(m^2)$ time. Now we show that for each pair of K we can decide in $\mathcal{O}(n^3\, m^3)$ time whether this pair expands in K.

The pairs of the form (f,g) and (ε, g) are easy to handle; there are at most m states f' s.t. $f \stackrel{a}{\rightarrow} f'$, and at most m states g' with $g \stackrel{a}{\Rightarrow} g'$, hence we need to check only $\mathcal{O}(m^2)$ pairs to verify the first (and consequently also the second) condition of Definition 4. Each such pair can be checked in constant time, because the set of all pairs $(f,g), (\varepsilon, g)$ which belong to $Cl(K)$ has been already computed at the beginning.

Now let us consider a pair of the form (Y, g). First we need to verify that for each $Y \stackrel{a}{\rightarrow} \alpha$ there is some $g \stackrel{a}{\Rightarrow} h$ s.t. $(\alpha, h) \in Cl(K)$. This requires $\mathcal{O}(n\, m)$ tests whether $\alpha \in L(\mathcal{A}_h)$. As the length of α is $\mathcal{O}(n)$ and the size of $\mathcal{A}_h$ is $\mathcal{O}(n\, m^2)$, each such test can be done in $\mathcal{O}(n^2\, m^2)$ time, hence we need $\mathcal{O}(n^3\, m^3)$ time in total. As for the second condition of Definition 4, we need to find out whether for each $g \stackrel{a}{\rightarrow} h$ there is some $X \stackrel{a}{\Rightarrow} \alpha$ s.t. $(\alpha, h) \in Cl(K)$. To do that, we simply test the emptiness of $L(\mathcal{A}_{(X,a)}) \cap L(\mathcal{A}_h)$. The size of the product automaton is $\mathcal{O}(n^3\, m^2)$ and we need to perform only $\mathcal{O}(m)$ such tests, hence the time $\mathcal{O}(n^3\, m^3)$ suffices.

Pairs of the form (Yf, g) are handled in a similar way; the first condition of Definition 4 is again no problem, as we are interested only in the '$\stackrel{a}{\rightarrow}$' moves of the left component. Now let $g \stackrel{a}{\rightarrow} g'$. An existence of a 'good' $\stackrel{a}{\Rightarrow}$ move of Yf can be verified by testing whether one of the following conditions holds:

- $L(\mathcal{A}_{(Y,a)}) \cdot \{f\} \cap L(\mathcal{A}_{g'})$ is nonempty.
- $Y \stackrel{a}{\Rightarrow} \varepsilon$ and there is some $f \stackrel{\tau}{\Rightarrow} f'$ s.t. $(f', g') \in Cl(K)$.
- $Y \stackrel{\tau}{\Rightarrow} \varepsilon$ and there is some $f \stackrel{a}{\Rightarrow} f'$ s.t. $(f', g') \in Cl(K)$.

All those conditions can be checked in $\mathcal{O}(n^3\, m^3)$ time (the required analysis has been in fact done above). As K contains $\mathcal{O}(n\, m^2)$ pairs, the total time which is needed to compute $Exp(K)$ is $\mathcal{O}(n^4\, m^5)$. □

As the BPA process E (introduced at the beginning of this section) is an element of $Const(\Delta)$, we have that $E \approx F$ iff $(E, F) \in \mathcal{B}$. To compute $\mathcal{B}$, we have to perform the computation of the expansion $\mathcal{O}(n\, m^2)$ times (see Theorem 3). This gives us the following main theorem:

Theorem 6. *Weak bisimilarity is decidable between BPA and finite-state processes in* $\mathcal{O}(n^5\, m^7)$ *time.*

The fact that weak bisimilarity is not a congruence w.r.t. sequential composition is rather unpleasant; any equivalence which is to be considered as 'behavioural' should have this property. We propose a solution to this problem by designing a natural refinement of weak bisimilarity called *termination-sensitive bisimilarity*. This relation distinguishes between the following 'basic phenomenons' of sequencing:

- *successful termination* of the process which is currently being executed. The system can then continue to execute the next process in the queue.
- *unsuccessful termination* of the executed process (deadlock). This models a severe error which causes the whole system to 'get stuck'.
- *entering an infinite internal loop* (livelock).

Termination-sensitive bisimilarity is a congruence w.r.t. sequencing, and it is also decidable between BPA and finite state processes in polynomial time. It can be proved by adapting the proof for weak bisimilarity. Formal definitions and proofs are omitted due to the lack of space—see [21] for details.

4 Normed BPP Processes

In this section we prove that weak bisimilarity is decidable in polynomial time between normed BPP and finite-state processes. The basic structure of our proof is similar to the one for BPA. The key is that the weak bisimulation problem can be decomposed into problems about the single constants and their interaction with each other. In particular, a normed BPP process is finite w.r.t. weak bisimilarity iff every single reachable process constant is finite w.r.t. weak bisimilarity. This does not hold for general BPP and thus our construction does not carry over to general BPP.

Even for normed BPP, we have to solve some additional problems. The bisimulation base and its closure are simpler due to the normedness assumption, but the 'symbolic' representation of BPP state-space is more problematic (see below). The set of states which are reachable from a given BPP state in one '$\stackrel{a}{\Rightarrow}$' move is no longer regular, but it can be in some sense represented by a CF-grammar. In our algorithm we use the facts that emptiness of a CF language is decidable in polynomial time, and that CF languages are closed under intersection with regular languages. Most proofs in this section are omitted due to the lack of space. See [21] for details.

Let E be a BPP process and F a finite-state process with the underlying systems Δ and Γ, respectively. We can assume w.l.o.g. that $E \in Const(\Delta)$. Elements of $Const(\Delta)$ are denoted by $X, Y, Z, \ldots$, elements of $Const(\Gamma)$ by $f, g, h, \ldots$ The set of all parallel expressions over $Const(\Delta)$ is denoted by $Const(\Delta)^{\otimes}$ and its elements by Greek letters $\alpha, \beta, \ldots$ The size of Δ is denoted by n, and the size of Γ by m.

In our constructions we represent certain subsets of $Const(\Delta)^{\otimes}$ by finite automata and CF grammars. The problem is that elements of $Const(\Delta)^{\otimes}$ are considered modulo commutativity; however, finite automata and CF grammars of course distinguish between different 'permutations' of the same word. As the classes of regular and CF languages are not closed under permutation, this problem is important. As we want to clarify the distinction between α and its possible 'linear representations', we define for each α the set $Lin(\alpha)$ as follows:

$$Lin(X_1 \| \cdots \| X_k) = \{X_{p(1)} \cdots X_{p(k)} \mid p \text{ is a permutation of the set } \{1, \cdots, k\}\}$$

For example, $Lin(X\|Y\|Z) = \{XYZ,\ XZY,\ YXZ,\ YZX,\ ZXY,\ ZYX\}$. We also assume that each $Lin(\alpha)$ contains some (unique) element called *canonical form* of $Lin(\alpha)$. It is not important how the canonical form is chosen; we need it just to make some constructions deterministic (for example, we can fix some linear order on process constants and let the canonical form of $Lin(\alpha)$ be the sorted order of constants of α).

Definition 5. *A relation K is* fundamental *iff it is a subset of* $(Const(\Delta) \cup \{\varepsilon\}) \times Const(\Gamma)$. *The greatest fundamental relation is denoted by $\mathcal{G}$. The* bisimulation base *for Δ and Γ, denoted $\mathcal{B}$, is defined as follows:*

$$\mathcal{B} = \{(X, f) \mid X \approx f\} \cup \{(\varepsilon, f) \mid \varepsilon \approx f\}$$

Definition 6. *Let K be a fundamental relation. The* closure *of K, denoted $Cl(K)$, is the least relation M which satisfies*

1. $K \subseteq M$
2. *if* $(X, g) \in K$, $(\beta, h) \in M$, *and* $f \approx g\|h$, *then* $(\beta\|X, f) \in M$
3. *if* $(\varepsilon, g) \in K$, $(\beta, h) \in M$, *and* $f \approx g\|h$, *then* $(\beta, f) \in M$

The family of $Cl(K)^i$ approximations is defined in the same way as in the previous section.

Lemma 3. *Let* $(\alpha, f) \in Cl(K)$, $(\beta, g) \in Cl(K)$, $f\|g \approx h$. *Then* $(\alpha\|\beta, h) \in Cl(K)$.

Again, the closure of the bisimulation base is the greatest weak bisimulation between processes of Δ and Γ.

Theorem 7. *Let* $\alpha \in Const(\Delta)^{\otimes}$, $f \in Const(\Gamma)$. *We have that* $\alpha \approx f$ *iff* $(\alpha, f) \in Cl(\mathcal{B})$.

The closure of any fundamental relation can in some sense be represented by a finite-state automaton, as stated in the next theorem.

Theorem 8. *Let K be a fundamental relation. For each $g \in Const(\Gamma)$ there is a finite-state automaton $\mathcal{A}_g$ of size $\mathcal{O}(n\,m)$ constructible in $\mathcal{O}(n\,m)$ time s.t. the following conditions hold:*

- *whenever $\mathcal{A}_g$ accepts an element of $Lin(\alpha)$, then $(\alpha, g) \in Cl(K)$*
- *if $(\alpha, g) \in Cl(K)$, then $\mathcal{A}_g$ accepts at least one element of $Lin(\alpha)$*

It is important to realize that if $(\alpha, g) \in Cl(K)$, then $\mathcal{A}_g$ does not necessarily accept *all* elements of $Lin(\alpha)$. Generally, $\mathcal{A}_g$ cannot be 'repaired' to do so (see the beginning of this section); however, there is actually no need for such 'repairs', because $\mathcal{A}_g$ has the following nice property:

Lemma 4. *Let K be a fundamental relation s.t. $\mathcal{B} \subseteq K$. If $\alpha \approx g$, then the automaton $\mathcal{A}_g$ of (the proof of) Theorem 8 constructed for K accepts all elements of $Lin(\alpha)$.*

The set of states which are reachable from a given $X \in Const(\Delta)$ in one '$\stackrel{a}{\Rightarrow}$' move is no longer regular, but it can, in some sense, be represented by a CF grammar.

Theorem 9. *For all $X \in Const(\Delta)$, $a \in Act(\Delta)$ there is a context-free grammar $G_{(X,a)}$ in 3-GNF of size $\mathcal{O}(n^4)$ constructible in $\mathcal{O}(n^4)$ time s.t. the following two conditions hold:*

- *if $G_{(X,a)}$ generates an element of $Lin(\alpha)$, then $X \stackrel{a}{\Rightarrow} \alpha$*
- *if $X \stackrel{a}{\Rightarrow} \alpha$, then $G_{(X,a)}$ generates at least one element of $Lin(\alpha)$*

The notion of expansion is defined in a different way (when compared to the one of the previous section).

Definition 7. *Let K be a fundamental relation. We say that a pair $(X, f) \in K$* expands *in K iff the following two conditions hold:*

- *for each $X \stackrel{a}{\rightarrow} \alpha$ there is some $f \stackrel{a}{\Rightarrow} g$ s.t. $\overline{\alpha} \in L(\mathcal{A}_g)$, where $\overline{\alpha}$ is the canonical form of $Lin(\alpha)$.*
- *for each $f \stackrel{a}{\rightarrow} g$ the language $L(\mathcal{A}_g) \cap L(G_{(X,a)})$ is non-empty.*

A pair $(\varepsilon, f) \in K$ expands in K iff $f \stackrel{a}{\rightarrow} g$ implies $a = \tau$, and for each $f \stackrel{\tau}{\rightarrow} g$ we have that $\varepsilon \in L(\mathcal{A}_g)$. The set of all pairs of K which expand in K is denoted by $Exp(K)$.

Theorem 10. *Let K be a fundamental relation. The set $Exp(K)$ can be computed in $\mathcal{O}(n^{11}\,m^8)$ time.*

Proof. First we compute the automata $\mathcal{A}_g$ of Theorem 8 for all $g \in Const(\Gamma)$. This takes $\mathcal{O}(n\,m^2)$ time. Then we compute the grammars $G_{(X,a)}$ of Theorem 9 for all $X \in Const(\Delta)$, $a \in Act$. This takes $\mathcal{O}(n^6)$ time. Now we show that it is decidable in $\mathcal{O}(n^{10}\,m^7)$ time whether a pair (X, f) of K expands in K.

The first condition of Definition 7 can be checked in $\mathcal{O}(n^3\,m^2)$ time, as there are $\mathcal{O}(n)$ transitions $X \stackrel{a}{\rightarrow} \alpha$, $\mathcal{O}(m)$ states g s.t. $f \stackrel{a}{\Rightarrow} g$, and for each such pair (α, g) we verify whether $\overline{\alpha} \in L(\mathcal{A}_g)$ where $\overline{\alpha}$ is the canonical form of $Lin(\alpha)$; this membership test can be done in $\mathcal{O}(n^2\,m)$ time, as the size of $\overline{\alpha}$ is $\mathcal{O}(n)$ and the size of $\mathcal{A}_g$ is $\mathcal{O}(n\,m)$.

The second condition of Definition 7 is more expensive. To test the emptiness of $L(\mathcal{A}_g) \cap L(G_{(X,a)})$, we first construct a pushdown automaton $\mathcal{P}$ which recognises this language. $\mathcal{P}$ has $\mathcal{O}(m)$ control states and its total size is $\mathcal{O}(n^5\, m)$. Furthermore, each rule $pX \overset{a}{\rightarrow} q\alpha$ of $\mathcal{P}$ has the property that $length(\alpha) \leq 2$, because $G_{(X,a)}$ is in 3-GNF. Now we transform this automaton to an equivalent CF grammar by a well-known procedure described e.g. in [17]. The size of the resulting grammar is $\mathcal{O}(n^5\, m^3)$, and its emptiness can be thus checked in $\mathcal{O}(n^{10}\, m^6)$ time (cf. [17]). This construction has to be performed $\mathcal{O}(m)$ times, hence we need $\mathcal{O}(n^{10}\, m^7)$ time in total.

Pairs of the form (ε, f) are handled in a similar (but less expensive) way. As K contains $\mathcal{O}(n\, m)$ pairs, the computation of $Exp(K)$ takes $\mathcal{O}(n^{11}\, m^8)$ time. □

The previous theorem is actually a straightforward consequence of Definition 7. The next theorem says that *Exp* really does what we need.

Theorem 11. *Let K be a fundamental relation s.t. $Exp(K) = K$. Then $Cl(K)$ is a weak bisimulation.*

Proof. Let $(\alpha, f) \in Cl(K)^i$. We prove that for each $\alpha \overset{a}{\rightarrow} \beta$ there is some $f \overset{a}{\Rightarrow} g$ s.t. $(\beta, g) \in Cl(K)$ and vice versa. By induction on i.

- $i = 0$. Then $(\alpha, f) \in K$, and we can distinguish the following two possibilities:
  1. $\alpha = X$
     Let $X \overset{a}{\rightarrow} \beta$. By Definition 7 there is $f \overset{a}{\Rightarrow} g$ s.t. $\overline{\beta} \in L(\mathcal{A}_g)$ for some $\overline{\beta} \in Lin(\beta)$. Hence $(\beta, g) \in Cl(K)$ due to the first part of Theorem 8.
     Let $f \overset{a}{\rightarrow} g$. By Definition 7 there is some string $w \in L(\mathcal{A}_g) \cap L(G_{(X,a)})$. Let $w \in Lin(\beta)$. We have $X \overset{a}{\Rightarrow} \beta$ due to the first part of Theorem 9, and $(\beta, g) \in Cl(K)$ due to Theorem 8.
  2. $\alpha = \varepsilon$
     Let $f \overset{a}{\rightarrow} g$. Then $a = \tau$ and $\varepsilon \in L(\mathcal{A}_g)$ by Definition 7. Hence $(\varepsilon, g) \in Cl(K)$ due to Theorem 8.
- **Induction step.** Let $(\alpha, f) \in Cl(K)^{i+1}$. There are two possibilities.
  I. $\alpha = X \| \gamma$ and there are r, s s.t. $(X, r) \in K$, $(\gamma, s) \in Cl(K)^i$, and $r \| s \approx f$.
     Let $X \| \alpha \overset{a}{\rightarrow} \beta$. The action '$a$' can be emitted either by X or by α. We distinguish the two cases.
     1) $X \| \gamma \overset{a}{\rightarrow} \delta \| \gamma$. As $(X, r) \in K$ and $X \overset{a}{\rightarrow} \delta$, there is some $r \overset{a}{\Rightarrow} r'$ s.t. $(\delta, r') \in Cl(K)$. As $r \| s \approx f$ and $r \overset{a}{\Rightarrow} r'$, there is some $f \overset{a}{\Rightarrow} g$ s.t. $r' \| s \approx g$. To sum up, we have $(\delta, r') \in Cl(K)$, $(\gamma, s) \in Cl(K)$, $r' \| s \approx g$, hence $(\delta \| \gamma, g) \in Cl(K)$ due to Lemma 3.
     2) $X \| \gamma \overset{a}{\rightarrow} X \| \rho$. As $(\gamma, s) \in Cl(K)^i$ and $\gamma \overset{a}{\rightarrow} \rho$, there is $s \overset{a}{\Rightarrow} s'$ s.t. $(\rho, s') \in Cl(K)$. As $r \| s \approx f$ and $s \overset{a}{\Rightarrow} s'$, there is $f \overset{a}{\Rightarrow} g$ s.t. $(r \| s') \approx g$. Due to Lemma 3 we obtain $(X \| \rho, g) \in Cl(K)$.
     Let $f \overset{a}{\rightarrow} g$. As $r \| s \approx f$, there are $r \overset{x}{\Rightarrow} r'$, $s \overset{y}{\Rightarrow} s'$ where $x = a \wedge y = \tau$ or $x = \tau \wedge y = a$ s.t. $r' \| s' \approx g$. As $(X, r) \in K$, $(\gamma, s) \in Cl(K)^i$, there are $X \overset{x}{\Rightarrow} \delta$, $\gamma \overset{y}{\Rightarrow} \rho$ s.t. (δ, r'), $(\rho, s') \in Cl(K)$. Clearly $X \| \gamma \overset{a}{\Rightarrow} \delta \| \rho$ and $(\delta \| \rho, g) \in Cl(K)$ due to Lemma 3.
  II. $(\alpha, r) \in Cl(K)^i$ and there is some s s.t. $(\varepsilon, s) \in K$ and $r \| s \approx f$.
     The proof can be completed along the same lines as above. □

Now we can approximate (and compute) the bisimulation base in the same way as in the previous section.

Theorem 12. *There is a $j \in \mathbb{N}$, bounded by $\mathcal{O}(n\,m)$, such that $\mathcal{B}^j = \mathcal{B}^{j+1}$. Moreover, $\mathcal{B}^j = \mathcal{B}$.*

Theorem 13. *Weak bisimilarity between normed BPP and finite-state processes is decidable in $\mathcal{O}(n^{12}\,m^9)$ time.*

5 Conclusions

We have proved that weak bisimilarity is decidable between BPA processes and finite-state processes in $\mathcal{O}(n^5\,m^7)$ time, and between normed BPP and finite-state processes in $\mathcal{O}(n^{12}\,m^9)$ time. It may be possible to improve the algorithm by re-using previously computed information, for example about sets of reachable states, but the exponents would still be very high. This is because the whole bisimulation basis is constructed. To get a more efficient algorithm, one could try to avoid this. Note however, that once we construct $\mathcal{B}$ (for a BPA/nBPP system Δ and a finite-state system Γ) and the automaton $\mathcal{A}_g$ of Theorem 1/Theorem 8 (for $K = \mathcal{B}$ and some $g \in \mathit{Const}(\Gamma)$), we can decide weak bisimilatity between a BPA/nBPP process α over Δ and a process $f \in \mathit{Const}(\Gamma)$ in time $\mathcal{O}(|\alpha|)$—it suffices to test whether $\mathcal{A}_f$ accepts α (observe that there is no substantial difference between $\mathcal{A}_f$ and $\mathcal{A}_g$ except for the initial state).

The technique of bisimulation bases has also been used for strong bisimilarity in [15, 16]. However, those bases are different from ours; their design and the way how they generate 'new' bisimilar pairs of processes rely on additional algebraic properties of strong bisimilarity (which is a full congruence w.r.t. sequencing, allows for unique decompositions of normed processes w.r.t. sequencing and parallelism, etc.). The main difficulty of those proofs is to show that the membership in the 'closure' of the defined bases is decidable in polynomial time. The main point of our proofs is the use of 'symbolic' representation of infinite subsets of BPA and BPP state-space.

We would also like to mention that our proofs can be easily adapted to other bisimulation-like equivalences, where the notion of 'bisimulation-like' equivalence is the one of [19]. A concrete example is termination-sensitive bisimilarity of Section 3. Intuitively, almost every bisimulation-like equivalence has the algebraic properties which are needed for the construction of the bisimulation base, and the 'symbolic' technique for state-space representation can also be adapted. See [19] for details.

References

1. J.C.M. Baeten, J.A. Bergstra, and J.W. Klop. Decidability of bisimulation equivalence for processes generating context-free languages. *JACM*, 40:653–682, 1993.
2. J.C.M. Baeten and W.P. Weijland. *Process Algebra*. Number 18 in Cambridge Tracts in Theoretical Computer Science. Cambridge University Press, 1990.
3. A. Bouajjani, J. Esparza, and O. Maler. Reachability analysis of pushdown automata: application to model checking. In *Proceedings of CONCUR'97*, volume 1243 of *LNCS*, pages 135–150. Springer, 1997.

4. O. Burkart, D. Caucal, and B. Steffen. An elementary decision procedure for arbitrary context-free processes. In *Proceedings of MFCS'95*, volume 969 of *LNCS*, pages 423–433. Springer, 1995.
5. O. Burkart and J. Esparza. More infinite results. *ENTCS*, 5, 1997.
6. D. Caucal. Graphes canoniques des graphes algébriques. *RAIRO*, 24(4):339–352, 1990.
7. I. Černá, M. Křetínský, and A. Kučera. Comparing expressibility of normed BPA and normed BPP processes. *Acta Informatica*, 36(3):233–256, 1999.
8. S. Christensen, Y. Hirshfeld, and F. Moller. Bisimulation is decidable for all basic parallel processes. In *Proceedings of CONCUR'93*, volume 715 of *LNCS*, pages 143–157. Springer, 1993.
9. S. Christensen, H. Hüttel, and C. Stirling. Bisimulation equivalence is decidable for all context-free processes. *Information and Computation*, 121:143–148, 1995.
10. J. Esparza. Petri nets, commutative context-free grammars, and basic parallel processes. In *Proceedings of FCT'95*, volume 965 of *LNCS*, pages 221–232. Springer, 1995.
11. J. Esparza. Decidability of model checking for infinite-state concurrent systems. *Acta Informatica*, 34:85–107, 1997.
12. J. Esparza and J. Knop. An automata-theoretic approach to interprocedural data-flow analysis. In *Proceedings of FoSSaCS'99*, volume 1578 of *LNCS*, pages 14–30. Springer, 1999.
13. J.F. Groote. A short proof of the decidability of bisimulation for normed BPA processes. *Information Processing Letters*, 42:167–171, 1992.
14. Y. Hirshfeld. Bisimulation trees and the decidability of weak bisimulations. *ENTCS*, 5, 1996.
15. Y. Hirshfeld, M. Jerrum, and F. Moller. A polynomial algorithm for deciding bisimilarity of normed context-free processes. *TCS*, 158:143–159, 1996.
16. Y. Hirshfeld, M. Jerrum, and F. Moller. A polynomial algorithm for deciding bisimulation equivalence of normed basic parallel processes. *MSCS*, 6:251–259, 1996.
17. J.E. Hopcroft and J.D. Ullman. *Introduction to Automata Theory, Languages, and Computation*. Addison-Wesley, 1979.
18. H. Hüttel and C. Stirling. Actions speak louder than words: Proving bisimilarity for context-free processes. In *Proceedings of LICS'91*, pages 376–386. IEEE Computer Society Press, 1991.
19. P. Jančar, A. Kučera, and R. Mayr. Deciding bisimulation-like equivalences with finite-state processes. In *Proceedings of ICALP'98*, volume 1443 of *LNCS*, pages 200–211. Springer, 1998.
20. A. Kučera. On effective decomposability of sequential behaviours. *TCS*. To appear.
21. A. Kučera and R. Mayr. Weak bisimilarity with infinite-state systems can be decided in polynomial time. Technical report TUM-I9830, Institut für Informatik, TU-München, 1998.
22. A. Kučera and R. Mayr. Simulation preorder on simple process algebras. In *Proceedings of ICALP'99*, LNCS. Springer, 1999. To apper.
23. R. Mayr. Process rewrite systems. *Information and Computation*. To appear.
24. R. Mayr. Weak bisimulation and model checking for basic parallel processes. In *Proceedings of FST&TCS'96*, volume 1180 of *LNCS*, pages 88–99. Springer, 1996.
25. R. Mayr. Strict lower bounds for model checking BPA. *ENTCS*, 18, 1998.
26. R. Milner. *Communication and Concurrency*. Prentice-Hall, 1989.
27. F. Moller. Infinite results. In *Proceedings of CONCUR'96*, volume 1119 of *LNCS*, pages 195–216. Springer, 1996.
28. R. Paige and R. Tarjan. Three partition refinement algorithms. *SIAM Journal of Computing*, 16(6):973–989, 1987.
29. D.M.R. Park. Concurrency and automata on infinite sequences. In *Proceedings* 5th *GI Conference*, volume 104 of *LNCS*, pages 167–183. Springer, 1981.
30. J. Stříbrná. Hardness results for weak bisimilarity of simple process algebras. *ENTCS*, 18, 1998.

Robust Satisfaction

Orna Kupferman[1] and Moshe Y. Vardi[2*]

[1] Hebrew University, The institute of Computer Science, Jerusalem 91904, Israel
Email: orna@cs.huji.ac.il, URL: http://www.cs.huji.ac.il/~orna
[2] Rice University, Department of Computer Science, Houston, TX 77251-1892, U.S.A.
Email: vardi@cs.rice.edu, URL: http://www.cs.rice.edu/~vardi

Abstract. In order to check whether an open system satisfies a desired property, we need to check the behavior of the system with respect to an arbitrary environment. In the most general setting, the environment is another open system. Given an open system M and a property ψ, we say that M *robustly satisfies* ψ iff for every open system M', which serves as an environment to M, the composition $M \| M'$ satisfies ψ. The problem of *robust model checking* is then to decide, given M and ψ, whether M robustly satisfies ψ. In this paper we study the robust-model-checking problem. We consider systems modeled by nondeterministic Moore machines, and properties specified by branching temporal logic (for linear temporal logic, robust satisfaction coincides with usual satisfaction). We show that the complexity of the problem is EXPTIME-complete for CTL and the μ-calculus, and is 2EXPTIME-complete for CTL*. We partition branching temporal logic formulas into three classes: universal, existential, and mixed formulas. We show that each class has different sensitivity to the robustness requirement. In particular, unless the formula is mixed, robust model checking can ignore nondeterministic environments. In addition, we show that the problem of classifying a CTL formula into these classes is EXPTIME-complete.

1 Introduction

Today's rapid development of complex and safety-critical systems requires reliable verification methods. In formal verification, we verify that a system meets a desired property by checking that a mathematical model of the system satisfies a formal specification that describes the property. We distinguish between two types of systems: *closed* and *open* [20]. (Open systems are called *reactive* systems in [20].) A closed system is a system whose behavior is completely determined by the state of the system. An open system is a system that interacts with its environment and whose behavior depends on this interaction. Thus, while in a closed system all the nondeterministic choices are internal, and resolved by the system, in an open system there are also external nondeterministic choices, which are resolved by the environment [19].

In order to check whether a closed system satisfies a desired property, we translate the system into a formal model, typically a state transition graph, specify the property

* Supported in part by the NSF grants CCR-9628400 and CCR-9700061, and by a grant from the Intel Corporation. Part of this work was done when this author was a Varon Visiting Professor at the Weizmann Institute of Science.

with a temporal-logic formula, and check formally that the model satisfies the formula. Hence the name *model checking* for the verification methods derived from this viewpoint [5, 33]. In order to check whether an open system satisfies a desired property, we need to check the behavior of the system with respect to an arbitrary environment [11]. In the most general setting, the environment is another open system. Thus, given an open system M and a specification ψ, we need to check whether for every (possibly infinite) open system M', which serves as an environment to M, the composition $M||M'$ satisfies ψ. If the answer is yes, we say that M *robustly satisfies* ψ. The problem of *robust model checking*, initially posed in [12], is to determine, given M and ψ, whether M robustly satisfies ψ.

Two possible views regarding the nature of time induce two types of temporal logics [27]. In *linear* temporal logics, time is treated as if each moment in time has a unique possible future. Thus, linear temporal logic formulas are interpreted over linear sequences and we regard them as describing a behavior of a single computation of a program. In *branching* temporal logics, each moment in time may split into various possible futures. Accordingly, the structures over which branching temporal logic formulas are interpreted can be viewed as infinite computation trees, each describing the behavior of the possible computations of a nondeterministic program. It turns out that traditional model-checking algorithms and tools are not suitable for the verification of open systems with respect to branching temporal logics [24]. In other words, it may be that while M satisfies ψ, it does not robustly satisfy ψ.

To see the difference between robust satisfaction and usual satisfaction, consider the open system M described below. The system M models a cash machine (ATM). At the

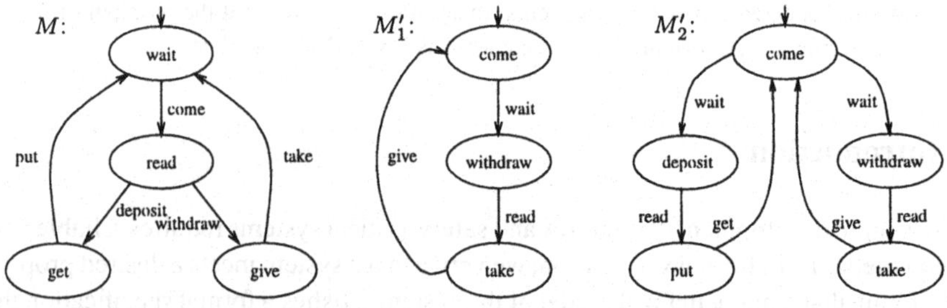

state labeled *wait*, M waits for costumers. When a costumer comes, M moves to the state labeled *read*, where it reads whether the costumer wants to *deposit* or *withdraw* money. According to the external choice of the costumer, M moves to either a *get* or *give* state, from which it returns to the *wait* state. An environment for the ATM is an infinite line of costumers, each with his depositing or withdrawing plans. Suppose that we want to check whether the ATM can always get money eventually; thus, whether it satisfies the temporal logic formula $\psi = AGEFget$. Verification algorithms that refer to M as a closed system, perform model checking in order to verify the correctness of the ATM. Since $M \models \psi$, they get a positive answer to this question. Nonetheless, it is easy to see that the ATM does not satisfy the property ψ with respect to all environments. For example, the composition of M with the environment M_1', in which all the costumers only withdraw money, does not satisfy ψ. Formally, M_1' never supplies to M the input *deposit*, thus M_1' disables the transition of M from the *read* state to the *get* state. Consequently, the composition $M||M_1'$ contains a single computation, in which *get* is not reachable.

A first attempt to solve the robust-model-checking problem was presented in [24], which suggested the method of *module checking*. In this algorithmic method we check, given an open system modeled as a finite state-transition graph, and a desired property specified as a temporal-logic formula, whether, no matter how an environment disables some of the system's transitions, the composition of the system with the environment satisfies the property. In particular, in the ATM example, the module-checking paradigm takes into consideration the fact that the environment can consistently disable the transition from the *read* state to the *get* state, and detects the fact that the ATM cannot always get money eventually. The model discussed in [24] is somewhat simplistic as it does not allow the system to have internal variables. This assumption is removed in [25], which considers module checking with *incomplete information*. In this setting, the system has internal variables, which the environment cannot read. While [24] considers arbitrary disabling of transitions, the setting in [25] is such that whenever two computations of the system differ only in the values of internal variables along them, the disabling of transitions along them coincide. While the setting in [25] is more general, it still does not solve the general robust-model-checking problem.

To see this, let us go back to the ATM example. Suppose that we want to check whether the ATM can either move from all the successors of the initial state to a state where it gets money, or it can move from all the successors of the initial state to a state where it gives money. When we regard M as a closed system, this property is satisfied. Indeed, M satisfies the temporal-logic formula $\varphi = AXEX get \vee AXEX give$. Moreover, no matter how we remove transitions from the computation tree of M, the trees we get satisfy either $AXEX get$ or $AXEX give$[1]. In particular, $M||M_1'$ satisfies $AXEX give$. Thus, if we follow the module-checking paradigm, the answer to the question is positive. Consider now the environment M_2'. The initial state of $M||M_2'$ has two successors. One of these successors has a single successor in which the ATM gives money and the second has a single successor in which the ATM gets money. Hence, $M||M_2'$ does not satisfy φ. Intuitively, while the module-checking paradigm considers only disabling of transitions, and thus corresponds to the composition of M with all deterministic environments, robust model checking considers all, possibly nondeterministic, environments. There, the composition of the system with an environment may not just disable some of the system's transitions, but may also, as in the example above, increase the nondeterminism of the system.

In this work we consider the problem of verification of open systems in its full generality and solve the robust-model-checking problem. Thus, given an open system M and a specification ψ, we study the problem of determining whether M robustly satisfies ψ. Both M and its environment are nondeterministic Moore machines. They communicate via input and output variables and they both may have private variables

[1] We assume that the composition of the system and the environment is *deadlock free*, thus every state has at least one successor.

and be nondeterministic. Our setting allows the environment to be infinite, and to have unbounded branching degree. Nevertheless, we show that if there is some environment M' for which $M||M'$ does not satisfy ψ, then there is also a finite environment M'' with a bounded branching degree (which depends on the number of universal requirements in ψ) such that $M||M''$ does not satisfy ψ.

We solve the robust-model-checking problem for branching temporal specifications. As with module checking with incomplete information, *alternation* is a suitable and helpful automata-theoretic mechanism for coping with the internal variables of M and M'. In spite of the similarity to the incomplete information setting, the solution the robust model-checking problem is more challenging, as one needs to take into consideration the fact that a module may have different reactions to the same input sequence, yet this is possible only when different nondeterministic choices have been taken along the sequence. Using *alternating tree automata*, we show that the problem of robust satisfaction is EXPTIME-complete for CTL and the μ-calculus, and is 2EXPTIME-complete for CTL*. The internal variables of M make the time complexity of the robust-model-checking problem exponential already in the size of M. The same complexity bounds hold for the problem of module checking with incomplete information [25]. Thus, on the one hand, the problem of robust model checking, which generalizes the problem of module checking with incomplete information, is not harder than the latter problem. On the other hand, keeping in mind that the system to be checked is typically a parallel composition of several components, which by itself hides an exponential blow-up [18], our results imply that checking verification of open systems with respect to branching temporal specifications is rather intractable.

Recall that not all specification formalisms are sensitive to the distinction between open and closed systems. The study of verification of open system has motivated the use of *universal temporal logic* [13] as a specification formalism. Formulas of universal temporal logics describe requirements that should hold in *all* computations of the system. These requirements may be either linear or branching. In both cases, the more behaviors the system has, the harder it is for the system to satisfy the requirements. Indeed, universal temporal logics induce the *simulation* order between systems [28, 6]. That is, a system M simulates a system M' if and only if all universal temporal logic formulas that are satisfied in M are satisfied in M' as well. It follows that traditional model-checking methods are applicable also for the verification of open systems with respect to universal properties. Indeed, since M simulates $M||M'$ for every M', satisfaction of a universal property in M implies its satisfaction in all the compositions of M with an environment.

One of the main advantages of branching temporal logics with respect to linear temporal logic is, however, the ability to mix universal and *existential* properties; e.g., in order to specify possibility properties like $AGEFp$. Existential properties describe requirements that should hold in *some* computations of the system. We show that non-universal properties can be partitioned into two classes, each with a different sensitivity to the distinction between open and closed systems. We say that a temporal-logic formula φ is *existential* if it imposes only existential requirements on the system, thus $\neg\varphi$ is universal. The formula φ is *mixed* if it imposes both existential and universal requirements, thus φ is neither universal nor existential. While universal formulas are insensitive to the system being open, we show that existential formulas are insensitive to the environment being nondeterministic. Thus, for such formulas, one can use the module-checking method. We study the problems of determining whether a given formula is universal or mixed, and show that they are both EXPTIME-complete. These result are relevant also in the contexts of modular verification [13] and backwards reasoning [17].

In the discussion, we compare robust model checking with previous work about verification of open systems as well as with the closely-related area of supervisory control [35, 3]. We also argue for the generality of the model studied in this paper and show that it captures settings in which assumptions about the environment are known, as well as settings with global actions and possible deadlocks.

2 Preliminaries

2.1 Trees and Automata

Given a finite set Υ, an *Υ-tree* is a set $T \subseteq \Upsilon^*$ such that if $x \cdot \upsilon \in T$, where $x \in \Upsilon^*$ and $\upsilon \in \Upsilon$, then also $x \in T$. When Υ is not important or clear from the context, we call T a tree. The elements of T are called *nodes*, and the empty word ϵ is the *root* of T. For every $x \in T$, the nodes $x \cdot \upsilon \in T$ where $\upsilon \in \Upsilon$ are the *children* of x. Each node $x \neq \epsilon$ of T has a *direction* in Υ. The direction of a node $x \cdot \upsilon$ is υ. We denote by $dir(x)$ the direction of node x. An Υ-tree T is a *full infinite tree* if $T = \Upsilon^*$. Unless otherwise mentioned, we consider here full infinite trees. A *path* η of a tree T is a set $\eta \subseteq T$ such that $\epsilon \in \eta$ and for every $x \in \eta$ there exists a unique $\upsilon \in \Upsilon$ such that $x \cdot \upsilon \in \eta$. The i'th *level* of T is the set of nodes of length i in T. Given two finite sets Υ and Σ, a *Σ-labeled Υ-tree* is a pair $\langle T, V \rangle$ where T is an Υ-tree and $V : T \to \Sigma$ maps each node of T to a letter in Σ. When Υ and Σ are not important or clear from the context, we call $\langle T, V \rangle$ a labeled tree.

Alternating tree automata generalize nondeterministic tree automata and were first introduced in [29]. An alternating tree automaton $\mathcal{A} = \langle \Sigma, Q, q_0, \delta, \alpha \rangle$ runs on full Σ-labeled Υ-trees (for an agreed set Υ of directions). It consists of a finite set Q of states, an initial state $q_0 \in Q$, a transition function δ, and an acceptance condition α (a condition that defines a subset of Q^ω).

For a set Υ of directions, let $\mathcal{B}^+(\Upsilon \times Q)$ be the set of positive Boolean formulas over $\Upsilon \times Q$; i.e., Boolean formulas built from elements in $\Upsilon \times Q$ using $\wedge$ and $\vee$, where we also allow the formulas **true** and **false** and, as usual, $\wedge$ has precedence over $\vee$. The transition function $\delta : Q \times \Sigma \to \mathcal{B}^+(\Upsilon \times Q)$ maps a state and an input letter to a formula that suggests a new configuration for the automaton. For example, when $\Upsilon = \{0, 1\}$, having $\delta(q, \sigma) = (0, q_1) \wedge (0, q_2) \vee (0, q_2) \wedge (1, q_2) \wedge (1, q_3)$ means that when the automaton is in state q and reads the letter σ, it can either send two copies, in states q_1 and q_2, to direction 0 of the tree, or send a copy in state q_2 to direction 0 and two copies, in states q_2 and q_3, to direction 1. Thus, unlike nondeterministic tree automata, here the transition function may require the automaton to send several copies to the same direction or allow it not to send copies to all directions.

A *run of an alternating automaton* $\mathcal{A}$ on an input Σ-labeled Υ-tree $\langle T, V \rangle$ is a tree $\langle T_r, r \rangle$ in which the root is labeled by q_0 and every other node is labeled by an element of $\Upsilon^* \times Q$. Unlike T, in which each node has exactly $|\Upsilon|$ children, the tree T_r may have nodes with many children and may also have *leaves* (nodes with no children). Thus, $T_r \subset \mathbb{N}^*$ and a path in T_r may be either finite, in which case it contains a leaf, or infinite. Each node of T_r corresponds to a node of T. A node in T_r, labeled by (x, q), describes a copy of the automaton that reads the node x of T and visits the state q. Note that many nodes of T_r can correspond to the same node of T; in contrast, in a run of a nondeterministic automaton on $\langle T, V\rangle$ there is a one-to-one correspondence between the nodes of the run and the nodes of the tree. The labels of a node and its children have to satisfy the transition function. Formally, $\langle T_r, r\rangle$ is a Σ_r-labeled tree where $\Sigma_r = \Upsilon^* \times Q$ and $\langle T_r, r\rangle$ satisfies the following:

1. $\epsilon \in T_r$ and $r(\epsilon) = (\epsilon, q_0)$.
2. Let $y \in T_r$ with $r(y) = (x, q)$ and $\delta(q, V(x)) = \theta$. Then there is a (possibly empty) set $S = \{(c_0, q_0), (c_1, q_1), \ldots, (c_{n-1}, q_{n-1})\} \subseteq \Upsilon \times Q$, such that:
   - S satisfies θ, and
   - for all $0 \leq i < n$, we have $y \cdot i \in T_r$ and $r(y \cdot i) = (x \cdot c_i, q_i)$.

For example, if $\langle T, V\rangle$ is a $\{0,1\}$-tree with $V(\epsilon) = a$ and $\delta(q_0, a) = ((0, q_1) \vee (0, q_2)) \wedge ((0, q_3) \vee (1, q_2))$, then the nodes of $\langle T_r, r\rangle$ at level 1 include the label $(0, q_1)$ or $(0, q_2)$, and include the label $(0, q_3)$ or $(1, q_2)$. Note that if $\theta = \mathbf{true}$, then y need not have children. This is the reason why T_r may have leaves. Also, since there exists no set S as required for $\theta = \mathbf{false}$, we cannot have a run that takes a transition with $\theta = \mathbf{false}$.

Each infinite path ρ in $\langle T_r, r\rangle$ is labeled by a word $r(\rho)$ in Q^ω. Let $inf(\rho)$ denote the set of states in Q that appear in $r(\rho)$ infinitely often. A run $\langle T_r, r\rangle$ is accepting iff all its infinite paths satisfy the acceptance condition. In *Büchi* alternating tree automata, $\alpha \subseteq Q$, and an infinite path ρ satisfies α iff $inf(\rho) \cap \alpha \neq \emptyset$. In *Rabin* alternating tree automata, $\alpha \subseteq 2^Q \times 2^Q$, and an infinite path ρ satisfies an acceptance condition $\alpha = \{\langle G_1, B_1\rangle, \ldots, \langle G_m, B_m\rangle\}$ iff there exists $1 \leq i \leq m$ for which $inf(\rho) \cap G_i \neq \emptyset$ and $inf(\rho) \cap B_i = \emptyset$. As with nondeterministic automata, an automaton accepts a tree iff there exists an accepting run on it. We denote by $\mathcal{L}(\mathcal{A})$ the language of the automaton $\mathcal{A}$; i.e., the set of all labeled trees that $\mathcal{A}$ accepts. We say that an automaton is *nonempty* iff $\mathcal{L}(\mathcal{A}) \neq \emptyset$.

Formulas of branching temporal logic can be translated to alternating tree automata [8,4]. Since the modalities of conventional temporal logics, such as CTL* and the μ-calculus, do not distinguish between the various successors of a node (that is, they impose requirements either on all the successors of the node or on some successor), the alternating automata that one gets by translating formulas to automata are of a special structure, in which whenever a state q is sent to direction v, the state q is sent to all the directions $v \in \Upsilon$, in either a disjunctive or conjunctive manner. Formally, following the notations in [15], the formulas in $\mathcal{B}^+(\Upsilon \times Q)$ that appear in the transitions of such alternating tree automata are members of $\mathcal{B}^+(\{\Box, \Diamond\} \times Q)$, where $\Box q$ stands for $\bigwedge_{v \in \Upsilon}(v, q)$ and $\Diamond q$ stands for $\bigvee_{v \in \Upsilon}(v, q)$. As we shall see in Section 3, this structure of the automata is crucial for solving the robust model-checking problem. We say that an alternating tree automaton is *symmetric* if it has the special structure described above.

2.2 Modules

A *module* is a tuple $M = \langle I, O, W, w^{in}, i^{in}, \rho, \pi \rangle$, where I is a finite set of Boolean input variables, O is a finite set of Boolean output variables (we assume that $I \cap O = \emptyset$), W is a (possibly infinite) set of states, $w^{in} \in W$ is an initial state, $i^{in} \in 2^I$ is an initial input, $\rho : W \times 2^I \rightarrow 2^W$ is a nondeterministic transition function, and $\pi : W \rightarrow 2^O$ is a labeling function that assigns to each state its output. We require that for all $w \in W$ and $\sigma \in 2^I$, the set $\rho(w, \sigma)$ is not empty. Intuitively, the module can always respond to external inputs, though the response might be to enter a "bad" state. Intuitively, M starts its execution in w^{in}, where it expect the input i^{in}. Whenever M is in state w and the input is $\sigma \subseteq I$, it moves nondeterministically to one of the states in $\rho(w, \sigma)$. A module is *open* if $I \neq \emptyset$. Otherwise, it is *closed*. The *degree* of M is the minimal integer k such that for all w and σ, the set $\rho(w, \sigma)$ contains at most k states. If for all w and σ the set $\rho(w, \sigma)$ contains exactly k states, we say that M is of *exact degree k*.

Let $M_1 = \langle I, O, W_1, w_1^{in}, i_1^{in}, \rho_1, \pi_1 \rangle$ and $M_2 = \langle O, I, W_2, w_2^{in}, i_2^{in}, \rho_2, \pi_2 \rangle$ be two modules such that $\pi_1(w_1^{in}) = i_2^{in}$ and $\pi_2(w_2^{in}) = i_1^{in}$. Note that the inputs of M_1 are the outputs of M_2 and vice versa. The *composition* of M_1 and M_2 is the closed module $M_1 \| M_2 = \langle \emptyset, I \cup O, W, w^{in}, \emptyset, \rho, \pi \rangle$, where

- $W = W_1 \times W_2$.
- $w^{in} = \langle w_1^{in}, w_2^{in} \rangle$.
- For every $\langle w_1, w_2 \rangle \in W$, we have $\rho(\langle w_1, w_2 \rangle, \emptyset) = \rho_1(w_1, \pi_2(w_2)) \times \rho_2(w_2, \pi_1(w_1))$.
- For every $\langle w_1, w_2 \rangle \in W$, we have $\pi(\langle w_1, w_2 \rangle) = \pi_1(w_1) \cup \pi_2(w_2)$.

Note that since we assume that for all $w \in W$ and $\sigma \in 2^I$, the set $\rho(w, \sigma)$ is not empty, the composition of M with M' is *deadlock free*, thus every reachable state has at least one successor. Note also that the restriction to M' that closes M does not effect the answer to the robust-model-checking problem. Indeed, if there is some M' such that $M \| M'$ is open and does not satisfy ψ, we can easily extend M' so that its composition with M would be closed and would still not satisfy ψ.

Every module $M = \langle I, O, W, w^{in}, i^{in}, \rho, \pi \rangle$ induces an *enabling tree* $\langle T, V \rangle$. The enabling tree of M is a full infinite $\{\top, \bot\}$-labeled $(W \times 2^I)$-tree, thus $T = (W \times 2^I)^*$. We define $dir(\epsilon)$ to be $\langle w^{in}, i^{in} \rangle$, and we label ϵ by $\top$. Intuitively, $\langle T, V \rangle$ indicates which behaviors of M are enabled. Consider a node $x \in T$ such that $dir(x) = \langle w, \sigma \rangle$. For every state $w' \in W$ and input $\sigma' \in 2^I$, we define $V(x.\langle w', \sigma' \rangle)$ as $\top$ if $w' \in \rho(w, \sigma)$, and as $\bot$ otherwise. Consider a node $x = \langle w_1, \sigma_1 \rangle, \langle w_2, \sigma_2 \rangle, \ldots, \langle w_m, \sigma_m \rangle \in T$. By the definition of V, the module M can traverse a computation $w^{in}, w_1, w_2, \ldots, w_m$ when it reads the input sequence $i^{in}, \sigma_1, \sigma_2, \ldots, \sigma_{m-1}$ iff all the prefixes y of x have $V(y) = \top$. Indeed, then and only then we have $w_1 \in \rho(w^{in}, i^{in})$, and $w_{i+1} \in \rho(w_i, \sigma_i)$ for all $1 \leq j \leq m - 1$.

Following the definition of a product between two modules, the enabling tree of $M_1 \| M_2$ is a $\{\top, \bot\}$-labeled $(W_1 \times W_2)$-tree. Intuitively, M_2 supplies to M_1 its input (and vice versa). Therefore, while the trees of M_1 are $(W_1 \times 2^I)$-trees, reflecting the fact that every state in M_1 may read $2^{|I|}$ different inputs and move to $|W_1|$ successors, the tree of $M_1 \| M_2$ is a $(W_1 \times W_2)$-tree, reflecting the fact that every state in $M_1 \| M_2$ may have $|W_1| \cdot |W_2|$ successors. Note that M_2 may be nondeterministic. Accordingly, a node associated with a state w of M_1 may have k successors that are labeled $\top$ in the enabling tree of M_1 and have $k' > k$ successors that are labeled $\top$ in the enabling tree of $M_1 \| M_2$. That is, M_2 can not only prune transitions of M_1; it can also split transitions of M_1.

Recall that the enabling tree of a module M is a full infinite $\{\top, \bot\}$-labeled $(W \times 2^I)$-tree. As we shall see in Section 3, the fact that the tree is full circumvents some technical difficulties. We now define when M satisfies a formula. For that, we prune from the full tree nodes that correspond to unreachable states of M. Since each state of M has at least one successor, every node in the pruned tree also has at least one successor. Consequently, we are able, in Section 3, to duplicate subtrees and go back to convenient full trees. For an enabling tree $\langle T, V \rangle$, the $\top$-*restriction* of $\langle T, V \rangle$ is the $\{\top\}$-labeled tree with directions in $(W \times 2^I)$ that is obtained from $\langle T, V \rangle$ by pruning subtrees with a root labeled $\bot$. For a module M, the *computation tree* of M is a $2^{I \cup O}$-labeled $(W \times 2^I)$-tree obtained from the $\top$-restriction of M's enabling tree by replacing the $\top$ label of a node with direction $\langle w, \sigma \rangle$ by the label $\pi(w) \cup \sigma$. Note that when M is closed, its computation tree is a W-tree. We say that M satisfies a branching temporal logic formula ψ over $I \cup O$ iff M's computation tree satisfies ψ. The problem of *robust model checking* is to determine, given M and ψ, whether for every M', the composition $M \| M'$ satisfies ψ (we assume that the reader is familiar with branching temporal logic. We refer here to the logics CTL, CTL* , and the μ-calculus [10, 22]).

3 Robust Model Checking

In this section we solve the robust-model-checking problem and study its complexity. Thus, given a module M and a branching temporal logic formula ψ, we check whether for every M', the composition $M \| M'$ satisfies ψ. We assume that M has finitely many states and allow M' to have infinitely many states. Nevertheless, we show that if some environment that violates ψ exists, then there exists also a violating environment with finitely many states and a bounded branching degree. For a branching temporal logic formula ψ, we denote by $\mathcal{E}(\psi)$ the number of existential subformulas (subformulas of the form $E\xi$) in ψ. It is known that $\mathcal{E}(\psi)$ bounds the branching degree required in order to satisfy ψ [10]. We now extend this result and show that, also in robust model checking, it suffices to consider environments of degree $\mathcal{E}(\psi)$. For an integer $k \geq 1$, let $[k] = \{1, \ldots, k\}$.

Theorem 1. *Consider a module M and a branching temporal logic formula ψ over $I \cup O$. Let $k = \max\{1, \mathcal{E}(\psi)\}$. If there exists M' such that $M \| M' \models \psi$, then there also exists M' of exact degree k such that $M \| M' \models \psi$.*

Proof (sketch): Assume that $M \| M' \models \psi$ for some M'. Thus, the computation tree $\langle T, V \rangle$ of $M \| M'$ satisfies ψ. In order for that to be true, each node in $\langle T, V \rangle$ has to satisfy a set of subformulas of ψ. Formally, there is a mapping V' of T to sets of subformulas of ψ such that $\psi \in V'(\epsilon)$, and for every $x \in T$, the set $V'(x)$ contains formulas that hold in x, such that the labeling along paths that start at x is enough to "justify" $V'(x)$. For example, if a node x is labeled by EXp, then at least one successor of x is labeled by p. Consider a node x. Some formulas in $V'(x)$ impose on the paths starting at x universal requirements. To satisfy these requirements, x need not have children (yet all the children that x does have, belong to paths that satisfy these universal requirements). In addition, some formulas in $V'(x)$ impose on the paths starting at x at most k existential requirements. Each such requirement needs to be satisfied by some path starting at x, yet it does not have to be satisfied by more than one such path. Also, it may be that the existential requirements impose particular values on the input variables in the successors of x, and different existential requirements may impose the same value. Accordingly, we can prune some of the paths that start at x and satisfy the formulas in $V'(x)$ with not more than k successors of x for each $\sigma \in 2^I$, or by a single successor, in the case $V'(x)$ contains no existential requirements. The pruned tree can therefore be obtained by taking the product of M with a module M'' of degree k where M'' is a suitable pruning of the infinite module obtained by unwinding M'. In order to get a module of exact degree k, we can then duplicate some of the subtrees of M''. (For the μ-calculus, the proof is considerably more complicated and uses techniques from [36].) □

In order to understand the difference between Theorem 1 and the classical "bounded-degree property" for branching temporal logic, recall that the theorem refers to the branching degree of the environment, rather than to that of the composition $M \| M'$. Consider, for example, a module M with an initial state that has two successors, one labeled p and one labeled $\neg p$. In order for M to satisfy the formula $\psi = EX(p \wedge q) \wedge EX(p \wedge \neg q)$, for an input variable q, a split of the state labeled p is required. Though $\mathcal{E}(\psi) = 2$, such a split results in a composition of branching degree 4. It can, however, be achieved by composing M with an environment M' of branching degree 2. Theorem 1 shows that, though we may sometimes need the branching degree of $M \| M'$ to be bigger than $\mathcal{E}(\psi)$, it is sufficient to compose M with an environment of branching degree $\mathcal{E}(\psi)$. We now use Theorem 1 to show that the robust-satisfaction problem for branching temporal logics can be reduced to the emptiness problem for alternating tree automata.

Theorem 2. *Consider a module M and branching temporal logic formula ψ over $I \cup O$. Let $\mathcal{A}_\psi$ be the symmetric alternating tree automaton that corresponds to ψ and let $k = \max\{1, \mathcal{E}(\psi)\}$. There is an alternating tree automaton $\mathcal{A}_{M,\psi}$ over 2^I-labeled $(2^O \times [k])$-trees such that*

1. *$\mathcal{L}(\mathcal{A}_{M,\psi})$ is empty iff M robustly satisfies $\neg\psi$.*
2. *$\mathcal{A}_{M,\psi}$ and $\mathcal{A}_\psi$ have the same acceptance condition.*
3. *The size of $\mathcal{A}_{M,\psi}$ is $O(|M| \cdot |\mathcal{A}_\psi| \cdot k)$.*

Proof (sketch): Before we describe $\mathcal{A}_{M,\psi}$, let us explain the difficulties in the construction and why alternation is so helpful solving them. The automaton $\mathcal{A}_{M,\psi}$ searches for a module M' of exact degree k for which $M \| M' \in \mathcal{L}(\mathcal{A}_\psi)$. The modules M and M' interact via the sets I and O of variables. Thus, M' does not know the state in which M is, and it only knows M's output. Accordingly, not all $\{\top, \bot\}$-labeled $(W \times W')$-trees are possible enabling trees of a product $M \| M'$. Indeed, $\mathcal{A}_{M,\psi}$ needs to consider only trees in which the behavior of M' is consistent with its incomplete information: if two nodes have the same output history (history according to M''s incomplete information), then either they agree on their label (which can be either $\bot$ or a set of input variables), or the two nodes are outcomes of two different nondeterministic choices that M' has taken along this input history. This consistency condition is non-regular and cannot be checked by an automaton [37]. It is this need, to restrict the set of candidate enabling trees to trees that meet some non-regular condition, that makes robust model checking in the branching paradigm so challenging. The solution is to consider, instead $(W \times W')$-trees, $(2^O \times [k])$-trees. Each node in such a tree may correspond to several nodes in a $(W \times W')$-tree, all with the same output history. Then, alternation is used in order to make sure that while all these nodes agree on their labeling, each of them satisfy requirements that together guarantee the membership in $\mathcal{A}_\psi$.

Let $M = \langle I, O, W, w^{in}, i^{in}, \rho, \pi \rangle$. For $w \in W$, $\sigma \in 2^I$, and $v \in 2^O$, we define

$$s(w, \sigma, v) = \{w' \mid w' \in \rho(w, \sigma) \text{ and } \pi(w') = v\}.$$

That is, $s(w, \sigma, v)$ contains all the states with output v that w moves to when it reads σ. The definition of the automaton $\mathcal{A}_{M,\psi}$ can be viewed as an extension of the product alternating tree automaton obtained in the alternating-automata theoretic framework for branching time model checking [4]. There, as we are concerned with model checking, there is a single computation tree with respect to which the formula is checked, and the automaton obtained is a 1-letter automaton. The difficulty here, as we are concerned with robust model checking, is that there are many computation trees to check, so a 1-letter automaton does not suffice. Let $\mathcal{A} = \langle 2^{I \cup O}, Q, q_0, \delta, \alpha \rangle$. We define $\mathcal{A}_{M,\psi} = \langle 2^I, Q', q_0, \delta', \alpha' \rangle$, where

- $Q' = \{q_0\} \cup (W \times Q)$. Intuitively, when the automaton is in state $\langle w, q \rangle$, it accepts all trees that are induced by an environment M' for which the composition with M with initial state w is accepted by $\mathcal{A}$ with initial state q. The initial state q_0 corresponds to the state $\langle w^{in}, q_0 \rangle$, yet it also checks that the first input is i^{in}.
- The transition function $\delta' : Q' \times 2^I \rightarrow \mathcal{B}^+((2^O \times [k]) \times Q')$ is defined as follows.
  - For all w, q, and σ, the transition $\delta'(\langle w, q \rangle, \sigma)$ is obtained from $\delta(q, \sigma \cup \pi(w))$ by replacing a conjunction $\Box q'$ by the conjunction

$$\bigwedge_{v \in 2^O} \bigwedge_{j \in [k]} \bigwedge_{w' \in s(w,\sigma,v)} (\langle v, j \rangle, \langle w', q' \rangle),$$

    and replacing a disjunction $\Diamond q'$ by the disjunction

$$\bigvee_{v \in 2^O} \bigvee_{j \in [k]} \bigvee_{w' \in s(w,\sigma,v)} (\langle v, j \rangle, \langle w', q' \rangle).$$

  - For the initial state q_0, we define $\delta'(q_0, i^{in}) = \delta'(\langle w^{in}, \psi \rangle, i^{in})$. For all $\sigma \neq i^{in}$, we define $\delta'(q_0, \sigma) = \textbf{false}$.

  Consider, for example, a transition from the state $\langle w, q \rangle$. Let $\sigma \in 2^I$ be such that $\delta(q, \sigma \cup \pi(w)) = \Box s \wedge \Diamond t$. The successors of w that are enabled with input σ should satisfy $\Box s \wedge \Diamond t$. Thus, all these successors should satisfy s and at least one successor should satisfy t. The state w may have several successors in $\rho(w, \sigma)$ with the same output $v \in 2^O$. These successors are indistinguishable by M'. Therefore, if M' behaves differently in such two successors, it is only because M' is in a different

state when it interacts with these successors. The number k bounds the number of states in $\rho(w, \sigma)$. Accordingly, M' can exhibit k different behaviors when it interacts with indistinguishable successors of w. For each $j \in [k]$, the automaton sends all the successors of w in $s(w, \sigma, v)$ to the same direction $\langle v, j \rangle$, where they are going to face the same future. Since $\delta(q, \sigma \cup \pi(w)) = \Box s \wedge \Diamond t$, a copy in state s is sent to all the successors, and a copy in state t is sent to some successor. Note that as M is deadlock free, the conjunctions and disjunctions in δ cannot be empty.
- α' is obtained from α by replacing every set participating in α by the set $W \times \alpha$.

□

We now consider the complexity bounds for various branching temporal logics that follow from our algorithm.

Theorem 3. *Robust model checking is*

(1) *EXPTIME-complete for CTL, μ-calculus, and the alternation-free μ-calculus.*
(2) *2EXPTIME-complete for CTL*.*

Proof (sketch): Consider a branching temporal logic formula ψ of length n. Let $\mathcal{A}_\psi$ be the symmetric alternating tree automaton that corresponds to ψ. By [8, 4], $\mathcal{A}_\psi$ is a Büchi automaton with $O(n)$ states for ψ in CTL or in the alternation-free μ-calculus, $\mathcal{A}_\psi$ is a parity automaton with $O(n)$ states and d sets in the acceptance condition for ψ in μ-calculus with alternation depth d, and $\mathcal{A}_\psi$ is a Rabin automaton with $2^{O(n)}$ states and 2 pairs in the acceptance condition for ψ in CTL*. In Theorem 2, we reduced the robust-model-checking problem of M with respect to $\neg\psi$ to the problem of checking the nonemptiness of the automaton $\mathcal{A}_{M,\psi}$, which is of size $|M| \cdot |\mathcal{A}_\psi| \cdot \max\{1, \mathcal{E}(\psi)\}$, and which has the same type and size of acceptance condition as $\mathcal{A}_\psi$. The upper bounds then follow from the complexity of the nonemptiness problem for the various automata [30, 38, 26].

For the lower bounds, one can reduce the satisfiability problem for a branching temporal logic to the robust-model-checking problem for that logic. To see this, note that, by the "bounded-degree property" of branching temporal logic, a search for a satisfying model for ψ can be reduced to a search for a satisfying $2^{I \cup O}$-labeling of a tree with branching degree $\max\{1, \mathcal{E}(\psi)\}$. Then, one can relate the choice of the labels to choices made by the environment. □

The *implementation complexity* of robust model checking is the complexity of the problem in terms of the module, assuming that the specification is fixed. As we discuss in Section 4, there are formulas for which robust model checking coincides with module checking with incomplete information. Since module checking with incomplete information is EXPTIME-hard already for CTL formulas of that type, it follows that the implementation complexity of robust model checking for CTL (and the other, more expressive, logics) is EXPTIME-complete.

In our definition of robust satisfaction, we allow the environment to have infinitely many states. We now claim that finite environments are strong enough. The proof is based on a "finite-model property" of tree automata, proven in [34] for nondeterministic tree automata and extended in [30, 25] to alternating tree automata. As we discuss in

Section 5, this result is of great importance in the dual paradigm of supervisory control, where instead of hostile environments we consider collaborative controllers.

Theorem 4. *Given a module M and a branching temporal logic formula ψ, if there is an infinite module M' of degree k such that $M||M'$ satisfies ψ, then there also exists a finite module M'' of degree k such that $M||M''$ satisfies ψ.*

The alternating-automata-theoretic approach to CTL and CTL* model checking is extended in [23] to handle Fair-CTL and Fair-CTL* [9]. Using the same extension, we can solve the problem of robust model checking also for handle modules augmented with fairness conditions.

4 Universal and Mixed Formulas

The study of verification of open system has motivated the use of universal temporal logic [13]. Formally, a formula ψ is *universal* iff for every module M, if M satisfies ψ, then for every M', the composition $M||M'$ also satisfies ψ. By the above definition, M satisfies a universal property ψ iff M robustly satisfies ψ. In this section we show that the set of non-universal properties can be further partitioned into two classes, each with a different sensitivity to the robustness of the satisfaction. In addition, we study the complexity of classifying a CTL formula to its sensitivity class. We say that a CTL formula ψ is *mixed* iff ψ imposes both universal and existential properties in a nontrivial way. Thus, ψ is mixed iff neither ψ nor $\neg\psi$ is universal. We first show that formulas that are not mixed are insensitive to the environment being nondeterministic.

Theorem 5. *Consider a module M and a specification ψ. If ψ is not mixed, then M robustly satisfies ψ iff $M||M' \models \psi$ for every deterministic M'.*

Proof (sketch): Clearly, if M robustly satisfies ψ, then $M||M' \models \psi$ for every deterministic M'. For the other direction, assume that ψ is not mixed and that $M||M' \models \psi$ for every deterministic M'. We prove that then, M robustly satisfies ψ. Thus, that $M||M' \models \psi$ for every possibly nondeterministic M'. We distinguish between two cases. If ψ is universal, then, as M simulates $M||M'$ for every (possibly nondeterministic) M', robust satisfaction coincides with usual satisfaction and we are done. If ψ is existential, assume that there is a nondeterministic M' such that $M||M'$ does not satisfy ψ. Let M'' be any deterministic module obtained from M' by removing transitions. Since M' simulates M'', the composition $M||M'$ simulates the composition $M||M''$ [13]. Therefore, as ψ is existential, it must be that $M||M''$ does not satisfy ψ as well. □

Thus, to robustly model check formulas that are not mixed, one can use the method of module checking with incomplete information [25]. We now study the problems of determining whether a given CTL formula is universal (or existential) or mixed, and show that they are all EXPTIME-complete.

Theorem 6. *Given a CTL formula ψ, checking whether ψ is universal is EXPTIME-complete.*

Proof (sketch): For a set $\mathcal{T}$ of trees and an integer k, we define $reshape(\mathcal{T}, k)$ as the set of trees obtained from trees in $\mathcal{T}$ by prunings or duplications of subtrees, so that each node has at most k successors. Given a CTL formula ψ, let $k = \max\{1, \mathcal{E}(\psi)\}$, and let $\mathcal{T}$ be the set of trees of branching degree k that satisfy ψ. It can be shown that the formula ψ is universal iff $reshape(\mathcal{T}) \subseteq \mathcal{T}$. Given ψ, let $\mathcal{A}_\psi$ be a nondeterministic Buchi automation for ψ; that is, $\mathcal{L}(\mathcal{A}_\psi) = \mathcal{T}$. By "reshaping" the transition function of $\mathcal{A}_\psi$, we can define a nondeterministic Büchi automaton $\mathcal{A}'_\psi$ such that $\mathcal{L}(\mathcal{A}_\psi) = reshape(\mathcal{T}, k)$. Then, ψ is universal iff $\mathcal{L}(\mathcal{A}'_\psi) \subseteq \mathcal{L}(\mathcal{A}_\psi)$. In order to check the latter, we check the nonemptiness of $\mathcal{L}(\mathcal{A}'_\psi) \cap \mathcal{L}(\mathcal{A}_{\neg\psi})$. Since both $\mathcal{A}'_\psi$ and $\mathcal{A}_{\neg\psi}$ are exponential in $|\psi|$, and the nonemptiness check is polynomial, the EXPTIME upper bound follows.

For the lower bound, we do a reduction from alternating linear-space Turing machines. Given a machine T, we construct a CTL formula ψ such that ψ is universal iff the machine T does not accept the empty tape. Typically, ψ is satisfied in a tree iff the tree does not represent an accepting computation tree of T on the empty tape. We can define ψ that is polynomial in T. One can then prove that the machine T rejects the empty tape iff $\psi = \mathbf{true}$, and that $\psi = \mathbf{true}$ iff ψ is universal. □

Theorem 7. *Given a CTL formula ψ, checking whether ψ is mixed is EXPTIME-complete.*

Proof (sketch): Since ψ is mixed iff both ψ and $\neg\psi$ are non-universal, the upper bound follows from Theorem 6. The lower bound is similar to the one in Theorem 6, only that now we prove that ψ is mixed iff the machine T accepts the empty tape. To prove this, we replace the second claim in the proof of Theorem 6 with the claim that $\psi = \mathbf{true}$ iff ψ is not mixed. □

5 Related Work and Discussion

Different researchers have considered the problem of reasoning about open systems. The distinction, in [20], between closed and open systems first led to the realization that *synthesis* of open systems corresponds to a search for a winning strategy in a *game* between the system and the environment [32], in which the winning condition is expressed in terms of a linear temporal logic formula. Transformation of the game-theoretic approach to model checking and adjustment of verification methods to the open-system setting started, for linear temporal logic, with the problem of *receptiveness* [7, 2, 14]. Essentially, the receptiveness problem is to determine whether every finite prefix of a computation of a given open system can be extended to an infinite computation that satisfies a linear temporal property irrespective of the behavior of the environment. In *module checking* [24], the setting is again game-theoretic: an open system is required to satisfy a branching temporal property no matter how the environment disables its transitions. Verification of open systems was formulated in terms of a game between agents in a multi-agent system in [1]. *Alternating-time temporal logic*, introduced there, enables path quantifiers to range over computations that a team of agents can force the system into, and thus enables the specification of multi-agent systems. In particular, ATL and ATL* are the alternating-time versions of CTL and CTL*, respectively.

Unlike [1], in which all the agents of the system are specified, our setting here assumes that only one agent, namely the system, is given. We ask whether there exists another agent, namely the environment, which is not yet known, such that the composition of the system and the environment violates a required property. Thus, while the outcome of the games that correspond to alternating temporal logic are computations, here the outcomes are trees[2]. The unknown environment may be nondeterministic, thus the branching structure of the trees is not necessarily a restriction of the branching structure of the system. Since the properties we check are branching, the latter point is crucial. As follows from the 2EXPTIME lower bounds for both ATL* model checking and CTL* robust model checking, verification of general properties of open systems is "robustly hard". Exceptions are universal properties, for which robust satisfaction coincides with usual satisfaction, as well as properties that can be specified in the logic ATL. Indeed, the logic ATL identifies a class of properties for open systems for which it suffices to solve iterated finite games, which can be done in linear time.

Robust satisfaction is closely related to *supervisory control* [35, 3]. Given a finite-state machine whose transitions are partitioned into controllable and uncontrollable, and a specification for the machine, the control problem requires the construction of a controller that chooses the controllable transitions so that the machine always satisfies the specification. Clearly, checking whether all the compositions $M||M'$ of a system M with an environment M' satisfies a property ψ is dual to checking whether there is a controller M' such that $M||M'$ satisfy the property $\neg\psi$. Thus, from a control-theory point of view, the results of this paper generalize known supervisory-control methods to the case where both the system and the controller are nondeterministic Moore machines. In particular, our results imply that nondeterministic controllers are more powerful than deterministic ones, and describe how to synthesize finite-state controllers.

Often, the requirement that M satisfies ψ in all environments is too restrictive, and we are really concerned in the satisfaction of ψ in compositions of M with environments about which some *assumptions* are known. In the *assume-guarantee* paradigm to verification, each specification is a pair $\langle\varphi, \psi\rangle$, and M satisfies $\langle\varphi, \psi\rangle$ iff for every M', if $M||M'$ satisfies φ, then $M||M'$ also satisfies ψ. When φ and ψ are given in linear temporal logic, M satisfies $\langle\varphi, \psi\rangle$ iff M satisfies the implication $\varphi \rightarrow \psi$ [31] (see also [21]). The situation is different in the branching paradigm. For universal temporal logic, M satisfies $\langle\varphi, \psi\rangle$ iff ψ is satisfied in the composition $M||M_\varphi$, of M with a module M_φ that embodies all the behaviors that satisfy φ [13, 23]. For general branching temporal logic, the above is no longer valid. Robust model checking can be viewed as a special case of the assume-guarantee setting, where φ is $\mathbf{true}$. Robust model checking, however, can be used to solve the general assume-guarantee setting. Indeed, M satisfies $\langle\varphi, \psi\rangle$ iff M robustly satisfies the implication $\varphi \rightarrow \psi$. Thus, while in the linear framework the assume-guarantee paradigm corresponds to usual model checking, robustness is required in the branching framework.

Since assumptions about the environment and its interaction with the systems are natural part of the specification in robust model checking, the model studied in this paper subsumes extensions that can be expressed in terms properties of the environment and its interaction with the system. For example, recall that our compositions here are deadlock free, thus deadlock is modeled by entering some "bad" state. In order to check that M satisfies a property ψ in all the compositions $M||M'$ in which this bad state is

[2] Game logic [1] considers games in which the output are trees, yet both players are known.

not reachable, we have to perform robust model checking of M with respect to the property $(AG\theta) \rightarrow \psi$, with $\theta = \neg bad$, assuming that the bad state is labeled by bad. In a similar way, we can specify in θ other global assumptions about the composition, and thus model settings that support handshaking or other forms of coordinations between processes, as well as more general global actions, as in [16].

References

1. R. Alur, T.A. Henzinger, and O. Kupferman. Alternating-time temporal logic. In *Proc. 38th IEEE Symposium on Foundations of Computer Science*, pages 100–109, 1997.
2. M. Abadi and L. Lamport. Composing specifications. *ACM Transactions on Programming Languages and Systems*, 15(1):73–132, 1993.
3. M. Antoniotti. *Synthesis and verification of discrete controllers for robotics and manufacturing devices with temporal logic and the Control-D system.* PhD thesis, New York University, New York, 1995.
4. O. Bernholtz, M.Y. Vardi, and P. Wolper. An automata-theoretic approach to branching-time model checking. In D. L. Dill, editor, *Proc. 6th CAV*, LNCS 818, pages 142–155, 1994.
5. E.M. Clarke and E.A. Emerson. Design and synthesis of synchronization skeletons using branching time temporal logic. In *Proc. Workshop on Logic of Programs*, LNCS 131, pages 52–71, 1981.
6. E.M. Clarke, O. Grumberg, and M.C. Browne. Reasoning about networks with many identical finite-state processes. In *Proc. 5th PODC*, pages 240–248, 1986.
7. D.L. Dill. *Trace theory for automatic hierarchical verification of speed independent circuits.* MIT Press, 1989.
8. E.A. Emerson and C. Jutla. Tree automata, Mu-calculus and determinacy. In *Proc. 32nd IEEE Symposium on Foundations of Computer Science*, pages 368–377, San Juan, October 1991.
9. E.A. Emerson and C.-L. Lei. Temporal model checking under generalized fairness constraints. In *Proc. 18th Hawaii International Conference on System Sciences*, North Holywood, 1985. Western Periodicals Company.
10. E.A. Emerson. Temporal and modal logic. *Handbook of Theoretical Computer Science*, pages 997–1072, 1990.
11. M.J. Fischer and L.D. Zuck. Reasoning about uncertainty in fault-tolerant distributed systems. In M. Joseph, editor, *Proc. Symp. on Formal Techniques in Real-Time and Fault-Tolerant Systems*, LNCS 331, pages 142–158, 1988.
12. O. Grumberg and D.E. Long. Model checking and modular verification. In *Proc. 2nd CONCUR*, LNCS 527, pages 250–265, 1991.
13. O. Grumberg and D.E. Long. Model checking and modular verification. *ACM Trans. on Programming Languages and Systems*, 16(3):843–871, 1994.
14. R. Gawlick, R. Segala, J. Sogaard-Andersen, and N. Lynch. Liveness in timed and untimed systems. In *Proc. 21st ICALP*, LNCS 820, pages 166–177, 1994.
15. E. Graedel and I. Walukiewicz. Guarded fixed point logic. In *Proc. 14th LICS*, 1999.
16. J.Y. Halpern and R. Fagin. Modelling knowladge and action in distributed systems. *Distributed Computing*, 3(4):159–179, 1989.
17. T.A. Henzinger, O. Kupferman, and S. Qadeer. From pre-historic to post-modern symbolic model checking. In *Proc. 10th CAV*, LNCS 1427, 1998.
18. D. Harel, O. Kupferman, and M.Y. Vardi. On the complexity of verifying concurrent transition systems. In *Proc. 8th CONCUR*, LNCS 1243, pages 258–272, 1997. Springer-Verlag.
19. C.A.R. Hoare. *Communicating Sequential Processes.* Prentice-Hall, 1985.

20. D. Harel and A. Pnueli. On the development of reactive systems. In K. Apt, editor, *Logics and Models of Concurrent Systems*, volume F-13 of *NATO Advanced Summer Institutes*, pages 477–498. Springer-Verlag, 1985.
21. B. Jonsson and Y.-K. Tsay. Assumption/guarantee specifications in linear-time temporal logic. In *Proc. TAPSOFT '95*, LNCS 915, pages 262–276, 1995.
22. D. Kozen. Results on the propositional μ-calculus. *Theoretical Computer Science*, 27:333–354, 1983.
23. O. Kupferman and M.Y. Vardi. On the complexity of branching modular model checking. In *Proc. 6th CONCUR*, LNCS 962, pages 408–422, 1995.
24. O. Kupferman and M.Y. Vardi. Module checking. In *Proc. 8th CAV*, LNCS 1102, pages 75–86, 1996.
25. O. Kupferman and M.Y. Vardi. Module checking revisited. In *Proc. 9th CAV*, LNCS 1254, pages 36–47, 1997.
26. O. Kupferman and M.Y. Vardi. Weak alternating automata and tree automata emptiness. In *Proc. 30th STOC*, pages 224–233, 1998.
27. L. Lamport. Sometimes is sometimes "not never" - on the temporal logic of programs. In *Proc. 7th POPL*, pages 174–185, 1980.
28. R. Milner. An algebraic definition of simulation between programs. In *Proc. 2nd International Joint Conference on Artificial Intelligence*, pages 481–489. British Computer Society, September 1971.
29. D.E. Muller and P.E. Schupp. Alternating automata on infinite trees. *Theoretical Computer Science*, 54,:267–276, 1987.
30. D.E. Muller and P.E. Schupp. Simulating alternating tree automata by nondeterministic automata: New results and new proofs of theorems of Rabin, McNaughton and Safra. *Theoretical Computer Science*, 141:69–107, 1995.
31. A. Pnueli. In transition from global to modular temporal reasoning about programs. In K. Apt, editor, *Logics and Models of Concurrent Systems*, volume F-13 of *NATO Advanced Summer Institutes*, pages 123–144. Springer-Verlag, 1985.
32. A. Pnueli and R. Rosner. On the synthesis of a reactive module. In *Proc. 16th POPL*, 1989.
33. J.P. Queille and J. Sifakis. Specification and verification of concurrent systems in Cesar. In *Proc. 5th International Symp. on Programming*, LNCS 137, pages 337–351, 1981.
34. M.O. Rabin. Weakly definable relations and special automata. In *Proc. Symp. Math. Logic and Foundations of Set Theory*, pages 1–23. North Holland, 1970.
35. P.J.G. Ramadge and W.M. Wonham. The control of descrete event systems. *IEEE Transactions on Control Theory*, 77:81–98, 1989.
36. R.S. Streett and E.A. Emerson. An automata theoretic decision procedure for the propositional mu-calculus. *Information and Computation*, 81(3):249–264, 1989.
37. J.W. Thatcher. Tree automata: an informal survey. In A.V. Aho, editor, *Currents in the theory of computing*, pages 143–172. Prentice-Hall, Englewood Cliffs, 1973.
38. M.Y. Vardi and P. Wolper. Automata-theoretic techniques for modal logics of programs. *Journal of Computer and System Science*, 32(2):182–221, April 1986.

Statecharts Via Process Algebra*

Gerald Lüttgen[1], Michael von der Beeck[2], and Rance Cleaveland[3]

[1] Institute for Computer Applications in Science and Engineering, NASA Langley Research Center, Hampton, VA 23681-2199, USA, luettgen@icase.edu
[2] Department of Computer Science, Munich University of Technology, Arcisstr. 21, D-80290 München, Germany, beeck@in.tum.de
[3] Department of Computer Science, State University of New York at Stony Brook, Stony Brook, NY 11794-4400, USA, rance@cs.sunysb.edu

Abstract. *Statecharts* is a visual language for specifying the behavior of *reactive systems*. The language extends *finite-state machines* with concepts of *hierarchy*, *concurrency*, and *priority*. Despite its popularity as a design notation for *embedded systems*, precisely defining its semantics has proved extremely challenging. In this paper, we present a simple *process algebra*, called *Statecharts Process Language* (SPL), which is expressive enough for encoding Statecharts in a structure-preserving and semantics-preserving manner. We also establish that the behavioral equivalence *bisimulation*, when applied to SPL, preserves Statecharts semantics.

1 Introduction

Statecharts is a visual language for specifying the behavior of *reactive systems* [7]. The language extends the notation of *finite-state machines* with concepts of (i) *hierarchy*, so that one may speak of a state as having sub-states, (ii) *concurrency*, thereby allowing the definition of systems having simultaneously active subsystems, and (iii) *priority*, so that one may express that certain system activities have precedence over others. Statecharts has become popular among engineers as a design notation for *embedded systems*, and commercially available tools provide support for it [10]. Nevertheless, precisely defining its semantics has proved extremely challenging, with a variety of proposals [6, 8, 9, 17, 18, 19, 22, 23] being offered for several dialects [26] of the language. The semantic subtlety of Statecharts arises from the language's capability for defining transitions whose enabledness disables other transitions. A Statechart may react to an *event* by engaging in an enabled transition, thereby performing a so-called *micro step*, which may generate new events that may in turn trigger new transitions while disabling others. When this chain reaction

* This work was supported by the National Aeronautics and Space Administration under NASA Contract No. NAS1-97046 while the first author was in residence at the Institute for Computer Applications in Science and Engineering (ICASE), NASA Langley Research Center, Hampton, VA 23681-2199, USA. The third author was supported by NSF grants CCR-9257963, CCR-9505662, CCR-9804091, and INT-9603441, AFOSR grant F49620-95-1-0508, and ARO grant P-38682-MA.

comes to a halt, one execution step – also referred to as a *macro step* – is complete. At a technical level, the difficulty for defining an operational semantics capturing the "macro-step" behavior of Statecharts arises from the fact that such a semantics should exhibit the following desirable properties: (i) the *synchrony hypothesis* [2], which guarantees that a reaction to an external event terminates before the next event enters the system, (ii) *compositionality*, which ensures that the semantics of a Statechart is defined in terms of the semantics of its components, and (iii) *causality*, which demands that the participation of each transition in a macro step must be causally justified. Huizing and Gerth showed that an operational semantics in which transitions are labeled purely by sets of events – i.e., the "observations" a user would make – cannot be given, if one wishes all three properties to hold [15]. In fact, the traditional semantics of Statecharts – as defined by Pnueli and Shalev [22] – satisfies the synchrony hypothesis and causality, but is not compositional. Other approaches, e.g. [17], have achieved all three goals, but at the expense of including complex information regarding causality in transition labels.

While not as well-established in practice, *process algebras* [1, 12, 21] offer many of the semantic advantages that have proved elusive in Statecharts. In general, these theories are operational, and place heavy emphasis on issues of compositionality through the study of *congruence relations*. Many of the behavioral aspects of Statecharts have also been studied for process algebras. For example, the synchrony hypothesis is related to the *maximal progress assumption* developed in *timed* process algebras [11, 28]. In these algebras, event transitions and "clock" transitions are distinguished, with only the latter representing the advance of time. Maximal progress then ensures that time may proceed only if the system under consideration cannot engage in internal computation. Clocks may therefore be viewed as "bundling" sequences of event transitions, which may be thought of as analogous to "micro steps," into a single "time step," which may be seen as a "macro step." The concept of priority has also been studied in process-algebraic settings [4], and the Statecharts hierarchy operator is related to the *disabling* operator of LOTOS [3].

In this paper, we present a new, *process-algebraic* semantics of Statecharts. Our approach synthesizes the observations above; specifically, we present a new process algebra, called *Statecharts Process Language* (SPL), and we show that it is expressive enough for embedding several Statecharts variants. SPL is inspired by Hennessy and Regan's *Timed Process Language* (TPL) [11] which extends Milner's CCS [21] by the concept of an abstract, global clock. Our algebra replaces the handshake communication of TPL by a *multi-event communication*, and introduces a mechanism to specify *priority* among transitions as well as a *hierarchy operator* [25]. The operational semantics of SPL uses SOS rules to define a transition relation whose elements are labeled with simple sets of events; then, using traditional process-algebraic results we show that SPL has a compositional semantic theory based on *bisimulation* [21]. We connect SPL with Statecharts by embedding the variant of the language considered by Maggiolo-Schettini et al. in [17]. More precisely, we define a compositional translation from Statecharts to SPL that preserves the macro-step semantics of the former. This result depends crucially on our treatment of the SPL macro-step transition relation as a *derived* one: the standard SPL transition relation becomes in essence a micro-step semantics. Thus, while our macro-step semantics cannot be compositional (cf. the result of Huizing and Gerth [15]), we obtain a compositional theory, in the form of a semantic congruence, at a lower, micro-step level. In addition to the usual benefits conferred by compositional reasoning, this semantics has a practical advantage: given the unavoidable complexity of inferring macro steps, actual users of Statecharts would benefit from a finer-grained semantics that helps them understand how the macro steps of their systems are arrived at.

2 Statecharts

Statecharts is a specification language for *reactive systems*, i.e., concurrent systems which are characterized by their ongoing interaction with their *environment*. They subsume finite state machines whose transitions are labeled by pairs of events, where the first component is referred to as *trigger* and may include *negated events*, and the second component is referred to as *action*. Intuitively, if the environment offers the events in the trigger, but not the negated ones, then the transition is triggered; it fires, thereby producing the events in the label's action. Concurrency is achieved by allowing Statecharts to be composed from more simple ones running in parallel, which may communicate via *broadcasting* events. Elementary, or *basic states* in Statecharts may also be hierarchically refined by injecting other Statecharts.

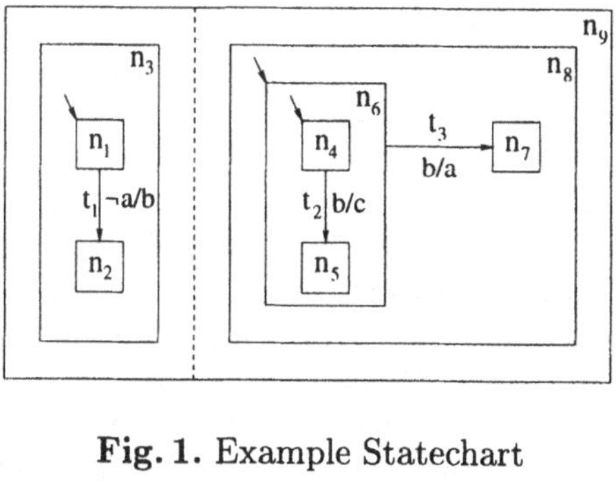

Fig. 1. Example Statechart

As an example, consider the Statechart depicted to the right. It consists of a so-called *and-state*, labeled by n_9, which denotes the parallel composition of the two Statecharts labeled by n_3 and n_8. Actually, n_3 and n_8 are the names of *or-states*, describing sequential state machines. The first consists of two states n_1 and n_2 that are connected via transition t_1 with label $\neg a/b$. The label specifies that t_1 is triggered by $\neg a$, i.e., by the absence of event a, and produces event b. States n_1 and n_2 are not refined further and, therefore, are referred to as basic states. Or-state n_8 is refined by or-state n_6 and basic state n_7, connected via a transition labeled by b/a. Or-state n_6 is further refined by basic states n_4 and n_5, and transition t_2 labeled by b/c. The variant of Statecharts considered here does not include *interlevel transitions* – i.e., transitions crossing borderlines of states – and *state references* – i.e., triggers of the form in_n, where n is a state name. Moreover, state hierarchy does not impose implicit priorities on transitions. The impact of altering our approach to accommodate these concepts is discussed in Sec. 6.

Table 1. States and transitions of Statecharts terms

$\mathsf{states}([n]) := \{n\}$	$\mathsf{states}([n : s; l; T]) := \{n\} \cup \bigcup\{\mathsf{states}(s_i) \mid 1 \le i \le k\}$
	$\mathsf{states}([n : s]) := \{n\} \cup \bigcup\{\mathsf{states}(s_i) \mid 1 \le i \le k\}$
$\mathsf{trans}([n]) := \emptyset$	$\mathsf{trans}([n : s; l; T]) := T \cup \bigcup\{\mathsf{trans}(s_i) \mid 1 \le i \le k\}$
	$\mathsf{trans}([n : s]) := \bigcup\{\mathsf{trans}(s_i) \mid 1 \le i \le k\}$

For our purposes, it is convenient to represent Statecharts not visually but by terms. This is also done in related work [16, 17, 24]; we closely follow [17]. Formally, let $\mathcal{N}$ be a countable set of names for Statecharts states, $\mathcal{T}$ be a countable set of names for Statecharts transitions, and Π be a countable set of Statecharts events. Moreover, we associate with every event $e \in \Pi$ its negated counterpart $\neg e$. We also lift negation to negated events by defining $\neg\neg e := e$. Finally, we write $\neg E$ for $\{\neg e \mid e \in E\}$. Then, the set of Statecharts terms is defined to be the least set satisfying the following rules.

1. **Basic state:** If $n \in \mathcal{N}$, then $s = [n]$ is a Statecharts term.
2. **Or-state:** If $n \in \mathcal{N}$, $s_1, \ldots, s_k$ are Statecharts terms, $k > 0$, $\rho = \{1, \ldots, k\}$, $T \subseteq \mathcal{T} \times \rho \times 2^{\Pi \cup \neg\Pi} \times 2^{\Pi} \times \rho$, and $1 \le l \le k$, then $s = [n : (s_1, \ldots, s_k); l; T]$ is a Statecharts term. Here, $s_1, \ldots, s_k$ are the sub-states of s, and T is the set of transitions between these states. Statechart s_1 is the default state of s, while s_l is the currently active state.
3. **And-state:** If $n \in \mathcal{N}$ and if $s_1, \ldots, s_k$ are Statecharts terms for $k > 0$, then $s = [n : (s_1, \ldots, s_k)]$ is a Statecharts term.

We refer to n as the *root* of s and write $\mathsf{root}(s) := n$. If $t = \langle \hat{t}, i, E, A, j \rangle \in T$ is a transition of or-state $[n : (s_1, \ldots, s_k); l; T]$, then we define $\mathsf{name}(t) := \hat{t}$, $\mathsf{out}(t) := s_i$, $\mathsf{ev}(t) := E$, $\mathsf{act}(t) := A$, and $\mathsf{in}(t) := s_j$. We write SC for the set of Statecharts terms, in which (i) all state names and transition names are mutually disjoint, (ii) no transition t produces an event that contradicts its trigger, i.e., $\mathsf{ev}(t) \cap \neg\mathsf{act}(t) = \emptyset$, and (iii) no transition t produces an event that is included in its trigger, i.e., $\mathsf{ev}(t) \cap \mathsf{act}(t) = \emptyset$. As a consequence of (i), states and transitions in Statecharts terms are uniquely referred to by their names. Therefore, we may identify a Statecharts state s and transition t with its name $\mathsf{root}(s)$ and $\mathsf{name}(t)$, respectively. The sets $\mathsf{states}(s)$ and $\mathsf{trans}(s)$ of all states and transitions of s are inductively defined on the structure of s, as depicted in Table 1, where $s = (s_1, \ldots, s_k)$. Finally, let us return to our example Statechart in Fig. 1, and present it as a Statecharts term $s_9 \in$ SC. We choose $\Pi := \{a, b, c\}$, $\mathcal{N} := \{n_1, n_2, \ldots, n_9\}$, and $\mathcal{T} := \{t_1, t_2, t_3\}$.

$$
\begin{array}{lll}
s_9 := [n_9 : (s_3, s_8)] & s_3 := [n_3 : (s_1, s_2); 1; \{\langle t_1, 1, \{\neg a\}, \{b\}, 2\rangle\}] & s_1 := [n_1] \\
s_2 := [n_2] & s_8 := [n_8 : (s_6, s_7); 1; \{\langle t_3, 6, \{b\}, \{a\}, 7\rangle\}] & s_7 := [n_7] \\
s_4 := [n_4] & s_6 := [n_6 : (s_4, s_5); 1; \{\langle t_2, 4, \{b\}, \{c\}, 5\rangle\}] & s_5 := [n_5]
\end{array}
$$

In the remainder of this section, we formally present the semantics of Statecharts terms as is defined in [17], which is a slight variant of the "traditional" semantics proposed by Pnueli and Shalev [22]. More precisely, this semantics differs from [22] in that it does not allow the step-construction function, which we present below, to fail. The semantics of a Statecharts term s is a transition system, whose states and transitions are referred to as configurations and macro steps, respectively. Configurations of s are usually sets $\mathsf{conf}(s)$ of names of states which are currently active [22]. We define $\mathsf{conf}(s)$ along the structure of s: (i) $\mathsf{conf}([n]) := \{n\}$, (ii) $\mathsf{conf}([n:(s_1,\dots,s_k);l;T]) := \{n\} \cup \mathsf{conf}(s_l)$, and (iii) $\mathsf{conf}([n:(s_1,\dots,s_k)]) := \{n\} \cup \bigcup\{\mathsf{conf}(s_i) \mid 1 \le i \le k\}$. However, for our purposes it is more convenient to use Statecharts terms for configurations, as every or-state contains a reference to its active sub-state. Consequently, the *default configuration* $\mathsf{default}(s)$ of Statecharts term s may be defined inductively as follows: (i) $\mathsf{default}([n]) := [n]$, (ii) $\mathsf{default}([n:(s_1,\dots,s_k);l;T]) := [n:(\mathsf{default}(s_1),\dots,\mathsf{default}(s_k));1;T]$, and (iii) $\mathsf{default}([n:(s_1,\dots,s_k)]) := [n:(\mathsf{default}(s_1),\dots,\mathsf{default}(s_k))]$. As mentioned before, a Statechart reacts to the arrival of some external events by triggering enabled micro steps, possibly in a chain-reaction–like manner, thereby performing a macro step. More precisely, a macro step comprises a maximal set of micro steps, or transitions, that are *triggered* by events offered by the environment or generated by other micro steps, that are mutually *consistent*, *compatible*, and *relevant*, and that obey *causality*. The Statecharts principle of *global consistency*, which prohibits an event to be present and absent in the same macro step, is subsumed by *triggered* and *compatible*. In the following, we formally introduce the above notions.

Table 2. Step-construction function

```
function step-construction(s, E); var T := ∅;
  while T ⊂ enabled(s, E, T) do choose t ∈ enabled(s, E, T) \ T; T := T ∪ {t} od;
return T
```

A transition $t \in \mathsf{trans}(s)$ is *consistent* with all transitions in $T \subseteq \mathsf{trans}(s)$, in signs $t \in \mathsf{consistent}(s,T)$, if t is not in the same parallel component as any transition in T. Formally, $\mathsf{consistent}(s,T) := \{t \in \mathsf{trans}(s) \mid \forall t' \in T.\, t \perp_s t'\}$. Here, we write $t \perp_s t'$, if $t = t'$, or if there exists an and-state $[n:(s_1,\dots,s_k)]$ in s, i.e., $n \in \mathsf{states}(s)$, such that $t \in \mathsf{trans}(s_i)$ and $t' \in \mathsf{trans}(s_j)$ for some $1 \le i,j \le k$ satisfying $i \ne j$. A transition $t \in \mathsf{trans}(s)$ is *compatible* to all transitions in $T \subseteq \mathsf{trans}(s)$, in signs $t \in \mathsf{compatible}(s,T)$, if no event produced by t appears negated in a trigger of a transition in T. Formally, $\mathsf{compatible}(s,T) := \{t \in \mathsf{trans}(s) \mid \forall t' \in T.\, \mathsf{act}(t) \cap \neg\mathsf{ev}(t') = \emptyset\}$. A transition $t \in \mathsf{trans}(s)$ is *relevant* for s, in signs $t \in \mathsf{relevant}(s)$, if the root of the source state of t is in the configuration of s. Formally, $\mathsf{relevant}(s) := \{t \in \mathsf{trans}(s) \mid \mathsf{root}(\mathsf{out}(t)) \in \mathsf{conf}(s)\}$. A transition $t \in \mathsf{trans}(s)$ is *triggered* by a set E of events, in signs $t \in \mathsf{triggered}(s,E)$, if the positive, but not the negative, trigger events of t are in E. Formally, $\mathsf{triggered}(s,E) := \{t \in \mathsf{trans}(s) \mid \mathsf{ev}(t) \cap \Pi \subseteq E \text{ and } \neg(\mathsf{ev}(t) \cap \neg\Pi) \cap E = \emptyset\}$. Finally, t is *enabled* in s regarding a set E of events and a set T of transitions, if $t \in \mathsf{enabled}(s, E, T)$, where $\mathsf{enabled}(s, E, T) := \mathsf{relevant}(s) \cap \mathsf{consistent}(s, T) \cap \mathsf{triggered}(s, E \cup \bigcup_{t \in T} \mathsf{act}(t)) \cap \mathsf{compatible}(s, T)$. Unfortunately, this formalism is still not rich enough to *causally* justify the triggering of each transition. The principle of *causality* may be introduced by computing macro steps, i.e., sets of transition names, using the nondeterministic *step-construction function* presented in Table 2. This function is adopted from [17], where also its soundness and completeness relative to the classical approach via the notion of *inseparability* of transitions [22] are stated. Note that the maximality of each macro step implements the synchrony hypothesis of Statecharts. The set of all macro steps that can be constructed using function *step-construction*, relative to a Statecharts term s and a set E of environment events, is denoted by $\mathsf{step}(s, E) \subseteq 2^{\mathcal{T}}$.

Table 3. Function update

$$\mathsf{update}([n], T') := [n] \quad \mathsf{update}([n : s], T') := [n : (\mathsf{update}(s_1, T_1), \ldots, \mathsf{update}(s_k, T_k))]$$

$$\mathsf{update}([n : s; l; T], T') := \begin{cases} [n : s; l; T] & \text{if } T' = \emptyset \\ [n : (s_1, \ldots, \mathsf{update}(s_l, T'), \ldots, s_k); l; T] & \text{if } \emptyset \neq T' \subseteq \mathsf{trans}(s_l) \\ [n : (s_1, \ldots, \mathsf{default}(s_m), \ldots, s_k); m; T] & \text{if } \emptyset \neq T' = \{\langle t', l, E, A, m \rangle\} \subseteq T \\ [n] & \text{otherwise} \end{cases}$$

For a set $T \in \mathsf{step}(s, E)$, Statecharts term s may evolve in a macro step to term $s' := \mathsf{update}(s, T)$ when triggered by the environment events in E and, thereby, produce the events $A := \bigcup \{\mathsf{act}(t) \mid t \in T\}$. We denote this macro step by $s \overset{E}{\underset{A}{\Longrightarrow}} s'$. The function update is defined in Table 3, where $s := (s_1, \ldots, s_k)$ and $T_i := T' \cap \mathsf{trans}(s_i)$, for $1 \leq i \leq k$. Observe that at most one transition of T may be enabled at the top-level of an or-state; thus, the "otherwise" case in Table 3 cannot occur in our context. Intuitively, $\mathsf{update}(s, T)$, when $T \subseteq \mathsf{trans}(s)$, re-defines the active states of s, when the transitions in T are executed.

3 Process-Algebraic Framework

Our process-algebraic framework is inspired by *timed process calculi*, such as Hennessy and Regan's TPL [11]. The *Statecharts Process Language* (SPL), which we intend to develop, includes a special action σ denoting the ticking of a global clock. SPL's semantic framework is based on a notion of transition system that involves two kinds of transitions, *action* transitions and *clock* transitions, modeling two different mechanisms of communication and synchronization in *concurrent* systems. The role of actions correspond to the one of events in Statecharts. A clock represents the progress of time, which manifests itself in a recurrent global synchronization event, the clock transition, in which all process components are forced to take part. However, action and clock transitions are not orthogonal concepts but are connected via the *maximal progress assumption* [11, 28]. Maximal progress implies that progress of time is determined by the *completion of internal computations* and, thus, mimics Statecharts' synchrony hypothesis. The key idea for embedding Statecharts terms in a timed process algebra is to represent a macro step as a sequence of micro steps that is enclosed by clock transitions, signaling the beginning and the end of the macro step, respectively. This sequence implicitly encodes causality and leads to a compositional Statecharts semantics. Unfortunately, existing timed process algebras are – in their original form – not suitable for embedding Statecharts. The reason is that Statecharts transitions may be labeled by *multiple* events and that some events may appear *negated*. The former feature implies that, in contrast to standard process algebras [1, 12, 21], processes may be forced to synchronize on more than one event simultaneously, and the latter feature is similar to mechanisms for handling priority [4]. Our framework must also include an operator similar to the *disabling operator* of LOTOS [3] for resembling state hierarchy [25].

Formally, let Λ be a countable set of *events* or *ports*, and let $\sigma \notin \Lambda$ be the distinguished *clock event* or *clock tick*. We define *input actions* to be of the form $\langle E, N \rangle$, where $E, N \subseteq \Lambda$, and *output actions* E to be subsets of Λ. In case of the input action $\langle \emptyset, \emptyset \rangle$, we speak of an *unobservable* or *internal* action, which is also denoted by $\bullet$. We let $\mathcal{A}$ stand for the set of all input actions. In contrast to CCS [21], the syntax of SPL includes two different operators for dealing with input and output actions, respectively. The *prefix operator* "$\langle E, N \rangle$." only permits prefixing with respect to input actions, which are instantly consumed in a single step. Output actions E are signaled to the environment of a process by attaching them to the process via the *signal* operator "$[E]\sigma(\cdot)$." They remain visible until the next clock tick σ occurs. The syntax of SPL is given by the following BNF

$$P \ ::= \ \mathbf{0} \mid X \mid \langle E, N \rangle.P \mid [E]\sigma(P) \mid P + P \mid P \rhd P \mid P \rhd_\sigma P \mid P|P \mid P \setminus L$$

where $L \subseteq \Lambda$ is a *restriction set*, and X is a *process variable* taken from some countable domain $\mathcal{V}$. We also allow the definition of *equations* $X \stackrel{\text{def}}{=} P$, where variable X is assigned to term P. If X occurs as a subterm of P, we say that X is *recursively* defined. We adopt the usual definitions for *open* and *closed* terms and *guarded* recursion, and refer to the closed and guarded terms as *processes* [21]. Moreover, we let $\mathcal{P}$, ranged over by P and Q, denote the set of all processes. Finally, the operators $\rhd$ and $\rhd_\sigma$, called *disabling* and *enabling* operator, respectively, allow us to model state hierarchy, as is illustrated below.

The operational semantics of an SPL process $P \in \mathcal{P}$ is given by a *labeled transition system* $\langle \mathcal{P}, \mathcal{A} \cup \{\sigma\}, \longrightarrow, P \rangle$, where $\mathcal{P}$ is the set of states, $\mathcal{A} \cup \{\sigma\}$ the alphabet, $\longrightarrow$ the transition relation, and P the start state. We refer to transitions with labels in $\mathcal{A}$ as *action transitions* and to those with label σ as *clock transitions*. For the sake of simplicity, we write (i) $P \xrightarrow[N]{E} P'$ instead of $\langle P, \langle E, N \rangle, P' \rangle \in \longrightarrow$ and (ii) $P \xrightarrow{\sigma} P'$ instead of $\langle P, \sigma, P' \rangle \in \longrightarrow$. We say that *$P$ may engage in a transition labeled by $\langle E, N \rangle$ or σ, respectively, and thereafter behave like process P'*. The transition relation is defined in Tables 4 and 5 using operational rules. In contrast to CCS [21], our framework does not provide a concept of output action transitions, such that "matching" input and output action transitions synchronize with each other and, thereby, simultaneously change states. Instead, output actions are attached to SPL processes via the signal operator. In order to present our communication mechanism, we need to introduce *initial output action sets*, $\overline{\mathbb{I}}(P)$, for $P \in \mathcal{P}$. These are defined as the *least* sets satisfying the equations in Table 4 (upper part). Intuitively, $\overline{\mathbb{I}}(P)$ collects all events which are initially offered by P.

Table 4. Initial output action sets & operational semantics (action transitions)

$$\overline{\mathbb{I}}([E]\sigma(P)) = E \qquad \overline{\mathbb{I}}(P+Q) = \overline{\mathbb{I}}(P) \cup \overline{\mathbb{I}}(Q) \qquad \overline{\mathbb{I}}(X) = \overline{\mathbb{I}}(P) \quad \text{if } X \stackrel{\text{def}}{=} P$$

$$\overline{\mathbb{I}}(P\,|\,Q) = \overline{\mathbb{I}}(P) \cup \overline{\mathbb{I}}(Q) \qquad \overline{\mathbb{I}}(P \setminus L) = \overline{\mathbb{I}}(P) \setminus L$$

$$\overline{\mathbb{I}}(P \rhd Q) = \overline{\mathbb{I}}(P) \cup \overline{\mathbb{I}}(Q) \qquad \overline{\mathbb{I}}(P \rhd_\sigma Q) = \overline{\mathbb{I}}(P)$$

$$\textsf{Act}\ \frac{-}{\langle E,N\rangle.P \xrightarrow[N]{E} P}$$

$$\textsf{Rec}\ \frac{P \xrightarrow[N]{E} P'}{X \xrightarrow[N]{E} P'}\ X \stackrel{\text{def}}{=} P \qquad \textsf{Sum1}\ \frac{P \xrightarrow[N]{E} P'}{P+Q \xrightarrow[N]{E} P'} \qquad \textsf{Par1}\ \frac{P \xrightarrow[N]{E} P'}{P\,|\,Q \xrightarrow[N]{E\setminus\overline{\mathbb{I}}(Q)} P'\,|\,Q}\ N \cap \overline{\mathbb{I}}(Q) = \emptyset$$

$$\textsf{En}\ \frac{P \xrightarrow[N]{E} P'}{P \rhd_\sigma Q \xrightarrow[N]{E} P' \rhd_\sigma Q} \qquad \textsf{Sum2}\ \frac{Q \xrightarrow[N]{E} Q'}{P+Q \xrightarrow[N]{E} Q'} \qquad \textsf{Par2}\ \frac{Q \xrightarrow[N]{E} Q'}{P\,|\,Q \xrightarrow[N]{E\setminus\overline{\mathbb{I}}(P)} P\,|\,Q'}\ N \cap \overline{\mathbb{I}}(P) = \emptyset$$

$$\textsf{Dis1}\ \frac{P \xrightarrow[N]{E} P'}{P \rhd Q \xrightarrow[N]{E} P' \rhd_\sigma Q} \qquad \textsf{Dis2}\ \frac{Q \xrightarrow[N]{E} Q'}{P \rhd Q \xrightarrow[N]{E} Q'} \qquad \textsf{Res}\ \frac{P \xrightarrow[N]{E} P'}{P \setminus L \xrightarrow[N\setminus L]{E} P' \setminus L}\ E \cap L = \emptyset$$

The semantics for action transitions, depicted in Table 4 (lower part), is set up such that $P \xrightarrow[N]{E} P'$ means: P can evolve to P', if the environment offers communications on all ports in E, but none on any port in N. More precisely, process $\langle E,N\rangle.P$ may engage in input action $\langle E,N\rangle$ and then behave like P. The *summation operator* + denotes *nondeterministic choice*, i.e., process $P+Q$ may either behave like P or Q. Process $P\,|\,Q$ stands for the *parallel composition* of P and Q according to an interleaving semantics with synchronization on common ports. Rule Par1 describes the interaction of process P with its environment Q. If P can engage in a transition labeled by $\langle E,N\rangle$ to P', then P and Q synchronize on the events in $E \cap \overline{\mathbb{I}}(Q)$, provided that Q does not offer a communication on a port in N, i.e., $N \cap \overline{\mathbb{I}}(Q) = \emptyset$ holds. In this case, $P\,|\,Q$ can engage in a transition labeled by $\langle E \setminus \overline{\mathbb{I}}(Q), N\rangle$ to $P'\,|\,Q$. Rule Par2 deals with the symmetric case, where the roles of P and Q are interchanged. The semantics of the *disabling* and *enabling operators* are tightly connected. Process $P \rhd Q$ may behave as Q, thereby permanently disabling P, or as $P \rhd_\sigma Q$. In the latter case only P may proceed, and Q is disabled until the next clock tick arrives. This allows for modeling Statecharts or-states, where process P is on a lower level than Q. The *restriction operator* $\setminus L$ encapsulates all ports in L. Rule Res states that process $P \setminus L$ can only engage in an action transition labeled by $\langle E, N\rangle$, if there is no event in E, which is restricted by L. Moreover, the events in L may be eliminated from N. Finally, process variable X, where $X \stackrel{\text{def}}{=} P$, is identified with a process that behaves as a distinguished solution of the equation $X = P$.

Table 5. Operational semantics (clock transitions)

$$\textsf{tNil}\ \frac{-}{\mathbf{0} \xrightarrow{\sigma} \mathbf{0}} \qquad \textsf{tAct}\ \frac{-}{\langle E,N\rangle.P \xrightarrow{\sigma} \langle E,N\rangle.P}\ \langle E,N\rangle \neq \bullet \qquad \textsf{tOut}\ \frac{-}{[E]\sigma(P) \xrightarrow{\sigma} P}$$

$$\textsf{tPar}\ \frac{P \xrightarrow{\sigma} P' \quad Q \xrightarrow{\sigma} Q'}{P \,|\, Q \xrightarrow{\sigma} P' \,|\, Q'}\ \bullet \notin \mathrm{I}(P \,|\, Q) \qquad \textsf{tSum}\ \frac{P \xrightarrow{\sigma} P' \quad Q \xrightarrow{\sigma} Q'}{P + Q \xrightarrow{\sigma} P' + Q'}$$

$$\textsf{tDis}\ \frac{P \xrightarrow{\sigma} P' \quad Q \xrightarrow{\sigma} Q'}{P \rhd Q \xrightarrow{\sigma} P' \rhd Q'} \qquad \textsf{tEn}\ \frac{P \xrightarrow{\sigma} P'}{P \rhd_\sigma Q \xrightarrow{\sigma} P' \rhd Q}$$

$$\textsf{tRes}\ \frac{P \xrightarrow{\sigma} P'}{P \setminus L \xrightarrow{\sigma} P' \setminus L}\ \bullet \notin \mathrm{I}(P \setminus L) \qquad \textsf{tRec}\ \frac{P \xrightarrow{\sigma} P'}{X \xrightarrow{\sigma} P'}\ X \stackrel{\text{def}}{=} P$$

The operational rules for clock transitions deal with the maximal progress assumption, i.e., if $\bullet \in \mathrm{I}(P) := \{\langle E,N\rangle \,|\, \exists P'.\, P \xrightarrow[N]{E} P'\}$, then a clock tick σ is inhibited. The reason that transitions other than labeled by $\bullet$ do not have pre-emptive power is that these only indicate the *potential* of progress, whereas $\bullet$ denotes *real* progress in our framework. Rule tNil states that inaction process $\mathbf{0}$ can idle forever. Similarly, process $\langle E,N\rangle.P$ may idle for clock σ, whenever $\langle E,N\rangle \neq \bullet$. The *signal operator* in $[E]\sigma(P)$, which offers communications on the ports in E to its environment, disappears as soon as the next clock tick arrives and, thereby, enables P. Time has to proceed equally on both sides of summation, parallel composition, and disabling, i.e., $P + Q$, $P \,|\, Q$, and $P \rhd Q$ can engage in a clock transition if and only if both P and Q can. The side condition of Rule tPar implements maximal progress and states that there is no pending communication between P and Q. The reason for the side condition in Rule tRes is that the restriction operator may turn observable input actions into the internal, unobservable input action $\bullet$ (cf. Rule tRes) and, thereby, may pre-empt the considered clock transition. Finally, Rule tEn states that a clock tick switches the enabling operator to the disabling operator.

The operational semantics for SPL possesses several pleasant algebraic properties which are known from various timed process algebras [11, 28], such as (i) the *idling* property, i.e., $\bullet \notin \mathbb{I}(P)$ implies $\exists P'. P \xrightarrow{\sigma} P'$, for all $P \in \mathcal{P}$, (ii) the *maximal progress* property, i.e., $\exists P'. P \xrightarrow{\sigma} P'$ implies $\bullet \notin \mathbb{I}(P)$, for all $P \in \mathcal{P}$, and (iii) the *time determinacy* property, i.e., $P \xrightarrow{\sigma} P'$ and $P \xrightarrow{\sigma} P''$ implies $P' = P''$, for all $P, P', P'' \in \mathcal{P}$. Moreover, the summation and parallel operators are *associative* and *commutative.* The well-known *behavioral equivalence* bisimulation [21] may be adapted to cater for SPL as follows. Other work can be used for establishing that it is a well-defined *congruence* for SPL [27].

Definition 1 (Bisimulation). Bisimulation equivalence, $\sim\, \subseteq \mathcal{P} \times \mathcal{P}$, *is the largest symmetric relation such that for $P \sim Q$ the following conditions hold.*

1. $\overline{\mathbb{I}}(P) \subseteq \overline{\mathbb{I}}(Q)$ *2. If* $P \xrightarrow[N]{E} P'$ *then* $\exists Q' \in \mathcal{P}.\, Q \xrightarrow[N]{E} Q'$ *and* $P' \sim Q'$.

4 Embedding of Statecharts

In this section we present an embedding of Statecharts in SPL, which is a mapping $[\![\cdot]\!]$ from Statecharts terms to processes defined by (mutually recursive) equations. Although SPL's semantics is defined on a "micro-step level," SPL allows us to encode the synchrony hypothesis of Statecharts by using maximal progress. More precisely, a macro step in Statecharts semantics corresponds to a sequence of SPL action transitions which is enclosed by clock transitions. These sequences implicitly contain the causal order inherent in a Statecharts macro step. Formally, we choose $\Pi \cup \neg\Pi$ for the set Λ of ports and $\mathcal{N} \cup \{\hat{n} \,|\, n \in \mathcal{N}\} \cup \mathcal{T}$ for the set $\mathcal{V}$ of process variables. We define the embedding $[\![\cdot]\!]$ inductively along the structure of Statecharts terms, where $\sum$ is the indexed version of $+$ satisfying $\sum_{i \in \emptyset} P_i := \mathbf{0}$.

1. If $s = [n]$, then $[\![s]\!] := n$ where $n \stackrel{\text{def}}{=} \hat{n} \stackrel{\text{def}}{=} \mathbf{0}$.
2. If $s = [n : (s_1, \ldots, s_k); l; T]$ and $n_i = \mathsf{root}(s_i)$, for $1 \leq i \leq k$, then $[\![s]\!] := n$, where $n \stackrel{\text{def}}{=} \hat{n}_l$ and $\hat{n}_i \stackrel{\text{def}}{=} n_i \rhd \sum\{\{\![t]\!\}\,|\, t \in T \text{ and } \mathsf{out}(t) = s_i\}$, together with the equations of $[\![s_1]\!], \ldots, [\![s_k]\!]$. Please see below for the translation $\{\![t]\!\}$ of t.
3. If $s = [n : (s_1, \ldots, s_k)]$, then $[\![s]\!] := n$ and $n \stackrel{\text{def}}{=} \hat{n} \stackrel{\text{def}}{=} \mathsf{root}(s_1) \,|\, \cdots \,|\, \mathsf{root}(s_k)$, together with the equations of $[\![s_1]\!], \ldots, [\![s_k]\!]$.

Semantically, a basic state corresponds to inaction process $\mathbf{0}$, whereas an or-state can either behave according to the embedding of the currently active state s_l, or it may exit s_l by engaging in a transition $t \in T$ with $\mathsf{out}(t) = s_l$. Observe that an or-state is mapped using the disabling operator. The translation of an and-state maps its component states to the parallel composition of the processes resulting from the translations of each of these states. The interesting part of the definition of $[\![\cdot]\!]$ is the translation $\{\![t]\!\}$ of a transition $\langle t, i, E, A, j\rangle$. In the following, E' stands for $E \cap \Pi$ and N' for $\neg(E \cap \neg\Pi) \cup \neg A$. We define $\{\![t]\!\} := \langle E', N'\rangle.t$ where $t \stackrel{\text{def}}{=} [A \cup (E \cap \neg\Pi)]\sigma(\hat{n}_j)$. The translation splits a transition $\langle t, i, E, A, j\rangle$ in two parts, one handling its trigger E and one executing its action A. In order for t to trigger, all positive events in E must be offered by the environment, and all negative events in E must be absent. However, there is one more thing we have to obey: *global consistency*. Especially, we must ensure that there is no previous transition t' in the same macro step, which has fired because of the absence of an event in A. Therefore, to prevent t from triggering, we include a distinguished event $\neg e$, where $e \in A$, in the set N' of $\llbracket t \rrbracket$, and we make sure that $\neg e$ is offered when t' triggers. Hence, $\llbracket t \rrbracket$ can evolve via a SPL transition labeled by $\langle E', N' \rangle$ to process t, whenever the trigger of t is satisfied according to Statecharts semantics and whenever global consistency is preserved. Process t signals that transition t has fired by offering the events in A as well as the already mentioned negated events $\neg e$ for $\neg e \in E \cap \neg\Pi$. These events are offered until the current macro step is completed, i.e., until a clock transition is executed. Thus, SPL's two-level semantics of action and clock transitions allows for broadcasting events using SPL's synchronization mechanism and SPL's maximal progress assumption.

Table 6. Embedding of the Example Statechart

$$
\begin{aligned}
\llbracket s_9 \rrbracket &= n_9 \stackrel{\text{def}}{=} n_3 \mid n_8 \\
\llbracket s_3 \rrbracket &= n_3 \stackrel{\text{def}}{=} \hat{n}_1 \\
\hat{n}_1 &\stackrel{\text{def}}{=} n_1 \rhd \langle \emptyset, \{a, \neg b\} \rangle . t_1 \\
t_1 &\stackrel{\text{def}}{=} [\{b, \neg a\}] \sigma(\hat{n}_2) \\
\llbracket s_1 \rrbracket &= n_1 \stackrel{\text{def}}{=} \hat{n}_1 \stackrel{\text{def}}{=} \mathbf{0} \\
\llbracket s_2 \rrbracket &= n_2 \stackrel{\text{def}}{=} \hat{n}_2 \stackrel{\text{def}}{=} \mathbf{0} \\
\llbracket s_8 \rrbracket &= n_8 \stackrel{\text{def}}{=} \hat{n}_6 \\
\hat{n}_6 &\stackrel{\text{def}}{=} n_6 \rhd \langle \{b\}, \{\neg a\} \rangle . t_3 \\
t_3 &\stackrel{\text{def}}{=} [\{a\}] \sigma(\hat{n}_7) \\
\llbracket s_6 \rrbracket &= n_6 \stackrel{\text{def}}{=} \hat{n}_4 \\
\hat{n}_4 &\stackrel{\text{def}}{=} n_4 \rhd \langle \{b\}, \{\neg c\} \rangle . t_2 \\
t_2 &\stackrel{\text{def}}{=} [\{c\}] \sigma(\hat{n}_5) \\
\llbracket s_4 \rrbracket &= n_4 \stackrel{\text{def}}{=} \hat{n}_4 \stackrel{\text{def}}{=} \mathbf{0} \\
\llbracket s_5 \rrbracket &= n_5 \stackrel{\text{def}}{=} \hat{n}_5 \stackrel{\text{def}}{=} \mathbf{0} \\
\llbracket s_7 \rrbracket &= n_7 \stackrel{\text{def}}{=} \hat{n}_7 \stackrel{\text{def}}{=} \mathbf{0}
\end{aligned}
$$

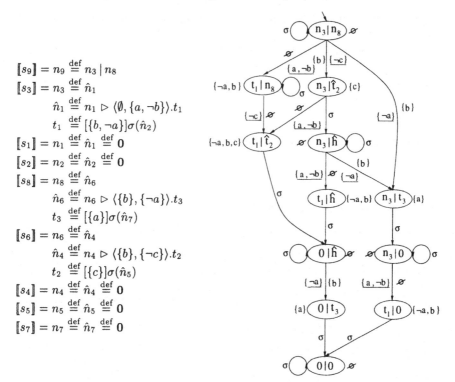

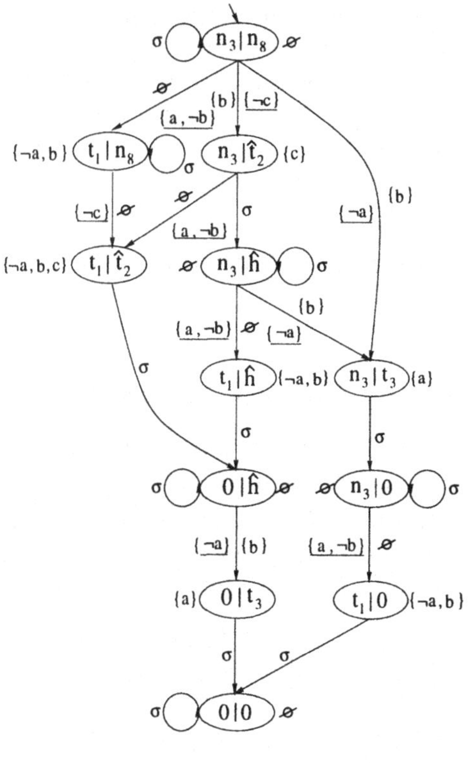

We now return to our introductory example by presenting its formal translation to SPL in Table 6, left-hand side. The embedding's operational semantics is depicted on the right-hand side of Table 6, where $\hat{t}_2 \stackrel{\text{def}}{=} t_2 \rhd_\sigma \langle \{b\}, \{\neg a\} \rangle . t_3$, and $\hat{h} \stackrel{\text{def}}{=} \mathbf{0} \rhd \langle \{b\}, \{\neg a\} \rangle . t_3$. Moreover, the initial output action set $\mathbb{I}(P)$, for some $P \in \mathcal{P}$, is denoted next to the ellipse symbolizing state P, and the sets N' appearing in the label of transitions are underlined in order to distinguish them from the sets E'. Let us have a closer look at the leftmost path of the transition system, passing the states $(n_3 \,|\, n_8)$, $(t_1 \,|\, n_8)$, $(t_1 \,|\, \hat{t}_2)$, $(\mathbf{0} \,|\, \hat{h})$, $(\mathbf{0} \,|\, t_3)$, and $(\mathbf{0} \,|\, \mathbf{0})$. The first three states are separated from the last three states by a clock transition. Hence, the considered sequence corresponds to two "potential" macro steps. We say "potential," since macro steps only emerge when composing our Statecharts embedding with an environment which triggers macro steps. The events needed to trigger the transitions and the actions produced by them can be extracted from a macro-step sequence as follows. For obtaining the trigger, consider all transition labels $\langle E, N \rangle$ occurring in the sequence, add up all events in components E, and include the negations of all positive events in components N. Regarding the generated actions, consider the set of positive events in the initial output action sets of the states preceding the clock transition which signals the end of the macro step. Thus, the first potential macro step of the example sequence is triggered by $\neg a$ and produces events b and c, whereas the second is triggered by b and produces a. The state names along a sequence also indicate, which transitions have fired. More precisely, whenever a state includes a variable $t \in \mathcal{T}$ at its top-level, transition t participates in the current macro step. Thus, for the first potential macro step transitions t_1 and t_2 are chosen, whereas the second consists of transition t_3 only. Note that t_3 is not enabled in states $(t_1 \,|\, n_8)$ or $(t_1 \,|\, \hat{t}_2)$, since event $\neg a$ is in their initial output action sets and $a \in \mathsf{act}(t_3)$. Hence, our embedding respects global consistency, which prohibits t_1 and t_3 to occur in the same macro step.

5 Semantic Correspondence

For formalizing the semantic relation between Statecharts terms and their SPL embeddings, we define a notion of SPL *macro steps* by combining several transitions to a single step, as outlined in the previous section. We write $P \overset{E}{\underset{A}{\Longrightarrow}} P'$ if $\exists P'' \in \mathcal{P}.\ (\mathsf{Env}_E \,|\, P) \setminus \Lambda \overset{\emptyset}{\underset{\emptyset}{\longrightarrow}}{}^{*} (\mathsf{Env}_E \,|\, P'') \setminus \Lambda \overset{\sigma}{\longrightarrow} (\mathbf{0} \,|\, P') \setminus \Lambda$ and $\mathbb{I}(P'') = A$, where $\mathsf{Env}_E \overset{\text{def}}{=} [E]\sigma(\mathbf{0})$. Intuitively, P is placed in context $(\mathsf{Env}_E \,|\, \cdot) \setminus \Lambda$, where Env_E models a single-step environment which offers the events in E until clock tick σ occurs. The following relation, which we refer to as *step correspondence*, provides the formal foundation for relating Statecharts and SPL macro steps.

Definition 2 (Step Correspondence). *A relation $\mathcal{R} \subseteq SC \times \mathcal{P}$ is a* step correspondence *if for all $\langle s, P \rangle \in \mathcal{R}$ and $E, A \subseteq \Pi$ the following conditions hold:*

1. *$\forall s' \in SC.\ s \overset{E}{\underset{A}{\Longrightarrow}} s'$ implies $\exists P' \in \mathcal{P}.\ P \overset{E}{\underset{A}{\Longrightarrow}} P'$ and $\langle s', P' \rangle \in \mathcal{R}$.*
2. *$\forall P' \in \mathcal{P}.\ P \overset{E}{\underset{A}{\Longrightarrow}} P'$ implies $\exists s' \in SC.\ s \overset{E}{\underset{A}{\Longrightarrow}} s'$ and $\langle s', P' \rangle \in \mathcal{R}$.*

s is step-correspondent *to P, if $\langle s, P \rangle \in \mathcal{R}$ for some step correspondence $\mathcal{R}$.*

Theorem 1 (Embedding). *Every $s \in SC$ is step-correspondent to $[\![s]\!]$.*

We close this section by returning to the behavioral relation $\sim$.

Theorem 2 (Preservation). *Let $P, Q \in \mathcal{P}$ such that $P \sim Q$, and suppose that $P \overset{E}{\underset{A}{\Rightarrow}} P'$. Then $\exists Q' \in \mathcal{P}.\, Q \overset{E}{\underset{A}{\Rightarrow}} Q'$ and $P' \sim Q'$.*

Now, we can state our desired result, namely that the behavioral equivalence *bisimulation*, when applied to SPL, preserves Statecharts semantics.

Corollary 1. *Let $E, A \subseteq \Pi$, $s \in SC$, and $P \in \mathcal{P}$ such that $[\![s]\!] \sim P$. Then*

1. $\forall s' \in SC.\ s \overset{E}{\underset{A}{\Longrightarrow}} s'$ *implies* $\exists P' \in \mathcal{P}.\, P \overset{E}{\underset{A}{\Rightarrow}} P'$ *and* $[\![s']\!] \sim P'$.
2. $\forall P' \in \mathcal{P}.\, P \overset{E}{\underset{A}{\Rightarrow}} P'$ *implies* $\exists s' \in SC.\ s \overset{E}{\underset{A}{\Longrightarrow}} s'$ *and* $[\![s']\!] \sim P'$.

6 Adaptability to Other Statecharts Variants

For Statecharts a variety of different semantics has been introduced in the literature [26]. In this section, we show how our approach can be adapted to these variants and, thereby, testify to its flexibility.

In the Statecharts variant examined in this paper, two features are left out which are often adopted in other variants. One feature concerns *inter-level transitions*, i.e., transitions which cross the "borderlines" of Statecharts states and, thus, permit a style of "goto"-programming. Unfortunately, when allowing inter-level transitions the syntax of Statecharts terms cannot be defined compositionally and, consequently, nor its semantics. The second feature left out is usually referred to as *state reference*, which permits the triggering of a transition to depend on the fact whether a certain parallel component is in a certain state. Such state references can be encoded in SPL's communication scheme by introducing special events in_n, for $n \in \mathcal{N}$, which are signaled by a process if it is in state n.

Another issue concerns the sensing of *internal* and *external events*. Usually, internal events are sensed within a macro step, but external events are not. Hence, events are *instantaneous*, i.e., an event exists only for the duration of the macro step under consideration. This is reflected in our signal operator which stops signaling events as soon as the next clock tick arrives. In the semantics of Statemate [8], an event is only sensed in the macro step following the one in which it was generated. This behavior can be encoded in our embedding by splitting every state $t \in \mathcal{T}$ into two states that are connected via a clock transition. The specific sensing of events in Statemate greatly simplifies the development of a *compositional* semantics [6].

The Statecharts concept of *negated events* forces transitions to be triggered only when certain events are absent. However, when permitting negated events in a macro-step semantics, one has to guarantee that the *effect of a transition is not contradictory to its cause.* Regarding this issue, one may distinguish two concepts: *global consistency* and *local consistency.* The first one prohibits a transition containing a negative trigger event $\neg e$ to be executed, if a micro step within the same macro step produces e. In our embedding, this is enforced by offering $\neg e$, whenever a transition triggers due to the absence of e. Moreover, $\neg e$ is included in the set of events which need to be absent in all Statecharts transitions producing e. When leaving out these events $\neg e$ in our embedding, we obtain the weaker notion of local consistency, i.e., once an event e is signaled in a micro-step, no following micro step of the same macro step may fire if its trigger contains $\neg e$. Local consistency implicitly holds in our embedding, since an event is always signaled until the next macro step begins.

In addition to encoding priorities between transitions via negated events, one may introduce an implicit priority mechanism along *state hierarchy*, as is done, e.g., in Statemate [10] but not in the Statecharts variant [17] considered in this paper. More precisely, a transition leaving an or-state may be given priority over any transition within this state, i.e., or-states can then be viewed as *pre-emptive interrupt* operators. SPL can easily be extended to capture this behavior.

7 Related Work

Achieving a compositional semantics for Statecharts is known to be a difficult task. The problems involved were systematically analyzed and investigated by Huizing and Gerth in the early nineties in the more general context of real-time reactive systems [15], for which three criteria have found to be desirable: (i) *responsiveness*, which corresponds to the synchrony hypothesis of Statecharts, (ii) *modularity*, which refers to the aspect of compositionality, and (iii) *causality*. Huizing and Gerth proved that these properties cannot be combined in a single-leveled semantics. In our approach the three properties hold on different levels: compositionality holds on the micro-step level – the level of SPL action transitions – whereas responsiveness and causality are guaranteed on the macro-step level – the level where sequences of SPL action transitions between global synchronizations, caused by clock ticks σ, are bundled together.

Uselton and Smolka [24, 25] and Levi [16] also focused on achieving a compositional semantics for Statecharts by referring to process algebras. In contrast to our approach, Uselton and Smolka's notion of transition system involves labels of the form $\langle E, \prec \rangle$, where E is a set of events, and $\prec$ is a transitive, irreflexive order on E encoding causality. Unfortunately, their semantics does not correspond, as intended, to the semantics of Pnueli and Shalev [22], as pointed out in [16, 17]. Levi repaired this shortcoming by modifying the domains of the arguments of $\prec$ to sets of events and by allowing empty steps to be represented explicitly.

Maggiolo-Schettini et al. considered a hierarchy of equivalences for Statecharts and studied congruence properties with respect to Statecharts operators [17]. For this purpose, they defined a compositional, operational macro-step semantics of Statecharts, which slightly differs from the one of Pnueli and Shalev, since it does not allow the step-construction function to fail. Their semantics is also expressed in terms of labeled transition systems, where labels consist of four-tuples which include information about causal orderings, global consistency, and negated events. The framework of Maggiolo-Schettini et al. serves well for the purpose of studying certain algebraic properties of equivalences on Statecharts, such as fully-abstractness results and axiomatizations [14, 15].

Another popular design language with a visual appeal like Statecharts and, moreover, a solid algebraic foundation is *Argos* [18]. However, the semantics of Argos – defined via SOS-rules as labeled transition systems – significantly differs from classical Statecharts semantics. For example, Argos is deterministic, abstracts from "non-causal" Statecharts by semantically identifying them with a *failure* state, and allows a single parallel component to fire more than once within a macro step.

Interfacing Statemate [10] to verification tools is a main objective in [19, 20]. The former work formalizes Statemate semantics in Z, while the latter work translates a subset of Statemate to the model-checking tool Spin [13].

8 Conclusions and Future Work

This paper presented a process-algebraic approach to defining a compositional semantics for Statecharts. Our technique translates Statecharts terms to terms in SPL which allows one to encode a "micro-step" semantics of Statecharts. The macro-step semantics may then be given in terms of a derived transition relation. We demonstrated the utility of our technique by formally embedding the Statecharts semantics of [17], which is a slight variant of Pnueli and Shalev's semantics [22], in SPL. Our approach also allows for interfacing Statecharts to existing verification tools and for the possibility of lifting behavioral equivalences from process algebras to Statecharts. We illustrated the viability of this last point by showing that bisimulation equivalence, which is a congruence for SPL, preserves Statecharts macro-step semantics.

Regarding future work, we plan to continue our investigation of behavioral equivalences for Statecharts in general, and "weak" equivalences in particular, by studying them for SPL. It may also be interesting to characterize the "Statecharts sub-algebra" of SPL. Moreover, we intend to implement SPL and our embedding in the *Concurrency Workbench of North Carolina* (CWB-NC) [5].

We would like to thank Peter Kelb, Ingolf Krüger, Michael Mendler, and the anonymous referees for many valuable comments and suggestions.

References

[1] J.C.M. Baeten and W.P. Weijland. *Process Algebra*, vol. 18 of *Cambridge Tracts in Theoretical Computer Science*. Cambridge University Press, 1990.

[2] G. Berry and G. Gonthier. The ESTEREL synchronous programming language: Design, semantics, implementation. *Science of Computer Programming*, 19:87–152, 1992.

[3] T. Bolognesi and E. Brinksma. Introduction to the ISO specification language LOTOS. *Computer Networks and ISDN Systems*, 14:25–59, 1987.

[4] R. Cleaveland, G. Lüttgen, and V. Natarajan. Priority in process algebra. In *Handbook of Process Algebra*. Elsevier, 1999.

[5] R. Cleaveland and S. Sims. The NCSU Concurrency Workbench. In *CAV '96*, vol. 1102 of *LNCS*, pages 394–397, 1996. Springer Verlag.

[6] W. Damm, B. Josko, H. Hungar, and A. Pnueli. A compositional real-time semantics of STATEMATE designs. In *Compositionality: The Significant Difference*, vol. 1536 of *LNCS*, pages 186–238, 1997. Springer Verlag.

[7] D. Harel. Statecharts: A visual formalism for complex systems. *Science of Computer Programming*, 8:231–274, 1987.

[8] D. Harel and A. Naamad. The STATEMATE semantics of Statecharts. *ACM Transactions on Software Engineering*, 5(4):293–333, 1996.

[9] D. Harel, A. Pnueli, J.P. Schmidt, and R. Sherman. On the formal semantics of Statecharts. In *LICS '87*, pages 56–64, 1987. IEEE Computer Society Press.

[10] D. Harel and M. Politi. *Modeling Reactive Systems with Statecharts: The STATEMATE Approach.* McGraw Hill, 1998.

[11] M. Hennessy and T. Regan. A process algebra for timed systems. *Information and Computation*, 117:221–239, 1995.

[12] C.A.R. Hoare. *Communicating Sequential Processes.* Prentice-Hall, 1985.

[13] G.J. Holzmann. The model checker Spin. *IEEE Transactions on Software Engineering*, 23(5):279–295, 1997.

[14] J.J.M. Hooman, S. Ramesh, and W.-P. de Roever. A compositional axiomatization of Statecharts. *Theoretical Computer Science*, 101:289–335, 1992.

[15] C. Huizing. *Semantics of Reactive Systems: Comparison and Full Abstraction.* PhD thesis, Eindhoven University of Technology, 1991.

[16] F. Levi. *Verification of Temporal and Real-Time Properties of Statecharts.* PhD thesis, University of Pisa-Genova-Udine, 1997.

[17] A. Maggiolo-Schettini, A. Peron, and S. Tini. Equivalences of Statecharts. In *CONCUR '96*, vol. 1119 of *LNCS*, pages 687–702, 1996. Springer Verlag.

[18] F. Maraninchi. Operational and compositional semantics of synchronous automaton compositions. In *CONCUR '92*, vol. 630 of *LNCS*, pages 550–564, 1992. Springer Verlag.

[19] E. Mikk, Y. Lakhnech, C. Petersohn, and M. Siegel. On formal semantics of Statecharts as supported by STATEMATE. In *Second BCS-FACS Northern Formal Methods Workshop*, 1997. Springer Verlag.

[20] E. Mikk, Y. Lakhnech, M. Siegel, and G.J. Holzmann. Verifying Statecharts with Spin. In *WIFT '98*, 1998. IEEE Computer Society Press.

[21] R. Milner. *Communication and Concurrency.* Prentice-Hall, 1989.

[22] A. Pnueli and M. Shalev. What is in a step: On the semantics of Statecharts. In *TACS '91*, vol. 526 of *LNCS*, pages 244–264, 1991. Springer Verlag.

[23] P. Scholz. *Design of Reactive Systems and their Distributed Implementation with Statecharts.* PhD thesis, Munich University of Technology, 1998.

[24] A.C. Uselton and S.A. Smolka. A compositional semantics for Statecharts using labeled transition systems. In *CONCUR '94*, vol. 836 of *LNCS*, pages 2–17, 1994. Springer Verlag.

[25] A.C. Uselton and S.A. Smolka. A process-algebraic semantics for Statecharts via state refinement. In *PROCOMET '94*. North Holland/Elsevier, 1994.

[26] M. v.d. Beeck. A comparison of Statecharts variants. In *FTRTFT '94*, vol. 863 of *LNCS*, pages 128–148, 1994. Springer Verlag.

[27] C. Verhoef. A congruence theorem for structured operational semantics with predicates and negative premises. *Nordic Journal of Computing*, 2:274–302, 1995.

[28] W. Yi. CCS + time = an interleaving model for real time systems. In *ICALP '91*, vol. 510 of *LNCS*, pages 217–228, 1991. Springer Verlag.

A Partial Order Event Model for Concurrent Objects

José Meseguer[1] and Carolyn Talcott[2]

[1] SRI International, Menlo Park CA, 94025 USA,
meseguer@csl.sri.com,
WWW home page: http://www.csl.sri.com/meseguer/meseguer.html
[2] Stanford University, Stanford CA, 94305, USA
clt@cs.stanford.edu,
WWW home page: http://www-formal.cs.stanford.edu/clt/

Abstract. The increasing importance and ubiquity of distributed and mobile object systems makes it very desirable to develop rigorous semantic models and formal reasoning techniques to ensure their correctness. The concurrency model of rewriting logic has been extensively used by a number of authors to specify, execute, and validate concurrent object systems. This model is a *true concurrency* model, associating an algebra of proof terms $\mathcal{T}^o_{\mathcal{R}}$ to the rewrite theory $\mathcal{R}$ specifying the desired system. The elements of $\mathcal{T}^o_{\mathcal{R}}$ are concurrent computations described as *proofs* modulo an equational theory of proof/computation equivalence. This paper builds a very intuitive alternate model $\mathcal{E}_{\mathcal{R}}$, also of a true concurrency nature, but based instead on the notion of concurrent *events* and a causality partial order between such events. The main result of the paper is the equivalence of these two models expressed as an isomorphism. Both models have straightforward extensions to similar models of infinite computations. The models are very general and can express both synchronous and asynchronous object computations. In the asynchronous case the Baker-Hewitt event model for actors appears as a special case of our model.

1 Introduction

The increasing importance and ubiquity of distributed and mobile object systems makes it very desirable to develop rigorous semantic models well-suited to object systems, and to use logically-based techniques to reason about their correctness. In the past few years, a number of authors have used rewriting logic [20, 23] to axiomatize concurrent object systems and have exploited its associated model of concurrency in a variety of ways (see for example [19, 21, 22, 26, 18, 29, 31, 30, 16, 17, 32, 28, 11] and other references in the survey paper [25]).

A concurrent object system axiomatized by a rewrite theory $\mathcal{R}$ has an associated semantic model $\mathcal{T}^o_{\mathcal{R}}$ whose elements are abstract concurrent computations corresponding to equivalence classes of proofs in the logic. Therefore, a computation is possible in the model if and only if it is provable in the logic. The model $\mathcal{T}^o_{\mathcal{R}}$ provides an *algebraic* abstract model of "true concurrency", in which equivalent descriptions of the same concurrent computation are identified. The purpose of this paper is to develop a different semantic model for concurrent object systems $\mathcal{E}_{\mathcal{R}}$ based on the notion of *events* having a partial order causality relationship between them. This second model is very natural and intuitive, and yields as special cases other known models such as the *event diagram* model for Actors [1,4]. The main result of this paper is that both models coincide, in the strong sense of being in fact isomorphic. This validates from an independent perspective the intuitive adequacy of the algebraic model proposed by rewriting logic, and provides a precise semantic bridge between different representations of concurrent object-based computations that, as we discuss further in the paper, can each be more useful in different kinds of applications. In particular, this equivalence of models yields an efficient decision procedure for proof equivalence.

Our model equivalence result for concurrent object systems should be placed within the broader context of related results for other kinds of concurrent systems showing semantic connections between the initial model $\mathcal{T}_{\mathcal{R}}$ of a rewrite theory $\mathcal{R}$ axiomatizing a concurrent system of a given kind, and well-known models of true concurrency. For example, Degano, Meseguer and Montanari [7] showed that, for a place/transition Petri net axiomatized by $\mathcal{R}$, the model $\mathcal{T}_{\mathcal{R}}$ is isomorphic to the commutative process model of Best and Devillers [2]. Similarly, for $\mathcal{R}$ a rewrite theory without additional equational axioms, Corradini, Gadducci and Montanari [5] showed that the computations in $\mathcal{T}_{\mathcal{R}}$ starting at a term t form, under natural assumptions, a prime algebraic domain. Another result by Laneve and Montanari [15] showed that for $\mathcal{R}$ the rewrite theory of the lambda calculus, the traditional model of parallel lambda calculus computation coincides with a quotient model $\mathcal{T}_{\mathcal{R}}/E$ modulo a few natural equations E. Yet another result by Carabetta, Degano and Gadducci [3] shows that, for $\mathcal{R}$ the rewrite theory of CCS, a quotient model $\mathcal{T}_{\mathcal{R}}/E$ under a few natural equations is equivalent to the proved transition causal model of Degano and Priami [9]. Therefore, our result should be seen as a further piece of evidence—involving an important class of concurrent systems—towards the longer-term research project of exploring the naturalness and expressiveness of rewriting logic as a semantic framework for concurrency (see references in [23,25]).

To better motivate the paper and make it more accessible, the rest of this introduction recapitulates in an informal style the main ideas of how rewriting logic axiomatizes concurrent object systems and yields a semantic model $\mathcal{T}_{\mathcal{R}}$ for a system so axiomatized—that for the case of object systems is further refined in this paper to the model $\mathcal{T}^{o}_{\mathcal{R}}$; we then sketch the basic ideas about the new isomorphic model based on a partial order of events.

1.1 Concurrent Objects in Rewriting Logic

We explain how concurrent object systems are axiomatized in rewriting logic, and how concurrent object-based computations correspond to proofs in the logic and yield an algebraic model of "true concurrency". In general, a rewrite theory is a pair $\mathcal{R} = ((\Omega, \Gamma), R)$, with (Ω, Γ) an equational specification with signature of operations Ω and a set of equational axioms Γ; and with R a collection of labelled rewrite rules. The equational specification describes the *static* structure of the system's state space as an algebraic data type. The *dynamics* of the system are described by the rules in R that specify local concurrent *transitions* that can occur in the system axiomatized by $\mathcal{R}$, and that can be applied *modulo* the equations Γ.

Let us then begin explaining how the state space of a concurrent object system can be axiomatized as the initial algebra of an equational theory (Ω, Γ). That is, we explain the key state-building operations in Ω and the equations Γ that they satisfy. The concurrent state of an object-oriented system, often called a *configuration*, has typically the structure of a *multiset* made up of objects and messages. Therefore, we can view configurations as built up by a binary multiset union operator which we can represent with empty syntax (i.e. juxtaposition) as $\_\,\_ : \mathbf{Conf} \times \mathbf{Conf} \longrightarrow \mathbf{Conf}$. (Following the conventions of mix-fix notation, we use _s to indicate argument positions.) The operator $\_\,\_$ is declared to satisfy the structural laws of associativity and commutativity and to have identity $\emptyset$. Objects and messages are singleton multiset configurations, and belong to subsorts $\mathbf{Object}, \mathbf{Msg} < \mathbf{Conf}$, so that more complex configurations are generated out of them by multiset union.

An *object* in a given state is represented as a term $\langle O : C \mid a_1 : v_1, \ldots, a_n : v_n \rangle$, where O is the object's name or identifier, C is its class, the a_i's are the names of the object's *attribute identifiers*, and the v_i's are the corresponding *values*. The set of all the attribute-value pairs of an object state is formed by repeated application of the binary union operator $\_,\_$ which also obeys structural laws of associativity, commutativity, and identity; i.e., the order of the attribute-value pairs of an object is immaterial. This finishes the description of some of the sorts, operators, and equations in the theory (Ω, Γ) axiomatizing the states of a concurrent object system. Particular systems will have additional operations and equations, specifying, for example, the data operations on attribute values. But the top level structure of the concurrent object system is always given by the multiset union operator.

The associativity and commutativity of a configuration's multiset structure make it very fluid. We can think of it as "soup" in which objects and messages float, so that any objects and messages can at any time come together and participate in a concurrent transition corresponding to a communication event of some kind. In general, the rewrite rules in R describing the dynamics of an object-oriented system can have the form

$$
\begin{aligned}
r(\bar{x}) :\quad & M_1 \ldots M_n \, \langle O_1 : F_1 \mid atts_1 \rangle \ldots \langle O_m : F_m \mid atts_m \rangle \\
\longrightarrow\ & \langle O_{i_1} : F'_{i_1} \mid atts'_{i_1} \rangle \ldots \langle O_{i_k} : F'_{i_k} \mid atts'_{i_k} \rangle \\
& \langle Q_1 : D_1 \mid atts''_1 \rangle \ldots \langle Q_p : D_p \mid atts''_p \rangle \\
& M'_1 \ldots M'_q \\
& \textit{if } C
\end{aligned}
$$

where r is the label, $\bar{x}$ is a list of the variables occurring in the rule, the Ms are message expressions, $i_1, \ldots, i_k$ are different numbers among the original $1, \ldots, m$, and C is the rule's condition. That is, a number of objects and messages can come together and participate in a transition in which some new objects may be created, others may be destroyed, and others can change their state, and where some new messages may be created. If two or more objects appear in the left-hand side, we call the rule *synchronous*, because it forces those objects to jointly participate in the transition. If there is only one object in the left-hand side, we call the rule *asynchronous*. For example, we can consider three classes of objects, `Buffer`, `Sender`, and `Receiver`. The buffer stores a list of numbers in its `q` attribute. Lists of numbers are built using an associative list concatenation operator, $\_\,.\,\_$ with identity `nil`, and numbers are regarded as lists of length one. The name of the object reading from the buffer is stored in its `reader` attribute. The sender and receiver objects store a number in a `cell` attribute that can also be empty (`mt`) and have also a counter (`cnt`) attribute. The sender stores also the name of the receiver in an additional attribute. The counter attribute is used to ensure that messages are received by the receiver in the same order as they are sent by the sender even though communication between the two parties is asynchronous. Each time the sender gets a new message from the buffer, it increments its counter. It uses the current value of the counter to tag the message sent to the receiver. The receiver only accepts a message whose tag is its current counter. It then increments its counter indicating that it is ready for the next message. Three typical rewrite rules for objects in these classes (where `E` and `N` range over natural numbers, `L` over lists of numbers, `L.E` is a list with last element `E`, and `(to Z : E from (Y,N))` is a message) are

```
read(X,Y,L,E,N) : < X : Buffer | q: L . E, reader: Y >
                  < Y : Sender | cell: mt, cnt: N >
               => < X : Buffer | q: L, reader: Y >
                  < Y : Sender | cell: E, cnt: N + 1 >

send(Y,Z,E,N) : < Y : Sender | cell: E, cnt: N, receiver: Z >
   => < Y : Sender | cell: mt, cnt: N > (to Z : E from (Y,N))

receive(Z,Y,E,N) : < Z : Receiver | cell: mt, cnt: N >
                     (to Z : E from (Y,N))
                  => < Z : Receiver | cell: E, cnt: N + 1 >
```

where the `read` rule is synchronous and the `send` and `receive` rules asynchronous. These rules are applied *modulo* the associativity and commutativity of the multiset union operator, and therefore allow both object synchronization and message sending and receiving events anywhere in the configuration, regardless of the position of the objects and messages. We can then consider the rewrite theory $\mathcal{R} = ((\Omega, \Gamma), R)$ axiomatizing the object system with these three classes and with R the three rules above (and perhaps other rules, such as one for the receiver to write its contents into another buffer object, that we omit).

Rewriting logic then gives an inference system to deduce, for a system axiomatized by a rewrite theory $\mathcal{R}$, which are all the finitary concurrent computations possible in such a system. Such computations are identified with *proofs* of the general form $\alpha : t \longrightarrow t'$ in the logic. In what follows, to simplify the exposition, we specialize the general inference system to the case of object systems. Idle, or "identity" computations, in which nothing changes in a state t, are denoted by t itself, and elementary rewrites corresponding to the application of a single rule are denoted by the appropriate substitution instance of the rule label; for example, `send(b,c,3,1)` is the rewrite in which sender object `b` sends to `c` the value `3` with counter `1`. More complex computations are then built up by parallel and sequential composition of elementary proofs, according to the following inference system, that specifies both the inferences and the new proof terms associated to proofs α and β.

1. **Congruence**: $\dfrac{\alpha : t \longrightarrow t' \quad \beta : u \longrightarrow u'}{\alpha\,\beta : t\,u \longrightarrow t'\,u'}$.
2. **Transitivity**: $\dfrac{\alpha : t_1 \longrightarrow t_2 \quad \beta : t_2 \longrightarrow t_3}{\alpha; \beta : t_1 \longrightarrow t_3}$.

For example, a buffer object a, and sender and receiver objects b and c can be involved in a concurrent computation in which b reads a value from a and sends it to c, and then, simultaneously, c receives it and b reads a second value from a. Suppose that we begin with the following initial configuration C_0

```
< a : Buffer | q: 7 . 9, reader: b >
< c : Receiver | cell: mt, cnt: 1 >
< b : Sender | cell: mt, cnt: 0, receiver : c >
```

Then, the above concurrent computation can be described by the proof term α, built up by repeated application of congruence and transitivity

```
(read(a,b,7,9,0) < c : Receiver | cell: mt, cnt: 1 >);
(send(b,c,9,1) < c : Receiver | cell: mt, cnt: 1 >
               < a : Buffer | q: 7, reader: b >);
(read(a,b,nil,7,1) receive(c,b,9,1))
```

and has as its final configuration C_1

```
< a : Buffer | q: nil, reader: b >
< b : Sender | cell: 7, cnt: 2, receiver : c >
< c : Receiver | cell: 9, cnt: 2 >
```

This is fine, but when do two different proofs describe the *same* concurrent computation? This typical "true concurrency" question is posed by approaches in which a more abstract description of concurrent computations is sought, and models characterizing such computations are built. For example, in our computation α we could replace the last step, namely `(read(a,b,nil,7,1) receive(c,b,9,1))`, by either of two different "interleaving" proof terms equivalent to it: one in which the `receive` happens after the `read`, and another in which they happen in opposite order.

Rewriting logic is in this sense a "true concurrency" approach. The abstract model giving a semantics to the concurrent computations of a system axiomatized by a rewrite theory $\mathcal{R} = ((\Omega, \Gamma), R)$ is denoted $\mathcal{T}_{\mathcal{R}}$ and is a quotient of the algebra of proof terms modulo some simple equations that express basic equivalences between proofs. The details of this algebraic construction for an arbitrary rewrite theory can be found in [20]. In this paper we give a detailed construction $\mathcal{T}^o_{\mathcal{R}}$ specializing to the case of object-oriented systems and further refining the $\mathcal{T}_{\mathcal{R}}$ model. We can give the flavor of these proof equivalences by pointing out that, firstly, all the equations Γ in $\mathcal{R}$ are also applied to proof terms (in particular, parallel composition of proofs in object systems is associative and commutative, because multiset union enjoys those axioms) and in addition parallel and sequential composition obey, among others, the following three equations that are sufficient for proving that our proof term α above is equivalent to the two proof terms in which the last parallel step has been replaced by interleavings.

1. *Associativity.* $(\alpha;\beta);\gamma = \alpha;(\beta;\gamma)$.
2. *Identities.* For each $\alpha : t \longrightarrow t'$, $\alpha; t' = \alpha$ and $t;\alpha = \alpha$.
3. *Functoriality.* $(\alpha_1;\beta_1)(\alpha_2;\beta_2) = (\alpha_1\ \alpha_2);(\beta_1\ \beta_2)$.

By identifying proof terms using equations such as these, the model $\mathcal{T}_{\mathcal{R}}$ characterizes abstract concurrent computations in an object-oriented system as equivalence classes $[\alpha]$ of proof terms α.

1.2 A Partial Order Event Model

The model $\mathcal{T}_{\mathcal{R}}$ is algebraic and enjoys many of the good properties of algebraic constructions, including parallel and sequential composition operations and being initial in a much broader category of models [20]. By contrast, other approaches to true concurrency build partial-order or Petri net-like models that are more topological in nature, e.g. [1, 2, 13, 8, 6]. They complement the description given by an algebraic construction by offering a more intuitive description of the abstract computations. The ideal result one would like to have is one of complete *equivalence* between two true concurrency models: one based on an algebraic construction, and another based on a topological construction. In this way, we can freely move back and forth between equivalent descriptions that may each have strong advantages for different purposes. For example, we can visualize a computation using its topological description and can do algebraic manipulations with a corresponding proof term. In addition, in the process of proving such an equivalence, the topological model typically becomes endowed with an important *algebraic* structure that can be very valuable.

For the case of rewrite theories $\mathcal{R}$ axiomatizing Place/Transition Petri nets, the equivalence of $\mathcal{T}_{\mathcal{R}}$ with a well-known combinatorial model of "true concurrency" for nets—namely the commutative processes of Best and Devillers [2]—has been shown in [7] . The main result of this paper is a similar equivalence of models for concurrent object systems, namely, an equivalence between an algebraic model $\mathcal{T}_{\mathcal{R}}^{o}$ of proof terms modulo a few natural equations—that specializes to object systems the general model $\mathcal{T}_{\mathcal{R}}$—and a very natural and intuitive partial order of events model $\mathcal{E}_{\mathcal{R}}$ that is fully general, in the sense of allowing both synchronous and asynchronous computations. Furthermore, the models $\mathcal{T}_{\mathcal{R}}^{o}$ and $\mathcal{E}_{\mathcal{R}}$ have a straightforward extension to models of *infinite* computations $\mathcal{T}_{\mathcal{R}}^{\infty}$ and $\mathcal{E}_{\mathcal{R}}^{\infty}$ that are again isomorphic. In fact, for rewrite theories $\mathcal{R}$ whose rules are all asynchronous and obey the actor locality laws (cf. [30]), $\mathcal{E}_{\mathcal{R}}^{\infty}$ specializes to the well-known partial order of events model of Hewitt and Baker for Actor systems [1, 4].

Mathematically, we prove our desired equivalence of models as an isomorphism $\mathcal{T}_{\mathcal{R}}^{o} \cong \mathcal{E}_{\mathcal{R}}$ of the corresponding algebraic structures. We can give the flavor of $\mathcal{E}_{\mathcal{R}}$ by showing in Figure 1 the partial order of events corresponding in $\mathcal{E}_{\mathcal{R}}$ to the equivalence class $[\alpha]$ of proof terms for the proof term α in our example. In $\mathcal{T}_{\mathcal{R}}^{o}$ the equivalence class of proofs $[\alpha]$ labels an arrow between the beginning and ending configurations C_0 and C_1. In $\mathcal{E}_{\mathcal{R}}$ the same arrow is instead labeled by the corresponding partial order of events. Note that the events are the elementary rewrites corresponding to applying each one of the rules, and the order between them is the *causality* relation between such events. Thus, the first `send` and the `receive` events are causally connected, but the second `read` and the `receive` are unrelated in the causal partial order.

The rest of the paper is organized as follows. §2 introduces *object theories*, a general class of rewrite theories capturing the essential aspects of concurrent object systems. §3 gives a detailed construction of the proof term model $\mathcal{T}_{\mathcal{R}}^{o}$ for an object rewrite theory $\mathcal{R}$, and studies some of its key properties. §4 then gives our main contribution, namely the construction of the partial order of events model $\mathcal{E}_{\mathcal{R}}$ for an object rewrite theory $\mathcal{R}$ and the isomorphism $\mathcal{T}_{\mathcal{R}}^{o} \cong \mathcal{E}_{\mathcal{R}}$. §5 extends this isomorphism to an isomorphism

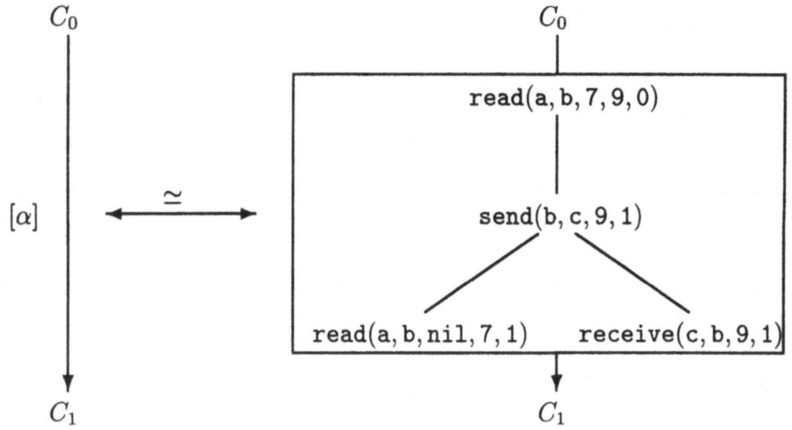

Fig. 1. Equivalence between proof and event partial order descriptions of a concurrent computation.

of corresponding models of infinite computations $\mathcal{T}_{\mathcal{R}}^{\infty} \cong \mathcal{E}_{\mathcal{R}}^{\infty}$. We finish with some concluding remarks.

2 Object Theories

In this section we describe the class of rewrite theories, called object theories, for which the event model is defined. We consider rewrite theories whose underlying equational logic is *membership equational logic* [24]. A membership equational logic specification has the form (Ω, Γ) where Ω contains a family K of *kinds*, a K-kinded family of operators of the form $f : k_1 \dots k_n \longrightarrow k_{n+1}$ with $k_i \in K$, and for each $k \in K$ a set of *sorts* S_k; and where Γ is a set of Horn clauses whose atomic formulas are either equations $t = t'$ between terms of kind k, or membership assertions $t : s$ between a term of kind k and a sort $s \in S_k$. Intuitively, terms with a kind but without a sort are "error" or "undefined" expressions. Subsort relations $s \leq s'$, and operator declarations $f : s_1 \dots s_n \longrightarrow s_{n+1}$ at the level of sorts can be seen as syntactic sugar for their corresponding Horn clauses.

For a rewrite theory $\mathcal{R} = ((\Omega, \Gamma), R)$ we first describe the constraints on the equational part (Ω, Γ), then we describe the constraints on the rules R. Object theories generalize the typical rewrite theories for object-oriented systems. It may be that not all multiset unions of configurations are meaningful. For example consider object systems in which legal configurations should not have different objects with the same identity. To account for this possibility, we postulate a subsort $\mathbf{Coh} < \mathbf{Conf}$ of "coherent" configurations characterizing the meaningful system states. The introduction of object identities is postponed until § 4.

2.1 Equational Part

(Ω, Γ) is a theory in membership equational logic such that there is a distinguished sort **Conf** of *configurations* with

(i) a binary operation _ _ on **Conf** that is associative and commutative with neutral element $\emptyset$, and such that there is no other operation in Ω that can take elements of sort **Conf** as arguments.
(ii) a subsort **Coh** < **Conf** called the *coherent configurations* satisfying the membership constraints $\emptyset : \mathbf{Coh}$ and $U\ V : \mathbf{Coh} \Rightarrow U : \mathbf{Coh} \wedge V : \mathbf{Coh}$.
(iii) for any set X of variables, the elements of sort **Conf** in the free algebra $T_{\Omega,\Gamma}(X)$ form a free multiset on the variables and "alien subterms" (that is, subterms whose top symbol is different from _ _ or $\emptyset$) of sort **Conf**, under the operation _ _ and neutral element $\emptyset$.

Remark. Since (in membership equational logic) for each sort S and terms t, t', whenever $\Gamma \vdash t : S$ and $\Gamma \vdash t = t'$, then $\Gamma \vdash t' : S$, we have, thanks to (i) and (ii), that for each $U, V, W : \mathbf{Conf}\ (U\ V)\ W : \mathbf{Coh}$ iff $U\ (V\ W) : \mathbf{Coh}$ and then U, V, W, $U\ V$, $V\ W$, and $U\ W$, all have sort **Coh**. Furthermore, thanks to (iii), if $\Gamma \vdash U\ V : \mathbf{Coh} \wedge U\ W : \mathbf{Coh} \wedge U\ V = U\ W$, then $\Gamma \vdash V = W$.

2.2 Rules Part

For a rewrite theory $\mathcal{R} = ((\Omega, \Gamma), R)$ whose equational part (Ω, Γ) satisfies the conditions of §2.1 we further require that rules only rewrite non-empty configurations and that ground coherent configurations are rewritten to coherent configurations. In particular we require the following.

(i) All rules have the form $r : Z \longrightarrow Z'$ if ψ , where ψ is a conjunction of equations, $Z, Z' : \mathbf{Conf}$, $Z \neq \emptyset$, and Z, Z' contain no variables of sort **Conf**.
(ii) For a ground term W such that $\Gamma \vdash W : \mathbf{Coh}$, if θ is a ground substitution and $r : Z \longrightarrow Z'$ if ψ is a rule in R such that $\theta(\psi)$ holds, that is $\Gamma \vdash \theta(\psi)$, and

$$W = U\ \theta(Z) \xrightarrow{U\ r(\theta)} U\ \theta(Z'),$$

then $\Gamma \vdash U\ \theta(Z') : \mathbf{Coh}$. That is, the rewrite rules in R always rewrite ground terms in **Coh** to other ground terms in **Coh**.

The above requirements for the equational and rules parts of an object theory are for example satisfied by the buffer, sender, and receiver object theory of Section 1, by taking as sort **Coh** of coherent configurations multisets of objects and messages that are actually sets and that satisfy natural object theory requirements such as uniqueness of object identifiers, as well as specific invariants such as messages originating from a given sender having a counter strictly smaller than that of the sender, and so on.

3 The Proof Term Model $\mathcal{T}^o_{\mathcal{R}}$

3.1 Algebra of Proof Terms

To each object theory $\mathcal{R}$ we associate an algebra of proof terms $\mathcal{T}^o_{\mathcal{R}}$. This proof term algebra is specified in *partial* membership equational logic [24]. Partial membership equational logic is the variant of membership equational logic in which, instead of kinds, we have a poset of sorts $(S, \leq)$ where each connected component C has a top element $\top_C$, operations $f : \top_{C_1} \ldots \top_{C_n} \longrightarrow \top_{C_{n+1}}$ are interpreted as partial functions, and Horn clauses have a partial interpretation.

The proof term algebra is given by means of the theory $\mathcal{P}_{\mathcal{R}} = (\Omega', \Gamma')$ where (noting that a specification in total membership algebra can always be regarded as a partial membership algebra specification) $(\Omega, \Gamma) \subseteq (\Omega', \Gamma')$ and the additional sorts and operations of Ω' are as follows.

(1) There is a sort $\mathbf{Prf}$ of proofs with $\mathbf{Coh} < \mathbf{Prf}$ such that
   - total operations $source, target : \mathbf{Prf} \to \mathbf{Coh}$ such that $source(U) = U$ and $target(U) = U$ for $U : \mathbf{Coh}$; given $\alpha : \mathbf{Prf}$ with $source(\alpha) = U$ and $target(\alpha) = V$ we use the notation $U \xrightarrow{\alpha} V$ to simultaneously state these three facts,
   - there is a partial operation $\_;\_ : \mathbf{Prf} \times \mathbf{Prf} \to \mathbf{Prf}$,
   - the partial operation $\_\ \_$ on $\mathbf{Coh}$ is extended to a partial operation $\_\ \_ : \mathbf{Prf} \times \mathbf{Prf} \to \mathbf{Prf}$.

(2) For each rule $r : Z \longrightarrow Z'$ if ψ in R with variables $x_1, \ldots, x_n$ of respective sorts $s_1, \ldots, s_n$, there is a partial operation $r : s_1, \ldots, s_n \to \mathbf{Prf}$ such that for any substitution θ, if $\theta(Z) : \mathbf{Coh}$ and $\theta(\psi)$ holds, then

$$r(\theta) : \mathbf{Prf},\quad source(r(\theta)) = \theta(Z),\quad target(r(\theta)) = \theta(Z'),$$

that is, $\theta(Z) \xrightarrow{r(\theta)} \theta(Z')$. We assume that the variables $x_1, \ldots, x_n$ of rule r are ordered; $r(\theta)$ then abbreviates the term $\theta(r(x_1, \ldots, x_n))$.

(3) If $U, U', V, V' : \mathbf{Coh}$ and $\alpha, \beta : \mathbf{Prf}$ such that $U \xrightarrow{\alpha} U'$, $V \xrightarrow{\beta} V'$, and $U\ V : \mathbf{Coh}$, then $U\ V \xrightarrow{\alpha\ \beta} U'\ V'$.

(4) If $U \xrightarrow{\alpha} V$ and $V \xrightarrow{\beta} W$, then $U \xrightarrow{\alpha;\beta} W$.

The additional equations of Γ' are the following.

A. Category

(id) $\quad Z \xrightarrow{\alpha} Z' \Rightarrow Z;\alpha = \alpha; Z' = \alpha$

(assoc(_;_)) $\quad Z_0 \xrightarrow{\alpha_0} Z_1 \wedge Z_1 \xrightarrow{\alpha_1} Z_2 \wedge Z_2 \xrightarrow{\alpha_2} Z_3$

$$\Rightarrow (\alpha_0; \alpha_1); \alpha_2 = \alpha_0; (\alpha_1; \alpha_2)$$

B. Partial Monoidal Structure

(funct) $Z_0 \xrightarrow{\alpha_0} Z_1 \xrightarrow{\alpha_1} Z_2 \wedge Z'_0 \xrightarrow{\alpha'_0} Z'_1 \xrightarrow{\alpha'_1} Z'_2 \wedge Z_0\ Z'_0 : \mathbf{Coh}$

$$\Rightarrow (\alpha_0; \alpha_1)\ (\alpha'_0; \alpha'_1) = (\alpha_0\ \alpha'_0); (\alpha_1\ \alpha'_1)$$

(id-$\emptyset$) $\emptyset\ \alpha = \alpha \quad \text{for} \quad \alpha : \mathbf{Prf}$

(comm(_ _)) $\bigwedge_{j<2} (Z_j \xrightarrow{\alpha_j} Z'_j) \wedge (Z_0\ Z_1) : \mathbf{Coh} \Rightarrow (\alpha_0\ \alpha_1) = (\alpha_1\ \alpha_0)$

(assoc(_ _)) $\bigwedge_{j<3} (Z_j \xrightarrow{\alpha_j} Z'_j) \wedge (Z_0\ Z_1\ Z_2) : \mathbf{Coh} \Rightarrow (\alpha_0\ \alpha_1)\ \alpha_2 = \alpha_0\ (\alpha_1\ \alpha_2)$

3.2 Basic Properties

A proof term is in sequential form if it is a sequence of basic rewrites, $U_i\ r_i(\theta_i)$. Lemma 1 says that any proof term is equivalent to one in sequential form.

Lemma 1 (Sequentialization). If α is a ground Ω'-term such that $Z \xrightarrow{\alpha} Z'$, then we can find $n \in \mathbf{Nat}$ and $\alpha_i, r_i(\theta_i) : \mathbf{Prf},\ U : \mathbf{Coh}$ for $1 \leq i \leq n$ such that $\alpha_i = U_i\ r_i(\theta_i)$ and $\Gamma' \vdash \alpha = \alpha_1; \ldots; \alpha_n$. Furthermore, the multiset of elementary rewrites $r_i(\theta_i)$ is independent of the particular sequentialization.

Lemma 2 says that in passing from (Ω, Γ) to (Ω', Γ'), no new coherent configurations are introduced (no junk) and no Γ-equivalence classes of coherent configurations are collapsed by Γ' (no confusion).

Lemma 2 (Protection). Let $\mathcal{R} = ((\Omega, \Gamma), R)$ be an object theory with $\mathbf{Coh}$ its sort of coherent configurations, and let $\mathcal{P}_\mathcal{R} = (\Omega', \Gamma')$ be its associated proof term theory. Then for each Ω'-ground term U such that $\Gamma' \vdash U : \mathbf{Coh}$ there exists an Ω-ground term V such that $\Gamma \vdash V : \mathbf{Coh}$ and $\Gamma' \vdash U = V$. Also, for U, V Ω-ground terms such that $\Gamma \vdash U, V : \mathbf{Coh}$, we have $\Gamma' \vdash U = V \Leftrightarrow \Gamma \vdash U = V$.

Definition 1 (The proof category $\mathcal{T}^o_\mathcal{R}$). If $\mathcal{R} = ((\Omega, \Gamma), R)$ is an object theory with coherent configuration sort $\mathbf{Coh}$ and proof term theory $\mathcal{P}_\mathcal{R}$ as given in §3.1 with proof sort $\mathbf{Prf}$, then $\mathcal{T}^o_\mathcal{R}$ is the partial monoidal category with set of objects $|\mathcal{T}^o_\mathcal{R}|$ the ground Ω-terms of sort $\mathbf{Coh}$ modulo the equations Γ, or equivalently, by lemma 2, Ω'-ground terms of sort $\mathbf{Coh}$ modulo the equations of Γ', and arrows the ground Ω'-terms of sort $\mathbf{Prf}$ modulo the equations Γ' on proofs.

We call $\mathcal{T}^o_\mathcal{R}$ partial monoidal because the operation _ _ satisfies the typical equations of a (symmetric) monoidal product but is only a partial operation. In other words, the specification $\mathcal{P}_\mathcal{R}$ has an initial partial algebra $\mathcal{T}_{\mathcal{P}_\mathcal{R}}$ [24] which has a partial monoidal category structure for terms of sort $\mathbf{Prf}$. The category $\mathcal{T}^o_\mathcal{R}$ is obtained by restricting $\mathcal{T}_{\mathcal{P}_\mathcal{R}}$ to just those partial monoidal category operations and forgetting the additional sorts and algebraic structure. Alternatively, we can regard $\mathcal{T}^o_\mathcal{R}$ as a natural refinement and specialization to object theories and coherent configurations of the initial model $\mathcal{T}_\mathcal{R}$ associated to a rewrite theory $\mathcal{R}$ [20].

4 Associating Event Partial Orders to Proofs

Assume $\mathcal{R} = ((\Omega, \Gamma), R)$ is an object theory as defined in §2, and $\mathcal{T}^o_{\mathcal{R}}$ the associated proof category as given in §3. We refine the notion of object theory to objects with identities by postulating two more requirements. We want to treat occurrences of rewrite rule applications in a proof as events, each of which can be uniquely identified (condition 1). Furthermore in order to be able to identify causal connections between events due for example to the asynchronous sending and receiving of messages, not only objects, but also messages need to have unique identity (condition 2).

Definition 2 (Object theory with identities).
1. Assume further that all occurrences of a rule application in a proof are necessarily distinct. Using Lemma 1 we can express this as follows. Let $Z_0 \xrightarrow{\alpha} Z_n$ be a proof, and let

$$Z_0 \xrightarrow{U_1\ r_1(\theta_1)} \dots \xrightarrow{U_n\ r_n(\theta_n)} Z_n$$

be any sequentialization of α. Then for $0 \leq i, j \leq n$, we have $\mathcal{T}^o_{\mathcal{R}} \models r_i(\theta_i) \neq r_j(\theta_j)$ whenever $i \neq j$. Note that this condition can easily be attained by suitable annotation of objects and messages. Because of this property, we can regard each rewrite rule application $r_i(\theta_i)$ as a distinct *event* in the overall computation formalized as a proof.

2. Assume given a set **Id** of identities, and a function $ids : |\mathcal{T}^o_{\mathcal{R}}| \to \mathbf{P}_\omega(\mathbf{Id})$ such that

(1) $Z : \mathbf{Coh}$ implies $ids(Z) = \emptyset$ iff $Z = \emptyset$,
(2) $Z\ Z' : \mathbf{Coh}$ implies $ids(Z) \cap ids(Z') = \emptyset$ and $ids(Z\ Z') = ids(Z) \cup ids(Z')$.

Definition 3 (Event ids, $ids(r(\theta))$). For basic rewrites $r(\theta)$ define

$$ids(r(\theta)) = ids(source(r(\theta))) \cup ids(target(r(\theta))).$$

Now we associate to each ground term α of **Prf** a structure $[\![\alpha]\!] = \langle \mathcal{E}_\alpha, <_\alpha \rangle$ where $\mathcal{E}_\alpha$ is the set of events (the rewrite rule applications) of α and $<_\alpha$ is a partial order on these events. The structure $[\![\alpha]\!]$ is defined by induction on the generation of α as follows.

Definition 4 ($[\![\alpha]\!]$).

(id) $[\![U]\!] = \langle \emptyset, \emptyset \rangle$
(rule) $[\![r(\theta)]\!] = \langle \{r(\theta)\}, \emptyset \rangle$
(par) $[\![\alpha\ \beta]\!] = [\![\alpha]\!]\ [\![\beta]\!] = [\![\alpha]\!] \cup [\![\beta]\!]$, where $\langle \mathcal{E}_\alpha, <_\alpha \rangle \cup \langle \mathcal{E}_\beta, <_\beta \rangle = \langle \mathcal{E}_\alpha \cup \mathcal{E}_\beta, <_\alpha \cup <_\beta \rangle$[1]
(seq) $[\![\alpha; \beta]\!] = [\![\alpha]\!]; [\![\beta]\!] = \langle \mathcal{E}_\alpha \cup \mathcal{E}_\beta, <_{\alpha;\beta} \rangle$, where
$<_{\alpha;\beta} = \mathrm{TC}(<_\alpha \cup <_\beta \cup \{e_0 < e_1 \mid e_0 \in \mathcal{E}_\alpha \wedge e_1 \in \mathcal{E}_\beta \wedge ids(e_0) \cap ids(e_1) \neq \emptyset\})$
and TC(_) is the transitive closure operation.

Note that the condition $ids(e_0) \cap ids(e_1) \neq \emptyset$ in the (seq) case above is precisely the point where causal connections between events appears. $e_0 < e_1$ just if e_0 precedes e_1 in some sequentialization and if they involve a common identity (object or message).

[1] Note that, by condition (1) and Lemma 1, $\mathcal{E}_\alpha$ and $\mathcal{E}_\beta$ are disjoint, and hence the union as defined is a partial order.

Definition 5 (The event category $\mathcal{E}_{\mathcal{R}}$). If $\mathcal{R} = ((\Omega, \Gamma), R)$ is an object theory with identities with coherent configuration sort **Coh** and proof term theory $\mathcal{P}_{\mathcal{R}}$ as given in §3.1 with proof sort **Prf**, then $\mathcal{E}_{\mathcal{R}}$ is the partial monoidal category with set of objects $|\mathcal{E}_{\mathcal{R}}|$ the ground Ω-terms of sort **Coh** modulo the equations Γ, and arrows $[\![\alpha]\!] : U \longrightarrow V$, for α a ground term of sort **Prf**, such that $U = \mathit{source}(\alpha)$, and $V = \mathit{target}(\alpha)$. The sequential composition and monoidal product operations on arrows are the operations $\_;\_$ and $\_\,\_$ given in definition 4 above.

The main result of this section is the isomorphism between the proof category $\mathcal{T}^o_{\mathcal{R}}$ and the category $\mathcal{E}_{\mathcal{R}}$ of event partial orders associated to an object theory $\mathcal{R}$.

Theorem 1 (Proof-event isomorphism). Let $\mathcal{R} = ((\Omega, \Gamma), R)$ be an object theory with identities. Then, the partial monoidal categories $\mathcal{T}^o_{\mathcal{R}}$ and $\mathcal{E}_{\mathcal{R}}$ are isomorphic.

Proof: The isomorphism is the identity on objects and associates to each proof equivalence class $[\alpha]$ the event partial order $[\![\alpha]\!]$. That this is indeed an isomorphism follows from lemmas 6 and 7 below. □

Before proving lemmas 6 and 7 we establish some basic properties of events and the causal ordering. Lemma 3 gives a simple characterization of the event partial order associated to a proof in sequential form.

Lemma 3. If $\alpha = U_1\ r_1(\theta_1); \ldots ; U_k\ r_k(\theta_k)$, then $[\![\alpha]\!] = \langle \{r_1(\theta_1), \ldots, r_k(\theta_k)\}, < \rangle$ where $<$ is the transitive closure of $\{(r_i(\theta_i), r_j(\theta_j)) \mid i < j, \mathit{ids}(r_i(\theta_i)) \cap \mathit{ids}(r_j(\theta_j)) \neq \emptyset\}$.

Lemma 4 shows that adjacent events may be permuted if not causally related, that is, if they act on different parts of the configuration.

Lemma 4. If $U_0\ r_0(\theta_0); U_1\ r_1(\theta_1) :$ **Prf** such that $\mathit{ids}(r_0(\theta_0)) \cap \mathit{ids}(r_1(\theta_1)) = \emptyset$ (i.e. $r_1(\theta_1)$ is minimal in $[\![U_0\ r_0(\theta_0); U_1\ r_1(\theta_1)]\!]$) then there are $V_0, V_1 :$ **Coh** such that $U_0\ r_0(\theta_0); U_1\ r_1(\theta_1) = V_1\ r_1(\theta_1); V_0\ r_0(\theta_0)$.

Lemma 5 shows that sequentializations of a proof correspond to linearizations of the associated event partial order.

Lemma 5. If $\alpha :$ **Prf**, then every linearization of $[\![\alpha]\!]$ corresponds to a sequentialization of α.

Proof: By induction on the generation of α using Lemma 4. □

Lemma 6 (proof-event.1). If $\Gamma' \vdash \alpha = \alpha'$, then $[\![\alpha]\!] = [\![\alpha']\!]$.

Proof: By induction on the proof of equality from the equational rules of §3. The logical rules are easy to check and it remains to check each of the equational axioms. The details are omitted here. □

Lemma 7 (proof-event.2). If $[\![\alpha]\!] = [\![\alpha']\!]$ and $\mathit{source}(\alpha) = \mathit{source}(\alpha')$, then $\Gamma' \vdash \alpha = \alpha'$.

5 Infinite Computations and Event Models

We extend the proof and event partial order models of object theories to infinite computations as follows.

Let $\mathcal{R}$ be an object rewrite theory with associated proof term theory $\mathcal{P}_{\mathcal{R}}$. An infinite proof (or computation) π is simply an infinite sequence $\{\alpha_i \mid i \in \mathbf{Nat}\}$ of ground proof terms $\alpha_i : \mathbf{Prf}$ such that $target(\alpha_i) = source(\alpha_{i+1})$ for $i \in \mathbf{Nat}$. We call these infinite proofs *proof paths* or just *paths*. A path π is in *sequential form* if each step $\pi(i)$ contains at most one event (its sequentialization has length 0 or 1).

We denote by $\mathbf{Prf}^{\infty}$ the set of proof paths. The initial segment at i of a path π, written $\pi{\downarrow}_i$, is the sequential composition of the path elements $\pi(j)$ for $j \leq i$. This is defined by induction on i as follows:

$$\pi{\downarrow}_0 = \pi(0)$$
$$\pi{\downarrow}_{i+1} = \pi{\downarrow}_i ; \pi(i+1)$$

We define *prefix* orderings $\_ \preceq \_$ on proof terms and on event partial orders as follows.

Definition 6. The prefix ordering on proof terms

$$\alpha \preceq \beta \Leftrightarrow (\exists \alpha' : \mathbf{Prf})(\Gamma' \vdash \alpha ; \alpha' = \beta).$$

Note that an alternative and equivalent representation of infinite computations is as an increasing infinite sequence of finite computations using the $\preceq$ ordering.

The corresponding prefix ordering on event partial orders is the initial segment relation given by

$$\langle \mathcal{E}, < \rangle \preceq \langle \mathcal{E}', <' \rangle$$
$$\Leftrightarrow \mathcal{E} \subseteq \mathcal{E}' \wedge (\forall e, e' \in \mathcal{E})(e < e' \Leftrightarrow e <' e') \wedge (\forall e \in \mathcal{E}, e' \in \mathcal{E}')(e' <' e \Rightarrow e' \in \mathcal{E}).$$

Lemma 8. For $U \xrightarrow{\alpha} V$ and $U \xrightarrow{\beta} W$, we have $\alpha \preceq \beta$ iff $[\![\alpha]\!] \preceq [\![\beta]\!]$.

Two paths π and π' are equivalent, written $\pi \cong \pi'$, iff for each initial segment of π there is an initial segment of π' that extends it, and conversely.

$$\pi \sqsubseteq \pi' \Leftrightarrow (\forall i \in \mathbf{Nat})(\exists j \in \mathbf{Nat})(\pi{\downarrow}_i \preceq \pi'{\downarrow}_j),$$
$$\pi \cong \pi' \Leftrightarrow \pi \sqsubseteq \pi' \wedge \pi' \sqsubseteq \pi.$$

We define $\mathcal{T}^{\infty}_{\mathcal{R}}$ to be the quotient of $\mathbf{Prf}^{\infty}$ under this equivalence relation.

The following lemma shows that, modulo the path equivalence relation, we can restrict attention to paths in sequential form.

Lemma 9. For any $\pi \in \mathbf{Prf}^{\infty}$ we can find $\pi' \in \mathbf{Prf}^{\infty}$ such that π' is in sequential form and $\pi \cong \pi'$.

The mapping from proofs to event partial orders extends to paths as follows. To assure the desired isomorphism, we associate with each path a pair consisting of the initial configuration of the path and the associated event partial order. For $\pi \in \mathbf{Prf}^{\infty}$,

$$[\![\pi]\!] = (source(\pi(0)), \bigcup_{i \in \mathbf{Nat}} [\![\pi{\downarrow_i}]\!])$$

that is,

$$[\![\pi]\!] = (U, \langle \mathcal{E}_\pi, <_\pi \rangle)$$

where $U = source(\pi(0))$ $\mathcal{E}_\pi = \bigcup_{i \in \mathbf{Nat}} \mathcal{E}_{\pi\downarrow_i} = \bigcup_{i \in \mathbf{Nat}} \mathcal{E}_{\pi(i)}$, and $<_\pi = \bigcup_{i \in \mathbf{Nat}} <_{\pi\downarrow_i}$. Thus, $e <_\pi e'$ just if $e, e' \in \mathcal{E}_i$ for some $i \in \mathbf{Nat}$ and $e <_{\pi\downarrow_i} e'$. Note that, by Lemma 8, $<_\pi$ is indeed a partial order.

We define $\mathcal{E}_{\mathcal{R}}^{\infty}$ to be the image of $\mathcal{T}_{\mathcal{R}}^{\infty}$ under this mapping. We now extend the isomorphism $\mathcal{T}_{\mathcal{R}}^{o} \cong \mathcal{E}_{\mathcal{R}}$ to a similar result for infinite computation paths and infinite partial orders.

Theorem 2. For any $\pi, \pi' : \mathbf{Prf}^{\infty}$, $\pi \cong \pi' \Leftrightarrow [\![\pi]\!] = [\![\pi']\!]$.

6 Concluding Remarks

This paper has shown the equivalence between the algebraic proof term model for object rewrite theories, and a natural partial order of events model for such theories. This equivalence takes the form of an effective isomorphism between the two models that can be used in practice in a number of ways. For example, proof terms describing concurrent computations that can be easily generated during an execution of an object rewrite theory can now be visualized by their corresponding partial order of events. Furthermore, proof equivalence can be decided in $O(n^2 \times m)$ time by comparing the associated event partial orders, where n is the number of events, and m the size of the biggest event as a term.

In a similar way, the equivalence between proof and partial order descriptions can be used to check the consistency of object-oriented design descriptions written in different diagrammatic notations. Work of Knapp and Wirsing [32] has shown how object theories can be associated to UML-like notations. It now becomes possible to check that a given sequence diagram — that is essentially a partial order of events description — for a system is consistent with such an object theory, by checking whether the corresponding proof term is a valid proof in the logic.

Using techniques similar to those used in [5], our partial order of events model could be cast in an event structure formulation by viewing the computations starting from an object-oriented configuration C as forming a prime algebraic domain under the order $\alpha \leq \beta$ iff there is a γ such that $\beta = \alpha; \gamma$. Due to space limitations, the details will appear in a full version of the paper.

Our event model also has interesting connections with other formalisms that should be further explored. For example, the distributed temporal logic of objects recently proposed by Denker [10] is explicitly based on our model, and the temporal logic proposed by Katz [14] assumes a partial order true concurrency model closely related to ours. The axiomatization of the actor model in temporal logic of Duarte [12] is being extended to also axiomatize actor event diagrams. Similarly, in the area of model checking there has been growing interest in partial order descriptions that can reduce the search space (see for example [27]).

Acknowledgements. We thank Grit Denker, Steven Eker, Narciso Martí-Oliet, Ugo Montanari and several anonymous referees for their valuable suggestions for improving previous versions of this paper. This research was partially supported by ONR grant N00014-94-1-0857, NSF grants CRR-9633419 and CCR-9633363, DARPA/Rome Labs grant AF F30602-96-1-0300, and DARPA/NASA contract NAS2-98073.

References

1. H. G. Baker and C. Hewitt. Laws for communicating parallel processes. In *IFIP Congress*, pages 987–992. IFIP, Aug. 1977.
2. E. Best and R. Devillers. Sequential and concurrent behavior in Petri net theory. *Theoretical Computer Science*, 55:87–136, 1989.
3. G. Carabetta, P. Degano, and F. Gadducci. CCS semantics via proved transition systems and rewriting logic. *Proc. 2nd Intl. Workshop on Rewriting Logic and its Applications*, ENTCS, North Holland, 1998.
4. W. D. Clinger. Foundations of actor semantics. AI-TR- 633, MIT Artificial Intelligence Laboratory, May 1981.
5. A. Corradini, F. Gadducci, and U. Montanari. Relating two categorical models of term rewriting. In J. Hsiang, editor, *Rewriting Techniques and Applications*, volume 914 of *Lecture Notes in Computer Science*, pages 225–240, 1995.
6. P. Darondeau and P. Degano. Causal trees. In M. Ausiello, G. DezaniCiancaglini and S. Ronchi Della Rocca, editors, *16th ICALP*, Lecture Notes in Computer Science, pages 234–248. Springer-Verlag, 1989.
7. P. Degano, J. Meseguer, and U. Montanari. Axiomatizing the algebra of net computations and processes. *Acta Informatica*, 33:641–667, 1996.
8. P. Degano and U. Montanari. Concurrent histories, a basis for observing distributed systems. *J. of Computer and System Sciences*, 34:422–461, 1987.
9. P. Degano and C. Priami. Proved trees. In *Proc. ICALP'92*, pages 629–640. Springer LNCS 623, 1992.
10. G. Denker. From rewrite theories to temporal logic theories: A distributed temporal logic extension of rewriting logic. In C. Kirchner and H. Kirchner, editors, *2nd International Workshop on Rewriting Logic and its Applications, WRLA'98*, volume 15 of *Electronic Notes in Theoretical Computer Science*, 1998. URL: `http://www.elsevier.nl/locate/entcs/volume15.html`.
11. G. Denker, J. Meseguer, and C. Talcott. Protocol specification and analysis in Maude. In N. Heintze and J. Wing, editors, *Workshop on Formal Methods and Security Protocols, 25 June 1998, Indianapolis, Indiana*, 1998.
12. C. H. C. Duarte. *Proof-theoretic Foundations for the Design of Extensible Software Systems*. PhD thesis, Imperial College, University of London, 1999.
13. U. Goltz and W. Reisig. The nonsequential behavior of petri nets. *Information and Computation*, 57:125–147, 1983.
14. S. Katz. Refinement with global equivalence proofs in temporal logic. In *DIMACS Series in Discrete Mathematics, Vol. 29*, pages 59–78. American Mathematical Society, 1997.
15. C. Laneve and U. Montanari. Axiomatizing permutation equivalence. *Mathematical Structures in Computer Science*, 6:219–249, 1996.

16. U. Lechner, C. Lengauer, F. Nickl, and M. Wirsing. How to overcome the inheritance anomaly. in *Proc.ECOOP'96*, Springer LNCS, 1996.
17. U. Lechner, C. Lengauer, and M. Wirsing. An object-oriented airport. In E. Astesiano, G. Reggio, and A. Tarlecki, editors, *Recent Trends in Data Type Specification, Santa Margherita, Italy, May/June 1994*, pages 351–367. Springer LNCS 906, 1995.
18. P. Lincoln, N. Martí-Oliet, and J. Meseguer. Specification, transformation, and programming of concurrent systems in rewriting logic. In G. Blelloch, K. Chandy, and S. Jagannathan, editors, *Specification of Parallel Algorithms*, pages 309–339. DIMACS Series, Vol. 18, American Mathematical Society, 1994.
19. J. Meseguer. A logical theory of concurrent objects. In *ECOOP-OOPSLA'90 Conference on Object-Oriented Programming, Ottawa, Canada, October 1990*, pages 101–115. ACM, 1990.
20. J. Meseguer. Conditional rewriting logic as a unified model of concurrency. *Theoretical Computer Science*, 96(1):73–155, 1992.
21. J. Meseguer. A logical theory of concurrent objects and its realization in the Maude language. In G. Agha, P. Wegner, and A. Yonezawa, editors, *Research Directions in Concurrent Object-Oriented Programming*, pages 314–390. MIT Press, 1993.
22. J. Meseguer. Solving the inheritance anomaly in concurrent object-oriented programming. In O. M. Nierstrasz, editor, *Proc. ECOOP'93*, pages 220–246. Springer LNCS 707, 1993.
23. J. Meseguer. Rewriting logic as a semantic framework for concurrency: a progress report. In *Proc. CONCUR'96, Pisa, August 1996*, pages 331–372. Springer LNCS 1119, 1996.
24. J. Meseguer. Membership algebra as a semantic framework for equational specification. In F. Parisi-Presicce, editor, *Workshop on Abstract Data Types, WADT'97*, volume 1376 of *Lecture Notes in Computer Science*, pages 18–61. Springer, 1998.
25. J. Meseguer. Research directions in rewriting logic. In U. Berger and H. Schwichtenberg, editors, *Computational Logic, NATO Advanced Study Institute, Marktoberdorf, Germany, July 29 – August 6, 1997*. Springer-Verlag, 1998.
26. J. Meseguer and X. Qian. A logical semantics for object-oriented databases. In *Proc. International SIGMOD Conference on Management of Data*, pages 89–98. ACM, 1993.
27. D. Peled. Combining partial order reductions with on-the-fly model-checking. *Formal Methods in System Design*, 8:39–64, 1996.
28. I. Pita and N. Martí-Oliet. A Maude specification of an object oriented database model for telecommunication networks. In J. Meseguer, editor, *Proc. First Intl. Workshop on Rewriting Logic and its Applications*, volume 4 of *Electronic Notes in Theoretical Computer Science*. Elsevier, 1996. URL: `http://www1.elsevier.nl/mcs/tcs/pc/volume4.htm`.
29. H. Reichel. An approach to object semantics based on terminal co-algebras. To appear in *Mathematical Structures in Computer Science*, 1995. Presented at *Dagstuhl Seminar on Specification and Semantics*, Schloss Dagstuhl, Germany, May 1993.
30. C. L. Talcott. An actor rewriting theory. In J. Meseguer, editor, *Proc. 1st Intl. Workshop on Rewriting Logic and its Applications*, volume 4 of *Electronic Notes in Theoretical Computer Science*, pages 360–383. North Holland, 1996. `http://www1.elsevier.nl/mcs/tcs/pc/volume4.htm`.
31. C. L. Talcott. Interaction semantics for components of distributed systems. In E. Najm and J.-B. Stefani, editors, *1st IFIP Workshop on Formal Methods for Open Object-based Distributed Systems, FMOODS'96*, 1996. Proceedings published in 1997 by Chapman & Hall.
32. M. Wirsing and A. Knapp. A formal approach to object-oriented software engineering. In J. Meseguer, editor, *Proc. First Intl. Workshop on Rewriting Logic and its Applications*, volume 4 of *Electronic Notes in Theoretical Computer Science*. Elsevier, 1996. URL: `http://www1.elsevier.nl/mcs/tcs/pc/volume4.htm`.

Partial Order Reduction for Model Checking of Timed Automata *

Marius Minea

Carnegie Mellon University, School of Computer Science
Pittsburgh, PA 15213-3891
marius+@cs.cmu.edu

Abstract. The paper presents a partial order reduction method applicable to networks of timed automata. The advantage of the method is that it reduces both the number of explored control states and the number of generated time zones. The approach is based on a local-time semantics for networks of timed automata defined by Bengtsson et al. [1998], and used originally for local reachability analysis. In this semantics, each component automaton executes asynchronously, in its own local time scale, which is tracked by an auxiliary reference clock. On communication transitions, the automata synchronize their time scales. We show how this model can be used to perform model checking for an extension of linear temporal logic, which can express timing relations between events. We also show how for a class of timed automata, the local-time model can be implemented using difference bound matrices without any space penalty, despite the need to represent local time. Furthermore, we analyze the dependence relation between transitions in the new model and give practical conditions for selecting a reduced set of transitions.

1 Introduction

Model checking [5] has emerged as a very successful automatic verification technique for finite-state systems. However, its application is still limited by the *state space explosion* problem. The number of possible states in a system grows exponentially with the number of component parts, quickly exceeding the current capabilities of verification tools. For timed systems, the complexity in the control space is increased by the timing information that needs to be maintained, since each untimed state can be reached at many different time instances.

Partial order reduction [8, 14, 15] is a well-established method to reduce the complexity of state space exploration in asynchronous systems. It explores a restricted number of interleavings for independent concurrent transitions, while preserving the verified property in the reduced model. However, in timed systems the implicit synchronization among transitions, caused by the passage of time, makes the application of this technique problematic. This paper shows how to perform partial order reduction for continuous-time systems modeled as timed

* This research was sponsored in part by the Semiconductor Research Corporation (SRC), the National Science Foundation (NSF), and the Defense Advanced Research Projects Agency (DARPA).

automata, while preserving properties specified in an extension of linear-time temporal logic augmented with explicit time constraints.

2 Timed Automata

2.1 Definition

Timed automata [1] are transition systems extended with real-valued clocks which advance at the same rate and can be reset on executing a transition. Both states and transitions are associated with temporal constraints on the clocks.

Definition 1. *A* clock *is a variable over the set* $\mathbb{R}^+$ *of nonnegative reals. A* clock valuation *for a set of clocks* $C = \{x_1, \cdots, x_n\}$ *is a function* $v : C \to \mathbb{R}^+$.

Definition 2. *An* atomic clock constraint *is an inequality of the form* $x \prec c$, $c \prec x$, *or* $x - y \prec c$, *where* x, y *are clocks,* $c \in \mathbb{Z}$ *is an integer and* $\prec \in \{<, \leq\}$. *A clock constraint is a conjunction of atomic clock constraints or the value* true. *The set of clock constraints over a set of clocks* C *is denoted by* $\mathcal{B}(C)$.

Definition 3. *A timed automaton is a tuple* $A = (S, S^0, C, E, I, \mu)$, *where*
• S *is a finite set of nodes (control states);* $S^0 \subseteq S$ *is the set of initial nodes*
• C *is a finite set of real-valued non-negative clocks*
• $E \subseteq S \times \mathcal{B}(C) \times 2^C \times S$ *is a finite set of edges. An edge* $e = \langle s, \psi, R, s' \rangle$ *has an* enabling condition ψ *and a set* R *of clocks that are reset on traversing the edge.*
• $I : S \to \mathcal{B}(C)$ *defines an* invariant condition *associated with each node*
• $\mu : S \to 2^P$ *labels each node with atomic propositions from a set* P

A satisfied enabling condition does not force the execution of a transition. An automaton can remain at the same node as long as the node invariant is satisfied.

We define a network of timed automata using a general parallel composition:

Definition 4. *Consider* n *timed automata* $A_i = (S_i, S_i^0, C_i, E_i, I_i, \mu_i)$, *and a synchronization function* $f : \prod_{i=1}^n (E_i \cup \{\epsilon\}) \to \{0, 1\}$ *(where* ϵ *is a symbol denoting a null edge). The network of timed automata* $A_1 \parallel A_2 \parallel \ldots \parallel A_n$ *is a timed automaton* $A = (S, S^0, C, E, I, \mu)$, *where:*
• $S = S_1 \times S_2 \times \ldots \times S_n$ *and* $S^0 = S_1^0 \times S_2^0 \times \ldots \times S_n^0$
• $C = C_1 \cup C_2 \cup \ldots \cup C_n$ *(assuming* $C_i \cap C_j = \emptyset$, *for* $i \neq j$*)*
• E *contains a family of edges (a transition) for each tuple with* $f(e_1, \cdots, e_n) = 1$. *For transition* a, *let* $e_i = \langle s_i, \psi_i, R_i, s_i' \rangle$ *if* $e_i \neq \epsilon$ *and* $active(a) = \{i \mid e_i \neq \epsilon\}$. *The edges of* a *have endpoints with* s_i *and* s_i' *given by* e_i *for* $i \in active(a)$, $s_j = s_j' \in S_j$ *arbitrary for* $j \notin active(a)$, $\psi = \bigwedge_{i \in active(a)} \psi_i$, *and* $R = \bigcup_{i \in active(a)} R_i$.
• $I(s) = \bigwedge_{i=1}^n I_i(s_i)$
• $\mu(s) = \bigcup_{i=1}^n \mu(s_i)$ *(assuming pairwise disjoint sets of atomic propositions* P_i*)*

A transition corresponds to the synchronous traversal of edges in several component automata. The synchronization function determines which automata execute (the *active set* of the transition) and which ones remain at their local state. This allows the modeling of many common synchronization paradigms, including pairwise communication. A transition with more than one automaton in its active set is called a *synchronization transition*, otherwise it is called *local*.

2.2 Semantics

Given a clock valuation v and $d \in \mathbb{R}^+$, $v+d$ is the valuation given by $(v+d)(x) = v(x) + d$, $\forall x \in C$. For $R \subseteq C$, $v[R \mapsto 0]$ is the clock valuation that is zero for clocks in R and agrees with v for all other clocks. The truth value of the clock constraint $\psi \in \mathcal{B}(C)$ for a clock valuation v is denoted by $\psi(v)$.

Definition 5. *A model of a timed automaton is a state-transition graph* $\mathcal{S}(A) = (\Sigma, \Sigma^0, \rightarrow)$, *where*
• $\Sigma = \{(s, v) \mid I(s)(v)\}$ *is the set of timed states satisfying the node invariant*
• $\Sigma^0 = \{(s^0, 0_C) \mid s^0 \in S^0\}$ *is the set of initial states, with* $0_C(x) = 0$, $\forall x \in C$
• $\rightarrow$ *is the transition relation defined as union of* delay *and* action *transitions:*
- $(s, v) \overset{d}{\leadsto} (s, v + d)$ *if* $d \in \mathbb{R}^+$, *and for all* $0 \leq d' \leq d$, $I(s)(v + d')$ *holds*
- $(s, v) \overset{a}{\rightarrow} (s', v[R \mapsto 0])$ *for* $a \in \mathcal{T}$ *(the set of transitions of* A*) if there exists an edge* $e = (s, \psi, R, s') \in a$, *such that* $\psi(v)$ *is true and* $I(s')(v[R \mapsto 0])$ *holds*

A *delay transition* models the elapse of time in the same control state, while maintaining the invariant. An *action transition* can be executed (instantaneously) if the clock valuation satisfies the enabling condition. Clocks in the set R are reset, the other clocks maintain their value.

We assume that node invariants contain only constraints of the form $x_i \prec c$, because the constraints $x_i - x_j \prec c$ or $c \prec x_i$ are not falsified by time passage, and can be incorporated into the enabling condition of incoming edges. Also, since clock constraints are convex, invariants must only be checked in the final state of a delay transition: $(s, v) \overset{d}{\leadsto} (s, v + d)$ if $d \in \mathbb{R}^+$ and $I(s)(v + d)$ holds.

Definition 6. *An* execution trace *of a timed automaton is a finite or infinite sequence* $\sigma = (s^0, 0_C) \rightarrow (s^1, v^1) \ldots \rightarrow (s^k, v^k) \ldots$ *starting from a state* $s^0 \in S^0$.

We denote by $\sigma(k) = (s^k, v^k)$ the kth state on the trace σ, by σ_k the finite prefix of σ ending at (s^k, v^k) and by σ^k the suffix of σ starting at the same state.

2.3 The Model Checking Problem

Several model checkers for timed automata exist. The KRONOS tool is a model checker for TCTL and timed μ-calculus [9], and UPPAAL [11] verifies properties in a timed modal logic. However, partial order approaches have been so far restricted to less expressive properties: Pagani [12, 13] performs deadlock detection, whereas Bengtsson et al. [2] check local reachability within one process.

We use an extension of LTL inspired from the timed temporal logic for nets (TNL) of [18], which has been used to verify time Petri nets. By allowing constraints on two clock differences, the logic permits reasoning about the time separation of two events, since the difference between two clocks corresponds to the difference between the execution times of the transitions that reset them.

The formulas of our logic, called LTL_Δ, are defined by the grammar:

$$\psi ::= \mathit{true} \mid p \mid x - y \prec c \mid \neg\varphi \mid \varphi_1 \wedge \varphi_2 \mid \varphi_1 \,\mathcal{U}\, \varphi_2$$

where $p \in P$ is an atomic proposition, $x, y \in C$ are clocks, $c \in Z$ and $\prec \in \{<, \leq\}$.

Definition 7. *Consider an infinite execution trace* $\sigma = (s^0, v^0) \rightarrow (s^1, v^1) \rightarrow \ldots \rightarrow (s^k, v^k) \rightarrow \ldots$ *The semantics of an* LTL_Δ *formula is defined as follows:*
- $(s, v) \models p$ *iff* $p \in \mu(s)$
- $(s, v) \models x - y \prec c$ *iff* $v(x) - v(y) \prec c$.
- $\sigma \models \varphi_a$ *iff* φ_a *is an atomic formula and* $(s^0, v^0) \models \varphi_a$
- $\sigma \models \neg\varphi$ *iff* $\sigma \models \varphi$ *does not hold*
- $\sigma \models \varphi_1 \wedge \varphi_2$ *iff* $\sigma \models \varphi_1$ *and* $\sigma \models \varphi_2$
- $\sigma \models \varphi_1 \mathcal{U} \varphi_2$ *iff* $\exists k \geq 0$ *such that* $\sigma^k \models \varphi_2$ *and* $\sigma^j \models \varphi_1$ *for all* $0 \leq j < k$
- $\mathcal{S}(A) \models \varphi$ *iff* $\sigma \models \varphi$ *for any infinite execution trace* σ *of* $\mathcal{S}(A)$.

Since control state and clock differences are preserved by time passage, all intermediate states traversed by a delay transition have the same truth value for any atomic subformulas in LTL_Δ. Thus, the given semantics based on transition endpoints corresponds to the intuitive meaning of continuous execution.

3 The Model Checking Approach

3.1 Effect of Transition Interleavings

The traditional reachability analysis algorithm for networks of timed automata explores all possible transition interleavings among the individual components. Partial order methods choose a representative from each set of equivalent interleavings, exploring only a reduced portion of the state space. However, in our model of time, clocks advance simultaneously in all automata, and different interleavings may produce different assignments to clock values. The independence of transitions in the underlying untimed system may not be preserved.

Consider the system of two automata in Fig. 1 and its exploration using timed zones [9]. From the initial state $\langle(s_1, s_2), x = y\rangle$, transition a leads to the state $\langle(s_1', s_2), x \leq y\rangle$ (since clock x is reset). Next, on executing b, clock y is reset, leading to state $\langle(s_1', s_2'), x \geq y\rangle$. If b is executed before a, the system reaches first the state $\langle(s_1, s_2'), x \geq y\rangle$, and then the state $\langle(s_1', s_2'), x \leq y\rangle$.

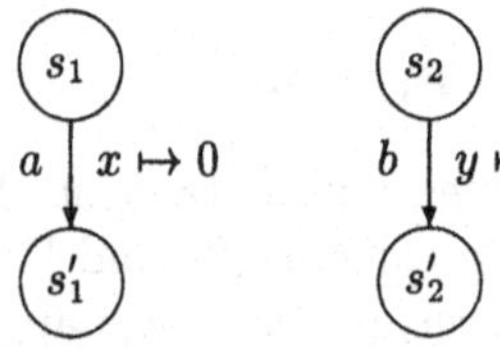

Fig. 1. Effect of transition interleavings

The two interleavings lead to the same control state, but to distinct clock zones and thus distinct states in the zone automaton. Hence, the transitions are not independent and usual partial order reduction techniques cannot be applied.

For a property insensitive to the ordering of x and y, both interleavings are still equivalent, leading to a timed state in the union of the two zones, $\langle(s_1', s_2'), x \geq y \vee x \leq y\rangle = \langle(s_1', s_2'), \mathit{true}\rangle$. Our goal is a partial order reduction method that produces a zone containing the timed states reachable by all transition interleavings, while exploring only one interleaving, and thus fewer states.

3.2 Related Work

Partial order reduction has been investigated by Yoneda and Schlingloff [18] for time Petri nets, which have earliest and latest firing times associated with transitions and are thus less expressive than timed automata. The logic used for specifications is similar to LTL_Δ, but the dependency relation between transitions uses run-time information about the time component of the current state. Lilius [10] improves on this technique by not storing the transition firing order in the timing constraints and reducing branching in the generated graph.

For timed automata, Pagani [12, 13] shows that in many cases timing introduces dependencies and reduces the amount of partial order reduction. The analysis is limited to deadlock detection. Dams et al. [6] handle some of these cases, generalizing the notion of independence and selecting at a state those transitions whose executions cover the result of exploring other interleavings.

Belluomini and Myers [3] use an event model with lower and upper time bounds associated to transitions. Timing information is represented in the form of partially ordered sets, reducing the number of generated time zones. However, their analysis does not reduce the number of explored transition interleavings.

The method from which we draw most is that of Bengtsson, Jonsson, Lilius and Wang [2]. They define a local-time semantics based on desynchronized execution of the component automata and local time delays, with additional reference clocks to model synchronization. In this model the same independence conditions as in the untimed case apply, and an algorithm is given to decide the reachability of a local control state.

3.3 Local-Time Model

We revisit the local-time model of Bengtsson et al. [2] using somewhat different notations and prove several results underlying its use in model checking.

Consider the interaction of action and delay transitions. The enabling of an action transition and the resulting state change depend only on the state of the participating automata. Hence, two action transitions with disjoint active sets are independent. On the other hand, a delay transition changes the state in *all* automata by incrementing the values of all clocks. It is therefore dependent on any action transition that also changes clock values (specifically, resets clocks).

However, one can view a global delay transition as a set of simultaneous transitions with equal delay in all component automata. This suggests that time-induced dependencies can be removed by separating a global delay transition into individual transitions for each component automaton, without requiring their simultaneity. To this effect, local passage of time is introduced as follows:

For a clock valuation v, $d \in \mathbb{R}$ and $i \in \overline{1,n}$, define the clock valuation $v +_i d$ by: $(v +_i d)(x) = v(x) + d$ for $x \in C_i$ and $(v +_i d)(x) = v(x)$ otherwise.

A *local delay transition* $\overset{d}{\rightsquigarrow}_i$ increments only the clocks in automaton A_i. We identify it with a pair $(d, i) \in \mathcal{T}_\Delta = \mathbb{R}^+ \times \overline{1,n}$, define $active(\overset{d}{\rightsquigarrow}_i) = \{i\}$ and denote $\mathcal{T}_l = \mathcal{T} \cup \mathcal{T}_\Delta$. For $i \in \overline{1,n}$, define the functions $delay_i : \mathcal{T}_l \mapsto \mathbb{R}^+$ as follows: $delay_i(\overset{d}{\rightsquigarrow}_i) = d$, $delay_i(\overset{d}{\rightsquigarrow}_j) = 0$ for $i \neq j$, and $delay_i(\overset{a}{\rightarrow}) = 0$ for $a \in \mathcal{T}$. They indicate the delay caused by a transition in a component automaton.

Definition 8. *The local-time model $\mathcal{L}(A)$ for a network of timed automata $A = A_1 \parallel A_2 \parallel \ldots \parallel A_n$ is a state-transition graph with state set Σ, initial state set Σ^0 and execution traces $\sigma = (s^0, v^0) \overset{\tau_1}{\rightarrow} (s^1, v^1) \ldots \overset{\tau_k}{\rightarrow} (s^k, v^k) \ldots$ starting from a state $(s^0, v^0) \in \Sigma^0$ and satisfying one of the following conditions for any $k \geq 1$:*
- *$\tau_k = (d, i) \in \mathcal{T}_\Delta$, $s^k = s^{k-1}$, $v^k = v^{k-1} +_i d$ and $\forall d' \in [0, d]. I_i(s_i^k)(v^k + d')$, or*
- *$\tau_k \in \mathcal{T}$, $(s^{k-1}, v^{k-1}) \overset{\tau_k}{\rightarrow} (s^k, v^k)$ and $\sum_{l=1}^{k-1} delay_i(\tau_l) = \sum_{l=1}^{k-1} delay_j(\tau_l)$ for all $i, j \in active(\tau_k)$*

The first case is a local delay transition $(s^{k-1}, v^{k-1}) \overset{d}{\rightsquigarrow}_i (s^k, v^k)$ in automaton A_i. In the second case, an action transition $(s^{k-1}, v^{k-1}) \overset{\tau_k}{\rightarrow} (s^k, v^k)$ is executed, under the additional constraint that the elapsed time (the sum of delays) is identical for all automata in the active set. (For a local action transition, this additional constraint is void). In both cases, the transition τ_k is said to be *enabled* after the execution of σ_{k-1}. Denote by $enabled(\sigma)$ and $enabled^*(\sigma)$ the set of transitions and transition sequences, respectively, that can follow a finite trace σ.

For a finite execution trace $\sigma = (s^0, v^0) \overset{\tau_1}{\rightarrow} (s^1, v^1) \ldots \overset{\tau_k}{\rightarrow} (s, v)$, define $time_i(\sigma) = t_0 + \sum_{l=1}^{k} delay_i(\tau_l)$, where $t_0 \in \mathbb{R}^+$ is an arbitrary timepoint at which the execution of σ starts. Then, $time_i(\sigma)$ (or $time_i$, when σ is implicit) denotes the timepoint reached in A_i after executing σ. The *local configuration* of A_i reached by σ is the tuple $cfg_i(\sigma) = (s_i, v_i, time_i)$, where v_i denotes the restriction of v to the clocks of A_i. The *global configuration* of A is the tuple $cfg(\sigma) = (cfg_1(\sigma), cfg_2(\sigma), \cdots, cfg_n(\sigma))$, also denoted $cfg(\sigma) = (s, v, time)$ with $time = (time_1, time_2, \cdots, time_n)$. The set of configurations is $\Sigma_C = \Sigma \times (\mathbb{R}^+)^n$.

Note that the enabling of an action transition is defined in terms of the trace executed so far. The following result shows that a configuration determines completely the subsequently enabled transitions. The proof follows directly from the definitions of parallel composition and the local-time model.

Proposition 1. *The following properties hold in the local-time model $\mathcal{L}(A)$ for finite execution traces σ and σ' and transition $\tau \in enabled(\sigma)$:*
- *if $cfg_i(\sigma) = cfg_i(\sigma')$ for all $i \in active(\tau)$, then $\tau \in enabled(\sigma')$ and $cfg_i(\sigma\tau) = cfg_i(\sigma'\tau)$ for all $i \in active(\tau)$*
- *$cfg_j(\sigma\tau) = cfg_j(\sigma)$ for all $j \notin active(\tau)$, where $\sigma\tau$ denotes the trace obtained by extending σ with the transition τ.*

Consequently, two finite execution traces leading to the same configuration have the same enabled transitions. For a configuration $\gamma \in \Sigma_C$ one can thus define $enabled(\gamma) = enabled(\sigma)$, where σ is an execution trace such that $cfg(\sigma) = \gamma$. Likewise, the successor configuration of γ by a transition $\tau \in enabled(\sigma)$ is defined as the configuration reached when extending the trace σ by transition τ: $succ_\tau(\gamma) = cfg(\sigma\tau)$. This is again independent of σ and we write $\gamma \overset{\tau}{\rightarrow} succ_\tau(\gamma)$.

We now prove the desired independence properties for transitions in $\mathcal{L}(A)$. In general, two transitions are called independent if neither disables the execution of the other, and the same state is reached by executing them in either order:

Definition 9. *Two transitions* τ_1 *and* τ_2 *are independent iff for any finite execution trace* σ *such that* $\tau_1, \tau_2 \in \mathit{enabled}(\sigma)$ *the following two conditions hold:*
- Enabledness: $\tau_2 \in \mathit{enabled}(\sigma\tau_1) \wedge \tau_1 \in \mathit{enabled}(\sigma\tau_2)$
- Commutativity: $\mathit{fin}(\sigma\tau_1\tau_2) = \mathit{fin}(\sigma\tau_2\tau_1) \wedge \mathit{enabled}^*(\sigma\tau_1\tau_2) = \mathit{enabled}^*(\sigma\tau_2\tau_1)$

where $\mathit{fin}(\sigma)$ *denotes the last state on the trace* σ.

Theorem 1. *Two (action or local delay) transitions* $\tau_1, \tau_2 \in \mathcal{T}_l$ *that involve disjoint sets of automata* ($\mathit{active}(\tau_1) \cap \mathit{active}(\tau_2) = \emptyset$) *are independent.*

Proof. For all $j \in \mathit{active}(\tau_2)$, we have $j \notin \mathit{active}(\tau_1)$, hence $\mathit{cfg}_j(\sigma\tau_1) = \mathit{cfg}_j(\sigma)$. Therefore, $\tau_2 \in \mathit{enabled}(\sigma) \Rightarrow \tau_2 \in \mathit{enabled}(\sigma\tau_1)$, and symetrically for τ_1. Also, since $\mathit{active}(\tau_1) \cap \mathit{active}(\tau_2) = \emptyset$, each local configuration is changed at most once, either by τ_1 or by τ_2, irrespective of their ordering. Therefore, $\mathit{cfg}(\sigma\tau_1\tau_2) = \mathit{cfg}(\sigma\tau_2\tau_1)$ and $\mathit{fin}(\sigma\tau_1\tau_2) = \mathit{fin}(\sigma\tau_2\tau_1)$. Since the enabled transitions are determined by the reached configuration, $\mathit{enabled}^*(\sigma\tau_1\tau_2) = \mathit{enabled}^*(\sigma\tau_2\tau_1)$. □

A finite trace σ in $\mathcal{L}(A)$ is called *synchronized* if $\mathit{time}_i(\sigma) = \mathit{time}_j(\sigma)$ for all $i, j \in \overline{1,n}$, i.e., if all automata have executed for the same amount of time, denoted by $\mathit{time}(\sigma)$. The following theorem relates the reachable state spaces of the standard and local-time models (cf. [2]):

Theorem 2. *Each state reachable in* $\mathcal{S}(A)$ *is also reachable in* $\mathcal{L}(A)$. *Moreover, each state reached by a synchronized trace in* $\mathcal{L}(A)$ *is also reachable in* $\mathcal{S}(A)$.

Proof. First, any trace in $\mathcal{S}(A)$ yields a trace in $\mathcal{L}(A)$ by replacing each global delay transition $\overset{d}{\rightsquigarrow}$ with the sequence of local delay transitions $\overset{d}{\rightsquigarrow}_1 \dots \overset{d}{\rightsquigarrow}_n$.

The reverse implication follows by induction on the number of action transitions in the trace σ_l of $\mathcal{L}(A)$. For the base case, if σ_l is synchonized and contains only local delay transitions, they sum up to the same total delay d. Then, $\mathit{fin}(\sigma_l)$ is reachable in $\mathcal{S}(A)$ by executing the global delay transition $\overset{d}{\rightsquigarrow}$.

For the induction step, let a be the action transition in σ_l executed at the latest timepoint, $t_a \leq t = \mathit{time}(\sigma_l)$. In every automaton, σ_l ends with local delay transitions totaling at least $t - t_a$. Removing this delay in every automaton yields a synchronized trace σ'_l with $\mathit{time}(\sigma'_l) = t_a$. In σ'_l, a is the last transition in all participating automata. Its removal yields a synchronized execution trace σ''_l with fewer action transitions. By the induction hypothesis, $\mathit{fin}(\sigma''_l)$ is reachable in $\mathcal{S}(A)$, and $\mathit{fin}(\sigma_l)$ is reachable from it by executing $\overset{a}{\rightarrow}$ followed by $\overset{t-t_a}{\rightsquigarrow}$. □

3.4 Local-Time Zone Automaton

An analogue of the zone automaton [9], which represents sets of timed states using clock constraints can be derived for the local-time model [2]. A *local-time zone* is a convex set of configurations $z \in \Sigma_C$ with the same control state. A transition is enabled in a zone iff it is enabled in some configuration in the zone: $\mathit{enabled}(z) = \{\tau \in \mathcal{T}_l \mid \exists \gamma \in z.\, \tau \in \mathit{enabled}(\gamma)\}$. The successor of a zone z by a transition $\tau \in \mathit{enabled}(z)$ is $\mathit{succ}_\tau(z) = \{\mathit{succ}_\tau(\gamma) \mid \gamma \in z \wedge \tau \in \mathit{enabled}(\gamma)\}$.

For the standard zone automaton, an exploration step consists of an action transition followed by a delay transition of arbitrary amount. For the local-time model, we combine an action transition with subsequent delay transitions in all automata belonging to its active set, and show:

Proposition 2. *For any finite execution trace σ, there exists a trace σ' with the same final configuration, which starts with a local delay transition in each component automaton, after which every subsequent action transition is followed by local delay transitions in all participating automata.*

Proof. A delay transition $\overset{d}{\rightsquigarrow}_i$ commutes with any other delay transition, and with action transitions a for which $i \notin active(a)$. Thus, a delay transition can be moved towards the beginning of the execution trace σ (merging consecutive delay transitions in the same automaton) until the preceding action transition involves the same automaton, or there are no preceding action transitions. □

Based on this result, we define the zone successor operation as follows:

$succ^Z_l(z, a) = \{\gamma_k \in \Sigma_C \mid \exists \gamma \in z, \exists d_{i_1}, \cdots, d_{i_k} \in \mathbb{R}^+ . \gamma \overset{a}{\rightarrow} \gamma' \overset{d_{i_1}}{\rightsquigarrow} \gamma_1 \ldots \overset{d_{i_k}}{\rightsquigarrow} \gamma_k\}$
where $active(a) = \{i_1, i_2, \cdots, i_k\}$. An initial local-time zone is the set of all configurations reachable from an initial state by a sequence of delay transitions:
$init^Z_l(s^0) = \{cfg(\sigma) \mid \exists d_{i_1}, \cdots, d_{i_n} \in \mathbb{R}^+ . \sigma = (s^0, 0_C) \overset{d_{i_1}}{\rightsquigarrow} (s^0, v^1) \ldots \overset{d_{i_n}}{\rightsquigarrow} (s^0, v^n)\}$
If $succ^\Delta_i(z) = \{\gamma' \mid \exists \gamma \in z, \exists d \in \mathbb{R}^+ . \gamma \overset{d}{\rightsquigarrow}_i \gamma'\}$ is the successor by an arbitrary local delay, then $init^Z_l(s^0) = (succ^\Delta_n \circ \ldots \circ succ^\Delta_1)(\gamma^0(s^0))$ and $succ^Z_l(z, a) = (succ^\Delta_{i_k} \circ \ldots \circ succ^\Delta_{i_1} \circ succ_a)(z)$, where $\circ$ denotes function composition.

Definition 10. *The local-time zone automaton $\mathcal{Z}_l(A)$ for a network of automata A is a tuple $(Z_l, Z^0_l, succ^Z_l)$, with $Z^0_l = \{init^Z_l(s^0) \mid s^0 \in S^0\}$ the set of initial local-time zones, $succ^Z_l$ the successor relation defined above, and Z_l the set of local-time zones reachable by successive application of $succ^Z_l$ from an initial zone.*

Together with Prop. 2, this definition implies directly the following:

Theorem 3. *A state is reachable in the model $\mathcal{L}(A)$ iff it belongs to a zone z which is reachable in the local-time zone automaton $\mathcal{Z}_l(A)$.*

3.5 Representation of Local-Time Zones

In [2], it is shown how local-time zones can be represented by difference bound matrices [7] using one additional variable per automaton. For a class of timed automata, we derive an improved representation which does not need additional space compared to the standard zone automaton.

The difference between two clocks is invariant to global delay transitions, but in the local-time model, it may be changed by a local delay transition if the clocks belong to different automata. However, since a transition $\overset{d}{\rightsquigarrow}_i$ increments both $time_i$ and the clocks in C_i, the value $time_i - v_i(x)$ is invariant to local delay transitions. Indeed, it represents the timepoint at which clock x was last reset.

Consider the new variables t_i for $i \in \overline{1,n}$ (the reference time in A_i) and t_x for all clocks $x \in C$ (the last reset time of x). Denote $T_i = \{t_x \mid x \in C_i\}$ for $i \in \overline{1,n}$, $T_i^+ = T_i \cup \{t_i\}$, $T = \{t_x \mid x \in C\} = \bigcup_{i=1}^n T_i$, and $T^+ = \bigcup_{i=1}^n T_i^+$. For a configuration $(s, v, time)$, define the valuation $\bar{v} : T^+ \to \mathbb{R}^+$ by $\bar{v}(t_i) = time_i$ for $i \in \overline{1,n}$ and $\bar{v}(t_x) = time_i - v(x)$ for $x \in C_i$. Conversely, $\bar{v}$ uniquely determines v and *time*, and $(s, \bar{v})$ is an alternate representation for a configuration.

Any atomic clock constraint appearing in the description of A can be rewritten as a difference constraint over T^+. In a difference constraint $x - y \prec c$, both clocks belong to the same automaton A_i, and $x - y = (t_i - t_x) - (t_i - t_y) = t_y - t_x$. Likewise, $x \prec c$ and $c \prec x$ are rewritten as $t_i - t_x \prec c$ and $t_x - t_i \prec -c$.

A *local-time clock zone* is the set of valuations belonging to a local-time zone. A zone is written as $\langle s, \psi_l \rangle$ with s the control state and ψ_l the clock zone.

Proposition 3. *A local-time clock zone can be written as a difference constraint over the variables in* T^+: $\psi_l = \bigwedge_{t_u, t_w \in T^+} t_u - t_w \prec c_{uw}$.

Proof. Initially, $t_x = t_i = t_0$, $\forall x \in C_i$, $i \in \overline{1,n}$. Thus, $\psi_l = \bigwedge_{t_u, t_w \in T^+} (t_u = t_w)$. For an action transition $(s, \bar{v}) \xrightarrow{a} (s', \bar{v}')$, we have $\bar{v}'(t_u) = \bar{v}(t_u)$ for $u \notin R_a$ and $\bar{v}'(t_x) = t_{i_x}$ for $x \in R_a$ (with $x \in C_{i_x}$). We denote this by $\bar{v}' = \bar{v}[t_x \mapsto t_{i_x}]_{x \in R_a}$ and extend the notation to clock zones. Also, the enabling condition ψ_a holds for $\bar{v}$ and the reference times in $T_a = \{t_i \mid i \in active(a)\}$ are equal. Thus, $succ_a(\psi_l) = \{\bar{v}' \mid (s, \bar{v}) \xrightarrow{a} (s', \bar{v}')\} = (\psi_l \wedge \psi_a \wedge \bigwedge_{t_i, t_j \in T_a} t_i = t_j)[t_x \mapsto t_{i_x}]_{x \in R_a} = [\exists_{X_a} . \psi_l \wedge \psi_a \wedge \bigwedge_{t_i, t_j \in T_a} t_i = t_j] \wedge \bigwedge_{x \in R_a} t_x = t_{i_x}$, with $X_a = \{t_x \mid x \in R_a\}$ and $\exists_{X_a}$ denoting quantification over all variables in X_a. Since difference constraints are closed under conjunction and quantification, $succ_a(\psi_l)$ is a difference constraint.

For a local delay transition $(s, \bar{v}) \overset{d}{\rightsquigarrow}_i (s, \bar{v}')$, we have $\bar{v}'(t_i) = \bar{v}(t_i) + d$ and $\bar{v}'(t_u) = \bar{v}(t_u)$ for all $t_u \in T^+ \setminus \{t_i\}$. Denote this by $\bar{v}' = \bar{v} +_i d$ and the successor of ψ_l after an arbitrary delay $\overset{d}{\rightsquigarrow}_i$ as $\psi_l \Uparrow^i = \{\bar{v}' \mid \exists \bar{v} \in \psi_l, \exists d \in \mathbb{R}^+ . \bar{v}' = \bar{v} +_i d\}$. We have $\psi_l \Uparrow^i = \exists d \in \mathbb{R}^+ . \psi_l[t_i - d/t_i] = \exists t_i' \in \mathbb{R}^+ . \psi_l[t_i'/t_i] \wedge t_i' - t_i \leq 0$, where $e[y/x]$ denotes substitution of y for x in e. Since $(s, \bar{v}) \overset{d}{\rightsquigarrow}_i (s, \bar{v}')$ iff $\bar{v}' = \bar{v} +_i d$ and $I_i(s_i)(\bar{v}')$ holds, we have $succ_i^{\Delta}(\psi_l) = \psi_l \Uparrow^i \wedge I_i(s_i)$, again a difference constraint.

Combining action and delay steps, we obtain the relation: $succ_l^Z(\psi_l, a) = ([\exists_{X_a} . \psi_l \wedge \psi_a \wedge \bigwedge_{t_i, t_j \in T_a} t_i = t_j] \wedge \bigwedge_{x \in R_a} t_x = t_{i_x}) \Uparrow^{i_1} \ldots \Uparrow^{i_k} \wedge \bigwedge_{i \in active(a)} I_i(s_i')$.

This representation of a local-time zone is monolithic and relates reset times of clocks to reference times in *all* automata, using n auxiliary reference times. For a certain class of networks, we prove the following simpler representation:

Proposition 4. *If each synchronization transition in a network of automata A resets at least one clock in each participating automaton, a local-time clock zone has the form* $\psi_l = \psi_\Delta(T) \wedge \bigwedge_{i=1}^n \psi_i(T_i, t_i)$, *where:*
- $\psi_\Delta(T) = \bigwedge_{t_x \neq t_y \in T} t_x - t_y \prec c_{xy}$, *with* $c_{xy} \in \mathbb{Z}$
- $\psi_i(T_i, t_i) = \bigwedge_{t_x \in T_i} (t_i - t_x \prec c_{ix} \wedge t_x - t_i \prec c_{xi})$ *with* $c_{ix}, c_{xi} \in \mathbb{Z}$

We call A a *sync-reset* network of automata. The term $\psi_\Delta(T)$ relates pairs of two reset times, while $\psi_i(T_i, t_i)$ relates t_i to the reset times in automaton A_i.

Proof. The initial zone is written as: $init_l^Z(s^0) = \bigwedge_{x,y\in C}(t_x = t_y) \wedge \bigwedge_{i=1}^{n} I_i(s_i^0)$. For $succ_l^Z$, the term $\psi_l \wedge \psi_a \wedge \bigwedge_{t_i,t_j \in T_a} t_i = t_j$ from Prop. 3 has the required form, save for $t_i = t_j$. Quantification over X_a adds constraints between t_i and t_z, for $i \in active(a), t_z \in T$. By assumption, for every $i \in active(a)$, a clock $x \in R_a \cap C_i$ is reset, yielding $t_x = t_i$. Hence, constraints on $t_i - t_z$ can be included in ψ_Δ as constraints on $t_x - t_z$. Finally, executing $\Uparrow^i$ for $i \in active(a)$ removes the equalities $t_i = t_j$, and adds constraints on $t_z - t_j$ with $z \notin C_j$. Likewise, these can be replaced with $t_z - t_y$ for $y \in R_a \cap C_j$, which are in the desired form. □

Clock constraints are usually represented as *difference-bound matrices* [7], which are indexed by clock variables and whose elements are *bounds*, i.e., pairs $(\prec, c)$ corresponding to an atomic clock constraint. The component ψ_Δ of a local-time zone can be represented as a DBM of dimension $|C|$ (the total number of clocks). Each constraint ψ_i requires $2 * |C_i|$ time bounds, for a total of $2 * |C|$, i.e., an additional row and column. Thus, ψ_l can be represented by a matrix of dimension $|C| + 1$, the same size as the DBM used in the standard algorithm. However, only the submatrices corresponding to individual automata (with reference time) and the submatrix for ψ_Δ (without reference times) are subject to DBM operations. The successor computation is done first on the submatrix corresponding to the active automata (after enforcing the synchronization constraints $t_i = t_j$). Strengthened constraints may lead to the recanonicalization of ψ_Δ and possibly of submatrices for other individual automata.

If an automaton in the network has synchronization transitions that do not reset clocks, an additional clock can be inserted into the automaton for this purpose. This transforms any network of automata into a sync-reset network, with potentially fewer than n additional time variables.

3.6 Preservation of LTL_Δ Formulas

Since in the local-time model $\mathcal{L}(A)$ the execution order of transitions is relaxed, $\mathcal{L}(A)$ accepts a richer set of behaviors than $\mathcal{S}(A)$. This section establishes restrictions on the local-time model which ensure that each of its traces is equivalent to a trace of the standard model with respect to a given LTL_Δ formula φ.

We extend LTL_Δ to the local-time model by defining the satisfaction of an atomic time constraint in a configuration: $(s, \bar{v}) \models x - y \prec c$ iff $\bar{v}(t_y) - \bar{v}(t_x) \prec c$. For $x \in C_i$ and $y \in C_j$ we have $\bar{v}(t_y) - \bar{v}(t_x) = (time_j - v(y)) - (time_i - v(x))$. Thus, in a synchronized configuration, the semantics is the same as in $\mathcal{S}(A)$. The transitions which affect the truth value of a formula are identified as follows:

Definition 11. *(Visibility) A transition $(s, v) \rightarrow (s', v')$ is* invisible *with respect to a specification φ if every atomic subformula of φ that has the same truth value in (s, v) and (s', v'). A transition which is not* invisible *is called* visible.

A transition in $\mathcal{L}(A)$ is visible if it connects two states which differ by at least one atomic proposition in the specification or it resets at least one clock in the specification, affecting the truth value of a difference constraint. Delay transitions are invisible, since they don't change the control state and don't reset clocks.

For a network of timed automata A and a formula φ in LTL_Δ denote by $\mathcal{F}^\varphi(A)$ the set of those traces of $\mathcal{L}(A)$ which satisfy the following properties:
• *Ordering* (**O**): Visible transitions occur in increasing order of their execution times. That is, in any trace $\sigma \in \mathcal{F}^\varphi(A)$, for visible transitions τ_k and τ_l with $k < l$, we have $time(\tau_k) \leq time(\tau_l)$ (where $time(\tau)$ is the execution time of τ).
• *Fairness* (**F**): Time progress is unbounded in all automata. That is, for any trace $\sigma \in \mathcal{F}^\varphi(A)$, $i \in \overline{1,n}$ and $M \in \mathbb{R}^+$, there exists $k \in \mathbb{N}$ with $time_i(\sigma_k) > M$.

Theorem 4. *Given an LTL_Δ formula φ, for any execution trace in $\mathcal{S}(A)$ there exists an execution trace in $\mathcal{F}^\varphi(A)$ with the same truth value for φ and vice versa.*

Proof. The direct implication is straightforward: from a trace σ in $\mathcal{S}(A)$ construct a trace σ_l in $\mathcal{L}(A)$ by replacing each global delay transition $\overset{d}{\leadsto}$ with the sequence of local delay transitions $\overset{d}{\leadsto}_1 \ldots \overset{d}{\leadsto}_n$. The trace σ_l satisfies **O**, since no action transitions are reordered, and **F**, since the same delay transitions are executed in each automaton. Because delay transitions are invisible, this transformation preserves the truth value of φ, and $\sigma \models \varphi$ iff $\sigma_l \models \varphi$.

For the reverse implication, we construct σ from σ_l by reordering all transitions in increasing order of their timepoints. The ordering condition **O** guarantees that no visible transitions are reordered, and the truth value of the formula is not changed. Delay transitions may be split and reordered so every action transition is preceded by equal delays in all automata. The fairness condition **F** guarantees that for all automata, local delay transitions with the needed amount exist in σ_l. Finally, all local delay transitions between two consecutive action transitions are merged into a global delay transition, resulting in a trace σ of $\mathcal{S}(A)$. □

Based on the above theorem, we proceed as follows: We first define a restricted local-time model $\mathcal{L}^\varphi(A)$ whose traces satisfy the ordering condition **O**. Next, we construct a zone automaton $\mathcal{Z}_l^\varphi(A)$ whose states are local-time *atoms*, i.e., sets of configurations with the same truth value for all atomic subformulas of φ. We show a correspondence between the traces of $\mathcal{L}^\varphi(A)$ and $\mathcal{Z}_l^\varphi(A)$, and then impose a fairness condition corresponding to **F** to ensure equivalence with the standard model. Finally, we apply a *maximization* to the atoms in $\mathcal{Z}_l^\varphi(A)$ to obtain an automaton $\mathcal{M}_l^\varphi(A)$ which is finite and therefore amenable to model checking.

To preserve the ordering of visible transitions, we introduce a new reference variable t_v, denoting the timepoint of the last visible transition executed. The domain of the valuation $\bar{v}$ is extended to include t_v. In the initial configuration, $\bar{v}(t_v) = 0$. The model $\mathcal{L}^\varphi(A)$ is defined in the same way as $\mathcal{L}(A)$, but with the additional restriction $\bar{v}(t_v) \leq time(a)$ for executing a visible transition a, and $\bar{v}'(t_v) = time(a)$ in the resulting configuration. Thus, each visible transition is executed at a later timepoint than the previous one, and condition **O** holds.

The zone successor formula for a visible transition becomes: $succ_a^v(\psi_l) = [\exists X_a \exists t_v . \psi \wedge \psi_a \wedge \bigwedge_{t_i,t_j \in T_a} t_i = t_j \wedge \bigwedge_{t_i \in T_a} t_v \leq t_i] \wedge \bigwedge_{t_i \in T_a} t_v = t_i \wedge \bigwedge_{x \in R_a} t_x = t_{i_x}$, For invisible transitions, the successor operation remains the same.

The ordering condition **O** can also be ensured without a new variable by a stronger condition on the traces of $\mathcal{L}^\varphi(A)$. This requires a visible transition to precede in time *all* action transitions which follow it in the execution trace and is enforced by the conjunct $\bigwedge_{j \notin active(a)} time(a) \leq t_j$.

In this case, the zone successor formula for visible transitions is written: $succ_a^v(\psi_l) = [\exists x_a . \psi \wedge \psi_a \wedge \bigwedge_{t_i, t_j \in T_a} t_i = t_j \wedge \bigwedge_{t_i \in T_a, t_j \notin T_a} t_i \leq t_j] \wedge \bigwedge_{x \in R_a} t_x = t_{i_x}$.

To perform model checking, we consider zones in which every configuration satisfies the same atomic subformulas of the specification φ (cf. [18]):

Definition 12. *(Atom) Given a timed automaton A and a LTL_Δ formula φ, an* atom *is a zone $\langle s, \psi_l \rangle$ such that $\bar{v}_1(t_y) - \bar{v}_1(t_x) \prec c \Leftrightarrow \bar{v}_2(t_y) - \bar{v}_2(t_x) \prec c$ for all $\bar{v}_1, \bar{v}_2 \in \psi_l$ and any constraint $x - y \prec c$ in φ.*

For each atomic clock constraint in φ, consider a new atomic proposition $q_k = t_{y_k} - t_{x_k} \prec_k c_k$. Thus, φ is reduced to a next-time free LTL formula φ_q. All configurations in an atom have the same truth value for all propositions q_k. The atoms comprising a zone $\langle s, \psi_l \rangle$ are given by the nonempty intersections between ψ_l and all constraints $t_{y_k} - t_{x_k} \prec_k c_k$, either in positive or negated form:

$$atoms^\varphi(\langle s, \psi_l \rangle) = \{\langle s, \phi \rangle \mid \phi = \psi_l \wedge \bigwedge_{k=1}^{m} q'_k,\ \phi \neq false,\ q'_k = q_k \text{ or } q'_k = \neg q_k\}.$$

Define transitions between atoms as follows: $z \overset{a}{\Rightarrow} z'$ if $a \in enabled(z)$ and $z' \in atoms^\varphi(succ_l^Z(z, a))$, and $z \overset{\xi}{\Rightarrow} z$ if at least one local state of z has the invariant $I_i(s_i) = true$. We obtain an *atom graph* for A and the formula φ:

Definition 13. *(Atom graph) The atom graph $A^\varphi(A)$ of a timed automaton A with respect to formula φ is a state-transition graph $(Z_l^\varphi, Z_l^0, \Rightarrow)$, with Z_l^0 the set of initial local-time zones, $\Rightarrow$ the atom transition relation and Z_l^φ the set of atoms reachable from Z_l^0 by repeated application of $\Rightarrow$.*

Then, our problem reduces to LTL model checking:

Proposition 5. *For each execution trace σ_l of $\mathcal{L}^\varphi(A)$, there is an atom sequence in $A^\varphi(A)$ that has the same truth value for φ_q as σ_l has for φ and vice versa.*

Proof. The proof is based on reordering transitions as in Prop. 2 (cf. also [18]), with $\overset{a}{\Rightarrow}$ transitions corresponding to series of action-delay transitions in $\mathcal{L}^\varphi(A)$. In addition, $\overset{\xi}{\Rightarrow}$ transitions correspond to delay transitions in automata which remain indefinitely at a state with the invariant *true*. Again, the ordering condition **O** ensures that the truth value of the formula is preserved. □

We now restrict the zone execution sequences such that the execution traces included herein satisfy the fairness condition **F**. Otherwise, the local-time model may contain traces that stop executing some automata and do not correspond to any trace in the standard model. The fairness condition **F** is violated if the execution trace does not make infinite time progress in some automaton, i.e., if the growth of a clock is always restricted by a state invariant. This cannot happen if any clock which is infinitely often limited by an invariant is reset infinitely often, allowing time to diverge. The fairness constraint can thus be written in terms of the structure of the automaton, $\bigwedge_{x \in C} \mathbf{GF}x.bounded \Rightarrow \mathbf{GF}x.reset$. The model checking problem on the initial network of automata is thus reduced to LTL model checking of a finite Kripke structure with a set of fairness constraints.

A stronger fairness constraint restricts the atom graph $A^{\varphi}(A)$ to zones that are *synchronizable*, i.e., contain at least one *synchronized* configuration (with $\bar{v}(t_i) = \bar{v}(t_j)$ for all $i, j \in \overline{1,n}$). This ensures that no more zones are explored in the local-time zone automaton than in the standard zone automaton, and the reduction is applied to a state space which is not larger than the original one. This guarantee comes at the expense of an additional check for the enabledness of transition $\stackrel{a}{\Rightarrow}$ in a given atom z, namely that $succ_l^Z(z,a)$ be synchronizable.

3.7 Building a Finite Model

The local-time zone automaton can be infinite, since difference bounds on clocks can become arbitrarily large. In [2], a finite quotient is shown to exist, but no method to compare local-time zones for equivalence is given. We show that, just as for the standard zone automaton, the actual value of time bounds does not affect the enabledness of transitions, once a certain value is exceeded. Hence, each local-time zone can be normalized to obtain a finite model.

We adapt the *maximization* (rounding) operation described, e.g., in [17] to the local-time model. Let $c_{\max}$ be the maximum absolute value of all constants in the automaton A and the formula φ. Adapting the region graph construction of [1], two valuations $\bar{v}$ and $\bar{v}'$ are called *region-equivalent* (written $\bar{v} \simeq_{\text{reg}} \bar{v}'$) if for any time variables $t_u, t_v \in T^+$, either $\lfloor \bar{v}(t_u) - \bar{v}(t_v) \rfloor = \lfloor \bar{v}'(t_u) - \bar{v}'(t_v) \rfloor$ or both differences have the same sign and are greater in absolute value than $c_{\max}$. Region equivalence extends to configurations by defining $(s, \bar{v}) \simeq_{\text{reg}} (s', \bar{v}')$ iff $s = s'$ and $\bar{v} \simeq_{\text{reg}} \bar{v}'$. *Regions* are the equivalence classes induced by $\simeq_{\text{reg}}$ on the set of configurations Σ_C. It is straightforward to show:

Lemma 1. *Let $\bar{v} \simeq_{\text{reg}} \bar{v}'$. Then:*
1. *If ψ is any constraint in A or in the specification φ, then $\bar{v} \in \psi$ iff $\bar{v}' \in \psi$.*
2. *For any clock set R, $\bar{v}[R \mapsto 0] \simeq_{\text{reg}} \bar{v}'[R \mapsto 0]$.*
3. *For $i \in \overline{1,n}$ and $d \geq 0$ there exists $d' \geq 0$ such that $\bar{v} +_i d \simeq_{\text{reg}} \bar{v}' +_i d'$.*

Since Lemma 1 covers all operations involved in executing a transition, the following property follows (cf. [1]):

Proposition 6. *Let $\gamma \simeq_{\text{reg}} \gamma'$ be two region-equivalent configurations in Σ_C.*
1. *If $\gamma \stackrel{a}{\to} \gamma_1$, there exists $\gamma_1' \simeq_{\text{reg}} \gamma_1$ such that $\gamma' \stackrel{a}{\to} \gamma_1'$.*
2. *If $\gamma \stackrel{d}{\rightsquigarrow}_i \gamma_1$, there exists $d' \in \mathbb{R}^+$ and $\gamma_1' \simeq_{\text{reg}} \gamma_1$ such that $\gamma' \stackrel{d'}{\rightsquigarrow}_i \gamma_1'$.*

The maximization of a zone z is the set of configurations which have some region-equivalent configuration in z: $\max(z) = \{\gamma' \in \Sigma_C \mid \exists \gamma \in z.\, \gamma \simeq_{\text{reg}} \gamma'\}$. A maximized zone is therefore a convex union of regions. It is easily seen that a maximized zone is obtained from the canonical representation of a zone by modifying all constraints involving constants $\pm c'$ with $c' > c_{\max}$: $t_u - t_v \prec -c'$ becomes $t_u - t_v < -c_{\max}$ and $t_u - t_v \prec c'$ becomes $t_u - t_v < \infty$ (trivially true). Furthermore, by point (1) of Lemma 1, a maximized atom is in turn an atom. Define $succ_l^M(z,a) = \max(succ_l^Z(z,a))$ and let $\mathcal{M}_l^{\varphi}(A)$ be the atom graph induced by $succ_l^M$ through repeated application from an initial zone. Since the constants in a maximized zone are bounded, it follows that $\mathcal{M}_l^\varphi(A)$ is finite.

By Prop. 6, the same transitions are enabled in every point of a region. Since a maximized atom is the closure of an atom with respect to region equivalence, this implies that the atom graph $\mathcal{A}^\varphi(A)$ and the maximized atom graph graph $\mathcal{M}^\varphi(A)$ are bisimilar. Putting the previous results together, we obtain the following theorem, which reduces our initial problem to LTL model checking with fairness constraints on a finite model:

Theorem 5. *The model $\mathcal{M}_l^\varphi(A)$ with the fairness constraint* $\mathbf{F}$ *is equivalent to the standard model $\mathcal{S}(A)$ with respect to the formula φ.*

3.8 Partial Order Reduction

Partial order reduction constructs only a representative part of the state space of a model, while preserving the verified property. This is done by exploring a subset of the enabled transitions at each states, instead of the entire set. Several criteria for choosing the subset of explored transitions have been developed. We follow the approach of Peled [14], in which the selected transitions are denoted as an *ample set* and have to satisfy the following conditions:
C0 *Emptiness:* $ample(s) = \emptyset$ iff $enabled(s) = \emptyset$.
C1 *Ample decomposition:* On any path from any state s, a transition in $ample(s)$ appears before the first transition dependent on a transition in $ample(s)$.
C2 *Invisibility:* If $ample(s) \neq enabled(s)$, all transitions in $ample(s)$ are invisible.
C3 *Cycle closing:* A transition enabled in every state of a cycle in the reduced state graph belongs to the ample set of some state on that cycle.

Having established the visible transitions in the model $\mathcal{M}_l^\varphi(A)$, one needs to determine the transition dependence relation. Bengtsson et al. [2] give a purely structural dependence relation, identical to that for untimed parallel composition: two transitions are independent if the two sets of automata involved in each of them are disjoint. This condition is sufficient for the local-time model $\mathcal{L}(A)$, as shown by Theorem 1. Since transitions in the zone automaton are composed of action and local delay transitions in the local-time model, the commutativity relation also follows for the zone automaton:

$$succ_l^Z(succ_l^Z(z,a),b) = succ_l^Z(succ_l^Z(z,b),a) \text{ if } active(a) \cap active(b) = \emptyset$$

However, in the local-time zone automaton, just like in the standard zone automaton, transitions which are both enabled in a zone may actually be enabled in different sets of configurations belonging to that zone.

Let A_1 and A_2 be two automata with clock sets $\{x, u\}$ and $\{y, v\}$, and consider a zone that is reached after executing two synchronization transitions, one resetting x and y, and the second resetting u and v. Thus, we have $t_x = t_y$ and $t_u = t_v$. Assume now that transition a in A_1 has enabling condition $x - u = t_u - t_x < 2$ and transition b in A_2 requires $y - v = t_v - t_y > 3$. Since $t_u - t_x = t_v - t_y$ due to the previous synchronizations, the two conditions cannot be satisfied simultaneously. Exploring either of $\stackrel{a}{\Rightarrow}$ and $\stackrel{b}{\Rightarrow}$ restricts the current local-time zone to a fragment where the other transition is no longer enabled.

Consequently, when selecting an ample set of transitions, one needs to check, just as for full state exploration, whether for every configuration in the current zone each of the explored automata is either be forced to execute an action transition or allows indefinite time progress. Otherwise, a potential deadlock exists. For a local transition, this check can be made statically by analyzing the invariant of the originating state together with the guard condition of the transition. This gives us a practical condition for the selection of an ample set:

Proposition 7. *In a sync-reset network of automata, a local transition in a process with a single clock does not disable transitions in other automata.*

Proof. Given local transition a in automaton A_i with a single clock x, the constraints in the enabling condition of a can be of the form $t_x - t_i \prec c$ and $t_i - t_x \prec c$. In a sync-reset network of automata, the representation of a local-time constraint links t_i only to clocks in the same automaton, i.e., to t_x. Therefore, the conjunction $\psi_l \wedge \psi_a$ does not induce stronger constraints on the other time variables and does not affect the enabledness of transitions in other automata. □

Based on the above results, we can use the ample set approach [14] to construct a reduced model for the automaton $\mathcal{M}_l^{\varphi}(A)$, and perform model checking by composing it with the tableau for the LTL formula [16].

4 Conclusions and Future Work

We have presented a method that allows the application of partial order reduction to systems modeled as a composition of timed automata. The method results in reduction in the state space, as well as in the number of clock zones that are generated for each control state. Compared to previous related work, this paper shows that partial order reduction can be used for model checking of properties described in a timed extension of linear temporal logic, rather than just for local reachability analysis. Furthermore, for a certain class of automata, we show that the local-time zones can be represented as efficiently as standard clock zones. We also analyze the dependence relation between transitions in the new model and give practical conditions for selecting an ample set.

An implementation of the presented algorithm is in progress, and we expect it to support the theoretical claims for efficiency improvement with experimental results. We also plan to extend the technique to models with other variants of synchronization, such as timed automata with deadlines. Of particular interest is a detailed comparison of the present approach with techniques developed for other timed models, such as time Petri nets and timed event level structures, and possible improvements that can result from here. Finally, we plan to explore how partial order reduction can be used for other finite quotient representations of timed automata, such as the region graph construction.

Acknowledgements. The author wishes to thank his advisor, Ed Clarke, for careful reading and suggestions on improving earlier drafts of this paper.

References

1. R. Alur and D. Dill. Automata for modeling real-time systems. In *Automata, Languages, and Programming. 17th Int. Colloquium Proc.*, *LNCS* v. 443, pp. 322-35, Coventry, UK, July 1990. Springer.
2. J. Bengtsson, B. Jonsson, J. Lilius, and Wang Yi. Partial order reductions for timed systems. In *CONCUR'98: Concurrency Theory. 8th Int. Conf. Proc.*, *LNCS* v. 1466, pp. 485-500, Nice, France, September 1998. Springer.
3. W. Belluomini and C. J. Myers. Verification of timed systems using POSETs. In *Computer Aided Verification. 10th Int. Conf., CAV'98. Proc.*, *LNCS* v. 1427, pp. 403-15, Vancouver, BC, Canada, June 1998. Springer.
4. *Computer Aided Verification. 2nd Int. Conf., CAV'90. Proc.*, *LNCS* v. 531, New Brunswick, NJ, USA, June 1990. Springer.
5. E. M. Clarke and E. A. Emerson. Design and synthesis of synchronization skeletons using branching time temporal logic. *Logic of Programs: Workshop, Yorktown Heights, NY*, *LNCS* v. 131, pp. 52-71. Springer, May 1981.
6. D. Dams, R. Gerth, B. Knaack, and R. Kuiper. Partial-order reduction techniques for real-time model checking. In *Proc. 3rd Int. Workshop on Formal Methods for Industrial Critical Systems*, pp. 157-69, Amsterdam, The Netherlands, May 1998.
7. D. L. Dill. Timing assumptions and verification of finite-state concurrent systems. *Proc. Int. Workshop Automatic Verification Methods for Finite State Systems.*, *LNCS* v. 407, pp. 197-212, Grenoble, June 1989. Springer.
8. P. Godefroid. Using partial orders to improve automatic verification methods. In [4], pp. 176-85.
9. T. A. Henzinger, X. Nicollin, J. Sifakis, and S. Yovine. Symbolic model checking for real-time systems. In *Proc. Seventh Ann. IEEE Symp. on Logic in Computer Science*, pp. 394-406, Santa Cruz, CA, USA, June 1992. IEEE Comp. Soc. Press.
10. J. Lilius. Efficient state space search for time Petri nets. In *Proc. MFCS'98 Workshop on Concurrency*, Brno, Czech Republic, August 1998. Elsevier.
11. K. G. Larsen, P. Pettersson, and Wang Yi. Compositional and symbolic model-cheking of real-time systems. In *Proc. 16th IEEE Real-Time Systems Symp.*, pp. 76-87, Pisa, Italy, Dec. 1995. IEEE Comp. Soc. Press.
12. F. Pagani. Partial orders and verification of real-time systems. In *Formal Techniques in Real-Time and Fault-Tolerant Systems. 4th Int. Symp. Proc.*, *LNCS* v. 1135, pp. 327-46, Uppsala, September 1996. Springer.
13. F. Pagani. *Ordres partiels pour la vérification de systèmes temps réel (Partial orders for verification of real-time systems)*. PhD thesis, Centre d'Études et de Recherches de Toulouse, September 1997.
14. D. Peled. All from one, one for all: on model checking using representatives. In *Computer Aided Verification. 5th Int. Conf., CAV'93. Proc.*, *LNCS* v. 697, pp. 409-23, Elounda, Greece, June 1993. Springer.
15. A. Valmari. A stubborn attack on state explosion. In [4], pp. 156-65.
16. M. Y. Vardi and P. Wolper. An automata-theoretic approach to automatic program verification. In *Proc. Symp. on Logic in Computer Science*, pp. 332-44, Cambridge, MA, USA, June 1986. IEEE Comp. Soc. Press.
17. H. Wong-Toi. *Symbolic Approximations for Verifying Real-Time Systems*. PhD thesis, Stanford University, December 1994.
18. T. Yoneda and B.-H. Schlingloff. Efficient verification of parallel real-time systems. *Formal Methods in System Design*, 11(2):197-215, August 1997.

On the Semantics of Place/Transition Nets

Rémi Morin and Brigitte Rozoy

L.R.I., Bât. 490, Université de Paris Sud, 91405 Orsay Cedex, France

Abstract. We present several interpretations of the behavior of P/T nets in terms of traces, event structures, and partial orders. Starting from results of Hoogers, Kleijn and Thiagarajan, we show how Petri nets determine local trace languages; these may be represented by local event structures in many ways, each method leading to a particular coreflection. One of these semantics is finally proved to be appropriate for the construction of a behavior preserving unfolding of Petri nets.

1 Introduction

Petri nets are a well-known and widely used model for concurrent systems. Several interpretations of their behavior have been studied in the literature since the seminal papers of Mazurkiewicz [13] and Nielsen, Plotkin and Winskel [17].

As far as Place/Transition nets are concerned, two different interpretations have been followed. On the one hand, Meseguer, Montanari and Sassone [14] have adopted the "individual token philosophy" for which it matters which token is used for the firing of a transition when several ones are available. In other words, tokens are provided with individual identities. In that way they succeeded to lift formally to level of Place/Transition nets the natural relationship between 1-safe nets and prime event structures [17]. This is based on a notion of unfolding by means of decorated occurrence nets which generalise Engelfriet's branching processes [3]. On the other hand, the "collective token philosophy" does not distinguish between tokens in the same place. In this direction, Hoogers, Kleijn and Thiagarajan have developped a *trace semantics* for Place/Transition nets in which the behavior of a net is essentially described by equivalence classes of multiset firing sequences [7]. Furthermore they have extended the strong relationship between 1-safe nets and prime event structures to the level of Place/Transition nets and *local event structures* [8]. Finally van Glabbeek and Plotkin [5] have also followed the "collective token philosophy" and presented an unfolding construction for Place/Transition nets which is proved to respect several behavioral equivalences.

In this paper, we adopt the second approach. We study some new semantics of Place/Transition nets and compare them to known ones with the help of examples. We present several interpretations of the behavior of Place/Transition nets in terms of traces, events structures and partial orders. We also introduce a new construction of unfolding for Petri nets which preserves and reflects the concurrent behaviors of the nets.

We first explain how the trace semantics of Hoogers, Kleijn and Thiagarajan [7] may be adapted to the simpler formalism of local trace languages [11].

Following the "collective token philosophy", the behaviors of a Petri net are described here by a set of equivalent firing sequences which form a local trace language. Moreover the local trace languages which correspond to a Petri net are characterized by a generalized regional criterion.

Next we come to the core of the paper. We recall that the model of local event structures was introduced by Hoogers, Kleijn and Thiagarajan [8] in order to lift to the level of Place/Transition nets the classical connection between 1-safe nets and prime event structures. Actually only a specific subclass of local event structures, required to satisfy a so-called "unique occurrence property" [8], was used as the framework of an event structure semantics of Petri nets. We give a positive answer to a question of Hoogers who suggested in his thesis a simpler "unique occurrence property", and asked whether another semantics of Petri nets is possible with the corresponding local event structures [6, p.139]. As described in Fig. 1, in order to represent the behavior of a Petri net by a local event structure, we first consider its corresponding local trace language and then apply some recent results which link local trace languages and local event structures [11]. We show that *the answer to Hoogers' question relies on a new more sophisticated method to translate the local trace languages of Petri nets into local event structures.* On the other hand, we prove that the corresponding partial orders of configurations admit a simple axiomatic characterization.

In fact, our main result gives a solution to a more generic problem which provides theoretically several other event structure semantics of Petri nets formalized by coreflections. Similarly to the approach of [17], each of these semantics determines a notion of unfolding. However, these constructions suffers from a major drawback which prevents any extension of the useful notion of finite partial unfolding [12,4]: they admit infinitely many places even if the underlying event structure is finite. That is why we propose a new unfolding construction which is similar to the classical notion of [17] — and different from the proposal of [5] — in that it essentially relies on an occurrence net. The main property of this construction is *a one-to-one correspondence between the multiset firing sequences of a net and those of its unfolding.*

However, for technical reasons, our study mainly concerns Petri nets without auto-concurrency, i.e. two instances of the same transition cannot occur concurrently at any marking. In the future, we should extend the results of [11] in order to avoid this restriction. Moreover, the approach followed in this paper might also be applied to some more general classes of Petri nets, e.g. with capacities, read arcs or inhibitor arcs [9,15].

2 Local Trace Languages

In recent years, several generalizations of the classical Mazurkiewicz traces have been proposed in order to describe a concurrency between actions which can depend on the context [7,18,10]. In [7], Hoogers, Kleijn and Thiagarajan have introduced *generalized trace languages* in order to lift the semantical theory of 1-safe nets to the level of Place/Transition nets. We show in this section how

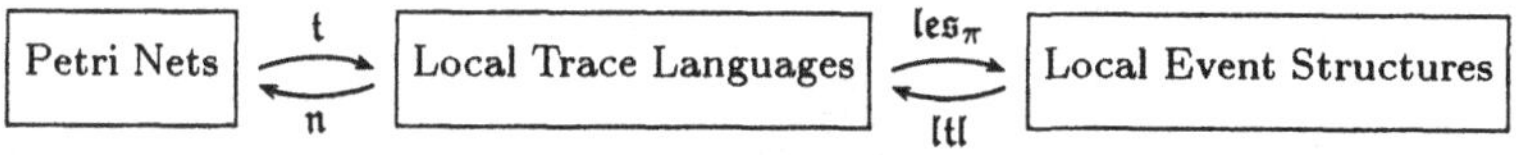

FIG. 1. Back and forth between Petri nets and local event structures.

their results adapt to the simpler formalism of local trace languages of [10]: any Petri net naturally determines a local trace language; moreover the local trace languages which correspond to a Petri net are characterized by a so-called regional criterion.

Basic Notions and Notations. We will use the following notations: for any (possibly infinite) alphabet Σ, and any words $u \in \Sigma^\star$, $v \in \Sigma^\star$, we write $u \leq v$ if u is a prefix of v, i.e. there is $z \in \Sigma^\star$ such that $u.z = v$; the empty word is denoted by ε. We write $|u|_a$ for the number of occurrences of $a \in \Sigma$ in $u \in \Sigma^\star$ and $\wp_f(\Sigma)$ denotes the set of finite subsets of Σ. We note $\mathcal{M}_f(\Sigma)$ the set of finite multisets over Σ and for any multisets p_1 and p_2 over Σ, we write $p_1 \subseteq p_2$ if $\forall a \in \Sigma$, $p_1(a) \leq p_2(a)$. For any $p \in \mathcal{M}_f(\Sigma)$, $\mathrm{Lin}(p) = \{u \in \Sigma^\star \mid \forall a \in \Sigma, |u|_a = p(a)\}$ is the set of linearisations of p. If $\lambda : \Sigma \rightharpoonup \Sigma'$ is a partial function from Σ to Σ', we also write $\lambda : \Sigma^\star \to \Sigma'^\star$ and $\lambda : \mathcal{M}_f(\Sigma) \to \mathcal{M}_f(\Sigma')$ to denote the naturally associated monoid morphisms.

DEFINITION 2.1. *A* Petri net *is a quadruple* $\mathcal{N} = (S, T, W, M_{in})$ *where*
- S *is a set of places and* T *is a set of transitions such that* $S \cap T = \emptyset$;
- W *is a map from* $(S \times T) \cup (T \times S)$ *to* $\mathbb{N}$, *called* weight function*;*
- M_{in} *is a map from* S *to* $\mathbb{N}$, *called* initial marking.

Given a Petri net $\mathcal{N} = (S, T, W, \mathrm{M}_{in})$, $\mathrm{Mar}_\mathcal{N}$ denotes the set of all markings of $\mathcal{N}$ that is to say functions $\mathrm{M} : S \to \mathbb{N}$; a multiset p of transitions is *enabled* at $\mathrm{M} \in \mathrm{Mar}_\mathcal{N}$ if $\forall s \in S$, $\mathrm{M}(s) \geq \sum_{t \in T} p(t).W(s,t)$; in this case, we note $\mathrm{M}\,[p\rangle\,\mathrm{M}'$ where $\mathrm{M}'(s) = \mathrm{M}(s) + \sum_{t \in T} p(t).(W(t,s) - W(s,t))$ and say that the transitions of p may be *fired* concurrently and lead to the marking M'. A *multiset firing sequence* consists of a sequence of markings $\mathrm{M}_0, \ldots, \mathrm{M}_n$ and a sequence of multisets of transitions $p_1, \ldots, p_n$ such that $\mathrm{M}_0 = \mathrm{M}_{in}$ and $\forall k \in [1, n]$, $\mathrm{M}_{k-1}\,[p_k\rangle\,\mathrm{M}_k$.

From Nets to Traces. The fundamental principle of generalized Mazurkiewicz traces is that the independence relation between actions in a given configuration depends on the sequence of actions that lead to this configuration.

DEFINITION 2.2. *A* local independence relation *on* Σ *is a non-empty subset* I *of* $\Sigma^\star \times \mathcal{M}_f(\Sigma)$. *The* trace equivalence $\sim$ *induced by* I *is the least equivalence on* $\Sigma^\star$ *such that*

TE_1*:* $\forall u, u' \in \Sigma^\star, \forall a \in \Sigma, u \sim u' \Rightarrow u.a \sim u'.a$;

TE_2*:* $\forall (u,p) \in I, \forall p' \subseteq p, \forall v_1, v_2 \in \mathrm{Lin}(p'), u.v_1 \sim u.v_2$.

A (local) trace *is an* $\sim$*-equivalence class* $[u]$ *of a word* $u \in \Sigma^\star$.

Consider for instance the Petri net $\mathcal{N}_1$ of Fig. 2. In the initial marking, transitions a and c can be fired concurrently whereas the step $\{a, b\}$ is not enabled; thus

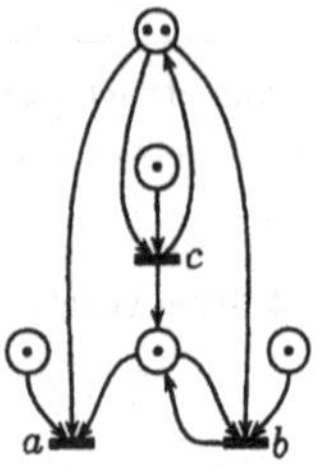

FIG. 2. $\mathcal{N}_1$

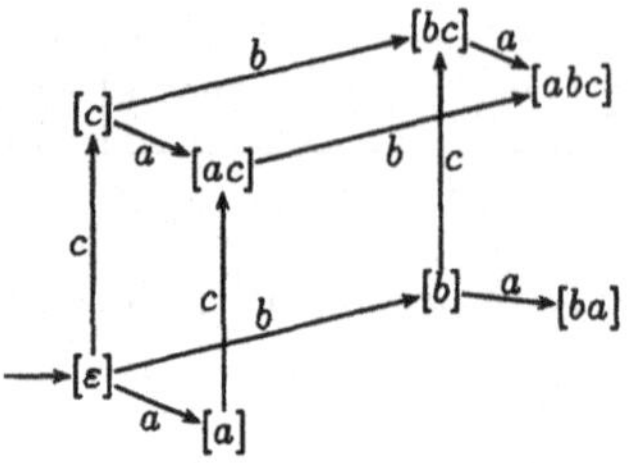

FIG. 3. $\mathcal{L}_1$

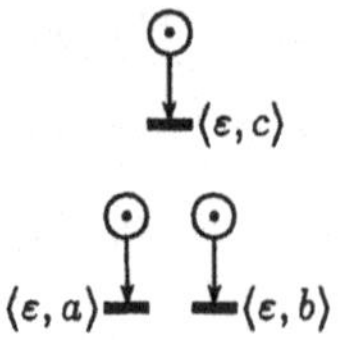

FIG. 4. $\mathcal{N}_1^o$

the associated local independence relation I should satisfy $(\varepsilon, \{a, c\}) \in I$ and $(\varepsilon, \{a, b\}) \notin I$. Yet, after c has fired, a and b become independent so $(c, \{a, b\}) \in I$. More generally, we adopt naturally the following definition.

DEFINITION 2.3. *Let $\mathcal{N} = (S, T, W, M_{in})$ be a Petri net. The associated local independence relation is $I_{\mathcal{N}} = \{(a_1...a_n, p) \in T^\star \times \mathcal{M}_f(T) \mid M_{in} [a_1\rangle M_1 ... [a_n\rangle M_n \wedge p$ is enabled at $M_n\}$.*

A local independence relation which represents a Petri net clearly satisfies some natural properties, formalized by Axioms LTL_1,..., LTL_4 of the next definition, which make it *complete* in the sense of [10]. Therefore it corresponds to a local trace language as defined in [11] and is naturally associated to a prefix-closed set of sequential observations defined by Axiom LTL_5 below.

DEFINITION 2.4. *A* local trace language *(LTL) over Σ is a structure $\mathcal{L} = (\Sigma, I, L)$ where $L \subseteq \Sigma^\star$ and I is a local independence relation on Σ such that*

LTL_1: $(u, p) \in I \wedge p' \subseteq p \Rightarrow (u, p') \in I$;
LTL_2: $(u, p) \in I \wedge p' \subseteq p \wedge v \in \mathrm{Lin}(p') \Rightarrow (u.v, p \setminus p') \in I$;
LTL_3: $u \sim u' \wedge (u, p) \in I \Rightarrow (u', p) \in I$;
LTL_4: $(u.a, \emptyset) \in I \Rightarrow (u, \{a\}) \in I$;
LTL_5: $u \in L \Leftrightarrow (u, \emptyset) \in I$.

The local trace language associated to a Petri net $\mathcal{N} = (S, T, W, M_{in})$ is $\mathrm{t}(\mathcal{N}) = (T, I_{\mathcal{N}}, L)$ where the local independence relation $I_{\mathcal{N}}$ is defined in Def. 2.3 and the set of sequential observations is $L = \{u \in T^\star \mid (u, \emptyset) \in I_{\mathcal{N}}\}$.

For instance, the local trace language corresponding to the Petri net $\mathcal{N}_1$ of Fig. 2 admits eight traces (depicted in Fig. 3) which are equivalence classes of sequential observations.

... and Back. Now, the characterization of the local trace languages which correspond to a Petri net is known to rely on a notion of region [7].

DEFINITION 2.5. *A* region *of $\mathcal{L} = (\Sigma, I, L)$ is a triple $\rho = (Mar_\rho, Pre_\rho, Post_\rho)$ where Mar_ρ, Pre_ρ and $Post_\rho$ are maps: $Mar_\rho : L \to \mathbb{N}$, $Pre_\rho : \Sigma \to \mathbb{N}$, $Post_\rho : \Sigma \to \mathbb{N}$ such that*

Reg_1: $\forall u \in L$: $Mar_\rho(u) = Mar_\rho(\varepsilon) + \sum_{a \in \Sigma} |u|_a.(Post_\rho(a) - Pre_\rho(a))$;
Reg_2: $\forall (u, p) \in I$: $Mar_\rho(u) \geq \sum_{a \in \Sigma} p(a).Pre_\rho(a)$.

A region ρ is *trivial* if $\forall a \in \Sigma$, $Pre_\rho(a) = Post_\rho(a) = 0$. We denote by $\mathcal{R}_\mathcal{L}$ the set of non trivial regions of $\mathcal{L}$. Note here that the map Mar_ρ is entirely specified by $Mar_\rho(\varepsilon)$, Pre_ρ, and $Post_\rho$.

THEOREM 2.6. *A local trace language $\mathcal{L} = (\Sigma, I, L)$ is the language of a Petri net iff it satisfies the following regional condition:*
$\forall u \in L$, $\forall p \in \mathcal{M}_f(\Sigma)$: $\left(\forall \rho \in \mathcal{R}_\mathcal{L}, Mar_\rho(u) \geq \sum_{a\in\Sigma} p(a).Pre_\rho(a)\right) \Rightarrow (u,p) \in I$.
Moreover, in this case, $\mathcal{L}$ is precisely the language of the Petri net $\mathfrak{n}(\mathcal{L}) = (\mathcal{R}_\mathcal{L}$, Σ, W, $M_{in})$ such that

$$
\begin{array}{rcl|rcl}
W : (\mathcal{R}_\mathcal{L} \times \Sigma) \cup (\Sigma \times \mathcal{R}_\mathcal{L}) & \to & \mathbb{N} & M_{in} : \mathcal{R}_\mathcal{L} & \to & \mathbb{N} \\
(\rho, a) & \mapsto & Pre_\rho(a) & \rho & \mapsto & Mar_\rho(\varepsilon) \\
(a, \rho) & \mapsto & Post_\rho(a) & & &
\end{array}
$$

Proof. We establish a correspondence between the local trace languages of Def. 2.4 and the generalized trace languages of [7] which satisfies axioms $(D1)$, $(D2)$, $(D3)$, $(PN1)$, $(PN2)$, and $(PN3)$ of that paper. Then the regional criterion $(PN4)$ of [7] adapts to the local trace languages and regions presented here. □

We should stress here that the Petri net $\mathfrak{n}(\mathcal{L})$ associated to a local trace language $\mathcal{L}$ appears to be the *maximal* Petri net whose language is $\mathcal{L}$ — in particular it admits an infinite number of places. This property will allow to prove that $\mathfrak{n}$ is the left adjoint of the translation $\mathfrak{t}$ (Def. 2.4 and Th. 2.9). However, some approaches, such as the synthesis problem of bounded nets [1], use a limited number of regions and places. This will be also the case in the third section of this paper devoted to a notion of unfolding for Petri nets.

Universality of the Constructions. We provide here Petri nets and local trace languages with behavior preserving morphisms and adapt again results of Hoogers' thesis [6] to obtain a coreflection between the two models [19].

DEFINITION 2.7. *[16, 8] A morphism from $\mathcal{N} = (S, T, W, M_{in})$ to $\mathcal{N}' = (S', T'$, W', $M'_{in})$ is a pair (α, β) of partial functions $\alpha : T \rightharpoonup T'$ and $\beta : S' \rightharpoonup S$ such that*
- *$\forall s' \in S'$, if $\beta(s')$ is defined then $M'_{in}(s') = M_{in}(\beta(s'))$,*
- *$\forall t \in T$, $\forall s' \in S'$, $W'(s', \alpha(t)) = W(\beta(s'), t) \wedge W'(\alpha(t), s') = W(t, \beta(s'))$, with the convention that $W(x,y) = 0$ if x or y is undefined.*

We denote by $\mathbb{PN}^+$ the category of Petri nets.

Local trace languages are equipped with morphisms which preserve independencies between actions.

DEFINITION 2.8. *A morphism λ from $\mathcal{L} = (\Sigma, I, L)$ to $\mathcal{L}' = (\Sigma', I', L')$ is a partial function $\lambda : \Sigma \rightharpoonup \Sigma'$ such that $\forall (u,p) \in I$, $(\lambda(u), \lambda(p)) \in I'$. We denote by $\mathbb{LTL}^+$ the category of local trace languages provided with these morphisms.*

The synthesis problem solved by Th. 2.6 can now be expressed in a categorical framework.

THEOREM 2.9. *The full subcategory $\mathbb{LTL}^+_{\mathbb{PN}}$ of local trace languages satisfying the regional condition of Th. 2.6 is coreflective into the category of Petri nets.*

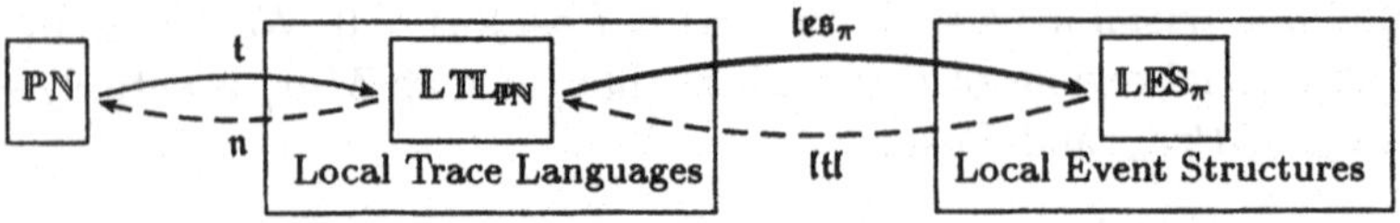

FIG. 5. From co-safe Petri nets ($\mathbb{PN}$) to π-singular local event structures ($\mathbb{LES}_\pi$).

Proof. The maps $\mathfrak{t} : \mathbb{PN}^+ \to \mathbb{LTL}^+_{\mathbb{PN}}$ and $\mathfrak{n} : \mathbb{LTL}^+_{\mathbb{PN}} \to \mathbb{PN}^+$ extend to adjoint functors which form a coreflection: $\mathbb{LTL}^+_{\mathbb{PN}} \hookrightarrow \mathbb{PN}^+$. This results from an isomorphism between $\mathbb{LTL}^+$ and the category of local trace languages of [6]. □

Connection with the Next Sections. In the rest of this paper, we exclude auto-concurrency from the behavior of Petri nets; thus we restrict our study to the class $\mathbb{LTL}$ of local trace languages $\mathcal{L} = (\Sigma, I, L)$ such that whenever $(u,p) \in I$ then p is a *set*. Consequently, and following [8] for the same technical reasons, we will only consider *co-safe* Petri nets, that is to say nets whose corresponding local trace language is without auto-concurrency.

3 Local Event Structures

In this section, we study some new connections between co-safe Petri nets and local event structures. Introduced in [8], this model is a powerful extension of classical event structures. Actually only a specific subclass of local event structures, required to satisfy a so-called "unique occurrence property", was used by Hoogers, Kleijn and Thiagarajan as the framework of an event structure semantics of co-safe Petri nets. The aim of this section is to give a positive answer to a question of Hoogers who suggested in his thesis [6, p.139] a simpler "unique occurrence property" and asked whether another semantics of Petri nets is possible with the corresponding local event structures.

In order to represent the behavior of a Petri net by a local event structure, we will first consider its corresponding local trace language defined in Section 1 (Fig. 1) and then use some recent results of [10,11] which establish a generic link between local trace languages and local event structures (Fig. 5). *In that way, Hoogers' suggestion and the local event structure semantics of [8] appear as two instances of a more abstract problem.* Then we observe that the semantics of [8] immediately results from Section 1 and the main result of [10] whereas Hoogers' question does not: as shown by an example, this latter needs a more sophisticated method to translate local trace languages of Petri nets into local event structures. On the other hand, the corresponding partial orders of configurations admit a simple axiomatic characterization (Th. 3.16).

In fact, our main result (Th. 3.11) gives a solution to the more abstract problem which provides theoretically several other event structures semantics of Petri nets, one of which will be studied in details in the third section.

Local Event Structures Are Co-Safe Petri Nets. Local event structures were defined in [8] as families of configurations of events provided with an enabling relation which specifies the local independencies between events.

DEFINITION 3.1. *A* local event structure *(LES) is a triple* $\mathcal{E} = (E, C, \vdash)$ *where* E *is a set of events,* $C \subseteq \wp_f(E)$ *is a set of finite subsets of events called configurations and* $\vdash \subseteq C \times \wp_f(E)$ *is an enabling relation such that*

LES_1: $(\emptyset \vdash \emptyset) \wedge (\forall e \in E, \exists c \in C, e \in c)$;
LES_2: $\forall c \in C$: $c \neq \emptyset \Rightarrow \exists e \in c, c \setminus \{e\} \vdash \{e\}$;
LES_3: $\forall c \in C, \forall p \in \wp_f(E)$: $c \vdash p \Rightarrow c \cap p = \emptyset$;
LES_4: $\forall c \in C, \forall p \in \wp_f(E), \forall p' \subseteq p$: $c \vdash p \Rightarrow (c \vdash p' \wedge c \cup p' \vdash p \setminus p')$.

LES_1 guarantees that the empty set is always a configuration and that the enabling relation is never empty. Also by LES_1, each event occurs in at least one configuration. LES_2 ensures that every non-empty configuration can be reached from the (initial) empty configuration. LES_3 implies that each event occurs at most once and by LES_4 each concurrent set can be split arbitrarily into subsets of concurrent events. To each local event structure $\mathcal{E}$ a set of finite sequential observations can be associated; these are called the paths of $\mathcal{E}$; formally, $\text{Paths}(\mathcal{E}) = \{e_1...e_n \in E^\star \mid \forall i \in [1, n], \{e_1, ..., e_{i-1}\} \vdash \{e_i\}\}$. As shown in [8], an event appears at most once along a path and each path u leads to a unique configuration $\text{Cfg}(u)$ defined by $\text{Cfg}(u) = \{e \mid |u|_e = 1\}$.

As noticed in [10], it is easy to associate to each local event structure $\mathcal{E}$ a local trace language $\mathfrak{ltl}(\mathcal{E})$ with the same sequential observations and a local independence relation faithfully representing the concurrency in $\mathcal{E}$.

DEFINITION 3.2. *Let* $\mathcal{E} = (E, C, \vdash)$ *be a local event structure. The local trace language* $\mathfrak{ltl}(\mathcal{E})$ *associated to* $\mathcal{E}$ *is* $\mathfrak{ltl}(\mathcal{E}) = (E, I, \text{Paths}(\mathcal{E}))$ *where* $I = \{(u, p) \in \Sigma^\star \times \wp_f(\Sigma) \mid u \in \text{Paths}(\mathcal{E})$ *and* $\text{Cfg}(u) \vdash p\}$.

With help of Th. 2.6 and similarly to a result of [8], we establish the following proposition which asserts that the trace language of a local event structure is the trace language of a co-safe Petri net; thus any local event structure $\mathcal{E}$ may be identified with the Petri net $\mathfrak{n} \circ \mathfrak{ltl}(\mathcal{E})$.

PROPOSITION 3.3. *For any local event structure* $\mathcal{E}$, $\mathfrak{ltl}(\mathcal{E})$ *is the local trace language of a co-safe Petri net.*

Now the question is to build a connection in the other direction, from co-safe Petri nets to local event structures. In the first section, we explained how each Petri net naturally specifies a local trace language. Thus we just have now to translate local trace languages into local event structures. This is however a much more difficult problem.

Abstracting Events from Local Traces. We briefly recall the method introduced in [10] in order to associate a local event structure to a given local trace language $\mathcal{L} = (\Sigma, I, L)$. The problem is to abstract events; they are actually identified as equivalence classes of *prime intervals* which are pairs $(u, a) \in \Sigma^\star \times \Sigma$ such that $u.a \in L$; we write $\text{Pr}(\mathcal{L})$ for the set of prime intervals of $\mathcal{L}$.

DEFINITION 3.4. *An* equivalence of prime intervals *of* $\mathcal{L}$ *is an equivalence* $\asymp_{\mathcal{L}}$ *over* $\text{Pr}(\mathcal{L})$ *which satisfies*

Ind: $(u, \{a, b\}) \in I \wedge a \neq b \Rightarrow (u, a) \asymp_{\mathcal{L}} (u.b, a)$ *[Independence]*

Cfl: $(u,a) \in \Pr(\mathcal{L}) \wedge (u',a) \in \Pr(\mathcal{L}) \wedge u \sim u' \Rightarrow (u,a) \asymp_{\mathcal{L}} (u',a)$ *[Confluence]*
Lab: $(u,a) \asymp_{\mathcal{L}} (v,b) \Rightarrow a = b$ *[Labeling]*
Occ: $u.a \leq v.a \wedge\ (u,a) \asymp_{\mathcal{L}} (v,a) \Rightarrow u = v$ *[Occurrence Separation]*

Let $\asymp_{\mathcal{L}}$ be an equivalence of prime intervals of $\mathcal{L}$. For any word $u \in L$, the set of events in u is denoted by $\mathrm{Eve}_{\asymp_{\mathcal{L}}}(u) = \{\langle v,b\rangle_{\mathcal{L}} \mid v.b \leq u\}$, where $\langle v,b\rangle_{\mathcal{L}}$ denotes the $\asymp_{\mathcal{L}}$-class of (v,b). As established in [10], $\asymp_{\mathcal{L}}$ determines a local event structure $\mathfrak{les}_{\asymp_{\mathcal{L}}}(\mathcal{L})$ defined as follows.

DEFINITION 3.5. *The local event structure* $\mathfrak{les}_{\asymp_{\mathcal{L}}}(\mathcal{L})$ *is the triple* $(E, C, \vdash)$ *where* $C = \{Eve_{\asymp_{\mathcal{L}}}(u) \mid u \in L\}$, $E = \bigcup C$, *and*

$$c \vdash \{e_1, ..., e_n\} \Leftrightarrow \begin{cases} \exists u \in \Sigma^\star, \exists a_1, ..., a_n \in \Sigma, (u, \{a_1, ..., a_n\}) \in I \\ Eve_{\asymp_{\mathcal{L}}}(u) = c \wedge\ \forall i \in [1,n], e_i = \langle u, a_i\rangle_{\mathcal{L}} \end{cases}$$

Punctuation, Singularity, and Symmetry. In order to translate any local trace language into a local event structure, one may simply choose an equivalence of prime intervals $\asymp_{\mathcal{L}}$ for each local trace language $\mathcal{L}$ and use the construction of Def. 3.6. In [11], such a choice of equivalences was called a *punctuation.*

DEFINITION 3.6. *A* punctuation *is a family of equivalences* $\pi = (\asymp_{\mathcal{L}})_{\mathcal{L} \in \mathbf{LTL}}$ *such that each* $\asymp_{\mathcal{L}}$ *is an equivalence of prime intervals of* $\mathcal{L}$ *(Def. 3.4) and for any isomorphism* $\lambda : \mathcal{L} \to \mathcal{L}'$ *in* $\mathbb{LTL}$: $(u,a) \asymp_{\mathcal{L}} (v,b) \Rightarrow (\lambda(u), \lambda(a)) \asymp_{\mathcal{L}'} (\lambda(v), \lambda(b))$. *We write* $\mathfrak{les}_\pi$ *for the translation from local trace languages to local event structures for which* $\mathcal{L}$ *maps to* $\mathfrak{les}_{\asymp_{\mathcal{L}}}(\mathcal{L})$ *(Def 3.5).*

We note here that the equivalences which constitute a punctuation are coherent with isomorphisms; this insures that isomorphic trace languages will be represented through $\mathfrak{les}_\pi$ by the same local event structure.

On the event structure side, each punctuation π determines a particular subclass of local event structures, called π-singular, for which the translation $\mathfrak{ltl}$ of Def 3.2 is a right inverse of $\mathfrak{les}_\pi$ (see Prop. 2.9 of [11]).

DEFINITION 3.7. *A local event structure* $\mathcal{E}$ *is* π-singular *w.r.t. a punctuation* $\pi = (\asymp_{\mathcal{L}})_{\mathcal{L} \in \mathbf{LTL}}$ *if* $\forall u_1.e, u_2.e \in \mathrm{Paths}(\mathcal{E})$, $(u_1,e) \asymp_{\mathfrak{ltl}(\mathcal{E})} (u_2,e)$.

Singularity is a central notion in this section because it is bound to the "unique occurrence properties" of [8,6]. Consider first the punctuation *History* $\pi^h = (\asymp^h_{\mathcal{L}})_{\mathcal{L} \in \mathbf{LTL}}$ such that each $\asymp^h_{\mathcal{L}}$ is the least equivalence over $\Pr(\mathcal{L})$ satisfying Ind (Def. 3.4) and Cjc (Conjunction):
Cjc : $(u,a) \in \Pr(\mathcal{L}) \wedge (u',a) \in \Pr(\mathcal{L}) \wedge \mathrm{Eve}_{\asymp_{\mathcal{L}}}(u) = \mathrm{Eve}_{\asymp_{\mathcal{L}}}(u') \Rightarrow (u,a) \asymp_{\mathcal{L}} (u',a)$
It is clear that $\asymp^h_{\mathcal{L}}$ satisfies Cfl, Lab and Occ; moreover *the local event structures satisfying the "unique occurrence property" of [8] are precisely the* π^h*-singular local event structures.* This is not surprising because History was inspired by the rule of identification of events used in [8]. For instance, the local trace language $\mathcal{L}_1$ — depicted in Fig. 2 — describes the behaviors of $\mathcal{N}_1$ of Fig. 3 and admits four events according to History: $\langle\varepsilon,a\rangle$, $\langle\varepsilon,b\rangle$, $\langle\varepsilon,c\rangle$, and $\langle b,a\rangle$.

Consider now the punctuation *Configuration* $\pi^c = (\asymp^c_{\mathcal{L}})_{\mathcal{L} \in \mathbf{LTL}}$ such that each $\asymp^c_{\mathcal{L}}$ is the least equivalence over $\Pr(\mathcal{L})$ which satisfies Ind and the following condition:

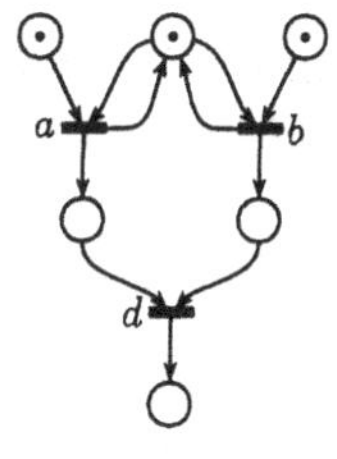

FIG. 6. $\mathcal{N}_2$

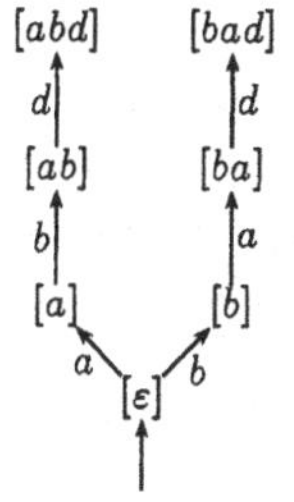

FIG. 7. $\mathcal{L}_2$

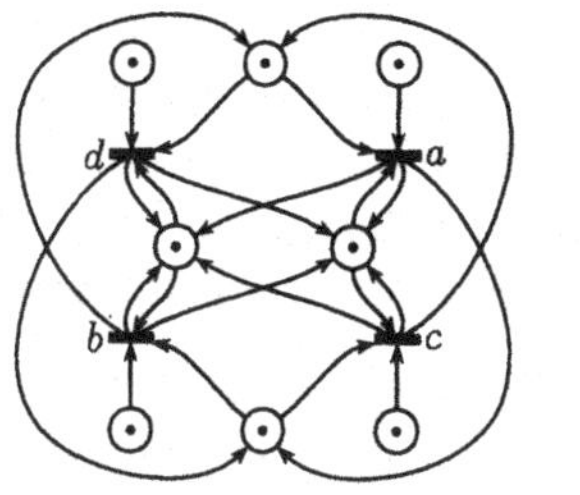

FIG. 8. $\mathcal{N}_3$

$$[(u,a) \in \Pr(\mathcal{L}) \wedge (u',a) \in \Pr(\mathcal{L}) \wedge \forall x \in \Sigma, |u|_x = |u'|_x] \Rightarrow (u,a) \asymp_{\mathcal{L}} (u',a).$$

This punctuation will play a central role here because *the local event structures satisfying the alternative "unique occurrence property" of [6, p. 139] are precisely the π^c-singular local event structures.* The point is that for any local event structure, if u and u' are two paths leading to a same configuration which enables a then (u,a) and (u',a) should be identified. For instance the local trace language $\mathcal{L}_2$ of Fig. 7 admits five events according to Configuration, but six events w.r.t. History. Note also that for any local trace language $\mathcal{L}$, $\asymp^h_{\mathcal{L}} \subseteq \asymp^c_{\mathcal{L}}$.

Now on the side of traces, an interesting subclass of local trace languages consists of those which are *symmetric* w.r.t. a punctuation:

DEFINITION 3.8. *A local trace language $\mathcal{L} = (\Sigma, I, L)$ is* π-symmetric *w.r.t. a punctuation π if $(u,p) \in I \wedge\ Eve_{\asymp_{\mathcal{L}}}(u) = Eve_{\asymp_{\mathcal{L}}}(u') \Rightarrow (u',p) \in I$.*

A crucial point here is that *any trace language describing a Petri net is π-symmetric w.r.t. any punctuation π* because if $\mathrm{Eve}_{\asymp_{\mathcal{L}}}(u) = \mathrm{Eve}_{\asymp_{\mathcal{L}}}(u')$ then u and u' contain the same actions and lead to the same marking.

From Co-Safe Petri Nets to π-Singular Local Event Structures. We now reach the core of this section. The problem is to connect the local trace languages of Petri nets with the π-singular local event structures w.r.t. a given punctuation π (Def. 3.7 and Fig. 5). For instance, such a connection w.r.t. the punctuation Configuration π^c would lead to a positive answer to Hoogers' question. Now as far as History π^h is concerned, this was achieved in [8] and results also immediately from the main result of [10] which asserts that any π^h-symmetric local trace language determines through $\mathfrak{les}_{\pi^h}$ a π^h-singular local event structure. Thus, we obtain the connection described in Fig. 5: the π^h-singular local event structure associated to a Petri net $\mathcal{N}$ is $\mathfrak{les}_{\pi^h} \circ \mathfrak{t}(\mathcal{N})$.

However, this is not always so easy, in particular with the Configuration punctuation π^c. Consider for instance the Petri net $\mathcal{N}_2$ of Fig. 6 and its corresponding local trace language $\mathcal{L}_2$ depicted in Fig. 7; we observe here $\mathfrak{les}_{\pi^c}(\mathcal{L}_2)$ is *not* π^c-singular: the point is that $(ab,d) \asymp^c_{\mathcal{L}_2} (ba,d)$ but

$$((\langle\varepsilon,a\rangle.\langle a,b\rangle, \langle ab,d\rangle) \not\asymp^c_{\mathfrak{ltl} \circ \mathfrak{les}_{\pi^c}(\mathcal{L}_2)} ((\langle\varepsilon,b\rangle.\langle b,a\rangle, \langle ba,d\rangle).$$

Therefore *the connection depicted in Fig. 5 does not hold with π^c.* However, in order to obtain a positive answer to Hoogers' question, we have to link co-safe Petri nets with π^c-singular local event structures. For that, we will exhibit a punctuation $\pi^\dagger$ for which the connection of Fig. 5 holds and such that π^c-singular local event structures are precisely $\pi^\dagger$-singular local event structures. In fact, we tackle here the more abstract problem which consists in translating local trace languages of co-safe Petri nets into π-singular local event structures. However, for technical reasons, we will restrict our study to *stable* punctuations.

DEFINITION 3.9. *A punctuation $\pi = (\asymp_{\mathcal{L}})_{\mathcal{L}\in\mathbb{LTL}}$ is* stable *if any local trace language $\mathcal{L} \in \mathbb{LTL}$ satisfies* Cjc*:*
$(u,a) \in \mathrm{Pr}(\mathcal{L}) \wedge (u',a) \in \mathrm{Pr}(\mathcal{L}) \wedge \; Eve_{\asymp_\mathcal{L}}(u) = Eve_{\asymp_\mathcal{L}}(u') \Rightarrow (u,a) \asymp_\mathcal{L} (u',a)$,
and if for any morphism $\lambda : \mathcal{L} \to \mathcal{L}'$ in $\mathbb{LTL}$:

$$(u,a) \asymp_\mathcal{L} (v,b) \Rightarrow (\lambda(u),\lambda(a)) \asymp_{\mathcal{L}'} (\lambda(v),\lambda(b)).$$

Note here that History and Configuration are stable. The crucial technical lemma of this section is the following.

LEMMA 3.10. *For any stable punctuation $\pi = (\asymp_\mathcal{L})_{\mathcal{L}\in\mathbb{LTL}}$, there is a stable punctuation $\pi^\dagger = \left(\asymp^\dagger_\mathcal{L}\right)_{\mathcal{L}\in\mathbb{LTL}}$ which satisfies the three following properties:*

- *$\pi^\dagger$-singular local event structures are exactly π-singular local event structures;*
- *for any local trace language $\mathcal{L}$, $\asymp^\dagger_\mathcal{L} \subseteq \asymp_\mathcal{L}$;*
- *if a local trace language $\mathcal{L}$ is $\pi^\dagger$-symmetric then $\mathfrak{les}_{\asymp^\dagger_\mathcal{L}}(\mathcal{L})$ is π-singular.*

Moreover, if $\mathcal{L}$ is the local trace language of a co-safe Petri net then $\asymp^\dagger_\mathcal{L}$ is the largest *equivalence of prime intervals of $\mathcal{L}$ such that $\mathfrak{les}_{\asymp^\dagger_\mathcal{L}}(\mathcal{L})$ is π-singular.*

Proof. For each local trace language $\mathcal{L}$, we consider $\asymp^\dagger_\mathcal{L}$ which is the least equivalence over $\mathrm{Pr}(\mathcal{L})$ such that

1. for any π-singular local event structure $\mathcal{E}$, for any paths $u.a$, $v.a$ of $\mathcal{E}$ and for any morphism $\lambda : \mathfrak{ltl}(\mathcal{E}) \to \mathcal{L}$: $(\lambda(u),\lambda(a)) \asymp^\dagger_\mathcal{L} (\lambda(v),\lambda(a))$;
2. $\asymp^\dagger_\mathcal{L}$ satisfies Cjc (Def. 3.9).

Because π is stable, $\asymp^\dagger_\mathcal{L} \subseteq \asymp_\mathcal{L}$ and $\pi^\dagger = \left(\asymp^\dagger_\mathcal{L}\right)_{\mathcal{L}\in\mathbb{LTL}}$ is a stable punctuation. We easily check that $\pi^\dagger$-singular local event structures are precisely π-singular local event structures. The difficult point is to prove that for any $\pi^\dagger$-symmetric local trace language $\mathcal{L}$, $\mathfrak{les}_{\pi^\dagger}(\mathcal{L})$ is $\pi^\dagger$-singular. For that we consider the morphism $\mathrm{Act}^\dagger_\mathcal{L}$ from $\mathfrak{ltl} \circ \mathfrak{les}_{\pi^\dagger}(\mathcal{L})$ to $\mathcal{L}$ such that $\langle u,a\rangle^\dagger$ maps to a and observe that $\mathrm{Act}^\dagger_\mathcal{L}$ extends to a bijection between $\mathrm{Pr}(\mathfrak{ltl} \circ \mathfrak{les}_{\pi^\dagger}(\mathcal{L}))$ and $\mathrm{Pr}(\mathcal{L})$. Then the technical point is to check that $(u,a) \asymp^\dagger_{\mathfrak{ltl}\circ\mathfrak{les}_{\pi^\dagger}(\mathcal{L})} (v,b) \Leftrightarrow (\mathrm{Act}^\dagger_\mathcal{L}(u), \mathrm{Act}^\dagger_\mathcal{L}(a)) \asymp^\dagger_\mathcal{L} (\mathrm{Act}^\dagger_\mathcal{L}(v), \mathrm{Act}^\dagger_\mathcal{L}(b))$; this implies that $\mathfrak{les}_{\pi^\dagger}(\mathcal{L})$ is $\pi^\dagger$-singular. Finally we consider $\mathcal{L} \in \mathbb{LTL}_{\mathbf{PN}}$ and $\asymp'_\mathcal{L}$ an equivalence of prime intervals of $\mathcal{L}$ such that $\mathfrak{les}_{\asymp'_\mathcal{L}}(\mathcal{L})$ is π-singular. Because $\mathcal{L} \in \mathbb{LTL}_{\mathbf{PN}}$, we can prove that the map $\mathrm{Act}'_\mathcal{L}$ for which each event $\langle u,a\rangle'$ of $\mathfrak{les}_{\asymp'_\mathcal{L}}(\mathcal{L})$ maps to a is a morphism from $\mathfrak{ltl} \circ \mathfrak{les}_{\asymp'_\mathcal{L}}(\mathcal{L})$ to $\mathcal{L}$. Now, due to Th 2.12 of [11], $\mathrm{Act}'_\mathcal{L}$ can be factorized by $\mathrm{Act}^\dagger_\mathcal{L}$: there exists a morphism λ from $\mathfrak{les}_{\asymp'_\mathcal{L}}(\mathcal{L})$ to $\mathfrak{les}_{\pi^\dagger}(\mathcal{L})$ such that $\mathrm{Act}'_\mathcal{L} = \mathrm{Act}^\dagger_\mathcal{L} \circ \lambda$. Furthermore, a simple induction on $|u|$ insures that $\lambda(\langle u,a\rangle') = \langle u,a\rangle^\dagger$. Therefore $\asymp'_\mathcal{L} \subseteq \asymp^\dagger_\mathcal{L}$. □

Consequently we obtain the main result of this section: each method to abstract events from traces determines a particular semantics of co-safe Petri nets.

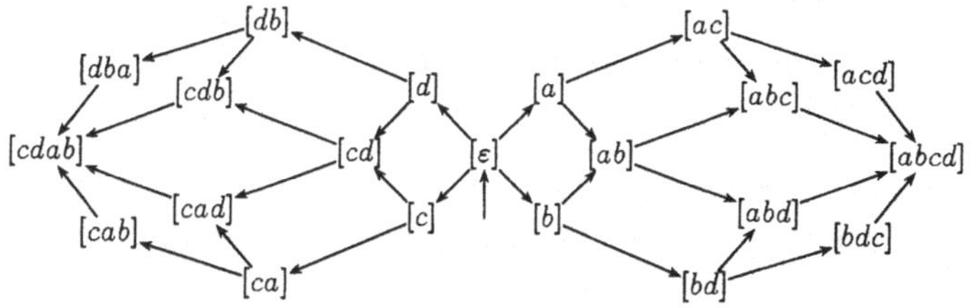

FIG. 9. $\mathcal{L}_3$

THEOREM 3.11. *Let π be a stable punctuation. Any co-safe Petri net $\mathcal{N}$ may be represented by the π-singular local event structure $\mathfrak{les}_{\pi^\dagger} \circ \mathfrak{t}(\mathcal{N})$ — by means of its corresponding local trace language $\mathfrak{t}(\mathcal{N})$ and the punctuation $\pi^\dagger$ of Lemma 3.10.*

EXAMPLE 3.12. We consider here again the punctuation Configuration π^c and the local trace language $\mathcal{L}_2$ of Fig. 7; according to Lemma 3.10, $\asymp^\dagger_{\mathcal{L}_2} \subseteq \asymp^c_{\mathcal{L}_2}$ and $\mathfrak{les}_{\asymp^\dagger_{\mathcal{L}_2}}(\mathcal{L}_2)$ is π^c-singular; therefore $(ab, d) \not\asymp^\dagger_{\mathcal{L}_2} (ba, d)$ whereas $(ab, d) \asymp^c_{\mathcal{L}_2} (ba, d)$. We should stress also here that the Petri net $\mathcal{N}_3$ of Fig. 8 and its corresponding trace language $\mathcal{L}_3$ depicted in Fig. 9 are represented by the π^c-singular local event structure $\mathfrak{les}_{\pi^c}(\mathcal{L}_3)$ which admits 4 events whereas $\mathfrak{les}_{\pi^h}(\mathcal{L}_3)$ admits 8 events; therefore π^h *and* π^c *do specify two different semantics of Petri nets.*

Universality of the Construction. We now apply the main result of [11] in order to formalize the connection of Th. 3.11 and Prop. 3.3 between π-singular local event structures and co-safe Petri nets in a categorical framework. We recall first the definition of behavior preserving morphisms of local event structures. A morphism η from $\mathcal{E} = (E, C, \vdash)$ to $\mathcal{E}' = (E', C', \vdash')$ is a partial function $\eta : E \rightharpoonup E'$ such that $\forall c \in C$, $\forall p \in \wp_f(E)$: $c \vdash p \Rightarrow \eta(c) \vdash' \eta(p)$. We denote by $\mathbb{LES}$ the category of local event structures provided with these morphisms. We also denote by $\mathbb{PN}$ the full subcategory of co-safe Petri nets and $\mathbb{LTL}_{\mathbb{PN}}$ the full subcategory of local trace languages which correspond to a co-safe Petri net.

COROLLARY 3.13. *Let π be a stable punctuation. The full subcategory $\mathbb{LES}_\pi$ of π-singular local event structures is coreflective into the category of co-safe nets.*

Proof. First, due to Th. 2.12 of [11] and Lemma 3.10, we have a coreflection $\mathfrak{ltl} : \mathbb{LES}_\pi \hookrightarrow \mathbb{LTL}_{\mathbb{PN}}$ whose right-adjoint is $\mathfrak{les}_{\pi^\dagger}$. Now the coreflection of Th. 2.9 induces obviously a coreflection $\mathbb{LTL}_{\mathbb{PN}} \hookrightarrow \mathbb{PN}$ when auto-concurrency is excluded from both models. □

Local Partial Orders versus π^c-Singular Local Event Structures. The preceding result gives, when applied with the punctuation Configuration π^c, a positive answer to Hoogers' question. Thus any co-safe Petri net may be faithfully represented by a π^c-singular local event structure. We now go further towards a more abstract model and give a simple axiomatic criterion for the partial orders associated to these particular local event structures.

DEFINITION 3.14. *Let $\mathcal{E} = (E, C, \vdash)$ be a local event structure; its associated partial order of configurations is $(C, \leq)$ where $\leq$ is the least transitive and reflexive relation on C such that $\mathsf{c} \vdash \{e\} \Rightarrow \mathsf{c} \leq \mathsf{c} \cup \{e\}$.*

Because the punctuation Configuration π^c is more general and also somehow simpler than History, we are able to characterize the partial orders of configurations of π^c-singular local event structures.

Let $(D, \leq)$ be a partially ordered set. If all elements of D are larger than a single element $x \in D$ then x is said to be the least element, denoted by $\perp$. For any $x, y \in D$, we write $x \prec y$ if $x < y$ and $\forall z \in D, x < z \leq y \Rightarrow y = z$. A chain from x to y is a sequence $z_0, ..., z_n$ in D such that $z_0 = x$, $z_n = y$ and $z_{i-1} \prec z_i$ for any $i \in [1, n]$. A prime interval of $(D, \leq)$ is a pair $[x, y]$ such that $x \prec y$; we write $[x, y] \prec [x', y']$ if $x \prec x'$, $y \prec y'$ and $x' \neq y$. Projectivity $\succ\!\prec$ is the symmetric and transitive closure of the relation $\prec$ over the prime intervals.

DEFINITION 3.15. *A partial order $(D, \leq)$ is a* local partial order *if it satisfies the following conditions:*

M: *D admits a least element $\perp$;*
F: *$\forall x \in D$, $\{y \in D \mid y < x\}$ is finite;*
R: *if $[x, y] \succ\!\prec [x', y']$ and $x = x'$ then $y = y'$;*
L: *for any chains $(x_i)_{i \in [0,n]}$ and $(y_j)_{j \in [0,m]}$ from $\perp$ to x and y respectively: $x = y$ iff for any $\succ\!\prec$-equivalence class of prime interval e,*

$$\mathrm{Card}\{i \in [1, n] \mid [x_{i-1}, x_i] \in e\} = \mathrm{Card}\{j \in [1, m] \mid [y_{j-1}, y_j] \in e\}.$$

Local partial orders satisfy some useful and classical properties. First, due to F, whenever $x \leq y$ there is a chain from x to y. Second, for any equivalence class of prime intervals e and any element x, let $n(x, e)$ denote the number of prime intervals in e in any chain from $\perp$ to x; this is well-defined because of L; as previously remarked by Droste [2], Axiom R implies that $n(x, e) \leq 1$: an equivalence class of prime intervals appears at most once along a chain.

THEOREM 3.16. *The partial order of configurations of any π^c-singular local event structure is a local partial order. Conversely, any local partial order is isomorphic to the partial order of configurations of a π^c-singular local event structure.*

4 Unfolding of Co-Safe Petri Nets

In the preceding section, we established that several identifications of events may be used to build different event structure semantics of Petri nets. From a theoretical point of view, each semantics determines a particular notion of unfolding; precisely, given a punctuation π, the unfolding of a net $\mathcal{N}$ could be defined as the Petri net describing the π-singular local event structure associated to $\mathcal{N}$ (Fig. 1, Th. 3.11, Prop. 3.3 and Fig. 5). *However this net admits an infinite number of places even if the local event structure $\mathsf{les}_\pi \circ \mathsf{t}(\mathcal{N})$ admits only a finite number of events.* That is why we present in this section a simplified unfolding construction. In this direction, we will use a new rule of identification of events which guarantees nice behavior preserving properties (Th. 4.8).

Recall first that Nielsen, Plotkin and Winskel [17] have established that the behavior of a 1-*safe* Petri net may be described by an (unfolded) *occurrence net* whose transitions correspond to the events of the *prime event structure* naturally associated to the 1-safe net. Analogously, the transitions of the unfolding of a co-safe Petri net will correspond to the events of its associated local event structure. Although this latter is *not* a prime event structure, we will extract from it notions of causality and conflict which will form the skeleton of the unfolding.

New Identification of Events. The punctuation *Unfolding* studied in this section is slightly more general than History; the main idea here is that if $u.a$ and $u'.a$ describe two sequential executions and if the events in u' only differ from those in u by occurrences of actions that are known to be independent with a, then the prime intervals (u,a) and (u',a) represent the same event.

DEFINITION 4.1. *For any local trace language* $\mathcal{L} = (\Sigma, I, L)$ *and any equivalence* $\asymp_{\mathcal{L}}$ *over* $\Pr(\mathcal{L})$*, the set of events independent with* $(u,a) \in \Pr(\mathcal{L})$ *is* $CoEve_{\asymp_{\mathcal{L}}}(u,a) = \{\langle v,b\rangle_{\mathcal{L}} \mid a \neq b \wedge (v,\{a,b\}) \in I \wedge (v,a) \asymp_{\mathcal{L}} (u,a)\}$.

The punctuation Unfolding $\pi^u = (\asymp^u_{\mathcal{L}})_{\mathcal{L}\in\mathbb{LTL}}$ *is such that each* $\asymp^u_{\mathcal{L}}$ *is the least equivalence over* $\Pr(\mathcal{L})$ *satisfying* Unf*:*

Unf*:* $(u,a) \in \Pr(\mathcal{L}) \wedge (u',a) \in \Pr(\mathcal{L}) \wedge Eve_{\asymp_{\mathcal{L}}}(u) \setminus CoEve_{\asymp_{\mathcal{L}}}(u,a) = Eve_{\asymp_{\mathcal{L}}}(u') \setminus CoEve_{\asymp_{\mathcal{L}}}(u,a) \Rightarrow (u,a) \asymp_{\mathcal{L}} (u',a)$ *[Unfolding]*

We easily check that the equivalences $\asymp^u_{\mathcal{L}}$ do exist and form a stable punctuation; in particular, Unf $\Rightarrow$ Ind $\wedge$ Cjc (Def. 3.4 and 3.9) thus $\asymp^h_{\mathcal{L}} \subseteq \asymp^u_{\mathcal{L}}$ for any local trace language $\mathcal{L}$. The following example shows that this inclusion may be strict.

EXAMPLE 4.2. We consider here again the local trace language $\mathcal{L}_1$ of Fig. 3; according to the punctuation Unfolding π^u, $(\varepsilon,a) \asymp^u_{\mathcal{L}_1} (b,a)$ whereas for History and Configuration these two prime intervals are *not* equivalent.

Now this new punctuation is easily proved to be stable (Def. 3.9) so it brings yet another event structure semantics for co-safe Petri nets (Corollary 3.13), this time with π^u-singular local event structures; moreover, similarly to History, the π^u-singular local event structure associated to a Petri net $\mathcal{N}$ is simply $\mathfrak{les}_{\pi^u} \circ \mathfrak{t}(\mathcal{N})$ (Def. 2.4 and 3.6). Formally, we have the following result.

LEMMA 4.3. $\mathfrak{les}_{\pi^u}$ *is the right-adjoint of the coreflection* $\mathfrak{ltl} : \mathbb{LES}_{\pi^u} \hookrightarrow \mathbb{LTL}_{\mathbb{PN}}$.

Construction of an Unfolded Petri Net. We consider here a co-safe Petri net $\mathcal{N} = (S, T, W, M_{in})$ and describe how the behavior of $\mathcal{N}$ may be faithfully represented by an *unfolded* net for which each transition is never fired twice along an execution. The basis of the construction of this unfolded net is the local event structure $\mathfrak{les}_{\pi^u}(\mathcal{L}) = (E, C, \vdash)$ associated to the local trace language $\mathcal{L}$ which describes the executions of $\mathcal{N}$. As formally established by Hoogers, Kleijn and Thiagarajan, any local event structure may be "completed" into a *prime event structure* [8]; here we only recall which causality and conflict relations over the events are naturally associated to $\mathfrak{les}_{\pi^u}(\mathcal{L})$: we note $e_1 \preceq e_2$ if

$\forall c \in C, e_2 \in c \Rightarrow e_1 \in c$ and $e_1 \sharp e_2$ if $\forall c \in C, e_2 \in c \Rightarrow e_1 \notin c$. We observe that $\preceq$ is a partial order over E, called *causal relation*, and $\sharp$ is a symmetric irreflexive relation, called *conflict relation*; moreover, $e_1 \sharp e_2 \preceq e_3 \Rightarrow e_1 \sharp e_3$, so the structure $(E, \preceq, \sharp)$ is a *prime event structure* [17,8]. It is well-known that any prime event structure may be represented by an occurrence net which will be use here as the skeleton of the unfolding of $\mathcal{N}$.

First, we introduce some supplementary notations. We write $e_1 \prec\!\!\!\cdot\; e_2$ if $e_1 \prec e_2$ and $e_1 \prec e_3 \preceq e_2 \Rightarrow e_3 = e_2$. The immediate conflict relation $\sharp_\mu$ is the symmetric binary relation over E such that $e_1 \sharp_\mu e_2$ if $e_1 \sharp e_2$ and for all events e_1' and e_2': $e_1' \preceq e_1 \wedge e_2' \preceq e_2 \wedge e_1' \sharp e_2' \Rightarrow (e_1 = e_1' \wedge e_2 = e_2')$. Now the occurrence net associated to the structure $(E, \preceq, \sharp)$ is simply the Petri net $\mathcal{N}^o = (S^o, E, W^o, \mathrm{M}^o_{in})$ such that the transitions are the events E, the places, called *conditions*, are $S^o = \{(e_1, e_2) \in E \times E \mid e_1 \prec\!\!\!\cdot\; e_2\} \cup \{\{e_1, e_2\} \subseteq E \mid e_1 \sharp_\mu e_2\} \cup \{(*, e_1) \mid e_1 \in E \wedge \forall e_2 \in E, e_2 \preceq e_1 \Rightarrow e_1 = e_2\}$, the weight function is given by

$$\begin{aligned} W^o : (S^o \times E) \cup (E \times S^o) &\to \mathbb{N} \\ (e_1, e_2), e' &\mapsto 1 \text{ if } e' = e_2, 0 \text{ otherwise} \\ \{e_1, e_2\}, e' &\mapsto 1 \text{ if } e' \in \{e_1, e_2\}, 0 \text{ otherwise} \\ (*, e_1), e' &\mapsto 1 \text{ if } e' = e_1, 0 \text{ otherwise} \\ e', (e_1, e_2) &\mapsto 1 \text{ if } e' = e_1, 0 \text{ otherwise} \\ e', \{e_1, e_2\} &\mapsto 0 \\ e', (*, e_1) &\mapsto 0 \end{aligned}$$

and in the initial marking M^o_{in} each condition contains one token except for the conditions $(e_1, e_2) \in S^o$ which are initially empty.

EXAMPLE 4.4. The occurrence net $\mathcal{N}_1^o$ associated to the Petri net $\mathcal{N}_1$ of Fig. 2 is depicted in Fig. 4. Obviously, this Petri net does not faithfully represent $\mathcal{N}_1$ because some of its behaviours are forbidden in $\mathcal{N}_1$; for instance the concurrent firing of the three transitions. That is why we will add some places to this occurrence net in order to restrict its behaviors and get a more faithfull representation.

Now in order to restrict the behaviors of the occurrence net $\mathcal{N}^o$, we simply add to $\mathcal{N}^o$ a copy of each place of $\mathcal{N}$ with the same initial marking; the connection between these new places and the events are given by the simple following weight function W^u in accordance with the weight function W of $\mathcal{N}$:

$$\begin{aligned} W^u : (S \times E) \cup (E \times S) &\to \mathbb{N} \\ s, \langle u, a \rangle &\mapsto W(s, a) \\ \langle u, a \rangle, s &\mapsto W(a, s) \end{aligned}$$

Thus the unfolding of $\mathcal{N}$ could be defined as $(S^o \cup S, E, W^o \cup W^u, \mathrm{M}^o_{in} \cup \mathrm{M}_{in})$. However, it may be the case that some places of $\mathcal{N}^o$ play the same role as some places of $\mathcal{N}$ and may be removed in order to get a simpler unfolding. Formally a condition $s^o \in S^o$ is *trivial* if there is a place $s \in S$ such that $\forall e \in E$, $W^o(s^o, e) = W^u(s, e) \wedge W^o(e, s^o) = W^u(e, s)$ and $\mathrm{M}^o_{in}(s^o) = \mathrm{M}_{in}(s)$. Clearly, the behavior of the unfolding does not change by removing such trivial conditions. Finally, we can sum up the definition of the unfolding of $\mathcal{N}$ in the following way.

DEFINITION 4.5. *The unfolding of the Petri net $\mathcal{N}$ is* $\mathrm{u}(\mathcal{N}) = (S', E, W', M'_{in})$ *where the set of places S' consists of the places of $\mathcal{N}$ and the non-trivial conditions of $\mathcal{N}^o$, the set of transitions E consists of the events of the local event structure $\mathcal{E} = \mathfrak{les}_{\pi^u} \circ \mathrm{t}(\mathcal{N})$ and the values of the weight function W' and the initial marking M'_{in} are those of $W^o \cup W^u$ and $M^o_{in} \cup M_{in}$ respectively.*

EXAMPLE 4.6. Continuing Example 4.4, we observe that the unfolding of the Petri net $\mathcal{N}_1$ of Fig. 2 is isomorphic to $\mathcal{N}_1$. This is different from the unfolding construction of [14] for which it matters which tokens are used for the firing of c in the initial marking. This is also different from the unfolding of [5] which admits an infinite number of events and does not rely on an occurrence net.

Main Properties of this Construction. We should stress first that the structure of the events in the underlying occurrence net is meaningful. Clearly, two events in conflict never appear together in the same firing sequence and two events in causal relation always appear in the corresponding order. Moreover we establish the following correspondence between the possible independency of two events of the unfolding $\mathrm{u}(\mathcal{N})$ and their structural concurrency within the underlying occurrence net.

LEMMA 4.7. *Two events e and e' of $\mathrm{u}(\mathcal{N})$ are neither in conflict nor in causal relation (w.r.t. the underlying occurrence net) iff the step $\{e, e'\}$ appears in a multiset firing sequence of $\mathrm{u}(\mathcal{N})$.*

This lemma is the basis of a strong connection between $\mathcal{N}$ and its unfolding.

THEOREM 4.8. *There is a bijection between the multiset firing sequences of $\mathcal{N}$ and those of its unfolding. More precisely, let φ be the map from the transitions E of the unfolding $\mathrm{u}(\mathcal{N})$ to the transitions T of $\mathcal{N}$ for which $\langle u, t\rangle$ maps to t:*

1. *for any multiset firing sequence $M'_{in} [p'_1\rangle M'_1 ... [p'_n\rangle M'_n$ in the unfolding $\mathrm{u}(\mathcal{N})$, there exists a (unique) multiset firing sequence $M_{in} [\varphi(p'_1)\rangle M_1 ... [\varphi(p'_n)\rangle M_n$ in $\mathcal{N}$.*
2. *for any multiset firing sequence $M_{in} [p_1\rangle M_1 ... [p_n\rangle M_n$ in $\mathcal{N}$, there is a* unique *multiset firing sequence $M'_{in} [p'_1\rangle M'_1 ... [p'_n\rangle M'_n$ in $\mathrm{u}(\mathcal{N})$ such that $\forall k \in [1, n]$, $\varphi(p'_k) = p_k$.*

This result guarantees that the unfolding construction offers a faithful representation of the concurrent executions of a Petri net. This extends a similar property of the classical unfolding of 1-safe nets [17]. However, because our study is restricted to co-safe Petri nets, the multiset firing sequences of a net and those of its unfolding do not admit auto-concurrency: they are all *step firing sequences* in the sense of [6].

The reader may have observed that the construction of the unfolding might have been proceeded with any other punctuation; however, one can easily check that the unfolding of Petri net $\mathcal{N}_1$ of Fig. 2 w.r.t. the punctuation History π^h would not satisfy the second property of the theorem above: the fact is that the sequence $b.a$ would correspond to two distinct sequences: $\langle \varepsilon, b\rangle.\langle b, a\rangle$ and

$\langle\varepsilon, b\rangle.\langle\varepsilon, a\rangle$ — whereas $(\varepsilon, a) \asymp^u_{\mathcal{L}_1} (b, a)$. In fact, Axiom Unf of Def. 4.1 is necessary to obtain an unfolding for which the second part of Th. 4.8 holds.

Finally, we observe that the unfolding construction is idempotent.

COROLLARY 4.9. *For any co-safe Petri net* $\mathcal{N}$, $\mathrm{u}(\mathrm{u}(\mathcal{N}))$ *is isomorphic to* $\mathrm{u}(\mathcal{N})$.

Acknowledgments. We are grateful to H.C.M. Kleijn for introducing us to Hoogers' question and for several motivating discussions on this subject. We are also grateful to E. Badouel and Ph. Darondeau for their advice to study the partial orders associated to Petri nets. Thanks to the suggestions of some anonymous referees, we were able to improve the presentation of our results.

References

1. Badouel E., Bernardinello L., Darondeau Ph.: *Polynomial algorithms for the synthesis of bounded nets.* CAAP, LNCS **915** (1995) 647–679
2. Droste M.: *Event structures and domains.* TCS **68** (1989) 37–47
3. Engelfriet J.: *Branching processes of Petri nets.* Act. Inf. **28** (1991) 575–591
4. Esparza J.: *Model Checking Using Net Unfoldings.* LNCS **668** (1993) 613–628
5. Van Glabbeek R.J., Plotkin G.D.: *Configuration Structures.* LICS **95** (1995) 199–209
6. Hoogers P.W.: *Behavioural aspects of Petri nets.* Thesis, av. at `ftp://ftp.wi.leidenuniv.nl/pub/CS/PhDTheses/hoogers-94.ps.gz` (Leiden University, 1994)
7. Hoogers P.W., Kleijn H.C.M., Thiagarajan P.S.: *A Trace Semantics for Petri Nets.* Information and Computation **117** (1995) 98–114
8. Hoogers P.W., Kleijn H.C.M., Thiagarajan P.S.: *An Event Structure Semantics for General Petri Nets.* Theoretical Computer Science **153** (1996) 129–170
9. Janicki R., Koutny M.: *Semantics of inhibitors nets.* Inf. and Comp. **123** (1995) 1–15
10. Kleijn H.C.M., Morin R., Rozoy B.: *Event Structures for Local Traces.* Electronic Notes in Theoretical Computer Science **16-2** (1998) – 16 p.
11. Kleijn H.C.M., Morin R., Rozoy B.: *Categorical Connections between Local Event Structures and Local Traces.* Technical Report 99-01, Univ. Leiden (1999) – 10 p.
12. MacMillan K.L.: *Using unfoldings to avoid the state explosion problem in the verification of asynchronous circuits.* CAV'92, LNCS **663** (1992) 165–177
13. Mazurkiewicz A.: *Concurrent program schemes and their interpretations.* Aarhus University Publication (DAIMI PB-78, 1977)
14. Meseguer J., Montanari U., Sassone V.: *On the Semantics of Petri Nets.* Concur'92, LNCS **630** (1992) 286–301
15. Montanari U., Rossi F.: *Contextual nets.* Act. Inf. **36** (1995) 545–596
16. Mukund M.: *Petri Nets and Step Transition Systems.* International Journal of Foundations of Computer Science **3** (1992) 443–478
17. Nielsen M., Plotkin G., Winskel G.: *Petri nets, events structures and domains, Part I.* Theoretical Computer Science **13** (1981) 85–108
18. Nielsen M., Sassone V., Winskel G.: *Relationships between Models of Concurrency.* LNCS **803** (1994) 425–475
19. Pierce B.C.: *Category Theory for Computer Scientists.* (The MIT Press, 1991)

Validating Firewalls in Mobile Ambients

Flemming Nielson, Hanne Riis Nielson,
René Rydhof Hansen, and Jacob Grydholt Jensen

Department of Computer Science, Aarhus University,
Ny Munkegade, DK-8000 Aarhus C, Denmark.

E-mail: {fn,hrn,rrh,grydholt}@daimi.au.dk
Web: http://www.daimi.au.dk/~fn/FlowLogic.html

Abstract. The ambient calculus is a calculus of computation that allows active processes (mobile ambients) to move between sites. A firewall is said to be protective whenever it denies entry to attackers not possessing the required passwords. We devise a polynomial time algorithm for rejecting proposed firewalls that are not guaranteed to be protective. This is based on a control flow analysis for recording what processes may turn up inside what other processes; in particular, we develop a syntax-directed system for specifying the acceptability of an analysis, we prove that all acceptable analyses are semantically sound, and we demonstrate that each process admits a least analysis.

1 Introduction

The ambient calculus is a calculus of computation that allows active processes (called mobile ambients) to move between sites; it thereby extends the notion of mobility found in Java (e.g. [8]) where only passive code may move between sites. The untyped calculus was introduced in [5] and a type system for a polyadic variant was presented in [6]. The calculus is molded on traditional process algebras (such as the π-calculus) but rather than focusing on communication (of values, channels, or processes) it focuses on the movement of processes between different sites; the sites correspond to administrative domains and are modelled using a notion of ambients. We refer to Section 2 for a review of the ambient calculus.

Since processes may evolve when moving around, it becomes harder to analyse what processes may turn up inside what other processes. In Section 3 we show how to adapt techniques from the static analysis of functional programs to develop a control flow analysis [9] for the ambient calculus. This takes the form of a syntax-directed system for specifying when the analysis results are acceptable; we then prove that all acceptable analyses are semantically sound (by means of a subject-reduction result); finally we demonstrate that each process not only admits an analysis but in fact admits a least analysis (by means of a Moore-family result) and we discuss the existence of polynomial-time algorithms.

In [5] the ambient calculus is used to model and study a firewall where only agents knowing the required passwords are supposed to enter; indeed, it is shown that all agents in a special form will in fact enter. However, it is at least as important to ensure that an attacker not knowing the required passwords cannot enter, since this would present a useful technique for screening a system against attackers. In Section 4 we use our analysis to present a polynomial-time procedure for rejecting a class of non-protective firewalls that do not ban illegal access to the internals of the non-protective firewall; this is based on identifying an attacker that is as hard to protect against as any other attacker (somewhat in the manner of hard problems for a given complexity class).

Because of space limitations we only present the analysis for the (Turing complete) core fragment of the calculus where communicaton is now allowed; however, the control flow analysis has been designed so that it scales up to the full language.

2 Mobile Ambients

Syntax. We follow the presentation in [5]; we make a syntactic distinction between capabilities (M) and namings (N) as is also implicit in the type system of [6]. The syntax of processes $P \in \mathbf{Proc}$, capabilities $M \in \mathbf{Cap}$ and namings $N \in \mathbf{Nam}$ is given by:

$P ::= (\nu n^{\mu})P$	restriction	$M ::= \mathsf{in}^{l^t} N$	enter N
$\mid\ \mathbf{0}$	inactivity	$\mid\ \mathsf{out}^{l^t} N$	exit N
$\mid\ P \mid P'$	composition	$\mid\ \mathsf{open}^{l^t} N$	open N
$\mid\ !P$	replication		
$\mid\ N^{l^a}[P]$	ambient	$N ::= n$	name
$\mid\ M.P$	movement		

To allow the analysis to deal with the α-conversion that is part of the semantics we distinguish between the *name* n introduced by a restriction operator (and that may be α-renamed) and the corresponding *stable name* $\mu \in \mathbf{SNam}$ (that cannot be α-renamed). One way to understand this distinction is to think of a name as an internet address (e.g. `daimi.au.dk`) and to think of a stable name as the corresponding absolute address (e.g. `130.225.16.40`); clearly it is possible for the internet address to change (e.g. from `daimi.aau.dk` to `daimi.au.dk`) without a similar change in the absolute address (e.g. `130.225.16.40`). Another way to understand the distinction between names and stable names is to regard the stable names as static representations of the names arising dynamically.

As is customary for the Flow Logic approach to control flow analysis [3, 4] we have also placed labels $l^a \in \mathbf{Lab}^a$ on ambients and labels $l^t \in \mathbf{Lab}^t$ on transitions – this is merely a convenient way of indicating "program points" and is useful when developing the analysis. The sets of names, stable names and labels are

$P \equiv P$	$P \mid Q \equiv Q \mid P$
$P \equiv Q \wedge Q \equiv R \Rightarrow P \equiv R$	$(P \mid Q) \mid R \equiv P \mid (Q \mid R)$
$P \equiv Q \Rightarrow Q \equiv P$	$!P \equiv P \mid !P$
$P \equiv Q \Rightarrow (\nu\, n^{\mu})P \equiv (\nu\, n^{\mu})Q$	$(\nu\, n^{\mu_n})(\nu\, m^{\mu_m})P \equiv (\nu\, m^{\mu_m})(\nu\, n^{\mu_n})P$
$P \equiv Q \Rightarrow P \mid R \equiv Q \mid R$	$\underline{\text{if}}\ n \neq m$
$P \equiv Q \Rightarrow\ !P \equiv\ !Q$	$(\nu\, n^{\mu})(P \mid Q) \equiv P \mid (\nu\, n^{\mu})Q$
$P \equiv Q \Rightarrow N^l[P] \equiv N^l[Q]$	$\underline{\text{if}}\ n \notin \underline{\text{fn}}(P)$
$P \equiv Q \Rightarrow M.\,P \equiv M.\,Q$	$(\nu\, n^{\mu})(m^l[P]) \equiv m^l[(\nu\, n^{\mu})P]$
$P \mid \mathbf{0} \equiv P$	$\underline{\text{if}}\ n \neq m$
$(\nu\, n^{\mu})\mathbf{0} \equiv \mathbf{0}$	$(\nu\, n^{\mu})P \equiv (\nu\, m^{\mu})(P\{n \leftarrow m\})$
$!\mathbf{0} \equiv \mathbf{0}$	$\underline{\text{if}}\ m \notin \underline{\text{fn}}(P)$ (α-renaming)

Table 1. Structural congruence.

left unspecified but are assumed to be non-empty; it is not essential thay they be mutually disjoint and we occasionally write $l \in \mathbf{Lab} = \mathbf{Lab}^a \cup \mathbf{Lab}^t$.

We write $\underline{\text{fn}}(P)$ for the set of *free names* of P and similarly for M and N. The *programs* of interest are ambients in the form $n_\star^{l_\star}[P_\star]$ where $n_\star \notin \underline{\text{fn}}(P_\star)$.

Example 1. Consider the following example from [5] for illustrating how an agent crosses a firewall using the prearranged passwords $\mathtt{k}$, $\mathtt{k}'$ and $\mathtt{k}''$:

$$\begin{array}{ll} \textit{Firewall}: & (\nu\, \mathtt{w}^{w})\mathtt{w}^{\mathrm{A}}[\mathtt{k}^{\mathrm{B}}[\mathsf{out}^1\mathtt{w}.\,\mathsf{in}^2\mathtt{k}'.\,\mathsf{in}^3\mathtt{w}] \mid \mathsf{open}^4\mathtt{k}'.\,\mathsf{open}^5\mathtt{k}''.P] \\ \textit{Agent}: & \mathtt{k}'^{\mathrm{C}}[\mathsf{open}^6\mathtt{k}.\,\mathtt{k}''^{\mathrm{D}}[Q]] \end{array}$$

The program of interest is $n_\star^{l_\star}[\textit{Firewall} \mid \textit{Agent}]$. We use typewriter font for names, italics for stable names, roman for ambient labels, and numbers for transition labels. □

Semantics. The semantics is given by a structural congruence relation $P \equiv Q$ and a reduction relation $P \to Q$ in the manner of the π-calculus. The congruence relation of Table 1 is a straightforward modification of a similar table in [5] with the exception that we have added the side condition "$\underline{\text{if}}\ n \neq m$" to the clause for $(\nu\, n^{\mu_n})(\nu\, m^{\mu_m})P$; in our setting it will be incorrect to have no side condition because the association between names and stable names must be maintained at all times. We write $P\{n \leftarrow m\}$ for the process that is as P but with all free occurrences of n replaced by m.

The reduction relation is given in Table 2 and is as in [5]; a pictorial representation of the three basic rules is given in Figure 1. It should be clear that the annotations in the syntax have no semantic consequences. We can make this precise as follows. Let $\mu_\bullet$ be a distinguished stable name, $l_\bullet^a$ a distinguished ambient label and $l_\bullet^t$ a distinguished transition label. Given a process P write $\lfloor P \rfloor$ for the process where all stable names are replaced by $\mu_\bullet$, all ambient labels by $l_\bullet^a$ and all transition labels by $l_\bullet^t$:

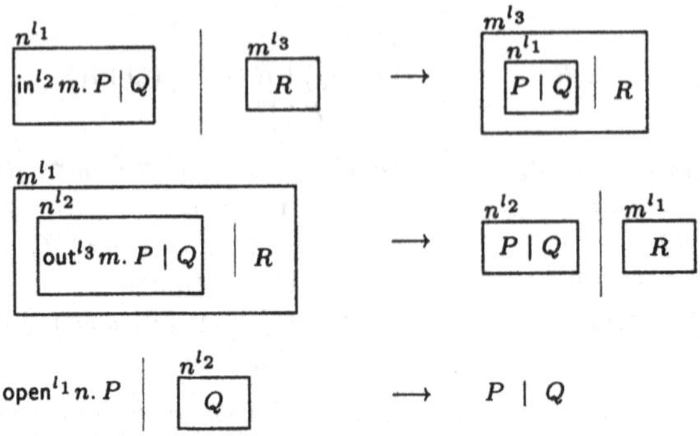

Fig. 1. Pictorial representation of the basic reduction rules.

Fact 1. $P \to^* Q \wedge \lfloor P \rfloor = \lfloor P' \rfloor \;\Rightarrow\; \exists Q' : P' \to^* Q' \wedge \lfloor Q \rfloor = \lfloor Q' \rfloor$.

The proof is by induction on the length of the derivation $P \to^* Q$; for the induction step $P \to^* R \to Q$ we proceed by induction on the shape of the inference of $R \to Q$. □

Example 2. We have the following sequence of reduction steps for $n_\star^{l_\star}[\mathit{Firewall} \mid \mathit{Agent}]$; in each step we have underlined the capability to be executed next and we have assumed that $\mathtt{w} \notin \underline{\mathrm{fn}}(Q)$.

$$
\begin{aligned}
& n_\star^{l_\star}[(\nu\,\mathtt{w}^w)\mathtt{w}^{\mathrm{A}}[\mathtt{k}^{\mathrm{B}}[\underline{\mathsf{out}^1\mathtt{w}}.\,\mathsf{in}^2\mathtt{k}'.\,\mathsf{in}^3\mathtt{w}] \mid \mathsf{open}^4\mathtt{k}'.\,\mathsf{open}^5\mathtt{k}''.\,P] \mid \mathtt{k}'^{\mathrm{C}}[\mathsf{open}^6\mathtt{k}.\,\mathtt{k}''^{\mathrm{D}}[Q]]] \\
\to\; & n_\star^{l_\star}[(\nu\,\mathtt{w}^w)(\mathtt{k}^{\mathrm{B}}[\underline{\mathsf{in}^2\mathtt{k}'}.\,\mathsf{in}^3\mathtt{w}] \mid \mathtt{w}^{\mathrm{A}}[\mathsf{open}^4\mathtt{k}'.\,\mathsf{open}^5\mathtt{k}''.\,P] \mid \mathtt{k}'^{\mathrm{C}}[\mathsf{open}^6\mathtt{k}.\,\mathtt{k}''^{\mathrm{D}}[Q]])] \\
\to\; & n_\star^{l_\star}[(\nu\,\mathtt{w}^w)(\mathtt{w}^{\mathrm{A}}[\mathsf{open}^4\mathtt{k}'.\,\mathsf{open}^5\mathtt{k}''.\,P] \mid \mathtt{k}'^{\mathrm{C}}[\mathtt{k}^{\mathrm{B}}[\mathsf{in}^3\mathtt{w}] \mid \underline{\mathsf{open}^6\mathtt{k}}.\,\mathtt{k}''^{\mathrm{D}}[Q]])] \\
\to\; & n_\star^{l_\star}[(\nu\,\mathtt{w}^w)(\mathtt{w}^{\mathrm{A}}[\mathsf{open}^4\mathtt{k}'.\,\mathsf{open}^5\mathtt{k}''.\,P] \mid \mathtt{k}'^{\mathrm{C}}[\underline{\mathsf{in}^3\mathtt{w}} \mid \mathtt{k}''^{\mathrm{D}}[Q]])] \\
\to\; & n_\star^{l_\star}[(\nu\,\mathtt{w}^w)\mathtt{w}^{\mathrm{A}}[\underline{\mathsf{open}^4\mathtt{k}'}.\,\mathsf{open}^5\mathtt{k}''.\,P \mid \mathtt{k}'^{\mathrm{C}}[\mathtt{k}''^{\mathrm{D}}[Q]]]] \\
\to\; & n_\star^{l_\star}[(\nu\,\mathtt{w}^w)\mathtt{w}^{\mathrm{A}}[\underline{\mathsf{open}^5\mathtt{k}''}.\,P \mid \mathtt{k}''^{\mathrm{D}}[Q]]] \\
\to\; & n_\star^{l_\star}[(\nu\,\mathtt{w}^w)\mathtt{w}^{\mathrm{A}}[P \mid Q]]
\end{aligned}
$$

The transition sequence shows that the firewall (which has the private name $\mathtt{w}$) sends out the pilot ambient named $\mathtt{k}$; since the agent knows the right passwords, and is in the right form, the pilot ambient can enter the agent and then guide it inside the firewall. □

$$
\begin{array}{ll}
P \to Q \Rightarrow (\nu\, n^\mu)P \to (\nu\, n^\mu)Q & n^{l_1}[\mathsf{in}^{l_2} m.\, P \mid Q] \mid m^{l_3}[R] \to m^{l_3}[n^{l_1}[P \mid Q] \mid R] \\
P \to Q \Rightarrow n^l[P] \to n^l[Q] & m^{l_1}[n^{l_2}[\mathsf{out}^{l_3} m.\, P \mid Q] \mid R] \to n^{l_2}[P \mid Q] \mid m^{l_1}[R] \\
P \to Q \Rightarrow P \mid R \to Q \mid R & \mathsf{open}^{l_1} n.\, P \mid n^{l_2}[Q] \to P \mid Q
\end{array}
$$

$$
P \equiv P' \;\wedge\; P' \to Q' \;\wedge\; Q' \equiv Q \Rightarrow P \to Q
$$

Table 2. Reduction relation.

3 Control Flow Analysis

Immediate constituents of ambients. The main aim of the analysis is to obtain the following information for each ambient: *(i)* which ambients *may* be immediately contained in it, and *(ii)* which transitions *may* it perform. An ambient will be identified by its label $l^a \in \mathbf{Lab}^a$ and a transition by its associated *stable capability* $\tilde{m} \in \mathbf{SCap}$; stable capabilities are given by

$$\tilde{m} ::= \mathsf{in}^{l^t}\,\mu \mid \mathsf{out}^{l^t}\,\mu \mid \mathsf{open}^{l^t}\,\mu$$

and correspond to capabilities except that names have been replaced by stable names. The analysis records this information in the following component:

$$I \in \mathbf{InAmb} = \mathbf{Lab}^a \to \mathcal{P}(\mathbf{Lab}^a \cup \mathbf{SCap})$$

When specifying the analysis we shall also use the "inverse" mapping I^{-1} : $(\mathbf{Lab}^a \cup \mathbf{SCap}) \to \mathcal{P}(\mathbf{Lab}^a)$ that returns the set of ambients in which the given ambient or transition might occur; formally $z \in I(l^a)$ if and only if $l^a \in I^{-1}(z)$. Later we shall write $I^+(l) \ni l'$ to mean that there exists $l_1, \cdots, l_n$ (for $n \geq 1$) such that $l = l_1$, $l' = l_n$, and $\forall i < n : I(l_i) \ni l_{i+1}$.

Stable names of ambients. Each occurrence of an ambient has a stable name and to keep track of this information the analysis also contains the following component:

$$H \in \mathbf{HNam} = \mathbf{Lab}^a \to \mathcal{P}(\mathbf{SNam})$$

Similarly to before we shall use the "inverse mapping" $H^{-1} : \mathbf{SNam} \to \mathcal{P}(\mathbf{Lab}^a)$ that returns the set of ambients that might have the given stable name; formally $\mu \in H(l^a)$ if and only if $l^a \in H^{-1}(\mu)$.

Naming environment. The association between free names and their stable names is expressed by a naming environment:

$$me \in \mathbf{MEnv} = \mathbf{Nam} \to_{\mathsf{fin}} \mathbf{SNam}$$

We shall write $me_\star$ for the initial naming environment for the program $n_\star^{l_\star}[P_\star]$ of interest and $dom(me_\star)$ for its finite domain.

Example 3. Consider the following analysis information (where the initial naming environment maps the names $\mathtt{k}$, $\mathtt{k}'$ and $\mathtt{k}''$ to k, k' and k'', respectively):

label	I	H
$l_\star$	$\{\mathrm{A}, \mathrm{B}, \mathrm{C}\}$	$\{\}$
A	$\{\mathsf{out}^1 w, \mathsf{in}^2 k', \mathsf{in}^3 w, \mathsf{open}^4 k', \mathsf{open}^5 k'', \mathsf{open}^6 k, \mathrm{A}, \mathrm{B}, \mathrm{C}, \mathrm{D}\}$	$\{w\}$
B	$\{\mathsf{out}^1 w, \mathsf{in}^2 k', \mathsf{in}^3 w\}$	$\{k\}$
C	$\{\mathsf{out}^1 w, \mathsf{in}^2 k', \mathsf{in}^3 w, \mathsf{open}^6 k, \mathrm{A}, \mathrm{B}, \mathrm{C}, \mathrm{D}\}$	$\{k'\}$
D	$\{\}$	$\{k''\}$

This shows that the ambient labelled A might perform transitions consuming any of the capabilities labelled 1–6 and that it might contain any of the ambients labelled A–D; in particular it might contain the ambient labelled C indicating that the agent might enter the firewall – and as shown in Example 2 this is indeed the case. □

3.1 The acceptability relation

The acceptability of the analysis is defined by the following four predicates (defined in Table 3 and explained below):

$(I,H) \models^{l}_{me} P$	for checking a process $P \in \mathbf{Proc}$;
$(I,H) \rhd_{me} M : \tilde{M}$	for translating a capability $M \in \mathbf{Cap}$ into a set $\tilde{M} \in \mathcal{P}(\mathbf{SCap})$ of stable capabilities;
$(I,H) \Vert\!\!\models_{me} N : \tilde{N}$	for decoding a naming $N \in \mathbf{Nam}$ into a set $\tilde{N} \in \mathcal{P}(\mathbf{SNam})$ of stable names;
$(I,H) \models^{l} \tilde{m}$	for checking a stable capability $\tilde{m} \in \mathbf{SCap}$.

The first part of Table 3 gives a simple syntax-directed definition of what it means for an analysis result (I,H) to be acceptable for the process P. The predicate is defined relative to the current naming environment me and the current label l of the enclosing ambient. The naming environment is updated whenever we pass through a restriction operator and the label is updated whenever we pass inside a new ambient. Note that the analysis cannot distinguish between whether a process occurs only once or many times: $!P$ and P are analysed in the same way (as are $P \mid P$ and P).

The clause for ambients $N^{l^a}[P]$ first checks the subprocess P using the appropriate naming environment and label. It then demands that the label of the ambient is recorded as being inside the current label. Finally, it demands that the stable name of the ambient is recorded as being a name of the ambient: Intuitively, $\tilde{N}$ is the singleton $\{me(n)\}$ when N is n; this is made precise by the third part of the table. As in Prolog, any free identifier on the righthandsides (like $\tilde{N}$) is assumed to be existentially quantified.

The clause for movement $M.P$ first checks the subprocess P using the appropriate naming environment and label. It then translates the capability M into the set of stable capabilities $\tilde{M}$ by replacing names with stable names: Intuitively $\tilde{M}$ is the singleton $\{\mathsf{in}^{l^t}\, me(n)\}$ when M is $\mathsf{in}^{l^t}\, n$ and similarly for the other capabilities; this is made precise by the second part of the table. Finally, each stable capability is analysed in turn as explained below.

The last part of Table 3 shows how to check stable capabilities against the analysis result (I,H). Figure 2 illustrates these clauses pictorially; the similarity

$(I,H) \models^{l}_{me} (\nu n^{\mu})P$	iff $(I,H) \models^{l}_{me[n\mapsto\mu]} P$
$(I,H) \models^{l}_{me} \mathbf{0}$	iff true
$(I,H) \models^{l}_{me} P \mid P'$	iff $(I,H) \models^{l}_{me} P \ \wedge\ (I,H) \models^{l}_{me} P'$
$(I,H) \models^{l}_{me} !P$	iff $(I,H) \models^{l}_{me} P$
$(I,H) \models^{l}_{me} N^{l^a}[P]$	iff $(I,H) \models^{l^a}_{me} P \ \wedge\ l^a \in I(l) \ \wedge$ $(I,H) \Vdash_{me} N : \tilde{N} \ \wedge\ \tilde{N} \subseteq H(l^a)$
$(I,H) \models^{l}_{me} M.P$	iff $(I,H) \models^{l}_{me} P \ \wedge$ $(I,H) \rhd_{me} M : \tilde{M} \ \wedge\ \forall \tilde{m} \in \tilde{M} : (I,H) \models^{l} \tilde{m}$
$(I,H) \rhd_{me} \mathsf{in}^{l^t} N : \tilde{M}$	iff $(I,H) \Vdash_{me} N : \tilde{N} \ \wedge\ \tilde{M} \supseteq \{\mathsf{in}^{l^t} \mu \mid \mu \in \tilde{N}\}$
$(I,H) \rhd_{me} \mathsf{out}^{l^t} N : \tilde{M}$	iff $(I,H) \Vdash_{me} N : \tilde{N} \ \wedge\ \tilde{M} \supseteq \{\mathsf{out}^{l^t} \mu \mid \mu \in \tilde{N}\}$
$(I,H) \rhd_{me} \mathsf{open}^{l^t} N : \tilde{M}$	iff $(I,H) \Vdash_{me} N : \tilde{N} \ \wedge\ \tilde{M} \supseteq \{\mathsf{open}^{l^t} \mu \mid \mu \in \tilde{N}\}$
$(I,H) \Vdash_{me} n : \tilde{N}$	iff $\tilde{N} \supseteq \{me(n)\}$
$(I,H) \models^{l} \mathsf{in}^{l^t} \mu$	iff $\mathsf{in}^{l^t} \mu \in I(l) \ \wedge$ $\forall l^a \in I^{-1}(\mathsf{in}^{l^t} \mu) : \forall l^{a'} \in I^{-1}(l^a) :$ $\forall l^{a''} \in I(l^{a'}) \cap H^{-1}(\mu) : l^a \in I(l^{a''})$
$(I,H) \models^{l} \mathsf{out}^{l^t} \mu$	iff $\mathsf{out}^{l^t} \mu \in I(l) \ \wedge$ $\forall l^a \in I^{-1}(\mathsf{out}^{l^t} \mu) : \forall l^{a'} \in I^{-1}(l^a) \cap H^{-1}(\mu) :$ $\forall l^{a''} \in I^{-1}(l^{a'}) : l^a \in I(l^{a''})$
$(I,H) \models^{l} \mathsf{open}^{l^t} \mu$	iff $\mathsf{open}^{l^t} \mu \in I(l) \ \wedge$ $\forall l^a \in I^{-1}(\mathsf{open}^{l^t} \mu) : \forall l^{a'} \in I(l^a) \cap H^{-1}(\mu) :$ $\forall l' \in I(l^{a'}) : l' \in I(l^a)$

Table 3. Control flow analysis.

between Figures 2 and 1 stresses the systematic way in which a control flow analysis may be developed from a formal semantics.

The clause for $\mathsf{in}^{l^t} \mu$ first ensures that the stable capability is properly recorded as part of the current ambient l. Then it ensures that all contexts l^a in which the capability could occur (and this clearly includes l) are properly recorded as being possible subambients of all sibling ambients $l^{a''}$ having the stable name μ. This involves quantifying over all possible parent ambients $l^{a'}$ and using the component H to obtain the stable name of $l^{a''}$.

The clause for $\mathsf{out}^{l^t} \mu$ follows a similar pattern. First it ensures that the stable capability is recorded as part of the current ambient l. Next it ensures that all contexts l^a in which the capability could occur (and again this includes l) are

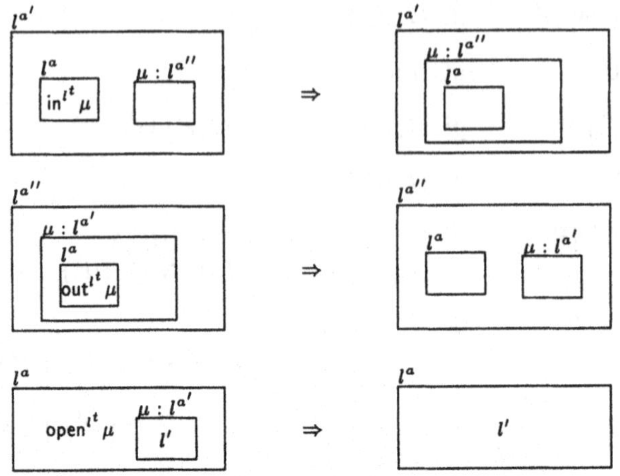

Fig. 2. Pictorial representation of the analysis of stable capabilities.

properly recorded as being possible ambients in all the possible grand parents $l^{a''}$ provided that the parent $l^{a'}$ has the stable name μ.

For the stable capability $\text{open}^{l^t}\mu$ we once again start by ensuring that it is properly recorded as part of the current ambient l. Then we consider all contexts l^a in which the capability could occur (and once more this includes l) and find all subambients $l^{a'}$ having the stable name μ; these are opened by ensuring that whatever is included in the subambient $l^{a'}$ also occurs in the parent ambient l^a.

It is crucial to observe that we need to consult all possible contexts l^a in which the capability could occur and not just the obvious candidate l. This is because, in order to establish semantic soundness, the analysis has to take into account that the current ambient might be dissolved by an open capability. (This fine point was the main difficulty that needed to be overcome when developing the analysis.)

Example 4. Let us check the condition $(I,H) \models^{A}_{me} \mathtt{k}^{B}[\mathsf{out}^1\mathtt{w}.\,\mathsf{in}^2\mathtt{k}'.\,\mathsf{in}^3\mathtt{w}]$ that arises when checking that the analysis information (I,H) of Example 3 correctly validates the program $n^{l_\star}_\star[\mathit{Firewall} \mid \mathit{Agent}]$ of Example 1; here the naming environment me maps $\mathtt{k}$, $\mathtt{k}'$, $\mathtt{k}''$ and $\mathtt{w}$ to k, k', k'' and w, respectively. First we decide to let $\tilde{N}$ be $\{k\}$. We then need to check that $(I,H) \Vdash_{me} \mathtt{k} : \{k\}$ (which follows from the choice of me), that $\{k\} \subseteq H(\text{B})$ (which follows from Example 3), that $\text{B} \in I(\text{A})$ (which once more follows from Example 3) and that $(I,H) \models^{B}_{me} \mathsf{out}^1\mathtt{w}.\,\mathsf{in}^2\mathtt{k}'.\,\mathsf{in}^3\mathtt{w}$ (see below).

To check that $(I,H) \models^{B}_{me} \mathsf{out}^1\mathtt{w}.\,\mathsf{in}^2\mathtt{k}'.\,\mathsf{in}^3\mathtt{w}$ we first decide to let $\tilde{M}$ be $\{\mathsf{out}^1 w\}$. We then need to check that $(I,H) \rhd_{me} \mathsf{out}^1\mathtt{w} : \{\mathsf{out}^1 w\}$ (which follows from $(I,H) \Vdash_{me} \mathtt{w} : \{w\}$), that $(I,H) \models^{B} \mathsf{out}^1 w$ (see below) and that $(I,H) \models^{B}_{me} \mathsf{in}^2\mathtt{k}'.\,\mathsf{in}^3\mathtt{w}$ (which amounts to twice repeating the checking illustrated for $\mathsf{out}^1\mathtt{w}$).

Finally, let us check that $(I,H) \models^{\mathrm{B}} \mathsf{out}^1 w$. First we check that $\mathsf{out}^1 w \in I(\mathrm{B})$ (using Example 3). For the second condition we have $l^a \in I^{-1}(\mathsf{out}^1 w) = \{\mathrm{A},\mathrm{B},\mathrm{C}\}$ and for each of the choices for l^a we have $l^{a'} \in I^{-1}(l^a) \cap H^{-1}(w) = \{\mathrm{A},\mathrm{C}\} \cap \{\mathrm{A}\} = \{\mathrm{A}\}$ so the parent ambient $l^{a'}$ of l^a will always be A. The grand parent of l^a is $l^{a''} \in I^{-1}(\mathrm{A}) = \{\mathrm{A},\mathrm{C}\}$ so the second condition amounts to checking that all of A, B and C are elements of both $I(\mathrm{A})$ and $I(\mathrm{C})$ and clearly this is the case. □

3.2 Properties of the analysis

In the terminology of data flow analysis [9] the above analysis is *flow-insensitive* since we ignore the order in which the capabilities occur; also it is *context-insensitive* (or *monovariant*) since a capability is analysed in the same way for all contexts in which it occurs.

Semantic correctness. Having specified what it means for an analysis result (I,H) to be acceptable the next step is to show that the notion of acceptability is semantically meaningful. We begin by establishing some auxiliary properties.

Fact 2. The analysis enjoys the following monotonicity properties:

(i) If $(I,H) \models^{l_1}_{me} P$ and $I(l_1) \subseteq I(l_2)$ then $(I,H) \models^{l_2}_{me} P$.
(ii) If $(I,H) \rhd_{me} M : \tilde{M}_1$ and $\tilde{M}_1 \subseteq \tilde{M}_2$ then $(I,H) \rhd_{me} M : \tilde{M}_2$.
(iii) If $(I,H) \Vdash_{me} N : \tilde{N}_1$ and $\tilde{N}_1 \subseteq \tilde{N}_2$ then $(I,H) \Vdash_{me} N : \tilde{N}_2$.
(iv) If $(I,H) \models^{l_1} \tilde{m}$ and $I(l_1) \subseteq I(l_2)$ then $(I,H) \models^{l_2} \tilde{m}$.

The proofs of *(ii)*, *(iii)* and *(iv)* are immediate; the proof of *(i)* is by structural induction. □

To express the next fact we shall write $me_1 =_P me_2$ to mean that me_1 and me_2 are equal on the free names of P and similarly for M and N.

Fact 3. The analysis only depends on the stable free names:

(i) If $me_1 =_P me_2$ and $(I,H) \models^{l}_{me_1} P$ then $(I,H) \models^{l}_{me_2} P$.
(ii) If $me_1 =_M me_2$ and $(I,H) \rhd_{me_1} M : \tilde{M}$ then $(I,H) \rhd_{me_2} M : \tilde{M}$.
(iii) If $me_1 =_N me_2$ and $(I,H) \Vdash_{me_1} N : \tilde{N}$ then $(I,H) \Vdash_{me_2} N : \tilde{N}$.

The proofs of *(iii)* and then *(ii)* are immediate; the proof of *(i)* is by structural induction. □

Lemma 1. If $P \equiv Q$ then $(I,H) \models^{l}_{me} P$ if and only if $(I,H) \models^{l}_{me} Q$.
The proof is by induction on the proof of $P \equiv Q$ and relies on Fact 3. □

We shall follow the approach from type systems and express the semantic correctness result as a *Subject Reduction Result*:

Theorem 1. If $(I,H) \models^{l}_{me} P$ and $P \to Q$ then $(I,H) \models^{l}_{me} Q$.

The proof is by induction on the transition $P \to Q$ and relies on Lemma 1 and Fact 2. □

As a consequence, if (I, H) is an acceptable analysis result for the program $n_\star^{l_\star}[P_\star]$ of interest (with respect to the initial naming environment $me_\star$) then it will continue being so for all the derivatives of the program.

Existence of analysis results. So far we have only shown how to check that a given pair (I, H) is indeed an acceptable analysis result; we have not studied *(i)* whether or not acceptable analysis results always exist, and if they do, *(ii)* whether or not there always is a least analysis result.

To obtain these results we shall show that the set of acceptable analysis results constitutes a *Moore family* (or has a model intersection property):

> A subset Y of a complete lattice $(L, \sqsubseteq)$ is a Moore family whenever $Y' \subseteq Y$ implies that $\sqcap Y' \in Y$.

By taking $Y' = \emptyset$ we see that a Moore family Y cannot be empty and by taking $Y' = Y$ we see that it always contains a least element; this will be essential for answering *(i)* and *(ii)* in the affirmative.

In our setting the complete lattice of interest is the set **InAmb** × **HNam** of pairs of mappings (I, H) and the ordering is the pointwise extension of the subset ordering. We then have:

Theorem 2. $\{(I, H) \mid (I, H) \models^{l}_{me} P\}$ is a Moore family for all l, me and P.
The proof shows that all of the sets

$$\{(I, H, \tilde{N}) \mid (I, H) \Vdash_{me} N : \tilde{N}\}$$
$$\{(I, H) \mid (I, H) \models^{l} \tilde{m}\},$$
$$\{(I, H, \tilde{M}) \mid (I, H) \rhd_{me} M : \tilde{M}\}$$

are Moore families and then proceeds by structural induction. □

By restricting the attention to a given program $n_\star^{l_\star}[P_\star]$ of size s one can devise an $O(s^5)$ algorithm for computing the least solution. Roughly the idea is as follows [7]. There are $O(s)$ places where conditions needs to be checked. Each condition can have length $O(s^3)$ because there are at most three nested quantifiers each ranging over $O(s)$ entities. Hence at most $O(s^4)$ basic conditions need to be checked. Since the height of each set of values is $O(s)$ this can be implemented in $O(s^5)$ basic steps using standard worklist algorithms. (We conjecture that a more sophisticated implementation will be able to achieve $O(s^4)$.)

4 Validating Firewalls

In the examples we have studied a notion of firewall given by its private name w and the passwords k, k′ and k″ used for entering it. One aspect of being a firewall is that agents in the *approved* form must be allowed to enter. For the firewall proposed in Example 1 the approved form is $\mathtt{k}'^{\mathrm{C}}[\mathsf{open}^6\mathtt{k}.\,\mathtt{k}''^{\mathrm{D}}[Q]]$ and in Example 2 we showed that agents in this form can indeed enter the firewall: $\mathit{Firewall} \mid \mathit{Agent} \rightarrow^* (\nu\, \mathtt{w}^w)\mathtt{w}^{\mathrm{A}}[P \mid Q]$ (assuming that $\mathtt{w} \notin \underline{\mathrm{fn}}(Q)$). As in [5] this can be strengthened to establish that $\mathit{Firewall} \mid \mathit{Agent}$ is observationally equivalent to $(\nu\, \mathtt{w}^w)\mathtt{w}^{\mathrm{A}}[P \mid Q]$ (assuming that $\mathtt{w} \notin \underline{\mathrm{fn}}(Q)$).

Another aspect of being a firewall, not dealt with in [5], is to ensure that processes not knowing the right passwords cannot enter. Due to the power of the ambient calculus this is not as trivial as it might appear at first sight. As an example, a process that does not initially know the passwords might nonetheless learn them by other means. As another example, the firewall might contain a trapdoor through which processes might be able to enter (see Example 5 below).

We define a process U to be *ignorant* whenever $\underline{\mathrm{fn}}(U) \cap \{\mathtt{k}, \mathtt{k}', \mathtt{k}''\} = \emptyset$. We then define a proposed firewall F to be *protective* whenever the semantics of Section 2 prevents it from allowing any ignorant process to enter.

Example 5. Consider the proposed firewall

$$\begin{aligned} \mathit{Firewall}' : (\nu\, \mathtt{w}^w)\mathtt{w}^{\mathrm{A}}[&\mathtt{k}^{\mathrm{B}}[\mathsf{out}^1\mathtt{w}.\,\mathsf{in}^2\mathtt{k}'.\,\mathsf{in}^3\mathtt{w}] \mid \mathsf{open}^4\mathtt{k}'.\,\mathsf{open}^5\mathtt{k}''.P \\ &\mid \mathtt{t}^{\mathrm{E}}[\mathsf{out}^7\mathtt{w}.\,\mathsf{in}^8\mathtt{w}.\,\mathsf{open}^9\mathtt{q}] \mid \mathsf{open}^{10}\mathtt{t}] \end{aligned}$$

that additionally contains a trapdoor $\mathtt{t}$. It is easy to check that

$$\mathit{Firewall}' \mid \mathit{Agent} \rightarrow^* (\nu\, \mathtt{w}^w)\mathtt{w}^{\mathrm{A}}[\cdots \mid P \mid Q]$$

using *Agent* of Example 1 (assuming that $\mathtt{w} \notin \underline{\mathrm{fn}}(Q)$). But now the ignorant process $\mathtt{q}^{\mathrm{F}}[\mathsf{in}^{11}\mathtt{t}.\,Q]$ can also enter as is shown by

$$\mathit{Firewall}' \mid \mathtt{q}^{\mathrm{F}}[\mathsf{in}^{11}\mathtt{t}.\,Q] \rightarrow^* (\nu\, \mathtt{w}^w)\mathtt{w}^{\mathrm{A}}[\cdots \mid P \mid Q]$$

(assuming that $\mathtt{w} \notin \underline{\mathrm{fn}}(Q)$) unlike what was intended. This means that *Firewall'* is not a protective firewall because it can be entered by a process not knowing the right passwords. □

The control flow analysis can be used to devise a test for whether or not a proposed firewall F is protective; the test is displayed in Table 4 and will be explained below. Since the control flow analysis is approximate also the test for protectiveness will be approximate; however, we shall ensure that *whenever the test is passed then no ignorant processes can enter.* We believe this to be typical of applications where software developed by subcontractors is *validated* before being embedded in the software system under construction.

Let us fix the distinct stable names w, k, k', k'', $\mu_\circ$, and $\mu_\diamond$, the distinct labels $l^a_\diamond$ and $l^t_\diamond$, and the distinct names $n_\circ$ and $n_\diamond$. Thanks to Fact 1 we may without loss of generality assume that a proposed firewall $F = (\nu\, \mathtt{w}^w)\mathtt{w}^{\mathrm{A}}[F']$ does not contain any of these distinguished symbols in the subprocess $\mathtt{w}^{\mathrm{A}}[F']$.

INPUT:	a proposed firewall $F = (\nu\, \mathtt{w}^w)\mathtt{w}^{\mathsf{A}}[F']$ without distinguished symbols in $\mathtt{w}^{\mathsf{A}}[F']$
OUTPUT:	"accept" or "reject"
METHOD:	construct $\breve{F}$ (see the text) construct $me_0 = me_\star\&(\breve{F}, \mu_\circ)$ construct $\hat{T}$ (see the text) find the least (I, H) such that $(I, H) \models^{l_\star}_{me_0} \breve{F} \mid \hat{T}$ if $\exists l : l^a_\diamond \in I^+(l) \wedge w \in H(l)$ then "reject" else "accept"

Table 4. Testing for protectiveness.

Given a process Q we shall write $\hat{Q}$ for the process where all stable names are replaced by $\mu_\diamond$, all ambient labels by $l^a_\diamond$, and all transition labels by $l^t_\diamond$. Write B for the process

$$\mathsf{in}^{l^t_\diamond} n_\diamond \mid \mathsf{out}^{l^t_\diamond} n_\diamond \mid \mathsf{open}^{l^t_\diamond} n_\diamond \mid \mathsf{in}^{l^t_\diamond} n_\circ \mid \mathsf{out}^{l^t_\diamond} n_\circ \mid \mathsf{open}^{l^t_\diamond} n_\circ$$

and define T to be

$$B \mid n_\diamond^{l^a_\diamond}[\, B \mid n_\circ^{l^a_\diamond}[\mathbf{0}]\,]$$

and note that this defines an ignorant process with $T = \hat{T}$.

Define the naming environment $me_\star$ by $me_\star(\mathbf{k}) = k$, $me_\star(\mathbf{k}') = k'$, $me_\star(\mathbf{k}'') = k''$, $me_\star(n_\circ) = \mu_\circ$, and $me_\star(n_\diamond) = \mu_\diamond$. For a naming environment me, a set X of names and a stable name μ define the naming environment $me\&(X, \mu)$ by

$$(me\&(X,\mu))(n) = \begin{cases} me(n) & \text{if } n \in dom(me) \\ \mu & \text{if } n \in X \setminus dom(me) \\ \text{undefined} & \text{if } n \notin X \cup dom(me) \end{cases}$$

and note that $(me\&(X,\mu))(n) = me(n)$ whenever $me(n)$ is defined. We shall allow to write $me\&(P,\mu)$ for $me\&(\underline{\mathrm{fn}}(P),\mu)$.

Given a proposed firewall $F = (\nu\, \mathtt{w}^w)\mathtt{w}^{\mathsf{A}}[F']$ we shall write $\breve{F}$ for the process $(\nu\, \mathtt{w}^w)\mathtt{w}^{\mathsf{A}}[F'']$ where F'' is like F' except that all stable names have been replaced by $\mu_\circ$. We have now defined all the notation used in the test displayed in Table 4. It operates on a proposed firewall F and outputs "accept" or "reject". It is clearly deterministic and given that the least (I, H) can be found in polynomial time it operates in polynomial time itself (in the size of F).

The correctness of the test hinges on the following key result; it shows that, from the point of view of the analysis, it is as hard to protect a firewall F against the process T as it is to protect the firewall F against any other ignorant process U:

Lemma 2. Let $F = (\nu\, \mathtt{w}^w)\mathtt{w}^{\mathsf{A}}[F']$ be a proposed firewall as demanded in Table 4 and let (I, H) be as in Table 4. If U is an ignorant process then

$$(I, H) \models^{l_\star}_{me'} \breve{F} \mid \hat{U}$$

where $me' = (me_\star \& (\breve{F}, \mu_\circ)) \& (\hat{U}, \mu_\diamond)$.

Proof. Write $me_0 = me_\star \& (\breve{F}, \mu_\circ)$ as in Table 4. By construction of (I, H) we have $(I, H) \models^{l_\star}_{me_0} \breve{F} \mid \hat{T}$ and using Fact 3 we get $(I, H) \models^{l_\star}_{me'} \breve{F} \mid \hat{T}$ from which $(I, H) \models^{l_\star}_{me'} \breve{F}$ and $(I, H) \models^{l_\star}_{me'} \hat{T}$ follows. By expansion of the latter we get

$$(I, H) \models^{l} \mathsf{in}^{l^t_\diamond} \mu, \qquad (I, H) \models^{l} \mathsf{out}^{l^t_\diamond} \mu, \qquad (I, H) \models^{l} \mathsf{open}^{l^t_\diamond} \mu \qquad (1)$$

for all $l \in \{l_\star, l^a_\diamond\}$ and $\mu \in \{\mu_\circ, \mu_\diamond\}$; we also obtain

$$l^a_\diamond \in I(l_\star), \qquad l^a_\diamond \in I(l^a_\diamond), \qquad \mu_\diamond \in H(l^a_\diamond), \qquad \mu_\circ \in H(l^a_\diamond) \qquad (2)$$

To conclude that $(I, H) \models^{l_\star}_{me'} \breve{F} \mid \hat{U}$ we need to prove that $(I, H) \models^{l_\star}_{me'} \hat{U}$ holds. Since U is ignorant this follows by Fact 3 from the following auxiliary result holding for arbitrary processes R:

$$(I, H) \models^{l}_{me} \hat{R}$$
for all l, me satisfying $l \in \{l_\star, l^a_\diamond\} \ \wedge \ \forall n \in \underline{\text{fn}}(R) : me(n) \in \{\mu_\circ, \mu_\diamond\}$

The proof of the auxiliary result is by structural induction on R and most of the cases are immediate so let us only consider the two interesting ones.

The case $\hat{R} = N^{l^a_\diamond}[\hat{R}_0]$: That $(I, H) \models^{l^a_\diamond}_{me} \hat{R}_0$ follows from the induction hypothesis; that $l^a_\diamond \in I(l)$ follows from (2); taking $\tilde{N} = \{\mu_\circ, \mu_\diamond\}$ we have $\tilde{N} \subseteq H(l^a_\diamond)$ from (2) and $(I, H) \Vert\!\!\models_{me} N : \tilde{N}$ is immediate.

The case $\hat{R} = \hat{M}.\,\hat{R}_0$: That $(I, H) \models^{l}_{me} \hat{R}_0$ follows from the induction hypothesis; taking $\tilde{M} = \{\mathsf{in}^{l^t_\diamond} \mu_\circ, \mathsf{out}^{l^t_\diamond} \mu_\circ, \mathsf{open}^{l^t_\diamond} \mu_\circ, \mathsf{in}^{l^t_\diamond} \mu_\diamond, \mathsf{out}^{l^t_\diamond} \mu_\diamond, \mathsf{open}^{l^t_\diamond} \mu_\diamond\}$ we have $\forall \tilde{m} \in \tilde{M} : (I, H) \models^{l} \tilde{m}$ from (1) and $(I, H) \rhd_{me} M : \tilde{M}$ is immediate. □

When F passes the test and U is an ignorant process we want to show that no subambient of U ever passes inside $\mathtt{w}$. Informally, this will take the form of assuming that $F \mid U \to^* R$ and guaranteeing that R contains no subambient $\mathtt{w}^{l^a_1}[\cdots \mathtt{u}^{l^a_2}[\cdots] \cdots]$ where $\mathtt{u}$ comes from U. Formalising this is somewhat tricky and we shall therefore avail ourselves of Fact 1 that allows us to arrange the labelling to suit our needs. Indeed, if $F \mid U \to^* R$ then $\breve{F} \mid \hat{U} \to^* R'$ for some R' such that $\lfloor R \rfloor = \lfloor R' \rfloor$.

Theorem 3. If F passes the test of Table 4 and U is an ignorant process and if $\breve{F} \mid \hat{U} \to^* R$ then R contains no subterm $n_1^{l^a_1}[\cdots n_2^{l^a_2}[\cdots] \cdots]$ where n_1 has stable name w and l^a_2 is $l^a_\diamond$.

Proof. Setting $me' = (me_\star \& (\breve{F}, \mu_\circ)) \& (\hat{U}, \mu_\diamond)$ and letting (I, H) be as in Table 4, it follows from Lemma 2 that $(I, H) \models^{l_\star}_{me'} \breve{F} \mid \hat{U}$. By Theorem 1 we also have $(I, H) \models^{l_\star}_{me'} R$. Suppose for the sake of contradiction that R does contain $n_1^{l^a_1}[\cdots n_2^{l^a_2}[\cdots] \cdots]$ where n_1 has stable name w and l^a_2 is $l^a_\diamond$. Then it follows from $(I, H) \models^{l_\star}_{me'} R$ that $l^a_\diamond \in I^+(l^a_1) \wedge w \in H(l^a_1)$ showing that the test could not have been passed. □

In summary, we have succeeded in using the control flow analysis to devise a polynomial time algorithm for ensuring that a proposed firewall is indeed protective; a web-based implementation is accessible via the Flow Logic webpage `http://www.daimi.au.dk/~fn/FlowLogic.html`.

Example 6. To test *Firewall* from Example 1 and *Firewall'* from Example 5 we need to be more precise about the subprocess P; in our tests we have used

$$!\mathsf{p}[\mathsf{in}\ \mathsf{p} \mid \mathsf{out}\ \mathsf{p} \mid \mathsf{open}\ \mathsf{p} \mid \mathsf{p}[\mathbf{0}]]$$

(omitting labels) as an example of an unrestricted internal process. Then *Firewall* passes the test because $H^{-1}(w) = \{\mathrm{A}\}$ and $I^+(\mathrm{A}) \not\ni l^a_\diamond$ but *Firewall'* fails the test because $H^{-1}(w) = \{\mathrm{A}\}$ and $I^+(\mathrm{A}) \ni l^a_\diamond$. □

5 Conclusion

It is well known that static techniques are needed for determining whether or not programs always evaluate in a permissible manner. Type systems have already been extensively used to study the properties of web-based languages and related calculi (e.g. [1,6]) but more "traditional" approaches [9] to static analysis have much to offer as well. In this paper we developed a control flow analysis for the ambient calculus building on recent developments for the pi-calculus [3,4].

The interplay between type systems and control flow analyses is not yet fully understood. While both type systems and control flow analyses can be proved semantically sound using a subject-reduction result, it would seem that only approaches based on control flow analyses admit least analyses for all processes. Indeed, often type systems (e.g. [1,6]) lack the corresponding notion of principal type, thereby making them harder to use in practice as there may be exponentially many types to consider before any conclusions can be drawn. In a subsequent paper we hope to use state-of-the-art techniques from data flow analysis to present an even stronger analysis than the control flow analysis developed here.

More importantly we demonstrated how a careful exploitation of the detailed operation of the control flow analysis allowed us to construct an attacker that was as hard to protect against as any other attacker; this is somewhat reminiscent of the identification of hard problems in a given complexity class. This allowed us to predict the operation of the firewall in conjunction with all ignorant attackers based on its operation in conjunction with the hard attacker; if it successfully protects against the hard attacker it will also protect against all other ignorant attackers.

This is a novel approach to the validation of software systems and we expect it to scale up to other calculi. Indeed, the basic machinery of the control flow analysis has already been developed for a number of calculi. Furhermore, by considering more powerful analyses expressed in the form of Flow Logics it is likely that one can reduce the gap between processes that know some of the passwords (hence are not ignorant) but still do not display them in the approved form.

References

1. M.Abadi: Secrecy by typing in security protocols. In Proceedings of Theoretical Aspects of Computer Software, volume 1281 of Lecture Notes in Computer Science, pages 611–638, Springer Verlag, 1997.
2. A.Aiken: Set constraints: Results, applications and future directions. In Proceedings of the Second Workshop on the Principles and Practices of Concurrent Programming, volume 874 of Lecture Notes in Computer Science, pages 171–179, Springer Verlag, 1994.
3. C.Bodei, P.Degano, F.Nielson, H.R.Nielson: Control Flow Analysis for the π-calculus. In Proceedings CONCUR'98, volume 1466 of Lecture Notes in Computer Science, pages 84–98, Springer Verlag, 1998.
4. C.Bodei, P.Degano, F.Nielson, H.R.Nielson: Static analysis of processes for no read-up and no write-down. In Proceedings FoSSaCS'99, volume 1578 of Lecture Notes in Computer Science, pages 120–134, Springer Verlag, 1999.
5. L.Cardelli, A.D.Gordon: Mobile Ambients. In Proceedings FoSSaCS'98, volume 1378 of Lecture Notes in Computer Science, pages 140–155, Springer Verlag, 1998.
6. L.Cardelli, A.D.Gordon: Types for Mobile Ambients. In Proceedings POPL'99, pages 79–92, ACM Press, 1999.
7. R.R.Hansen, J.G.Jensen: Flow Logics for Mobile Ambients. M.Sc.-thesis, 1999.
8. T.Jensen, D.LeMétayer, T.Thorn: Security and Dynamic Class Loading in Java: a formalisation. Report IRISA, Rennes, 1997.
9. F.Nielson, H.R.Nielson, C.Hankin: *Principles of Program Analysis*, Springer Verlag, 1999.

On Coherence Properties in Term Rewriting Models of Concurrency

Thomas Noll

Lehrstuhl für Informatik II, Aachen University of Technology
Ahornstr. 55, D–52056 Aachen, Germany
noll@@informatik.rwth-aachen.de

Abstract. This paper introduces a generic and uniform approach to integrate different design languages for distributed systems in verification tools. It is based on Meseguer's Rewriting Logic, hence transitions between the states of the respective system are modeled as (conditional) term rewriting steps modulo an equational theory. We argue that, for reasons of efficiency, it is intractable to admit arbitrary equations, and propose to employ rewriting modulo associativity and commutativity instead, using oriented versions of the equations. Furthermore the question is raised under which conditions this implementational restriction is complete. To this aim we define a coherence property which guarantees that every transition which is possible in the (fully equational) semantics can also be computed using the oriented equations, and we show that this property can be verified by testing the joinability of finitely many conditional critical pairs between transition rules and oriented equations.

1 Introduction

Because of the inherent complexity of distributed systems, tools for supporting their development become more and more indispensable. During the last years several prototypes have been developed, e.g. the *Edinburgh Concurrency Workbench* (see [5]), the *Concurrency Factory* ([4]), *Spin* ([16]), *Truth* ([17]), and the symbolic model checker *SMV* ([15]). Most of the tools are tailored for a specific syntactic and semantic setting, such as CCS with transition system semantics and μ–calculus model checking.

In order to ease the task of changing the design language accepted by the *Concurrency Workbench of North Carolina* (CWB–NC; [13]), the *Process Algebra Compiler* (PAC–NC; [14]) has been developed. Given the description of the syntax and operational semantics of a design language like CCS, it generates ML source code implementing a frontend which allows the CWB–NC to analyze systems specified in this language. However, since the semantics is specified in terms of structural operational rules, the semantic scope of this tool is restricted to (labeled) transition systems.

We want to add a further degree of freedom by allowing also the semantic domain of the design language to be specified. This goal can be achieved by employing Meseguer's *Rewriting Logic* (cf. [10]). This approach aims at a separate description of the static and of the dynamic aspects of a distributed system. More exactly, it distinguishes the laws describing the structure of the states of the system from the rules which specify its possible transitions. The two aspects are respectively formalized as a set of equations E and as a (conditional) rewrite system R. Both structures operate on states, represented as (equivalence classes of) Σ–terms where Σ is the signature of the design language under consideration. Since a single transition may comprise several independent rewriting steps, concurrent behavior can explicitly be modeled.

Rewriting Logic has successfully been applied to specify various languages and semantic domains; an overview can be found in [11]. Among others, Viry gives very natural specifications of CCS (see [18]) and of the π–calculus ([20]). However, since (conditional) term rewriting modulo arbitrary equational theories is generally too complex or even undecidable, it is hard to implement this approach directly. Instead, following the ideas of Viry in [18, 19], we propose to decompose E into a set of directed equations ER and into a set AC expressing associativity and commutativity of certain binary operators in Σ. If ER is terminating modulo AC, then rewriting by R modulo E can be implemented by a combination of normalizing by ER and rewriting by R, both modulo AC.

Since both ER and AC are contained in E, this approach is obviously sound, that is, a transition computed by the implementation is also possible in the (fully equational) semantics. However, the reverse implication, i.e. the completeness, does not always hold. In this paper we present sufficient conditions under which this property can be guaranteed. We show that the language specification has to match certain coherence properties which can be tested by inspecting a finite set of conditional critical pairs between rules of R and ER.

The remainder of this paper is organized as follows. In Sect. 2 we collect the fundamental definitions and results dealing with rewriting. Sect. 3 presents our specification formalism and its implementation, whose completeness properties are investigated in Sect. 4. Finally, Sect. 5 concludes with some remarks.

2 Preliminaries

2.1 Abstract Reduction Systems

A *reduction system* is a pair $(A, \to)$ where A is a set and $\to\ \subseteq A \times A$ is a binary relation on A. The symmetric, the transitive, the reflexive–transitive closure and the inverse of $\to$ is denoted by $\leftrightarrow$, $\to^+$, $\to^*$, and $\leftarrow$, respectively. An element $a \in A$ is called *$\to$-reducible* if there exists $b \in A$ such that $a \to b$, otherwise *$\to$-irreducible*. If $a \to^* b$, then b is called a *$\to$-successor* of a. An irreducible successor of a is called a $\rightarrow$*-normal form* of a; this is indicated by $a \rightarrow\!\!\!\!| b$. The relation $\rightarrow$ is called *confluent* if every pair of successors of some element of A possesses a common successor. It is called *terminating* if there is no infinite descending chain of the form $a_0 \rightarrow a_1 \rightarrow \ldots$.

It is well-known that in a confluent reduction system normal forms are unique if they exist. In particular, in a *convergent* (i.e., confluent and terminating) reduction system every element possesses a unique normal form.

2.2 Term Rewriting Systems

A *signature* Σ is a finite set of symbols, called *operators*, in which with every operator a natural number is associated, called its *rank*. For every $n \geq 0$, $\Sigma^{(n)}$ is the set of operators of rank n. We write $f^{(n)}$ to indicate that $f \in \Sigma^{(n)}$. Let X be an additional set of symbols, called *variables*. The set of Σ*-terms over* X is denoted by $T_\Sigma(X)$. For every $t \in T_\Sigma(X)$, $Pos(t) \subseteq \{1,2,\ldots\}^*$ and $X(t) \subseteq X$ are the set of all *positions* and the set of all variables contained in t, respectively. If every variable occurs at most once in t, then t is called *linear*. In particular, it is called *ground* if $X(t) = \emptyset$. Every position $w \in Pos(t)$ uniquely identifies a *subterm* of t, denoted by $t|_w$, where $t|_\varepsilon = t$. The tree $t[w \leftarrow s] \in T_\Sigma(X)$ is obtained from t by *replacing* the subtree at $w \in Pos(t)$ by the tree $s \in T_\Sigma(X)$. A *substitution* is a mapping $\sigma : X \rightarrow T_\Sigma(X)$; we identify it with its homomorphic extension to (tuples of) terms and denote the set of all substitutions by Sub.

A *term rewriting system (TRS)* is a set of rules $R \subseteq T_\Sigma(X)^2$, each represented as $l \rightarrow r$, where $l \notin X$ and $X(r) \subseteq X(l)$. As usual, the *rewrite relation* induced by R, $\xrightarrow{R} \subseteq T_\Sigma(X)^2$, is the smallest relation which comprises R and which is closed under arbitrary substitutions and contexts: for every $s,t \in T_\Sigma(X)$, $s \xrightarrow{R} t$ iff there exists $l \rightarrow r \in R$, $w \in Pos(s)$, and $\sigma \in Sub$ such that $s|_w = l\sigma$ and $t = s[w \leftarrow r\sigma]$. Here, $l\sigma$ is called an R*-redex* of s.

2.3 Term Rewriting Modulo Equational Theories

For a set E of *equations*, each of the form $u = v$, the same notations as above are used where usually the symmetric relation $\xleftrightarrow{E}$ is considered. Given $t \in T_\Sigma(X)$, $[t]_E$ denotes the *congruence class* of t modulo E, that is, $[t]_E = \{s \in T_\Sigma(X) \mid s \xleftrightarrow{E}^* t\}$.

For any relations $\xrightarrow{R}$ and $\xrightarrow{S}$ on $T_\Sigma(X)$, let

$$\xrightarrow{R/S} = \xrightarrow{S}^* \xrightarrow{R} \xrightarrow{S}^*$$

where juxtaposition denotes the composition of relations. In particular, when $\xrightarrow{S}$ is the replacement relation $\xleftrightarrow{E}$ induced by a set E of equations, we obtain the notion of *term rewriting modulo E*: for any $s, t \in T_\Sigma(X)$,

$$[s]_E \xrightarrow{[R]_E} [t]_E \quad \text{iff} \quad s \xrightarrow{R/E} t \quad \text{iff} \quad s \xleftrightarrow{E}{}^* \xrightarrow{R} \xleftrightarrow{E}{}^* t.$$

If $\xrightarrow{[R]_E}$ is confluent (terminating, convergent), then R is respectively said to be *confluent (terminating, convergent) modulo E*.

3 Rewriting Logic

In this section we give the syntax and semantics of the rewriting framework which we use to specify concurrent systems. It is essentially based on Meseguer's *Rewriting Logic*, as presented in [10], which exploits the observation that several models for concurrency have the notion of state and transition in common. However, they differ in their distributed structure (e.g. interleaving vs. true concurrency).

Rewriting Logic is intended to serve as a unifying mathematical model and uses notions from rewrite systems over equational theories. It aims at a separate description of the static and of the dynamic aspects of a distributed system. More exactly, it distinguishes the laws describing the structure of the states of the system from the rules which specify its possible transitions. The two aspects are respectively formalized as a set of equations E and as a (conditional) TRS R. Both structures operate on states, represented as (equivalence classes of) Σ-terms where Σ is the signature of the design language under consideration. Since a single transition may comprise several independent rewriting steps, concurrent behavior can explicitly be modeled.

Our aim is to use this approach as the formal basis of our compiler generator which, given the definition of a design language, automatically derives corresponding parsing and semantic functions which can be used as a frontend for verification tools such as *Truth* ([17]). However, since (conditional) term rewriting modulo arbitrary equational theories is generally too complex or even undecidable, we decompose E into a set of directed equations ER (in other words, a TRS) and into a set AC expressing associativity and commutativity of certain binary operators in Σ. If ER is terminating modulo AC, then rewriting by R modulo E can be implemented by a combination of normalizing by ER and rewriting by R, both modulo AC. We will see later under which assumptions this combination behaves as expected.

3.1 Syntax

Extending the notion of a rewrite theory in *Rewriting Logic* ([10]), we define the syntax of our specification formalism as follows.

Definition 1. *An* oriented rewrite theory (ORT for short) *is a quadruple* $\mathfrak{T} = (\Sigma, AC, ER, R)$ *where*

- Σ *is a signature with* $\Sigma_{AC} \subseteq \Sigma^{(2)}$ *being the set of* AC–symbols,
- $AC = \{x + (y + z) = (x + y) + z \mid + \in \Sigma_{AC}\} \cup \{x + y = y + x \mid + \in \Sigma_{AC}\}$,
- $ER \subseteq T_\Sigma(X)^2$ *is a finite TRS convergent modulo* AC, *and*
- $R \subseteq (T_\Sigma(X)^2)^+$ *is a finite set of* (conditional) transition rules, *each represented as* $\dfrac{c_1 \to c_1', \ldots, c_k \to c_k'}{l \to r}$, *where*
  - $l \notin X$,
  - l *is linear and* $\xrightarrow{ER/AC}$*-irreducible,*
  - $X(c_i) \subseteq X(l)$ *for every* $i \in \{1, \ldots, k\}$, *and*
  - $X(r) \subseteq X(l) \cup \bigcup_{i=1}^{k} X(c_i')$.

Thus we have two kinds of rules: rules in *ER* are always given an equational interpretation, that is, the oriented rewrite theory defines equivalence classes modulo $AC \cup ER$. The convergence of *ER* is required to provide unique normal forms. In contrast, rules in R describe transitions between states of the system under consideration. Here, the conditions accommodate for the fact that the behavior of a complex system may depend on the behavior of its components.

Our definition of an ORT differs from the one in [19] with regard to the following aspects:

- Viry does not take conditions in the transition rules into account, which is crucial for many applications.
- We do not necessarily assume the transition relation induced by an ORT to be congruent. This will be justified later.

In the following example we apply our formalism to the well–known CCS process algebra, the *Calculus of Communicating Systems* presented in [12], where we consider only the finite part. The exposition is inspired by the approach described in [18]. In particular, the additional binary operator $\{.\}.$ is used to simulate the (action) labels of the transition steps, which are not provided in the formal definition of the transition rules.

Example 2. The ORT $\mathfrak{T} = (\Sigma, AC, ER, R)$ is given by the following components, using the set $X = \{\alpha, x, x', y, y', z\}$ of variables:

- $\Sigma = \{\mathsf{nil}^{(0)}, .^{(2)}, +^{(2)}, \|^{(2)}, \bar{\ }^{(1)}, \tau^{(0)}, \{\ \}^{(2)}\} \cup A$, where $A = \{a^{(0)}, b^{(0)}, \ldots\}$ is a set of actions (restrictions and relabelings are omitted for simplification),
- $\Sigma_{AC} = \{+, \|\}$,
- $ER = \{z + \mathsf{nil} \to z,\ z \parallel \mathsf{nil} \to z,$
  $z + z \to z,\ \bar{\bar{\alpha}} \to \alpha\}$, and

– $R = \left\{ \frac{}{\alpha.x \to \{\alpha\}x}, \quad \frac{x \to \{\alpha\}x'}{x + y \to \{\alpha\}x'}, \quad \frac{x \to \{\alpha\}x'}{x \parallel y \to \{\alpha\}(x' \parallel y)}, \frac{x \to \{\alpha\}x', y \to \{\bar{\alpha}\}y'}{x \parallel y \to \{\tau\}(x' \parallel y')} \right\}.$

Since the equations in AC and ER identify certain states, the state space of the resulting system as well as the number of rewrite rules is reduced. For example, the symmetric counterparts of the "+" and of the first "||" rule in R are not required above because $\Sigma_{AC} = \{+, \parallel\}$.

3.2 Semantics

The (operational) semantics of ORTs expresses that a concurrent system whose current state is represented by the term s (or some equivalent thereof) can evolve to the state t provided that there exists a transition rule whose left–hand side matches s modulo AC and ER and whose conditions are fulfilled. This intuitive notion is formally described as follows where $\mathfrak{T} = (\Sigma, AC, ER, R)$ denotes an arbitrary ORT.

Definition 3. *The* semantic transition relation *of* $\mathfrak{T}$, $\xrightarrow{\mathfrak{T}^s} = \bigcup_{n \in \mathbb{N}} \xrightarrow{\mathfrak{T}^s_n}$ *with* $\xrightarrow{\mathfrak{T}^s_n} \subseteq T_\Sigma(X)^2$ *for every* $n \in \mathbb{N}$, *is inductively given by*

$$\xrightarrow{\mathfrak{T}^s_0} = \emptyset$$

$$\xrightarrow{\mathfrak{T}^s_{n+1}} = \{(s,t) \mid ex. \frac{C}{l \to r} \in R, \sigma \in Sub \text{ s.t. } s \xleftrightarrow{E}^* l\sigma, r\sigma \xleftrightarrow{E}^* t, \text{ and } c\sigma \xrightarrow{\mathfrak{T}^s_n} c'\sigma \text{ for every } c \to c' \in C\}$$

where $E = AC \cup ER$, *and where* $n \in \mathbb{N}$ *is called the* depth *of the transition.*

Note that the transition relation induced by an ORT is closed under substitutions, like in the (conditional) TRS case. However, in contrast to the original definition in [10], it is not closed under contexts since the left–hand side l of the rule has to be matched against the whole term s. In other words, its symmetric, reflexive, and transitive closure is not necessarily a congruence relation. Otherwise, in the CCS example above a transition of the form

$$a.b.\mathsf{nil} \xrightarrow{\mathfrak{T}^s} \{b\}a.\mathsf{nil}$$

would be possible which should clearly be forbidden.

However, if the congruence property is desired then it can be achieved by adding an appropriate rule for every operator. For example, the congruence with respect to an operator $f \in \Sigma^{(n)}$ can be expressed by the rule

$$\frac{x_1 \to y_1, \ldots, x_n \to y_n}{f(x_1, \ldots, x_n) \to f(y_1, \ldots, y_n)}.$$

3.3 Implementation

As explained above, the transition relation is to be implemented by exploiting the decomposition of E into AC and ER: before applying a rule from R, the current term is transformed into $\xrightarrow{ER}$-normal form modulo AC.

Definition 4. *The* implementational transition relation *of* $\mathfrak{T}$, $\xrightarrow{\mathfrak{T}^i} = \bigcup_{n\in\mathbb{N}} \xrightarrow{\mathfrak{T}^i_n}$ *with* $\xrightarrow{\mathfrak{T}^i_n} \subseteq T_\Sigma(X)^2$ *for every* $n \in \mathbb{N}$, *is inductively given by*

$$\xrightarrow{\mathfrak{T}^i_0} = \emptyset$$

$$\xrightarrow{\mathfrak{T}^i_{n+1}} = \{(s,t) \mid s \xrightarrow{ER/AC}\text{-irreducible, ex. } \frac{C}{l \to r} \in R, \sigma \in Sub \text{ s.t.}$$
$$s \xleftrightarrow{AC}{}^* l\sigma, r\sigma \xrightarrow{ER/AC}\!\!\!\!| \, t, \text{ and}$$
$$c\sigma \xrightarrow{ER/AC}\!\!\!\!| \xrightarrow{\mathfrak{T}^i_n} |\!\!\!\!\xleftarrow{ER/AC} c'\sigma \text{ for every } c \to c' \in C\}.$$

Note that $\xrightarrow{\mathfrak{T}^i}$ relates $\xrightarrow{ER/AC}$-normal forms only.

If, in addition to Definition 1, every rule $\dfrac{c_1 \to c'_1, \ldots, c_k \to c'_k}{l \to r} \in R$ satisfies the following requirements, then the implementational transition relation is decidable: for every $i \in \{1,\ldots,k\}$,

(i) c_i is a proper subterm of l, i.e., there exists $p \in Pos(l)\backslash\{\varepsilon\}$ such that $l|_p = c_i$, and
(ii) no ER–rule is applicable to any instance of a non–variable subterm of c'_i (modulo AC), i.e., for every $g \to d \in ER$ and every $p \in Pos(c'_i)$ with $c'_i|_p \notin X$, there exists no $\sigma \in Sub$ such that $(c'_i|_p)\sigma \xleftrightarrow{AC}{}^* g\sigma$. (In Sect. 4 we will see that this property is decidable.)

Under these assumptions the normalizing reductions in Definition 4 starting in $c\sigma$ and in $c'\sigma$ can obviously be omitted since both terms are $\xrightarrow{ER/AC}$-irreducible. Hence it is possible to compute the set of all direct $\xrightarrow{\mathfrak{T}^i}$-successors of a given term $s \in T_\Sigma(X)$ in $\xrightarrow{ER/AC}$-normal form using the following informal algorithm. For every rule $\dfrac{C}{l \to r} \in R$ whose left–hand side l matches s modulo AC (using the substitution $\sigma \in Sub$, i.e., $s \xleftrightarrow{AC}{}^* l\sigma$), every condition $c \to c' \in C$ has to be verified. This is done by recursively computing every $d \in T_\Sigma(X)$ with $c\sigma \xrightarrow{\mathfrak{T}^i} d$, and by extending σ such that $c'\sigma$ matches d modulo AC. If this is possible for every condition, then $t = r\sigma$ is a direct $\xrightarrow{\mathfrak{T}^i}$-successor of s.

Note that our CCS specification (see Example 2) satisfies both requirements. Property (i) above is essential to guarantee the termination of the recursive evaluation of the conditions. It would be violated if we added the CCS rule

$$\frac{x[P/\mathsf{fix}\,P.x] \to \{\alpha\}x'}{\mathsf{fix}\,P.x \to \{\alpha\}x'}$$

which describes the "unwinding" of fixpoints.

Condition (ii) prevents problems which arise from the fact that irreducible terms may become reducible under certain substitutions. The following example illustrates this situation.

Example 5. Let $\mathfrak{T} = (\Sigma, AC, ER, R)$ be given by $\Sigma = \{f^{(2)}, a^{(0)}, b^{(0)}, c^{(0)}\}$, $\Sigma_{AC} = \emptyset$, $ER = \{f(b,c) \to b\}$, and $R = \left\{\frac{}{a \to f(b,c)}, \frac{x \to f(y,z)}{f(x,y) \to z}\right\}$.

According to Definition 4, the term $s = f(a,b)$ possesses the direct $\xrightarrow{\mathfrak{T}^i_2}$-successor $t = c$: under the substitution $\sigma = [x/a, y/b, z/c]$, we have $s = f(x,y)\sigma$, $x\sigma = a \xrightarrow{R} f(b,c) \xrightarrow{ER} b$ (and hence $x\sigma \xrightarrow{\mathfrak{T}^i_1} b$), $f(y,z)\sigma = f(b,c) \xrightarrow{ER} b$, and $z\sigma = c = t$.

On the other hand, the above algorithm would proceed as follows.

(i) The left–hand side $f(x,y)$ is matched against $s = f(a,b)$, using the substitution $\tau = [x/a, y/b]$.
(ii) As above, the recursive evaluation of the condition's left–hand side ($x\tau = a \xrightarrow{R} f(b,c) \xrightarrow{ER} b$) yields $x\tau \xrightarrow{\mathfrak{T}^i} b$.
(iii) However, the instantiated right–hand side $f(y,z)\tau = f(b,z)$ is $\xrightarrow{ER}$-irreducible. Thus, the validity of the condition $x \to f(y,z)$ can only be shown by "guessing" the substitution $[z/c]$, which is of course impossible. Clearly, requirement (ii) above disables such situations.

However, for the remainder of the paper the above properties are not required.

The following observation can easily be proved using induction on n and exploiting the fact that $\xrightarrow{ER/AC}{}^* \subseteq \xleftrightarrow{E}{}^*$.

Lemma 6. *For every* $n \in \mathrm{N}$,

(i) $\xrightarrow{\mathfrak{T}^s_n} \subseteq \xrightarrow{\mathfrak{T}^s_{n+1}}$,
(ii) $\xrightarrow{\mathfrak{T}^i_n} \subseteq \xrightarrow{\mathfrak{T}^i_{n+1}}$, *and*
(iii) $\xrightarrow{\mathfrak{T}^i_n} \subseteq \xrightarrow{\mathfrak{T}^s_n}$.

In particular, part (iii) of this lemma yields the fact that the implementational relation $\xrightarrow{\mathfrak{T}^i}$ is *sound* with respect to the semantic relation $\xrightarrow{\mathfrak{T}^s}$:

Corollary 7. *For every* $s, t \in T_\Sigma(X)$, $s \xrightarrow{ER/AC}\!\!| \xrightarrow{\mathfrak{T}^i} t$ *implies* $s \xrightarrow{\mathfrak{T}^s} t$.

4 Completeness of the Implementation

Now we will consider completeness, i.e., we have to discuss the question whether every transition which is possible in the (fully equational) semantics can also be computed using the oriented equations.

4.1 Level Coherence

The following definition presents a sufficient criterion. It expresses that the choice of the $\xrightarrow{ER/AC}$-normal form must not restrict the potential for transitions. Note that similar properties have been investigated in the setting of term rewriting modulo equational theories (cf. [8]).

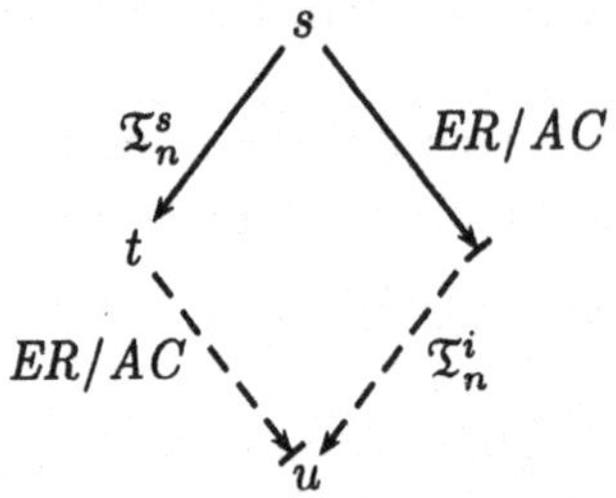

Definition 8. *$\mathfrak{T}$ is* level–coherent *iff, for every $n \in \mathrm{N}$ and every $(s,t) \in \xrightarrow{\mathfrak{T}^s_n}$, there exists $u \in T_\Sigma(X)$ such that $s \xrightarrow{ER/AC}\!| \xrightarrow{\mathfrak{T}^i_n} u$ and $t \xrightarrow{ER/AC}\!| u$. This property is illustrated by the diagram on the right–hand side.*

Obviously the property of level coherence implies the *completeness* of $\xrightarrow{\mathfrak{T}^i}$ with respect to $\xrightarrow{\mathfrak{T}^s}$.

Corollary 9. *If $\mathfrak{T}$ is level-coherent, then for every $(s,t) \in \xrightarrow{\mathfrak{T}^s}$ there exists $u \in T_\Sigma(X)$ such that $s \xrightarrow{ER/AC}\!| \xrightarrow{\mathfrak{T}^i} u$ and $t \xrightarrow{ER/AC}\!| u$.*

Note that level coherence is generally not necessary for the completeness of $\xrightarrow{\mathfrak{T}^i}$ with respect to $\xrightarrow{\mathfrak{T}^s}$. If we have, for example, $R = \left\{\frac{}{a \to c}, \frac{a \to c}{b \to c}\right\}$ and $ER = \{a \to b\}$, then it is impossible to close the following "peak" by means of a $\mathfrak{T}^i_1$–step: $c \xleftarrow{\mathfrak{T}^s_1} a \xrightarrow{ER/AC}\!| b$. Instead, the proper conditional rule $\frac{a \to c}{b \to c}$ must be applied:

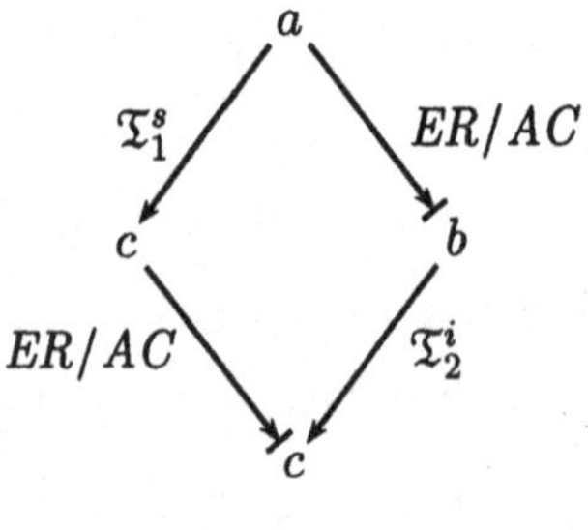

In principle such situations could be taken into account by introducing a more general notion of coherence, abstracting from the level index n, which would complicate many of the subsequent definitions and theorems. However, with the applications we have in mind it suffices to consider level coherence since normally the TRS *ER* is "simplifying"; that is, it reduces the depth of the reduction. This holds e.g. for the CCS specification in Example 2.

4.2 Conditional Critical Pairs

We will now give a necessary and sufficient criterion which guarantees the level coherence of an ORT and, thus, the completeness of the implementation. Note that Definition 8 potentially describes infinitely many critical situations since at the root position arbitrary terms s are admitted. However we observe the following possibilities for simplifications, leading to a finite collection of critical pairs to be investigated:

- Since reductions at non–overlapping positions of the start term are independent (see the proof of Theorem 12), only proper matchings between rules of R and ER have to be considered. (This is comparable to the confluence analysis of ordinary TRS.)
- Since rules in R can only be applied at the root position, only occurrences of ER–redexes in left–hand sides of R–rules have to be regarded (and not vice versa).
- Instead of taking into account all possible instances, it suffices to analyze those critical pairs which are obtained by applying "most general" unifiers. Now it becomes important that the AC–unification problem is decidable and finitary; that is, one can always decide whether two given terms $s, t \in T_\Sigma(X)$ are AC–equivalent under some substitution and, if they are, determine a finite minimal complete set of substitutions σ such that $s\sigma \xleftrightarrow{AC}{}^{*} t\sigma$ (see [1] for details). We denote this set by $MCU_{AC}(s,t)$.
- The potential infinity which arises from the conditions is captured by representing the corresponding dependences symbolically.

In order to simplify the representation we assume without loss of generality that the variables in R and ER are disjoint.

Definition 10. *Let* $\mathfrak{T} = (\Sigma, AC, ER, R)$ *be an ORT. The set of* conditional critical pairs *of* $\mathfrak{T}$, $CCP(\mathfrak{T}) \subseteq T_\Sigma(X)^2 \times \mathfrak{P}(T_\Sigma(X))$, *is given by*

$$CCP(\mathfrak{T}) = \{(s,t,C\sigma) \mid ex.\ \frac{C}{l \to r} \in R, p \in Pos(l)\ with\ l|_p \notin X, g \to d \in ER, and\ \sigma \in MCU_{AC}(l|_p, g)\ s.t.\ s = r\sigma\ and\ t = (l[p \leftarrow d])\sigma\}.$$

Note that since R and ER are assumed to be finite, $CCP(\mathfrak{T})$ is finite as well.

4.3 Level Joinability of Critical Pairs

Now we define a property which is related to the notion of *shallow joinability* of critical pairs in conditional TRS. In contrast to the latter, which does not guarantee the confluence (at least for join systems; cf. [2]), our condition characterizes the level coherence of an ORT and, thus, assures the completeness of the implementation.

Definition 11. *Let* $(s,t,C) \in CCP(\mathfrak{T})$ *and* $n \in \mathrm{N}$. *A substitution* $\sigma \in Sub$ *is called* (C,n)–feasible *iff* $c\sigma \xrightarrow{ER/AC}\!\!| \xrightarrow{\mathfrak{T}^i_n} |\!\!\xleftarrow{ER/AC} c'\sigma$ *for every* $c \rightarrow c' \in C$. *A critical pair* $(s,t,C) \in CCP(\mathfrak{T})$ *is called* level–joinable *iff, for every* $n \in \mathrm{N}$ *and every* (C,n)*–feasible substitution* $\sigma \in Sub$,

$$s\sigma \xrightarrow{ER/AC}\!\!| \xleftrightarrow{\mathfrak{T}^i_{n+1}} |\!\!\xleftarrow{ER/AC} t\sigma.$$

$CCP(\mathfrak{T})$ *is called* level–joinable *iff every critical pair* $(s,t,C) \in CCP(\mathfrak{T})$ *is.*

Theorem 12. $\mathfrak{T}$ *is level–coherent iff* $CCP(\mathfrak{T})$ *is level–joinable.*

Proof. We start with the "only if" part. Let $\mathfrak{T}$ be level–coherent, $(s,t,C) \in CCP(\mathfrak{T})$, $n \in \mathrm{N}$, and let $\sigma \in Sub$ be a (C,n)–feasible substitution. By Definition 10, there exist $\dfrac{c_1 \rightarrow c'_1, \ldots, c_k \rightarrow c'_k}{l \rightarrow r} \in R$, $g \rightarrow d \in ER$, $p \in Pos(l)$ with $l|_p \notin X$, and $\tau \in MCU_{AC}(l|_p, g)$ such that $s = r\tau$, $t = (l[p \leftarrow d])\tau$, and $C = \{c_i\tau \rightarrow c'_i\tau \mid 1 \leq i \leq k\}$. The (C,n)–feasibility of σ implies that $c_i\tau\sigma \xrightarrow{ER/AC}\!\!| \xrightarrow{\mathfrak{T}^i_n} |\!\!\xleftarrow{ER/AC} c'_i\tau\sigma$, and hence (using Lemma 6(iii)) $c_i\tau\sigma \xrightarrow{\mathfrak{T}^s_n} c'_i\tau\sigma$ for every $i \in \{1,\ldots,k\}$. This allows us to conclude that $l\tau\sigma \xrightarrow{\mathfrak{T}^s_{n+1}} r\tau\sigma = s\sigma$.

Since $\tau \in MCU_{AC}(l|_p, g)$, since term rewriting relations are closed under both substitutions and contexts, and since l is linear, we have $l\tau\sigma \xleftrightarrow{AC}{}^* (l[p \leftarrow g])\tau\sigma \xrightarrow{ER} (l[p \leftarrow d])\tau\sigma = t\sigma$, and hence $s\sigma \xleftarrow{\mathfrak{T}^s_{n+1}} l\tau\sigma \xrightarrow{ER/AC} t\sigma$. According to Definition 8, we obtain the joinability as follows:

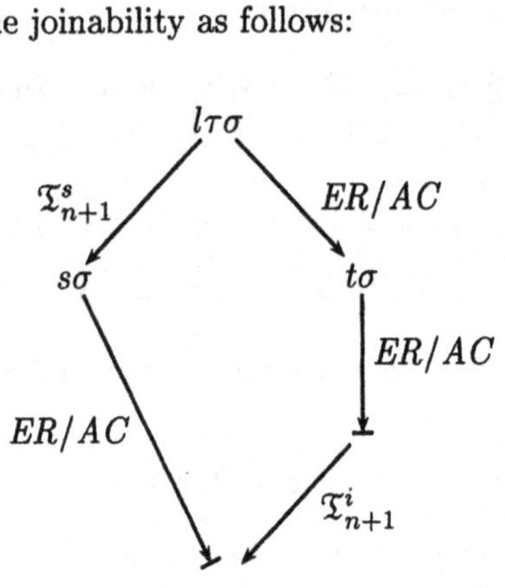

To prove the "if" part, let $CCP(\mathfrak{T})$ be level–joinable. By complete induction on n we establish the level coherence of $\mathfrak{T}$ by showing that, for every $s,t,u \in$

$T_\Sigma(X)$ and every $n \in \mathrm{N}$,

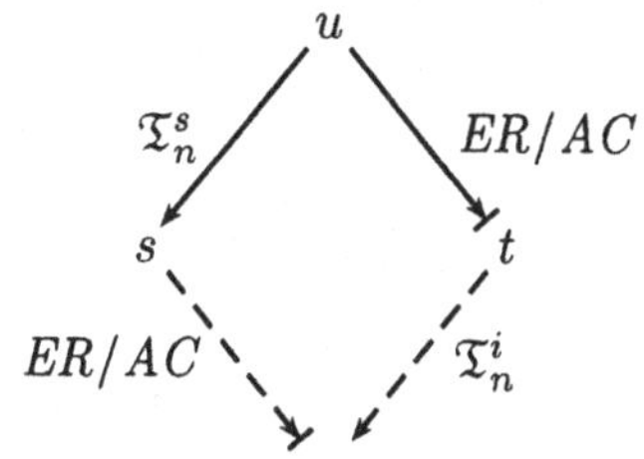

$n = 0$: Since $\mathfrak{I}_0^s = \emptyset$ (see Definition 3), the proposition holds trivially.

$n \rightsquigarrow n+1$: Since $u \xrightarrow{\mathfrak{I}_{n+1}^s} s$, there exist $\dfrac{C}{l \to r} \in R$ and $\sigma \in Sub$ such that $u \xleftrightarrow{E}{}^* l\sigma$, $r\sigma \xleftrightarrow{E}{}^* s$, and $c\sigma \xrightarrow{\mathfrak{I}_n^s} c'\sigma$ for every $c \to c' \in C$. By induction hypothesis,

$$c\sigma \xrightarrow{ER/AC}\!\!| \xrightarrow{\mathfrak{I}_n^i} |\!\!\xleftarrow{ER/AC} c'\sigma \qquad (*)$$

such that, using Lemma 6(iii),

$$c\sigma \xrightarrow{ER/AC}\!\!| \xrightarrow{\mathfrak{I}_n^s} |\!\!\xleftarrow{ER/AC} c'\sigma \qquad (**)$$

for every $c \to c' \in C$.

We proceed by showing that $l\sigma \xrightarrow{ER/AC}\!\!| \xrightarrow{\mathfrak{I}_{n+1}^i} |\!\!\xleftarrow{ER/AC} s$ using complete induction on m where $m \in \mathrm{N}$ is the number of steps of the longest $\xrightarrow{ER/AC}$-reduction starting in $l\sigma$. (Note that ER is assumed to be terminating modulo AC.) Since $u \xleftrightarrow{E}{}^* l\sigma$ and $u \xrightarrow{ER/AC}\!\!| t$, the convergence of ER modulo AC then enables us to conclude that $l\sigma \xrightarrow{ER/AC}\!\!| t$, closing the above diagram as desired.

$m = 0$: Here, $l\sigma$ is $\xrightarrow{ER/AC}$-irreducible. Hence, using $(*)$, $l\sigma \xrightarrow{\mathfrak{I}_{n+1}^i} |\!\!\xleftarrow{ER/AC} r\sigma$ where $r\sigma \xleftrightarrow{E}{}^* s$ again implies $r\sigma \xrightarrow{ER/AC}\!\!|\,|\!\!\xleftarrow{ER/AC} s$ by convergence.

$m \rightsquigarrow m+1$: Let $l\sigma$ be $\xrightarrow{ER/AC}$-reducible, that is, there exists $t' \in T_\Sigma(X)$ such that $l\sigma \xleftrightarrow{AC}{}^* \xrightarrow{ER} t'$. This implies that there are $p \in Pos(t')$ and $g \to d \in ER$ with $l\sigma \xleftrightarrow{AC}{}^* t'[p \leftarrow g\sigma]$ and $t'|_p = d\sigma$. (Note that we can reuse the above substitution σ since R and ER are assumed to be variable–disjoint.)

If $\dfrac{C}{l \to r}$ and $g \to d$ are applied at overlapping positions, i.e., $g\sigma \xleftrightarrow{AC}{}^* (l|_q)\sigma$ for some $q \in Pos(l)$ with $l|_q \notin X$, then $(r\sigma, t', C\sigma)$ is an instance of a conditional critical pair: there exist $\tau \in MCU_{AC}(l|_q, g)$ and $\tau' \in Sub$ such that $(r\sigma, (l[q \leftarrow d])\tau, C\tau) \in CCP(\mathfrak{I})$, $r\sigma \xleftrightarrow{AC}{}^* r\tau\tau'$, $t' \xleftrightarrow{AC}{}^* (l[q \leftarrow d])\tau\tau'$, $c\sigma \xleftrightarrow{AC}{}^* c\tau\tau'$, and $c'\sigma \xleftrightarrow{AC}{}^* c'\tau\tau'$ for every

$c \to c' \in C$. Since $(*)$ holds for every $c \to c' \in C$, $\tau\tau'$ is a feasible substitution. Hence the level joinability (Definition 11) and the fact that $s \xleftrightarrow{E}^{*} r\sigma \xleftrightarrow{AC}^{*} r\tau\tau'$ yields, as desired,

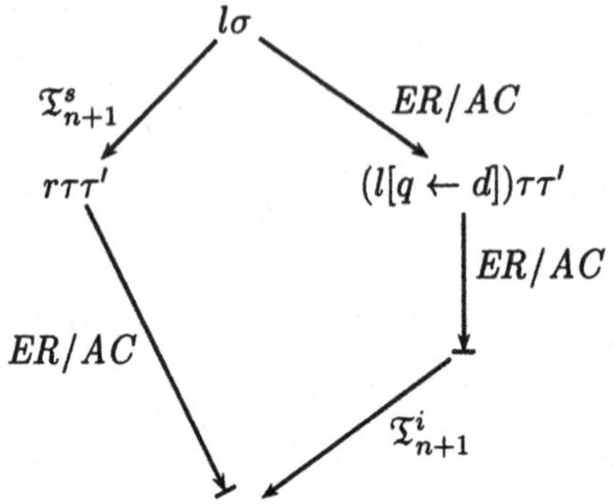

Otherwise $g \to d$ is applied below $\frac{C}{l \to r}$, i.e., for some $x \in X(l)$ and some $p \in Pos(x\sigma)$, $g\sigma \xleftrightarrow{AC}^{*} (x\sigma)|_p$.

Since l is linear, the $\xrightarrow{ER}$–successor t' of $l\sigma$ can be represented as $t' \xleftrightarrow{AC}^{*} l\sigma'$ where $\sigma' \in Sub$ is given by $x\sigma' = (x\sigma)[p \leftarrow d\sigma]$ and $y\sigma' = y\sigma$ for every $y \in X \setminus \{x\}$. Now, on the one hand, we have seen above $(**)$ that

$$c\sigma \xrightarrow{ER/AC}\!\!| \xrightarrow{\mathfrak{T}^s_n} |\!\!\xleftarrow{ER/AC} c'\sigma$$

for every $c \to c' \in C$. On the other hand, the definition of σ' yields $c\sigma \xrightarrow{ER/AC}^{*} c\sigma'$ and $c'\sigma \xrightarrow{ER/AC}^{*} c'\sigma'$ such that the $\xrightarrow{ER/AC}\!\!|$–normal forms coincide pairwise. Hence we have

$$c\sigma' \xrightarrow{ER/AC}\!\!| \xrightarrow{\mathfrak{T}^s_n} |\!\!\xleftarrow{ER/AC} c'\sigma'$$

for every $c \to c' \in C$. Thus, by Definition 3, $l\sigma' \xrightarrow{\mathfrak{T}^s_{n+1}} r\sigma'$. Since $r\sigma \xrightarrow{ER/AC}^{*} r\sigma'$ as well, the above diagram can be closed as follows, using the induction hypothesis for m.

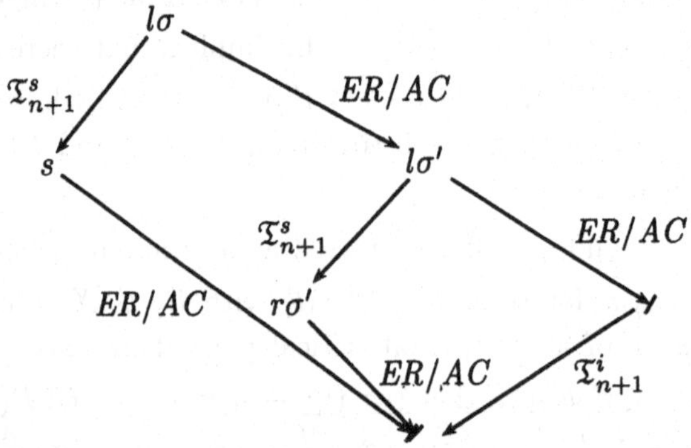

Example 13. It is possible to show that the CCS specification in Example 2 is level–coherent. For example, the rules $\dfrac{x \rightarrow \{\alpha\}x'}{x \parallel y \rightarrow \{\alpha\}(x' \parallel y)} \in R$ and $z \parallel \mathsf{nil} \rightarrow z \in ER$ yield the two conditional critical pairs

$$(\{\alpha\}(x' \parallel \mathsf{nil}), z, \{z \rightarrow \{\alpha\}x'\}), (\{\alpha\}(x' \parallel z), z, \{\mathsf{nil} \rightarrow \{\alpha\}x'\}) \in CCP(\mathfrak{T}).$$

(Choose $\sigma = [x/z, y/\mathsf{nil}]$ in Definition 10 to obtain both.)

Regarding the second pair, it is obvious that there does not exist any $(\{\mathsf{nil} \rightarrow \{\alpha\}x'\}, n)$–feasible substitution since the term **nil** is $\xrightarrow{\mathfrak{T}^i}$–irreducible. Hence, the critical pair is trivially level–joinable.

Now let $\sigma \in Sub$ be a $(\{z \rightarrow \{\alpha\}x'\}, n)$–feasible substitution for the first pair. According to Definition 11, there exists $u \in T_\Sigma(X)$ such that $z\sigma \xrightarrow{ER/AC}\!\!| \xrightarrow{\mathfrak{T}^i_n} u |\!\!\xleftarrow{ER/AC} (\{\alpha\}x')\sigma$. Then the level joinability can be established as follows. On the one hand, using the oriented equation $z \parallel \mathsf{nil} \rightarrow z \in ER$,

$$(\{\alpha\}(x' \parallel \mathsf{nil}))\sigma \xrightarrow{ER/AC} (\{\alpha\}x')\sigma \xrightarrow{ER/AC}\!\!| u.$$

On the other hand, by $\xrightarrow{\mathfrak{T}^i_n} \subseteq \xrightarrow{\mathfrak{T}^i_{n+1}}$ (Lemma 6(ii)),

$$z\sigma \xrightarrow{ER/AC}\!\!| \xrightarrow{\mathfrak{T}^i_{n+1}} u.$$

5 Conclusion

In this paper we have proposed a variant of Meseguer's *Rewriting Logic* as a semantic framework in which different design languages for distributed systems can easily be described. We have demonstrated its appropriateness by giving a natural specification of the well–known CCS process algebra. We have argued that, for an efficient implementation of this approach, the underlying principle of (conditional) term rewriting modulo arbitrary equational theories is intractable. Instead we have proposed to employ rewriting modulo AC, using oriented versions of the equations, and we have shown that this implementation is always sound. With regard to completeness, we have investigated the property of level coherence as a sufficient criterion which can be verified by testing the level joinability of finitely many conditional critical pairs.

Currently we are developing a prototype version of a compiler generator which, given the specification of a design language in terms of oriented rewrite theories, builds a corresponding *Haskell* frontend for our *Truth* verification tool. For the actual term rewriting steps, it employs the *ELAN* system (cf. [6]). From the point of view of memory efficiency, the prototype is very successful regarding the state–space reduction. However its run–time performance leaves much to be desired, due to the string–based *Haskell/ELAN* interface and to the interpreting implementation of *ELAN*. Future releases will hopefully overcome these drawbacks by compiling the rewrite rules. It could also be sensible to use other languages such as *CafeOBJ* ([3]) or *Maude* ([9]).

In addition, the coherence test has still to be implemented. But if this test fails for a given oriented rewrite theory, then the user is left alone with the information that one of the conditional critical pairs is not level–joinable, which causes the incoherence of the specification. He or she gets no hint on which rules should be added in order to assure the completeness of the implementation. We are therefore further seeking for coherence completion strategies which can be used to determine such rules. Because of the similarity between the joinability and the confluence property of conditional TRS, completion algorithms for the latter (see e.g. [7]) could be a good starting point.

References

1. F. Baader and J. Siekmann. Unification theory. In *Handbook of Logic in Artificial Intelligence and Logic Programming*, volume 2, pages 41–125. Oxford University Press, 1994.
2. J.A. Bergstra and J.W. Klop. Conditional rewrite rules: Confluence and termination. *Journal of Computer and System Sciences*, 32(3):323–326, 1986.
3. CafeOBJ home page. `http://caraway.jaist.ac.jp/cafeobj/`.
4. The Concurrency Factory. `http://www.cs.sunysb.edu/~concurr/`.
5. The Edinburgh Concurrency Workbench. `http://www.dcs.ed.ac.uk/home/cwb/`.
6. ELAN home page. `http://www.loria.fr/ELAN/`.
7. H. Ganzinger. A completion procedure for conditional equations. *Journal of Symbolic Computation*, 11(1–2):51–82, 1991.
8. J.–P. Jouannaud and C. Kirchner. Completion of a set of rules modulo a set of equations. *SIAM Journal on Computing*, 15(4):1155–1194, 1986.
9. The Maude system. `http://maude.csl.sri.com/`.
10. J. Meseguer. Conditional rewriting logic as a unified model of concurrency. *Theoretical Computer Science*, 96(1):73–155, April 1992.
11. J. Meseguer. Rewriting logic as a semantic framework for concurrency: a progress report. In *Seventh International Conference on Concurrency Theory (CONCUR'96)*, volume 1119 of *Lecture Notes in Computer Science*, pages 331–372. Springer–Verlag, August 1996.
12. R. Milner. *Communication and Concurrency*. International Series in Computer Science. Prentice–Hall, 1989.
13. The Concurrency Workbench of North Carolina. `http://www.csc.ncsu.edu/eos/users/r/rance/WWW/cwb-nc.html`.
14. The Process Algebra Compiler of North Carolina. `http://www.csc.ncsu.edu/eos/users/s/stsims/WWW/pac/pac-nc.html`.
15. Model checking at CMU. `http://www.cs.cmu.edu/~modelcheck/`.
16. Spin. `http://netlib.bell-labs.com/netlib/spin/whatispin.html`.

17. Truth home page.
`http://www-i2.informatik.rwth-aachen.de/Forschung/MCS/Truth/`.
18. P. Viry. Rewriting: An effective model of concurrency. In *Proceedings of PARLE'94 - Parallel Architectures and Languages Europe*, volume 817 of *Lecture Notes in Computer Science*, pages 648–660. Springer–Verlag, 1994.
19. P. Viry. Rewriting modulo a rewrite system. Technical Report TR–95–20, Università di Pisa, Dipartimento di Informatica, December 1995.
20. P. Viry. A rewriting implementation of pi–calculus. Technical Report TR–96–30, Università di Pisa, Dipartimento di Informatica, March 1996.

Synchronous Structures

David Nowak, Jean-Pierre Talpin, and Paul Le Guernic

INRIA-Rennes – IRISA, Campus de Beaulieu, F-35042 Rennes Cdex

Abstract. Synchronous languages have been designed to ease the development of reactive systems, by providing a methodological framework for assisting system designers from the early stages of requirement specifications to the final stages of code generation or circuit production. Synchronous languages enable a very high-level specification and an extremely modular design of complex reactive systems. We define an order-theoretical model that gives a unified mathematical formalization of all the above aspects of the synchronous methodology (from relations to circuits). The model has been specified and validated using a theorem prover as part of the certified, reference compiler of a synchronous programming language.

1 Introduction

Synchronous languages, such as SIGNAL [2], LUSTRE [9] and ESTEREL [4] have been designed to ease the development of reactive systems. The synchronous hypothesis provides a deterministic notion of concurrency where operations and communications are instantaneous. In a synchronous language, concurrency is meant as a logical way to decompose the description of a system into a set of elementary communicating processes. Interaction between concurrent components is conceptually performed by broadcasting events. Synchronous languages enable a very high-level specification and an extremely modular design of complex reactive systems by structurally decomposing them into elementary processes. The use of synchronous languages provides a methodological framework for assisting the users from the early stages of requirement specifications to the final stages of code generation or circuit production while obeying compliance to expressed and implied safety requirements. In that context, the synchronous language SIGNAL is particularly interesting, in that it allows the specification of (early) relational properties of systems which can then be progressively refined in order to obtain an executable specification. All the stages of this design process can easily be modeled and understood in isolation. The purpose of our presentation is to define a mathematical model which gives a unified formalization of all the aspects of a synchronous methodology and which contains each of them in isolation. The model uses basic notions of set-theory and order-theory. It has been specified and validated using the Coq proof assistant [7]. This implementation is part of a certified, reference compiler of the SIGNAL language. It completes and extends the results of [12] on the definition of a co-inductive trace semantics of SIGNAL in Coq.

Influential Analogy. In 1545, the great Italian mathematician Gerolamo Cardano wrote an important and influential treatise on Algebra: "Ars Magna" [5] in which the first complete expression for the solution of a general cubic equation was put forward. Cardano noticed that, in the case of some equation with three real solutions, he was forced to take at a certain stage the square root of a negative number. The imaginary numbers were borned. Analogically, we generalize the classical notion of signal ([2, 3, 10]) with imaginary signals. This extension has no material counterpart. It is used to compute intermediate results. For instance, the temporal abstractions of signals (called clocks) have necessary a greatest lower bound but do not always have a (real) least upper bound. In that case, we need to define an imaginary least upper bound. This axiomatization allows to extend the notion of classical clocks (a clock is a temporal abstraction of a signal) with imaginary clocks and define a boolean lattice of clocks. In this lattice-theoretical model, temporal relations between signals always have a solution. If the solution contains imaginary signals, this means that the system has no real solution in the classical model and that it does not thus form an executable specification.

Plan We first introduce the synchronous language SIGNAL in the section 2. In the section 3, we abstract the notion of control dependence in a mathematical structure that we call a synchronous structure. Within this structure we formalize the notions of signals, clocks and instants, and their relations. We define some internal operations on signals and clocks, prove their algebraic properties, prove that the set of clocks forms a boolean lattice, and define a Cartesian closed category of signals with product and coproduct. In the section 4, we add a valuation function and a data dependency relation to synchronous structure. In the section 5, we briefly expose the outcome of our model for the compilation of programs written in the synchronous language SIGNAL.

2 Overview of Signal

SIGNAL is an equational synchronous programming language. A SIGNAL program is modularly organized into processes consisting of simultaneous equations on signals. In SIGNAL, an equation is an elementary and instantaneous operation on input signals which defines an output. A signal is a sequence of values defined over a totally ordered set of instants. At any given instant, a signal x is either present or absent.

$P ::= (P \mid P')$	parallel composition
$\mid\ P/\mathbf{x}$	restriction
$\mid\ R(x1, \cdots, xn)$	instantaneous relation
$\mid\ z := x$ when y	selection
$\mid\ z := x$ default y	deterministic merge
$\mid\ y := x\$$ init v	delay

In SIGNAL, a process P is either an equation or the synchronous composition $P \mid P'$ of processes. Parallel composition $P \mid P'$ synchronizes the events produced by P and P'. P/x masks the signal x in the process P i.e. x is a local signal of the process P. An instantaneous relation $R(x_1, \cdots, x_n)$ forces the signals $x_1, \cdots, x_n$ to be synchronous and theirs instantaneous values to satisfy the relation R. A delay $y := x\$$ init v (called "shift register" in [2]) stores the value v' of x and outputs the previous value v of x to y. A deterministic merge $z := x$ default y outputs the value of x to z (if x is present) or the value of y (if x is absent). A selection (or down-sampling) $z := x$ when y outputs x to z when y is present and true. When all the inputs of an equation are absent, a transition takes place but no value is given to its output.

A set of equations can be encapsulated as a new reusable process. It consists of an interface providing parameters, input and output signals with their types. The pervasive operators when, default and \$ of SIGNAL offer a flexible mean for progressively specifying reactive systems, from the early specification of system properties or requirements to late executable programs. To illustrate this process, let us consider the design of a simple replenishable tank where *capacity* is an integer parameter, $fill$ is an input signal of type event (a subtybe of boolean onlay inhabited by *true*), and *empty* is an output signal of type boolean.

```
process tank = {integer capacity} (? event fill ! boolean empty)
  (| synchro (when (zn = 0), fill)
   | zn := n$ init 0
   | n := (capacity when fill) default (zn − 1)
   | empty := when (n = 0) default (not fill) |) / n, zn
```

This program uses an extended and more intuitive syntax of SIGNAL [11] that can be translated into the SIGNAL-kernel described in this overview. synchro is a process that forces its input signals to be synchronous. The following table illustrates an execution of the process *tank* with a capacity equal to 3.

fill	t				t				t	
zn	0	3	2	1	0	3	2	1	0	3
n	3	2	1	0	3	2	1	0	3	2
empty	f			t	f			t	f	

It is easy to observe that without its first equation synchro {when $(zn = 0), fill\}$ the process *tank* would not form an executable specification. In this case, it would be impossible to relate the clock of the signal *fill* with the other clocks of the program. But it would be a correct sub-specification that could be composed with another specification to remove the non-determinism. In this paper, we show how to deal with non-determinism using imaginary signals.

3 Control Dependence

In this section, we focus on a characterization of control dependencies, i.e., the temporal relations between events or the dates of events relative to some reference of time, not the value of events. Let us informally depict a synchronization scenario between two sequences of events (i.e. sets of ordered events). They exchange (dotted) synchronization messages using an asynchronous medium for their communications. This involve a synchronization relation between events. The natural structure of time of the whole system is that of a partial pre-order. In this section, we will abstract the notions involved in this example.

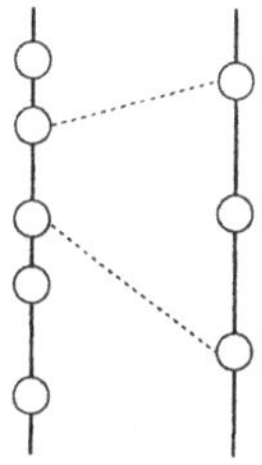

3.1 Synchronous Structure

We define a synchronous structure as an ordered set (its elements are called *events*) with a particular equivalence relation $\sim$. Intuitively, $x \sim y$ means that x and y are synchronous, that is to say the events x and y must occur simultaneously. The order relation $\leq$ is the temporal causality between two events: $x < y$ means that x must occur strictly before y.

Definition 1. *$(\mathcal{E}, \ll)$ is a synchronous structure iff $\mathcal{E}$ is a non empty set (of events) and $\ll$ is a preorder on $\mathcal{E}$ such that:*

$$\forall x \in \mathcal{E}, \{y \in \mathcal{E} \mid y \leq x\} \text{ is finite, where } \begin{array}{l} x \sim y \Leftrightarrow_{def} x \ll y \wedge y \ll x \\ x < y \Leftrightarrow_{def} x \ll y \wedge x \not\sim y \\ x \leq y \Leftrightarrow_{def} x < y \vee x = y \end{array}$$

For instance, the left part of the figure 1 depicts eight events which define a synchronous structure. To give easier explanations, the events are numbered from 1 to 8. Dotted lines represent the equivalence relation $\sim$ and bold lines represent the strict order relation $<$ as a Hasse diagram: $x < y$ iff there is a sequence of connected bold line segments moving downwards from x to y.

The preorder $\ll$ mixes the synchronicity relation and the temporal causality relation. It defines a notion of time for the whole system. We will explain this structure in more details after introducing the notion of signal. The right part of the figure depicts the preorder relation $\ll$ between events as a Hasse diagram where synchronous events are grouped in one node. From the fact that $\leq$ is well founded, we deduce that $\ll$ is a well founded preorder.

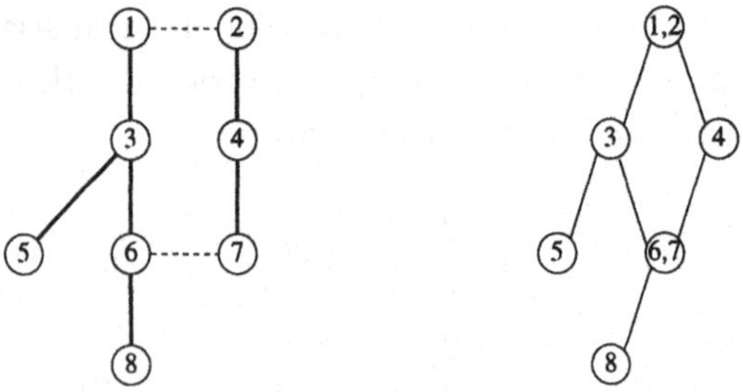

Fig. 1. Events and associated preorder

The following proposition comes directly from the definition of a synchronous structure. In the example, it guarantees that the events numbered 1 and 8 cannot be synchronous.

Proposition 1. $\forall (x, y_1, y_2, z) \in \mathcal{E}^4, x \ll y_1 \wedge y_1 < y_2 \wedge y_2 \ll z \Rightarrow \neg x \sim z$

We say that an event x is *covered by* an event y, and write $x \prec y$, iff $x < y$ and there is no event z satisfying $x < z < y$. From the fact that $\leq$ is well founded, we can deduce the following proposition. This proposition is important to guarantee a discrete model of synchronous programming.

Proposition 2. $\forall (x,y) \in \mathcal{E}^2, x < y \Rightarrow \exists z \in \mathcal{E}, z \prec y$

Indeed, $(\mathcal{E}, \leq)$ is not dense because $\leq$ is well founded.

3.2 Signal, Clock and Instant

In this subsection we define the objects of the model and their relations. First, we formalize the notion of *signal.* Usually, a (real) signal is a totally ordered set of events. This total order implies that two different events cannot be synchronous. We generalize this definition to enable partially ordered sets of events to be (imaginary) signals. A signal just have to satisfy the property that two different events cannot be synchronous. In the subsection 3.4, this relaxed condition is used to define internal operations.

Definition 2. *Let X be a subset of $\mathcal{E}$. X is a* signal *iff it satisfies the following axiom:*

$$\forall (x,y) \in X^2, x \sim y \Rightarrow x = y \tag{1}$$

Let $\mathcal{S}$ be the set of signals. For instance, in the figure 1, $\{1,3,5,8\}$ and $\{2,6,8\}$ are in $\mathcal{S}$. A *real signal* is then a particular case of signal which is totally preordered by $\ll$. For instance, in the figure 1, $\{1,3,5\}$, $\{2,6,8\}$ and $\emptyset$ are real signals but not $\{1,3,5,8\}$. An *imaginary signal* is a signal which is not a real signal. An imaginary signal enables to represent the lack of synchronization constraints in a sub-specified reactive system. In SIGNAL, a sub-specification is a correct specification that cannot be executed because of non-determinism. It needs to be composed with another specification to remove the non-determinism. Let X be a signal. From the axiom 1 we deduce that $\ll$ is antisymmetric on X and then is an order relation on X. X is totally ordered by $\ll$ iff X is a real signal. From the proposition 2, we deduce proposition 3. Then, we define a preorder relation $\preceq$ on $\mathcal{S}$ (definition 3, see, for instance, the figure 2). The preorder $\preceq$ gives rise to an equivalence relation $\widehat{=}$ (definition 4, we say that X and Y are *synchronous* iff $X \widehat{=} Y$).

Proposition 3. $\forall X \in \mathcal{S}, \forall (x,y) \in X^2, x < y \Rightarrow \exists z \in X, z \prec y$

Definition 3. *For all signals X and Y, $X \preceq Y$ iff $\forall x \in X, \exists y \in Y, x \sim y$*

Definition 4. *For all signals X and Y, $X \widehat{=} Y$ iff $X \preceq Y$ and $Y \preceq X$.*

In order to study the temporal relations between signals, we define the equivalence classes of signals by $\widehat{=}$.

Definition 5. *The* set of clocks $\mathcal{C}$ *is the quotient of $\mathcal{S}$ by $\widehat{=}$.*

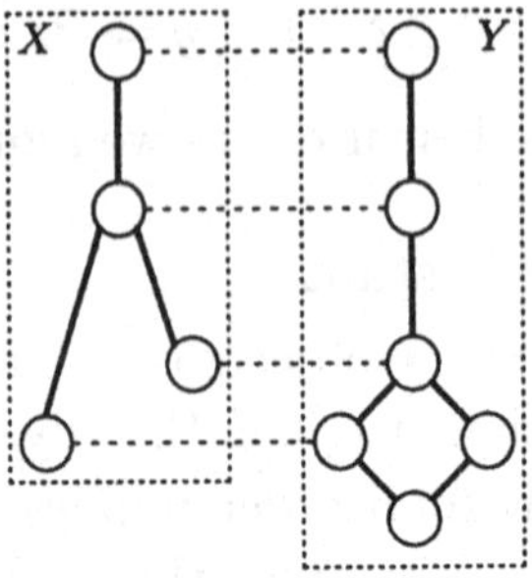

Fig. 2. Preordered signals ($X \preceq Y$)

For any signal X, we write $\widehat{X}$ its equivalence class that we call its *clock*. $\widehat{\emptyset}$ is called the *null clock*. The clock of a real (resp. imaginary) signal is a real (resp. imaginary) clock. The preorder $\preceq$ on $\mathcal{S}$ gives rise to an order $\sqsubseteq$ on $\mathcal{C}$.

Definition 6. *For all signals X and Y, $\widehat{X} \sqsubseteq \widehat{Y}$ iff $X \preceq Y$.*

We define the equivalence classes of events by $\sim$. Intuitively, these classes will represent the notion of logical instant.

Definition 7. *The* set of instants $\mathcal{I}$ *is the quotient of $\mathcal{E}$ by $\sim$.*

For any event x, we write $\widetilde{x}$ its equivalence class that we call its *instant*. The preorder $\ll$ gives rise to an order $\rhd$ on $\mathcal{I}$.

Definition 8. *For all event x and y, $\widetilde{x} \rhd \widetilde{y}$ iff $x \ll y$.*

Intuitively, it is clear that a clock should be related to a set of instants and conversely. We show that the set of clocks $\mathcal{C}$ and the powerset of $\mathcal{I}$ are isomorphic. Let $\mathcal{P}(\mathcal{I})$ be the powerset of $\mathcal{I}$. Using the Axiom of Choice, we prove the following theorem.

Theorem 1. *$(\mathcal{C}, \sqsubseteq)$ and $(\mathcal{P}(\mathcal{I}), \subseteq)$ are isomorphic.*

Let I be a set of instants. By definition of an instant, I is a set of disjoint sets of events. The Axiom of Choice is then necessary to "choose" a single event from each element of I. Then we can construct a signal and take its clock which is then the associated clock of I. Therefore there is a function f from $\mathcal{P}(\mathcal{I})$ to $\mathcal{C}$. We show that this function is invertible and f and f^{-1} are increasing. This is sufficient to prove that f is an isomorphism.

3.3 Trace

We can link this order-theoretic approach to our trace semantics of SIGNAL developped in [12]. Let $i \in \mathcal{I}$ be an instant. $t_X(i)$ is the event at the intersection of X and i if it exists. Or else it is the special value $\perp \notin \mathcal{E}$ if the intersection is empty. t_X is called the *trace of* X.

$$t_X : \mathcal{I} \longrightarrow \mathcal{E} \cup \{\perp\}$$
$$i \longmapsto \begin{cases} x \text{ if } X \cap i = \{x\} \\ \perp \text{ else} \end{cases}$$

The following lemma guarantees that this definition is correct.

Lemma 1. *For any signal X, for any instant i, for any event x of X, $X \cap i = \emptyset \vee \exists x \in X, X \cap i = \{x\}$ and $X \cap \widetilde{x} = \{x\}$.*

The two approaches are linked by the logical property: $X = Y \Leftrightarrow t_X = t_Y$.

3.4 Operations on Signals and Clocks

In this subsection we define some operations on signals and clocks which denote the control part of the instructions of SIGNAL [2]. Let X and Y be signals. First we define the selection of a signal at the clock of another signal.

Definition 9 (Selection). *For all signals X and Y,*

$$X \otimes Y =_{def} \{x \in X | \exists y \in Y, x \sim y\}$$

$\otimes$ is an internal operation on $\mathcal{S}$ i.e. for all signals X and Y, $X \otimes Y$ is a signal. For instance, in the left part of the figure 3, the selection of the signal Y at the clock of X is depicted. Although X and Y are imaginary signals, the result $Y \otimes X$ is a real signal in this example. The operator $\otimes$ on $\mathcal{S}$ gives rise to the greatest lower bound operator $\sqcap$ on $\mathcal{C}$.

Definition 10. *For all signals X and Y, $\widehat{X} \sqcap \widehat{Y} =_{def} \widehat{X \otimes Y}$.*

Proposition 4. *$C \sqcap D$ is the greatest lower bound of clocks C and D.*

We define the merge of two signals with priority to the left event.

Definition 11 (Deterministic Merge). *For all signals X and Y,*

$$X \oplus Y =_{def} X \cup \{y \in Y | \neg\exists x \in X, x \sim y\}$$

$\oplus$ is an internal operation on $\mathcal{S}$ i.e. for all signals X and Y, $X \oplus Y$ is a signal. For instance, in the right part of the figure 3, the deterministic merge of the signals X and Y is depicted. Although X and Y are real signals, their deterministic merge $X \oplus Y$ is an imaginary signal because its events are not totally preordered by $\ll$. The operator $\oplus$ on $\mathcal{S}$ gives rise to the least upper bound operator $\sqcup$ on $\mathcal{C}$.

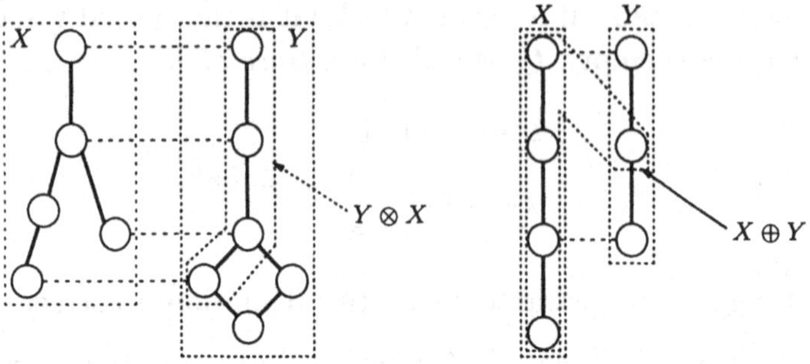

Fig. 3. Examples of selection and deterministic merge

Definition 12. *For all signals X and Y,* $\widehat{X} \sqcup \widehat{Y} =_{def} \widehat{X \oplus Y}$.

Proposition 5. $C \sqcup D$ *is the least upper bound of clocks C and D.*

Every couple of clocks $\{C, D\}$ has a least upper bound $C \sqcup D$ and a greatest lower bound $C \sqcap D$. Therefore $(\mathcal{C}, \sqcap, \sqcup)$ is a lattice[1]. From the isomorphism between $\mathcal{P}(\mathcal{I})$ and $\mathcal{C}$, we deduce that the lattice $(\mathcal{C}, \sqcap, \sqcup)$ is boolean i.e. it is complete, distributive and there exists a null element $\widehat{\emptyset}$ and a universal element $\top$. Let f be the morphism from $\mathcal{P}(\mathcal{I})$ to $\mathcal{C}$. The universal element $\top$ is equal to $f(\mathcal{I})$. We define the operator $\setminus$ on clocks which is the counterpart of the operator $\setminus$ on sets of instants which subtracts a set from an other. Let f be the morphism from $\mathcal{C}$ to $\mathcal{P}(\mathcal{I})$: $C \setminus D =_{\mathrm{def}} f^{-1}(f(C) \setminus f(D))$. The complementary of a signal X is a "chosen" signal $\overline{X}$ (using the Axiom of Choice) of clock $\top \setminus \widehat{X}$. Algebraic properties of these operations on signals and clocks are summarized in the figure 4. They are easily proved by case analysis using the trace semantics. We just have to translate the signal operators $\otimes$ and $\oplus$ into the trace semantics. We define an operator . on traces such that $t_{X \otimes Y} = t_X . t_Y$ and an operator $+$ on traces such that $t_{X \oplus Y} = t_X + t_Y$.

$$t_X . t_Y : \mathcal{I} \longrightarrow \mathcal{E} \cup \{\perp\} \qquad i \longmapsto \begin{cases} t_X(i) : t_X(i) \neq \perp, t_Y(i) \neq \perp \\ \perp \quad \text{othewise} \end{cases}$$

$$t_X + t_Y : \mathcal{I} \longrightarrow \mathcal{E} \cup \{\perp\} \qquad i \longmapsto \begin{cases} t_X(i) : t_X(i) \neq \perp \\ t_Y(i) \quad \text{othewise} \end{cases}$$

Note that $\oplus$ is not distributive to the right with respect to $\otimes$. Indeed, if $t_X(i) = x$, $t_Y(i) = \perp$ and $t_Z(i) = z$ then $((t_X . t_Y) + t_Z)(i) = z$ and $((t_X + t_Z).(t_Y + t_Z))(i) = x$.

[1] This is not true for real clocks as they do not always have a real least upper bounds

$$
\begin{array}{cc}
X \otimes Y \preceq X & X \preceq X \oplus Y \\
X \otimes Y \preceq Y & Y \preceq X \oplus Y \\
X \otimes Y \widehat{=} Y \otimes X & X \oplus Y \widehat{=} Y \oplus X \\
X \otimes (Y \otimes Z) = (X \otimes Y) \otimes Z & X \oplus (Y \oplus Z) = (X \oplus Y) \oplus Z \\
X \otimes (Y \oplus Z) = (X \otimes Y) \oplus (X \otimes Z) & X \oplus (Y \otimes Z) = (X \oplus Y) \otimes (X \oplus Z) \\
(X \oplus Y) \otimes Z = (X \otimes Z) \oplus (Y \otimes Z) &
\end{array}
$$

Fig. 4. Algebraic properties of $\otimes$ and $\oplus$

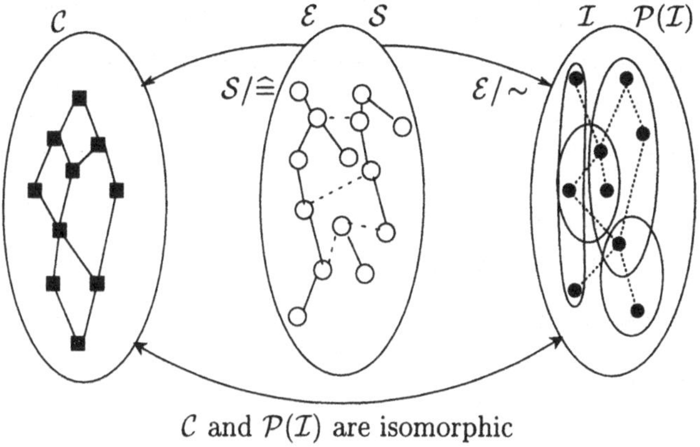

Fig. 5. Summary of the results presented so far

3.5 The Category of Signals

Another way to study temporal relations between signals is to define a category of signals in which a morphism describes the temporal relation between two signals. Suppose that X and Y are two signals such that $X \preceq Y$. Then, for any event $x \in X$, there exists an event $y \in Y$ such that $x \sim y$, by definition of $\preceq$. This event y is unique by definition of a signal. Hence, we can define a total function $[Y]_X$, called *signal morphism*, from X to Y:

$$
\begin{aligned}
[Y]_X : X &\longrightarrow Y \\
x &\longmapsto y \text{ such that } x \sim y
\end{aligned}
$$

For all signals X and Y such that $X \preceq Y$,

1. $[Y]_X$ is injective: $\forall (x, x') \in X^2, [Y]_X(x) = [Y]_X(x') \Rightarrow x = x'$
2. $[Y]_X$ is strictly monotonic: $\forall (x, x') \in X^2, x < x' \Rightarrow [Y]_X(x) < [Y]_X(x')$
3. $[Y]_X$ is bijective (with $[Y]_X^{-1} = [X]_Y$) iff $X \widehat{=} Y$.

The identity $[X]_X$ is a signal morphism and signal morphisms can be composed: for all signals X, Y and Z such that $X \preceq Y \preceq Z$, $[Z]_X = [Z]_Y \circ [Y]_X$. The set of signals and the set of morphisms define a small (preorder) category Sig with product $\otimes$ and coproduct $\oplus$. More precisely, let X and Y be two objects (i.e. signals) of the category Sig. The product object $X \otimes Y$ and the two projections $[X]_{X\otimes Y}$ and $[Y]_{X\otimes Y}$ are a product of X and Y. These data satisfy the property that, for any object Z and all morphisms $f : Z \longrightarrow X$ and $g : Z \longrightarrow Y$, there exists a unique morphism $\langle f, g\rangle : Z \longrightarrow X \otimes Y$ such that the left-diagram of figure 6 commutes. The coproduct object $X \oplus Y$ and the two injections $[X \oplus Y]_X$ and $[X \oplus Y]_Y$ are a coproduct of X and Y. These data satisfy the property that, for all object Z and all morphisms $f : X \longrightarrow Z$ and $g : Y \longrightarrow Z$, there exists a unique morphism $[f, g] : X \oplus Y \longrightarrow Z$ such that the right-diagram of figure 6 commutes.

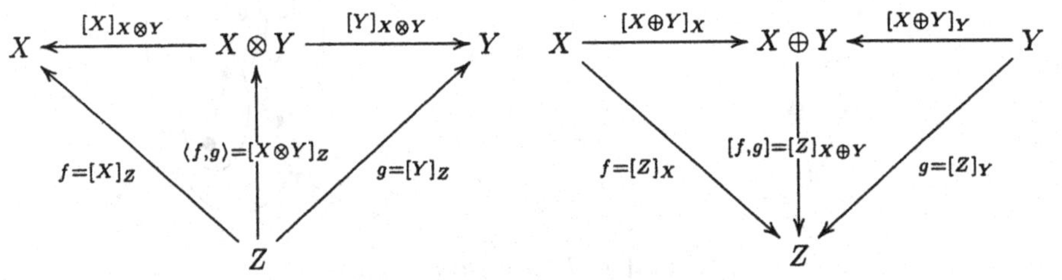

Fig. 6. Morphisms for product and coproduct objects

The signal $\emptyset$ is the unique initial object of the category Sig i.e. for any object X of Sig there exists a unique morphism $[X]_\emptyset : \emptyset \longrightarrow X$. And the coproduct $\oplus$ is defined for each ordered pair of objects of Sig. Hence the category Sig has finite coproducts. It is also possible to construct a terminal object. Let C be the clock corresponding to the set of all instants $\mathcal{I}$. Let $Y \in C$ be a signal of clock C. This signal Y is a terminal object i.e. for any object X of Sig there exists a unique morphism $[Y]_X : X \longrightarrow Y$. And the product $\otimes$ is defined for each ordered pair of objects of Sig. Hence the category Sig has finite products. Let $Y \Rightarrow Z$ be the object $\overline{Y} \oplus Z$ and $\mathsf{Apply}_{Y,Z} : (Y \Rightarrow Z) \otimes Y \longrightarrow Z$ be the morphism $[Z]_{(Y\Rightarrow Z)\otimes Y}$. $\mathsf{Apply}_{Y,Z}$ is correctly defined because $(Y \Rightarrow Z) \otimes Y \preceq Z$:

$$(Y \Rightarrow Z)\otimes Y = (\overline{Y}\oplus Z)\otimes Y = (\overline{Y}\otimes Y)\oplus(Z\otimes Y) = \emptyset\oplus(Z\otimes Y) = (Z\otimes Y) \preceq Z$$

In addition, $(Y \Rightarrow Z) \otimes Y = (Z \otimes Y)$. Therefore $\mathsf{Apply}_{Y,Z} = [Z]_{Z\otimes Y}$. Sig is Cartesian closed i.e. for all objects Z and each morphism $f : X \otimes Y \longrightarrow Z$ there exists a unique morphism $\lambda(f) = [Y \Rightarrow Z]_X : X \longrightarrow (Y \Rightarrow Z)$ such that the following diagram[2] commutes. The proof consists in proving that $X \preceq (Y \Rightarrow Z)$ i.e. $\lambda(f) = [Y \Rightarrow Z]_X$ is correctly defined. We

[2] $\forall f : X \longrightarrow Y, f : X' \longrightarrow Y', f \otimes g =_{\text{def}}: X \otimes X' \longrightarrow Y \otimes Y' = \langle f \circ [X]_{X\otimes X'}, g \circ [X']_{X\otimes X'}\rangle$

conjecture that the category Sig can be related to the category of event strutures ([15]) through functors.

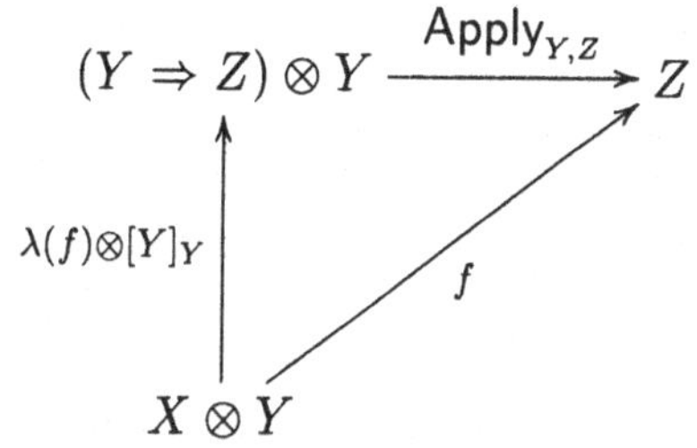

4 Data Dependence

In this section, we complete our notion of partial ordered time to deal with data dependence.

4.1 Valuated Synchronous Structure

We associate a valuation function v and a data dependency relation $\rightarrow$ to synchronous structure.

Definition 13. *Let $\mathcal{D}$ be a set. $(\mathcal{E}, \ll, v : \mathcal{E} \longrightarrow \mathcal{D}, \rightarrow)$ is a* valuated synchronous structure *iff $(\mathcal{E}, \ll)$ is a synchronous structure, v a function from $\mathcal{E}$ to $\mathcal{D}$ and $\rightarrow$ is a partial order included in $\ll$ i.e. :*

$$\forall (x, y) \in \mathcal{E}^2, x \rightarrow y \Rightarrow x \ll y \tag{2}$$

The definition 8 of the partial order $\triangleright$ on instants and the axiom 2 guarantee that the value of an event cannot depend on the value of a future event. The data dependencies of an event come only from past or present values of other events. A signal is said *of domain $D \subseteq \mathcal{D}$* iff all its events x are such that $v(x) \in D$. Let $\ll_{\mathrm{v}}$ be the transitive closure of the union of the relations $\ll$ and $\rightarrow$ The preorder $\ll_{\mathrm{v}}$ defines a notion of time which takes into account the synchronicity relation, control dependencies and data dependencies. We define a preorder relation $\preceq_{\mathrm{v}}$ on $\mathcal{W}$.

Definition 14. $X \preceq_v Y$ *iff* $X \preceq Y \wedge \forall x \in X, v(x) = v([Y]_X(x))$

The preorder $\preceq_{\mathrm{v}}$ gives rise to an equivalence relation $\widehat{=}_{\mathrm{v}}$. Intuitively, $X \widehat{=}_{\mathrm{v}} Y$ means that X and Y are synchronous and provide same values in same order.

Definition 15. $X \widehat{=}_v Y$ *iff* $X \preceq Y \wedge Y \preceq X$

4.2 Scheduling Specification

We define a ternary relation, called *conditional dependency.* Intuitively, $X \overset{C}{\rightarrow} Y$ states that, at the instants of the clock C, there are dependencies $\rightarrow$ from an event of X to an event of Y in the same instant. In this relation, we are only interested in instantaneous dependencies. Practically this relation is used to schedule the computation that have to be done in the same logical instant. A set of conditional dependencies is called a *scheduling specification.*

Definition 16. $X \xrightarrow{\widehat{Z}} Y \Leftrightarrow_{def} \forall x \in X \otimes Z, \exists y \in Y, x \sim y \wedge x \rightarrow y$

The following theorem enables to compute the transitive closure of a scheduling specification.

Theorem 2. *For all signals X, Y and Z, for all clocks C and D,*

$$X \xrightarrow{C} Y \wedge Y \xrightarrow{D} Z \Rightarrow X \xrightarrow{C \sqcap D} Z \qquad X \xrightarrow{C} Y \wedge X \xrightarrow{D} Y \Rightarrow X \xrightarrow{C \sqcup D} Y$$

In the figure 7, the diagram on the left depicts a scheduling specification involving local variables. These are hidden in the diagram on the right, using the theorem 2.

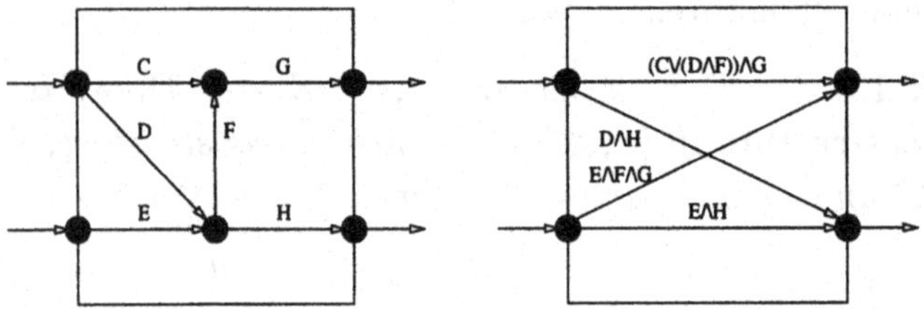

Fig. 7. Abstraction of Scheduling Specifications

5 Compilation of SIGNAL

First, we define the functions $^t.$ and $[.]$ which compute the sub-signal of the true valued events of a signal of domain $\{false, true\}$, and its clock.

Definition 17. *Let X be a signal of domain $\{false, true\}$.*

$$^tX = \{x \in X | v(x) = true\} \qquad [X] = \widehat{^tX}$$

The delay enables to move forward the valuation of a real signal[3]. The value of an event of a delayed real signal is the value of the previous event if it exists. In the other case, a default value is given. $\mathsf{Pre}(u, X, Y)$ states that Y is the delayed signal of X initialized with u.

$$\mathsf{Pre}(u, X, Y) \Leftrightarrow_{def} X \hat{=} Y \wedge \forall y \in Y, \begin{cases} y \text{ minimal element of } Y \Rightarrow v(y) = u \\ \exists y^- \in Y, y^- \prec y \Rightarrow v(y) = v([X]_Y(y^-)) \end{cases}$$

[3] It would make no sense to apply delay to imaginary signal.

We define a predicate that constrains a set of signals to be synchronous and to satisfy a predicate between their values at every instant. In SIGNAL, it is called an *instantaneous relation.*

Definition 18 (Instantaneous Relation). *Let $X_1, \cdots, X_n$ be n signals and P be a predicate on $\mathcal{D}^n$.*

$$R^n_P(X_1, \cdots, X_n) \Leftrightarrow_{def}$$
$$X_1 \widehat{=} \cdots \widehat{=} X_n \wedge \forall (x_1, \ldots, x_n) \in X_1 \times \ldots \times X_n, x_1 \sim \ldots \sim x_n \Rightarrow P(x_1, \ldots, x_n)$$

The denotational semantics of SIGNAL in this model is given in figure 8. The symbol := is not only denoted by $\widehat{=}_{\mathrm{v}}$ but also by dependence relations from the signals involved in the right part of an equation to the signal of the left part at the clock of the latter signal.

$$\begin{aligned}
[\![\mathtt{v}]\!] &\in \mathcal{D} \\
[\![\mathtt{f}]\!] &\in \mathcal{D} \times \cdots \times \mathcal{D} \longrightarrow \mathcal{D} \\
[\![\mathtt{x}]\!]_\rho &= \rho(\mathtt{x}) \in \mathcal{S} \\
[\![\mathtt{y:=f(x1,\ldots,xn)}]\!]_\rho &= F^n_{[\![\mathtt{f}]\!]}([\![\mathtt{x1}]\!]_\rho, \cdots, [\![\mathtt{xn}]\!]_\rho) \wedge \bigwedge_{i=1}^n [\![\mathtt{xi}]\!]_\rho \xrightarrow{\widehat{[\![\mathtt{y}]\!]_\rho}} [\![\mathtt{y}]\!]_\rho \\
[\![\mathtt{z:=x\ when\ y}]\!]_\rho &= [\![\mathtt{z}]\!]_\rho \widehat{=}_{\mathrm{v}} [\![\mathtt{x}]\!]_\rho \otimes {}^t[\![\mathtt{y}]\!]_\rho \wedge [\![\mathtt{x}]\!]_\rho \xrightarrow{\widehat{[\![\mathtt{x}]\!]_\rho} \sqcap [\![\mathtt{y}]\!]_\rho} [\![\mathtt{z}]\!]_\rho \\
[\![\mathtt{z:=x\ default\ y}]\!]_\rho &= [\![\mathtt{z}]\!]_\rho \widehat{=}_{\mathrm{v}} [\![\mathtt{x}]\!]_\rho \oplus [\![\mathtt{y}]\!]_\rho \wedge [\![\mathtt{x}]\!]_\rho \xrightarrow{\widehat{[\![\mathtt{x}]\!]_\rho}} [\![\mathtt{z}]\!]_\rho \wedge [\![\mathtt{y}]\!]_\rho \xrightarrow{\widehat{[\![\mathtt{y}]\!]_\rho} \setminus \widehat{[\![\mathtt{x}]\!]_\rho}} [\![\mathtt{z}]\!]_\rho \\
[\![\mathtt{y:=x\$\ init\ v}]\!]_\rho &= \mathsf{Pre}([\![\mathtt{v}]\!], [\![\mathtt{x}]\!]_\rho, [\![\mathtt{y}]\!]_\rho) \\
[\![P_1 \mid P_2]\!]_\rho &= [\![P_1]\!]_\rho \wedge [\![P_2]\!]_\rho \\
[\![P/\mathtt{x}]\!]_\rho &= \exists X \in \mathcal{S}, [\![P]\!]_{\rho, \mathtt{x} \mapsto X} \\
[\![(?\mathtt{x}_{1,..m} ! \mathtt{y}_{1,..n})P]\!]_\rho &= \lambda(x_{1,..m}, y_{1,..n}).[\![P]\!]_{\rho, \mathtt{x}_1 \mapsto x_1, \cdots, \mathtt{x}_m \mapsto x_m, \mathtt{y}_1 \mapsto y_1, \cdots, \mathtt{y}_n \mapsto y_n}
\end{aligned}$$

Fig. 8. The denotational semantics of SIGNAL

Endochrony refers to the Ancient Greek: "$\varepsilon\nu\delta o$", and literally means "time defined from the inside". An endochronous specification defines a reactive system where "time defined from the inside" translates into the property that the production of its outputs only depends on the presence of its inputs. An endochronous system reacts to inputs by having an activation clock computable from that of its inputs. This activation clock directs the execution of the program. By contrast with the classical synchronous programming model, in which the activation clock of a system is not always definable, it is always possible to manipulate real or imaginary clocks in our model (because the set of clock $\mathcal{C}$ is a complete lattice) and eventually to compute a real (endochronous) signal. Hierarchization is the implementation of the property of endochrony for the compilation of SIGNAL programs. It is the medium used in SIGNAL for compiling the parallelism specified using synchronous composition. It consists of organizing the computation of signals as a tree that defines a correct scheduling of computations into tasks. Each node of the tree consists of synchronous signals. It denotes the task of computing them when the clock is active. Each relation of a node with a sub-tree represents a sub-task of smaller clock.

6 Related Works

There are several ways to characterize the essentials of the synchronous paradigm. In [12], we introduce a co-inductive semantics of SIGNAL. A theorem library is developed and enable to express and prove not only liveness and safety properties of a synchronous program but also its correctness and its completeness. But it is not powerful enough to deal with more theoretical aspect of synchronous programming such as dependencies. The semantics of a synchronous language can be described in a better way with Symbolic Transition System (STS) [13]. This is a formalism on which fundamental questions can be investigated. But it manipulates the absence of a signal as a special value. This is not consistent with reality: the absence of a signal has to be inferred by the program (endochrony). In [3], STS is extended with preorders and partial orders to model causality relations, schedulings and communications. This pre-order theoretic model is put into practice in the design of BDL ([14]), a synchronous specification language that uses families of pre-orders to specify systems. In [6], the problem of characterizing synchrony without using a special symbol for absence is addressed in terms of multiple onput-output sequential machines. In [8], the language SIGNAL has been modelled in interaction categories ([1]) where processes are morphisms and objects are types of processes.

7 Conclusion

We have defined a unified model which formalizes all aspects of the development of a reactive system using the underlying programming methodology of synchronous languages (from relations to circuits). This model uses basic notions of set-theory and order-theory and has been specified and validated using the COQ theorem prover. This implementation is part of the development of a certified SIGNAL compiler.

References

1. Samson Abramsky. Interaction Categories (Extended Abstract). In G. L. Burn, Simon J. Gay, and M. D. Ryan, editors, *Theory and Formal Methods 1993: Proceedings of the First Imperial College Department of Computing Workshop on Theory*

and Formal Methods, pages 57–70. Springer-Verlag Workshops in Computer Science, 1993.

2. Albert Benveniste, Paul Le Guernic, and Christian Jacquemot. Synchronous programming with events and relations : the Signal language and its semantics. *Science of Computer Programming*, 16:103–149, 1991.
3. Albert Benveniste, Paul Le Guernic, and Pascal Aubry. Compositionality in dataflow synchronous languages: specification & code generation. Research report 3310, INRIA, 1997.
4. Grard Berry and Georges Gonthier. The ESTEREL synchronous programming language: design, semantics, implementation. *Science of Computer Programming*, 19:87–152, 1992.
5. Gerolamo Cardano. *Ars Magna*. 1545.
6. P. Caspi. Clocks in dataflow languages. *Theoretical Computer Science*, 94(1):125–140, March 1992.
7. Bruno Barras et al. *The Coq Proof Assistant Reference Manual - Version 6.2*. INRIA, Rocquencourt, May 1998.
8. Simon J. Gay and Raja Nagarajan. Modelling SIGNAL in Interaction Categories. In Geoffrey L. Burn, Simon J. Gay, and Mark D. Ryan, editors, *Theory and Formal Methods 1993: Proceedings of the First Imperial College Department of Computing Workshop on Theory and Formal Methods*. Springer-Verlag Workshops in Computer Science, 1993.
9. N. Halbwachs, P. Caspi, P. Raymond, and D. Pilaud. The synchronous dataflow programming language Lustre. *Proc. of the IEEE*, 79(9):1305–1320, September 1991.
10. Nicolas Halbwachs. *Synchronous Programming of Reactive Systems*. Kluwer Academic Pub., 1993.
11. D. Nowak, J.P. Talpin, and T. Gautier. Un systme de modules avanc pour Signal. Technical Report 3176, Irisa / Inria-Rennes, 1997.
12. David Nowak, Jean-Ren Beauvais, and Jean-Pierre Talpin. Co-inductive Axiomatization of a Synchronous Language. In *Proceedings of Theorem Proving in Higher Order Logics (TPHOLs'98)*, number 1479 in LNCS, pages 387–399. Springer Verlag, September 1998.
13. Amir Pnueli, Natarajan Shankar, and Eli Singerman. Fair Synchronous Transition Systems and their Liveness Proofs. Technical Report SRI-CSL-98-02, The Weizmann Institute of Science and SRI International, 1998.
14. J.-P. Talpin, A. Benveniste, B. Caillaud, C. Jard, Z. Bouziane, and H. Canon. Bdl, a language of distributed reactive objects. In *Proceedings of the International Symposium on Object-Oriented Real-Time Distributed Computing*. IEEE press, april 1998.
15. G. Winskel. Event structures. In W. Brauer, W. Reisig, and G. Rozenberg, editors, *Petri Nets: Applications and Relationships to Other Models of Concurrency, Advances in Petri Nets 1986, Part II; Proceedings of an Advanced Course*, Bad Honnef, September 1986, volume 255 of *Lecture Notes in Computer Science*, pages 325–392. Springer-Verlag, 1987.

Weakest-Congruence Results for Livelock-Preserving Equivalences

Antti Puhakka and Antti Valmari

Tampere University of Technology, Software Systems Laboratory,
PO Box 553, FIN-33101 Tampere, FINLAND,
email: anpu@cs.tut.fi, ava@cs.tut.fi

Abstract. A behavioural equivalence is a *congruence*, if a system is guaranteed to remain equivalent when any one of its component processes is replaced by an equivalent component processes. An equivalence is *weaker* than another equivalence if the latter makes at least the same distinctions between systems as the former. An equivalence *preserves a property*, if no equivalence class contains one system that has that property and another system that lacks the property. Congruences that preserve such properties as deadlocks or livelocks are important in automatic verification of systems, and knowledge of the weakest such congruences is useful for designing verification algorithms. A simple denotational characterisation of the weakest deadlock-preserving congruence has been published in 1995. In this article simple characterisations are given to the weakest livelock-preserving congruence, and to the weakest congruence that preserves all livelocking traces. The results are compared to Hoare's failures-divergences equivalence in the CSP theory.

1 Introduction

In this article we investigate *weakest congruences* for process-algebraic systems. A process algebra consists of a language for defining systems, and a semantic theory that defines one or more *equivalences* for the behaviours of systems. The language contains operators with which processes can be constructed and combined to form larger processes. An equivalence is a *congruence*, if and only if the replacement of a component process of a larger process with an equivalent component process always yields a result that is equivalent with the original larger process. Whether or not an equivalence is a congruence may depend on the set of operators that are allowed when constructing processes. An equivalence "$\simeq_1$" is *weaker* than another equivalence "$\simeq_2$" if and only if $P \simeq_2 Q$ implies $P \simeq_1 Q$.

The research on weakest congruence results may have its origin in Robin Milner's remark in p. 206 of [8]: "Hoare's failures equivalence ... is important, because it appears to be the weakest equivalence which never equates a deadlocking agent with one which does not deadlock." Milner probably required that the equivalence must be a congruence, because otherwise the weakest equivalence would be the trivial one that has precisely two equivalence classes: the processes (that is, Milner's agents) that deadlock, and those that do not. In [10] it was proven that Milner's guess was not precisely correct. The weakest deadlock-preserving congruence depends on the set of allowed process composition operators. Furthermore, assuming a reasonable choice of operators, it is the same as Hoare's failures equivalence only in the absence of so-called *divergence*. From now on we will call Hoare's equivalence *CSP-equivalence* to avoid confusion with some other important types of "failures" and "failures equivalences" that have appeared in the literature.

Another interesting weakest congruence result was proven in [5], where the so-called *nondivergent failures divergences equivalence* (*NDFD-equivalence*) was shown to be the weakest congruence that preserves the validity of formulae written in classic Manna-Pnueli linear time temporal logic [7] from which the "next state" operator "$\bigcirc$" has been removed. This logic is extremely important in verification of concurrent systems. Furthermore, if the congruence has to preserve also deadlocks, then the weakest congruence is the *Chaos-free failures divergences* (*CFFD*) *equivalence*. Because the Manna-Pnueli logic is state-based and process-algebraic equivalences are action-based, these results required an interpretation of the logic in an action-based setting. This can be done in more than one way. An alternative interpretation that is perhaps more relevant for practical verification than the original one was given in [11] (more easily found in [12] pp. 498–499).

Some researchers have tried to find the weakest congruence that preserves the results of certain kinds of tests. The solution with a fair way of testing was given by Brinksma, Rensink and Vogler in [2], and Leduc came to the conclusion that with another view to testing, the NDFD-equivalence is the solution [6].

Some equivalences investigated in weakest congruence research have their origin in [1].

In this article we are interested in weakest congruences that distinguish between diverging and non-diverging systems. Divergence is an important phenomenon, because it corresponds to livelock, and has perhaps been the biggest stumbling block in the quest of natural deadlock-preserving congruences. We also compare our results to the well-known CSP-equivalence.

Although the motivation of this article is mostly theoretical, weakest congruence results have also practical significance for automatic verification. One powerful way of fighting the well-known state explosion problem in automatic verification is *compositional LTS construction*, in which some reduction algorithm is applied to an LTS before using it as a component of a larger system. One way of guaranteeing that this approach produces correct results is to ensure that the reduction algorithm preserves some equivalence that is a congruence and that preserves the property in question. For instance, any reduction algorithm that preserves the weakest deadlock-preserving congruence can be used in compositional analysis of deadlocks.

Section 2 gives the earlier definitions, etc. that we will rely on in this article. The weakest congruence that preserves divergence traces is given in Section 3, and the weakest congruence that distinguishes between a diverging and non-diverging system in Section 4. Section 5 is devoted to an analysis of CSP-equivalence from the point of view of weakest congruences, and the paper ends with a Conclusions section.

2 Background

Let A^* denote the set of finite and A^ω infinite strings of elements of a set A. The empty string is denoted with ε, and it is an element of A^*, but not of A^ω. That a (finite or infinite) string σ is a prefix of a string ρ is denoted with $\sigma \leq \rho$, and $\sigma < \rho$ means that $\sigma \leq \rho \wedge \sigma \neq \rho$. The length of the string σ is denoted with $|\sigma|$.

The behaviour of a process consists of executing *actions*. There are two kinds of actions: *visible* and *invisible*. The invisible actions are denoted with a special symbol τ. The behaviour of a process is often represented as a *labelled transition system*. It is a directed graph whose edges are labelled with action names, with one state distinguished as the initial state of the process.

Definition 1. *A* labelled transition system, *abbreviated* LTS, *is a four-tuple* $(S, \Sigma, \Delta, \hat{s})$, *where*

- S *is the set of* states,
- Σ, *the* alphabet, *is the set of the* visible actions *of the process; we assume that* $\tau \notin \Sigma$,
- $\Delta \subseteq S \times (\Sigma \cup \{\tau\}) \times S$ *is the set of* transitions, *and*
- $\hat{s} \in S$ *is the* initial state.

An LTS is *finite* if and only if its S and Σ are finite.

The following notation is useful for talking about the execution of a process starting at some given state. The "$-x\rightarrow$"-notation requires that all actions along the execution path are listed, while the τ-actions are skipped in the "$=x\Rightarrow$"-notation.

Definition 2. *Let* $(S, \Sigma, \Delta, \hat{s})$ *be an LTS,* $s, s' \in S$, $a, a_1, a_2, \ldots, a_n, \ldots \in \Sigma \cup \{\tau\}$, *and* $b_1, b_2, \ldots, b_n, \ldots \in \Sigma$.

- $s -a\rightarrow s'$ *is an abbreviation for* $(s, a, s') \in \Delta$.
- $s -a_1a_2\cdots a_n\rightarrow s'$ *means that there are* $s_0, s_1, \ldots, s_n \in S$ *such that* $s_0 = s$, $s_n = s'$ *and* $s_{i-1} -a_i\rightarrow s_i$ *whenever* $1 \leq i \leq n$.
- $s -a_1a_2\cdots a_n\rightarrow$ *means that there is* s' *such that* $s -a_1a_2\cdots a_n\rightarrow s'$.
- $s -a_1a_2a_3\cdots\rightarrow$ *means that there are* $s_0, s_1, s_2, \ldots$ *such that* $s_0 = s$ *and* $s_{i-1} -a_i\rightarrow s_i$ *whenever* $1 \leq i$.
- *restr*$(a_1a_2\cdots a_n, A)$, *the* restriction *of* $a_1a_2\cdots a_n$ *to* A, *is the result of the removal of those* a_i *from* $a_1a_2\cdots a_n$ *that are not in* A. *The restriction of an infinite string is defined similarly.*
- $s{=}\varepsilon{\Rightarrow}s'$ *means that* $s{-}\tau^n{\rightarrow}s'$ *for some* $n \geq 0$, *where* τ^n *denotes the sequence of* n τ*-symbols.*

- $s = b_1 b_2 \cdots b_n \Rightarrow s'$ *means that there are* $s_0, s_1, \ldots, s_n \in S$ *such that* $s_0 = s$, $s_n = s'$ *and* $s_{i-1} = b_i \Rightarrow s_i$ *whenever* $1 \leq i \leq n$. *That is,* $s = b_1 b_2 \cdots b_n \Rightarrow s'$ *if and only if there is* $\sigma \in (\Sigma \cup \{\tau\})^*$ *such that* $s - \sigma \rightarrow s'$ *and* $restr(\sigma, \Sigma) = b_1 b_2 \cdots b_n$.
- $s = b_1 b_2 b_3 \cdots \Rightarrow$ *means that there are* $s_0, s_1, s_2, \ldots$ *such that* $s_0 = s$ *and* $s_{i-1} = b_i \Rightarrow s_i$ *whenever* $1 \leq i$.

The semantic equivalences that we will discuss will use the following abstract sets extracted from an LTS. The *traces* of an LTS are the sequences of visible actions generated by any finite execution that starts in the initial state. An infinite execution that starts in the initial state generates either an *infinite trace* or a *divergence trace*, depending on whether the number of visible actions in the execution is infinite.

Definition 3. *Let* $L = (S, \Sigma, \Delta, \hat{s})$ *be an LTS.*

- $Tr(L) = \{\sigma \in \Sigma^* \mid \hat{s} = \sigma \Rightarrow\}$ *is the set of the* traces *of* L.
- $Inftr(L) = \{\xi \in \Sigma^\omega \mid \hat{s} = \xi \Rightarrow\}$ *is the set of the* infinite traces *of* L.
- $Divtr(L) = \{\sigma \in \Sigma^* \mid \exists s : \hat{s} = \sigma \Rightarrow s \wedge s - \tau^\omega \rightarrow\}$, *where* τ^ω *denotes an infinite sequence of* τ*-actions, is the set of the* divergence traces *of* L.

It is obvious that $Divtr(L) \subseteq Tr(L)$ and, furthermore, if $\xi \in Inftr(L)$ and $\sigma < \xi$, then $\sigma \in Tr(L)$. If an LTS (or just its set of states) is finite, then its infinite traces are determined by its ordinary traces, as was shown in [13], for instance.

Proposition 1. *Let* $(S, \Sigma, \Delta, \hat{s})$ *be an LTS. If* S *is finite, then*
$Inftr(L) = \{\xi \in \Sigma^\omega \mid \forall \sigma : (\sigma < \xi \Rightarrow \sigma \in Tr(L))\}$.

We will later define some additional abstract sets. *Tr*, *Divtr* and *Inftr* are actually functions that take an LTS as input. Any collection of such functions can be used to define a semantic model of, and an equivalence between, LTSs as is shown below. Please notice that we will talk about an equivalence between two LTSs only if the LTSs have the same alphabet.

Definition 4. *Let* $f_1, f_2, \ldots, f_k$ *be any unary functions that take an LTS as their arguments.*

- *The* semantic model *of an LTS* L induced by $f_1, f_2, \ldots, f_k$ *is the* k*-tuple* $(f_1(L), f_2(L), \ldots, f_k(L))$.
- *Assume that the LTSs* L *and* L' *have the same alphabet. The* equivalence induced by $f_1, f_2, \ldots, f_k$ *is the equivalence "*$\simeq$*" defined as*
$L \simeq L' \iff f_1(L) = f_1(L') \wedge f_2(L) = f_2(L') \wedge \cdots \wedge f_k(L) = f_k(L')$.
We will call it the f_1-f_2-…-f_k-equivalence.

Almost every process algebra contains some *parallel composition operator*. In this article we use the version which forces precisely those component processes to participate in the execution of a visible action that have that action in their alphabets. The invisible action is always executed by one component process at a time. We first define the product of LTSs as the LTS that satisfies the above description and has the Cartesian product of component state sets as its set of states, and then define parallel composition by picking the part of the product that is reachable from the initial state of the product.

Definition 5. *Let* $L_1 = (S_1, \Sigma_1, \Delta_1, \hat{s}_1)$ *and* $L_2 = (S_2, \Sigma_2, \Delta_2, \hat{s}_2)$ *be LTSs. Their* product *is the LTS* $(S', \Sigma, \Delta', \hat{s})$ *such that the following hold:*

- $S' = S_1 \times S_2$
- $\Sigma = \Sigma_1 \cup \Sigma_2$
- $((s_1, s_2), a, (s_1', s_2')) \in \Delta'$ *if and only if either*
  - $a \in (\Sigma_1 \cup \{\tau\}) - \Sigma_2$ *and* $(s_1, a, s_1') \in \Delta_1 \wedge s_2' = s_2$, *or*
  - $a \in (\Sigma_2 \cup \{\tau\}) - \Sigma_1$ *and* $(s_2, a, s_2') \in \Delta_2 \wedge s_1' = s_1$, *or*
  - $a \in \Sigma_1 \cap \Sigma_2$ *and* $(s_1, a, s_1') \in \Delta_1$ *and* $(s_2, a, s_2') \in \Delta_2$.
- $\hat{s} = (\hat{s}_1, \hat{s}_2)$

The parallel composition $L_1 || L_2$ *is the LTS* $(S, \Sigma, \Delta, \hat{s})$ *such that*

- $S = \{ s \in S' \mid \exists \sigma \in \Sigma^* : \hat{s} =\sigma\Rightarrow s \}$
- $\Delta = \Delta' \cap (S \times (\Sigma \cup \{\tau\}) \times S)$

The following formulae describe the traces, etc. of a parallel composition as functions of the traces, etc. of its component processes. Their proofs are omitted because they basically consist of dull systematic checking against the definitions given above. Similar formulae can be found in the literature, for instance in [13].

Proposition 2. *Let* $L_1 = (S_1, \Sigma_1, \Delta_1, \hat{s}_1)$ *and* $L_2 = (S_2, \Sigma_2, \Delta_2, \hat{s}_2)$ *be LTSs.*

- $Tr(L_1||L_2) =$
  $\{ \sigma \in (\Sigma_1 \cup \Sigma_2)^* \mid restr(\sigma, \Sigma_1) \in Tr(L_1) \wedge restr(\sigma, \Sigma_2) \in Tr(L_2) \}$
- $Divtr(L_1||L_2) =$
  $\{ \sigma \in Tr(L_1||L_2) \mid restr(\sigma, \Sigma_1) \in Divtr(L_1) \vee restr(\sigma, \Sigma_2) \in Divtr(L_2) \}$
- $Inftr(L_1||L_2) = \{ \xi \in (\Sigma_1 \cup \Sigma_2)^\omega \mid$
  $restr(\xi, \Sigma_1) \in Inftr(L_1) \wedge restr(\xi, \Sigma_2) \in Tr(L_2) \cup Inftr(L_2) \vee$
  $restr(\xi, \Sigma_1) \in Tr(L_1) \wedge restr(\xi, \Sigma_2) \in Inftr(L_2) \}$

Another operator that is almost invariably found in process algebras in one form or another is *hiding*.

Definition 6. *Let* $L = (S, \Sigma, \Delta, \hat{s})$ *be an LTS, and* A *any set of action names. The LTS* **hide** A **in** L *is the LTS* $(S, \Sigma', \Delta', \hat{s})$ *such that the following hold:*

- $\Sigma' = \Sigma - A$
- $(s, a, s') \in \Delta'$ *if and only if*
  $a = \tau \wedge \exists b \in A : (s, b, s') \in \Delta$, *or* $a \notin A \wedge (s, a, s') \in \Delta$.

The traces, etc. of also **hide** A **in** L are functions of the traces, etc. of L.

Proposition 3. *Let* $L = (S, \Sigma, \Delta, \hat{s})$ *be an LTS, and let* Σ' *be the alphabet of* **hide** A **in** L.

- $Tr(\textbf{hide } A \textbf{ in } L) = \{\, \sigma \in \Sigma'^* \mid \exists \rho \in Tr(L) : \sigma = restr(\rho, \Sigma') \,\}$
- $Divtr(\textbf{hide}A\textbf{in}L) = \{\, \sigma \in \Sigma'^* \mid \exists \zeta \in Divtr(L) \cup Inftr(L) : \sigma = restr(\zeta, \Sigma') \,\}$
- $Inftr(\textbf{hide } A \textbf{ in } L) = \{\, \xi \in \Sigma'^\omega \mid \exists \zeta \in Inftr(L) : \xi = restr(\zeta, \Sigma') \,\}$

An equivalence "$\simeq$" is a *congruence* with respect to a process operator $op(L_1, \ldots, L_n)$ if and only if $L_1 \simeq L_1' \wedge \cdots \wedge L_n \simeq L_n'$ implies $op(L_1, \ldots, L_n) \simeq op(L_1', \ldots, L_n')$. We can reason from the above formulae that the *Tr-Divtr-Inftr*-equivalence is a congruence with respect to hiding and "$||$" [13]. Namely, if $Tr(L) = Tr(L')$, $Divtr(L) = Divtr(L')$, and $Inftr(L) = Inftr(L')$, then

$$\begin{aligned} Divtr(\textbf{hide } A \textbf{ in } L) &= \{\, \sigma \in \Sigma'^* \mid \exists \zeta \in Divtr(L) \cup Inftr(L) : \sigma = restr(\zeta, \Sigma') \,\} \\ &= \{\, \sigma \in \Sigma'^* \mid \exists \zeta \in Divtr(L') \cup Inftr(L') : \sigma = restr(\zeta, \Sigma') \,\} \\ &= Divtr(\textbf{hide } A \textbf{ in } L'), \end{aligned}$$

where Σ is the common alphabet of L and L', and $\Sigma' = \Sigma - A$. Similar reasoning applies to $Tr(\textbf{hide } A \textbf{ in } L)$ and $Inftr(\textbf{hide } A \textbf{ in } L)$. One can also immediately show with the same technique that $Tr(L_1||L_2) = Tr(L_1'||L_2')$ etc., given that $Tr(L_1) = Tr(L_1')$, etc.

In general, if $f_1(op(L_1, \ldots, L_n))$, $f_2(op(L_1, \ldots, L_n))$, $\ldots$, $f_k(op(L_1, \ldots, L_n))$ can be represented as functions of $f_1(L_1)$, $f_2(L_1)$, $\ldots$, $f_k(L_1)$, $\ldots$, $f_1(L_n)$, $f_2(L_n)$, $\ldots$, $f_k(L_n)$, then the equivalence induced by f_1, f_2, $\ldots$, f_k is a congruence with respect to op.

3 The Weakest Divergence-Trace-Preserving Congruence

The weakest divergence-trace-preserving congruence is the equivalence that preserves all divergence traces of a process, and is the weakest congruence with respect to hiding and "$||$" that has this property. In this section we will define *eventually nondivergent infinite traces* and then show that the equivalence induced by them together with traces and divergence traces is the weakest divergence-trace-preserving congruence. Eventually nondivergent infinite traces are those infinite traces, of whose prefixes only finitely many are divergence traces.

Definition 7. *Let* $L = (S, \Sigma, \Delta, \hat{s})$ *be an LTS. The set of the* eventually nondivergent infinite traces *of* L *is* $Enditr(L) = Inftr(L) - Divcl(L)$, *where*
$Divcl(L) = \{\, \xi \in \Sigma^\omega \mid \forall \sigma : (\sigma < \xi \Rightarrow \exists \sigma' : \sigma \leq \sigma' < \xi \wedge \sigma' \in Divtr(L)) \,\}$.

The abbreviation *Divcl* stands for "divergence closure". The definition immediately implies that $Enditr(L) \subseteq Inftr(L)$.

Proposition 4. *Let* L, L_1 *and* L_2 *be LTSs, and let* Σ_1, Σ_2 *and* Σ' *be the alphabets of* L_1, L_2 *and* **hide** A **in** L.

1. $Divtr(\textbf{hide}\ A\ \textbf{in}\ L) =$
   $\{\sigma \in \Sigma'^* \mid \exists \zeta \in Divtr(L) \cup Enditr(L) : \sigma = restr(\zeta, \Sigma')\}$
2. $Enditr(\textbf{hide}\ A\ \textbf{in}\ L) =$
   $\{\xi \in \Sigma'^\omega \mid \exists \zeta \in Enditr(L) : \xi = restr(\zeta, \Sigma')\} - Divcl(\textbf{hide}\ A\ \textbf{in}\ L)$
3. $Enditr(L_1 || L_2) = \{\xi \in (\Sigma_1 \cup \Sigma_2)^\omega \mid$
   $restr(\xi, \Sigma_1) \in Enditr(L_1) \wedge restr(\xi, \Sigma_2) \in Tr(L_2) \cup Enditr(L_2) \vee$
   $restr(\xi, \Sigma_1) \in Tr(L_1) \wedge restr(\xi, \Sigma_2) \in Enditr(L_2)\} - Divcl(L_1 || L_2)$
4. *The Tr-Divtr-Enditr-equivalence is a congruence with respect to hiding and "||".*

Proof. Because Proposition 3 gives that $Divtr(\textbf{hide}\ A\ \textbf{in}\ L) = \{\sigma \in \Sigma'^* \mid \exists \zeta \in Divtr(L) \cup Inftr(L) : \sigma = restr(\zeta, \Sigma')\}$, and because $Enditr(L) \subseteq Inftr(L)$, to prove 1 it suffices to show that whatever the strings in $Inftr(L) - Enditr(L)$ contribute to $Divtr(\textbf{hide}\ A\ \textbf{in}\ L)$ would be in the latter set anyway. These strings have arbitrarily long prefixes that are divergence traces of L. Let $\zeta \in Inftr(L) - Enditr(L)$, and $\sigma = restr(\zeta, \Sigma')$. If σ is infinite, then it is ruled out by the condition $\sigma \in \Sigma'^*$ in the right hand side of 1. Otherwise, ζ has a finite prefix ζ_1 such that $\sigma = restr(\zeta_1, \Sigma')$. Because $\zeta \in Inftr(L) - Enditr(L)$, ζ has a prefix ζ_2 such that $\zeta_1 \leq \zeta_2$ and $\zeta_2 \in Divtr(L)$. We have $\sigma = restr(\zeta_1, \Sigma') \leq restr(\zeta_2, \Sigma') \leq restr(\zeta, \Sigma') = \sigma$, so σ is included due to the part "$\exists \zeta \in Divtr(L) : \ldots$".

In a similar way one can show that if $\zeta \in Inftr(L) - Enditr(L)$, then $restr(\zeta, \Sigma') \in Divcl(\textbf{hide}\ A\ \textbf{in}\ L)$ or $restr(\zeta, \Sigma')$ is finite. This implies 2. The part 3 is proven similarly, and 4 follows from the previous parts and Propositions 2 and 3. □

The next two propositions show that any equivalence that preserves the divergence traces and is a congruence with respect to "||" and hiding must preserve also ordinary traces and eventually nondivergent infinite traces.

Proposition 5. *Let "$\simeq$" be a congruence with respect to "||" such that $L \simeq L'$ implies $Divtr(L) = Divtr(L')$. Then $L \simeq L'$ implies $Tr(L) = Tr(L')$.*

Proof. Let $L \simeq L'$, and let Σ be the common alphabet of L and L'. Let $\sigma = a_1 a_2 \cdots a_n \in \Sigma^*$, and let Test_1 be the LTS which has Σ as its alphabet, and whose states, transitions and initial state are as is shown in Figure 1. We have $\sigma \in Tr(L) \iff \sigma \in Divtr(L || \mathsf{Test}_1) \iff \sigma \in Divtr(L' || \mathsf{Test}_1) \iff \sigma \in Tr(L')$, where the first and last logical equivalences are due to the structure of Test_1, and the middle one follows from the congruence requirement and that the equivalence preserves divergence traces. □

Proposition 6. *Let "$\simeq$" be a congruence with respect to "||" and hiding such that $L \simeq L'$ implies $Divtr(L) = Divtr(L')$. Then $L \simeq L'$ implies $Enditr(L) = Enditr(L')$.*

Proof. Let $L \simeq L'$, let Σ be the common alphabet of L and L', and $\xi = a_1 a_2 a_3 \cdots \in Enditr(L)$. Because $\xi \in Enditr(L)$, it has a prefix $a_1 a_2 \cdots a_n$ such

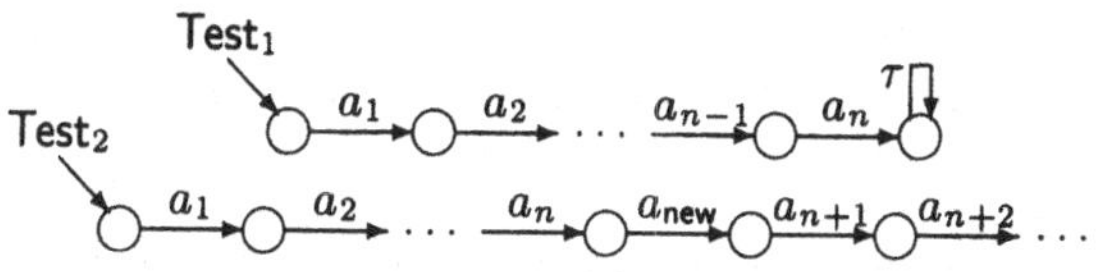

Fig. 1. Two LTSs used in proofs

that every divergence trace of L is either a prefix of $a_1 a_2 \cdots a_{n-1}$, or not a prefix of ξ. Let a_{new} be a symbol that is not in Σ, and let Test_2 be the LTS that has $\Sigma \cup \{a_{\mathsf{new}}\}$ as its alphabet, and whose states and transitions are as is shown in Figure 1. We see that $a_{\mathsf{new}} \in \mathit{Divtr}(\textbf{hide}\,\Sigma\,\textbf{in}\,(L||\mathsf{Test}_2))$. The congruence requirement implies that $a_{\mathsf{new}} \in \mathit{Divtr}(\textbf{hide}\,\Sigma\,\textbf{in}\,(L'||\mathsf{Test}_2))$. From this we can conclude that either $\xi \in \mathit{Inftr}(L')$, or $\mathit{Divtr}(L')$ contains some ρ such that $a_1 a_2 \cdots a_n \leq \rho < \xi$. In the latter case $\mathit{Divtr}(L) = \mathit{Divtr}(L')$ would imply that ρ is a divergence trace of L that is a prefix of ξ but not a prefix of $a_1 a_2 \cdots a_{n-1}$, which is in contradiction with the choice of n. So we see that $\xi \in \mathit{Inftr}(L')$ and, furthermore, $\xi \notin \mathit{Divcl}(L')$. These imply that $\xi \in \mathit{Enditr}(L')$. In conclusion, $\mathit{Enditr}(L) \subseteq \mathit{Enditr}(L')$. By replacing the roles of L' and L we see that also $\mathit{Enditr}(L') \subseteq \mathit{Enditr}(L)$. □

Putting the results of this section together gives the following theorem.

Theorem 1. *The Tr-Divtr-Enditr-equivalence is the weakest congruence with respect to "||" and hiding that preserves all divergence traces.*

Proof. The two preceding propositions say that any congruence that preserves divergence traces implies the *Tr-Divtr-Enditr*-equivalence. On the other hand, because this equivalence is a congruence, it is the required weakest congruence. □

The proof of Proposition 6 used an infinite LTS. Therefore, if we make the *a priori* assumption that all LTSs are finite, then the proposition cannot any more be used, at least not without a new proof. However, Proposition 5 remains valid in such a situation, because its proof did not assume infinite LTSs to be available. This fact allows us to show that Theorem 1 holds also if all LTSs are assumed to be finite, and even if hiding is removed from the set of operators with respect to which the equivalence must be a congruence.

Theorem 2. *The Tr-Divtr-Enditr-equivalence is the weakest congruence between finite LTSs with respect to "||" and hiding that preserves all divergence traces. The claim remains valid if "and hiding" is removed.*

Proof. Let "$\simeq$" preserve the divergence traces and be a congruence with respect to "||". By Proposition 5 it preserves also the traces. Proposition 1 implies that if $L \simeq L'$, then $\mathit{Inftr}(L) = \{\xi \in \Sigma^\omega \mid \forall \sigma : (\sigma < \xi \Rightarrow \sigma \in \mathit{Tr}(L))\} = \{\xi \in \Sigma^\omega \mid \forall \sigma : (\sigma < \xi \Rightarrow \sigma \in \mathit{Tr}(L'))\} = \mathit{Inftr}(L')$, so $\mathit{Enditr}(L) = \mathit{Inftr}(L) - \mathit{Divcl}(L) = \mathit{Inftr}(L') - \mathit{Divcl}(L') = \mathit{Enditr}(L')$. □

The key message of the proof is that if the LTSs are finite, then the *Tr-Divtr-Enditr*-equivalence collapses to the *Tr-Divtr*-equivalence.

4 The Weakest Divergence-Preserving Congruence

In the previous section we started with the requirement that the congruence must preserve all divergence traces. In this section our starting point is weaker: we assume only that the congruence preserves the one bit of information that tells if the process can diverge or not. As a result, we will end up with a strictly weaker equivalence.

The equivalence will be built from those traces and infinite traces that do not have divergence traces as their prefixes, and from those divergence traces that do not have divergence traces as their proper prefixes.

Definition 8. *Let L be an LTS and Σ its alphabet.*

- $diverges(L) = \mathsf{True}$ *if and only if* $Divtr(L) \neq \emptyset$. *Otherwise* $diverges(L) = \mathsf{False}$.
- *If* $X \subseteq \Sigma^*$, *then* $minimals(X) = \{\, \sigma \in X \mid \forall \rho : (\rho < \sigma \Rightarrow \rho \notin X) \,\}$
- *The set of the* mimimal divergence traces *of L is* $Mindiv(L) = minimals(Divtr(L))$
- *The set of the* extended divergence traces *of L is* $Divext(L) = \{\, \zeta \in \Sigma^* \cup \Sigma^\omega \mid \exists \rho : \rho \leq \zeta \wedge \rho \in Mindiv(L) \,\}$
- *The set of the* nondivergent traces *of L is* $Ndtr(L) = Tr(L) - Divext(L)$
- *The set of the* nondivergent infinite traces *of L is* $Ndinftr(L) = Inftr(L) - Divext(L)$

In analogy with the previous section, we need to show that the *Ndtr-Mindiv-Ndinftr*-equivalence is a congruence. We just present the formulae that give $Ndtr(\textbf{hide}\ A\ \textbf{in}\ L)$, etc. as functions of $Ndtr(L)$, etc., and skip their (boring) proof.

Proposition 7. *Let L be an LTS, and* $\Sigma' = \Sigma - A$.

- $Ndtr(\textbf{hide}\ A\ \textbf{in}\ L) =$
  $\{\, \sigma \in \Sigma'^* \mid \exists \rho \in Ndtr(L) : \sigma = restr(\rho, \Sigma') \,\} - Divext(L)$
- $Mindiv(\textbf{hide}\ A\ \textbf{in}\ L) =$
  $minimals(\{\, \sigma \in \Sigma'^* \mid \exists \zeta \in Mindiv(L) \cup Ndinftr(L) : \sigma = restr(\zeta, \Sigma') \,\})$
- $Ndinftr(\textbf{hide}\ A\ \textbf{in}\ L) =$
  $\{\, \xi \in \Sigma'^\omega \mid \exists \zeta \in Ndinftr(L) : \xi = restr(\zeta, \Sigma') \,\} - Divext(L)$
- $Ndtr(L_1 || L_2) =$
  $\{\, \sigma \in (\Sigma_1 \cup \Sigma_2)^* \mid restr(\sigma, \Sigma_1) \in Ndtr(L_1) \wedge restr(\sigma, \Sigma_2) \in Ndtr(L_2) \,\}$
- $Mindiv(L_1 || L_2) = minimals(\{\, \sigma \in (\Sigma_1 \cup \Sigma_2)^* \mid$
  $restr(\sigma, \Sigma_1) \in Mindiv(L_1) \wedge restr(\sigma, \Sigma_2) \in Ndtr(L_2) \cup Mindiv(L_2) \vee$
  $restr(\sigma, \Sigma_1) \in Ndtr(L_1) \wedge restr(\sigma, \Sigma_2) \in Mindiv(L_2) \,\})$

– $Ndinftr(L_1||L_2) = \{\, \xi \in (\Sigma_1 \cup \Sigma_2)^\omega \mid$
$restr(\xi, \Sigma_1) \in Ndinftr(L_1) \wedge restr(\xi, \Sigma_2) \in Ndtr(L_2) \cup Ndinftr(L_2) \vee$
$restr(\xi, \Sigma_1) \in Ndtr(L_1) \wedge restr(\xi, \Sigma_2) \in Ndinftr(L_2) \,\}$

That the *Ndtr-Mindiv-Ndinftr*-equivalence is the weakest *diverges*()-preserving congruence is a direct consequence of the following three propositions.

Proposition 8. *Let "$\simeq$" be a congruence with respect to "$||$" such that $L \simeq L'$ implies that $diverges(L) = diverges(L')$. Then $L \simeq L'$ implies $Mindiv(L) = Mindiv(L')$.*

Proof. Let $\sigma = a_1 a_2 \cdots a_n \in Mindiv(L)$. Let Test_3 have the same alphabet as L, and let its other components be as is shown in Figure 2. Then $diverges(L||\mathsf{Test}_3) = \mathsf{True}$, so also $diverges(L'||\mathsf{Test}_3) = \mathsf{True}$, from which we can reason that σ has a prefix ρ such that $\rho \in Mindiv(L')$. By repeating the argument with the roles of L and L' exchanged we see that ρ has a prefix σ' such that $\sigma' \in Mindiv(L)$. Due to the definition of *Mindiv* we have $\sigma' = \sigma$, so $\rho = \sigma$ and we get $Mindiv(L) = Mindiv(L')$. □

Fig. 2. Two more LTSs used in proofs

Proposition 9. *Let "$\simeq$" be a congruence with respect to "$||$" such that $L \simeq L'$ implies that $diverges(L) = diverges(L')$. Then $L \simeq L'$ implies $Ndtr(L) = Ndtr(L')$.*

Proof. Let $\sigma = a_1 a_2 \cdots a_n \in Ndtr(L)$. We use again Test_1 from Figure 1. We see that $diverges(L||\mathsf{Test}_1) = \mathsf{True}$, so also $diverges(L'||\mathsf{Test}_1) = \mathsf{True}$. This means that either σ has a prefix that is a divergence trace of L', or $\sigma \in Ndtr(L')$. Proposition 8 and the definition of $Ndtr(L)$ rule out the former possibility. Thus $\sigma \in Ndtr(L')$. So $Ndtr(L) \subseteq Ndtr(L')$. By symmetry also $Ndtr(L') \subseteq Ndtr(L)$. □

Proposition 10. *Let "$\simeq$" be a congruence with respect to "$||$" and hiding such that $L \simeq L'$ implies that $diverges(L) = diverges(L')$. Then $L \simeq L'$ implies $Ndinftr(L) = Ndinftr(L')$.*

Proof. Let $\xi = a_1 a_2 a_3 \cdots \in Ndinftr(L)$. It is the time to use Test_4 from Figure 2. Clearly **hide** Σ **in** $(L||\mathsf{Test}_4)$ diverges, so also **hide** Σ **in** $(L'||\mathsf{Test}_4)$ must diverge. This is possible only if either some prefix of ξ is a divergence trace of L', or $\xi \in Ndinftr(L')$. Like before, Proposition 8 and the definition of $Ndinftr(L)$ rule out the former possibility, so $\xi \in Ndinftr(L')$. Like before, that $\xi \in Ndinftr(L')$ implies $\xi \in Ndinftr(L)$ follows now from symmetry. □

Theorem 3. *The Ndtr-Mindiv-Ndinftr-equivalence is the weakest congruence with respect to "$\|$" and hiding that preserves the existence of divergence traces.*

We again used an infinite LTS, namely in the proof of Proposition 10. Like in the previous section, this proposition can be replaced by another one if we restrict ourselves to finite LTSs. As a matter of fact, the weaker assumption that the LTSs are *finitely branching* can be used as well.

Definition 9. *An LTS $(S, \Sigma, \Delta, \hat{s})$ is* finitely branching, *if and only if for each $s \in S$ and $a \in \Sigma \cup \{\tau\}$, the set $\{ s' \mid (s, a, s') \in \Delta \}$ is finite.*

Proposition 11. *If L is a finitely branching LTS, then*
$Ndinftr(L) = \{ \xi \in \Sigma^\omega \mid \forall \rho : (\rho < \xi \Rightarrow \rho \in Ndtr(L)) \}.$

Proof. The direction "$\xi \in Ndinftr(L) \Rightarrow$" is obvious. For the opposite direction, let $\xi = a_1 a_2 a_3 \cdots \in \Sigma^\omega$ such that $\forall \sigma : (\sigma < \xi \Rightarrow \sigma \in Ndtr(L))$. We show next that the system has an infinite execution $s_0 -b_1\rightarrow s_1 -b_2\rightarrow \cdots$, where $s_0 = \hat{s}$, such that $restr(b_1 b_2 \cdots, \Sigma) \leq \xi$. We do that by inductively demonstrating, for each $n \geq 0$, the existence of transitions $s_0 -b_1\rightarrow s_1 -b_2\rightarrow \cdots -b_n\rightarrow s_n$ and an infinite set E_n of arbitrarily long finite executions, such that the executions start with $s_0 -b_1\rightarrow \cdots -b_n\rightarrow s_n$, the traces of the executions are prefixes of ξ, and $s_0 = \hat{s}$.

A suitable E_0 is obtained by picking, for each $i \geq 0$, any execution that has $a_1 a_2 \cdots a_i$ as its trace. Of course, s_0 is chosen to be $\hat{s}$.

Because E_n contains an infinite number of arbitrarily long executions, it contains infinitely many arbitrarily long executions that are longer than n. The $(n+1)$th transition of any such execution is labelled either with τ, or with the a_k such that $restr(b_1 \cdots b_n, \Sigma) = a_1 \cdots a_{k-1}$. Because the LTS is finitely branching, there are only finitely many τ- and a_k-transitions that start in s_n. Thus infinitely many arbitrarily long members of E_n must share the same $(n+1)$th transition $s_n -b_{n+1}\rightarrow s_{n+1}$ (where $b_{n+1} = \tau$ or $b_{n+1} = a_k$). The set of those members can be chosen as E_{n+1}. This concludes the induction proof.

Our starting point included the assumption that all prefixes of ξ are nondivergent traces. Therefore, $restr(b_1 b_2 \cdots, \Sigma)$ cannot be finite, because otherwise the infinite execution $s_0 -b_1\rightarrow s_1 -b_2\rightarrow \cdots$ would generate a divergence trace that is a prefix of ξ. As a consequence, $restr(b_1 b_2 \cdots, \Sigma) = \xi$, so $\xi \in Ndinftr(L)$. □

Thus the *Ndtr-Mindiv-Ndinftr*-equivalence collapses to the *Ndtr-Mindiv*-equivalence if the LTSs are finitely branching.

Theorem 4. *The Ndtr-Mindiv-equivalence is the weakest congruence between finite LTSs with respect to "$\|$" and hiding that preserves the existence of divergence traces. The claim remains valid if "and hiding" is removed, and/or "finite" is replaced with "finitely branching".*

5 Comparison to CSP-Equivalence

Readers that are familiar with the CSP theory have certainly noticed that the *Ndtr-Mindiv-Ndinftr*-equivalence of the previous section has a striking similarity with the well-known failures-divergences equivalence of [3, 4, 9], and we will soon make this similarity explicit. As we hinted in the introduction, the meanings that this equivalence assigns to the terms "failure" and "divergence" are different from the meanings used elsewhere, so we prefer to call this equivalence *CSP-equivalence* to avoid confusion.

One important goal in the original definition of CSP-equivalence was to derive the meanings of recursive process equations directly — without first converting the processes to LTSs. In this way the need for an operational semantics was avoided. An analysis of the fixed points of process equations made such a definition possible. However, the definition had the consequence that no information of the behaviour of a process after it has executed a divergence trace is preserved by CSP-equivalence. A process that has executed a divergence trace is called *Chaos* in CSP literature. We assume in this section that all LTSs are finitely branching, because otherwise CSP-equivalence would not be a congruence ([9] p. 200).

CSP-equivalence can be defined in the LTS framework as follows ([9] p. 191). We use the additional concept of *stable failure*. A stable failure of an LTS is a pair consisting of a trace of that LTS and a subset of its alphabet. It is possible to execute that trace such that the LTS ends up in a state where it can execute neither invisible actions (the state is thus *stable*), nor any actions from the given subset. Stable failures or related concepts are important in equivalences that preserve deadlock information and are congruences with respect to "||".

Definition 10. *Let* $L = (S, \Sigma, \Delta, \hat{s})$ *be a finitely branching LTS.*

- $\mathit{sfail}(L) = \{ (\sigma, A) \in \Sigma^* \times 2^\Sigma \mid \exists s \in S : \hat{s} {=}\sigma{\Rightarrow} s \wedge \forall a \in A \cup \{\tau\} : \neg(s {-}a{\rightarrow}) \}$ *is the set of the* stable failures *of* L.
- $\mathit{CSPdivtr}(L) = \{ \sigma \in \Sigma^* \mid \exists \rho : \rho \leq \sigma \wedge \rho \in \mathit{Divtr}(L) \}$
- $\mathit{CSPfail}(L) = \mathit{sfail}(L) \cup (\mathit{CSPdivtr}(L) \times 2^\Sigma)$
- CSP-equivalence *is the CSPfail-CSPdivtr-equivalence.*

The following proposition implies that CSP-equivalence implies the *Ndtr-Mindiv*-equivalence.

Proposition 12. *Let* L *be a finitely branching LTS.*

1. $\mathit{Ndtr}(L) = \{ \sigma \mid (\sigma, \emptyset) \in \mathit{CSPfail}(L) \wedge \sigma \notin \mathit{CSPdivtr}(L) \}$
2. $\mathit{Mindiv}(L) = \mathit{minimals}(\mathit{CSPdivtr}(L))$

Proof. Let $\sigma \in \mathit{Ndtr}(L)$, and consider an arbitrary execution that produces σ. Because $\mathit{Ndtr}(L) \cap \mathit{Divtr}(L) = \emptyset$, any continuation of that execution with τ-transitions eventually leads to a stable state. Thus $(\sigma, \emptyset) \in \mathit{sfail}(L) \subseteq \mathit{CSPfail}(L)$. The definitions of $\mathit{Ndtr}(L)$ and $\mathit{CSPdivtr}(L)$ give that $\sigma \notin \mathit{CSPdivtr}(L)$. On the other hand, if $(\sigma, \emptyset) \in \mathit{CSPfail}(L)$ and $\sigma \notin \mathit{CSPdivtr}(L)$, then $(\sigma, \emptyset) \in \mathit{sfail}(L)$, so $\sigma \in \mathit{Tr}(L)$ and $\sigma \in \mathit{Ndtr}(L)$. Part 1 has now been proven. Part 2 follows easily from the definitions. □

In the opposite direction, clearly $CSPdivtr(L) = \{\sigma \in \Sigma^* \mid \exists \rho : \rho \leq \sigma \land \rho \in Mindiv(L)\}$. However, *CSPfail*$(L)$ cannot be obtained from *Ndtr*(L), *Mindiv*(L) and *Ndinftr*(L), and the *Ndtr-Mindiv-Ndinftr*-equivalence does not imply CSP-equivalence. The first two LTSs in Figure 3 prove this.

Fig. 3. Two pairs of *Ndtr-Mindiv-Ndinftr*-equivalent but not CSP-equivalent LTSs

It would be interesting to find a small strengthening to the starting point of the construction of the weakest divergence-preserving congruence such that the result would be precisely CSP-equivalence. Unfortunately, the task seems difficult. What *Ndtr-Mindiv-Ndinftr*-equivalence misses from CSP-equivalence is clearly related to the deadlock properties of LTSs. Unfortunately, it was shown in [10] that any equivalence that preserves the possibility of deadlocking and is a congruence with respect to "$||$" must preserve *sfail*(L). CSP-equivalence preserves only those stable failures whose trace part has no divergence trace as a prefix. As a consequence, the requirement of deadlock-preservation would strengthen the equivalence too much. It seems that the requirement must somehow be formulated such that it, like CSP-semantics, does not say anything about the behaviour after executing a minimal divergence trace.

A seemingly promising possibility would be to seek for the weakest "any-lock"-preserving congruence, that is, the weakest congruence that distinguishes systems that can stop executing visible actions from those that cannot. There are two ways in which a system can stop executing visible actions: deadlock (the system cannot execute anything), and livelock or divergence (the system executes infinitely many invisible actions). The congruence needs not distinguish between these two reasons, unless the congruence requirement indirectly forces such a distinction to be possible. CSP-equivalence implies this congruence, because CSP-equivalence is itself a congruence, $CSPdivtr(L) = \emptyset$ if and only if L has no divergences, and, in the case that L has no divergences, $(\sigma, \Sigma) \in CSPfail(L)$ if and only if σ is a trace that leads to a deadlock.

The proofs of Propositions 8, 9 and 10 can be carried through with the any-lock-preserving congruence by first adding a new action a_{new} to the alphabets of Test_3, Test_1 and Test_4, and attaching a local a_{new}-loop (that is, the transition $s{-}a_{\mathsf{new}}{\rightarrow}s$) to each state of Test_3, Test_1 and Test_4. Furthermore, the LTSs L_3 and L_4 in Figure 3 are *Ndtr-Mindiv-Ndinftr*-equivalent but not any-lock-equivalent. As a consequence, the any-lock-preserving congruence is strictly stronger than the *Ndtr-Mindiv-Ndinftr*-equivalence. Indeed, with the Test_5 in Figure 4 one can prove that the any-lock-preserving congruence must preserve all stable failures $(a_1 a_2 \cdots a_n, \{b_1, \ldots, b_k\})$ such that no prefix of $a_1 a_2 \cdots a_n b$ is a divergence trace, where b is any element of $\{b_1, \ldots, b_k\}$. This is slightly less than what we want — we would like to allow $a_1 a_2 \cdots a_n b$ (although not $a_1 a_2 \cdots a_n$) to diverge.

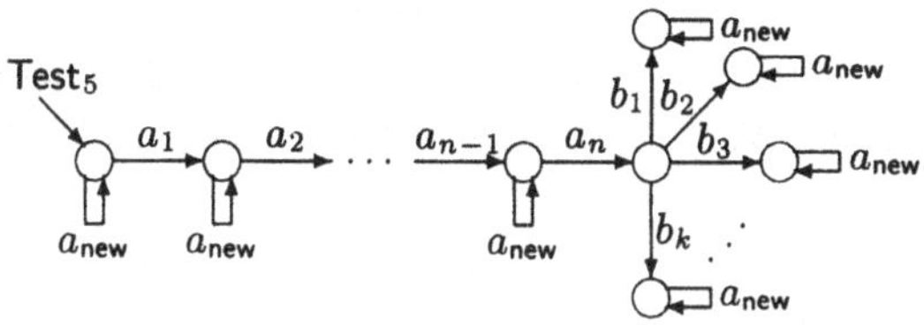

Fig. 4. An LTS for testing failures

Unfortunately, the weakest any-lock-preserving congruence seems to be strictly weaker than CSP-equivalence. Namely, CSP-equivalence distinguishes between L'_3 = **hide** $\{a_{\mathsf{new}}\}$ **in** L_3 and L'_4 = **hide** $\{a_{\mathsf{new}}\}$ **in** L_4, but it seems that the weakest any-lock-preserving congruence does not. Both of these two LTSs can stop executing visible actions by first executing a and then diverging. This remains true if the LTSs are put into an environment that hides a or eventually offers a. If the environment refuses a before offering it, then the systems can stop executing visible actions if and only if the environment can. It is thus difficult to think of an environment that would make it possible to distinguish between L'_3 and L'_4, when the only thing that can be observed is the ability of stopping executing visible actions.

Theorem 9.3.1 (iii) of [9] characterises CSP-equivalence as the weakest "immediate-any-lock-preserving" congruence, that is, as the weakest congruence that distinguishes systems that can deadlock or diverge *before executing any visible actions* from those that cannot. This result is not fully satisfactory from our point of view, because its proof uses a somewhat unusual "relational renaming" operator that can convert a transition to two transitions with different labels (see the errata that is in the www page of the book), whereas most of the other weakest congruence results rely only on ordinary parallel composition and hiding. Nevertheless, the result gives a characterisation of CSP-equivalence as the weakest congruence that satisfies a simple condition with respect to a well-known (large) set of operators.

6 Conclusions

We proved that the weakest livelock-preserving congruence is the *Ndtr-Mindiv-Ndinftr*-equivalence, and the weakest congruence that preserves all traces that lead to a livelock is the *Tr-Divtr-Enditr*-equivalence. We proved that Hoare's well-known CSP-equivalence implies the *Ndtr-Mindiv-Ndinftr*-equivalence but is not the same. As an attempt to give CSP-equivalence a characterisation as a weakest congruence we investigated the "any-lock"-preserving congruence, and found it to be (apparently strictly) between the *Ndtr-Mindiv-Ndinftr*-equivalence and CSP-equivalence.

An interesting topic for future research would be to check how sensitive the results in this article are to the particular choice of operators. We state as a hypothesis that the *Tr-Divtr-Enditr* and *Ndtr-Mindiv-Ndinftr*-equivalences remain congruences when other common process operators, such as choice, renaming, etc., are taken into account.

Acknowledgements The comments by the anonymous referees were useful in finishing this paper. The work of A. Puhakka, the main author, was funded by the TISE graduate school and the Academy of Finland, project "Specification and Verification of Distributed Systems with Synchronous Actions". The work of A. Valmari was funded by the Academy of Finland, project "Software Verification with CFFD-Semantics".

References

1. Bergstra, J. A., Klop, J. W. & Olderog, E.-R.: "Failures without Chaos: A New Process Semantics for Fair Abstraction". *Formal Description of Programming Concepts III*, North-Holland 1987, pp. 77–103.
2. Brinksma, E., Rensink, A. & Vogler, W.: "Fair Testing". *Proc. CONCUR '95, Concurrency Theory*, Lecture Notes in Computer Science 962, Springer-Verlag 1995, pp. 313–327.
3. Brookes, S. D. & Roscoe, A. W.: "An Improved Failures Model for Communicating Sequential Processes". *Proc. NSF-SERC Seminar on Concurrency*, Lecture Notes in Computer Science 197, Springer-Verlag 1985, pp. 281–305.
4. Hoare, C. A. R.: *Communicating Sequential Processes.* Prentice-Hall 1985, 256 p.
5. Kaivola, R. & Valmari, A.: "The Weakest Compositional Semantic Equivalence Preserving Nexttime-less Linear Temporal Logic". *Proc. CONCUR '92, Third International Conference on Concurrency Theory*, Lecture Notes in Computer Science 630, Springer-Verlag 1992, pp. 207–221.
6. Leduc, G.: "Failure-based Congruences, Unfair Divergences and New Testing Theory". *Proc. Protocol Specification, Testing and Verification*, XIV, Chapman & Hall, London (1995), pp. 252–267.
7. Manna, Z. & Pnueli, A.: *The Temporal Logic of Reactive and Concurrent Systems, Volume I: Specification.* Springer-Verlag 1992, 427 p.
8. Milner, R.: *Communication and Concurrency.* Prentice-Hall 1989, 260 p.
9. Roscoe, A. W.: *The Theory and Practice of Concurrency.* Prentice-Hall 1998, 565 p.
10. Valmari, A.: "The Weakest Deadlock-Preserving Congruence". *Information Processing Letters* 53 (1995) 341–346.
11. Valmari, A.: "Failure-based Equivalences Are Faster Than Many Believe". *Proc. Structures in Concurrency Theory 1995*, Springer-Verlag "Workshops in Computing" series, 1995, pp. 326–340.
12. Valmari, A.: "The State Explosion Problem". *Lectures on Petri Nets I: Basic Models*, Lecture Notes in Computer Science 1491, Springer-Verlag 1998, pp. 429–528.
13. Valmari, A. & Tienari, M.: "Compositional Failure-Based Semantic Models for Basic LOTOS". *Formal Aspects of Computing* (1995) 7: 440–468.

Proof-Checking Protocols Using Bisimulations

Christine Röckl and Javier Esparza

Technische Universität München,
Fakultät für Informatik, D-80290 München,
{roeckl,esparza}@in.tum.de

Abstract. We report on our experience in using the Isabelle/HOL theorem prover to mechanize proofs of observation equivalence for systems with infinitely many states, and for parameterized systems. We follow the direct approach: An infinite relation containing the pair of systems to be shown equivalent is defined, and then proved to be a weak bisimulation. The weak bisimilarity proof is split into many cases, corresponding to the derivatives of the pairs in the relation. Isabelle/HOL automatically proves simple cases, and guarantees that no case is forgotten. The strengths and weaknesses of the approach are discussed.

1 Introduction

Observation equivalence (or weak bisimilarity) is a natural notion of behavioural equivalence; it has been extensively studied and applied in the literature (see, for instance, [9, 10, 18]). There exist two general ways of showing that a system is observationally equivalent to its specification. One, semantically oriented, way is to follow the definition: exhibit a relation containing as a pair the system and its specification, and prove that it is a bisimulation. Conceptually, the difficult part is to exhibit the relation, while proving that the relation is indeed a bisimulation reduces to a (usually large) number of simple checks of the form "for each derivative there exists a matching derivative". In order to apply this method the only requirement is to have a good intuition about the system. The second, syntactically oriented, way is to use algebraic (equational) reasoning: the system is proved to be observationally equivalent to the specification by exhibiting a (usually long) chain of equalities starting with the system and ending with the specification. In this case no bisimulation relation has to be guessed but deep insight into the proof system is required.

Both ways can be applied to systems with a finite or infinite state space. In the second case, however, they cannot be completely automatized due to well known undecidability results. Still, algebraic techniques have been used to verify various infinite systems, including a variation of the ABP (see, for instance, [6, 4]), and tool support has been developed. In contrast, and to the best of our knowledge, the semantic way has not yet been mechanized for infinite systems, even though it could be very useful: In order to establish that a relation is a bisimulation, a large number of cases may have to be considered, even if the relation can be partitioned into uniformly representable infinite subsets. For instance, the infinite-state version of the alternating bit protocol (ABP) considered in [Mil89], yields a relation consisting of the union of 12 groups of pairs, leading to 94 proof obligations of the form 'the strong transition $\xrightarrow{\mu}$ on one side can be matched by a weak transition $\stackrel{\hat{\mu}}{\Longrightarrow}$ on the other side'. Proofs by hand consider only a few cases (Milner considers 6), leaving the others as obvious or similar. This procedure is of course prone to errors; in fact, Milner remarks that his proof of the ABP is on the verge of what is tractable by hand, and he proposes to mechanize the procedure with the help of theorem provers.

In this paper we follow Milner's suggestion. We report on our experience with the mechanization in Isabelle/HOL [16, 15] of three examples concerning communication protocols, including Milner's proof of the ABP. We examine to what extent the theorem prover is able to verify bisimulation relations automatically, and at what point the user has to provide additional information. We shall see that Isabelle automatically finds derivatives that are reachable by transition sequences of length 0 or 1, while longer transition sequences and their derivatives have to be specified by the user.

We have chosen examples from the area of communication protocols for several reasons. First, they are often used as 'real-life' examples to demonstrate the expressiveness and merits, or weaknesses, of concurrent frameworks and their behavioural or algebraic equivalences [3, 9, 1, 12, 19, 11]. Second, they often serve as benchmarks for proof environments, especially for algebraic proof systems in theorem provers. Examples of mechanizations are presented in [14, 13, 6, 4, 5, 8]. Finally, the bisimulation approach seems to be particularly suitable for communication protocols (see the discussion at the end of the paper).

We model both communication protocols and their specifications in terms of *labelled transition systems* [17], using a concurrent normal form that usually suffices for reactive systems (see also [6]): a finite number of sequential, yet nondeterministic, value-passing processes (i.e., processes with infinite summation) is connected by parallel composition, with an embracing restriction hiding the internal actions. Note that other models like communicating automata, for instance, would have done as well. Note further that the systems and their specifications need not even be described within the same model. Also, our experiments do not rely especially on Isabelle/HOL; we could as well have applied any other generic prover offering higher order logic, like PVS or Coq.

The paper is organized as follows: In Section 2 we give a short overview of the features of Isabelle/HOL we are going to exploit in our case study. The process algebraical framework is introduced in Section 3. The main part of this work is Section 4 where we establish bisimulation relations for three examples, and discuss their proofs in Isabelle: we show that a channel that loses or duplicates messages is observationally equivalent to a channel that further — detectably — garbles messages but applies a filter to discard the garbled ones before delivery. The channels are assumed to be of arbitrary length, thus both systems are infinite-state (Section 4.1). The second example is a mechanization of Milner's correctness proof for the Alternating Bit Protocol (ABP) [9] in Isabelle/HOL. Also this example is infinite-state as, again, the channels are assumed to be of arbitrary length (Section 4.2). The last example deals with a specification of a Sliding Window Protocol (SWP) in terms of a parallel composition of several channels applying the ABP [14]. It is a parameterized system (the parameter being essentially the window size) containing infinite-state components. Due to the compositionality of observation equivalence we can replace the ABP components with their specifications (i.e., with one-place buffers), and thus obtain a parameterized system of finite-state components (Section 4.3).

2 Isabelle/HOL

Using the generic theorem prover Isabelle [16] we conduct all proofs in its instantiation HOL for higher-order logic [15]. Proofs in Isabelle are based on unification, and are usually conducted in a backward resolution style: the user formulates the goal he/she intends to prove, and then — in interaction with Isabelle — continuously reduces it to simpler subgoals until all of the subgoals have been accepted by the tool. Upon this the goal can be stored in Isabelle's database as a theorem. Isabelle offers various tactics, most of them applying to single subgoals. The basic tactics allow the user to instantiate a theorem from Isabelle's database so that its conclusion can be applied to transform a current subgoal into instantiations of its premises. Further there exist automatic tactics using the basic tactics to prove given subgoals according to different heuristics. These heuristics have in common that a provable goal is always transformed into a set of provable subgoals; 'unsafe' rules (rules that might yield unprovable subgoals) are only applied if none of the resulting subgoals has to be reported to the user as currently unproved. Besides these *classical tactics* Isabelle offers *simplification tactics* based on algebraic transformations. The most general and powerful tactic is `Auto_tac` which interweaves classical and simplification tactics, and reasons about all subgoals of the current proof state simultaneously.

Isabelle's instantiation for higher-order logic offers a number of modules, called *theories*, including among others frameworks for arithmetics, sets, and lists. These modules include databases with basic theorems that have already been proved, and that can be referenced by the user when working with the modules. The arithmetic module of Isabelle/HOL is of particular interest, as it allows the user to give inductive definitions. Note that transition systems are defined inductively, i.e., as the *least set* satisfying certain axioms and rules. Isabelle then automatically generates various forms of the transition rules including case exhaustions stating all reasons that may have led to a given transition; e.g., "$P \parallel Q \xrightarrow{\tau} R$ because $P \xrightarrow{\tau} P'$ and $R = P' \parallel Q$, or ...". Further it generates tactics for structural induction. Although Isabelle/HOL offers a framework for coinduction, we do not make use of it, sticking to the original definition of observation equivalence given in terms of a predicate over binary relations.

Isabelle is a *generic* theorem prover, i.e., the user can define theories of his/her own. Such theory modules consist of two parts: in a definition part new types, constants, and rules for the constants are introduced; in a second part theorems are proved and added to the theorem database. A preamble to the definition part lists the theories upon which the new module is based; these can include built-in as well as user-defined theories.

3 Transition Systems and Observation Equivalence

Reactive systems generally consist of a finite set of *process components* which, according to their *states*, send and receive data along *connections*[1] between them (see also [6]).

Let Λ be a countably infinite set of *signals* (visible actions) ranged over by $a, b, \ldots$. Further, in order to transmit messages of some (unspecified) type α we introduce a countably infinite set of *connections* $(\alpha)Con$ ranged over by $c1, c2, \ldots$. The type of the connections is parameterized over the type variable α, thus we use $(\alpha)Con$ instead of simply writing Con. The set of *visible labels*, $(\alpha)\mathcal{L}$ is given by all inputs a and $c(v)$, and outputs $\bar{a}$ and $\bar{c}\langle v\rangle$ of signals and messages along the connections. We use $Act \stackrel{\text{def}}{=} \Lambda \cup \{\bar{a} \mid a \in \Lambda\}$ to denote the set of inputs and outputs on signals.

(States of) *systems* are defined inductively in terms of (states of) their *components* P_i — given in terms of constant identifiers, and representing the basic units — and parallel compositions between them. We use the term *processes* to refer to components and systems equally. Formally, process components and states have the following syntax:

$$\begin{array}{lcl} (\alpha)\mathcal{PC} & ::= & P_1(\tilde{x}_1) \quad | \ldots | \quad P_n(\tilde{x}_n), \\ (\alpha)\mathcal{S} & ::= & (\alpha)\mathcal{PC} \quad | \quad (\alpha)\mathcal{S} \parallel (\alpha)\mathcal{S}, \end{array}$$

where the $\tilde{x}_i$ are variables of type α. In the Isabelle formalization of the protocols, components are sequential yet possibly nondeterministic processes, where each state is denoted by a constant of its own.

We use a Plotkin-style transition semantics [17] given in terms of a *strong transition relation* $\stackrel{\mu}{\longrightarrow} \subseteq (\alpha)\mathcal{S} \times ((\alpha)\mathcal{L} \cup \{\tau\}) \times (\alpha)\mathcal{S}$. The transition rules are defined inductively via axioms for the components in $(\alpha)\mathcal{PC}$ (describing the input and output as well as the silent behaviour of the components), and rules for parallel composition including communication, where $\mu \in (\alpha)\mathcal{L} \cup \{\tau\}$, $a \in \Lambda$, $c \in (\alpha)Con$, and $v \in \alpha$:

$$\frac{P \stackrel{\mu}{\longrightarrow} P'}{P \parallel Q \stackrel{\mu}{\longrightarrow} P' \parallel Q}\,\text{P1} \qquad \frac{P \stackrel{a}{\longrightarrow} P' \quad Q \stackrel{\bar{a}}{\longrightarrow} Q'}{P \parallel Q \stackrel{\tau}{\longrightarrow} P' \parallel Q'}\,\text{C1} \qquad \frac{P \stackrel{c(v)}{\longrightarrow} P' \quad Q \stackrel{\bar{c}\langle v\rangle}{\longrightarrow} Q'}{P \parallel Q \stackrel{\tau}{\longrightarrow} P' \parallel Q'}\,\text{C3}$$

The rules P2, C2, and C4 are symmetric versions of P1, C1, and C3.

Note that the transition rules are defined inductively, i.e., the transition relation is defined as the least set satisfying the given axioms and rules (cf. Section 2). This implies that besides the constructive rules such as P1, P2, or

[1] We do not use the term *channel* here, in order to avoid ambiguities with the channels used by the protocols.

C1, one can give analysis rules telling for a given transition how it can be derived. Let, e.g., a be a signal, and let P and Q be processes. Then,

$$\frac{P \parallel Q \xrightarrow{a} R}{(\exists P' \,.\, R = P' \parallel Q \;\wedge\; P \xrightarrow{a} P') \;\vee\; (\exists Q' \,.\, R = P \parallel Q' \;\wedge\; Q \xrightarrow{a} Q')}$$

is such a case analysis. The general analysis rule comprising silent steps is more complicated, as it further considers possible communications between the components.

We model inputs in an early style, i.e., input rules are of the form $\forall w.\, P \xrightarrow{c(w)} P'(w)$. When formalizing the transition systems in Isabelle the user need not state the $\forall$-quantification explicitely but applies a formal parameter for which the quantification is then automatically provided by the prover. Note that Isabelle is able to distinguish between constants and formal parameters.

In order to abstract from internal activities of the systems (like, e.g., communications, idle loops, or the processing of data), we use a *weak transition relation* $\overset{\mu}{\Longrightarrow} \subseteq (\alpha)\mathcal{S} \times ((\alpha)\mathcal{L} \cup \{\tau, \epsilon\}) \times (\alpha)\mathcal{S}$ which allows for arbitrarily many τ-steps before and after each transition. As usual $\overset{\epsilon}{\Longrightarrow} \overset{\text{def}}{=} (\xrightarrow{\tau})^*$ denotes the reflexive transitive closure of (strong) internal steps. Below we give the rules for introducing ϵ, lifting strong transitions to weak ones, and for the expansion of weak transitions by silent steps, where $\nu \in (\alpha)\mathcal{L} \cup \{\tau\}$, and $\mu \in (\alpha)\mathcal{L} \cup \{\tau, \epsilon\}$,

$$\frac{}{P \overset{\epsilon}{\Longrightarrow} P}\,\text{E} \quad \frac{P \overset{\tau}{\Longrightarrow} P'}{P \overset{\epsilon}{\Longrightarrow} P'}\,\text{TE} \quad \frac{P \xrightarrow{\nu} P'}{P \overset{\nu}{\Longrightarrow} P'}\,\text{SW} \quad \frac{P \overset{\epsilon}{\Longrightarrow} P';\; P' \overset{\mu}{\Longrightarrow} P'';\; P'' \overset{\epsilon}{\Longrightarrow} P'''}{P \overset{\mu}{\Longrightarrow} P'''}\,\text{EX}$$

Further there exist weak versions of the rules P1, P2, C1, C2, C3, and C4.

We use the common abbreviation $\hat{\mu}$ to denote ϵ if $\mu = \tau$, and μ if μ is a visible label.

Isabelle allows a mixfix representation of constants (like actions, processes, or transitions). Exploiting this we can write `P -[c<v>]-> P'` for an output of value v along channel `c`, or `P -[c#{v}]-> P'(v)` for an input of `v`, or `P -[tau]-> P'` for a silent transition. We write `P =[u]=> P'` for a weak transition with label u.

So far we have not distinguished between *internal* and *external* signals or connections. In order to avoid an additional restriction operator, for which we would have to formalize additional rules in Isabelle, we defer the matter of interface to the definition of bisimularity. Let $A \subseteq \Lambda$ be a set of signals, and $C \subseteq (\alpha)Con$ a set of connections of type α. Then we define $(\alpha)\mathcal{L}|_{(A,C)}$ to contain all possible inputs and outputs on signals in A, as well as on connections in C.

Definition 1 (Observation Equivalence). A relation $\mathcal{R} \subseteq (\alpha)\mathcal{S} \times (\alpha)\mathcal{S}$ is a *(weak) bisimulation wrt.* $(A, C) \in \Lambda \times (\alpha)Con$, if for all $(P, Q) \in \mathcal{R}$ and $\mu \in (\alpha)\mathcal{L}|_{(A,C)} \cup \{\tau\}$, the following holds:

- If $P \xrightarrow{\mu} P'$, for some P', there exists a Q' s.t. $Q \overset{\hat{\mu}}{\Longrightarrow} Q'$ and $(P', Q') \in \mathcal{R}$.
- If $Q \xrightarrow{\mu} Q'$, for some Q', there exists a P' s.t. $P \overset{\hat{\mu}}{\Longrightarrow} P'$ and $(P', Q') \in \mathcal{R}$.

Two processes P and Q are ***observation equivalent wrt.*** (A,C), written $P \approx_{(A,C)} Q$, if there exists a weak bisimulation containing (P,Q).

Let $\setminus$ be the restriction operator from CCS, extended to be applicable both to signals and connections. Though we find it convenient not to have a restriction operator on $(\alpha)\mathcal{S}$, we are still close to the original notion of observation equivalence. Let $\mathcal{R}$ be a bisimulation wrt. (A,C). Then

$$\mathcal{R}' \stackrel{\text{def}}{=} \{(P\setminus(\Lambda - A, (\alpha)\mathcal{C}on - C), Q\setminus(\Lambda - A, (\alpha)\mathcal{C}on - C) \mid (P,Q) \in \mathcal{R}\}$$

is a weak bisimulation in the sense of [9]. This is due to observation equivalence being a congruence wrt. restriction. Further every bisimulation in the sense of [9] is a bisimulation wrt. $(\Lambda, (\alpha)\mathcal{C}on)$. This implies that two systems are observationally equivalent in the usual sense iff they are so in our sense, allowing us to adapt congruence properties and proof techniques without having to prove them explicitly in our framework.

4 A Case Study

In all of the following examples the sets of labels that are visible to the observer either consist of signals, or of messages sent alog channels. We can thus project observation equivalence wrt. (A,C) either to observation equivalence wrt. A, written $\approx_A$, or to observation equivalence wrt. C, written $\approx_C$.

All proofs in this section follow a uniform pattern: The user sets up the systems and their specifications by giving their states and transition rules in one module. From this definition Isabelle computes sets of rules that can be used to reason about the transitions in a constructive as well as in an analysing style; further Isabelle generates schemes for structural induction (cf. Sections 2, 3). Another module contains as a predicate the criterion for a relation to be a bisimulation. The bisimulation relation itself is defined and proved in a third module which relies on the two previous modules. Often a further module is necessary to provide additional theorems about the data types used to model the transition systems (e.g., about insertion into or deletion from the finite lists representing communication channels).

The proof that a relation is a bisimulation usually falls into the following parts: in a separate theorem for each label μ, we prove symbolically for every $(P,Q) \in \mathcal{R}$ that if $P \stackrel{\mu}{\longrightarrow} P'$, for some P', there exists a Q' s.t. $Q \stackrel{\hat{\mu}}{\Longrightarrow} Q'$ and $(P',Q') \in \mathcal{R}$; and similarly for Q. Then the main theorem of each theory stating that $\mathcal{R}$ is a bisimulation is instantiated according to Definition 1 and reduced to the above proof obligations. This, including the necessary swapping of quantifiers, is done automatically by Isabelle.

In the following we present three examples demonstrating the generality of the approach: in the first example we compare two infinite-state systems that are both non-deterministic; the second example deals with quite a large composed system containing non-deterministic components and a small deterministic specification; in the third example we consider a composed system which is parameterized wrt. the number of its components. The three examples are closely related, e.g., the system studied in our third example contains as its components the second one which can be replaced by its specification due to the compositionality of observation equivalence. It should be noted that, unlike in many proof mechanizations in theorem provers, our Isabelle proofs are not much different from proofs as one would perform them by hand. Yet, what is different is the emphasis put on different parts of the proofs. Whereas in proofs by hand one has to be careful not to forget about any strong transition, Isabelle automatically takes care of this. On the other hand, an Isabelle user has to spend a lot of time interacting with the tool in order to prove very simple theorems about the data structures manipulated during a transition. In both cases the weak transitions have to be found by the person conducting the proof. However, presenting them to Isabelle is not much more time-consuming than writing them down on a piece of paper, and often it even suffices to provide the prover with a scheme so to enable it to generate the transitions automatically.

4.1 Faulty channels of unbounded size

Our first example is taken from [19]. It is of interest to us as it compares two indeterminate infinite-state systems operating on similar data structures. Most of the resulting proof obligations refer to strong transitions in weak disguise.

Consider two channels of unbounded capacity, say K and L. We model their contents by finite lists of arbitrary length. Both may lose or duplicate messages, but K is further able to garble data. This is reflected by an additional bit attached to each message in K.

```
"L(s)                -[ci#{x}]->       L(x # s)"                (* accept *)
"L(s @ a # t)        -[tau]->          L(s @ t)"                (* lose *)
"L(s @ a # t)        -[tau]->          L(s @ a # a # t)"        (* dupl *)
"L(s @ [x])          -[co<x>]->        L(s)"                    (* deliver *)

"K(s)                -[ci#{x}]->       K((x, True) # s)"        (* accept *)
"K(s @ a # t)        -[tau]->          K(s @ t)"                (* lose *)
"K(s @ a # t)        -[tau]->          K(s @ a # a # t)"        (* dupl *)
"K(s @ (x, b) # t)   -[tau]->          K(s @ (x, False) # t)"   (* garble *)
"K(s @ [(x, b)])     -[cf<(x, b)>]->   K(s)"                    (* deliver *)
```

A filter attached to K delivers correctly transmitted messages and discards garbled ones. We consider a version of a filter that discards garbled messages immediately when it receives them, but may arbitrarily lose or duplicate a message once it has accepted it. A filter that delivers all messages it has accepted would not lead to a system observationally equivalent to L: a state in which a message has just been transferred to the filter could not match a losing action by L.

```
"Filter         -[cf#{(x, True)}]->    FF{x, 0}"                (* accept *)
```

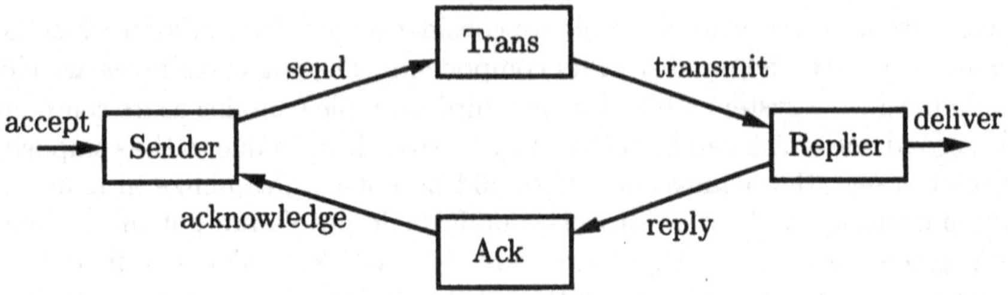

Fig. 1. Components of the ABP

```
"Filter          -[cf#{(x, False)}]->  Filter"         (* discard *)
"FF{x, 0}        -[co<x>]->            Filter"         (* deliver *)
"FF{x, Suc n}    -[co<x>]->            FF{x, n}"       (* deliver *)
"FF{x, 0}        -[tau]->              Filter"         (* lose *)
"FF{x, Suc n}    -[tau]->              FF{x, n}"       (* lose *)
"FF{x, n}        -[tau]->              FF{x, Suc n}"   (* dupl *)
```

By demonstrating that the relation `BS_Filter`, given below, is a bisimulation relation wrt. `{ci, co}`, we can conclude that `L` ≈ (`K` || `Filter`)\\{`cf`}.

```
"BStr == {(P, Q) . (EX s      .  P = K(s) || Filter
                              &  Q = L(map fst (filter snd s)))}"
"BSdl == {(P, Q) . (EX s x n .  P = K(s) || FF{x,n}
                              &  Q = L((map fst (filter snd s))
                                    @ (replicate (Suc n) x)))}"

"BS_Filter == BStr Un BSdl"
```

In the relation `BS_Filter`, `L` contains lists which are obtained from those stored in `K` by first eliminating all garbled messages (`filter snd s`; garbled messages are tagged with a `False` bit), and then projecting all elements of the resulting list to their first components (`map fst`). The list $\mathtt{x}^{\mathtt{n+1}}$, denoted by `replicate (Suc n) x`, models the `n` + 1 copies of message `x` stored in the filter.

Proving that `BS_Filter` is a bisimulation is not difficult, yet one has to take care of the lists of messages in the channels. As mentioned above most of the involvement by the user goes into theorems telling, for instance, how `map fst (filter snd s)` looks like if an element has been lost from `s`. Provided with these theorems, however, Isabelle proves by one single application of `Auto_tac` that `BS_filter` is a bisimulation. In particular, the user does not have to find the weak transitions.

The proof script contains less than 300 lines, and has been set up within a few hours only.

4.2 The Alternating Bit Protocol

The Alternating Bit Protocol (ABP), introduced in [2], is a well-established benchmark for proof methodologies implemented in theorem provers (see, for instance, [14, 4, 13, 5]). It turns unreliable channels into reliable communication lines. We consider an infinite-state variant in which the channels can hold arbitrarily many messages. The model as well as the outline of the proof follow [9].

The behaviour of the ABP can be specified in terms of a one-place buffer. This allows to abstract from the data to be transmitted. Note however that including them would not further complicate the proof, neither by hand nor in a theorem prover.

```
"accBuff -[accept]->  delBuff"                  (* accept a message *)
"delBuff -[deliver]-> accBuff"                  (* deliver a message *)
```

The ABP is designed around two faulty channels — one *transmitting* the messages and the other returning *acknowledgements* — a *sender*, and a *replier* module. A schematic view is given in Figure 1.

The channels, `Trans` and `Ack`, may both lose or duplicate but never swap messages. They behave exactly like channel `L` in the previous Section. We refer to the input and output connections of `Trans` as `cs` and `ct`, and to those of `Ack` as `cr` and `ca`, respectively.

The *sender* module continuously accepts messages from the environment, transmits them over the channel `Trans`, and waits for an acknowledgement along `Ack`, before accepting a new message. If an acknowledgement does not arrive within a certain time the sender assumes that the message has been lost and resends it. Yet, as the message may only have been delayed, the sender tags all messages with a bit so that new messages can be distinguished from old ones. An alternating bit suffices for this purpose, as the messages may not be swapped.

```
"Accept(b)  -[accept]->  Send(b)"               (* accept a message *)
"Send(b)    -[cs<b>]->   Sending(b)"            (* send message *)
"Sending(b) -[tau]->     Send(b)"               (* timeout *)
"Sending(b) -[ca#{b}]->  Accept(~b)"            (* correct acknowledge *)
"Sending(b) -[ca#{~b}]-> Sending(b)"            (* old acknowledge *)
```

After having delivered a message to the environment, the *replier* module repeatedly transmits tagged acknowledgements to the sender until a new message arrives.

```
"Deliver(b)  -[deliver]-> Reply(b)"             (* deliver a message *)
"Reply(b)    -[cr<b>]->   Replying(b)"          (* acknowledge *)
"Replying(b) -[tau]->     Reply(b)"             (* timeout *)
"Replying(b) -[ct#{~b}]-> Deliver(~b)"          (* receive new message *)
"Replying(b) -[ct#{b}]->  Replying(b)"          (* receive old message *)
```

The bisimulation relation, `BS_ABP`, is the union of the relations `BSaccept`, in which the processes are potentially able to accept a new message, and `BSdeliver`, in which the processes may deliver the current message. In every channel there are at most two types of messages or acknowledgements: those that are currently being delivered, and possibly copies of the previous ones. The finite lists are thus either of the form x^n, or $x^n y^m$; in Isabelle this is expressed by using the `replicate` operator defined in the built-in module for finite lists.

```
"BSaccept == {(P, Q) . Q = accBuff &
      ((EX b n p   .  P = Accept(~b) || Trans(replicate n b) ||
                         Ack(replicate p b) || Reply(b))
     | (EX b n p   .  P = Accept(~b) || Trans(replicate n b) ||
                         Ack(replicate p b) || Replying(b))
     | (EX b m p q .  P = Send(~b) || Trans(replicate m (~b)) ||
                         Ack(replicate p b @ replicate q (~b)) ||
                         Reply(~b))
     | (EX b m p q .  P = Send(~b) || Trans(replicate m (~b)) ||
                         Ack(replicate p b @ replicate q (~b)) ||
                         Replying(~b))
     | (EX b m p q .  P = Sending(~b) || Trans(replicate m (~b)) ||
                         Ack(replicate p b @ replicate q (~b)) ||
                         Reply(~b))
     | (EX b m p q .  P = Sending(~b) || Trans(replicate m (~b)) ||
                         Ack(replicate p b @ replicate q (~b)) ||
                         Replying(~b)))}"

"BSdeliver == {(P, Q) . Q = delBuff &
      ((EX b m p   .  P = Send(~b) || Trans(replicate m (~b)) ||
                         Ack(replicate p b) || Deliver(~b))
     | (EX b m p   .  P = Sending(~b) || Trans(replicate m (~b)) ||
                         Ack(replicate p b) || Deliver(~b))
     | (EX b m n p .  P = Send(~b) ||
                         Trans(replicate m (~b) @ replicate n b) ||
                         Ack(replicate p b) || Reply(b))
     | (EX b m n p .  P = Send(~b) ||
                         Trans(replicate m (~b) @ replicate n b) ||
                         Ack(replicate p b) || Replying(b))
     | (EX b m n p .  P = Sending(~b) ||
                         Trans(replicate m (~b) @ replicate n b) ||
                         Ack(replicate p b) || Reply(b))
     | (EX b m n p .  P = Sending(~b) ||
                         Trans(replicate m (~b) @ replicate n b) ||
                         Ack(replicate p b) || Replying(b)))}"

"BS_ABP == BSaccept Un BSdeliver"
```

To show that `BS_ABP` is indeed a bisimulation wrt. {`accept`, `deliver`} we follow our usual scheme. As a typical example consider the case where the ABP performs a strong `accept` transition. We have to prove the obligation, if `(P, Q)` ∈ `BS_ABP` and `P -[accept]-> P'`, then there exists a `Q'` s.t. `Q =[accept]=> Q'`, and `(P', Q')` ∈ `BS_ABP`. Out of the six subrelations in `BSaccept`, differing in the shape of `P`, Isabelle automatically extracts the first two as those in which `P` can do an `accept`. It remains to show that in both cases the resulting process `P'` fits the shape of `P` in the third and fourth subrelations of `BSdeliver`, the

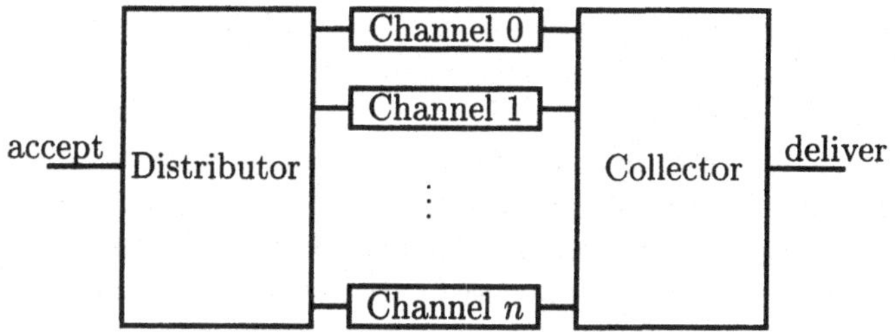

Fig. 2. A Specification of the SWP

difficulty being that the lists of messages in the channels look differently from those in `BSaccept`. However, once provided with the necessary theorems about finite lists, Isabelle manages to complete the proof fully automatically.

Another interesting example is the reverse case where `Q -[accept]-> Q'`, and `P =[accept]=> P'`. For the third through sixth case of `BSaccept` the user has to provide suitable sequences of weak transitions leading to the acceptance of a new message. In all of the cases we can apply the following scheme: remove all messages from `Trans` and `Ack` (that this is possible can be shown by an induction on the length of the lists stored in the channels), then have the replier transmit an acknowledgement to the sender, and finally execute the `accept` transition.

For the invisible transitions of the ABP we essentially have to show that they yield derivatives that still lie within `BSaccept` or `BSdeliver`, respectively. As for each of the processes there are several possibilities, we examine each of the twelve cases separately. Note, moreover, that a simultaneous treatment of all the cases may exceed the capacity of the prover, as also the hypothetical cases like "component `Ack` communicates with component `Trans`" have to be considered, resulting in an exponential blow-up of cases. Again, provided with the necessary theorems about lists, Isabelle proves the cases fully automatically.

The proof script contains about 800 lines. As a large part of it consists of theorems about the finite lists used in the channels, some experience with theorem provers is necessary to set up the proofs. The bisimulation part itself contains a bit more than 400 lines, and can be set up within a few days by a user experienced both in the bisimulation proof method and theorem proving. Notice, however, that this is only possible if the concept of the proof has already been clear before Isabelle is brought into play.

4.3 A Specification of the Sliding Window Protocol

In [14] a specification of the Sliding Window Protocol (SWP) is presented, given by the parallel composition of n communication lines which in turn use the ABP on faulty channels. Figure 2 gives a schematical view of the system. Incoming messages are cyclically distributed to the communication lines by a *distributor* module, and are recollected and delivered by a *collector* module. The system specifies the behaviour of an SWP with input and output windows of equal size.

A far simpler specification consists of an $(n+2)$-place buffer, if n is the number of parallel channels in our implementation (the distributor and the collector contribute with one place each).

We now reason that an implementation using n copies of the ABP and an $(n+2)$-place buffer are observation equivalent. Applying the compositionality of observation equivalence, we can use one-place buffers instead of the n copies of the ABP. We split the proof into three parts, again exploiting the compositionality of observation equivalence.

This time we cannot abstract from data, as the system need not deliver a message before accepting a new one. We have to guarantee that messages are not swapped. The one-place buffers EB[2] are thus of the following form, where ci_i and co_i are denoted by ci{i} and co{i}, respectively:

```
"EB{i,None} -[ci{i}#{x}]-> EB{i,Some x}"                    (* accept *)
"EB{i,Some x} -[co{i}<x>]-> EB{i,None}"                     (* deliver *)
```

Distributor and collector possess parameters h and l telling to which of the buffers a message is to be sent, or from which one it is to be taken. Each such transition increments the parameters by 1 modulo the number of buffers, yielding the cyclic behaviour of distributor and collector.

```
"D{n, h, None} -[ca#{x}]-> D{n, h, Some x}"                  (* accept *)
"D{n, h, Some x} -[ci{h}<x>]-> D{n, h {+n} 1, None}"         (* distr *)

"C{n, l, None} -[co{l}#{x}]-> C{n, l {+n} 1, Some x}"        (* collect *)
"C{n, l, Some x} -[cd<x>]-> C{n, l, None}"                   (* deliver *)
```

In the sequel we discuss the three parts of the proof:

(1) We need a finite representation of the n one-place buffers put in parallel. They can be described by a single component AB containing an array with n elements, one for the place of each buffer.

```
"[| i < length xs ; xs!i = None |] ==>
   AB{xs} -[ci{i}#{x}]-> AB{xs[i := Some x]}"                (* accept *)
"[| i < length xs ; xs!i = Some x |] ==>
   AB{xs} -[co{i}<x>]-> AB{xs[i := None]}"                   (* deliver *)
```

The first rule reads as follows: for all positions numbered $0 \le$ i $< n$ (with n = length xs), if the position is empty (xs!i = None), then AB can read a value on connection ci{i} and store it in place i. The second rule is the corresponding rule for destructive output.

In order to show that a parallel composition of n one-place buffers and an array buffer of size n are observation equivalent wrt. $\{$ci{i} $\mid 0 \le$ i $< n\} \cup \{$co{i} $\mid 0 \le$ i $< n\}$, we show that for every array of length n, there is a bisimulation containing (EB{(n), None} || AB{$\texttt{None}^{\texttt{n}}$}, AB{$\texttt{None}^{\texttt{n+1}}$}):

```
"BS_BuffInduct == {(P, Q) .
                   (EX xs x .  P = (EB{(length xs), x}) || AB{xs} &
                               Q = AB{xs @ [x]})}"
```

[2] EB stands for 'element buffer', as opposed to the 'array buffer' AB introduced later.

Exploiting the compositionality of observation equivalence, we can conclude by induction on the number of parallel components that

$$\texttt{EB\{0, None\}} \parallel \ldots \parallel \texttt{EB\{(n - 1), None\}} \approx_{\texttt{\{ci\{i\},co\{i\}\}}} \texttt{AB\{None}^{\texttt{n}}\texttt{\}}.$$

(2) As a consequence of (1) our implementation reduces to a system consisting of a distributor, a collector, and an array buffer AB of size n. We proceed by comparing this system to another system given by a variation of an n-place buffer BP, and two barrier one-place buffers, one attached to its 'front' (FB), and another attached to its 'back' (BB). Internally the n-place buffer is organized like the array buffer, yet it possesses only one input and one output connection, and stores and retrieves messages in a cyclic order; the parameters h and l indicate where to store messages and from where to retrieve them.

```
"cs!h = None    ==>    BP{h, l, cs} -[ci#{x}]->
        BP{h {+length(cs)} 1, l, cs[h := Some(x)]}"          (* accept *)
"cs!l = Some(x)    ==>    BP{h, l, cs} -[co<x>]->
        BP{h, l {+length(cs)} 1, cs[l := None]}"             (* deliver *)

"FB{None} -[ca#{x}]-> FB{Some x}"                             (* accept *)
"FB{Some x} -[ci<x>]-> FB{None}"                              (* deliver *)

"BB{None} -[co#{x}]-> BB{Some x}"                             (* accept *)
"BB{Some x} -[cd<x>]-> BB{None}"                              (* deliver *)
```

We can show that D{n, h, None} || AB{None$^{\texttt{n}}$} || C{n, l, None} and FB{None} || BP{h, l, None$^{\texttt{n}}$} || BB{None} are observation equivalent wrt. {ca, cd} by exhibiting the following bisimulation relation:

```
"BS_SWP == {(P, Q) . (EX xs x y h l .
            h < length xs & l < length xs &
            P = (FB{x}) || (BP{h, l, xs}) || BB{y} &
            Q = (D{length xs, h, x}) || (AB{xs}) || C{length xs, l, y})}"
```

(3) To complete the proof we have to show that BP behaves like an n-place buffer, nB, modelled as follows:

```
"length(s) < n ==> nB{n, s}      -[ci#{x}]-> nB{n, s @ [x]}"
                  "nB{n, x # s} -[co<x>]->  nB{n, s}"
```

A list None$^k \circ s \circ$ None^l stored in BP is reflected in nB by a list $\hat{s}$ (see BS1_1 below); $\hat{s}$ is obtained from s by mapping all elements Some x to x (s does not contain None anyway); a list $s_1 \circ$ None$^k \circ s_2$ in BP is reflected in nB by $\hat{s}_2 \circ \hat{s}_1$ (see BS1_2 below). The bisimulation relation looks as follows:

```
"BS1_1  ==  {(P, Q) . (EX n cs h l .
             P = nB{n, map the cs}
           & Q = BP{h, l, replicate l None @ cs @ replicate (n - h) None}
           & list_all (% x . x ~= None) cs
           & h < n
           & l + length cs = h)}"
```

```
"BS1_2  ==  {(P, Q) . (EX n cs1 cs2 h l .
              P = nB{n, map the cs2 @ map the cs1}
            & Q = BP{h, l, cs1 @ replicate (l - h) None @ cs2}
            & list_all (% x . x ~= None) cs1
            & list_all (% x . x ~= None) cs2
            & length cs1 = h
            & h <= l & l < n
            & l + length cs2 = n)}"

"BS_nBuffer == BS1_1 Un BS1_2"
```

The proof script for the three bisimulations verifying the SWP contains about 600 lines, and has been set in less than two weeks. The proofs of (1) and (2) are rather straightforward, as P and Q behave similarly. Isabelle deduces the proofs automatically without the user having to split them into single theorems covering the obligations. Note that the weak transitions are directly derivable from strong ones applying the rules TE and SW (see Section 3), thus need not be given by the user. Also, almost no additional results about the data types have to be provided by the user. The most challenging part concerning the mechanization is (3), as here cyclic structures are mapped to linear lists. The corresponding theorems make up for nearly two thirds of the proof. For these proofs certain expertise in theorem proving is indispensable.

5 Discussion

In the previous section we have presented a mechanization of the verification of communication protocols in a process-algebraical framework based on exhibiting bisimulation relations. In this section we discuss several questions about our approach, going from the more general (is a bisimulation framework appropriate?) to the more concrete (how to further improve our techniques).

Are bisimulation techniques suitable? Bisimulations are often argued to be too discriminating for many practical applications. In the area of communication protocols this seems to be a lesser problem. Due to the rather deterministic behaviour of the specifications of communication protocols, observation and fair testing equivalence [12] — and sometimes even trace equivalence [9] — coincide, no matter how the implementations of the protocols look like. Thus, in these cases one can profit from the bisimulation proof methodology to show the, usually less discriminating, notions of testing or trace equivalence.

Bisimulation equivalence does not preserve liveness properties since, e.g., one cannot infer from Q being divergence-free that necessarily P is so too, even if they are weakly bisimilar. However, in the area of communication protocols nondeterminism often models probabilistic choice (a message can be lost with a certain probability), and so it is often reasonable to assume that the system does not remain in any τ-loop indefinitely (as for instance in the ABP). Under this assumption bisimulation peserves liveness properties. It should also be mentioned that our approach can be extended to stronger bisimulation-like equivalences preserving liveness properties (see, for instance, [20]).

Comparison with algebraic techniques. Algebraic techniques are generally considered more elegant. Furthermore, it has been claimed that they succeed in cases where it is hard to find a bisimulation relation [7]. Their main drawback with respect to our approach is that they require deep insight into a proof system for bisimulation; on the contrary, exhibiting a bisimulation relation requires only good intuition about the system. A further point is that algebraic techniques usually require the transformation of a system into its normal form applying expansion, which in the presence of parallel compositions leads to an explosion of the size of the process. The degree of the explosion corresponds to the number of proof obligations the system produces in a bisimulation relation. A direct approach has the advantage that the explosion problem can be attacked by splitting the proof obligations, as we have done it in Section 4.2.

Keeping bisimulations manageable. Keeping the size of relations manageable is an important problem of our approach. Notice, for instance, that the description of BS_ABP already takes almost a page. A first solution is to use the compositionality of observation equivalence. In our third example we were able to replace the ABP channels by one-place buffers. Without this the bisimulation would have been unmanageable. Furthermore — though we have not used them — there exist various 'up to' techniques that can be exploited to reduce the size of the relations [10, 18]. 'Up to' techniques combine the direct approach with algebraic reasoning.

Dealing with data structures. Although our approach does not require to master a proof system for bisimulation, it still requires a lot of expertise in theorem proving, as usually the systems to be verified manipulate data. Proving simple facts about the data structures of a system (list, stacks, etc.) may amount to more than half of the interaction with the theorem prover. At this point the user has to decide whether to perform the full proof, or whether to provide certain necessary theorems as unproved axioms. Usually the properties of the data stuctures are rather straightforward, and so the proof does not lose much credibility.

Improving the approach. In our case study we have modelled all transition systems from scratch, slightly modifying the definition of observation equivalence in order to avoid a restriction (or hiding) operator, and projecting the labels either on the signals or on the use of connections. A formalization of a framework for further proofs would of course have to contain a restriction operator, and to implement a general definition of observation equivalence. Further there would have to be a transition rule lifting the user-defined axioms to the transition systems. We have implemented such a prototypical framework — yet still with restriction integrated in the definition of bisimilarity — and have transferred some simple proofs from Section 4 to it.

Acknowledgements: We would like to thank Daniel Hirschkoff and two anonymous referees for very helpful comments.

References

1. J. Baeten and W. Weijland. *Process Algebra.* Cambridge University Press, 1990.
2. K. A. Bartlett, R. A. Scantlebury, and P. T. Wilkinson. A note on reliable full-duplex transmission over half-duplex links. *Comm. of the ACM*, 12(5):260–261, May 1969.
3. J. A. Bergstra and J. W. Klop. Verification of an alternating bit protocol by means of process algebra. In *Mathematical Methods of Specification and Synthesis of Software Systems '85*, volume 215 of *LNCS*. Springer, 1985.
4. M. Bezem and J. F. Groote. A formal verification of the alternating bit protocol in the calculus of constructions. Logic Group Preprint Series 88, Dept. of Philosophy, Utrecht University, 1993.
5. E. Gimenez. An application of co-inductive types in Coq: Verification of the alternating bit protocol. In *Proc. TYPES'95*, volume 1158 of *LNCS*, pages 135–152. Springer, 1996.
6. J. F. Groote and J. G. Springintveld. Focus points and convergent process operators. Logic Group Preprint Series 142, Dept. of Philosophy, Utrecht University, 1995.
7. J. F. Groote and J. G. Springintveld. Algebraic verification of a distributed summation algorithm. Technical Report CS-R9640, CWI, Amsterdam, 1996.
8. T. Hardin and B. Mammass. Proving the bounded retransmission protocol in the pi-calculus. In *Proc. INFINITY'98*, 1998.
9. R. Milner. *Communication and Concurrency.* Prentice-Hall, 1989.
10. R. Milner and D. Sangiorgi. The problem of weak bisimulation up-to. In *Proc. CONCUR'92*, volume 630 of *LNCS*. Springer, 1992.
11. K. Namjoshi. A simple characterization of stuttering bisimulation. In *Proc. FSTTCS'97*, volume 1346 of *LNCS*, pages 284–296. Springer, 1997.
12. V. Natarajan and R. Cleaveland. Divergence and fair testing. In *Proc. ICALP'95*, volume 944 of *LNCS*, pages 648–659. Springer, 1995.
13. T. Nipkow and K. Slind. I/O automata in Isabelle/HOL. In *Proc. TYPES'94*, volume 996 of *LNCS*, pages 101–119. Springer, 1994.
14. K. Paliwoda and J. Sanders. The sliding-window protocol. Technical Report PRG-66, Programming Research Group, Oxford University, March 1988.
15. L. C. Paulson. Isabelle's object-logics. Technical Report 286, University of Cambridge, Computer Laboratory, 1993.
16. L. C. Paulson. *Isabelle: a generic theorem prover*, volume 828 of *LNCS*. Springer, 1994.
17. G. Plotkin. Structural operational semantics. Technical report, DAIMI, Aarhus University, 1981.
18. D. Sangiorgi. On the proof method for bisimulation. In *Proc. MFCS'95*, volume 969 of *LNCS*, pages 479–488. Springer, 1995.
19. J. L. A. Snepscheut. The sliding-window protocol revisited. *Formal Aspects of Computing*, 7:3–17, 1995.
20. D. Walker. Bisimulation and divergence. *Information and Computation*, 85(2):202–241, 1990.

Event Structures as Presheaves —Two Representation Theorems

Glynn Winskel

BRICS*, University of Aarhus, Denmark

Abstract. The category of event structures is known to embed fully and faithfully in the category of presheaves over pomsets. Here a characterisation of the presheaves represented by event structures is presented. The proof goes via a characterisation of the presheaves represented by event structures when the morphisms on event structures are "strict" in that they preserve the partial order of causal dependency.

1 Introduction

Presheaves have been advanced as a model of nondeterministic processes which supports a notion of bisimulation and as well extends to higher order [7, 12, 5, 4, 3, 10, 13, 6, 2]. At the start of this work, the paper [7] showed that the category of (labelled) event structures embedded fully and faithfully in the category of presheaves over pomsets; the embedding arises canonically from the fact that pomsets can be regarded as event structures. The paper [7] gave several grounds for viewing the presheaf category as consisting of generalised event structures.

Clearly some presheaves were not obtained from event structures, among them those presheaves which were not "rooted" in the sense of not having a unique starting state. The empty presheaf is not rooted. It allows no computation, not even the empty pomset. At the other extreme, the terminal presheaf, which assigns a singleton set to each pomset, supports all computational behaviour (like the "chaos" of CSP); although rooted it cannot correspond to an event structure, seen most quickly by noticing that all morphisms from pomsets to event structures are mono a state of affairs not reflected in the presheaf category for the terminal object. Other presheaves not corresponding to event structures could be nevertheless understood within broader classes of models such as certain categories of Petri nets. But the precise boundary was unclear; there remained the question of precisely which presheaves over pomsets arose from event structures.

This paper uncovers the conditions that characterise those presheaves represented by event structures (Theorem 17). The proof involves first showing an analogous result for a stricter class of morphisms on event structures (Theorem 9). A condition central to both theorems is one equivalent to saying that the presheaves should be *separated* with respect to a simple Grothendieck topology.

* **Basic Research in CS**, Centre of the Danish National Research Foundation.

2 Event Structures and Pomsets

We will work with labelled event structures, and throughout this paper we assume a fixed set of labels L.

A *(labelled) event structure* [11] is a structure $(E, \leq, Con, l)$ consisting of a set E, of *events* which are partially ordered by $\leq$, the *causal dependency relation*, a nonempty *consistency relation Con* consisting of finite subsets of events, and a *labelling function* $l : E \to L$, which satisfy

$$\{e' \mid e' \leq e\} \text{ is finite},$$
$$\{e\} \in Con,$$
$$Y \subseteq X \in Con \Rightarrow Y \in Con,$$
$$X \in Con \ \&\ e \leq e' \in X \Rightarrow X \cup \{e\} \in Con,$$

for all events e, e' and their subsets X, Y. Events $e, e' \in E$ are *concurrent* (causally independent) iff ($e \not\leq e'$ & $e' \not\leq e$ & $\{e, e'\} \in Con$). A *configuration* of E is a subset $x \subseteq E$ which is

- *downwards-closed:* $\forall e, e'.\ e' \leq e \in x \Rightarrow e' \in x$, and
- *consistent:* $\forall X.\ X$ finite & $X \subseteq x \Rightarrow X \in Con$.

An event e determines a *prime* configuration $[e] = \{e_1 \in E \mid e_1 \leq e\}$ consisting of all its causal predecessors and the event itself.

We restrict attention to label-preserving morphisms on event structures over the common labelling set L (the fibres of [11]). Let $E = (E, \leq, Con, l), E' = (E', \leq', Con', l')$ be event structures over L. A *morphism* from E to E' consists of a function $f : E \to E'$ on events which preserves labels (*i.e.* $l = l' \circ f$) such that if x is a configuration of E, then its image fx is a configuration of E' and if for $e_1, e_2 \in x$ their images are equal, *i.e.* $f(e_1) = f(e_2)$, then $e_1 = e_2$. We can equivalently describe a morphism of event structures from E to E' as a function $f : E \to E'$ such that

$$\forall e \in E.\ [f(e)] \subseteq f[e]\ \&$$
$$\forall X \in Con.\ [fX \in Con'\ \&\ (\forall e_1, e_2 \in X.\ f(e_1) = f(e_2) \Rightarrow e_1 = e_2)].$$

We say a morphism $f : E \to E'$ of event structures is *strict* iff $[f(e)] = f[e]$.

It is easy to check that the function composition of two morphisms of event structures is a morphism so that we obtain a category.

Definition 1. We write $\mathbf{E}$ for the associated category of event structures, writing $\mathbf{E}_s$ for the subcategory with strict morphisms.

In event structures a configuration, to be thought of as a computation path, carries more structure than simply a string of actions. A configuration inherits the shape of a pomset from the causal dependency and labelling of the event structure. Pomsets [9] are partial orders of labelled events and so can be identified with special event structures where all finite subsets of events are consistent.

Definition 2. Say a pomset is *prime* when it has a top event e with respect to the causal dependency relation $\leq$—so its set of events is $[e]$.

Morphisms from pomsets to event structures are 1-1 functions which send downwards-closed sets to downwards-closed sets. Thus a morphism from pomset P to pomset Q may not only extend P by extra events but also relax the causal dependency relation; two events causally related in P may have images no longer causally related in Q—of course this cannot occur for a strict morphism which would force P to be a pomset prefix of Q.

We separate the forms of morphism corresponding to the different ways one pomset can extend another.

Definition 3. Define **Pom** to be the full subcategory of event structures **E** with objects *finite* pomsets. Define $\mathbf{Pom}_s$ to be the subcategory of **Pom** where all morphisms are strict morphisms.

An epimorphism in **Pom** is called an *augmentation* (following [9], though note the switch of direction relative to *loc. cit.*).

It is clear that all isomorphisms in **Pom** are augmentations (and strict) and that restricting to augmentation morphisms also yields a subcategory of **Pom**.

Proposition 4. *In* **Pom**, *any morphism* $f : P \to Q$ *factors uniquely to within isomorphism as a composition* $f = P \xrightarrow{a} Q_0 \xrightarrow{j} Q$ *where* a *is an augmentation and* j *is a strict morphism.*

Such *augment-strict* factorisations play a central role in the proof of the second representation theorem.

3 Presheaf models

Here a presheaf over an (essentially small) category $\mathbb{P}$ is thought of as standing for a nondeterministic process whose computation paths have the shape of objects of $\mathbb{P}$; according to this view the morphisms of $\mathbb{P}$ express how one path shape extends to another. In this paper $\mathbb{P}$ will be either **Pom** or $\mathbf{Pom}_s$.

The objects (presheaves) of $\widehat{\mathbb{P}}$ consist of functors $\mathbb{P}^{op} \to \mathbf{Set}$, to the category of sets. The morphisms of $\widehat{\mathbb{P}}$ are natural transformations between functors. A presheaf $X : \mathbb{P}^{op} \to \mathbf{Set}$ can be thought of as specifying for a typical object P the set $X(P)$ of computation paths of shape P. It acts on a morphism $j : P \to Q$ in $\mathbb{P}$ to give a function $X(j) : F(Q) \to F(P)$ saying how Q-paths restrict to P-paths.

Notation: Let X be a presheaf over a category $\mathbb{P}$. Let $j : P \to Q$ be a morphism in $\mathbb{P}$. As is usual, we will frequently write $y \cdot j$ for $X(j)(y)$, the restriction of $y \in X(Q)$ along $j : P \to Q$, a morphism in $\mathbb{P}$. Note that the functoriality of X ensures that $(y \cdot k) \cdot j$, which we will most often write as $y \cdot k \cdot j$, equals $y \cdot (k \circ j)$, when $j : P \to Q$, $k : Q \to R$ and $y \in X(R)$.

Definition 5. Let X be a presheaf over $\mathbb{P}$. Define its *category of elements* $els(X)$ to be the category consisting of: *objects* (P, x) where P is an object of $\mathbb{P}$ and $x \in X(P)$; *morphisms* $j : (P, x) \to (Q, y)$ whenever $j : P \to Q$ in $\mathbb{P}$ and $x = y \cdot j$.

The Yoneda embedding $\mathcal{Y} : \mathbb{P} \to \widehat{\mathbb{P}}$ expresses how to regard a path P as the presheaf $\mathbb{P}[-, P]$, such presheaves being called *representables.* The category of presheaves $\widehat{\mathbb{P}}$ is the free colimit completion of $\mathbb{P}$: for any functor $F : \mathbb{P} \to \mathcal{E}$ where $\mathcal{E}$ has all small colimits, there is a functor $Lan_{\mathcal{Y}}(F) : \widehat{\mathbb{P}} \to \mathcal{E}$, unique to within isomorphism, such that

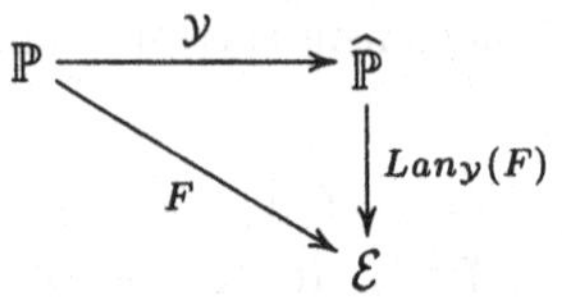

commutes. In particular, as presheaf categories have all small colimits we can instantiate $\mathcal{E}$ to a presheaf category $\widehat{\mathbb{Q}}$. The functor $Lan_{\mathcal{Y}}(F)$ (a left Kan extension) can be described explicitly (see *e.g.* [8]) as that functor such that

$$Lan_{\mathcal{Y}}(F)(X) = colim_{(P,x) \in els(X)} F(P)$$

for any $X \in \widehat{\mathbb{P}}$; its action on morphisms is determined by the universal property of colimits.

Colimits in Set: Colimits of presheaves are given pointwise in terms of colimits in **Set** for which we can make use of an explicit construction of colimits (see *e.g.* [1]):

Proposition 6. *Let $\mathbb{I}$ be a small category. Let $D : \mathbb{I} \to$ **Set** be a functor (called a diagram of shape $\mathbb{I}$ in **Set**). Then, D has a colimit in **Set** given explicitly as the cone consisting of the set C and functions $\gamma_i : D(i) \to C$, for $i \in \mathbb{I}$, described as follows. The set C is the set of equivalence classes*

$$C = \biguplus_{i \in I} D(i) \;/ \sim$$

where $\sim$ is the least equivalence relation on the set $\biguplus_{i \in I} D(i)$ for which

$$(i, x) \sim (j, y) \text{ if } D(f)(x) = y \text{ , for some } f : i \to j \text{ in } \mathbb{I} \text{ .}$$

The function $\gamma_i : D(i) \to C$, where $i \in \mathbb{I}$, takes $x \in D(i)$ to the equivalence class $\{(i, x)\}_\sim$.

As colimits are unique to within isomorphism, we can and shall assume that all the colimits in **Set** we consider are given explicitly as in Proposition 6.

4 The Problem

There is a *canonical functor* c from the category of event structures **E** to the category of presheaves $\widehat{\mathbf{Pom}}$. The functor c takes an event structure E of **E** to the presheaf $\mathbf{E}[-, E]$; in detail, $c(E)$ is the presheaf which for each path object P yields the set of paths $\mathbf{E}[P, E]$ from P into E. The functor c takes a morphism $f : E \to E'$ in $\mathbf{E}$ to the natural transformation $\mathbf{E}[-, f] : \mathbf{E}[-, E] \to \mathbf{E}[-, E']$ whose component at an object P of **Pom** is the function $\mathbf{E}[P, E] \to \mathbf{E}[P, E']$ taking p to $f \circ p$—intuitively, a path $p : P \to E$ in E is taken to a path $f \circ p : P \to E'$ in E'.

Because the inclusion functor $\mathbf{Pom} \hookrightarrow \mathbf{E}$ is dense,

Theorem 7. *[7] The canonical functor* $c : \mathbf{E} \to \widehat{\mathbf{Pom}}$ *is full and faithful.*

The canonical functor $c_s : \mathbf{E}_s \to \widehat{\mathbf{Pom}_s}$ is defined analogously, but with respect to strict morphisms on event structures and pomsets, and analogously:

Theorem 8. *The canonical functor from* $c_s : \mathbf{E}_s \to \widehat{\mathbf{Pom}_s}$ *is full and faithful.*

The problem addressed in this paper is the characterisation of those presheaves which correspond to event structures with respect to the canonical embeddings. These amount to representation theorems; a presheaf X over **Pom** is said to be *represented by* an event structure E in $\mathbf{E}$ iff $X \cong \mathbf{E}[-, E]$. It turns out that characterising the presheaves in $\widehat{\mathbf{Pom}}$ which are represented by event structures in $\mathbf{E}$ involves first characterising those presheaves in $\widehat{\mathbf{Pom}_s}$ represented by event structures in $\mathbf{E}_s$, the strict case.

5 Representation Theorem—Strict Morphisms

This section is devoted to showing our first representation theorem:

Theorem 9. *A presheaf* $X \in \widehat{\mathbf{Pom}_s}$ *is isomorphic to* $\mathbf{E}_s[-, E]$ *for some event structure* E *iff* X *is nonempty and satisfies the conditions*

(Mono) *For all* $j_1, j_2 : P \to Q$ *in* $\mathbf{Pom}_s$*, where* P *is prime,*

$$\forall x \in X(Q).\ x \cdot j_1 = x \cdot j_2 \Rightarrow j_1 = j_2.$$

(Separated) *For all* $x, x' \in X(Q)$ *where* Q *is a pomset, if* $(\forall j : P \to Q$ *in* $\mathbf{Pom}_s$ *with* P *prime,* $x \cdot j = x' \cdot j)$ *then* $x = x'$.

Remark 10. The *empty presheaf* assigns the emptyset to each pomset, even the empty pomset, and so cannot be represented by any event structure which will always have the empty configuration. As we will see the condition "Mono" expresses that morphisms from pomsets into event structures are mono. In fact "Mono" is equivalent to the corresponding condition where P is not restricted to be prime. The condition "Separated" is equivalent to saying that the presheaf X is *separated* with respect to the Grothendieck topology (see *e.g.* [8]) with basis consisting of collections $\{k_i : P_i \to Q \mid i \in I\}$ of jointly surjective morphisms. Note that "Separated" implies that any nonempty presheaf X is *rooted* in the sense that $X(\emptyset)$, the set assigned to the empty pomset $\emptyset$, is a singleton; because there are no prime pomsets mapping into the empty pomset.

It is easy to show the "only if" half of the theorem.

Lemma 11. *Let E be an event structure in $\mathbf{E}_s$. Let X be the presheaf $\mathbf{E}_s[-, E]$. Then X is nonempty and satisfies the conditions "Mono" and "Separated".*

Proof. *"Mono":* Let $x \in \mathbf{E}_s[Q, E]$ and $j_1, j_2 : P \to Q$ morphisms in $\mathbf{Pom}_s$. For the presheaf $\mathbf{E}_s[-, E]$ obtained via the hom-functor, $x \cdot j_1 = x \cdot j_2$ means $x \circ j_1 = x \circ j_2$, so $j_1 = j_2$ as x is 1-1 and thus mono.
"Separated": Suppose $x, x' \in \mathbf{E}_s[Q, E]$ have the property that $x \cdot j = x' \cdot j$ for all $j : P \to Q$ in $\mathbf{Pom}_s$, from a prime pomset P. But this implies $x \circ j = x' \circ j$ for all inclusions $j : [e] \hookrightarrow Q$ where e is an event of Q. Hence, x and x' agree on all events of Q and so are equal. □

To show the converse, "if" direction, of Theorem 9 we construct an event structure from a nonempty presheaf satisfying the "Mono" and "Separated" conditions. We do this by forming a colimit in $\mathbf{E}_s$. Not all colimits exist in $\mathbf{E}_s$. However if a nonempty presheaf X satisfies the "Mono" condition we can construct a colimit as follows.

Lemma 12. *Let X be a nonempty presheaf over $\mathbf{Pom}_s$ which satisfies the "Mono" condition. Then the colimit $colim_{(Q,x)\in els(X)} Q$ exists in $\mathbf{E}_s$. Its events E can be taken to be the colimit in* **Set**

$$\biguplus_{(Q,x)\in els(X)} Q \,/ \sim$$

where $\sim$ is the least equivalence relation such that

$$((Q,x),q) \sim ((Q',x'),q') \text{ if } \exists k : Q \to Q' \text{ in } \mathbf{Pom}_s.\ x = x'.k \ \&\ k(q) = q' \ ,$$

when the components of the colimiting cone in $\mathbf{E}_s$, at $(Q,x) \in els(X)$, are given by maps

$$\gamma_{Q,x} : Q \to E \text{ with } q \mapsto \{((Q,x),q)\}_\sim \ .$$

The causal dependency and consistency relations on E satisfy:

- *$e \leq e'$ iff there are $q \leq q'$ in Q for some pomset Q and $x \in X(Q)$ such that $\gamma_{Q,x}(q) = e$ and $\gamma_{Q,x}(q') = e'$,*
- *$C \in Con$ iff there is $S \subseteq Q$ for some pomset Q and $x \in X(Q)$ such that $C = \gamma_{Q,x} S$.*

Proof. Write $((Q,x),q) \sim_1 ((Q',x'),q')$ iff $\exists k : Q \to Q'.\ x = x' \cdot k \ \&\ k(q) = q'$. By definition, the relation $\sim$ is the symmetric transitive closure of $\sim_1$.

Suppose that $((Q,x),q) \sim_1 ((Q',x'),q')$ and that $i : [q] \hookrightarrow Q$ and $i' : [q'] \hookrightarrow Q'$ are the associated inclusion morphisms in $\mathbf{Pom}_s$. Then "restricting" along i and i' we obtain

$$\begin{array}{ccc} ((Q,x),q) & \sim_1 & ((Q',x'),q') \\ \sim_1 & & \sim_1 \\ (([q],x\cdot i),q) & \sim_1 & (([q'],x'\cdot i'),q') \ . \end{array}$$

Recalling that morphisms are strict we see that for $x \in X(Q), x' \in X(Q')$,

$$(([q], x \cdot i), q) \sim_1 (([q'], x' \cdot i'), q') \text{ iff } \exists k : [q] \cong [q']. \quad x \cdot i = x' \cdot i' \cdot k \,.$$

Thus a $\sim_1$-chain establishing $((Q, x), q) \sim ((Q', x'), q')$ restricts to a $\sim_1$-chain involving only prime pomsets. Noting that the $\sim_1$ relation is already symmetric and transitive when only prime pomsets are involved, we obtain

$$\begin{aligned} ((Q, x), q) \sim ((Q', x'), q') &\text{ iff } (([q], x \cdot i), q) \sim_1 (([q'], x' \cdot i'), q') \\ &\text{ iff } \exists k : [q] \cong [q']. \quad x \cdot i = x' \cdot i' \cdot k \end{aligned}$$

where $i : [q] \hookrightarrow Q$ and $i' : [q'] \hookrightarrow Q'$ are the inclusion morphisms.

It follows that each $\gamma_{Q,x} : Q \to E$ is 1-1. Suppose $q, q' \in Q$ and $((Q, x), q) \sim ((Q, x), q')$. Then we obtain

$$k : [q] \cong [q'] \;\&\; x \cdot i = x' \cdot i' \cdot k$$

where $i : [q] \hookrightarrow Q$ and $i' : [q'] \hookrightarrow Q'$. But X is assumed to satisfy the "Mono" condition. Hence $i = i' \circ k$ so that $q = i(q) = i' \circ k(q) = i'(q') = q'$, making $\gamma_{Q,x}$ a 1-1 function.

As $\sim_1$ respects causal predecessors $[-]$, defining the causal dependency and consistency relations as above yields an event structure and ensures that each $\gamma_{Q,x}$ is a morphism in $\mathbf{E}_s$. Together $\gamma_{Q,x}$, where $(Q, x) \in els(X)$, form a cone in $\mathbf{E}_s$, which is colimiting because it is so in **Set**. □

Prime pomsets distribute through the colimits of Lemma 12:

Lemma 13. *Let X be a nonempty presheaf over $\mathbf{Pom}_s$ satisfying the "Mono" condition. Let P be a prime pomset. The canonical map from the colimiting cone,*

$$\varphi_P : colim_{(Q,x) \in els(X)} \mathbf{E}_s[P, Q] \to \mathbf{E}_s[P, colim_{(Q,x) \in els(X)} Q] \,,$$

acting so

$$\varphi_P : \{((Q, x), j)\}_\sim \mapsto \gamma_{Q,x} \circ j \,,$$

is an isomorphism, where $\gamma_{Q,x}$ where $(Q, x) \in els(X)$, is the colimiting cone to $colim_{(Q,x) \in els(X)} Q$.

Proof. Write E for the event structure obtained as the colimit $colim_{(Q,x) \in els(X)} Q$ in Lemma 12. We first check that φ_P is well-defined. In the explicit presentation of the colimit $C = colim_{(Q,x) \in els(X)} \mathbf{E}_s[P, Q]$ in **Set** the equivalence relation $\sim$ is generated by $\sim_1$ where

$$((Q, x), j) \sim_1 ((Q', x'), j') \text{ iff } \exists k : Q \to Q' \text{ in } \mathbf{Pom}_s.\ x = x' \cdot k \;\&\; k \circ j = j'.$$

Thus, if $((Q, x), j) \sim_1 ((Q', x'), j')$, then there is $k : (Q, x) \to (Q', x')$ in $els(X)$. So, as E, γ is a cone, we directly obtain $\gamma_{Q,x} = \gamma_{Q',x'} \circ k$. Thus

$$\varphi_P(((Q, x), j)) = \gamma_{Q,x} \circ j = \gamma_{Q',x'} \circ k \circ j = \gamma_{Q',x'} \circ j' = \varphi_P(((Q', x'), j')) \,.$$

Hence φ_P is well-defined as a function. We require in addition that φ_P is 1-1 and onto.

"onto": Suppose $f : P \to E$ in $\mathbf{E}_s$. As a prime pomset, P is $[p]$ for some event p. The image $f(p)$, in E, is an equivalence class $\{((Q,x),q)\}_\sim$, choosing any representative $((Q,x),q)$, where $(Q,x) \in els(X)$ and $q \in Q$. Because morphisms are strict $[p] \cong [q]$, so f must factor through $\gamma_{Q,x}$ for some $j : P \to Q$ in $\mathbf{Pom}_s$:

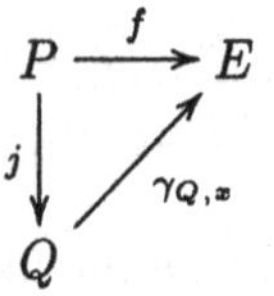

But now $\varphi_P(\{((Q,x),j)\}_\sim) = \gamma_{Q,x} \circ j = f$.

"1-1": Again, as P is prime it has the form $[p]$ for some $p \in P$. First note that any equivalence class $c \in colim_{(Q,x)\in els(x)} \mathbf{E}_s[P,Q]$ has a representative of the form $(([q],x),j),j)$ where $j : [p] \cong [q]$. To see this note that for any representative $((Q,y),l)$,

$$((Q,y),l) \sim_1 (([q], y \cdot i), l_0)$$

where $q = l(p)$ and l factors as $[p] \overset{l_0}{\cong} [q] \overset{i}{\hookrightarrow} Q$.

Thus assuming that $\varphi_P(c) = \varphi_P(c')$ for $c, c' \in colim_{(Q,x)\in els(X)} \mathbf{E}_s[P,Q]$, there are representatives $(([q],x),j)$ and $(([q'],x'),j')$ where $j : [p] \cong [q]$ and $j' : [p] \cong q']$ for which

$$\gamma_{[q],x} \circ j = \gamma_{[q'],x'} \circ j' . \tag{1}$$

Consequently, $\gamma_{[q],x}(q) = \gamma_{[q'],x'}(q')$, from which we obtain $((([q],x),q) \sim (([q'],x'),q')$ in E. But now (just as in the proof of Lemma 12) we derive the existence of an isomorphism k such that

$$k : [q] \cong [q'] \;\&\; x = x' \cdot k . \tag{2}$$

As $k : ([q],x) \cong ([q'],x')$ is a morphism in $els(X)$ and E, γ is a cone, we see that

$$\gamma_{[q],x} = \gamma_{[q'],x'} \circ k .$$

Hence, by (1),

$$\gamma_{[q'],x'} \circ j' = \gamma_{[q],x} \circ j = \gamma_{[q'],x'} \circ k \circ j ,$$

ensuring $j' = k \circ j$ from the injectivity of $\gamma_{[q'],x'}$. With (2), this yields $(([q],x),j) \sim_1 (([q'],x'),j')$ in C, making $c = c'$. Hence φ_P is 1-1. □

It is well-known that a presheaf is the colimit of its representables and that colimits in categories of presheaves are obtained pointwise [8]. With our explicit treatment of colimits in **Set** we obtain an explicit isomorphism:

Lemma 14. *Let X be a presheaf over $\mathbf{Pom}_s$. Let $P \in \mathbf{Pom}_s$. Then*

$$\psi_P : X(P) \cong colim_{(Q,x)\in els(X)} \mathbf{Pom}_s[P,Q] ,$$

where $\psi_P(z) = \{((P,z), 1_P)\}_\sim$.

Now we can prove the "if" half of the first representation theorem:

Lemma 15. *Suppose X, a nonempty presheaf over $\mathbf{Pom}_s$, satisfies the "Mono" and "Separated" conditions. Let E be the event structure obtained as the colimit $colim_{(Q,x)\in els(X)}Q$ in $\mathbf{E}_s$ (cf. Lemma 12). Then, there is a natural isomorphism*

$$\theta : X \cong \mathbf{E}_s[-, E]$$

which has components $\theta_Q : X(Q) \to \mathbf{E}_s[Q, E]$, at pomset Q, given by

$$\theta(x) = \gamma_{Q,x}$$

for $x \in X(Q)$. [We adopt the notation of Lemma 12 where $\gamma_{Q,x} : Q \to E$ is the component of the colimiting cone at $(Q, x) \in els(X)$.]

Proof. We first check that θ is a natural transformation. Suppose $j : Q \to Q'$ in $\mathbf{Pom}_s$. We require the following naturality square to commute:

$$\begin{array}{ccc} X(Q) & \xrightarrow{\theta_Q} & \mathbf{E}_s[Q, E] \\ {\scriptstyle X(j)}\uparrow & & \uparrow{\scriptstyle -\circ j} \\ X(Q') & \xrightarrow{\theta_{Q'}} & \mathbf{E}_s[Q', E] \end{array} .$$

I.e., letting $x' \in X(Q')$, we require $\gamma_{Q,x'\cdot j} = (\gamma_{Q',x'}) \circ j$. But this is a direct consequence of E, γ forming a cone.

For θ to be a natural isomorphism we need that each θ_Q, at a pomset Q, is 1-1 and onto:
"onto": Supposing $f : Q \to E$ the image of Q must be consistent in E. Hence, by the way the consistency relation is defined on E in Lemma 12, the map f must factor as

$$\begin{array}{ccc} Q & \xrightarrow{f} & E \\ {\scriptstyle j}\downarrow & \nearrow{\scriptstyle \gamma_{Q_0,x_0}} & \\ Q_0 & & \end{array}$$

for some $(Q_0, x_0) \in els(X)$. Take $x = x_0 \cdot j \in X(Q)$. Then, $f = \gamma_{Q_0,x_0} \circ j = \gamma_{Q,x}$ because E, γ is a cone and $(Q, x) \xrightarrow{j} (Q_0, x_0)$ in $els(X)$. Hence $\theta_Q(x) = f$.
"1-1": Suppose $\theta_Q(x) = \theta_Q(x')$ for $x, x' \in X(Q)$. Then, for any $j : P \to Q$ with P prime, $\theta_P(x \cdot j) = \theta_P(x' \cdot j)$ by naturality. Thus because X is "Separated", it is sufficient to show that θ_P is 1-1 for each prime pomset P. However, each component θ_P, when P is a prime pomset, arises as the composition of isomorphisms

$$\begin{aligned} X(P) &\overset{\psi_P}{\cong} colim_{(Q,x)\in els(X)}\mathbf{Pom}_s[P, Q] && \text{cf. Lemma 14,} \\ &= colim_{(Q,x)\in els(X)}\mathbf{E}_s[P, Q] && \text{as } \mathbf{Pom}_s \hookrightarrow \mathbf{E}_s \text{ is full,} \\ &\overset{\varphi_P}{\cong} \mathbf{E}_s[P, E] && \text{cf. Lemma 13.} \end{aligned}$$

□

As a corollary of Lemmas 11 and 15 we obtain the first representation theorem (Theorem 9) whose statement heads this section.

6 Representation Theorem—Nonstrict Morphisms

Our aim now is to characterise those presheaves over **Pom** represented by event structures in **E**.

Notation: We make heavy use of the augment-strict factorisation of Proposition 4 and it is helpful to adopt the convention that arrows $\twoheadrightarrow$ stand for augmentations while $\rightarrowtail$ stand for strict morphisms.

The statement of the second representation theorem involves a "confluence" condition on the category of elements of a presheaf.

Confluence Conditions: We will be interested in presheaves $Y \in \widehat{\mathbf{Pom}}$ for which the category of elements $els(Y)$ satisfies the *confluence condition*:

Letting $a : P \twoheadrightarrow Q$ and $f : P \to R$, in $els(Y)$,

$$\text{if} \quad \begin{array}{ccc} (R,z) & & \\ {\scriptstyle f}\uparrow & & \\ (P,x) & \xrightarrow[a]{} & (Q,y) \end{array} \qquad \text{then} \quad \begin{array}{ccc} (R,z) & \overset{a'}{\twoheadrightarrow} & (S,w) \\ {\scriptstyle f}\uparrow & & \uparrow{\scriptstyle f'} \\ (P,x) & \underset{a}{\twoheadrightarrow} & (Q,y) \end{array} \quad \text{commutes,}$$

for some (S,w) in $els(Y)$ with $a' : R \twoheadrightarrow S$ and $f' : Q \to S$.

We can summarise the confluence condition in the diagram:

$$\begin{array}{ccc} \cdot & \overset{a'}{\dashrightarrow} & \cdot \\ {\scriptstyle f}\uparrow & & \vdots\,{\scriptstyle f'} \\ \cdot & \underset{a}{\twoheadrightarrow} & \cdot \end{array}$$

Remark 16. By specialising f in the confluence condition to an augmentation we obtain a condition which we likewise summarise as the confluence diagram:

$$\begin{array}{ccc} \cdot & \overset{a'}{\dashrightarrow} & \cdot \\ {\scriptstyle f}\uparrow & & \vdots\,{\scriptstyle f'} \\ \cdot & \underset{a}{\twoheadrightarrow} & \cdot \end{array}$$

Note that in this case the morphism f' will also be an augmentation just because it is a second factor of an epimorphism.

The remainder of the paper is devoted to showing the second representation theorem:

Theorem 17. *A presheaf $Y \in \widehat{\mathbf{Pom}}$ is isomorphic to $\mathbf{E}[-, E]$ for some event structure E iff Y is nonempty and satisfies the conditions*

(Mono) *For all $j_1, j_2 : P \to Q$ in* **Pom**, *where P is prime,*

$$\forall y \in Y(Q).\ y \cdot j_1 = y \cdot j_2 \Rightarrow j_1 = j_2.$$

(Separated) *For all $y, y' \in Y(Q)$ where Q is a pomset, if $(\forall j : P \to Q$ in* **Pom**, *with P prime, $y \cdot j = y' \cdot j)$ then $y = y'$.*

(Confluent) *The confluence condition above holds of $els(Y)$.*

The proof of Theorem 17 uses the first representation theorem (Theorem 9) characterising which presheaves in $\widehat{\mathbf{Pom}_s}$ are represented by event structures E in $\mathbf{E}_s$. The proof has three main parts Sections 6.1, 6.2 and 6.3.

In Section 6.1 the extension of the obvious inclusion functor $\mathbf{Pom}_s \hookrightarrow \mathbf{Pom}$ to a colimit-preserving functor $L : \widehat{\mathbf{Pom}_s} \to \widehat{\mathbf{Pom}}$ is characterised (Lemma 18).

The next stage, presented in Section 6.2, is to relate the two canonical embeddings $c_s : \mathbf{E}_s \to \widehat{\mathbf{Pom}_s}$ and $c : \mathbf{E} \to \widehat{\mathbf{Pom}}$ in the diagram

$$\begin{array}{ccc} \mathbf{E}_s & \hookrightarrow & \mathbf{E} \\ {\scriptstyle c_s}\downarrow & & \downarrow{\scriptstyle c} \\ \widehat{\mathbf{Pom}_s} & \xrightarrow{L} & \widehat{\mathbf{Pom}} \end{array}$$

which is shown to commute up to isomorphism (Lemma 20). It follows that the presheaves in $\widehat{\mathbf{Pom}}$ represented by event structures in $\mathbf{E}$ are, to within isomorphism, the images under L of those presheaves in $\widehat{\mathbf{Pom}_s}$ represented by event structures in $\mathbf{E}_s$.

Finally, in Section 6.3, it is shown that, to within isomorphism, the images in $\widehat{\mathbf{Pom}}$ under L of presheaves in $\widehat{\mathbf{Pom}_s}$ are those which satisfy the "Confluent" condition, and that the "Mono" and "Separated" conditions transfer via L to the corresponding conditions in $\widehat{\mathbf{Pom}}$ (Lemma 21). This yields the second representation theorem (Theorem 17).

6.1 The Functor L

To within isomorphism, there is a colimit-preserving function $L : \widehat{\mathbf{Pom}_s} \to \widehat{\mathbf{Pom}}$ such that

$$\begin{array}{ccc} \mathbf{Pom}_s & \xhookrightarrow{I} & \mathbf{Pom} \\ {\scriptstyle \mathcal{Y}_s}\downarrow & & \downarrow{\scriptstyle \mathcal{Y}} \\ \widehat{\mathbf{Pom}_s} & \xrightarrow{L} & \widehat{\mathbf{Pom}} \end{array}$$

commutes to within isomorphism. The functor L may be obtained as the left-Kan extension, so $Lan_{\mathcal{Y}_s}(\mathcal{Y} \circ I)(X) = colim_{(P,x) \in els(X)} \mathcal{Y}(P)$ for $X \in \widehat{\mathbf{Pom}_s}$. By exploiting the augment-strict factorisation (Proposition 4) we give a more workable characterisation.

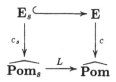

Lemma 18. *Let* $X \in \widehat{\mathbf{Pom}_s}, Q \in \mathbf{Pom}$. *Define*

$$L(X)(Q) = \{\{(P,x,a)\}_{\simeq} \mid x \in X(P) \ \&\ a : Q \twoheadrightarrow P)\}$$

where $(P,x,a) \simeq (P',x',a')$ *iff* $\exists k : P \cong P.\ x = x' \cdot k \ \&\ k \circ a = a'$. *For* $f : Q \to Q'$, *define* $L(X)(f) : L(X)(Q') \to (LX)(Q)$ *to act so*

$$\{(P',x',a')\}_{\simeq} \mapsto \{(P,x' \cdot i, a)\}_{\simeq}$$

where $i : P \rightarrowtail P'$ *and* $a : Q \twoheadrightarrow P$ *are an augment-strict factorisation* $i \circ a = a' \circ f$:

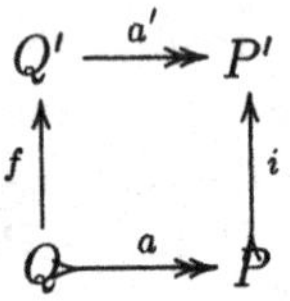

Then, $L(X)$ *is a presheaf over* **Pom** *such that* $L(X) \cong Lan_{\mathcal{Y}_S}(\mathcal{Y} \circ I)(X)$.

Proof. As colimits of presheaves are obtained pointwise, from the explicit description of colimits in **Set**, Proposition 6, we see

$$Lan_{\mathcal{Y}_s}(\mathcal{Y} \circ I)(X)(Q) = colim_{(P,x) \in els(X)} \mathbf{Pom}_s[Q,P] = \biguplus_{(P,x) \in els(X)} \mathbf{Pom}[Q,P]/\sim$$

where $\sim$ is the least equivalence relation such that

$$((P,x),f) \sim ((P',x'),f') \text{ iff } \exists k : P \rightarrowtail P'.\ x = x' \cdot k \ \&\ k \circ f = f'\ .$$

It follows that for each $((P',x'),f') \in \biguplus_{(P,x) \in els(X)} \mathbf{Pom}[Q,P]$

$$((P,x' \cdot i),a) \sim ((P',x'),f')$$

where an augment-strict factorisation of f is:

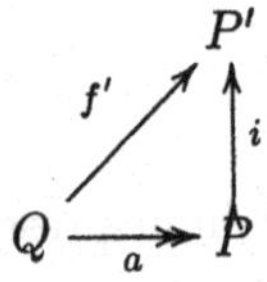

The isomorphism $L(X)(Q) \cong Lan_{\mathcal{Y}_s}(\mathcal{Y} \circ I)(X)(Q)$ is a direct consequence.

Via the isomorphism we obtain a colimiting cone with vertex $L(X)(Q)$; it has components $\gamma_{P,x} : \mathbf{Pom}_s[Q,P] \to L(X)(Q)$ for $(P,x) \in els(X)$ given by $\gamma_{P,x}(g) = \{(P_0, x \cdot i_0, a_0)\}_{\simeq}$ where g has augment-strict factorisation:

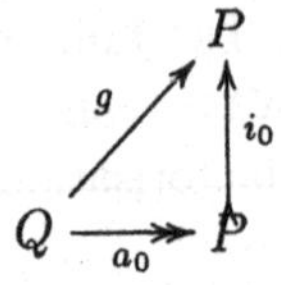

We require that the isomorphism is natural in Q. To show this it is sufficient to verify that with respect to $f : Q \to Q'$ in **Pom** the map $L(X)(f)$ is the (necessarily unique) mediating map from the colimiting cone $L(X)(Q'), \gamma'$ to the cone $L(X)(Q), \gamma \circ f$, *i.e.* for all $(P, x) \in els(X)$,

$$L(X)(f) \circ \gamma'_{P,x} = \gamma_{P,x} \circ f \ .$$

The verification relies on augment-strict factorisation being unique to within isomorphism. □

Remark 19. Let X be a presheaf over $\mathbf{Pom}_s$. In the special case when f is an augmentation $a_0 : Q \twoheadrightarrow Q'$, $L(X)(a_0) : \{(P', x', a')\}_{\simeq} \mapsto \{(P', x', a' \circ a_0)\}_{\simeq}$.

6.2 Relating Non-Strict and Strict

The next lemma relates the two canonical embeddings $c_s : \mathbf{E}_s \to \widehat{\mathbf{Pom}_s}$ and $c : \mathbf{E} \to \widehat{\mathbf{Pom}}$.

Lemma 20. *Let E be an event structure in $\mathbf{E}_s$. Then, $L \circ c_s(E) \cong c(E)$.*

Proof. We require that $L(\mathbf{E}_s[-, E]) \cong \mathbf{E}[-, E]$. From the definition of L,

$$L(\mathbf{E}_s[-, E])(Q) = \{\{(P, x, a)\}_{\simeq} \mid x : P \rightarrowtail E \ \&\ a : Q \twoheadrightarrow P\}$$

where $(P, x, a) \simeq (P', x', a')$ iff $\exists k : P \cong P'.\ k \circ a = a' \ \&\ x = x' \cdot k$. Thus elements of $L(\mathbf{E}_s[-, E])(Q)$ are in 1-1 correspondence with factorisations (to within isomorphism) of morphisms in $\mathbf{E}[Q, E]$. As such factorisations are unique, we obtain the isomorphism

$$\alpha_E : L(\mathbf{E}_s[-, E])(Q) \cong \mathbf{E}[Q, E] \text{ where } \{(P, x, a)\}_{\simeq} \mapsto x \circ a \ .$$

To check that the isomorphism α_Q is natural in Q, we require for $f : Q \to Q'$ that the naturality square

$$\begin{array}{ccc}
\{\{(P, x, a)\}_{\simeq} \mid x : P \rightarrowtail E \ \&\ a : Q \twoheadrightarrow P\} & \xrightarrow{\alpha_Q} & \mathbf{E}[Q, E] \\
{\scriptstyle L(\mathbf{E}_s[-,E])(f)} \uparrow & & \uparrow {\scriptstyle -\circ f} \\
\{\{(P', x', a')\}_{\simeq} \mid x' : P' \rightarrowtail E \ \&\ a' : Q' \twoheadrightarrow P'\} & \xrightarrow{\alpha_{A'}} & \mathbf{E}[Q', E]
\end{array}$$

commutes. However, by definition

$$L(\mathbf{E}_s[-, E])(f)(\{(P', x', a')\}_{\simeq}) = \{(P, x' \circ i, a)\}_{\simeq}$$

where $a : Q \twoheadrightarrow P$ and $i : P \rightarrowtail P'$ provide an augment-strict factorisation of $a' \circ f$:

$$\begin{array}{ccccc}
Q' & \xrightarrow{a'} & P' & \xrightarrow{x'} & E \\
{\scriptstyle f} \uparrow & & \uparrow {\scriptstyle i} & & \\
Q & \xrightarrow{a} & P & &
\end{array}$$

Clearly, $\alpha_Q(\{(P, x' \circ i, a)\}_{\simeq}) = x' \circ i \circ a = x' \circ a' \circ f = \alpha'_Q(\{P', x', a'\}_{\simeq}) \circ f$, so the naturality square commutes, as required. □

6.3 Transfer of Conditions via L

We characterise, to within isomorphism, those presheaves which are images under $L : \widehat{\mathbf{Pom}_s} \to \widehat{\mathbf{Pom}}$ as those which are "Confluent" and see how the key conditions of the first representation theorem transfer across L.

Lemma 21. *(i) Let $Y \in \widehat{\mathbf{Pom}}$. Then Y satisfies the "Confluent" condition iff $Y \cong L(X)$ for some $X \in \widehat{\mathbf{Pom}_s}$.*

(ii) Let $X \in \widehat{\mathbf{Pom}_s}$. Then, X satisfies the "Mono" and "Separated" conditions iff $L(X)$ satisfies the "Mono" and "Separated" conditions

Proof. (i) *"if":* Suppose that in $els(L(X))$

$$f : (Q,q) \to (R,r) \text{ and } b : (Q,q) \to (Q',q')$$

where $q \in L(X)(Q)$, $r \in L(X)(R)$ and $q' \in L(X)(Q')$ and $f : Q \to R$ and $b : Q \twoheadrightarrow Q'$ in **Pom**. Assume f factorises as

$$f = Q \overset{a_1}{\twoheadrightarrow} P_1 \overset{i}{\rightarrowtail} P.$$

Then, from the definition of $L(X)$,

$$r = \{(P,x,a)\}_{\simeq},\ q' = \{(P',x',a')\}_{\simeq} \text{ and } r = \{(P_1, x{\cdot}i, a_1)\}_{\simeq} = \{(P',x',a'{\circ}b)\}_{\simeq}$$

for some $a : R \twoheadrightarrow P$ with $x \in X(P)$, and $a' : Q' \twoheadrightarrow P'$ with $x' \in X(P')$.

Because $(P_1, x \cdot i, a_1) \simeq (P', x', a' \circ b)$ there is an isomorphism $j : P_1 \cong P'$ making $(P_1, x \cdot i, 1_{P_1}) \simeq (P', x', 1_{P'})$. Summarising all the facts in a diagram in $els(L(X))$ we obtain the two commuting squares

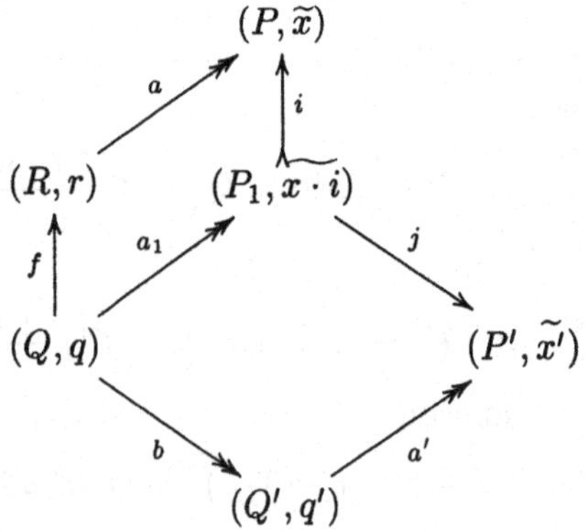

where $\widetilde{x} = \{(P,x,1_P)\}_{\simeq}$, $\widetilde{x'} = \{(P',x',1_{P'})\}_{\simeq}$ and $\widetilde{x \cdot i} = \{(P_1, x \cdot i, 1_{P_1})\}_{\simeq}$. In particular, noting the isomorphism $j : (P_1, \widetilde{x \cdot i}) \cong (P', \widetilde{x'})$, we see the "Confluent" condition is satisfied in $els(L(X))$.

"only if": To show the converse, we show how given $Y \in \widehat{\mathbf{Pom}}$ which is "Confluent" there is a presheaf $ext(Y) \in \widehat{\mathbf{Pom}_s}$ such that $L(ext(Y)) \cong Y$. The presheaf $ext(Y)$ consists of just the *extreme* elements of Y, those elements of Y which are not restrictions of elements with respect to any augmentations other than isomorphisms:

- $ext(Y)(P) = \{y \in Y(P) \mid \forall a : P \twoheadrightarrow Q, y' \in Y(Q)).\ \ y = y' \cdot a \Rightarrow a \text{ is iso.}\}$ for pomsets P.
- $ext(Y)(j)$ is the restriction of $Y(j)$, for morphisms $j : P \rightarrowtail P'$ in $\mathbf{Pom}_s$; that $ext(Y)(j)$ is well-defined, *i.e.* that if $y' \in ext(Y)(P')$, then $Y(j)(y') \in ext(Y)(P)$, follows directly from Y being "Confluent" and the uniqueness up to isomorphism of factorisation.

It is now clear that $ext(Y) \in \widehat{\mathbf{Pom}_s}$. We require that $L(ext(Y)) \cong Y$. By definition $L(ext(Y)(Q) = \{\{(P, y, a\}_{\simeq} \mid y \in ext(Y)(P) \ \&\ a : Q \twoheadrightarrow P\}$. Defining $\delta(\{P, y, a)\}_{\simeq}) = y \cdot a$ yields a function $\delta : L(ext(Y))(Q) \to Y(Q)$ which is seen to be well-defined directly from the definition of $\simeq$.
δ is 1-1: Suppose $\delta(\{(P, Y, a)\}_{\simeq}) = \delta(\{(P', y', a')\}_{\simeq})$. Then $y_0 =_{def} y \cdot a = y' \cdot a'$. As Y is assumed "Confluent" we obtain a commuting diagram

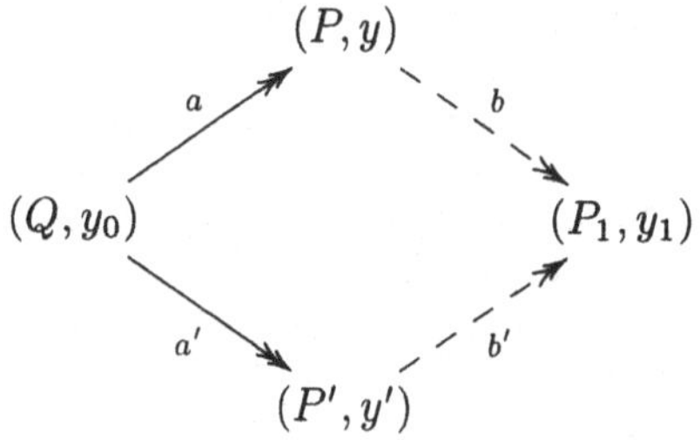

in $els(Y)$. However (P, y) and (P', y') are extreme elements of Y. Hence b and b' are isomorphisms making $(P, y, a) \simeq (P', y', a')$.
δ is onto: Suppose $y \in Y(Q)$. Because pomsets in **Pom** are finite, any chain

$$(Q, y) \overset{a_1}{\twoheadrightarrow} (Q, y) \overset{a_2}{\twoheadrightarrow} \cdots \overset{a_n}{\twoheadrightarrow} (Q_n, y_n) \overset{a_{n+1}}{\twoheadrightarrow} \cdots$$

is $els(Y)$ must eventually only involve isomorphisms, *i.e.* for some n for all $m \geq n$, each augmentation a_m is an isomorphism. Taking $a = a_{n-1} \circ \cdots \circ a_1$ there is an extreme element y_n for which $\delta(\{(Q_n, a, y_n)\}_{\simeq}) = y_n \cdot a = y$.

It follows that, to within isomorphism, the images of L are precisely those presheaves Y of $\widehat{\mathbf{Pom}}$ which are "Confluent".
(ii) We now show that the "Mono" and "Separated" conditions transfer via L. We first observe that for $X \in \widehat{\mathbf{Pom}_s}$,

$$ext(L(X)) \cong X$$

because extreme elements of $L(X)$, of the form $\{(P, x, 1_P)\}_{\simeq}$, are in 1-1 correspondence with $x \in X(P)$.
"if": Assuming $L(X)$ is "Mono" and "Separated", the "Mono" and "Separated" conditions can be also seen to hold in the restriction $ext(L(X))$, which is isomorphic to X.
"only if": Assuming X is "Mono" and "Separated" entails that $X \cong \mathbf{E}_s[-, E]$ for some event structure E. By Lemma 20, $L(X) \cong \mathbf{E}[-, E]$. Now, just as in the proof of Lemma 11, $\mathbf{E}[-, E]$ and so $L(X)$ satisfies "Mono" (because morphism from pomsets to event structure in $\mathbf{E}$ are mono) and "Separated" (because morphisms are determined by their actions on events.) □

We now obtain, as a corollary:

Proof of **Theorem 17**: By Lemma 20, a nonempty presheaf $Y \in \widehat{\mathbf{Pom}}$ is represented by some event structure in $\mathbf{E}$ iff $Y \cong L(X)$ for some nonempty $X \in \widehat{\mathbf{Pom}_s}$ which is "Mono" and "Separated". But Lemma 21 says that the latter properties hold of X iff $Y \cong L(X)$ is "Mono", "Separated" and "Confluent". □

References

1. F. Borceux. *Handbook of categorical algebra, vol. 1.* CUP, 1994.
2. G. L. Cattani. *Presheaf models for concurrency.* PhD thesis, University of Aarhus, 1999. Forthcoming.
3. G. L. Cattani, M. Fiore, and G. Winskel. A theory of recursive domains with applications to concurrency. In *Proceedings of LICS '98*, pages 214–225, IEEE Press, 1998.
4. G. L. Cattani, I. Stark, and G. Winskel. Presheaf models for the π-calculus. In *Proceedings of CTCS '97*, LNCS 1290, pages 106–126, 1997.
5. G. L. Cattani and G. Winskel. Presheaf models for concurrency. In *Proceedings of CSL' 96*, LNCS 1258, pages 58–75, 1997.
6. M. Fiore, G. L. Cattani and G. Winskel. Weak bisimulation and open maps. To appear in LICS'99.
7. A. Joyal, M. Nielsen, and G. Winskel. Bisimulation from open maps. *Information and Computation*, 127:164–185, 1996.
8. S. Mac Lane and I. Moerdijk. *Sheaves in geometry and logic: A First Introduction to Topos Theory.* Springer-Verlag, 1992.
9. Pratt, V.R., Modelling concurrency with partial orders. International Journal of Parallel Programming, 15,1, p.33-71, Feb. 1986.
10. A. J. Power, G. L. Cattani and G. Winskel. A categorical axiomatics for bisimulation. In *Proceedings of CONCUR'98*, LNCS 1466, pages 591–596, 1998.
11. G. Winskel and M. Nielsen. Models for concurrency. In *Handbook of logic in computer science, Vol. 4*, Oxford Sci. Publ., pages 1–148. Oxford Univ. Press, 1995.
12. G. Winskel. A presheaf semantics of value-passing processes. In *Proceedings of CONCUR'96*, LNCS 1119, pages 98–114, 1996.
13. G.Winskel. A Linear Metalanguage for Concurrency. In *Proceedings of AMAST'98*, LNCS, 1999.

Subtyping and Locality in Distributed Higher Order Processes[†]
(extended abstract)

Nobuko Yoshida and Matthew Hennessy

COGS, University of Sussex

Abstract. This paper studies one important aspect of distributed systems, *locality*, using a calculus of distributed higher-order processes in which not only basic values or channels, but also parameterised processes are transferred across distinct locations. An integration of the subtyping of λ-calculus and IO-subtyping of the π-calculus offers a tractable tool to control the locality of channel names in the presence of distributed higher order processes. Using a local restriction on channel capabilities together with a subtyping relation, locality is preserved during reductions even if we allow new receptors to be dynamically created by instantiation of arbitrary higher-order values and processes. We also show that our method is applicable to more general constraints, based on local and global channel capabilities.

1 Introduction

There have been a number of attempts at adapting traditional process calculi, such as CCS and CSP, so as to provide support for the modelling of certain aspects of distributed systems, such as *distribution* of resources and *locality*, [3, 10, 20, 25, 30]. Most of these are based on first-order extensions of the π-calculus [21]; first-order in the sense that the data exchanged between processes are from simple datatypes, such as basic values or channel names. There are various proposals for implementing the transmission of higher-order data using these first-order languages, mostly based on [27]. However these translations, as we will explain in Section 6, do not preserve the distribution and locality of the source language. Consequently we believe that higher-order extensions of the π-calculus should be developed in their own right, as formal modelling languages for distributed systems.

In this paper we design such a language and examine one important aspect of distributed systems, namely *locality*. The language is a simple integration of the call-by-value λ-calculus and the π-calculus [21], together with primitives for distribution and spawning of new code at remote sites. The combination of dynamic channel creation inherited from π-calculus and transmission of higher-order programs inherited from λ-calculus offers us direct descriptions of various distributed computational structures. As such, it has much in common with the core version of Facile [2, 9, 19], CML [8] and LLinda [22], and can be regarded as an extension of Blue-calculus [5] to a higher-order term passing.

A desirable feature of some distributed systems is that every channel name is associated with a *unique* receptor, which is called *receptiveness* in [28]; another property called *locality* where new receptors are not created by received channels, has also been studied in [3, 4, 20, 34] for an asynchronous version of the π-calculus [16]. The combination of these constraints provides a model of a realistic distributed environment, which

[†] Supported by EPSRC GR/K60701 and CONFER II. E-mail:{nobuko,matthewh}@cogs.susx.ac.uk.

Term:	$P,Q,\ldots ::=$	$V \mid PQ \mid$
		$u?(\tilde{x}{:}\tilde{\tau}).P \mid u!\langle\tilde{V}\rangle P \mid (\nu a{:}\sigma)P \mid *P \mid P\|Q \mid \mathbf{0}$
Value:	$V,W,\ldots ::=$	$u \mid \lambda(x{:}\tau).P$
Identifier:	$u,v,\ldots ::=$	$l \mid a \mid x$
Literal:	$l,l',\ldots ::=$	`true` $\mid$ `false` $\mid () \mid 0 \mid 1 \mid \ldots$

Figure 1. Syntax of $\pi\lambda$

regards a receptor as an object or a thread existing in a unique name space. A generalisation is also proposed in Distributed Join-calculus where not only single receptor but also several receptors with the same input channel are allowed to exist in the same location [10]; in this paper we call this more general condition *locality of channels*. In distributed object-oriented systems, objects with a given ID reside in a specific location even if multiple objects with the same ID are permitted to exist for efficiency reasons, as found in, e.g. CONCURRENT AGGREGATES [7]; This locality constraint should be obeyed even in the presence of parameterised object passing, which is recently often found in practice [11].

In this paper we show that, in a distributed higher-order process language, locality of channels can be enforced by a typing system with subtyping. The essential idea is to control the *input capability* of channels, guaranteeing at any one time this capability resides at exactly one location. As discussed in Section 3, ensuring locality in higher order processes is much more difficult than in systems which only allows name passing. However, using our typing system we only have to *static* type-check each local configuration to guarantee the required global invariance, namely *locality of channels*.

The main technical novelty of our work is an extension of the input/output type system of [14, 24] to a higher-order setting where the order theoretic property of subtyping relation, *finite-bounded completeness*, plays a pivotal role for a natural integration with arrow types. The framework will be generally applicable for other purposes where similar global constraints should be guaranteed using static local type checking.

The paper is organised as follows: Section 2 studies a call-by-value higher-order π-calculus, $\pi\lambda$, with subtyping. Section 3 introduces a distributed version of $\pi\lambda$, called $\mathrm{D}\pi\lambda$, and illustrates the difficulty of enforcing locality in $\mathrm{D}\pi\lambda$. Section 4 proposes a new typing system to ensure the locality. Section 5 discusses applications of our type discipline; extendibility of our system to more general global/local channel constraints studied in [30] in a higher-order setting, a multiple higher-order replication theorem extending [24], and the type checking. Section 6 concludes with discussions and related work. Due to space limitation, we leave the detailed explanations and proofs to the full version [35].

2 A Higher-order π-calculus with IO-subtyping

SYNTAX The syntax of $\pi\lambda$ is given in Figure 1. It uses an infinite set of *names* or *channels* $\mathbf{N}$, ranged over by $a,b,\ldots$, and an infinite set of *variables* $\mathbf{V}$, $x,y,\ldots$. We often use $X,Y,\ldots$ for variables over higher order terms explicitly. The syntax is a mixture of

Reduction Rules:

$$(\beta) \quad (\lambda(x{:}\tau).P)V \longrightarrow P\{V/x\} \qquad (\text{app}_l)\ \frac{P \longrightarrow P'}{PV \longrightarrow P'V}$$

$$(\text{com}) \quad u?(\tilde{x}{:}\tilde{\tau}).P \,|\, u!\langle\tilde{V}\rangle Q \longrightarrow P\{\tilde{V}/\tilde{x}\} \,|\, Q \qquad (\text{app}_r)\ \frac{Q \longrightarrow Q'}{PQ \longrightarrow PQ'}$$

$$(\text{par})\ \frac{P \longrightarrow P'}{P\,|\,Q \longrightarrow P'\,|\,Q} \qquad (\text{res})\ \frac{P \longrightarrow P'}{(\nu a{:}\sigma)P \longrightarrow (\nu a{:}\sigma)P'} \qquad (\text{str})\ \frac{P \equiv P' \longrightarrow Q' \equiv Q}{P \longrightarrow Q}$$

Structure Equivalence:

- $P \equiv Q$ if $P \equiv_\alpha Q$.
- $P|Q \equiv Q|P \quad (P|Q)|R \equiv P|(Q|R) \quad P|\mathbf{0} \equiv P \quad *P \equiv P|*P$
- $(\nu a)\mathbf{0} \equiv \mathbf{0} \quad (\nu a)(\nu b)P \equiv (\nu b)(\nu a)P \quad (\nu a)P|Q \equiv (\nu a)(P|Q)$ if $a \notin \mathsf{fn}(Q)$

Figure 2. Reduction for $\pi\lambda$

a call-by-value λ-calculus and the π-calculus. From the former there are values, consisting of basic values and abstractions, together with application. From the latter we have input and output on communication channels, dynamic channel creation, iteration and the empty process. We use the standard notational conventions; for example ignoring trailing occurrences of $\mathbf{0}$ and omitting type annotations unless they are relevant. We use $\mathsf{fn}(P)$/$\mathsf{fv}(P)$ to denote the sets of *free names/variables* respectively, and typically write $\lambda().P$ for a *thunk* of P, $\lambda(x{:}\mathtt{unit}).P$ assuming $x \notin \mathsf{fv}(P)$.

REDUCTION The reduction semantics of $\pi\lambda$ is given in Figure 2. The main reduction rules are β-reduction, (β), and communication, (com). The final contextual rule, (str), uses a structural rules from the π-calculus. We use $\longrightarrow\!\!\!\!\rightarrow$ to denote multi-step reductions.

EXAMPLE 2.1. (sq-server) Suppose that in the language we have a literal sq for squaring natural numbers. For a given name a let $\mathbf{sq}(a)$ represent the expression $*a?(y,z).\, z!\langle \mathsf{sq}(y)\rangle$, which we write as

$$\mathbf{sq}(a) \Longleftarrow *a?(y,z).\, z!\langle \mathsf{sq}(y)\rangle$$

This receives a value on y to be processed together with a return channel z to which the processed data is to be sent. It then processes the squaring data and then returns it along the return channel.

A sq-server, **sqServ**, is a process which on requests sends to the client the code for squaring values, which the client can initialise locally.

$$\mathbf{sqServ} \Longleftarrow *\mathrm{req}?(r).\, r!\langle \lambda(x).\, \mathbf{sq}(x)\rangle$$

Here the process receives a request on the channel req, in the form of a return channel r, to which the abstraction $\lambda(x).\, \mathbf{sq}(x)$ is sent. A client can now download this code and initialise it by a local channel a which will act as the request channel for data processing:

$$\mathbf{Client} \Longleftarrow (\nu r)\, \mathrm{req}!\langle r\rangle.\, r?(X).\, (\nu a)(\, X a \,|\, a!\langle 1,c_1\rangle \,|\, a!\langle 2,c_2\rangle \,|\, a!\langle 3,c_3\rangle \,|\, \cdots) \qquad \square$$

Type:

Term Type:	ρ	::=	$\texttt{proc} \mid \tau$
Value Type:	τ	::=	$\texttt{unit} \mid \texttt{bool} \mid \texttt{nat} \mid \tau \to \rho \mid \sigma$
Channel Type:	σ	::=	$\langle S_{\mathrm{I}}, S_{\mathrm{O}} \rangle$ with $S_{\mathrm{I}} \geq S_{\mathrm{O}}$, $S_{\mathrm{I}} \neq \bot$ and $S_{\mathrm{O}} \neq \top$.
Sort Type:	S	::=	$\bot \mid \top \mid (\tilde{\tau})$

Ordering:

(base)	$\texttt{proc} \leq \texttt{proc}$, $\texttt{nat} \leq \texttt{nat}$, $S \leq S$, etc.
$(\bot, \top)$	$\bot \leq S \quad S \leq \top$
(vec)	$\forall i.\ \tau_i \leq \tau'_i \ \Rightarrow\ (\tilde{\tau}) \leq (\tilde{\tau}')$
$(\to)$	$\tau \geq \tau',\ \rho \leq \rho' \ \Rightarrow\ \tau \to \rho \leq \tau' \to \rho'$
(chan)	$\sigma_i = \langle S_{i\mathrm{I}}, S_{i\mathrm{O}} \rangle,\ S_{1\mathrm{I}} \leq S_{2\mathrm{I}},\ S_{1\mathrm{O}} \geq S_{2\mathrm{O}} \ \Rightarrow\ \sigma_1 \leq \sigma_2.$

Abbreviations:

(input only)
$$(\tilde{\tau})^{\mathrm{I}} \stackrel{\mathrm{def}}{=} \langle (\tilde{\tau}), \bot \rangle$$
(output only)
$$(\tilde{\tau})^{\mathrm{O}} \stackrel{\mathrm{def}}{=} \langle \top, (\tilde{\tau}) \rangle$$
(input/output)
$$(\tilde{\tau})^{\mathrm{IO}} \stackrel{\mathrm{def}}{=} \langle (\tilde{\tau}), (\tilde{\tau}) \rangle$$

Figure 3. Types for $\pi\lambda$

IO-TYPES We use as types for $\pi\lambda$ a simplification of the input/output capabilities of [14] (in turn a *strict* generalisation of [24][1]). They are defined in Figure 3, where we assume a given set of base types, such as nat and bool, and a type for processes, proc. Value types may then be constructed from these types using the constructor $\to$, as in the λ-calculus. Here in addition we may also use channel types, ranged over by σ. These take the form $\langle S_{\mathrm{I}}, S_{\mathrm{O}} \rangle$, a pair consisting of an *input sort* S_{I} and an *output sort* S_{O}; these input/output sorts are in turn either a vector of value types or $\top$, denoting the highest capability, or $\bot$, denoting the lowest. The representation of IO-types as a tuple [14, 15] makes the definition of the subtyping relationship, also given in Figure 3, more natural when we integrate with arrow types of the λ-calculus; the ordering of input types is covariant, whereas that of output types is contravariant. The condition on channel types, $S_{\mathrm{I}} \geq S_{\mathrm{O}}$ is necessary to ensure that a receiver always takes fewer capabilities than specified by the outside environment, while a sender always send more capabilities than specified. Then IO-types in [24] are represented as a special case of our IO-types; to denote them, we introduce the abbreviations in Figure 3. Note that $(\tilde{\tau})^{\mathrm{IO}} \leq (\tilde{\tau})^{\mathrm{I}} \leq \langle \top, \bot \rangle$ and $(\tilde{\tau})^{\mathrm{IO}} \leq (\tilde{\tau})^{\mathrm{O}} \leq \langle \top, \bot \rangle$.[2]

The subtyping relation over types defined in Figure 3 is partial order and *finite bounded complete*, FBC, (cf. [14]). The partial meet operator $\sqcap$ and partial join operator $\sqcup$ can be also defined directly following [14]. For the base and arrow types, we define $\sqcap$/$\sqcup$ as the standard meet/join operators w.r.t. $\leq$. For channel types, we use the following definition:

$$\text{(vec)} \quad (\tilde{\tau}) \sqcup (\tilde{\tau}') \stackrel{\mathrm{def}}{=} (\tilde{\tau}'') \text{ with } \tau''_i = \tau_i \sqcup \tau'_i \quad \text{and} \quad (\tilde{\tau}) \sqcap (\tilde{\tau}') \stackrel{\mathrm{def}}{=} (\tilde{\tau}'') \text{ with } \tau''_i = \tau_i \sqcap \tau'_i$$

[1] Our general form of IO-types gives more typable terms than [24] even if we restrict our language to the pure polyadic π-calculus. See Example 2.5 in [35].

[2] Note also $\langle \top, \bot \rangle \neq \top$ because the former is a type for a channel which is only used as a value (i.e. empty capability), while the latter is the top of sort types. The side coditions $S_{\mathrm{I}} \neq \bot$ and $S_{\mathrm{O}} \neq \top$ ensure to avoid mismatching arity constraints.

Common Typing Rules:

ID: $\Gamma, u{:}\tau \vdash u : \tau$ SUB: $\dfrac{\Gamma \vdash P : \rho \quad \rho \leq \rho'}{\Gamma \vdash P : \rho'}$

Functional Typing Rules:

CONST: $\Gamma \vdash 1 : \mathtt{nat}$ etc.

ABS: $\dfrac{\Gamma, x{:}\tau \vdash P : \rho}{\Gamma \vdash \lambda(x{:}\tau).P : \tau \to \rho}$

APP: $\dfrac{\Gamma \vdash P : \tau \to \rho \quad \Gamma \vdash Q : \tau}{\Gamma \vdash PQ : \rho}$

Process Typing Rules:

IN: $\dfrac{\Gamma \vdash u : (\tilde{\tau})^{\mathrm{I}} \quad \Gamma, \tilde{x}{:}\tilde{\tau} \vdash P : \mathtt{proc}}{\Gamma \vdash u?(\tilde{x}{:}\tilde{\tau}).P : \mathtt{proc}}$ NIL: $\Gamma \vdash \mathbf{0} : \mathtt{proc}$

OUT: $\dfrac{\Gamma \vdash u : (\tilde{\tau})^{\mathrm{O}} \quad \Gamma \vdash V_i : \tau_i \quad \Gamma \vdash P : \mathtt{proc}}{\Gamma \vdash u!\langle \tilde{V} \rangle P : \mathtt{proc}}$ REP: $\dfrac{\Gamma \vdash P : \mathtt{proc}}{\Gamma \vdash *P : \mathtt{proc}}$

RES: $\dfrac{\Gamma, a{:}\sigma \vdash P : \mathtt{proc}}{\Gamma \vdash (\nu a{:}\sigma)P : \mathtt{proc}}$ PAR: $\dfrac{\Gamma \vdash P : \mathtt{proc} \quad \Gamma \vdash Q : \mathtt{proc}}{\Gamma \vdash P \mid Q : \mathtt{proc}}$

Figure 4. Typing System for $\pi\lambda$.

(chan) $\langle S_{\mathrm{I}}, S_{\mathrm{O}} \rangle \sqcup \langle S'_{\mathrm{I}}, S'_{\mathrm{O}} \rangle \stackrel{\text{def}}{=} \langle S_{\mathrm{I}} \sqcup S'_{\mathrm{I}}, S_{\mathrm{O}} \sqcap S'_{\mathrm{O}} \rangle$ and
$\langle S_{\mathrm{I}}, S_{\mathrm{O}} \rangle \sqcap \langle S'_{\mathrm{I}}, S'_{\mathrm{O}} \rangle \stackrel{\text{def}}{=} \langle S_{\mathrm{I}} \sqcap S'_{\mathrm{I}}, S_{\mathrm{O}} \sqcup S'_{\mathrm{O}} \rangle$ if $S_{\mathrm{I}} \geq S'_{\mathrm{O}}$ and $S'_{\mathrm{I}} \geq S_{\mathrm{O}}$; else undefined.

For sort types (but not value, term or channel types) we can ensure that both $\sqcap$ and $\sqcup$ are total; in all cases of $S \sqcap S'$ (respectively $S \sqcup S'$) not covered by the above clasues, then we set $S \sqcap S' = \bot$ (respectively $S \sqcup S' = \top$).

THE IO TYPING SYSTEM *Type environments*, ranged over by $\Gamma, \Delta, \ldots$, are functions from a finite subset of $\mathbf{N} \cup \mathbf{V}$ to the set of value types. We use the following notation:

(1) $\mathsf{dom}(\Gamma)$ denotes $\{u \mid u{:}\tau \in \Gamma\}$ and Γ/A denotes $\{u{:}\tau \in \Gamma \mid u \notin A\}$.
(2) $\Gamma, u{:}\tau$ means $\Gamma \cup \{u{:}\tau\}$, together with the assumption $u \notin \mathsf{dom}(\Gamma)$.
(3) $\Delta \leq \Gamma$ means $\Delta(u) \leq \Gamma(u)$ for all $u \in \mathsf{dom}(\Gamma)$.

Then we define the partial meet operator $\sqcap$ and the partial join operator $\sqcup$ as:

$\Gamma \sqcap \Delta \stackrel{\text{def}}{=} \Gamma/\mathsf{dom}(\Delta) \cup \Delta/\mathsf{dom}(\Gamma) \cup \{u{:}(\Delta(u) \sqcap \Gamma(u)) \mid u \in \mathsf{dom}(\Gamma) \cap \mathsf{dom}(\Delta)\}$ and
$\Gamma \sqcup \Delta \stackrel{\text{def}}{=} \{u{:}(\Delta(u) \sqcup \Gamma(u)) \mid u \in \mathsf{dom}(\Gamma) \cap \mathsf{dom}(\Delta)\}$

Typing Assignments are formulas $P : \rho$ for any term P and any type ρ. We write $\Gamma \vdash P : \rho$ if the formula $P : \rho$ is provable from a typing function Γ using the Typing System given in Figure 4. This is divided in two parts. The first is inherited from the λ-calculus, while the second is a simple adaptation of the IO-typing system from [14, 24].

EXAMPLE 2.2. (typed sq server) We now revisit Example 2.1. In the definition of $\mathbf{sq}(a)$ a pair of values are input, a natural number and a channel respectively, and this channel will be used to transmit a natural number. So we type it as:

$$\mathbf{sq}(a) \Longleftarrow *a?(y{:}\mathtt{int}, z{:}(\mathtt{int})^{\mathrm{O}}).\ z!\langle \mathsf{sq}(y) \rangle$$

Syntax:

System:	$M, N, \ldots$	$::=$	$P \mid N \| M \mid (\nu a{:}\sigma)N$	
Term:	$P, Q, \ldots$	$::=$	$\mathtt{Spawn}(P) \mid \cdots$	from Figure 1
Value:	as in Figure 1			

Distributed Reduction Rules:

$$\text{(spawn)} \quad (\cdots Q \,|\, \mathtt{Spawn}(P)) \longrightarrow (\cdots Q) \,\|\, P$$

$$(\text{com}_s) \quad (u?(\tilde{x}{:}\tilde{\tau}).P \,|\, \cdots) \,\|\, (u!\langle \tilde{V} \rangle Q \,|\, \cdots) \longrightarrow (P\{\tilde{V}/\tilde{x}\} \,|\, \cdots) \,\|\, (Q \,|\, \cdots)$$

$$(\text{par}_s)\ \frac{M \longrightarrow M'}{M \,\|\, N \longrightarrow M' \,\|\, N} \quad (\text{res}_s)\ \frac{N \longrightarrow N'}{(\nu a{:}\sigma)N \longrightarrow (\nu a{:}\sigma)N'} \quad (\text{str}_s)\ \frac{N \equiv N' \longrightarrow M' \equiv M}{N \longrightarrow M}$$

Figure 5. Syntax and Distributed Reduction in D$\pi\lambda$

where the process only receives the output capability on the return channel z, which is guaranteed by the assigned type $(\mathtt{int})^{\mathtt{O}}$ to z. Then we have $\Gamma \vdash \mathbf{sq}(a) : \mathtt{proc}$ for any typing function Γ such that $\Gamma(a) \leq (\mathtt{int}, (\mathtt{int})^{\mathtt{O}})^{\mathtt{I}}$. Now by ABS in Figure 4, we have:

$$\vdash \quad \lambda(x{:}(\mathtt{int},(\mathtt{int})^{\mathtt{O}})^{\mathtt{I}}).\mathbf{sq}(x) : (\mathtt{int},(\mathtt{int})^{\mathtt{O}})^{\mathtt{I}} \to \mathtt{proc}$$

which means that should x be instantiated by a channel whose capability is *dominated by* $(\mathtt{int},(\mathtt{int})^{\mathtt{O}})^{\mathtt{I}}$, then it becomes a safe process. □

This simple typing system satisfies the following standard subject reduction theorem.

THEOREM 2.3. (Subject Reduction) *If* $\Gamma \vdash P : \rho$ *and* $P \longrightarrow\!\!\!\!\to P'$, *then* $\Gamma \vdash P' : \rho$.

3 Locality of Channels in Distributed Higher Order π-calculus

DISTRIBUTED HIGHER ORDER π-CALCULUS The extended syntax for distributed processes is given by in Figure 5. Intuitively $N \| M$ represents two systems N, M running at two physically distinct locations, while the process $\mathtt{Spawn}(P)$ creates a new location at which the process P is launched. The reduction semantics of the previous section is extended to the new language, D$\pi\lambda$, in a straightforward manner, outlined in Figure 5. The structural equivalence of systems is defined by changing "$|$" to "$\|$" and P, Q, R to M, N, N' in Figure 2. The first two rules are the most important, namely spawning of a process at a new location (spawn) and communication between physically distinct locations, (com$_s$).

DEFINING LOCALITY We require that every input channel name is associated with a unique location. This is violated in, for example,

$$a?(y).\,P \;\;\|\;\; (a?(z).\,Q \,|\, b?(x_1).\,R_1 \,|\, b?(x_2).\,R_2)$$

because the name a can receive input at two distinct locations. Note however that the name b is located uniquely, although at that location a call can be serviced in two different ways.

A formal definition of this concept (or rather its complement), *locality error*, is given in Figure 6, using a predicate on systems, $N \xrightarrow{lerr}$. Intuitively this should be read as saying:

Input Predicate:

$$a?(\tilde{x}).\,P\downarrow a^{\mathrm{I}} \qquad \frac{P\downarrow a^{\mathrm{I}}}{(P|Q)\downarrow a^{\mathrm{I}}} \qquad \frac{Q\downarrow a^{\mathrm{I}}}{(P|Q)\downarrow a^{\mathrm{I}}} \qquad \frac{P\downarrow a^{\mathrm{I}} \quad a\neq b}{(\nu b)P\downarrow a^{\mathrm{I}}} \qquad \frac{P\downarrow a^{\mathrm{I}}}{*P\downarrow a^{\mathrm{I}}}$$

$$\frac{M\downarrow a^{\mathrm{I}}}{(N\,||\,M)\downarrow a^{\mathrm{I}}} \qquad \frac{N\downarrow a^{\mathrm{I}}}{(N\,||\,M)\downarrow a^{\mathrm{I}}} \qquad \frac{N\downarrow a^{\mathrm{I}} \quad a\neq c}{(\nu c)N\downarrow a^{\mathrm{I}}}$$

Locality Error:

$$\frac{N\downarrow a^{\mathrm{I}} \quad M\downarrow a^{\mathrm{I}}}{(N\,||\,M)\stackrel{lerr}{\longrightarrow}} \qquad \frac{N\stackrel{lerr}{\longrightarrow}}{(N\,||\,M)\stackrel{lerr}{\longrightarrow}} \qquad \frac{N\stackrel{lerr}{\longrightarrow}}{(M\,||\,N)\stackrel{lerr}{\longrightarrow}} \qquad \frac{N\stackrel{lerr}{\longrightarrow}}{(\nu c)N\stackrel{lerr}{\longrightarrow}}$$

Figure 6. Locality Error

Local Distributed Rules:

SPAWN: $\dfrac{\Gamma\vdash P:\texttt{proc}}{\Gamma\vdash \texttt{Spawn}(P):\texttt{proc}}$ INTRO: $\dfrac{\Gamma\vdash P:\texttt{proc}}{\Gamma\vdash_1 P}$ PAR$_l$: $\dfrac{\Gamma\vdash_1 N \quad \Delta\vdash_1 M \quad \Gamma\asymp_1\Delta}{\Gamma\sqcap\Delta\vdash_1 N\,||\,M}$ RES$_l$: $\dfrac{\Gamma,a{:}\sigma\vdash_1 M}{\Gamma\vdash_1 (\nu a{:}\sigma)M}$

Figure 7. Local Distributed Typing Rules

in the system N there is a runtime error, namely there is some name a which is ready to receive input at two distinct locations. The definition uses an input predicate $P\downarrow a^{\mathrm{I}}$ which is satisfied when P can immediately perform input on name a.

Now let us say a channel type σ is *local* if σ has an input capability, i.e $\sigma=\langle(\tilde{\tau}),S_{\mathrm{O}}\rangle$. We also call u is *local under* Γ if $\Gamma(u)$ is local.

DEFINITION 3.1. (system composable) Γ_1 and Γ_2 are *composable*, written by $\Gamma_1\asymp\Gamma_2$, if $\Gamma_1\sqcap\Gamma_2$ is defined, and Γ_1 and Γ_2 are *system-composable*, written by $\Gamma_1\asymp_1\Gamma_2$, if $\Gamma_1\asymp\Gamma_2$ and $u{:}\langle S_{i\mathrm{I}},S_{i\mathrm{O}}\rangle\in\Gamma_i$ $(i=1,2)$ implies $S_{1\mathrm{I}}=\top$ or $S_{2\mathrm{I}}=\top$. □

Intuitively this means that if a channel a is local in Γ_1, then it must not be local in another environment Γ_2.

The typing system for distributed systems given in Figure 7 is simply in form of $\Gamma\vdash_1 N$ where Γ is again the same environment. The most essential rule is PAR$_l$; this says that $N_1\,||\,N_2$ is typable with respect to Δ if Δ can be written as $\Gamma_1\sqcap\Gamma_2$, where $\Gamma_1\asymp_1\Gamma_2$ and N_i is typable with respect to Γ_i. If terms are system composable, then we have no immediate locality error since $P\downarrow a^{\mathrm{I}}$ and $\Gamma\vdash_1 P:\texttt{proc}$ imply $\Gamma\vdash a:(\tilde{\tau})^{\mathrm{I}}$ for some $\tilde{\tau}$. That is:

THEOREM 3.2. (Type Safety) $\Gamma\vdash_1 N$ *implies* $N\stackrel{lerr}{\not\longrightarrow}$.

It is however easy to see that system composability as defined above is not closed under reduction: indeed, we easily have $N\stackrel{lerr}{\not\longrightarrow}$ and $N\longrightarrow\!\!\!\!\!\rightarrow N'$ does *not* imply $N'\stackrel{lerr}{\not\longrightarrow}$.

DIFFICULTIES IN PRESERVING LOCALITY IN D$\pi\lambda$ There are basically two reasons why locality is not preserved after communication. The first is the use of a name received from another location as an input subject. Take $a?(x).\,P\,|\,b!\langle a\rangle \parallel b?(y).\,y?(z).\,Q$. Then it is easy to check that this can be typed with PAR_l in Figure 7. However after one reduction step, the communication along b, we obtain $a?(x).\,P \parallel a?(z).\,Q$, which is no longer typable. It is not difficult to exclude such terms which do not involve term passing by a simple syntactic condition or typing systems as studied in [2, 4, 20, 28]. The second, which is more complicated, concerns the parameterisations of processes and the instantiation of variables which occur in outgoing values. The presence of higher-order passing makes the problem subtle, as seen in the next example.

EXAMPLE 3.3. Let V denote the value $\lambda().\mathbf{sq}(a)$ in the slightly modified system

$$a?(x).\,P\,|\,b!\langle V\rangle \parallel b?(Y).\,Y\,()$$

This is a typable configuration; nevertheless, after the transmission of the value V to the new site and a reduction we get a system which violates our locality conditions. Next consider a similar code where V denotes $\lambda(x).\mathbf{sq}(x)$.

$$a?(x).\,P\,|\,b!\langle V\rangle \parallel b?(Y).\,(Y\,c)\,|\,c?(x).\,Q$$

This does not destroy locality while it creates a new receptor $\mathbf{sq}(c)$. However the following system which sends the same value as the above to b disturbs locality.

$$d?(X).\,X()\,|\,b!\langle V\rangle \parallel b?(Y).\,d!\langle\lambda().(Y\,c)\rangle\,|\,c?(x).\,Q \qquad \square$$

Certain values are *sendable* in that their transfer from location to location will never lead to a locality error. For example, the first value $\lambda().\mathbf{sq}(a)$ is immediately not sendable, although $\lambda(x).\mathbf{sq}(x)$ will be sendable, because it contains no free occurrence of input channels. However the algebra of *sendable* and non-*sendable* terms is not straightforward; in the third system, V is transmitted along b across locations, where it is used to dynamically construct a new dangerous value $\lambda().(V\,c)$; this is then transmitted across locations via d and when it is run we obtain once more a locality error.[3] We need a new set of *sendable*/non-*sendable* types and a typing system which controls the formation of values and ensures that in every occurrence of $b!\langle V\rangle$, where the term V can be exported to a new location, it can only evaluate to a value of *sendable* type.

4 Type Inference System for Locality

LOCAL TYPING SYSTEM We add a new type constructor $\mathbf{s}(\rho)$ for sendable terms; the formation rules and ordering are given in Figure 8. The side condition of arrow types simply avoids, as we will see, a sendable term having a non-sendable subterm; e.g. if either P or Q is non-sendable, then $P\,Q$ will automatically be non-sendable. A similar side condition on arrow types can be found in the *passive types* in [23].

The first extra ordering ensures that $\leq$ is a preorder. The second says that the constructor $\mathbf{s}(\,)$ preserves subtyping and the third that all sendable values are also values. In conjunction with (id), this rule implies that sendability is idempotent, $\mathbf{s}(\mathbf{s}(\rho)) \simeq \mathbf{s}(\rho)$ with $\simeq \stackrel{\text{def}}{=} \leq \cap \geq$. Similarly with (lift), we have: $\mathbf{s}(\mathbf{s}(\tau) \to \mathbf{s}(\rho)) \simeq \mathbf{s}(\tau) \to \mathbf{s}(\rho)$. Based on this ordering, we formalise *sendable types* as follows.

[3] See [35] for further non-trivial examples of higher-order processes.

Types:

Term Type:	ρ	::=	$\pi \mid \tau$
Process Type:	π	::=	$\texttt{proc} \mid \mathbf{s}(\texttt{proc})$
Value Type:	τ	::=	$\texttt{unit} \mid \texttt{nat} \mid \texttt{bool} \mid \sigma \mid \mathbf{s}(\tau)$ with $\tau \neq \sigma$
		$\mid$	$\tau \to \rho$ with $\rho \leq \mathbf{s}(\rho') \Rightarrow \tau \leq \mathbf{s}(\tau')$
Channel Type:	σ	::=	$\langle S_{\mathrm{I}}, S_{\mathrm{O}} \rangle$ with $S_{\mathrm{I}} \geq S_{\mathrm{O}}$, $S_{\mathrm{I}} \neq \bot$ and $S_{\mathrm{O}} \neq \top \quad \mid \quad \mathbf{s}(\langle \top, S_{\mathrm{O}} \rangle)$
Sort Type:	S	::=	as in Figure 3.

Ordering: All rules from Figure 3 and

(trans)	$\rho_1 \leq \rho_2 \quad \rho_2 \leq \rho_3 \;\Rightarrow\; \rho_1 \leq \rho_3$	(id)	$\mathbf{s}(\rho) \leq \mathbf{s}(\mathbf{s}(\rho))$
(mono)	$\rho \leq \rho' \;\Rightarrow\; \mathbf{s}(\rho) \leq \mathbf{s}(\rho')$	(lift)	$\mathbf{s}(\tau) \to \mathbf{s}(\rho) \leq \mathbf{s}(\mathbf{s}(\tau) \to \mathbf{s}(\rho))$
(sendable)	$\mathbf{s}(\rho) \leq \rho$		

Figure 8. Locality types for $\mathrm{D}\pi\lambda$

DEFINITION 4.1. Let Sble, the set of sendable types, be the least set of types which includes all types of the form $\mathbf{s}(\rho)$, and for which τ, $\rho \in$ Sble implies $\tau \to \rho \in$ Sble. We say ρ is *sendable* if $\rho \in$ Sble. □

The main properties of the set of sendable types is given in the following proposition:

PROPOSITION 4.2. (downwards closed) (1) Sble *is downwards closed with respect to subtyping:* $\rho' \leq \rho$ *and* $\rho \in$ Sble *implies* $\rho' \in$ Sble, (2) $\rho \in$ Sble *if and only if* $\rho \simeq \mathbf{s}(\rho)$, *and* (3) $\rho \in$ Sble *if and only if* $\rho \leq \mathbf{s}(\rho')$ *for some* ρ'.

The last statement of this Proposition is particularly relevant; in our revised typing system a value can only be exported to a new site if it can be assigned a type in Sble.

The essential order theoretic property, FBC, is also preserved on this subtyping relation. We extend the definition of $\sqcap$ and $\sqcup$ in § 2.3 to the sendable types as follows.

$$\mathbf{s}(\rho_1) \sqcap \mathbf{s}(\rho_2) = \mathbf{s}(\rho_1 \sqcap \rho_2) \quad \text{and} \quad \mathbf{s}(\rho_1) \sqcup \mathbf{s}(\rho_2) = \mathbf{s}(\rho_1 \sqcup \rho_2)$$
$$\mathbf{s}(\rho_1) \sqcap \rho_2 = \mathbf{s}(\rho_1 \sqcap \rho_2) \quad \text{and} \quad \mathbf{s}(\rho_1) \sqcup \rho_2 = \rho_1 \sqcup \rho_2 \quad \text{with } \rho_2 \neq \mathbf{s}(\rho_2').$$

The new type inference system, with judgements of the form $\Gamma \vdash_1 P : \rho$, is given in Figure 9 and uses the notion of *sendable type environments*, which only use sendable types.

DEFINITION 4.3. A typing environment Δ is *sendable*, written $\Delta \vdash_1$ SBL, if $u : \tau \in \Delta$ implies $\tau \in$ Sble or $a : \sigma \in \Delta$ implies $\sigma \in \langle \top, S_{\mathrm{O}} \rangle$. □

In Figure 9 the **Send Rules** determine which values can be exported to other locations, either by spawning or by communication. All constants and output capabilities on channels are automatically sendable. The crucial rule is TERM_l, which says that in a general term is sendable only if it can be derived from a sendable type environment. For processes, first we can create a process by spawn only if it is sendable. In OUT_d we require that values which will be sent across locations to have sendable types. However if the transmission is only done in the same location, this condition should be relaxed; in OUT_l, the message is guaranteed to be transmitted within the same location since name a has an

Send Rules:

$$\text{CONST}_l: \frac{\Gamma \vdash_1 l : \mathtt{nat}}{\Gamma \vdash_1 l : \mathtt{s(nat)}} \quad \text{etc.} \qquad \text{CHAN}_l: \frac{\Gamma \vdash_1 a : \langle T, S\rangle}{\Gamma \vdash_1 a : \mathtt{s}(\langle T, S\rangle)}$$

$$\text{TERM}_l: \frac{\Delta \vdash_1 P : \rho \quad \Delta \vdash_1 \mathtt{SBL} \quad \Delta \geq \Gamma}{\Gamma \vdash_1 P : \mathtt{s}(\rho)} \qquad \text{SPAWN}_l: \frac{\Gamma \vdash_1 P : \mathtt{s(proc)}}{\Gamma \vdash_1 \mathtt{Spawn}(P) : \mathtt{s(proc)}}$$

Common Rules: as in Figure 4.

Functional Rules: as in Figure 4.

Process Rules:

$$\text{OUT}_d: \frac{\Gamma \vdash_1 u : (\mathtt{s}(\tilde{\tau}))^{\mathtt{O}} \quad \Gamma \vdash_1 V_i : \mathtt{s}(\tau_i) \quad \Gamma \vdash_1 P : \pi}{\Gamma \vdash_1 u!\langle\tilde{V}\rangle P : \pi} \qquad \text{OUT}_l: \frac{\Gamma \vdash_1 u : \langle(\tilde{\tau}'), (\tilde{\tau})\rangle \quad \Gamma \vdash_1 V_i : \tau_i \quad \Gamma \vdash_1 P : \mathtt{proc}}{\Gamma \vdash_1 u!\langle\tilde{V}\rangle P : \mathtt{proc}}$$

NIL,REP,PAR,RES as in Figure 4 with proc replaced by π, and IN the same as in Figure 4.

Local Distributed Rules: PAR_l and RES_l as in Figure 7 and INTRO as in Figure 7 with $\vdash$ replaced by $\vdash_1$ in INTRO.

Figure 9. Locality Typing System for D$\pi\lambda$

input capability. Note also an input process has always the non-sendable type proc.

EXAMPLE 4.4. (Sq-server) In the following, we offer a non-trivial example of the use of sendability in typing. Recall Examples 2.1 and 2.2, and let us define

$$\sigma = (\mathtt{int}, (\mathtt{int})^{\mathtt{O}})^{\mathtt{I}} \qquad \tau = \sigma \to \mathtt{proc} \qquad \sigma' = (\mathtt{int}, (\mathtt{int})^{\mathtt{O}})^{\mathtt{IO}}$$

First we note $\lambda(x{:}\sigma).\, \mathbf{sq}(x)$ has a sendable type $\mathtt{s}(\sigma \to \mathtt{proc})$; the derivation is similar to that in Example 2.2, followed by an application of TERM_l. Then **SqServ** is typed as:

$$\mathsf{req} : ((\tau)^{\mathtt{O}})^{\mathtt{I}} \vdash_1 *\mathsf{req}?(r : (\tau)^{\mathtt{O}}).r!\langle\lambda(x{:}\sigma).\, \mathbf{sq}(x)\rangle : \mathtt{proc}$$

Next for **Client**, first let us define its body as $P \equiv (\ X\,a\ \ |a!\langle 1, c_1\rangle | \cdots)$. To accept $\lambda(x{:}\sigma).\, \mathbf{sq}(x)$ from the server and create $\mathbf{sq}(a)$ by applying a to $\lambda(x{:}\sigma).\, \mathbf{sq}(x)$, a will be used with both input/output capabilities in P. Hence P is typed as: $X{:}\tau,\ a{:}\sigma' \vdash_1 P : \mathtt{proc}$. Now define $\Gamma = \mathsf{req} : ((\tau)^{\mathtt{O}})^{\mathtt{O}}, r : (\tau)^{\mathtt{IO}}$. Then by applying RES and IN, we have:

$$\Gamma \vdash_1 r?(X : \tau).\, (\nu a{:}\sigma')P : \mathtt{proc}$$

To output r through "req", r should have a sendable type. Then, by $(\tau)^{\mathtt{IO}} \leq (\tau)^{\mathtt{O}}$, we have:

$$\text{CHAN}_l \quad \frac{\Gamma \vdash_1 r : (\tau)^{\mathtt{O}}}{\Gamma \vdash_1 r : \mathtt{s}((\tau)^{\mathtt{O}})}$$

The type of the channel "req" in the client is inferred as $\Gamma \vdash_1 \mathsf{req} : (\mathtt{s}((\tau)^{\mathtt{O}}))^{\mathtt{O}}$ by $\mathtt{s}((\tau)^{\mathtt{O}}) \leq (\tau)^{\mathtt{O}}$ as well as the contravariance of output capability. Combining these three, we infer:

$$\mathsf{req} : ((\tau)^{\mathtt{O}})^{\mathtt{O}} \vdash_1 (\nu r : (\tau)^{\mathtt{IO}})\mathsf{req}!\langle r\rangle r?(X : \tau).\, (\nu a{:}\sigma')P \equiv \mathbf{Client} : \mathtt{proc}$$

Finally since $\{\text{req}:((\tau)^{\text{O}})^{\text{I}}\} \asymp_1 \{\text{req}:((\tau)^{\text{O}})^{\text{O}}\}$, both systems are system composable.

$$\text{req}:((\tau)^{\text{O}})^{\text{IO}} \vdash_1 \mathbf{SqServ} \parallel \mathbf{Client}$$

Observe that:

(1) The sendable type $\mathtt{s}(\sigma \to \mathtt{proc})$ of $\lambda(x{:}\sigma).\,\mathbf{sq}(x)$ makes it possible to create a new server $\mathbf{sq}(a)$ in the client side.

(2) r is declared with both input and output capabilities in the **Client**. The **Client** itself uses the input capability but, because of the type of "req" it only sends the output capability to **SqServ**. This form of communication is essential to represent a continuation passing style programming in the π-calculus as studied in [16, 24, 27, 28]. Moreover it demonstrates the need for non-trivial subtyping on channels. □

One can refer to [35] for more examples which show that our systems eliminates various forms of behaviour which destroy the locality, like those in Example 3.3.

SUBJECT REDUCTION We now prove locality is preserved under reduction.

LEMMA 4.5. (1) (algebra on environments) $\Gamma_1 \vdash \text{SBL}$, *and* $\Gamma_2 \vdash \text{SBL}$ *imply* $\Gamma_1 \sqcap \Gamma_2 \vdash \text{SBL}$ *and* $\Delta_1 \asymp_1 \Delta_2$ *and* $\Delta_1 \asymp_1 \Delta_3$ *implies* $\Delta_1 \asymp_1 \Delta_2 \sqcap \Delta_3$.
(2) If $\rho \in$ Sble then $\Gamma \vdash_1 P : \rho$ *implies there exists* Δ *s.t.* $\Delta \geq \Gamma$, $\Delta \vdash_1 \text{SBL}$, *and* $\Delta \vdash_1 P : \rho$.
(3) (substitution) *Suppose* $\Gamma, x{:}\tau \vdash_1 P : \rho$ *and* $\Gamma \vdash_1 V : \tau$. *Then* $\Gamma \vdash_1 P\{V/x\} : \rho$.

The second property, which is needed to prove (3), is the most important. In the type system there are many different ways of inferring a sendable type, for example using CONST$_l$, CHAN$_l$, SUB or APP. However we can regard all sendable types as being inferred in a uniform manner by an application of TERM$_l$.[4]

The main lemma requires the order-theoretic property, FBC, of our subtyping relation, together with Lemma 4.5.

LEMMA 4.6. **(Main Lemma)** *Suppose* $\Gamma_1, x{:}\tau' \vdash_1 P : \pi$ *and* $\Gamma_2 \vdash_1 V : \mathtt{s}(\tau)$ *with* $\Gamma_1 \asymp_1 \Gamma_2$ *and* $\tau' \geq \mathtt{s}(\tau)$. *Then there exists* Δ *such that:* (1) $\Gamma_2 \leq \Delta$ with $\Delta \vdash_1 \text{SBL}$ and $\Delta \vdash_1 V : \mathtt{s}(\tau)$, (2) $\Gamma_1 \sqcap \Delta \vdash_1 P\{V/x\} : \pi$, and (3) $\Gamma_1 \sqcap \Delta \sqcap \Gamma_2 = \Gamma_1 \sqcap \Gamma_2$ with $\Gamma_1 \sqcap \Delta \asymp_1 \Gamma_2$.

The non-trivial case of the proof of the subject reduction property is when a value is sent to a different location by (com$_s$) rule: suppose $\Gamma_1 \vdash_1 a?(x{:}\tau).\,P$, $\Gamma_2 \vdash_1 a!\langle V\rangle.\,Q$ and $\Gamma_1 \asymp_1 \Gamma_2$. Then we must show:

$$\Gamma_1 \sqcap \Gamma_2 \vdash_1 a?(x{:}\tau).\,P \parallel a!\langle V\rangle.\,Q \text{ implies } \Gamma_1 \sqcap \Delta \vdash_1 P\{V/x\} \text{ and } \Gamma_2 \vdash_1 Q$$

with $\Gamma_1 \sqcap \Delta \asymp_1 \Gamma_2$ and $\Gamma_1 \sqcap \Delta \sqcap \Gamma_2 = \Gamma_1 \sqcap \Gamma_2$ for some Δ. By the main lemma, we can take Δ as a sendable environment such that $\Delta \vdash_1 V : \mathtt{s}(\tau)$. Now we establish:

THEOREM 4.7. **(Subject Reduction Theorem)**

$$\textit{If } \Gamma \vdash_1 N \textit{ and } N \longrightarrow\!\!\!\!\rightarrow M, \textit{ then } \Gamma \vdash_1 M.$$

[4]The proof of the second property relies directly on the constraint on the construction of arrow types. Relaxing this constraint would allow us to type more terms as sendable. Typical examples take the form $(\lambda xy.x)PQ$, with P being sendable and Q non-sendable. Indeed such terms may be exported between locations without violating locality constraints. But inventing a typing system which allows this behaviour is a topic for further research (cf. [23]).

See [35] for the proofs. Combining Theorem 3.2 and Theorem 4.7, we now have:

COROLLARY 4.8. (Type Safety) $\Gamma \vdash_1 N$ *and* $N \longrightarrow\!\!\!\!\rightarrow M$ *imply* $M \not\xrightarrow{lerr}$.

5 Further Development

5.1 Generalisation to Global/Local Subtyping

Our typing system has a static view of the role of channels. From the point of view of a given location they can only be used for input *locally* whereas there is *global* access to its output capability. A more general view is proposed in [30], whereby the input/output capabilities of each channel can be designated to be either global or be restricted to being local. Here we show that our typing system can be adapted to this framework.

In this extension channel types are labelled by one of the *locality modes*, $\{\mathtt{GG},\mathtt{LG},\mathtt{GL},\mathtt{LL}\}$, ranged over by $m, m', \ldots$. Their meaning is as follows:

(1) GG – a channel is allowed to be used as the input and output subjects anywhere.
(2) GL (resp. LG) – a channel is used as the input (resp. output) subject anywhere, while as the output (resp. input) subject only inside this location.
(3) LL – a channel is used as the input and output subjects only in this location.

A partial order on this set is given by a reflexive closure of $\mathtt{GG} \leq m$ with $m = \mathtt{LG},\mathtt{GL}$ and $m \leq \mathtt{LL}$. Then the syntax of channel type is extended to $m\langle S_\mathrm{I}, S_\mathrm{O}\rangle$ where m denotes how the channel is used as the subject while $S_\mathrm{I}, S_\mathrm{O}$ stand for the types of objects which it carries.

In the revised system judgements take the form: $\Gamma \vdash_g P : \rho$. First we replace CHAN$_l$ in Figure 9 with a more general rule which indicates when channel capabilities may be transferred between locations:

$$\frac{\Gamma \vdash_g a : \mathtt{LL}\langle S_\mathrm{I}, S_\mathrm{O}\rangle}{\Gamma \vdash_g a : \mathtt{s}(\mathtt{GG}\langle \top, \bot\rangle)} \quad \frac{\Gamma \vdash_g a : \mathtt{LG}\langle S_\mathrm{I}, S_\mathrm{O}\rangle}{\Gamma \vdash_g a : \mathtt{s}(\mathtt{GG}\langle \top, S_\mathrm{O}\rangle)} \quad \frac{\Gamma \vdash_g a : \mathtt{GL}\langle S_\mathrm{I}, S_\mathrm{O}\rangle}{\Gamma \vdash_g a : \mathtt{s}(\mathtt{GG}\langle S_\mathrm{I}, \bot\rangle)} \quad \frac{\Gamma \vdash_g a : \mathtt{GG}\langle S_\mathrm{I}, S_\mathrm{O}\rangle}{\Gamma \vdash_g a : \mathtt{s}(\mathtt{GG}\langle S_\mathrm{I}, S_\mathrm{O}\rangle)}$$

In general a capability can only be transmitted if it has the form $\mathtt{GG}\langle S_\mathrm{I}, S_\mathrm{O}\rangle$ for some $S_\mathrm{I}, S_\mathrm{O}$. As an example if this is GL, then it is prohibited from being used as the output subject in an other location; hence it can only be sent as the capability $\langle S_\mathrm{I}, \bot\rangle$, with the mode GG.

The input/output rules require minor modifications:

$$\text{IN}_{gl}: \frac{\Gamma \vdash_g u : \mathtt{LL}(\tilde{\tau})^\mathrm{I} \quad \Gamma, \tilde{x}{:}\tilde{\tau} \vdash_g P : \mathtt{proc}}{\Gamma \vdash_g u?(\tilde{x}{:}\tilde{\tau}).P : \mathtt{proc}} \quad \text{OUT}_{gl}: \frac{\Gamma \vdash_g u : \mathtt{LL}((\tilde{\tau}'),(\tilde{\tau})) \quad \Gamma \vdash_g V_i : \tau_i \quad \Gamma \vdash_g P : \mathtt{proc}}{\Gamma \vdash_g u!\langle \tilde{V}\rangle P : \mathtt{proc}} \quad \text{OUT}_{gd}: \frac{\Gamma \vdash_g u : \mathtt{LL}(\mathtt{s}(\tilde{\tau}))^\mathrm{O} \quad \Gamma \vdash_g V_i : \mathtt{s}(\tau_i) \quad \Gamma \vdash_g P : \mathtt{proc}}{\Gamma \vdash_g u!\langle \tilde{V}\rangle P : \mathtt{proc}}$$

Finally to compose systems we need a more general definition of *system composable*: two environments Γ_1 and Γ_2 are composable, denoted by $\Gamma_1 \asymp_g \Gamma_2$, if $\Gamma_1 \asymp \Gamma_2$ and if $u{:}m_i\langle S_{i\mathrm{I}}, S_{i\mathrm{O}}\rangle \in \Gamma_i$ $(i = 1,2)$, then (1) $m_i = \mathtt{LL}$ implies $S_{j\mathrm{I}} = \top$ and $S_{j\mathrm{O}} = \bot$, (2) $m_i = \mathtt{LG}$ implies $S_{j\mathrm{I}} = \top$, and (3) $m_i = \mathtt{GL}$ implies $S_{j\mathrm{O}} = \bot$ with $i \neq j$. This leads to the final change to the typing rules:

$$\text{PAR}_g: \quad \frac{\Gamma \vdash_g M \quad \Delta \vdash_g N \quad \Gamma \asymp_g \Delta}{\Gamma \sqcap \Delta \vdash_g M \| N}$$

THEOREM 5.1. (Subject Reduction) *If* $\Gamma \vdash_g N$ *and* $N \longrightarrow\!\!\!\!\rightarrow M$, *then* $\Gamma \vdash_g M$.

Type Safety Theorem could easily be established for this typing system, by introducing a *tagged* version of the language (cf. [14, 24, 30]).

5.2 *Behavioral Equivalence*

Typing systems impose constraints on the communication structure of processes and various authors, for example [24, 28, 33] have used this to define relativised behavioural equivalences. These have proved useful for example in studying the properties of translations between languages [3, 24, 33]. This technique can also be applied to D$\pi\lambda$, thereby opening up the possibility of obtaining interesting relativised behavioural theories for higher-order processes. Let $\approx_\Gamma$ (resp. $\sim_\Gamma$) denote a typed weak (resp. strong) barbed reduction-closed congruence defined by input/output predicates and reduction-closure property as in [2, 17, 24, 28, 33]. Various properties of $\approx_\Gamma$ and $\sim_\Gamma$, proved for variations of the π-calculus, can easily be generalised to D$\pi\lambda$; a simple example is closure under β-reduction (see page 10 in [2]). We can also prove various distributed equations, such as: $(P\,|\,\mathtt{Spawn}(Q))\,\|\,R \approx_\Delta (P\,\|\,Q)\,\|\,R \sim_\Delta P\,\|\,(Q\,|\,R)$ and $P\,\|\,(a!\langle\tilde{V}\rangle\,|\,Q) \sim_\Delta (P\,|\,a!\langle\tilde{V}\rangle)\,\|\,Q$.

We also have the following multiple higher-order strong replication theorem which is not valid in D$\pi\lambda$ without types, but valid in the local D$\pi\lambda$ studied in Section 4.

PROPOSITION 5.2. *Let us define* $R \stackrel{\text{def}}{=} *a?(\tilde{x}).\,R_1\,|\cdots|\,*a?(\tilde{x}).\,R_n$ *with* R_i *sendable. Then we have:* $(\nu a)(R\,\|\,P\,\|\,Q) \;\sim_\Gamma\; (\nu a)(R\,|\,P)\,\|\,(\nu a)(R\,|\,Q)$

Note we do not require any side condition for P and Q (cf. [24]). The proof is by observing that P and Q may only export the sendable value V via a since it is impossible that the name a is local in either P or Q. We can then apply the standard reasoning framework from [24, 27, 28]. See [35] for proofs. Note also that this proposition can not be derived in the framework of [28] since a is neither a linear nor an ω-receptive name.

Such theorems will be useful for reasoning about object-oriented systems where templates are shared among locations. Further extension of typed equivalences studied in π-calculus (e.g. [28, 33]) to distributed higher-order processes is an interesting research topic we intend to pursue.

5.3 *Type Checking*

For a practical use of a typing system, it is essential that we can check the well-typedness of a system N against a global type environment Γ. For this purpose, we can construct an equivalent typing system to $\vdash_1$ without TERM$_l$, to obtain a syntax directed system (in type reconstruction we use the partial meet operator to obtain a sendable type). Using this, we can easily obtain an algorithm to check the typability of P against Γ, as well as an algorithm to compute ρ such that $\Gamma \vdash_1 P{:}\rho$ along the line of [31]. Once processes in each location are type-checked, the algorithm which computes typability of a global system $\Gamma \vdash_1 N$ is simply obtained by decidability of $\Gamma \asymp_1 \Delta$. See [35] for details.

6 Discussion and Related Work

We have proposed a local subtyping system for a simple higher-order distributed process language D$\pi\lambda$ and we used it to show that a global safety condition can be guaranteed by static type-checking of each local configuration. Our typing system does not require additional information on the resources available at different locations to ensure that higher-order processes can be safely passed between locations without violating locality constraints on channels. The notion of *sendable* values and the corresponding *sendable* types plays an essential role in our typing system. Other schemes for restricting capabilities of higher-order terms by types may also be found in various different contexts; for example, in reference types [23], agent migration [26], an implementation of network protocols [18], and a location-based Linda language [22].

The distributed component of $D\pi\lambda$ is rather primitive, but we believe that the type inference system can easily be adapted to languages where, for example, locations can be named and dynamically generated as in [3, 14], or where there is more significant interplay between the concurrent and the functional language primitives, as in Facile [9, 19]. However extensions of our capability based typing systems to more advanced distributed primitives, such as hierarchical location spaces [32], process mobility [6, 10, 29], and cryptographic constructs [1, 13] will be more challenging. Since in our language we inherit the standard subtyping of the λ-calculus, it is also possible to consider the introduction of richer subtyping relations, for example those based on records, recursive types, or polymorphic types into type systems for distributed languages.

It has been argued that in some sense there is no need for higher-order constructs in π-calculus based languages. For example in [27] there is a concise translation of processes using higher order values into the π-calculus. However, as we will now examine in the context of $D\pi\lambda$, certain information is lost in such translations.

The basic idea of the translation in [27] is to replace the transmission of an abstraction with the transmission of a newly generated *trigger*. An application to the abstraction is then replaced by a transmission of the data to the trigger, which provides a copy of the abstraction body to process the data. Using this idea **sqServ** is replaced by

$$[\![\mathbf{sqServ}]\!] \Leftarrow * \mathrm{req}?(r).\,(\nu\,\mathrm{tr})\,(r!\langle\mathrm{tr}\rangle \mid \mathbf{S}_{\mathrm{tr}}) \quad \text{with } \mathbf{S}_{\mathrm{tr}} \Leftarrow * \mathrm{tr}?(x).\,\mathbf{sq}(x)$$

Here when a request is received, a new trigger is generated, and then returned to the client. Associated with the trigger is a trigger server, $\mathbf{S}_{\mathrm{tr}}$ which receives data on the trigger and then executes the body, namely $z!\langle\mathbf{sq}(x)\rangle$. Suppose we have the following $\mathbf{Client}_2$ who may already have a square server, for example for faster parallel evaluation.

$$\mathbf{Client}_2 \Leftarrow (\nu ar)(\mathrm{req}!\langle r\rangle.\,r?(X).\,X\,a \mid \mathbf{sq}(a) \mid a!\langle 1,c_1\rangle \mid a!\langle 2,c_2\rangle \mid \cdots)$$

Then the client is replaced by

$$[\![\mathbf{Client}_2]\!] \Leftarrow (\nu ar)(\mathrm{req}!\langle r\rangle.\,r?(\mathrm{tr}).\,\mathrm{tr}!\langle a\rangle \mid \mathbf{sq}(a) \mid a!\langle 1,c_1\rangle \mid a!\langle 2,c_2\rangle \mid \cdots)$$

The application in $\mathbf{Client}_2$ is replaced by a transmission of a to the trigger, which was received in response to the request. However there exists an essential difference between $\mathbf{sqServ} \parallel \mathbf{Client}_2$ and its translation $[\![\mathbf{sqServ}]\!] \parallel [\![\mathbf{Client}_2]\!]$. In the former, the new receptor $\mathbf{sq}(a)$ is created in the client location, whereas in the latter $\mathbf{sq}(a)$ is created on the server side:

$$\begin{array}{rl} \mathbf{sqServ} \parallel \mathbf{Client}_2 \longrightarrow\!\!\!\!\rightarrow \mathbf{sqServ} & \parallel\ (\nu a)(\mathbf{sq}(a) \mid \mathbf{sq}(a) \mid \cdots) \\ [\![\mathbf{sqServ}]\!] \parallel [\![\mathbf{Client}_2]\!] \longrightarrow\!\!\!\!\rightarrow (\nu a)(\mathbf{sq}(a) \mid [\![\mathbf{sqServ}]\!] & \parallel\ \mathbf{sq}(a) \mid a!\langle 1,c_1\rangle \cdots) \end{array}$$

This disturbs the locality on the channel a. Actually we can check that for all Γ, we have:

$\Gamma \not\vdash_1$ $[\![\mathbf{Client}_2]\!] \parallel [\![\mathbf{sqServ}]\!]$, since a should be used as input capability in the server side to create a new $\mathbf{sq}(a)$. But $\mathbf{sqServ} \parallel \mathbf{Client}_2$ is typable, as seen in Example 4.4. This example shows that it would be difficult to adapt the translation technique in [27] so that the local typing structure is preserved; it provides at least one reason why higher-order distributed calculi are worthy of investigation independently. Moreover, as argued in [8, 9, 19, 22, 29], many practical applications call for parameterised higher-order process passing, which may be difficult to represent directly without functional constructions, even in languages which support migration of the processes; their presence leads to a natural and powerful programming style as seen in the above literature.

Preserving the locality of channels has been studied extensively for the π-calculus, [3, 4, 20, 34, 10]. For example the (untyped) *local π-calculus* [4, 20, 34] is simply defined with the following input restriction rule

$$a?(x).\,P \qquad \text{if } x \text{ does not appear as a free input subject in } P$$

If we consider the subset of Dπλ, where the abstraction mechanisms are omitted and only a single location is used then the typing system automatically enforces this restriction on well-typed terms. However it would be wrong to generalise this restriction to higher-order terms by imposing the constraint:

$$\lambda x.P \qquad \text{if } x \text{ does not appear as a free input subject in } P$$

This is too strong; new receptors can never be created by β-reduction, hence Example 4.4 would no longer be typable. Moreover this idea does not work if we wish to control higher-order variable, as seen in Example 3.3 (see also [35]).

A locality condition similar to ours is used in [10] in describing various kinds of encodings in Distributed Join-Calculus. Our approach is more general; we have a formal typing system for arbitrary higher-order process passing and instantiation which ensures locality of receptors, although new receptors can be created inside the same location.

ACKNOWLEDGEMENTS: We thank James Riely, Kohei Honda and the members of Wednesday Study Group of Sussex University for comments and discussions, and anonymous referees for useful comments.

References

1. Abadi, M. and Gordon, A., The Spi-calculus, Computer and Communications Security, pp.36–47, ACM Press, 1997.
2. Amadio, R., Translating Core Facile, ECRC Research Report 944-3, 1994.
3. Amadio, R., An asynchronous model of locality, failure, and process mobility. INRIA Report 3109, 1997.
4. Boreale, M., On the Expressiveness of Internal Mobility in Name-Passing Calculi, *CONCUR'96*, LNCS 1119, pp.163–178, Springer-Verlag, 1996.
5. Boudol, G., The π-Calculus in Direct Style, *POPL'98*, pp.228–241, ACM Press, 1998.
6. Cardelli, L. and Gordon, A., Typed Mobile Ambients, *POPL'99*, pp.79–92, ACM Press, 1999.
7. Chien, A., Concurrent Aggregates, MIT Press, 1993.
8. Ferreira, W., Hennessy, M. and Jeffrey, M., A Theory of Weak Bisimulation for Core CML, *ICFP*, pp.201–212, ACM Press, 1996. The full version appeared in *J. Func. Pro.*, 8(5):447–491,1998.
9. Giacalone, A., Mistra, P. and Prasad, S., Operational and Algebraic Semantics for Facile, *ICALP'90*, LNCS 443, pp.765–780, Springer-Verlag, 1990.

10. Fournet, C., Gonthier, G., Lévy, J.-J., Maranget, L., and Rémy, D., A Calculus for Mobile Agents, *CONCUR'96*, LNCS 1119, pp.406–421, Springer-Verlag, 1996.
11. Sun Microsystems Inc., Java home page. http://www.javasoft.com/, 1995.
12. Hartonas, C. and Hennessy, M., Full Abstractness for a Functional/Concurrent Language With Higher-Order Value-Passing, *Information and Computation*, Vol. 145, pp.64–106, 1998.
13. Heintze, N. and Riecke, J., The SLam Calculus: Programming with Secrecy and Integrity, *POPL'98*, pp.365-377. ACM Press, 1998.
14. Hennessy, M. and Riely, J., Resource Access Control in Systems of Mobile Agents, CS Report 02/98, University of Sussex, http://www.cogs.susx.ac.uk, 1998.
15. Honda, K., Composing Processes, *POPL'96*, pp.344-357, ACM Press, 1996.
16. Honda, K. and Tokoro, M., An Object Calculus for Asynchronous Communication. *ECOOP'91*, LNCS 512, pp.133–147, Springer-Verlag 1991.
17. Honda, K. and Yoshida, N., On Reduction-Based Process Semantics. *TCS*, pp.437–486, No.151, North-Holland, 1995.
18. Jeffrey, A. and Wakeman, I., SafetyNet. Available from: http://klee.cs.depaul.edu/an/, 1998.
19. Leth, L. and Thomsen, B., Some Facile Chemistry, ERCC Technical Report, ERCC-92-14, 1992.
20. Merro, M. and Sangiorgi, D., On asynchrony in name-passing calculi, *ICALP'98*, LNCS 1443, pp.856–867, Springer-Verlag, 1998.
21. Milner, R., Parrow, J.G. and Walker, D.J., A Calculus of Mobile Processes. *Information and Computation*, 100(1), pp.1–77, 1992.
22. De Nicola, R., Ferrari, G. and Pugliese, R., Klaim: a Kernel Language for Agents Interaction and Mobility, IEEE Trans. on Software Engineering, Vol.24(5), 1998.
23. O'Hearn, P., Power, J., Takeyama, M., and Tennent, D., Syntactic Control of Interference Revised, *MFPS'97*, *ENCS*, Elsevier, 1997.
24. Pierce, B.C. and Sangiorgi. D, Typing and subtyping for mobile processes. *MSCS*, 6(5):409–454, 1996.
25. Pierce, B. and Turner, D., Pict: A Programming Language Based on the Pi-calculus, Indiana University, CSCI Technical Report, 476, March, 1997.
26. Riely, J. and Hennessy, M., Trust and Partial Typing in Open Systems of Mobile Agents, CS Technical Report, University of Sussex, 04/98, Available at: http://www.cogs.susx.ac.uk, 1998.
27. Sangiorgi, D., *Expressing Mobility in Process Algebras: First Order and Higher Order Paradigms*. Ph.D. Thesis, University of Edinburgh, 1992.
28. Sangiorgi, D., The name discipline of uniform receptiveness, *ICALP'97*, LNCS 1256, pp.303–313, 1997.
29. Sekiguchi, T. and Yonezawa, A., A calculus with code mobility, IFIP, pp.21–36, Chapman & Hall, 1997.
30. Sewell, P., Global/Local Subtyping and Capability Inference for a Distributed π-calculus, *ICALP'98*, LNCS 1443, pp.695–706, Springer-Verlag, 1998.
31. Vasconcelos, V. and Honda, K., Principal Typing Scheme for Polyadic π-Calculus. *CONCUR'93*, LNCS 715, pp.524-538, Springer-Verlag, 1993.
32. Vitek, J. and Castagna, G., A Calculus of Secure Mobile Computations, Available at: http://cuiwww.unige.ch/~ jvitek, 1999.
33. Yoshida, N., Graph Types for Monadic Mobile Processes, *FST/TCS'16*, LNCS 1180, pp. 371–386, Springer-Verlag, 1996. Full version as LFCS Technical Report, ECS-LFCS-96-350, 1996.
34. Yoshida, N., Minimality and Separation Results on Asynchronous Mobile Processes: representability theorems by concurrent combinators. *CONCUR'98*, pp.131–146, LNCS 1466, Springer-Verlag, 1998. Full version as CS Report 05/98, University of Sussex, Available at: http://www.cogs.susx.ac.uk, 1998.
35. The full version of this paper. CS Technical Report 01/99, University of Sussex, Available at: http://www.cogs.susx.ac.uk, 1999.

Author Index

Lecture Notes in Computer Science

For information about Vols. 1–1584
please contact your bookseller or Springer-Verlag

Vol. 1587: J. Pieprzyk, R. Safavi-Naini, J. Seberry (Eds.), Information Security and Privacy. Proceedings, 1999. XI, 327 pages. 1999.

Vol. 1589: J.L. Fiadeiro (Ed.), Recent Trends in Algebraic Development Techniques. Proceedings, 1998. X, 341 pages. 1999.

Vol. 1590: P. Atzeni, A. Mendelzon, G. Mecca (Eds.), The World Wide Web and Databases. Proceedings, 1998. VIII, 213 pages. 1999.

Vol. 1592: J. Stern (Ed.), Advances in Cryptology – EUROCRYPT '99. Proceedings, 1999. XII, 475 pages. 1999.

Vol. 1593: P. Sloot, M. Bubak, A. Hoekstra, B. Hertzberger (Eds.), High-Performance Computing and Networking. Proceedings, 1999. XXIII, 1318 pages. 1999.

Vol. 1594: P. Ciancarini, A.L. Wolf (Eds.), Coordination Languages and Models. Proceedings, 1999. IX, 420 pages. 1999.

Vol. 1595: K. Hammond, T. Davie, C. Clack (Eds.), Implementation of Functional Languages. Proceedings, 1998. X, 247 pages. 1999.

Vol. 1596: R. Poli, H.-M. Voigt, S. Cagnoni, D. Corne, G.D. Smith, T.C. Fogarty (Eds.), Evolutionary Image Analysis, Signal Processing and Telecommunications. Proceedings, 1999. X, 225 pages. 1999.

Vol. 1597: H. Zuidweg, M. Campolargo, J. Delgado, A. Mullery (Eds.), Intelligence in Services and Networks. Proceedings, 1999. XII, 552 pages. 1999.

Vol. 1598: R. Poli, P. Nordin, W.B. Langdon, T.C. Fogarty (Eds.), Genetic Programming. Proceedings, 1999. X, 283 pages. 1999.

Vol. 1599: T. Ishida (Ed.), Multiagent Platforms. Proceedings, 1998. VIII, 187 pages. 1999. (Subseries LNAI).

Vol. 1600: M. Wooldridge, M. Veloso (Eds.), Artificial Intelligence Today. VIII, 489 pages. 1999. (Subseries LNAI).

Vol. 1601: J.-P. Katoen (Ed.), Formal Methods for Real-Time and Probabilistic Systems. Proceedings, 1999. X, 355 pages. 1999.

Vol. 1602: A. Sivasubramaniam, M. Lauria (Eds.), Network-Based Parallel Computing. Proceedings, 1999. VIII, 225 pages. 1999.

Vol. 1603: J. Vitek, C.D. Jensen (Eds.), Secure Internet Programming. X, 501 pages. 1999.

Vol. 1604: M. Asada, H. Kitano (Eds.), RoboCup-98: Robot Soccer World Cup II. XI, 509 pages. 1999. (Subseries LNAI).

Vol. 1605: J. Billington, M. Diaz, G. Rozenberg (Eds.), Application of Petri Nets to Communication Networks. IX, 303 pages. 1999.

Vol. 1606: J. Mira, J.V. Sánchez-Andrés (Eds.), Foundations and Tools for Neural Modeling. Proceedings, Vol. I, 1999. XXIII, 865 pages. 1999.

Vol. 1607: J. Mira, J.V. Sánchez-Andrés (Eds.), Engineering Applications of Bio-Inspired Artificial Neural Networks. Proceedings, Vol. II, 1999. XXIII, 907 pages. 1999.

Vol. 1608: S. Doaitse Swierstra, P.R. Henriques, J.N. Oliveira (Eds.), Advanced Functional Programming. Proceedings, 1998. XII, 289 pages. 1999.

Vol. 1609: Z. W. Raś, A. Skowron (Eds.), Foundations of Intelligent Systems. Proceedings, 1999. XII, 676 pages. 1999. (Subseries LNAI).

Vol. 1610: G. Cornuéjols, R.E. Burkard, G.J. Woeginger (Eds.), Integer Programming and Combinatorial Optimization. Proceedings, 1999. IX, 453 pages. 1999.

Vol. 1611: I. Imam, Y. Kodratoff, A. El-Dessouki, M. Ali (Eds.), Multiple Approaches to Intelligent Systems. Proceedings, 1999. XIX, 899 pages. 1999. (Subseries LNAI).

Vol. 1612: R. Bergmann, S. Breen, M. Göker, M. Manago, S. Wess, Developing Industrial Case-Based Reasoning Applications. XX, 188 pages. 1999. (Subseries LNAI).

Vol. 1613: A. Kuba, M. Šámal, A. Todd-Pokropek (Eds.), Information Processing in Medical Imaging. Proceedings, 1999. XVII, 508 pages. 1999.

Vol. 1614: D.P. Huijsmans, A.W.M. Smeulders (Eds.), Visual Information and Information Systems. Proceedings, 1999. XVII, 827 pages. 1999.

Vol. 1615: C. Polychronopoulos, K. Joe, A. Fukuda, S. Tomita (Eds.), High Performance Computing. Proceedings, 1999. XIV, 408 pages. 1999.

Vol. 1616: P. Cointe (Ed.), Meta-Level Architectures and Reflection. Proceedings, 1999. XI, 273 pages. 1999.

Vol. 1617: N.V. Murray (Ed.), Automated Reasoning with Analytic Tableaux and Related Methods. Proceedings, 1999. X, 325 pages. 1999. (Subseries LNAI).

Vol. 1618: J. Bézivin, P.-A. Muller (Eds.), The Unified Modeling Language. Proceedings, 1998. IX, 443 pages. 1999.

Vol. 1619: M.T. Goodrich, C.C. McGeoch (Eds.), Algorithm Engineering and Experimentation. Proceedings, 1999. VIII, 349 pages. 1999.

Vol. 1620: W. Horn, Y. Shahar, G. Lindberg, S. Andreassen, J. Wyatt (Eds.), Artificial Intelligence in Medicine. Proceedings, 1999. XIII, 454 pages. 1999. (Subseries LNAI).

Vol. 1621: D. Fensel, R. Studer (Eds.), Knowledge Acquisition Modeling and Management. Proceedings, 1999. XI, 404 pages. 1999. (Subseries LNAI).

Vol. 1622: M. González Harbour, J.A. de la Puente (Eds.), Reliable Software Technologies – Ada-Europe'99. Proceedings, 1999. XIII, 451 pages. 1999.

Vol. 1623: T. Reinartz, Focusing Solutions for Data Mining. XV, 309 pages. 1999. (Subseries LNAI).

Vol. 1625: B. Reusch (Ed.), Computational Intelligence. Proceedings, 1999. XIV, 710 pages. 1999.

Vol. 1626: M. Jarke, A. Oberweis (Eds.), Advanced Information Systems Engineering. Proceedings, 1999. XIV, 478 pages. 1999.

Vol. 1627: T. Asano, H. Imai, D.T. Lee, S.-i. Nakano, T. Tokuyama (Eds.), Computing and Combinatorics. Proceedings, 1999. XIV, 494 pages. 1999.

Col. 1628: R. Guerraoui (Ed.), ECOOP'99 - Object-Oriented Programming. Proceedings, 1999. XIII, 529 pages. 1999.

Vol. 1629: H. Leopold, N. García (Eds.), Multimedia Applications, Services and Techniques - ECMAST'99. Proceedings, 1999. XV, 574 pages. 1999.

Vol. 1631: P. Narendran, M. Rusinowitch (Eds.), Rewriting Techniques and Applications. Proceedings, 1999. XI, 397 pages. 1999.

Vol. 1632: H. Ganzinger (Ed.), Automated Deduction – Cade-16. Proceedings, 1999. XIV, 429 pages. 1999. (Subseries LNAI).

Vol. 1633: N. Halbwachs, D. Peled (Eds.), Computer Aided Verification. Proceedings, 1999. XII, 506 pages. 1999.

Vol. 1634: S. Džeroski, P. Flach (Eds.), Inductive Logic Programming. Proceedings, 1999. VIII, 303 pages. 1999. (Subseries LNAI).

Vol. 1636: L. Knudsen (Ed.), Fast Software Encryption. Proceedings, 1999. VIII, 317 pages. 1999.

Vol. 1637: J.P. Walser, Integer Optimization by Local Search. XIX, 137 pages. 1999. (Subseries LNAI).

Vol. 1638: A. Hunter, S. Parsons (Eds.), Symbolic and Quantitative Approaches to Reasoning and Uncertainty. Proceedings, 1999. IX, 397 pages. 1999. (Subseries LNAI).

Vol. 1639: S. Donatelli, J. Kleijn (Eds.), Application and Theory of Petri Nets 1999. Proceedings, 1999. VIII, 425 pages. 1999.

Vol. 1640: W. Tepfenhart, W. Cyre (Eds.), Conceptual Structures: Standards and Practices. Proceedings, 1999. XII, 515 pages. 1999. (Subseries LNAI).

Vol. 1642: D.J. Hand, J.N. Kok, M.R. Berthold (Eds.), Advances in Intelligent Data Analysis. Proceedings, 1999. XII, 538 pages. 1999.

Vol. 1643: J. Nešetřil (Ed.), Algorithms – ESA '99. Proceedings, 1999. XII, 552 pages. 1999.

Vol. 1644: J. Wiedermann, P. van Emde Boas, M. Nielsen (Eds.), Automata, Languages, and Programming. Proceedings, 1999. XIV, 720 pages. 1999.

Vol. 1645: M. Crochemore, M. Paterson (Eds.), Combinatorial Pattern Matching. Proceedings, 1999. VIII, 295 pages. 1999.

Vol. 1647: F.J. Garijo, M. Boman (Eds.), Multi-Agent System Engineering. Proceedings, 1999. X, 233 pages. 1999. (Subseries LNAI).

Vol. 1648: M. Franklin (Ed.), Financial Cryptography. Proceedings, 1999. VIII, 269 pages. 1999.

Vol. 1649: R.Y. Pinter, S. Tsur (Eds.), Next Generation Information Technologies and Systems. Proceedings, 1999. IX, 327 pages. 1999.

Vol. 1650: K.-D. Althoff, R. Bergmann, L.K. Branting (Eds.), Case-Based Reasoning Research and Development. Proceedings, 1999. XII, 598 pages. 1999. (Subseries LNAI).

Vol. 1651: R.H. Güting, D. Papadias, F. Lochovsky (Eds.), Advances in Spatial Databases. Proceedings, 1999. XI, 371 pages. 1999.

Vol. 1652: M. Klusch, O.M. Shehory, G. Weiss (Eds.), Cooperative Information Agents III. Proceedings, 1999. XI, 404 pages. 1999. (Subseries LNAI).

Vol. 1653: S. Covaci (Ed.), Active Networks. Proceedings, 1999. XIII, 346 pages. 1999.

Vol. 1654: E.R. Hancock, M. Pelillo (Eds.), Energy Minimization Methods in Computer Vision and Pattern Recognition. Proceedings, 1999. IX, 331 pages. 1999.

Vol. 1656: S. Chatterjee, J.F. Prins, L. Carter, J. Ferrante, Z. Li, D. Sehr, P.-C. Yew (Eds.), Languages and Compilers for Parallel Computing. Proceedings, 1998. XI, 384 pages. 1999.

Vol. 1661: C. Freksa, D.M. Mark (Eds.), Spatial Information Theory. Proceedings, 1999. XIII, 477 pages. 1999.

Vol. 1662: V. Malyshkin (Ed.), Parallel Computing Technologies. Proceedings, 1999. XIX, 510 pages. 1999.

Vol. 1663: F. Dehne, A. Gupta. J.-R. Sack, R. Tamassia (Eds.), Algorithms and Data Structures. Proceedings, 1999. IX, 366 pages. 1999.

Vol. 1664: J.C.M. Baeten, S. Mauw (Eds.), CONCUR'99. Concurrency Theory. Proceedings, 1999. XI, 573 pages. 1999.

Vol. 1666: M. Wiener (Ed.), Advances in Cryptology – CRYPTO '99. Proceedings, 1999. XII, 639 pages. 1999.

Vol. 1668: J.S. Vitter, C.D. Zaroliagis (Eds.), Algorithm Engineering. Proceedings, 1999. VIII, 361 pages. 1999.

Vol. 1671: D. Hochbaum, K. Jansen, J.D.P. Rolim, A. Sinclair (Eds.), Randomization, Approximation, and Combinatorial Optimization. Proceedings, 1999. IX, 289 pages. 1999.

Vol. 1678: M.H. Böhlen, C.S. Jensen, M.O. Scholl (Eds.), Spatio-Temporal Database Management. Proceedings, 1999. X, 243 pages. 1999.

Vol. 1684: G. Ciobanu, G. Păun (Eds.), Fundamentals of Computation Theory. Proceedings, 1999. XI, 570 pages. 1999.

Vol. 1685: P. Amestoy, P. Berger, M. Daydé, I.S. Duff, V. Frayssé, L. Giraud, D. Ruiz (Eds.), Euro-Par'99. Parallel Processing. Proceedings, 1999. XXXII, 1503 pages. 1999.

Vol. 1688: P. Bouquet, L. Serafini, P. Brézillon, M. Benerecetti, F. Castellani (Eds.), Modeling and Using Context. Proceedings, 1999. XII, 528 pages. 1999. (Subseries LNAI).

Vol. 1689: F. Solina, A. Leonardis (Eds.), Computer Analysis of Images and Patterns. Proceedings, 1999. XIV, 650 pages. 1999.